AQA
A-level

AB
910
438491

910

£35-99

...phy

...el and AS

...rth Edition

...olm Skinner

...aul Abbiss

...hil Banks

...elen Fyfe

...Whittaker

...on. This means that we have checked that it broadly covers the specification and we are satisfied with the overall quality. Full details of our approval process can be found on our website.

We approve textbooks because we know how important it is for teachers and students to have the right resources to support their teaching and learning. However, the publisher is ultimately responsible for the editorial control and quality of this book.

Please note that when teaching the *AQA A-level Geography* course, you must refer to AQA's specification as your definitive source of information. While this book has been written to match the specification, it cannot provide complete coverage of every aspect of the course.

A wide range of other useful resources can be found on the relevant ... s of our website: www.aqa.org.uk.

D1354518

DYNAMIC
LEARNING

HODDER
EDUCATION
AN HACHETTE UK COMPANY

The Publishers would like to thank the following for permission to reproduce copyright material.

Photo credits

00.1 (part opener), © David Coleman/Alamy Stock Photo; 00.2 (part opener) © AndreyKrav/Thinkstock/iStockphoto/Getty Images; 00.3 (part opener) © Google Earth, © Infoterra Ltd & Bluesky Image, © Getmapping plc; 1.07, © Alexandra Kobalenko/All Canada Photos/Corbis; 1.09, © David Coleman/Alamy Stock Photo; 1.22, © Richard Croft; 1.37, © Gavriel Jecan/CORBIS; 1.42, © NASA; 1.51, © John Foster/Masterfile/Corbis; 2.07, © bodikp/Thinkstock/iStock/Getty Images; 2.08, © Jupiterimages/Thinkstock/iStock/Getty Images; 2.15, © Goddard Space Flight Center/NASA; 2.18, © Ann and Steve Toon/Alamy Stock Photo; 2.19a, © Robert Miramontes/Thinkstock/iStock/Getty Images; 2.19b, © Charles O. Cecil/Alamy Stock Photo; 2.20, © Bruce M. Herman/Science Photo Library; 2.25, © imageBROKER/Alamy Stock Photo; 2.29, © pum.eva/Thinkstock/iStock/Getty Images; 2.32, © blickwinkel/Alamy Stock Photo; 2.33, © lucky-photographer/Thinkstock/iStock/Getty Images; 2.34, © Patrick Burke/Thinkstock/iStock/Getty Images; 2.40a, NASA image created by Jesse Allen, using Landsat data provided by the United States Geological Survey; 2.40b, © NASA/United States Geological Survey; 2.41, © Peter Banos/Alamy Stock Photo; 2.42, © M. ou Me. Desjeux, Bernard/Corbis; 2.43b–2.45, © Ian Whittaker; 2.47a, © Tiziana and Gianni Baldizzone/Corbis; 2.47b, © Peter Giovannini/imageBROKER; 2.49, © George Steinmetz/Corbis; 2.50, © Michel Gounot/Godong/Corbis; 2.51, © G. R. 'Dick' Roberts/NSIL/Getty Images; 2.52a, © United States Geological Survey; 2.52b, © United States Geological Survey; 3.10, © John Worrall/Alamy Stock Photo; 3.2, © B.A.E. Inc./Alamy Stock Photo; 3.7a & b, © Phil Banks; 3.23, © imageBROKER/Alamy Stock Photo; 3.27a, © Mr. Nut/Alamy Stock Photo; 3.27b, © Robert Quinlan/Alamy Stock Photo; 3.30 & 3.32, © Ian Whittaker; 3.33, © NOAA; 3.37, © Luke Farmer/Alamy Stock Photo; 3.42–46, © Phil Banks; 3.51, © Pevensey Bay Coastal; 3.52, © Pevensey Bay Coastal; 3.56, © Universal Images Group North America LLC/Alamy Stock Photo; 3.58, © Timothy G. Laman/National Geographic/Getty Images; 3.60, © Neil Cooper/Alamy Stock Photo; 3.61, © Ian Whittaker; 4.07a, Adapted from © NASA; 4.07b, © NASA; 4.15, © Danita Delimont/Getty Images; 4.16, © Galen Rowell/Corbis; 4.18, © Scott Rylander Travel/Alamy Stock Photo; 4.28, © Matthew Williams-Ellis/robertharding/Corbis; 4.32, © Julia Katherine Rich/Fotolibra; 4.33, © Patrick Dieudonne/robertharding/Getty Images; 4.35, © Michael Dutton/Alamy Stock Photo; 4.41, © Cultura Creative (RF)/Alamy Stock Photo; 4.44, © Erin Paul Donovan/Alamy Stock Photo; 4.48, © Timothy Epp/Alamy Stock Photo; 4.52, © Dr. Marli Miller/Visuals Unlimited/Corbis; 4.58, © Bryan and Cherry Alexander Photography; 4.60, © Vladimir Romanovsky; 4.62, © Design Pics Inc/Alamy Stock Photo; 4.64, © Nigel Wilkins/Alamy Stock Photo; 4.66, © AllCanadaPhotos.com/Corbis; 4.68, © Bryan and Cherry Alexander Photography; 4.70, © Emma Wood/Alamy Stock Photo; 4.71, © NASA; 4.72, © NASA; 4.75, © Ian Whittaker; 4.76 a, b & c, © Mauri Pelto/Google Earth; 5.17, © Purple Pilchards/Alamy Stock Photo; 5.19, © Jon Helgason/123RF.COM; 5.24, © Jon Sheer, All Rights Reserved/Getty Images; 5.26, © U.S. Navy/Handout/Getty Images; 5.28, © Luke Lee/Contributor/Getty Images; 5.32, © Chris Graythen / Stringer/Getty Images; 5.34, © JAY DIRECTO/Stringer/Getty Images; 5.41, © TED ALJIBE/AFP/Getty Images; 5.42, © Pool AVENTURIER/LOVINY/Gamma-Rapho via Getty Images); 5.44, © Pacific Press Service/Alamy Stock Photo; 5.45, © robertharding/Alamy Stock Photo; 6.10, © Wayne Lynch/Getty Images; 6.12–13, © Phil Banks; 6.16, © Robin Bush/Getty Images; 6.20, © BrianLasenby/Thinkstock/iStock/Getty Images; 6.21, 6.26, 6.28, 6.29, © Phil Banks; 6.32, © John Morrison/Alamy Stock Photo; 6.45, © Universal Group North America LLC/Alamy Stock Photo; 6.47, © Universal Images Group North America LLC / Alamy Stock Photo; 6.48, © dpa picture alliance/Alamy Stock Photo; 6.56, © imageBROKER/Alamy Stock Photo; 7.3, © Alan Gignoux/Alamy Stock Photo; 7.11, © Prisma Bildagentur AG/Alamy Stock Photo; 7.13, © A.P.S. (UK)/Alamy Stock Photo; 7.16, © Jack Hinds/Alamy Stock Photo; 7.26, © ITAR-TASS Photo Agency/Alamy Stock Photo; 7.27, © Ian Dagnall/Alamy Stock Photo; 7.29, © Kin Cheung/AP/Press Association Images; 7.36, © EDU Vision/Alamy Stock Photo; 7.37, © United Nations Division for Sustainable Development; 07.42a, © wanderworldimages/Alamy Stock Photo; 07.42b, © Kevin Schafer/Alamy Stock Photo; 7.45, © staphy/Thinkstock/iStock/Getty Images; 7.47, © kkaplin/Thinkstock/iStock/Getty Images; 7.48, © Greenpeace; 8.3, © Steve Taylor ARPS/Alamy; 8.4, © Roger Cracknell 01/classic/Alamy; 8.5, © Alistair Laming/Alamy; 8.6, © dpa picture alliance archive/Alamy; 8.7, © Sven Schermer/123RF.COM; 8.8, Image by Wordsmith One for Transition Town Totnes' Clone-stopping Campaign'; 8.9, © Jim Wileman/Alamy; 8.10, Copyright - Transition Network/Designer - Trucie Mitchell; 8.11, Used with permission from Project for Public Places. Website: http://www.pps.org/reference/grplacefeat/; 8.13, © Amoret Tanner Collection/The Art Archive; 8.14, © Helen Fyfe; 8.15, © Education Images/UIG/ Getty Images; 8.16, © jvdwolf/123RF.COM; 8.19, © bloodua/123RF.COM; 8.20, © Pembrokeshire Coast National Park Authority; 8.21, © Science & Society Picture Library/Contributor/Getty Images; 8.22, © DEA PICTURE LIBRARY/De Agostini/Getty Images; 8.24, © Dr Jon Anderson, http://www.spatialmanifesto.com/methodology-projects/biomapping; 8.25, © Sean Pavone/Alamy Stock Photo; 8.26, © Helen Fyfe; 8.27, © classicpaintings/Alamy Stock Photo; 8.28, © eye35/Alamy Stock Photo; 8.29, © Alan Finlayson/123RF.COM; 8.30, © Luz Martin/Alamy Stock Photo; 8.31, From LSE Library's collections, BOOTH/E/1/5; 8.34, © Helen Fyfe; 8.35, © bombaert/123RF.COM; 8.45, © Becky Bettesworth (Award winning Devon artist, Becky Bettesworth lives in Torquay and creates vintage style travel posters and seaside prints available to buy at www.beckybettesworth.co.uk); 8.46, © Jamie Gladden/Alamy Stock Photo; 8.47, © Adam Dant 2014 All Rights Reserved; 8.48, © stevenjfrancis/Thinkstock/iStock/Getty Images; 8.49, © Helen Fyfe; 08.50a, © macfromlondon/123RF.COM; 08.50b, © Topical Press Agency/Hulton Archive/Getty Images; 8.51, © Helen Fyfe; 8.52, © S. R. Gaiger/Topical Press Agency/Hulton Archive/Getty Images); 9.4, Courtesy Windows On Our World; 9.6, © epa european pressphoto agency b.v./Alamy Stock Photo; 9.7, © Caroline Knowles; 9.8, © PIUS UTOMI EKPEI/AFP/Getty Images; 9.9, © Trinity Mirror/Mirrorpix/Alamy Stock Photo; 9.10 & 9.15, © Helen Fyfe; 9.17, © Adeel Halim/Bloomberg/Getty Images; 9.20, © lazyllama/Fotolia; 9.27a & b, NASA image by Marit Jentoft-Nilsen, based on Landsat-7 data.; 9.29, © Tony McConnell/Science Photo Library; 9.32, © John S Lander/LightRocket via Getty Images; 9.33, © tjs11/Thinkstock/iStock/Getty Images; 9.34, © Lou Linwei/Alamy Stock Photo; 9.38, © Zoonar RF/Thinkstock/iStock/Getty Images; 9.39, © Duncan Chard/Bloomberg/Getty Images; 9.41, © Cultura RM/Alamy Stock Photo; 9.46, © trekandshoot/Thinkstock/iStock/Getty Images; 9.47, © Saxon Holt/Alamy Stock Photo; 9.48, © CN/Lee Jae-Won LJW/REUTERS/; 9.49, © Wendy Connett/Alamy Stock Photo; 9.53, © ZUMA Press, Inc./Alamy Stock Photo; 9.54, © Lourens Smak/Alamy Stock Photo; 9.55, © Bethany Clarke/Getty Images; 9.57a & b, © Paul Talling/www.derelictlondon.com; 9.61, © Joanne Underhill/Alamy Stock Photo; 9.62, © Cultura RM Exclusive/Atli Mar Hafsteinsson/Getty Images; 9.63, © Starcevic/Thinkstock/iStock/Getty Images; 9.65a, © SJ Images/Alamy Stock Photo; 9.65b, © TravelStockCollection - Homer Sykes/Alamy Stock Photo; 9.66 & 9.67, © Londonmapper.org.uk; 9.70, © Dinodia Photos/Alamy Stock Photo; 9.71, © Dinodia Photos/Alamy Stock Photo; 10.00 (running head), © Paul Abbiss; 10.12, © All Rights Reserved. Dean Swope, 2009; 10.13, © Tim Gainey/Alamy Stock Photo; 10.16, © Yuri Kravchenko/Alamy Stock Photo; 10.18, © Kathy Wright/Alamy Stock Photo; 10.19, © Universal Images Group via Getty Images; 10.21, © Jim Holmes; 10.32, © Sanjeev Verma/Hindustan Times/Getty Images; 10.35, © The Image Works/TopFoto; 10.36, © Henrik Kettunen/Alamy Stock Photo; 10.37, © World Heart Federation; 10.38, © epa european pressphoto agency b.v./Alamy Stock Photo; 10.39, © STAN HONDA/AFP/Getty Images; 10.46, © Irene Abdou/Alamy Stock Photo; 10.47, © Klaus Lang/Alamy Stock Photo; 10.55, © Italian Navy/dpa/Press Association Images; 10.56, © Jeff J Mitchell/Getty Images; 10.60, © frans lemmens/Alamy Stock Photo; 10.75, © Wongalea/Alamy Stock Photo; 10.84, © Chris Fredriksson/Alamy Stock Photo; 10.90, © David Smith/Alamy Stock Photo; 10.94, © PearlBucknall/Alamy Stock Photo; 11.7a, © Paul Andrew Lawrence/Alamy Stock Photo; 11.10, © Thomas Moore/Alamy Stock Photo; 11.12, © World Energy Council; 11.32, © Cambridge Aerial Photography/Alamy Stock Photo; 11.36, © Images of Africa Photobank/Alamy Stock Photo; 11.38, © Design Pics Inc/Alamy Stock Photo; 11.41, © David Wootton/Alamy Stock Photo; 11.42, © Kevin Foy/Alamy Stock Photo; 11.45b, © Snorri Gunnarsson/Alamy Stock Photo; 11.54, Peter Byrne/PA Archive Press Association Images; 11.56, © Milesy/Alamy Stock Photo; 11.57, © Aurora Photos/Alamy Stock Photo; 11.58, © MediaWorldImages/Alamy Stock Photo; 11.68, © Tim Graham/Alamy Stock Photo; 11.69, © Agencia Brasil/Alamy Stock Photo; 11.74, © ABP/George Esiri GE/ Reuters; 11.75, © George Osodi/AP/Press Association Images; 11.77, © Pablo Lopez Luz/Barcroft Media; 11.79, © Eduardo Verdugo/AP/Press Assocation Images; 12.24, © Google Earth; 12.26, © Eric Fischer; 12.30, © Google Earth © Infoterra Ltd & Bluesky Image; © Getmapping plc.

Acknowledgements

3.50, 8.53, 12.31, 12.35: © Crown copyright 2016 Ordnance Survey. Licence number 100036470.

Every effort has been made to trace all copyright holders, but if any have been inadvertently overlooked, the Publishers will be pleased to make the necessary arrangements at the first opportunity.

Although every effort has been made to ensure that website addresses are correct at time of going to press, Hodder Education cannot be held responsible for the content of any website mentioned in this book. It is sometimes possible to find a relocated web page by typing in the address of the home page for a website in the URL window of your browser.

Hachette UK's policy is to use papers that are natural, renewable and recyclable products and made from wood grown in sustainable forests. The logging and manufacturing processes are expected to conform to the environmental regulations of the country of origin.

Orders: please contact Bookpoint Ltd, 130 Park Drive, Milton Park, Abingdon, Oxon OX14 4SE. Telephone: (44) 01235 827720. Fax: (44) 01235 400454. Email education@bookpoint.co.uk Lines are open from 9 a.m. to 5 p.m., Monday to Saturday, with a 24-hour message answering service. You can also order through our website: www.hoddereducation.co.uk

ISBN: 978 1 4718 5869 7

© Malcolm Skinner, Paul Abbiss, Phil Banks, Helen Fyfe and Ian Whittaker 2016

First published in 2016 by
Hodder Education,
An Hachette UK Company
Carmelite House
50 Victoria Embankment
London EC4Y 0DZ

www.hoddereducation.co.uk

Impression number   10 9 8 7 6 5 4 3 2

Year   2020 2019 2018 2017 2016

Cover photo © Andrzej Wojcicki/Science Photo Library

Illustrations by Barking Dog Art Design & Illustration

Typeset in ITC Berkeley Oldstyle Std Book 11/14 pt by Aptara Inc.

Printed in Italy

A catalogue record for this title is available from the British Library.

Contents

Introduction

This textbook covers the subject content for the AQA specification in both AS and A-level Geography, as it is laid out in the specification document. The new curriculum enables students to explore the world, its issues and their place in it. Within this, you will be able to explore current issues of local, national and global importance in contexts specific to different parts of the world. The book forms the backbone for studies of AQA Geography but should be supplemented by reference to up-to-date sources of information including newspapers, television, periodicals aimed at post-16 geography students and, of course, the internet.

The following are key features of the content of this book:

- Concepts are clearly explained, and related issues are explored and analysed.
- Relevant, up-to-date and detailed case studies are provided.
- A variety of stimulus material is provided, including maps, graphs, diagrams and photographs.
- Key terms for each topic are placed in boxes within the relevant chapter and are to be found in bold type within the text. These are the terms from the specification that it is essential you understand.
- Within the text you will find key questions which will enable you to think about the text as you read through the material and will help you to understand and process what you are reading, particularly when you read independently.
- Opportunities for fieldwork are indicated.
- The geographical skills that you will require at both AS and A-level are covered in a separate chapter. In the other chapters, there will be a focus on skills at certain points where you can use particular skills in the context of the course material.

- At the end of each chapter you will find review questions which will help you consider and process the information in the chapter you have just read. There will also be a number of practice exercises which will be written to AS or A-level standard.
- Opportunities for further reading will be indicated at the conclusion of each chapter.

Aims and assessment objectives

During your time studying this specification, you should aim to develop the following:

- your knowledge of locations, places, processes and environments, at all geographical scales from local to global across the specification as a whole
- an in-depth understanding of the selected core and non-core processes in physical and human geography at a range of temporal and spatial scales, and of the concepts which illuminate their significance in a range of locational contexts
- an ability to recognise and analyse the complexity of people–environment interactions at all geographical scales, and appreciate how these underpin understanding of some of the key issues facing the world today
- an understanding of, and an ability to apply, the concepts of place, space, scale and environment, that underpin both the National Curriculum and GCSE; at this higher level, developing a more nuanced understanding of these concepts
- an understanding of specialised concepts relevant to the core and non-core content. These must include the concepts of **causality, systems, equilibrium, feedback, inequality, representation, identity, globalisation, interdependence, mitigation, adaptation, sustainability, risk, resilience and thresholds**

- a better understanding of the ways in which values, attitudes and circumstances have an impact on the relationships between people, place and environment, and develop the knowledge and ability to engage, as citizens, with the questions and issues arising
- your confidence and competence in selecting, using and evaluating a range of quantitative and qualitative skills and approaches and applying them as an integral part of your studies. This should include observing, collecting and analysing geo-located data.

Like other geography specifications, AQA has three assessment objectives. Students should be able to:

- Demonstrate knowledge and understanding of places, environments, concepts, processes, interactions and change, at a variety of scales.
- Apply knowledge and understanding in different contexts to interpret, analyse and evaluate geographical information and issues.
- Use a variety of relevant quantitative, qualitative and fieldwork skills in order to investigate questions and issues, interpret, analyse and evaluate data and evidence, and construct arguments and draw conclusions.

Schemes of assessment

AS level

This course is designed to be taken over one or two years. It is a linear qualification which means that in order to achieve an award, you must complete all assessments at the end of the course and in the same series, that is, at the end of either one or two years.

The components of the assessment are as follows:

Component 1 (Physical geography and people and the environment) is a written paper of 1 hour and 30 minutes with two sections:

Section A Questions will be set on (a) Water and carbon cycles **or** (b) Coastal systems and landscapes **or** (c) Glacial systems and landscapes. You will have to answer questions from **one** subject; **either** (a), (b) **or** (c).

Section B You will have to answer questions from **either** Hazards **or** Contemporary urban environments.

In both sections, question types will consist of multiple choice, short answer and those questions which will be marked with levels of response, particularly answers requiring extended prose.

Component 2 (Human geography and geography fieldwork investigation) is also a written paper of 1 hour and 30 minutes with two sections:

Section A Changing places

Section B Questions will be set on your fieldwork investigation and geographical skills.

In both sections, question types will consist of multiple choice, short answer and those questions which will be marked with levels of response, particularly answers requiring extended prose.

A-level

This course is designed to be taken over two years. It is a linear qualification which means that in order to achieve an award you must complete all assessments at the end and in the same series, that is, after two years.

The components of the assessment are as follows:

Component 1 (Physical geography) is a written paper of 2 hours and 30 minutes with three sections:

Section A Water and carbon cycles

Section B either Hot desert systems and landscapes **or** Coastal systems and landscapes **or** Glacial systems and landscapes. You will have to answer questions on **one** of these topics.

Section C either Hazards **or** Ecosystems under stress. You will have to answer questions on **one** of these topics.

In all three sections, question types will consist of multiple choice, short answer and those marked with levels of response, particularly answers requiring extended prose.

Component 2 (Human geography) is a written paper of 2 hours and 30 minutes with three sections:

Section A Global systems and global governance

Section B Changing places

Section C either Contemporary urban environments **or** Population and the environment **or** Resource security

In all three sections, question types will consist of multiple choice, short answer and those marked with levels of response, particularly answers requiring extended prose.

Component 3 (Geographical investigation) You must complete an individual investigation which must include data collected in the field. The investigation must be based on a question or issue defined and developed by you and must relate to any part of the specification content. In length, it must be between 3,000 and 4,000 words and will be marked by your teachers, your whole centre then being moderated by AQA.

Making connections

When you are studying one aspect of geography covered by this book (for example, Water and carbon cycles or Hazards or Population and the environment), it is important that you recognise that the topic has connections to other aspects of geography. You will be expected to demonstrate your knowledge and understanding of these connections. What you should aim to do essentially is to 'think like a geographer'.

If we take the example of Chapter 1, which covers Water and carbon cycles, connections could be made between the following:

Water cycle:
- how the cycle operates in arid areas
- the connection within urban drainage
- glacial budgets

Climate change:
- desertification
- sea level change and coastal management
- frequency of tropical storms

Command words used in the examinations

One of the major challenges in any examination is interpreting the demands of the questions. Thorough revision is essential, but an awareness of what is expected from you is also required. Too often candidates attempt to answer the question which they think is there rather than the one that has actually been set.

Correct interpretation, therefore, of the **command words** is vital to your success. In AQA Geography examination papers a variety of command words are used. Some demand more of the candidate than others; some require a simple task to be performed; others require greater thought and a longer response. The information given below offers advice on the command words used in the examinations:

Analyse...

This requires a student to break down concepts, information and/or issues to convey an understanding of them by finding connections and causes and/or effects. Such questions sometimes carry a large number of marks where students will be expected to write a relatively long piece of prose.

Annotate...

You are required to add to a diagram, image or graphic material a number of words that describe and/or explain features, rather than just identify them (which is **labelling**).

Calculate...

You are required to work out the value of something.

Define...

You have to state the precise meaning of an idea or concept. Such questions usually require a relatively short answer, the size of the mark allocation being a useful indicator as to the length of answer required. The use of an example is often helpful.

Describe...

This requires you to give an account of something which may be an entity, an event, a feature, a pattern, a distribution or a process. Explanations are not required. A good example would be when you are asked to describe a landform. In this case, asking the following questions will be useful in forming an answer:
- What does it look like?
- What is it made from?
- How big is it? (Include figures if possible.)
- What is its relationship to other features which may be near it? (field relationship)

When presented with a graph or table, you may be asked to '**describe the changes in...**' Good use of accurate adverbs is required here – words such as 'rapidly', 'steeply', 'gently', slightly', 'greatly'.

Discuss...

This is often used in questions which carry a large number of marks and require a lengthy piece of prose. Students are expected to build up an argument about an issue, presenting more than one side of the argument. You should present arguments for and against, making good use of evidence and appropriate examples, and express an opinion of the merits of each side. In short, you should construct a written debate. It can be used in a variety of contexts such as: '**Discuss the extent to which…**' where a judgement about the validity of the evidence or the outcome of an issue is clearly requested; '**Discuss the varying/various attitudes to…**' where you are required to debate the variety of views that might exist.

Evaluate… /Assess… /Critically…

These command words ask you to consider several options and arguments, weigh them up and then come to a conclusion as to their importance/success/worth. **Assess** asks for a statement of the overall quality or value of the feature or issue being considered. **Evaluate** asks you to give an overall statement of value. In both cases, your own judgement is requested, together with a justification for that judgement. In some cases, the term **critically** often occurs, for example, 'Critically evaluate…'. In this case you are being asked to look at an issue or problem from the point of view of a critic with a particular focus on the strengths and weaknesses of the points of view being expressed. You should question not only the evidence itself but also its source and how it was collected. Such questions often carry a large number of marks where students will be expected to write a relatively long piece of prose.

Examine

You must consider carefully and provide a detailed account of the indicated topic.

Explain… /Why… /Suggest reasons for…

Here you are required to set out the causes of a phenomenon and/or the factors which influence its form/nature. This usually requires an understanding of processes. Explanation is a higher level skill than description and will usually be awarded a greater mark weighting.

Interpret…

This means that you are required to ascribe a meaning to something presented in the question.

Justify…

This command word asks you to give reasons for the validity of a view, idea or why some action should

be taken. This will probably involve discussing and discounting alternative views or actions. Decision making, for example, is not always straightforward. All options when decisions are made have positive and negative aspects. The options that are rejected may have some good elements and, equally, a chosen option will not be perfect in every respect. When options have been rejected you should give an overall statement of why the negatives outweigh the positives and, conversely, for a chosen option, why the positives outweigh the negatives. You should also be able to explain all of this review process.

Outline… /Summarise…

You are required to provide a brief account of relevant information. This will require a relatively short answer, the mark allocation being a useful indicator as to the length of answer expected.

State…

This command word asks for a brief answer to a simple task. The mark allocation (which will be low) is a useful indicator as to the length of answer required.

To what extent…

This means that you must form and express a view as to the merit or validity of a view or statement after examining the evidence available and/or different sides of an argument.

Skills required in producing a piece of extended prose (essay writing)

For many students, this represents the most difficult part of any examination. But it is also an opportunity to demonstrate your strengths. Before you start to write an essay you must have a plan of what you will write, either in your head or on paper. All such pieces of writing must have a beginning (the introduction), a middle section (argument) and an ending (conclusion).

The introduction

This does not have to be long – a few sentences should be sufficient. It may define the terms set out in the question, set the scene for the argument to follow or provide a brief statement of the idea or viewpoint to be developed in the main part of the answer.

The argument

This is the main part of the answer. It should consist of a series of paragraphs, each developing one point and following on logically from the previous one. Try to avoid paragraphs that list information without any depth, and avoid writing all you know about a particular topic without any link to the question set. Make good use of exemplar material, naming real places and make them count by giving accurate details specific to those locations. You should aim to demonstrate good, and often critical, knowledge and understanding of concepts.

The conclusion

The conclusion, like the opening, should not be too long. It should reiterate the main points as stated in the introduction, but now supported by the evidence and facts given in the argument. It must address the command word/s given in the question such as 'assess', or 'evaluate'.

Should you produce essay plans in the examination?

If you produce an essay plan at all, it must be brief, taking only 2 or 3 minutes to write on a separate piece of paper. The plan must reflect the above formula and make sure that you stick to it. Be logical, and only write an outline, retaining the examples in your head to use at the most appropriate point in your answer.

Other important points

Always keep an eye on the time. Make sure that you write clearly and concisely. Do not give confused answers, endlessly long sentences or pages of prose with no paragraphs. Above everything, carefully read the question and answer the question as it has been set.

How are questions marked?

Most questions at both AS and A-level are marked using a levels of response mark scheme. Such a scheme is broken down into levels, each of which has a descriptor. At each level the descriptor denotes the average performance expected for that level. Questions at both AS and A-level are marked with schemes giving two or three levels, however questions at A-level that require a longer essay-type response will be marked using four levels (assuming a 20-mark maximum for the question). One typical four-level scheme is set out

below, although it is entirely possible that schemes may exhibit slight variations from that shown depending upon the wording and the set-up of the question. Once published, you should always consult past papers and their mark schemes to see how descriptors are used in different contexts.

Level 1 (1–5 marks)

Level 1 answers are likely to show the following characteristics:
- will present a very limited and/or unsupported evaluative conclusion that is loosely based upon knowledge and understanding which is applied to the context of the question
- there will be very limited analysis and evaluation in the application of knowledge and understanding. The answer lacks clarity and coherence in this respect
- there will be very limited and rarely logical evidence of links between knowledge and understanding to the application of knowledge and understanding in different contexts
- there will be very limited relevant knowledge and understanding shown of place(s) and environments
- knowledge and understanding of key concepts and processes will be very isolated
- there will be very limited awareness of scale and temporal change which is rarely integrated where appropriate
- there may be a number of inaccuracies within the text.

Level 2 (6–10 marks)

Level 2 answers are likely to show the following:
- some sense of an evaluative conclusion partially based upon knowledge and understanding which is applied to the context of the question
- shows some partially relevant analysis and evaluation in the application of knowledge and understanding
- reveals some evidence of links between knowledge and understanding to the application of knowledge and understanding in different contexts
- shows some partially relevant knowledge and understanding of place(s) and environments
- reveals some knowledge and understanding of key concepts, processes and interactions and change
- shows some awareness of scale and temporal change which is sometimes integrated where appropriate
- there may be a few inaccuracies in the text.

Level 3 (11–15 marks)

Level 3 answers are likely to show the following:

- shows clear evaluative conclusions that are based on knowledge and understanding which is applied in the context of the question
- reveals generally clear, coherent and relevant analysis and evaluation in the application of knowledge and understanding
- shows generally clear evidence of links between knowledge and understanding to the application of knowledge and understanding in different contexts
- shows generally clear and relevant knowledge and understanding of place(s) and environments
- reveals generally clear and accurate knowledge and understanding of key concepts and processes
- shows generally clear awareness of scale and temporal change which is integrated where appropriate.

Level 4 (16–20 marks)

Level 4 answers are likely to show the following:

- reveals detailed evaluative conclusions that are rational and firmly based on knowledge and understanding which is applied to the context of the question
- makes a detailed, coherent and relevant analysis and evaluation in the application of knowledge and understanding throughout
- shows full evidence of links between knowledge and understanding to the application of knowledge and understanding in different contexts
- reveals a detailed, highly relevant and appropriate knowledge and understanding of place(s) and environments and used throughout the answer
- shows, throughout the answer, a full and accurate knowledge and understanding of key concepts and processes
- has a detailed awareness of scale and temporal change which is well integrated where appropriate.

Part 1

Physical Geography

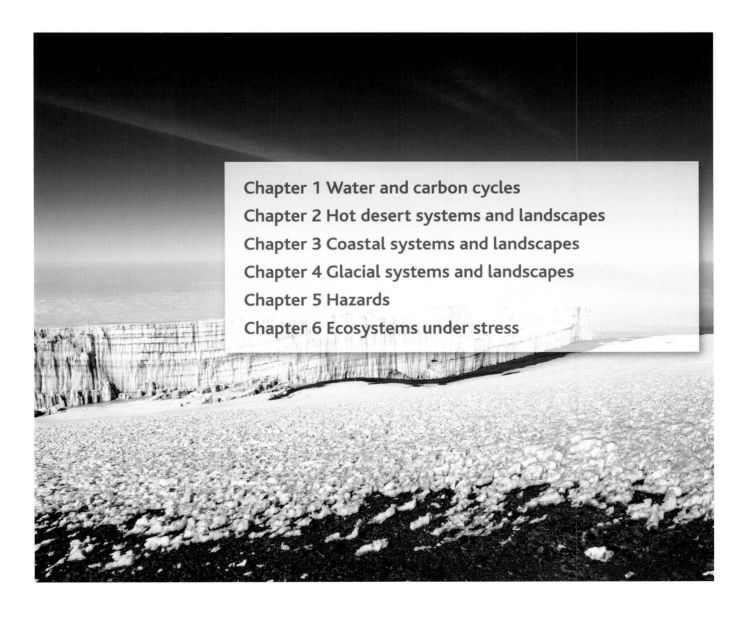

Chapter

1

Water and carbon cycles

The cycling of water has obvious and significant implications for the health and prosperity of society. The availability and quantity of water is vital to life on Earth and helps to tie together the Earth's lands, oceans and atmosphere into an integrated physical system. Added to this is the fact that water vapour is the most important greenhouse gas and is a major driving factor in determining climate. The global water cycle is driven by many complex processes and interactions at a variety of scales; these are often poorly understood and badly represented in model predictions.

Carbon is everywhere: in the oceans, in rocks and soils, in all forms of life and in our atmosphere. Without carbon, life would not exist as we know it. The well-being and functioning of our planet depends on carbon and how it cycles through the Earth's system. The carbon cycle plays a key role in regulating the Earth's global temperature and climate by controlling the amount of another greenhouse gas, carbon dioxide, in the atmosphere.

Both the water and carbon cycles (and other aspects of geography) can be studied by considering them as systems.

In this chapter you will study:
- systems frameworks and their application
- the water cycle
- the carbon cycle
- water, carbon, climate and life on Earth

1.1 Systems frameworks and their application

Because the Earth is highly complex, geographers have attempted to simplify aspects of it so that relationships between components can be better understood. These simplifications are called models (for example, the water cycle or the demographic transition model). One type of model that is widely used, particularly in physical geography, is the **system**.

A system is an assemblage of interrelated parts that work together by way of some driving process. They are a series of **stores** or **components** that have **flows** or **connections** between them. There are three types of property: **elements, attributes** and **relationships**. Elements are the things that make up the system of interest. Attributes are the perceived characteristics of the elements. Relationships are descriptions of how the various elements (and their attributes) work together to carry out some kind of process. Most systems share the same common characteristics:

- They have a structure that lies within a **boundary**.
- They are generalisations of reality, removing incidental detail that obscures fundamental relationships.
- They function by having **inputs** and **outputs** of material (energy and/or matter) that is processed within the components causing it to change in some way.

- They involve the flow of material between components.

Systems can be classified as:

- **Isolated systems**: these have no interactions with anything outside the system boundary. There is no input or output of energy or matter. Many controlled laboratory experiments are this type of system and they are rare in nature.
- **Closed systems**: these have transfers of energy both into and beyond the system boundary but not transfer of matter (Figure 1.1).
- **Open systems**: these are where matter and energy can be transferred from the system across the boundary into the surrounding environment. Most ecosystems are examples of open systems (Figure 1.2).

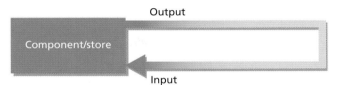

Figure 1.1 A closed system

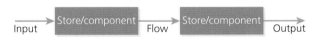

Figure 1.2 An open system, for example, a drainage basin hydrological system

When there is a balance between the inputs and outputs then the system is said to be in a state of **dynamic equilibrium.** This means that the stores stay the same. If, however, one of the elements of the system changes, for example one of the inputs increases without any corresponding change in the outputs, then the stores change and the equilibrium is upset. This is called feedback. There are two types of feedback:

- **positive feedback** where the effects of an action (for example, an increase in carbon dioxide) are amplified or multiplied by subsequent knock-on or secondary effects (Figure 1.3)
- **negative feedback** where the effects of an action (for example, the increased use of fossil fuels) are nullified by its subsequent knock-on effects (Figure 1.4).

Key question

Can you think of other examples of feedback in geography?

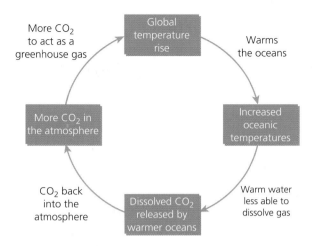

Figure 1.3 Example of positive feedback in a system

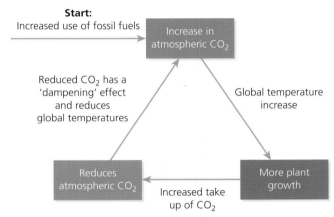

Figure 1.4 Example of negative feedback in a system

The Earth can be studied using a systems approach; indeed the Earth as a whole could be considered a closed system. Energy comes into the system in the form of solar energy. This is balanced by radiant energy lost by the Earth. It could be related to the concept of 'spaceship Earth' which is a term usually expressing concern over the use of limited resources available on Earth and encouraging everyone on it to act as a harmonious crew working towards the greater good.

At the global level the Earth has four major subsystems, including the atmosphere, lithosphere, hydrosphere and biosphere. Each of these can be considered to be an open system that forms part of a chain; a **cascading system**.

Interlocking relationships among the atmosphere, lithosphere, hydrosphere and biosphere have a profound effect on the Earth's climate and climate change.

1.2 The water cycle

Key terms

Atmospheric water – Water found in the atmosphere; mainly water vapour with some liquid water (cloud and rain droplets) and ice crystals.

Cryospheric water – The water locked up on the Earth's surface as ice.

Hydrosphere – A discontinuous layer of water at or near the Earth's surface. It includes all liquid and frozen surface waters, groundwater held in soil and rock and atmospheric water vapour.

Oceanic water – The water contained in the Earth's oceans and seas but not including such inland seas as the Caspian Sea.

Terrestrial water – This consists of groundwater, soil moisture, lakes, wetlands and rivers.

'Water is life's matter, mother and medium.'
Albert Szent-Gyorgyi, 1937 Nobel Prize acceptance speech

Water on planet Earth

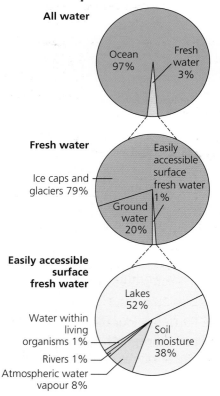

Figure 1.5 The distribution of the world's water

Water on or close to the Earth's surface is called the **hydrosphere**. Scientists have attempted many times to estimate the total amount of water in the hydrosphere. There is general agreement that it amounts to some 1.338×10^9 km^3. It is thought that approximately 97 per cent of this is **oceanic water**. Fresh water, which makes up the remaining 3 per cent, is locked up in land ice, glaciers and permafrost (**cryospheric water**), groundwater, lakes, soil, wetland, rivers, biomass (**terrestrial water**) and **atmospheric water**.

12,900 km^3 of water vapour are found in the atmosphere. This amounts to a global average of 26 kg/m^2 of water for each column of air on the surface of the Earth. There are large variations in this, however. Although atmospheric water only makes up 0.4 per cent of all water, it has a profound effect on our lives at present.

The amount of water in these stores is in a state of dynamic equilibrium with changes at a range of timescales from diurnal to geological. Changing amounts of atmospheric water in the future could be a major cause and/or important effect of climate change.

Oceanic water

The oceans dominate the amount of available water. Its exact amount is unknown with figures varying from 1,320,000,000 to 1,370,000,000 km^3 with an average depth of 3,682 m. That difference is greater than the sum of all the rest of the water put together. They cover approximately 72 per cent of the planet's surface (3.6×10^8 km^2). They are customarily divided into several principal oceans and smaller seas. Although the ocean contains 97 per cent of the Earth's water, oceanographers have stated that only 5 per cent has been explored.

Oceanic water tastes salty because it contains dissolved salts. These salts allow it to stay as liquid water below 0°C. They are alkaline with an average pH of about 8.14. The pH has fallen from about 8.25 in the last 250 years and it seems destined to continue falling. This change in the pH is linked to the increase in atmospheric carbon and may have a profound influence on marine ecosystems.

Cryospheric water

The cryosphere is those portions of the Earth's surface where water is in solid form. Figure 1.6 shows the five locations of cryospheric water.

2 + 2 = 4

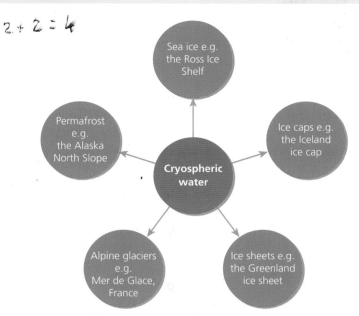

Figure 1.6 The locations of cryospheric water

Sea ice: Much of the Arctic Ocean is frozen; the amount of which grows in winter and shrinks in summer. The same is true of the waters surrounding Antarctica. Sea ice forms when water in the oceans is cooled to temperatures below freezing. Sea ice does not raise sea level when it melts, because it forms from ocean water. It is closely linked with our planet's climate, so scientists are concerned about its recent decline.

Figure 1.7 Chunks of broken sea ice in Yelverton Bay, Ellesmere Island, Canada

Ice shelves are platforms of ice that form where ice sheets and glaciers move out into the oceans. Ice shelves exist mostly in Antarctica and Greenland, as well as in the Arctic near Canada and Alaska. Icebergs are chunks of ice that break off glaciers and ice shelves and drift in the oceans. They raise sea level only when they first leave land and push into the water, but not when they melt in the water.

Ice sheets: An ice sheet is a mass of glacial land ice extending more than 50,000 km². The two major ice sheets on Earth today cover most of Greenland and Antarctica. During the last ice advance, ice sheets also covered much of North America, northern Europe and Argentina.

Figure 1.8 Mountains rising out of part of the Greenland ice sheet

Together, the Antarctic and Greenland ice sheets contain more than 99 per cent of the freshwater ice on Earth. The Antarctic Ice Sheet extends almost 14 million km², roughly the area of the United States and Mexico combined. It contains 30 million km³ of ice. The Greenland Ice Sheet extends about 1.7 million km², covering most of the island of Greenland.

Ice sheets form in areas where snow that falls in winter does not melt entirely over the summer. Over thousands of years, the layers of snow pile up into thick masses of ice, growing thicker and denser as the weight of new snow and ice layers compresses the older layers. Ice sheets are constantly in motion, slowly flowing downhill under their own weight. Near the coast, most of the ice moves through relatively fast-moving outlets called ice streams. This type of glacier is significant in the Antarctic where they can be up to 50 km wide, 2 km thick and hundreds of kilometres long. As long as an ice sheet accumulates the same mass of snow as it loses to the sea, it remains stable.

Ice sheets contain enormous quantities of frozen water. If the Greenland Ice Sheet melted, scientists estimate that sea level would rise about six metres. If the Antarctic Ice Sheet melted, sea level would rise by about 60 m.

Ice caps are thick layers of ice on land that are smaller than 50,000 km². They are usually found in mountainous areas. Ice caps tend to be dome-shaped and are centred over the highest point of an upland area. They flow outwards, covering almost everything in their path and becoming the major source for many glaciers.

Ice caps occur all over the world, from the polar regions to mountainous areas such as the Himalayas, the Rockies, the Andes and the Southern Alps of New Zealand. The Furtwangler Glacier on Kilimanjaro, at 60,000 m², is Africa's only remaining ice cap. It is melting rapidly and may soon disappear.

Figure 1.9 The Furtwangler Ice Sheet. The last ice sheet in Africa

Alpine glaciers are thick masses of ice found in deep valleys or in upland hollows. Most valley glaciers are fed by ice from ice caps or smaller corrie glaciers. These glaciers are particularly important in the Himalayas where about 15,000 Himalayan glaciers form a unique reservoir which supports perennial rivers such as the Indus, Ganges and Brahmaputra which, in turn, are the lifeline of millions of people in South Asian countries (Pakistan, Nepal, Bhutan, India and Bangladesh). Frozen ground and permafrost ring the Arctic Ocean. Glaciers, snow and ice cover the nearby land, including a thick sheet of snow and ice covering Greenland.

Permafrost is defined as ground (soil or rock and included ice or organic material) that remains at or below 0°C for at least two consecutive years. The thickness of permafrost varies from less than one metre to more than 1,500 m. Most of the permafrost existing today formed during cold glacial periods and has persisted through warmer interglacial periods, including the Holocene (the last 10,000 years). Some relatively shallow permafrost (30 to 70 m) formed during

the second part of the Holocene (the last 6,000 years) and some during the Little Ice Age (from 400 to 150 years ago). Subsea permafrost occurs at close to 0°C over large areas of the Arctic continental shelf, where it formed during the last glacial period on the exposed shelf landscapes when sea levels were lower. Permafrost is found beneath the ice-free regions of the Antarctic continent and also occurs beneath areas in which the ice sheet is frozen to its bed.

The permafrost has begun to melt as climate warms. This melting is releasing large amounts of carbon dioxide and methane, potentially affecting global climates.

Terrestrial water

Terrestrial water may be considered as falling into four broad classes:

- surface water
- groundwater
- soil water
- biological water.

Surface water is the free-flowing water of rivers as well as the water of ponds and lakes.

- **Rivers** act as both a store and a transfer of water; they are streams of water within a defined channel. They transfer water from the ground, from soils and from the atmosphere to a store. That store may be wetlands, lakes or the oceans. Rivers make up only a small percentage (0.0002 per cent) of all water, covering just 1,000,000 km² with a volume of 2,120 km³. One river alone, the Amazon in South America, is the largest river by discharge of water in the world, averaging a discharge of about 209,000 m³/s, greater than the next seven largest independent rivers combined. It drains an area of about 7,050,000 km² and accounts for approximately one fifth of the world's total river flow. The portion of the river's drainage basin in Brazil alone is larger than any other river's basin. The Amazon enters Brazil with only one fifth of the flow it finally discharges into the Atlantic Ocean, yet already has a greater flow at this point than the discharge of any other river.
- **Lakes** are collections of fresh water found in hollows on the land surface. They are generally deemed a lake if they are greater than two hectares in area. Any standing body of water smaller than this is termed a pond.

The majority of lakes on Earth are freshwater, and most lie in the Northern Hemisphere at higher latitudes. Canada has an estimated 31,752 lakes larger than 3 km² and an estimated total number of at least 2 million. Finland has 187,888 lakes 500 m² or larger, of which 56,000 are large (10,000 m²).

The largest lake is the Caspian Sea at 78,200 km³. It is a remnant of an ancient ocean and is about 5.5 million years old. It is generally fresh water, though becomes more saline in the south where there are few rivers flowing into it. The deepest lake in the world is Lake Baikal in Siberia with a mean depth of 749 m and a deepest point at 1,637 m.

- **Wetlands:** The **Ramsar Convention** defines wetlands as 'areas of marsh, fen, peatland or water, whether natural or artificial, permanent or temporary, with water that is static or flowing where there is a dominance by vegetation'.

They are areas where water covers the soil, or is present either at or near the surface of the soil all year or for varying periods of time during the year, including during the growing season. Water saturation determines how the soil develops and the types of plant and animal communities living in and on the soil. Wetlands may support both aquatic and terrestrial species. The prolonged presence of water creates conditions that favour the growth of specially adapted plants and promotes the development of characteristic wetland soils.

Wetlands vary widely because of regional and local differences in soils, topography, climate, hydrology, water chemistry, vegetation and other factors, including human disturbance. They are found from the polar regions to the tropics and on every continent except Antarctica.

Figure 1.10 The location of the Pantanal wetlands (shown in dark green)

The Pantanal of South America is often referred to as the world's largest freshwater wetland system. It extends through millions of hectares of central-western Brazil, eastern Bolivia and eastern Paraguay. It is a complex system of marshlands, flood plains, lagoons and interconnected drainage lines. It also provides economic benefits by being a huge area for water purification and groundwater discharge and recharge, climate stabilization, water supply, flood abatement, and an extensive, transport system, among numerous other important functions.

Wetlands are the main ecosystem in the Arctic. These peatlands, rivers, lakes, and shallow bays cover nearly 60 per cent of the total surface area. Arctic wetlands store enormous amounts of greenhouse gases and are critical for global biodiversity.

Groundwater is water that collects underground in the pore spaces of rock. Scientists have set a lower level for groundwater at a depth of 4,000 m but it is known that there are large quantities of water below that. A very deep borehole in the Kola Peninsula in Northern Russia found huge quantities of hot mineralised water at a depth of 13 km.

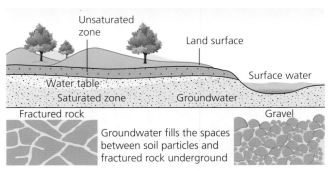

Figure 1.11 Groundwater

The depth at which soil pore spaces or fractures and voids in rock become completely saturated with water is called the water table. Groundwater is recharged from, and eventually flows to, the surface. Natural discharge often occurs at springs and seeps, and can form oases or wetlands. The amount of groundwater is reducing rapidly due to extensive extraction for use in irrigating agricultural land in dry areas.

Soil water is that which is held together with air in unsaturated upper weathered layers of the Earth. It is of fundamental importance to many hydrological, biological and biogeochemical processes. It affects weather and climate, run-off potential and flood control, soil erosion and slope failure, reservoir management, geotechnical engineering and water quality. Soil moisture is a key variable in controlling the exchange of water and heat energy between the land surface and the atmosphere through evaporation and plant transpiration. As a result, soil moisture plays an important role in the development of weather patterns and the production of precipitation.

Biological water constitutes the water stored in all the biomass. It varies widely around the globe depending on the vegetation cover and type. Areas of dense rainforest store much more water than deserts. The role of animals as a water store is minimal.

Trees take in water via their roots. This is either transported or stored in the trunk and branches of the tree. The water is lost by the process of transpiration through stomata in the leaves. This storage provides a reservoir of water that helps maintain some climatic environments. If the vegetation is destroyed, this store is lost to the atmosphere and the climate can become more desert-like. Many plants are adapted to store water in large quantities. Cacti are able to gather water via their extensive root system and then very slowly use it until the next rainstorm. The baobab tree stores water, but it is thought that this is to strengthen the structure of the tree rather than to be used in tree growth.

Atmospheric water

Atmospheric water exists in all three states. The most common atmospheric water exists as a gas: water vapour. This is clear, colourless and odourless and so we take its presence for granted. This atmospheric water vapour is important as it absorbs, reflects and scatters incoming solar radiation, keeping the atmosphere at a temperature that can maintain life. The amount of water vapour that can be held by air depends upon its temperature. Cold air cannot hold as much water vapour as warm air. This results in air over the poles being quite dry, whereas air over the tropics is very humid.

A small increase in water vapour will lead to an increase in atmospheric temperatures. This becomes positive feedback as a small increase in global temperature would lead to a rise in global water vapour levels, thus further enhancing the atmospheric warming.

Cloud is a visible mass of water droplets or ice crystals suspended in the atmosphere. Cloud formation is the

result of air in the lower layers of Earth's atmosphere becoming saturated due to either or both of two processes: cooling of the air and an increase in water vapour. When the cloud droplets grow they can eventually fall as rain.

Factors driving the change in magnitude of water stores

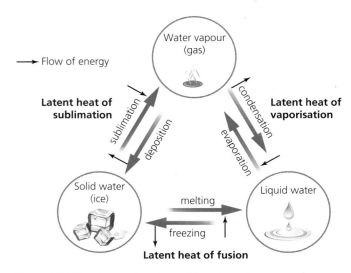

Figure 1.12 The phases of water and the phase changes

Water exists on Earth in three forms: liquid water, solid ice and gaseous water vapour. Figure 1.12 shows the processes that occur as water changes from one state to another. Energy, in the form of latent heat, is either absorbed or released depending on the process. This is particularly important in atmospheric processes such as cloud or precipitation formation.

Evaporation

Evaporation occurs when energy from solar radiation hits the surface of water or land and causes liquid water to change state from a liquid to a gas (water vapour). The rate of evaporation depends upon several factors:

- the amount of solar energy
- the availability of water (for example there is more evaporation from a pond than from a grassy field).
- the humidity of the air; the closer the air is to saturation point, the slower the rate of evaporation
- the temperature of the air; warmer air can hold more water vapour than cold air.

Linked to this is the fact that all terrestrial plants lose water through **transpiration**. This is where water is transported from the roots of a plant to its leaves and then lost through pores on the leaf surface. Leaves also intercept rain as it falls, and this water can be evaporated before it reaches the soil.

As water evaporates it uses energy in the form of latent heat and so cools its surroundings.

Condensation

As air cools it is able to hold less water vapour. This means that if it is cooled sufficiently then it will get to a temperature at which it becomes saturated. This is known as the dew point temperature. Excess water in the air will then be converted to liquid water in the process of **condensation**. The water molecules need something to condense on. These can be tiny particles (smoke, salt, dust, etc.) that are collectively called condensation nuclei, or surfaces (leaves, grass stems, windows, etc.) that are below the dew point temperature. If the surface is below freezing point then water vapour sublimates, changing directly from gas to solid in the form of hoar frost.

Condensation is the direct cause of all forms of precipitation.

- It takes place when the temperature of air is reduced to dew point but its volume remains constant. This occurs when:
  - warm moist air passes over a cold surface
  - on a clear winter's night heat is radiated out to space and the ground gets colder, cooling the air directly in contact with it.
- It also occurs when the volume of air increases but there is no addition of heat (adiabatic cooling). This happens when air rises and expands in the lower pressure of the upper atmosphere. This can occur when:
  - air is forced to rise over hills (relief or orographic effect)
  - masses of air of different temperatures and densities meet. The less dense warm air rises over the denser cold air (frontal effect).
  - localised warm surfaces heat the air above. This expands, becomes less dense and rises (convectional effect).

Key questions

Rain, drizzle, snow, sleet and hail are all forms of precipitation. What are the key differences in their formation?

What are the similarities and differences in the formation of dew and hoar frost?

Cryospheric processes

Chapter 4 discusses in detail the two main **cryospheric processes** of **accumulation** and **ablation** (melting).

It is thought that there have been five major glacial periods in Earth's history. The most recent started 2.58 million years ago and continues today; it is called the **Quaternary** glaciation. During this time there were:

- glacial periods when, due to the volume of ice on land, sea level was approximately 120 m lower than present and continental glaciers covered large parts of Europe, North America, and Siberia. This represents an interruption in the global hydrological cycle
- interglacial periods when global ablation exceeds accumulation and the hydrological cycle as we know it today returns.

Over the past 740,000 years there have been eight such glacial cycles.

Permafrost is formed when air temperatures are so low that they freeze any soil and groundwater present. It rarely occurs under ice because the temperatures are not low enough.

Drainage basin systems

Key terms

Condensation – The process by which water vapour changes to liquid water.

Cryospheric processes – Those processes that affect the total mass of ice at any scale from local patches of frozen ground to global ice amounts. They include accumulation (the build-up of ice mass) and ablation (the loss of ice mass).

Drainage basin – This is an area of land drained by a river and its tributaries. It includes water found on the surface, in the soil and in near-surface geology.

Evaporation – The process by which liquid water changes to a gas. This requires energy, which is provided by the sun and aided by wind.

Evapotranspiration – The total output of water from the drainage basin directly back into the atmosphere.

Groundwater flow – The slow movement of water through underlying rocks.

Infiltration – The downward movement of water from the surface into soil.

Key terms

Interception storage – The precipitation that falls on the vegetation surfaces (canopy) or human-made cover and is temporarily stored on these surfaces. Intercepted water either can be evaporated directly to the atmosphere, absorbed by the canopy surfaces or ultimately transmitted to the ground surface.

Overland flow – The tendency of water to flow horizontally across land surfaces when rainfall has exceeded the infiltration capacity of the soil and all surface stores are full to overflowing.

Percolation – The downward movement of water within the rock under the soil surface. Rates vary depending on the nature of the rock.

Run-off – All the water that enters a river channel and eventually flows out of the drainage basin.

Saturated – This applies to any water store that has reached its maximum capacity.

Stemflow – The portion of precipitation intercepted by the canopy that reaches the ground by flowing down stems, stalks or tree bole.

Storm and rainfall event – An individual storm is defined as a rainfall period separated by dry intervals of at least 24 hours and an individual rainfall event is defined as a rainfall period separated by dry intervals of at least 4 hours (Hamilton and Rowe, 1949).

Throughfall – The portion of the precipitation that reaches the ground directly through gaps in the vegetation canopy and drips from leaves, twigs and stems. This occurs when the canopy-surface rainwater storage exceeds its storage capacity.

Throughflow – The movement of water down-slope through the subsoil under the influence of gravity. It is particularly effective when underlying permeable rock prevents further downward movement.

Transpiration – The loss of water from vegetation through pores (stomata) on their surfaces.

Water balance – The balance between inputs (precipitation) and outputs (run-off, evapotranspiration, soil and groundwater storage) in a drainage basin.

A useful way of looking at **drainage basins** is to consider them as cascading systems. They are a series of open systems that link together so that the output of one is the input of the next.

Drainage basins

A drainage basin (or catchment area) is the area that supplies a river with its supply of water (see Figure 1.13). This includes water found below the water table as well as soil water and any surface flow. Drainage basins are separated from one another by high land called a watershed.

The input to the system is the precipitation. The nature, intensity and longevity of the precipitation have a direct bearing on what happens when the water hits the ground. In general, however, on a hill slope the following happens:

● Precipitation lands on the bare surface or, more likely, vegetation cover. This vegetation provides a

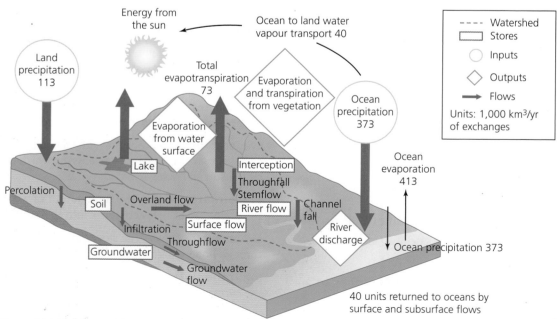

Figure 1.13 The drainage basin hydrological cycle

store: **interception store**. This occurs on the leaves/ branches of vegetation. Some vegetation is a better interception store than others. The density of the vegetation cover also plays a part. Studies show that forests made up of needle-leaf trees captured 22 per cent of the rainfall, while broad-leaf deciduous forests intercepted 19 per cent. This difference may be due to the density of the vegetation cover rather than the structure of the leaves. Some tropical rainforests intercept as much as 58 per cent of the rainfall.

- A lot of the water captured by vegetation surfaces is evaporated back into the atmosphere.
- Water making its way from the leaves of trees to the ground drips or falls by the process of **throughfall** from one interception store to another until it eventually reaches the ground. Water also flows down the stems of grasses, etc., and in very heavy storms it can flow straight down the trunks of trees. This is called **stemflow**.
- On reaching the ground, water soaks into the soil by the process of **infiltration**. The rate of this movement is called the **infiltration rate**. This movement of water into the **soil** is controlled by gravity, capillary action and soil porosity. Of these factors soil porosity is most important. A soil's porosity is controlled by its texture, structure and organic content. Coarse textured soils have larger pores and fissures than fine-grained soils and therefore allow for more water flow. Pores and fissures found in soils can be made larger through a number of factors that enhance internal soil structure. For example, the burrowing of worms and other organisms and penetration of plant roots can increase the size and number of macro- and micro-channels within the soil.

Fieldwork opportunities

There is an opportunity here for an investigation based on the infiltration rates of different surfaces.

The rate of infiltration normally declines rapidly during the early part of a **rainstorm event** and reaches a constant value after several hours of rainfall. A number of factors are responsible for this phenomenon:
- The filling of small pores on the soil surface with water reduces the ability of capillary forces to actively move water into the soil.

- As the soil moistens, the clay particles absorb water causing them to expand. This expansion reduces the size of soil pores.
- Raindrop impact breaks large soil clumps into smaller particles. These particles then clog soil surface pores reducing the movement of water into the soil.
- **Soil storage** is the amount of water stored in the soil. Soils consist of solid particles with pore spaces between them. Those pore spaces can be filled with air but also with water. The amount of pores in a soil is different for different types of soil. The pores in a clay soil account for 40 to 60 per cent of the volume. In fine sand this can be 20 to 45 per cent.
- Vegetation requires water to survive. Most plants remove water from the soil and store it in the structure of the plant. This is called **vegetation storage**. Plants lose water back to the atmosphere through stomata on their leaves. This is called **transpiration.**
- If rainfall intensity is greater than the infiltration rate then the soil has reached **infiltration capacity** and the soil will be **saturated**. Water will build up on the surface as **surface storage** which is usually in the form of puddles. While common in a man-made environment their natural occurrence is quite rare. Water usually disappears into the ground at a greater rate than it rains (that is, the infiltration rate is greater than the rate of precipitation). Water can only build up on the surface after a long period of rain, an intense rainstorm or on an impermeable surface (either man-made, for example an impacted footpath; or natural, for example a bare rock or frozen surface).
- Much of this surface storage then **evaporates** back into the atmosphere and is lost to the drainage basin. It is difficult to separate **evaporation** from transpiration and so the total amount outputted from the system in this way is referred to as **evapotranspiration.** When the surface stores are full then **overland flow** or **sheet flow** will begin on slopes. This is very fast flow, rapidly reaching the nearest channel.
- Any lateral movement of soil water downslope is called **throughflow** and eventually reaches the nearest channel. This tends to be much slower than overland flow. Generally speaking, the more vegetated an area, the faster the rate of throughflow because it is aided by root channels in the soil.

- Following infiltration water moves vertically down through the soil and unsaturated rock by the process of **percolation.** This can then be held in pore spaces in the rocks as **groundwater.** It then passes slowly into the zone of saturated rock where it can move vertically and laterally by the process of **groundwater flow.** This is very slow movement. It can feed rivers through long periods of drought. Some rocks are able to store a lot of water, especially if they are porous. These are called aquifers.

The sum of all these movements and stores adds up to form the drainage basin hydrological cycle. All of these flows lead water to the nearest river (or deep groundwater store). The river then transfers water by **channel flow.** The amount of water that leaves the drainage basin in this way is called **run-off.**

The rate of movement of all this water varies as shown in Figure 1.14. The fastest movement of water is along the surface. There are relatively few obstacles slowing it down. Urban surfaces have especially fast water movement because they are often designed to move the water quickly by having strategically placed slopes and very smooth surfaces. Water moves through soil at a slower rate. Again this varies greatly. Under woodland there are many channels created by roots as well as burrowing animals and these allow

relatively free movement. Clay soils retain water, hindering any movement. They can dry out from the surface down before they allow horizontal transfer.

Once in the rocks below the soil the rate of transfer slows considerably. Groundwater can be held for millennia.

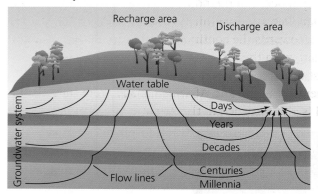

Figure 1.15 The varying timescales of water movements

Transport within the global water cycle

There is a net transport of some 38 units from ocean to land with about the same amount returning by the rivers to the ocean. However, the amount of precipitation over the continents is almost three times as high, indicating a considerable recirculation of water over land. This recirculation has a marked annual cycle as well as having large variations between continents. The recirculation is larger during the summer and for tropical land areas.

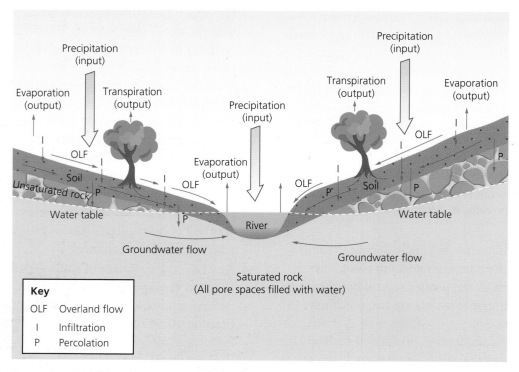

Figure 1.14 The hill slope drainage system

The hydrological cycle of the world's oceans interacts differently with that of the continents. Most of the water from the Pacific Ocean recirculates between different parts of the Pacific itself and there is little net transport towards land. The pattern of water exchange between ocean and land is different in the Atlantic and Indian Oceans. Two thirds of the total net transport of water towards the continents comes from the Atlantic Ocean, with the rest essentially from the Indian Ocean. Most of the continental water for North and South America, Europe and Africa emanates from the Atlantic and is also returned to the Atlantic by rivers.

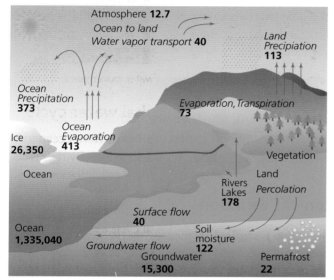

Figure 1.16 The varying amounts of water cycle movements (in thousand km³/year for exchanges and thousand km³ for storage)

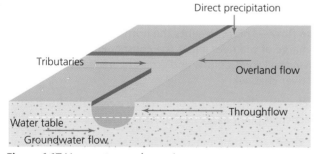

Figure 1.17 How water reaches a river

Water balance

Within a drainage basin, the balance between inputs (precipitation) and outputs (run-off, evapotranspiration, soil and groundwater storage) is known as the **water balance** or budget. Rivers only occur if the stores are able to release water, there is direct precipitation or there is overland flow into the river. As one moves downstream, larger rivers are also fed by their tributary streams.

Discharge levels rise and fall, often showing an annual pattern (called the river's **regime**). They also vary in the short term following heavy rainfall. The water balance can be shown using the following formula:

$$\text{Precipitation (P)} = \text{discharge (Q)} + \text{evapotranspiration (E)} \pm \text{changes in storage (S)}$$

The two most important parts of this relationship are precipitation and **'potential' evapotranspiration**. Evapotranspiration is closely related to the prevailing temperature. The warmer it is, the higher the evapotranspiration. It often happens that the temperature, and so the atmosphere's ability to hold water vapour, is greater than the amount of water available. Potential evapotranspiration is the amount of water that could be evaporated or transpired (or both) from an area if there was sufficient water available. The relationship between precipitation and potential evapotranspiration for a place over a year is illustrated by a soil moisture graph (Figure 1.18).

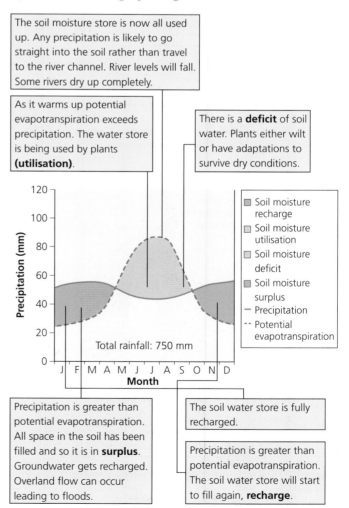

Figure 1.18 Soil water budget graph for eastern England

Soil moisture graphs vary considerably depending on the climate. Figure 1.19 shows the soil moisture budget for an equatorial area. In this case, the rainfall has two marked maxima. Because the temperatures vary very little throughout the year the potential evapotranspiration stays relatively constant. The high rainfall fills the soil stores rapidly. In the short times between the rainfall maxima, soil water does not go into deficit and so rivers and plants have a source of water all year round. There is a high potential for flooding between February and June and again in August to November.

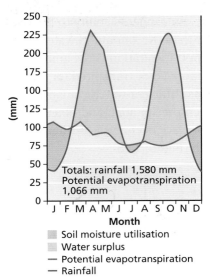

Figure 1.19 Soil moisture budget for Yaoundé, Cameroon

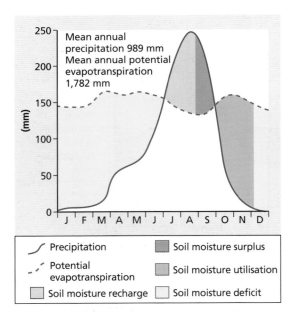

Figure 1.20 Soil moisture budget for Navrongo, northern Ghana (11°N)

Source: AQA

Figure 1.20 however, is for a tropical wet/dry climate. During the rainy season (July to October) the soils are recharged until there is surplus. This surplus does not last long when the rains stop. There is a short period of utilisation but from December to July the soil is dry. River levels fall with many drying up altogether. Vegetation has adapted to this seasonal water supply by evolving characteristics that enable it to survive drought. Humans have adapted traditionally by following migrating herds of animals that have themselves migrated following the rains north and south.

Key question

Can you think of other areas that might have a contrasting soil moisture budget to those illustrated in Figures 1.19 and 1.20?

Run-off variation

River flow is studied by measuring the **discharge** of a river. This is defined as the volume of water passing a measuring point in a given time. It is calculated by

Key terms

Bankfull – the maximum discharge that a river channel is capable of carrying without flooding.

Base flow – This represents the normal day-to-day discharge of the river and is the consequence of slow moving soil throughflow and groundwater seeping into the river channel.

Discharge – The amount of water in a river flowing past a particular point expressed as m^3s^{-1} (cumecs).

Lag time – the time between the peak rainfall and peak discharge.

Peak discharge – the point on a flood hydrograph when river discharge is at its greatest.

Storm flow – discharge resulting from storm precipitation involving both overland flow, throughflow and groundwater flow.

Storm hydrograph – A graph of discharge of a river over the time period when the normal flow of the river is affected by a storm event.

multiplying the cross-sectional area of the river by its velocity at the measuring point. It is measured in m³/sec, or cumecs.

Discharge is the combined result of the many climatological and geographical factors which interact within a drainage basin. This means knowledge of discharge is very important in the assessment and management of water resources (including irrigation provision), the design of water-related structures (reservoirs, bridges, flood banks, urban drainage schemes, sewage treatment works, etc.) and flood warning and alleviation schemes. The information can also help in developing hydroelectric power and protecting both the ecological health of watercourses and wetlands and their amenity and recreational value.

With climate change expected to impact very unevenly on river discharge patterns, keeping records of river flows is the key to identifying, quantifying and interpreting hydrological trends. This will help in the development of more effective ways of dealing with future flood and drought episodes.

The UK has a dense gauging station network of around 1,500 measuring stations supported by secondary and temporary monitoring sites. Such a large number is needed because the UK contains a multiplicity of mostly small river basins and is diverse in terms of its climate, topography, geology, land use and patterns of water usage.

Figure 1.22 Brant Broughton Gauging Station on the River Brant in Lincolnshire, England

There are many ways to measure the discharge of a river. Figure 1.22 shows a weir of known shape. The amount of water flowing over the weir is proportional to the depth of water at the weir. The depth of the water flowing over the weir can be converted into discharge using a simple equation.

River regimes

A river regime can be defined as *'the variability in its discharge throughout the course of a year in response to precipitation, temperature, evapotranspiration and drainage basin characteristics.'*

There are almost 1,500 river systems within the UK and they vary greatly, being extremely sensitive to climatic variation, land use changes and water abstraction. They are also greatly influenced by the landscapes through which they flow. UK rivers range from mountain torrents draining areas receiving up to five metres of rain a year, to much more placid groundwater-fed streams supported by much lower rainfall in parts of southeast England.

UK rainfall is fairly evenly distributed throughout the year with a slight autumn/winter maximum, particularly in the west. However, seasonal variations in temperature and sunshine amounts mean that evapotranspiration losses are heavily concentrated in the summer. This means that UK rivers have a marked seasonality in their discharge with a late summer/early autumn minimum (See Figure 1.23, page 16).

In other parts of the world river regimes differ widely. The Mohawk River flows out of the mountains of upper New York State. Here the extremely snowy, cold and long winters and drier, hot summers lead to a different pattern of river discharge (Figure 1.24, page 16).

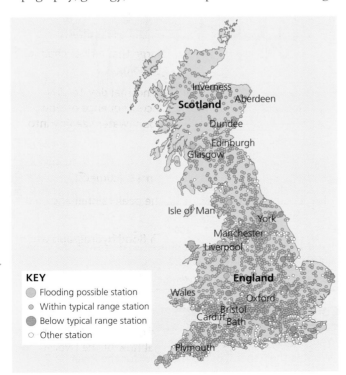

KEY
- Flooding possible station
- Within typical range station
- Below typical range station
- Other station

Figure 1.21 The UK gauging station network

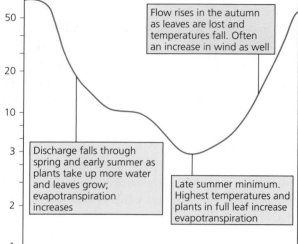

Figure 1.23 Sketch of the regime of the River Avon, Bath 1999

Highest discharge in winter. Slightly higher rainfall, much less evapotranspiration

Flow rises in the autumn as leaves are lost and temperatures fall. Often an increase in wind as well

Discharge falls through spring and early summer as plants take up more water and leaves grow; evapotranspiration increases

Late summer minimum. Highest temperatures and plants in full leaf increase evapotranspiration

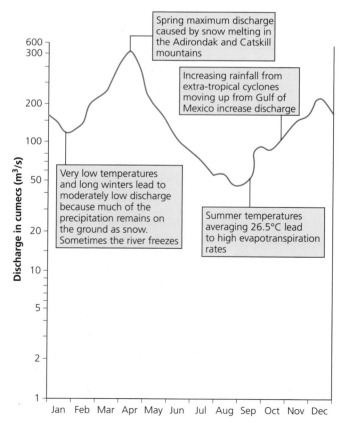

Figure 1.24 Sketch of the regime of the Mohawk River (New York State) USA

Spring maximum discharge caused by snow melting in the Adirondak and Catskill mountains

Increasing rainfall from extra-tropical cyclones moving up from Gulf of Mexico increase discharge

Very low temperatures and long winters lead to moderately low discharge because much of the precipitation remains on the ground as snow. Sometimes the river freezes

Summer temperatures averaging 26.5°C lead to high evapotranspiration rates

Key questions

What do you think the regimes of the following rivers look like and why?
- **River Nile**
- **River Ganges**

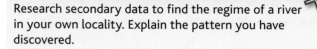

Skills focus

Research secondary data to find the regime of a river in your own locality. Explain the pattern you have discovered.

The geological characteristics of individual catchments have a major influence on flow patterns. Figure 1.25 shows daily flows for neighbouring catchments in the Thames basin. Although they experience almost identical climatic conditions, the flow regime for the Lambourn, which is supplied by springs from the underlying chalk, is much less variable than that for the Ock which drains an impermeable clay catchment.

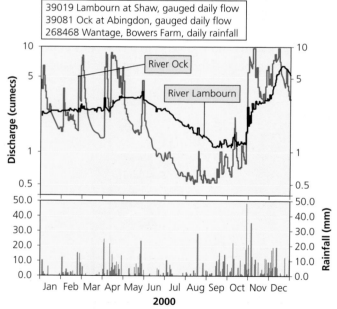

39019 Lambourn at Shaw, gauged daily flow
39081 Ock at Abingdon, gauged daily flow
268468 Wantage, Bowers Farm, daily rainfall

Figure 1.25 The variation in run-off for two rivers with the same rainfall pattern but different underlying geology

Source: Centre for Ecology & Hydrology

Skills focus

You should be able to describe trends from graphs. What is the general pattern of flow for the River Lambourn?

Rainfall events mean that the pattern of river flow is not smooth. This is illustrated by the hydrograph for the River Ock in Figure 1.25. Each time it rains the River Ock responds quickly, giving the hydrograph a jagged appearance. Each sudden rise and then subsequent fall in discharge can be closely studied using a storm (or flood) hydrograph. Storm hydrographs are graphs of discharge over the time period when the normal flow of a river is affected by a storm event.

The storm (flood) hydrograph

We have already seen that a hydrograph is a graph of river discharge against time. A **storm hydrograph** is the graph of the discharge of a river leading up to and following a storm or rainfall event. They are important because they can predict how a river might respond to a rainstorm. This can help in managing the river.

The storm hydrograph starts with the base flow. The river is fed by throughflow of soil water and groundwater. The slow movement of this water means that the changes in discharge are small. As storm water enters the drainage basin the river begins to be fed by much more fast-moving water. The discharge rises, as shown by the rising limb of the storm hydrograph. It eventually reaches a peak discharge, the highest flow in the channel for that event. The time taken from the peak rainfall to the peak discharge is called the **lag time**. The discharge begins to fall as shown by the receding limb of the graph. When all the storm water has passed through, the river returns to its base flow.

Although all storm hydrographs have the same common elements, they are not all the same shape. Hydrographs that have a short lag time, high peak discharge, steep rising and falling limbs are described as being 'flashy'. Others are a lot more subdued with gentle rising and falling limbs, long lag times and low peak discharge. This shape is determined by both physical and human factors. Physical factors:

- Drainage basins that are more circular in shape lead to more flashy hydrographs than those that are long and thin because each point in the drainage basin is roughly equidistant from the measuring point on a river.

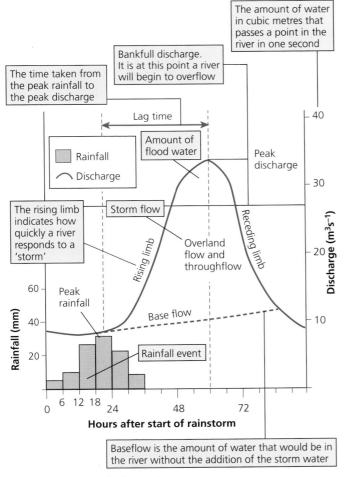

Figure 1.26 A storm hydrograph

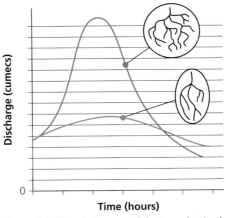

Figure 1.27 The influence of drainage basin shape on storm hydrographs: Water takes less time to reach the river in a round drainage basin than an elongated one

- Drainage basins with steep sides tend to have flashier hydrographs than gently sloped river basins. This is because water flows more quickly on the steep slopes, whether as throughflow or overland flow and so gets to the river more quickly (Figure 1.28, page 18).

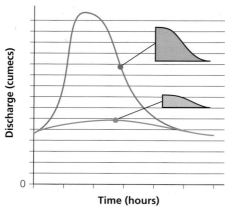

Figure 1.28 The influence of steepness on storm hydrographs: In steep sided drainage basins (or the upper reaches of a river) water gets to the river more quickly than in an area of gentle slopes

- Basins that have a high drainage density (that is, they have a lot of surface streams acting as tributaries to the main river) have flashy hydrographs. All the water arrives at the measuring station at the same time.

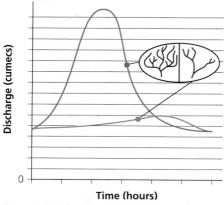

Figure 1.29 The influence of drainage density on flood hydrographs: Drainage basins with high drainage density have flashier river flows

- If the drainage basin is already saturated by antecedent rainfall then overland flow increases because infiltration capacity has been reached. Since overland flow is the fastest of the transfers the lag time is reduced. Again, peak discharge is higher resulting in a flashy hydrograph.
- If the soil or rock type within the river basin is impermeable (for example, clay soils or shale rocks) overland flow will be higher. Throughflow and infiltration will also be reduced, meaning a flashy hydrograph. The same can be said of surfaces baked hard by the sun during a long period of dry weather or frozen surfaces resulting from cold weather.

On the other hand, drainage basins underlain by sandstones have a subdued hydrograph because the water soaks into this porous rock.

- Thick vegetation cover in drainage basins will have a significant effect on a storm hydrograph. Vegetation intercepts the precipitation, holding the water on its leaves; this slows the movement of rainwater to the ground and so to river channels. Water is also lost due to evaporation and transpiration from vegetation surfaces reducing how much gets to the river. This subdues the storm hydrograph, increasing lag time and reducing peak discharge.
- The amount and intensity of precipitation can affect storm hydrographs. Heavy storms with a lot of water entering the drainage basin over a short time result in higher discharge. The type of precipitation can also have an impact. The lag time is likely to be greater if the precipitation is snow rather than rain. This is because snow takes time to melt before the water enters the river channel. When there is rapid melting of snow the peak discharge could be high.
- Large drainage basins catch more precipitation and so have a higher peak discharge compared to smaller basins. Smaller basins generally have shorter lag times because precipitation does not have as far to travel.

Human factors:
- Deforestation reduces interception rates allowing rainwater to hit the surface directly. The lack of vegetation roots reduces the infiltration rate into the soil. These both lead to rapid overland flow and flashy hydrographs. Deforestation also exposes the soil to greater rates of erosion, which leads to sedimentation of the channel. This reduces the **bankfull** capacity of a river and can lead to a greater chance of flooding.
- Afforestation has the opposite effect making it a useful flood prevention measure.
- Agriculture has a variety of effects, among which are:
  - Ploughing breaks up the topsoil and allows greater infiltration, subduing hydrographs. This can be enhanced by contour ploughing where furrows are created that run directly down slope, then they can act as small stream channels and lead

to flashier hydrographs. Ploughing wet soils can cause impermeable smears in the subsoil called plough pans. These inhibit percolation leading to greater surface flows.

- – Terracing on hillsides stops movement of water downhill and subdues hydrographs.
- – Grass crops increase infiltration and lead to subdued hydrographs.
- – Large numbers of animals on small areas can impact soils leading to overland flows.
- Growth of urban areas and other large impermeable surfaces such as roads lead to flashy hydrographs (see Chapter 9). This is exacerbated by the very fact that settlements have been built on floodplains. This urban growth leads to the expansion of built-up, impermeable surfaces such as roads, car parks, shopping centres, etc. Most settlements are designed to transfer water as quickly as possible away from human activity to the nearest river. This is achieved through road camber, building design and drainage systems. In many cities in the UK there has been a continued loss of front gardens in favour of paved drives. Due to the growing number of two-/three-car families, an area of vegetated garden equivalent to 300 ha/year was lost in London between 1998 and 2006.

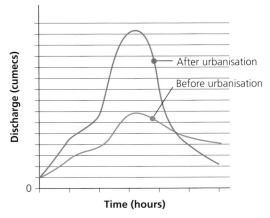

Figure 1.30 The influence of land use change on flood hydrographs: Impermeable surfaces shaped to get rid of water quickly, combined with a dense network of smooth drains, means that water gets to the river very quickly. The river itself can also be altered, for example, to move the water rapidly away from the urban area

- Some soft engineering flood management schemes attempt to reduce flashiness in a river's hydrograph. Afforestation increases interception and infiltration. This slows down the progress of water to the river channel and subdues any changes in discharge.
- Water abstraction reduces the base flow and so more water must reach the channel before it reaches bankfull capacity.

Examples of how land use changes can affect the water cycle

Deforestation

Tropical South America contains the world's largest continuous tropical forest and savannah ecosystems. This region is environmentally important not only because of traditional ecological measures, such as its high biodiversity, but also because it generates more than a quarter of the world's river discharge. It has undergone explosive development and deforestation in the last 50 years as national and international demand for cattle feed (mostly soy), beef and sugar cane for ethanol, have increased. Already about 10 per cent of the rainforest in this large region has been converted to cattle pasture and agriculture.

Deforestation and forest degradation result in a complex set of changes to streams of all sizes. When forests are removed, the new vegetation generally has fewer leaves and shallower roots. This means it uses less water than the forest it replaces. The result is that less water evaporates from the land surface to be returned to the atmosphere; more water runs off of the land and stream flow is increased. The amount of change that occurs depends on local conditions including the amount of rainfall, how much of a watershed is deforested, topography, soils and the land use after deforestation. Studies have shown that there is little effect with less than 20 per cent of a basin deforested but a large increase with 50 to 100 per cent of a basin deforested. These changes occur at the local scale, but rivers of all sizes are affected when deforestation is extensive.

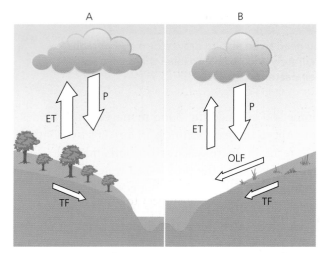

Figure 1.31a The effects of deforestation on the water cycle – localised deforestation

Before deforestation (A) much of the high precipitation (P) is returned to the atmosphere by evapotranspiration (ET). Overland flow is minimal. Most of the water that reaches the forest floor infiltrates into the soil and travels slowly to the river by throughflow, maintaining a steady flow in the river.

After deforestation (B), although the precipitation stays the same, the evapotranspiration is lower because the replacement vegetation has smaller leaves and roots and is less dense. Overland flow and throughflow occur because of the lack of vegetation. This leads to increased discharge and flashiness. This can cause localised flooding

This suggests several important points about the climate, land surface and water cycle:

- If deforestation does not cause decreased rainfall via atmospheric feedbacks, discharge will likely be significantly increased throughout the entire southern Amazon (Figure 1.31a).
- If rainfall does decrease via atmospheric feedbacks the resulting decrease in river discharge may be greater than the changes without feedbacks (Figure 1.31b).
- Changes in water resources caused by atmospheric feedbacks will not be limited to those catchment areas where deforestation has occurred but will be spread unevenly throughout the whole Amazon basin by atmospheric circulation.

Soil drainage

Subsurface drainage removes excess water from the soil profile. It is carried out usually through a network of perforated tubes installed 60–120 cm below the soil surface. These tubes are commonly called 'tiles' because they were originally made from short lengths of clay pipes known as tiles. Water would seep into

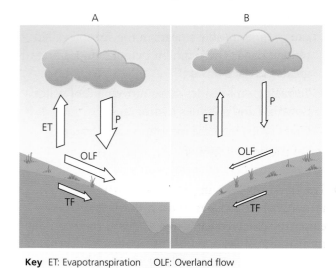

Key ET: Evapotranspiration   OLF: Overland flow
P: Precipitation   TF: Throughflow

Figure 1.31b The effects of deforestation on the water cycle – extensive deforestation

Where deforestation is extensive, positive feedback can occur in the basin hydrological system. In A, because evapotranspiration is low, much of the water leaves the area in the river channel rather than being recycled continuously between the forest and the atmosphere. Once the water has left the area there is less water vapour available in the atmosphere for precipitation and so precipitation levels fall. Less water gets to the river channel and the flow is reduced

the small spaces between the tiles and drain away. The most common type of 'tile' now is corrugated plastic tubing with small perforations to allow water entry. When the water table in the soil is higher than the tile, water flows into the tubing, either through holes in the plastic tube or through the small cracks between adjacent tiles. This lowers the water table to the depth of the tile over the course of several days. Drain tiles allow excess water to leave the field, but once the water table has been lowered to the elevation of the tiles, no more water flows through the tiles. In most years in the UK, drain tiles have not flowed between June and October.

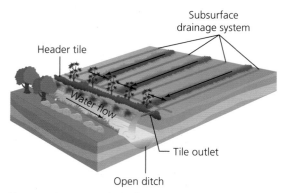

Figure 1.32 A typical tile drainage set-up

Although costly, agricultural drainage is very good for moderately to poorly drained soils. It increases the productivity of the field and helps to improve the efficiency of growers. A study by Ohio State University demonstrated that for every dollar spent on drainage the grower got back between $1.20 and $1.90 when growing corn and soybeans.

Some of the advantages of draining marginal farmland:

- The build-up of an improved soil structure, making it more friable and easier to work. It also makes it easier to achieve greater root penetration, enabling roots to travel faster and further.
- Improved aeration, which makes conditions more favourable for microorganisms to thrive. This increases the rate at which organic matter is broken down into humus and plant nutrients are mineralised into an available form. It also provides the necessary supply of air for root cell respiration.
- The increased aeration increases the ease with which the soil can be warmed. This can make possible earlier sowing of seeds, with greater likelihood of improved germination.
- Heavy machinery can work on the land without danger of compaction (and so leading to increased overland flow).
- Larger numbers of animals can be allowed to graze the land, once again without compacting the soil.

Some of the disadvantages:

- The insertion of drains artificially increases the speed of throughflow in the soil. Much more water reaches watercourses more quickly than before drainage. This can increase the likelihood of flooding and increase the range of flows in rivers. It is interesting to note that before the drainage of many floodplains in the UK from the eighteenth century onwards, rivers were more navigable than today; the annual flow regime was much more even.
- The dry topsoil can be subject to wind erosion if not properly protected. Soil loss by wind erosion has mainly been documented for sandy and peaty soils in the eastern and middle counties of England, especially arable fields in the East Midlands and East Anglia. Generally the area of England and Wales subject to wind erosion is small, although those fields that are affected are more severely eroded than by water erosion. The proportion of a field subject to wind erosion is likely to be greater than that subject to water erosion. Available estimates suggest that the mean wind erosion rate is of the order of 0.1 to 2 tonnes/ha/year, although maximum values for fields can be one or two orders of magnitude higher.
- Another major concern with regard to land drainage is nitrate loss. It can lead to eutrophication. Water draining from fields finds its way into local watercourses. There it enriches ponds, etc., with nitrogen or phosphorus. It causes algae and higher

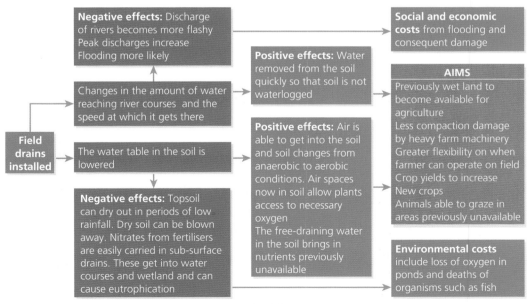

Figure 1.33 System diagram showing the effects of soil drainage

forms of plant life to grow too fast. This disturbs the balance of organisms present in the water and the quality of the water concerned.

One way to overcome some of these problems is to use controlled drainage. This keeps the water table high during the off-season when crops are not growing. The high water table increases the rate of denitrification (a process that converts nitrate to harmless nitrogen gas (N_2) as soon as the saturated soil warms up in the spring) and reduces nitrate loss to the environment.

Water abstraction

Problems can occur when the demand for water exceeds the amount available during a certain period. They happen frequently in areas with low rainfall and high population density, and/or in areas with intensive agricultural or industrial activity.

In many areas of Europe, groundwater is the dominant source of fresh water. In a number of places water is being pumped from beneath the ground faster than it is being replenished through rainfall. The result is sinking water tables, empty wells, higher pumping costs and, in coastal areas, the intrusion of saltwater from the sea which degrades the groundwater. This saline intrusion is widespread along the Mediterranean coastlines of Italy, Spain and Turkey, where the demands of tourist resorts are the major cause of over-abstraction. In Malta, most groundwater can no longer be used for domestic consumption or irrigation because of saline intrusion, and the country has resorted to expensive desalination plants.

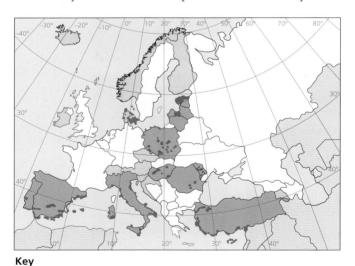

Key
- Salt water intrusion
- Groundwater overexploitation
- No groundwater overexploitation
- Data available
- No data
- Outside data coverage

Figure 1.34 Salt water intrusion in Europe

Source: European Environment Agency

Sinking water tables can also make rivers less reliable, since many river flows are maintained in the dry season by springs that dry up when water tables fall. Groundwater also helps sustain surface reservoirs of water such as lakes and wetlands that are often highly productive ecosystems and resources for tourism as well as leisure activities. These, too, are threatened by over-abstraction of groundwater.

Irrigation is the main cause of groundwater over-exploitation in agricultural areas. Examples include the Greek Argolid plain of eastern Peloponnesus where it is common to find boreholes 400 m deep contaminated by seawater intrusion. In Italy, over-exploitation of the Po River in the region of the Milan aquifer has led to a 25–40 m decrease in groundwater levels over the last 80 years. In Spain, more than half of the abstracted groundwater volume is obtained from areas facing over-exploitation problems.

Water abstraction from the chalk of southern England

The water within the chalk aquifer of southern England is replenished by rainfall that lands on the exposed chalk hills of the North and South Downs and the Chilterns. Normally recharge takes place during the winter months when potential evapotranspiration is low and soil moisture deficits are negligible. Groundwater amounts vary seasonally, with levels rising from autumn through winter into spring. During the summer months, potential evapotranspiration generally exceeds rainfall, soil moisture deficits build up, and little, if any, percolation takes place.

In the summer water still leaves the chalk from springs as well as by abstraction from boreholes. This pattern is not constant, since rainfall varies both over time and location. Rivers fed by groundwater from chalk aquifers can have intermittent sections. These streams, often referred to as 'bournes', are a natural characteristic of chalk downlands. The positions of the springs feeding these rivers differ throughout the year, being at a greater altitude in winter and spring. If there are one or more dry winters when the effective rainfall available for recharge is low then these rivers can dry up altogether.

Some of the most acute problems with over-abstraction have been found in chalk stream systems, where up to 95 per cent of the flow is derived from underground aquifers. The catchments of chalk streams provide underground reservoirs of generally high quality groundwater which can be abstracted for public supply. Abstraction for public water supply and industry has dramatically reduced the flow in many chalk streams and, in some cases, completely dried up sections of these important rivers, particularly during dry summers when public demand is at its highest. This also has an economic impact on local communities, resulting from the inability to fish, enjoy river views due to encroaching vegetation or undergo other recreational activities.

Water abstraction in the London Basin

Figure 1.35 shows the subsurface geology of the London Basin. The chalk layers form a syncline beneath the London area with the uplands of the Chilterns to the NW and the North Downs to the SE. Precipitation on these exposed chalk hills soaks into the porous chalk where it is stored and released naturally at springs where it is in contact with either Greensand or Palaeogene rocks.

Throughout history, in London, water has been abstracted from wells and boreholes that penetrate down to the chalk. During the nineteenth century and first part of the twentieth century, the chalk-basal

sands aquifer had been increasingly exploited, as a result of increased industrialisation and the associated development of groundwater sources. At the peak of abstraction in the 1960s, groundwater levels beneath central London had dropped to 88 m below sea level, creating a large depression in the water table. A smaller cone of depression also developed to the east beneath the River Roding in Essex.

Since the mid-1960s, industries, such as brewing, in central London relocated or were closed down. Economic activity turned more to service industries and commerce than heavy industry. The subsequent reduction in abstraction resulted in groundwater levels recovering by as much as 3 m/year in places by the early 1990s, leading to a gradual rebound of the water table. This then has posed the threat of rising groundwater to structures in the London Basin such as London Underground and building foundations. This led to the implementation of the General Aquifer Research, Development and Investigation Team (GARDIT) strategy to control water levels. As a result of careful management of both abstraction and artificial recharge the rise in groundwater that the GARDIT strategy was designed to arrest had largely been achieved by 2000, so this year provides a useful baseline year for comparisons.

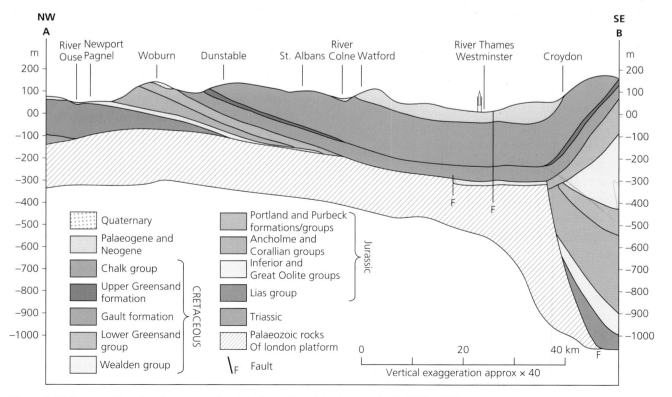

Figure 1.35 Cross section showing the confined chalk aquifer of the London Basin (BGS, 1996)

The differences in groundwater levels for January 2000 and January 2014:

- Groundwater levels in west London have risen due to limited abstraction in this area, in the order of four to eight metres since 2000 which has levelled off in recent years.
- In central and east London groundwater levels have fallen in the order of 5–7 m since 2000 as a result of increased abstraction.
- Groundwater levels have fallen more than 2 m across much of south London, with falls of up to 12 m concentrated around the many large public water supply abstractions.
- In east London, where there are chalk outcrops around the River Thames from Greenwich to Woolwich, there is a risk of saline intrusion. When groundwater levels near the river are lower than the water level in the River Thames, saline river water can enter the chalk aquifer.

1.3 The carbon cycle

Carbon (C) (from the Latin *carbo* meaning coal) is one of the most chemically versatile of all the elements. Carbon forms more compounds than any other element and scientists predict that there are more than ten million different carbon compounds in existence today on Earth. Carbon is found in all life forms in addition to sedimentary rocks, diamonds, graphite, coal and petroleum (oil and natural gas).

Carbon follows a certain route on Earth, called the carbon cycle. It is the complex processes carbon undergoes as it is transformed from organic carbon (the form found in living organisms such as plants and trees) to inorganic carbon and back again. Through following the carbon cycle we can also study energy flows on Earth, because most of the chemical energy needed for life is stored in organic compounds as bonds between carbon atoms and other atoms.

Carbon atoms move through the carbon cycle in many different forms. Some important examples of carbon compounds include:

- carbon dioxide (CO_2), a gas found in the atmosphere, soils and oceans
- methane (CH_4), a gas found in the atmosphere, soils and oceans and sedimentary rocks
- calcium carbonate ($CaCO_3$), a solid compound found in calcareous rocks, oceans and in the skeletons and shells of ocean creatures
- hydrocarbons – solids, liquids or gases usually found in sedimentary rocks
- bio-molecules – complex carbon compounds produced in living things. Proteins, carbohydrates, fats and oils, and DNA are examples of bio-molecules.

Of all these forms of carbon, we study CO_2 in most detail because it is thought that this has a profound effect on climate. It is also difficult to separate a *natural* carbon cycle from one that is affected by human activity. Human activity and associated emissions of carbon dioxide (anthropogenic CO_2) fundamentally affect the carbon cycle and so affect climate.

Origins of carbon on Earth

The primary source of carbon/CO_2 is the Earth's interior. It was stored in the mantle when the Earth formed. It escapes from the mantle at constructive and destructive plate boundaries as well as hot-spot volcanoes. Much of the CO_2 released at destructive margins is derived from the metamorphism of carbonate rocks subducting with the ocean crust. Some of the carbon remains as CO_2 in the atmosphere, some is dissolved in the oceans, some carbon is held as biomass in living or dead and decaying organisms, and some is bound in carbonate rocks. Carbon is removed into long-term storage by burial of sedimentary rock layers, especially coal and black shales (these store organic carbon from undecayed biomass) and carbonate rocks like limestone (calcium carbonate).

The major stores of carbon

A gigatonne of carbon dioxide equivalent (GtC) is the unit used by the United Nations climate change panel, the Intergovernmental Panel on Climate Change (IPCC), to measure the amount of carbon in various stores. 1 Gt amounts to 10^9 tonnes (1 billion tonnes).

Transfer (flux) of carbon within the cycle is measured in gigatonnes of carbon per year (GtC/years).

The lithosphere

The Earth's **lithosphere** includes the crust and the uppermost mantle; this constitutes the hard and rigid outer layer of the Earth. The uppermost part of the lithosphere, the layer that chemically reacts to the atmosphere, hydrosphere and **biosphere** through the soil forming process is called the pedosphere.

Carbon is stored in the lithosphere in both inorganic and organic forms. Inorganic deposits of carbon in the lithosphere include fossil fuels like coal, oil, and natural gas, oil shale (kerogens) and carbonate-based sedimentary deposits like limestone. Organic forms of carbon in the lithosphere include litter, organic matter and humic substances found in soils.

Carbon in the lithosphere is distributed between these stores:
- marine sediments and sedimentary rocks contain up to 100 million GtC
- soil organic matter contains between 1,500 and 1,600 GtC

- fossil fuel deposits of coal, oil and gas contain approximately 4,100 GtC
- peat, which is dead but undecayed organic matter found in boggy areas contains approximately 250 GtC.

The hydrosphere

The ocean plays an important part in the carbon cycle. Attempts to collate measurements of the amount of carbon in the oceans have been made by the Global Ocean Data Analysis project (GLODAP) using data from research ships, commercial ships and buoys. The measurements come from deep and shallow waters from all the oceans. There is some variation in the results and figures can only be an approximation.

The oceanic stores can be divided into three:
- the surface layer (euphotic zone) where sunlight penetrates so that photosynthesis can take place contains approximately 900 GtC.
- the intermediate (twilight zone) and the deep layer of water contain approximately 37,100 GtC
- living organic matter (fish, plankton, bacteria, etc.) amount to approximately 30 GtC and dissolved organic matter 700 GtC.

This gives a total for oceanic carbon of between 37,000 GtC to 40,000 GtC.

When organisms die, their dead cells, shells and other parts sink into deep water. Decay releases carbon dioxide into this deep water. Some material sinks right to the bottom, where it forms layers of carbon-rich sediments. Over millions of years, chemical and physical processes may turn these sediments into rocks. This part of the carbon cycle can lock up carbon for millions of years. It is estimated that this sedimentary layer could store up to 100 million GtC.

The biosphere

This is defined as the total sum of all living matter. For our purposes we are going to consider the terrestrial biosphere as being separate from the oceanic biosphere. The total amount of carbon stored in the terrestrial **biosphere** has been estimated to be 3,170 GtC. The distribution of this carbon depends upon the ecosystem as shown in Figure 1.36 (page 26).

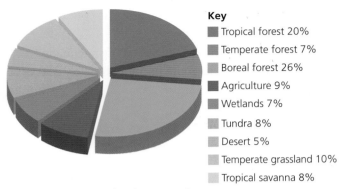

Key
- Tropical forest 20%
- Temperate forest 7%
- Boreal forest 26%
- Agriculture 9%
- Wetlands 7%
- Tundra 8%
- Desert 5%
- Temperate grassland 10%
- Tropical savanna 8%

Figure 1.36 Terrestrial carbon stores by ecosystem

The main stores of carbon in the terrestrial biosphere are as follows:

- **Living vegetation:** at the global level, 19 per cent of the carbon in the Earth's biosphere is stored in plants. Unlike the oceans, much of this carbon is stored directly in the tissues of the plants. Although the exposed part of the plant is the most visible, the below-ground biomass (the root system) must also be considered. The amount of carbon in the biomass varies from between 35 and 65 per cent of the dry weight. The amount varies depending on the location and the vegetation type. It is estimated that half of the carbon in forests occurs in high-latitude forests, and a little more than one third occurs in low-latitude forests. The two largest forest reservoirs of carbon are the vast expanses in Russia, which hold roughly 25 per cent of the world's forest carbon, and the Amazon basin, which contains about 20 per cent.

- **Plant litter:** this is defined as fresh, undecomposed, and easily recognisable (by species and type) plant debris. This can be anything from leaves, cones, needles, twigs, bark, seeds/nuts, etc. The type of litter is directly affected by the type of ecosystem. Leaf tissues account for about 70 per cent of litter in forests, but woody litter tends to increase with forest age. In grasslands there is very little above ground perennial tissue so the annual fall of litter is very low.

- **Soil humus:** this originates from litter decomposition. Humus is a thick brown or black substance that remains after most of the organic litter has decomposed. It gets dispersed throughout the soil by soil organisms such as earthworms. In all forests, tropical, temperate and boreal together, approximately 31 per cent of the carbon is stored in the biomass and 69 per cent in the soil. In tropical forests, approximately 50 per cent of the carbon is stored in the biomass and 50 per cent in the soil. Altogether the world's soils hold more carbon (2,500 GtC) than the vegetation. Soil carbon can be either organic (1,550 GtC) or inorganic carbon (950 GtC). The inorganic carbon component consists of carbon itself as well as carbonate materials such as calcite, dolomite and gypsum. The amount of carbon found in living plants and animals is comparatively small relative to that found in soil (560 GtC). The soil carbon pool is approximately 3.1 times larger than the atmospheric pool of 800 GtC. Only the ocean has a larger carbon store.

- **Peat:** this is an accumulation of partially decayed vegetation or organic matter that is unique to natural areas called peatlands or mires. Peat forms in wetland conditions, where almost permanent water saturation obstructs flows of oxygen from the atmosphere into the ground. This creates low oxygen anaerobic conditions that slow down rates of plant litter decomposition. Peatlands cover over four million km² or 3 per cent of the land and freshwater surface of the planet; they occur on all continents, from the tropical to boreal and Arctic zones and from sea level to high alpine conditions. It is estimated that peat stores more than 250 GtC worldwide.

- **Animals:** these play a small role in the storage of carbon. They are, however, very important in the generation of movement of carbon through the carbon cycle.

The atmosphere

Carbon has been in the atmosphere from early in the Earth's history. Atmospheric carbon dioxide levels have reached very high values in the deep past, possibly topping over 7,000 ppm (parts per million) in the Cambrian period around 500 million years ago. Its lowest concentration has probably been over the last 2 million years during the Quaternary glaciation when it sank to about 180 ppm.

Today, carbon dioxide is a trace gas in the Earth's atmosphere. Estimates of the overall amount of carbon stored in the atmosphere vary from 720 GtC to 800 GtC. It makes up about 0.04 per cent (400 ppm) of the atmosphere; this low concentration belies its importance to the planet and all life on it. Due to human activities, the present concentration of

CO_2 in the Earth's atmosphere is higher than it has been for at least 800,000 years, and, in all likelihood, the highest in the past 20 million years. Despite its relatively small concentration, CO_2 is a potent **greenhouse gas** and plays a vital role in regulating the Earth's surface temperature. The recent phenomenon of global warming has been attributed primarily to increasing industrial CO_2 emissions into Earth's atmosphere.

Atmospheric carbon has been measured at the Mauna Loa Observatory (MLO) on Hawaii since 1958. The undisturbed air, remote location and minimal influences of vegetation and human activity at MLO are ideal for monitoring atmospheric constituents. The observatory is part of the American National Oceanic and Atmospheric Administration (NOAA). The measurements show that the global annual mean concentration of CO_2 in the atmosphere has increased markedly since the Industrial Revolution, from 280 ppm to 317.7 ppm in March 1958 to 400.3 ppm as of February 2015. This increase is largely attributed to human derived (anthropogenic) sources, particularly the burning of fossil fuels and deforestation.

Figure 1.37 The NOAA atmospheric observatory station, Mauna Loa, Hawaii

A graph of this change has been named after the scientist who first started this research; it is called the Keeling Curve.

Keeling was one of the first scientists to gather evidence that linked fossil fuel emissions to rising levels of carbon dioxide. Keeling's research has been backed up by other readings from around the world; for example, the CO_2 trapped in ice cores from Antarctica and Greenland can be used to give a 'proxy'

measure of the CO_2 in the atmosphere at the time that snow was laid down.

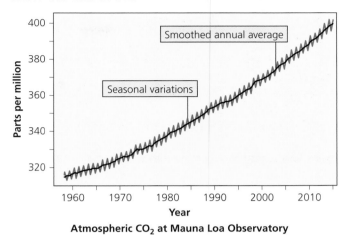

Atmospheric CO_2 at Mauna Loa Observatory

Figure 1.38 The Keeling curve of the changing concentration of atmospheric CO_2

Source: NOAA

The daily average at Mauna Loa first exceeded 400 ppm on 10 May 2013. It is currently rising at a rate of approximately 2 ppm/year and accelerating.

Movement of carbon

Carbon moves from one store to another in a continuous cycle (see Figure 1.39, page 28)

This cycle consists of several carbon stores as described above. The processes by which the carbon moves between these stores are known as transfers or fluxes. If more carbon enters a store than leaves it, that store is considered a **net carbon sink**. If more carbon leaves a store than enters it, that store is considered a **net carbon source**.

The geological component

The geological component of the carbon cycle is where it interacts with the rock cycle in the processes of **weathering**, burial, subduction and volcanic eruptions.

In the atmosphere, carbon dioxide is removed from the atmosphere by dissolving in water and forming carbonic acid:

$$CO_2 + H_2O \longrightarrow H_2CO_3 \text{ (carbonic acid)}$$

As this weakly acidic water reaches the surface as rain, it reacts with minerals at the Earth's surface, slowly dissolving them into their component ions through the process of chemical weathering. These component ions are carried in surface waters like streams and rivers, eventually to the ocean, where they settle out as minerals like calcite, a form of calcium carbonate ($CaCO_3$).

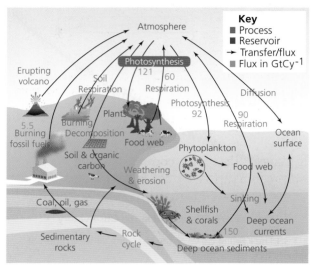

Figure 1.39 The carbon cycle

Key question

What landforms may be created by this process?

Calcium carbonate is also precipitated from calcium and bicarbonate ions in seawater by marine organisms like foraminifera, coccoliths or molluscs. When these creatures die, their skeletons sink to the bottom of the oceans where they collect as sediment. Burial by overlying layers of sediment can eventually turn these sediments into sedimentary limestone. Coral also extracts $CaCO_3$ from seawater. These creatures live and eventually die in the same location. Dead coral is built upon by later generations of live coral and so it too becomes buried. The carbon is now stored below the sea floor in layers of limestone. Tectonic uplift can then expose this buried limestone. One example of this occurs in the Himalayas where some of the world's highest peaks are formed of material that was once at the bottom of the ocean.

Tectonic forces cause plate movement to push the sea floor under continental margins in the process of subduction. The carbonaceous sea-floor deposits are pushed deep into the Earth where they heat up, eventually melt, and can rise back up to the surface through volcanic eruptions or in seeps, vents and CO_2-rich hot springs. In this way CO_2 returns to the atmosphere. Weathering, burial, subduction and volcanism control atmospheric carbon dioxide concentrations over time periods of hundreds of millions of years.

Photosynthesis

Tiny marine plants (phytoplankton) in the sunlit surface waters (the euphotic zone) of the oceans, as well as all terrestrial plants, photosynthetic algae and bacteria, turn the carbon into organic matter by the process of **photosynthesis**. They use energy from sunlight to combine carbon dioxide from the atmosphere with water to form carbohydrates. These carbohydrates store energy. Oxygen is a by-product that is released into the atmosphere.

$$\text{carbon dioxide} + \text{water} + \text{sunlight} \longrightarrow \text{carbohydrate} + \text{oxygen}$$

$$CO_2 + H_2O + \text{sunlight} \longrightarrow CH_2O + O_2$$

Respiration

Plants (and photosynthetic algae and bacteria) then use some of the stored carbohydrates as an energy source to carry out their life functions by the process of **respiration**. Some of the carbohydrates remain as biomass (the bulk of the plant, etc.). Consumers such as animals and bacteria get their energy from this excess biomass. Oxygen from the atmosphere is combined with carbohydrates to liberate the stored energy. Water and carbon dioxide are by-products.

$$\text{oxygen} + \text{carbohydrate} \longrightarrow \text{energy} + \text{water} + \text{carbohydrate}$$

$$O_2 + CH_2O \longrightarrow \text{energy} + H_2O + CO_2$$

Note: Photosynthesis and respiration are essentially the opposite of one another. Photosynthesis removes CO_2 from the atmosphere and replaces it with O_2. Respiration takes O_2 from the atmosphere and replaces it with CO_2. However, these processes are not in balance. Not all organic matter is oxidized. Some is buried in sedimentary rocks. The result is that over geologic time, there has been more oxygen put into the atmosphere and carbon dioxide removed by photosynthesis than the reverse.

Decomposition

Decomposition is more difficult to define than photosynthesis. In broad terms, decomposition includes physical, chemical and biological mechanisms that transform organic matter into increasingly stable forms. This broad definition includes physical break-up of organic material by wet-dry, shrink-swell, hot-cold and other cycles.

Animals, wind and even other plants can also cause this fragmentation. Leaching and transport in water is another important physical mechanism. Chemical transformations include oxidation and condensation. Biological mechanisms involve feeding and digestion aided by the catalytic effect of enzymes. The decomposition process is carried out by **decomposers** whose special role is to break down the cells and tissues in dead organisms into large biomolecules and then break those biomolecules down into smaller molecules and individual atoms. Decomposition ensures that the important elements of life – carbon, hydrogen, oxygen, nitrogen, phosphorus, sulphur, magnesium – can be continually recycled into the soil and made available for life. For example, a plant cannot make its DNA molecules unless it has a supply of nitrogen, phosphorus and sulphur atoms from the soil in addition to the carbon, hydrogen and oxygen atoms it obtains through photosynthesis. For this reason, plant growth is limited by the availability of nitrogen, phosphorus, magnesium and sulphur atoms in addition to the availability of carbon dioxide, water and light energy.

Oceanic carbon pumps

Water is able to dissolve CO_2. There is a negative correlation between the temperature of the water and the amount of CO_2 that can be dissolved. This leads to **vertical deep mixing**, a term used to describe the most important movement of CO_2 in the oceans. It occurs when warm water in oceanic surface currents is carried from the warm tropics to the cold polar regions. Here the water is cooled, making it dense enough to sink below the surface layer, sometimes all the way to the ocean bed. When cold water returns to the surface and warms up again, it loses carbon dioxide to the atmosphere. In this fashion, vertical circulation ensures that carbon dioxide is constantly being exchanged between the ocean and the atmosphere. This vertical circulation also acts as an enormous **carbon pump**, giving the ocean a lot more carbon than it would have if this surface water was not being constantly replenished.

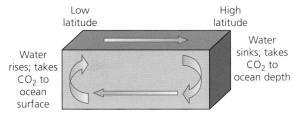

Figure 1.40 The oceanic carbon pump: The concept of vertical deep mixing, where carbon dioxide is transported from the ocean surface to the ocean depths by sinking cold water in the high latitudes. If brought to the surface (for instance, by upwelling) the cold water will warm up and release some of its carbon dioxide to the atmosphere

Living things in the ocean move carbon from the atmosphere into surface waters then down into the deeper ocean and eventually into rocks. This action of organisms moving carbon in one direction is often called a **biological pump**. Carbon gets incorporated into marine organisms as organic matter or structural calcium carbonate. When organisms die, their dead cells, shells and other parts sink into deep water. Decay releases carbon dioxide into this deep water. Some material sinks right to the bottom, where it forms layers of carbon-rich sediments. Over millions of years, chemical and physical processes may turn these sediments into rocks. This part of the carbon cycle can lock up carbon for millions of years.

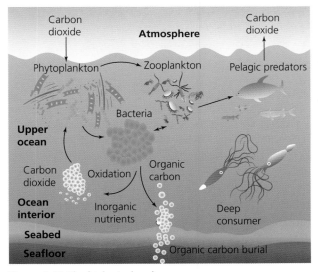

Figure 1.41 The biological carbon pump

Combustion

Combustion occurs when any organic material is reacted (burned) in the presence of oxygen to give off the products of carbon dioxide, water and energy. The organic material can be any vegetation or fossil fuel such as natural gas (methane), oil or coal. Organic materials contain at least carbon and hydrogen and may include oxygen. If other elements are present they also ultimately combine with oxygen to form a variety of pollutant molecules such as sulphur oxides and nitrogen oxides.

Biomass combustion

This is the burning of living and dead vegetation. It includes human-induced burning as well as naturally occurring fires. It happens most in:

- the boreal (northern) forests in Alaska, Canada, Russia, China and Scandinavia
- savannah grasslands in Africa
- tropical forests in Brazil, Indonesia, Colombia, Ivory Coast, Thailand, Laos, Nigeria, Philippines, Burma and Peru
- temperate forests in the US and western Europe
- agricultural waste after harvests in the US and western Europe.

Figure 1.42 The Moderate Resolution Imaging Spectroradiometer (MODIS) on NASA's Terra satellite shows fires around the world in 2010

Source: NASA

The relationship between forests and carbon dioxide emissions is complex. Forests have a life cycle: trees die after severe fire, setting the stage for new growth to begin. If a forest fully replaces itself, there will be no net carbon change over that life cycle. The fire consumes only about 10 to 20 per cent of the carbon and immediately emits it into the atmosphere. It kills trees but doesn't consume them. So, new trees grow (storing carbon), old trees decompose (emitting carbon) and the organic layer of the soil accumulates (storing carbon). This balance between simultaneous production and decomposition determines whether

the forest is a net source or sink. Left alone, terrestrial and atmospheric carbon stays more or less in balance. However, increasingly large and/or more frequent fires, possibly made worse by warming temperatures and changing precipitation levels, can change that carbon balance.

Every year, fires burn 3 to 4 million km^2 of the Earth's land surface area, and release more than a billion tonnes of carbon into the atmosphere in the form of carbon dioxide. However, massive old-growth northern latitude forests are also considered a carbon 'sink' because older trees are repositories of decades or centuries of carbon; their heavy canopy blocks sunlight from reaching the forest floor, slowing decomposition of the forest litter.

Volcanic activity

According to the United States Geological Survey (USGS) *'the carbon dioxide released in recent volcanic eruptions has never caused detectable global warming of the atmosphere.'* This is probably because:

- The warming effect of emitted CO_2 is counterbalanced by the large amount of sulphur dioxide that is given out. **Conversion of this sulphur dioxide to sulphuric acid, which forms fine droplets,** increases the reflection of radiation from the Sun back into space, cooling the Earth's lower atmosphere.
- The amounts of carbon dioxide released have not been enough to produce detectable global warming. For example, all studies to date of global volcanic CO_2 emissions indicate that present-day sub-aerial and submarine volcanoes have released less than 1 per cent of the CO_2 released currently by human activities. It has been proposed that intense volcanic release of carbon dioxide in the deep geologic past did cause a large enough increase in atmospheric CO_2 to cause a rise in atmospheric temperatures and possibly some mass extinctions, though this is a topic of scientific debate.

Hydrocarbon extraction and burning – cement manufacture

Dead plants or animals turn into fossil fuels following burial. The pressure from multiple layers of sediment leads to an anoxic (oxygen free) environment that allows for decomposition to take place without oxygen. When this is combined with heat from the Earth, the carbon in sugar molecules is rearranged to form other compounds. The type of material that is buried helps to determine what the final product will be. Animal remains tend to form petroleum (crude oil) while plant matter is more likely to form coal and natural gas. When these fossil fuels are extracted from the ground and then burnt, carbon dioxide and water are released into the atmosphere.

Cement manufacture contributes CO_2 to the atmosphere when calcium carbonate is heated, producing lime and carbon dioxide. CO_2 is also produced by burning the fossil fuels that provide the heat for the cement manufacture process.

It is estimated that the cement industry produces around 5 per cent of global anthropogenic CO_2 emissions, of which 50 per cent is produced from the chemical process itself, and 40 per cent from burning fuel to power that process. The amount of CO_2 emitted by the cement industry is more than 900 kg of CO_2 for every 1,000 kg of cement produced.

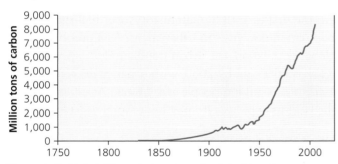

Figure 1.43 Global CO_2 emissions from fossil fuel burning, 1751 to 2006

Source: British Geological Survey

In 2013, global CO_2 emissions due to fossil fuel use and cement production were 36 GtC. This is 61 per cent higher than 1990 (the Kyoto Protocol reference year) and 2.3 per cent higher than 2012. CO_2 emissions were dominated by China (28 per cent), the USA (14 per cent) and India (7 per cent) with growth in all of these countries. The European Union (EU) also contributed 10 per cent but their 28 states were in an overall 1.8 per cent decline.

The 2013 CO_2 emissions (fossil fuel and cement production only) breakdown is: coal (43 per cent), oil (33 per cent), gas (18 per cent), cement (5.5 per cent) and gas flaring from oil wells (0.6 per cent).

Farming practices

When soil is ploughed, the soil layers invert, air mixes in, and soil microbial activity dramatically increases. It results in soil organic matter being broken down much more rapidly, and carbon is lost from the soil into the atmosphere. In addition to the effect on soil from ploughing, emissions from the farm tractors increases carbon dioxide levels in the atmosphere.

The largest source of carbon emissions within agriculture is enteric fermentation – when methane (CH_4) is produced by livestock during digestion and released via belches. This accounted in 2011 for 39 per cent of the sector's total greenhouse gas outputs. Emissions from animals increased by 11 per cent between 2001 and 2011.

Greenhouse gases resulting from biological processes in rice paddies that generate methane make up 10 per cent of total agricultural emissions, while the burning of tropical grasslands accounts for 5 per cent.

According to the United Nations Food and Agriculture Organization (FAO), in 2011, 44 per cent of agriculture-related greenhouse gas outputs occurred in Asia, followed by the Americas (25 per cent), Africa (15 per cent), Europe (12 per cent) and Oceania (4 per cent). This regional distribution has been constant over the last decade.

Land use change

CO_2 emissions that result from land use change (mainly **deforestation**) account for up to 30 per cent of anthropogenic CO_2 emissions.

Deforestation

Most deforestation is driven by the need for extra agricultural land. Often subsistence farmers will clear a few hectares to feed their families by cutting down trees and burning them in a process known as 'slash and burn' agriculture.

Logging operations also remove forest. Loggers, some of them acting illegally, also build roads to access more and more remote forests, which in turn leads to further deforestation. Forests are also cut as a result of growing urban sprawl. Not all deforestation is intentional. Some is caused by a combination of human and natural factors like wildfires and subsequent overgrazing, which may prevent the re-establishment of young trees.

The FAO estimates that about 13 million ha, an area roughly equivalent to the size of Greece, of the world's forests are cut down and converted to other land uses every year.

At the same time, planting of trees has resulted in forests being established or expanded on to abandoned agricultural land. This has reduced the net loss of total forest area. In the period 1990 to 2000 the world is estimated to have suffered a net loss of 8.9 million ha of forest each year, but in the period 2000 to 2005 this was reduced to an estimated 7.3 million ha/year. This means that the world lost about three per cent of its forests in the period 1990 to 2005; at present we are losing about 200 km² of forest each day. The world's rainforests could completely vanish in a hundred years at the current rate of deforestation.

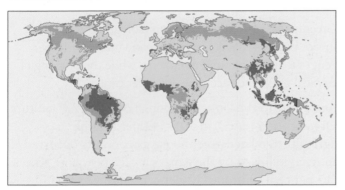

Key

■ >0.50% decrease per year   ■ >0.50% increase per year
■ Change rate between −0.50 and 0.50% per year

Figure 1.44 Countries with large net changes in forest area, 2000–2005

Source: FAO

Skills focus

You should be able to read and interpret atlas maps. Think about how you might describe the distribution of those areas that have a greater than 0.5 per cent decrease in forest area per year.

When forests are cleared for conversion to agriculture or pasture, a large proportion of the above-ground biomass may be burned, rapidly releasing most of its carbon into the atmosphere. Some of the wood may be used as wood products and so preserved for a longer time. Forest clearing also accelerates the decay of dead wood, litter and below-ground organic carbon.

Figure 1.45 shows how deforestation changes the carbon cycle but it is not the only way it affects climate. Forest soils are moist, but without the shade from tree cover they quickly dry out. Trees also help maintain the water cycle by returning water vapour back into the atmosphere through transpiration. Without trees to fill these roles, many former forestlands can quickly become barren deserts.

Urban growth

For the first time in human history, over half the world's population now lives in urban areas. As a proportion of global population, the urban population is expected to reach 60 per cent by 2030, with urban areas growing at a rate of 1.3 million people every week. As cities grow, the land use changes from either natural vegetation or

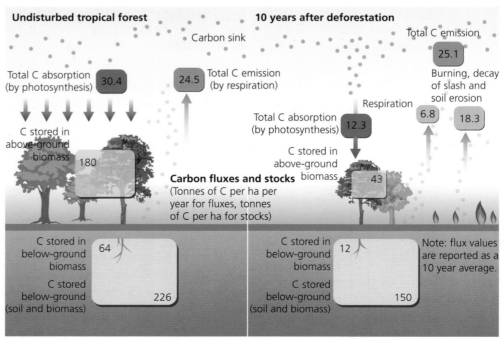

Figure 1.45 The effects of deforestation on the carbon cycle.

agriculture to one which is built up. The CO_2 emissions resulting from energy consumption for transport, industry and domestic use, added to the CO_2 emitted in the cement manufacture required for all the buildings and infrastructure, have increased. In 2012, cities (metropolitan areas above 0.5 million) were responsible for around 47 per cent of global carbon emissions. Under a business-as-usual scenario, this share of emissions is predicted to increase slightly to 49 per cent by 2030. Similar to population and economic output, the distribution of emissions is highly concentrated, with the 21 highest emitting cities contributing 10 per cent of global energy-related carbon emissions, 64 cities contributing 20 per cent and 139 cities contributing 30 per cent.

Average emissions per capita (tonnes of CO_2)

North America	15.6
Eastern Europe & Ex-Soviet states	9.9
Rest of Asia & Oceania	7.5
Western Europe	6.5
China	6.3
Middle East & North Africa	5.9
Latin America & Caribbean	3.3
South East Asia	3.3
India	2.8
Sub-Saharan Africa	1.8

Figure 1.46 Average carbon emissions per capita of cities above 0.5 million by region in 2012

An even greater concentration can be observed for carbon emissions growth from 2012 to 2030. In total, cities are projected to be responsible for 56 per cent of the global increase in carbon emissions during that period, with 10 cities contributing 10 per cent of global emissions growth, 28 cities contributing 20 per cent and 193 cities contributing 50 per cent.

Carbon sequestration

The process of **carbon sequestration** involves capturing CO_2 from the atmosphere and putting it into long-term storage. There are two primary types of carbon sequestration:

- **Geologic sequestration:** CO_2 is captured at its source (for example, power plants or industrial processes) and then injected in liquid form into stores underground. These could be depleted oil and gas reservoirs, thin, uneconomic coal seams, deep salt formations and the deep ocean. This is still at the experimental stage.

  The ocean is very capable of absorbing much more additional carbon than terrestrial systems simply because of its sheer size. An advantage of ocean carbon sequestration is that the carbon sequestered is quite literally 'sunk' within weeks or months of being captured from the air/water. Once in the deep ocean it is in a circulation system commonly measured in thousands of years. By the time this carbon reaches the seabed it has entered the Earth's geological cycle.

- **Terrestrial or biologic sequestration:** this involves the use of plants to capture CO_2 from the atmosphere and then to store it as carbon in the stems and roots of the plants as well as in the soil. The aim is to develop a set of land management practices that maximises the amount of carbon that

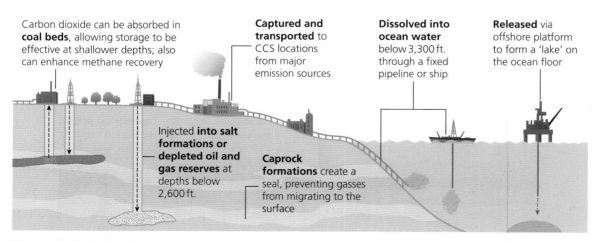

Carbon dioxide can be absorbed in **coal beds**, allowing storage to be effective at shallower depths; also can enhance methane recovery

Captured and transported to CCS locations from major emission sources

Dissolved into ocean water below 3,300 ft. through a fixed pipeline or ship

Released via offshore platform to form a 'lake' on the ocean floor

Injected **into salt formations or depleted oil and gas reserves** at depths below 2,600 ft.

Caprock formations create a seal, preventing gasses from migrating to the surface

Figure 1.47 Geologic carbon sequestration
Source: IPCC

remains stored in the soil and in plant material for the long term. Most authorities also believe that the enrichment of plant ecosystems is a positive environmental action with many associated benefits, including the enrichment of wildlife.

There are disadvantages to terrestrial sequestration. For instance, a forest planted to capture carbon might lose that carbon back to the air in a catastrophic forest fire or if the forest suffers disease or infestation. Land-based sequestration plantations are slow growing and require active monitoring and management for the lifetime of the plantation, usually many decades. The carbon within those systems is never removed permanently from the atmospheric system.

Effects of the changing carbon budget

It is clear that CO_2 levels have varied through time. It is certain that 100 million years ago carbon dioxide values were many times higher than now though the exact value is in doubt. 500 million years ago atmospheric CO_2 was some 20 times higher than present values. It dropped, then rose again 200 million years ago to a maximum of 4 to 5 times present levels. It then followed a slow decline until recent pre-industrial time.

The only way to calculate the effects of the changing levels of carbon is to build a computer model. A number of such models have been built over the years. They can have between 50 and 100 interacting equations describing all the different processes of the carbon cycle. The result of this is that models only predict possibilities, not probabilities.

Impact on the land

The impact of increasing atmospheric CO_2 on the land has been subject to intense research. Unfortunately the results are unclear because the study has, so far, been over a relatively short time period. This is coupled to the fact that there are so many other variables that could have an impact on the land and the atmosphere. These are both human and physical variables. Some of the impacts are summed up in Figure 1.49.

Note: Current research estimates that permafrost in the Northern Hemisphere holds 1,672 GtC. If just 10 per cent of this permafrost were to thaw, it could release enough extra carbon dioxide to the atmosphere to raise temperatures an additional 0.7 °C by 2100.

Impact on the oceans

Many of the observed physical and chemical changes in the ocean are consistent with increasing atmospheric CO_2 and a warming climate. Unfortunately, because of the complex nature of the

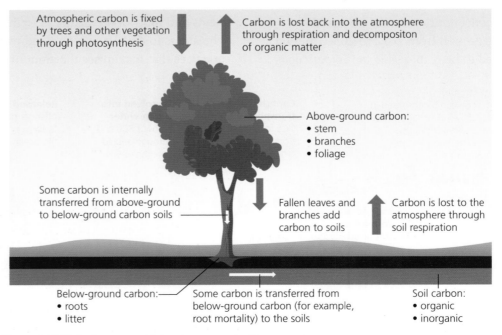

Figure 1.48 Biologic or terrestrial carbon sequestration

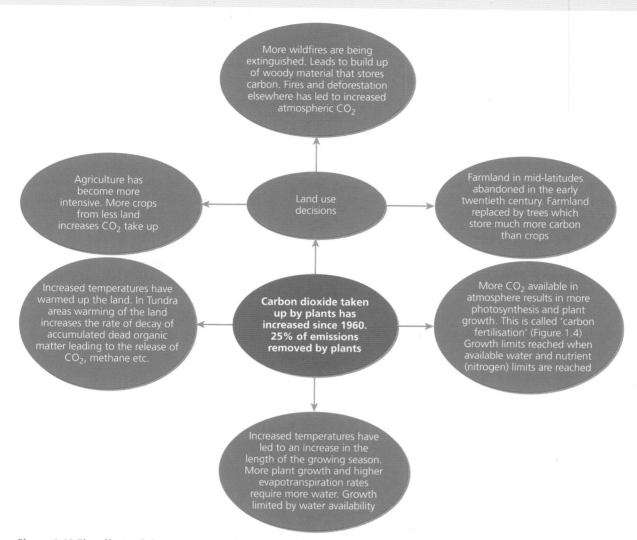

Figure 1.49 The effects of changes in the carbon cycle on the land.

chemistry and biochemistry of the oceans and its inhabitants, many of the causative links to climate change are still not well understood. It is difficult to predict the precise rate, magnitude and direction of change of CO_2 uptake and how that affects acidity, salinity, storminess and nutrient enrichment. It is even more difficult to map these effects at a local scale.

Ocean acidification

About 30 per cent of the CO_2 that has been released into the atmosphere has diffused into the ocean through direct chemical exchange. Dissolving carbon dioxide in the ocean creates carbonic acid. This makes the slightly alkaline ocean become a little less alkaline. Since 1750, the pH of the ocean's surface has dropped by 0.1, a 30 per cent change in acidity.

The impact of ocean acidification on ecosystems is largely unknown, with the exception of coral reefs. It affects marine organisms in two ways. Firstly, carbonic acid reacts with carbonate ions in the water to form bicarbonate. However, those same carbonate ions are what animals like coral and many planktonic species need to create their calcium carbonate shells. With less carbonate available, the animals need to expend more energy to build these shells. As a result, the shells end up being thinner and more fragile.

Coral reefs provide food and livelihood security for some 500 million people worldwide. Significant reef loss and the consequent fall in marine biodiversity threatens the survival of coastal communities through reduced food availability and a reduced capacity of coastlines to buffer the impact of sea level rise, including increased storm surges.

Polar and sub-polar marine ecosystems are projected to become so low in carbonate ions within this century that waters may actually become corrosive to unprotected shells and skeletons of organisms currently living there.

In 2007, farmed oyster larvae off the coast of Oregon and Washington in the USA began dying by the millions. It was found that these losses were directly linked to ocean acidification. The rise in acidity and subsequent oyster crash took a significant toll on coastal communities – from 2005 to 2009 lost production cost millions of dollars in lost sales. Further research showed evidence of acidic seawater rising up from the ocean depths and that the water rising from the deep ocean today holds CO_2 absorbed approximately 30 to 50 years ago. Benoit Eudeline, chief hatchery scientist for Taylor Shellfish Farms, the largest US producer of farmed shellfish, likened the current situation to 'sitting on a ticking time bomb'.

A more optimistic viewpoint is that the more acidic seawater is, the better it dissolves calcium carbonate rocks (chalk and limestone). Over time this reaction will allow the ocean to soak up excess CO_2 because the more acidic water will dissolve more rock, release more carbonate ions and increase the ocean's capacity to absorb CO_2.

Ocean warming

Warmer oceans, a product of the climate change, could decrease the abundance of phytoplankton, which grow better in cool, nutrient-rich waters. This could limit the ocean's ability to take carbon from the atmosphere through the biological carbon pump and lessen the effectiveness of the oceans as a carbon sink. On the other hand, CO_2 is essential for plant and phytoplankton growth. An increase in CO_2 could increase their growth by fertilising those few species of phytoplankton and ocean plants (like sea grasses) that take carbon dioxide directly from the water.

Ocean warming also kills off the symbiotic algae which coral needs in order to grow, leading to bleaching and eventual death of reefs.

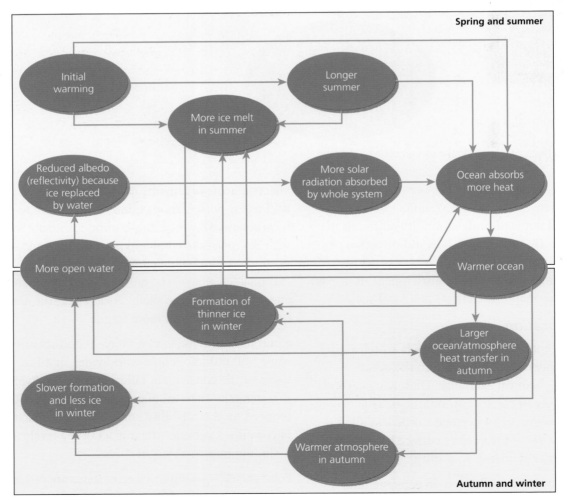

Figure 1.50 A systematic view of the oceanic carbon cycle showing the positive feedback effects of oceanic warming

Melting sea ice

In the last 35 years satellites monitoring sea ice in the Arctic have measured its retreat at 40 per cent. When sea ice melts, it is not just an indicator of a warming climate but also part of a feedback mechanism because the highly reflective ice is replaced by more heat absorbent water. When it starts to melt the ocean is able to absorb more sunlight, which in turn amplifies the warming that caused it to melt in the first place.

Sea ice also provides a unique habitat for algae that appear in more concentrated forms and with more fat content in the ice. The loss of those ice-bound algae affects marine predators all the way up the food chain, from krill and fish to seals, walruses and polar bears. Added to this is the fact that animals like the polar bear that rely on sea ice to get to their main food source of seals can no longer travel upon it.

Figure 1.51 A polar bear standing on an ice floe

Ocean salinity

There has been an observed decrease in salinity in the deep North Atlantic. This is probably caused by higher levels of precipitation and higher temperatures. The precipitation leads to higher river run-offs that eventually reach the sea. The higher temperatures are causing melting of the Greenland ice sheet and many alpine glaciers. This too will lead to an increase in fresh water reaching the oceans. These changes have been linked to a possible slowing down of the large-scale oceanic circulation in the North-East Atlantic. This in turn will have an effect on the climate of North West Europe.

Key question

What effects might the disruption of the North Atlantic Drift have on the climate of NW Europe?

Sea level rise

Although global sea levels have been more or less constant for the last 5,000 years they are subject to change. Studies of coastal landforms show that they have been much lower in the past than they are today. The last glacial retreat led to a worldwide rise in sea levels about 10,000 years ago. This increase was caused by the melting of land-locked freshwater ice.

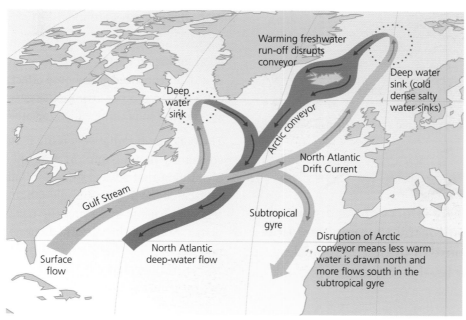

Figure 1.52 The effect of changing ocean salinity on the Atlantic Ocean currents

Research now indicates that sea levels worldwide have been rising at a rate of 3.5 mm/year since the early 1990s. Causes of this change include:

- **Melting of terrestrial ice:** recently, persistently higher temperatures have led to an increased rate of summer melting as well as a drop in snowfall in the shorter winters. This imbalance results in a significant net gain in water entering the oceans from rivers against evaporation from the ocean. Added to this is the fact that the massive ice sheets of Antarctica and Greenland are moving more quickly towards the oceans due to the increased amount of meltwater lubricating their bases.

- **Thermal expansion:** when water heats up, it expands. About half of the past century's rise in sea level is thought to be attributable to warmer oceans having a greater volume and so occupying more space. Accurate measurements of this phenomenon have not yet been possible.

If the Earth continues to warm up then we can expect the oceans to rise between 0.8 and 2 m by 2100. This is not an exact science and Figure 1.53 shows a range of predictions for rising sea levels.

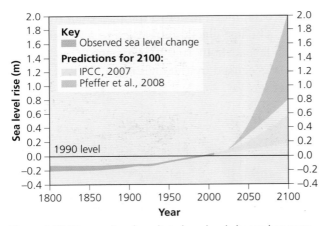

Figure 1.53 Observed and predicted sea level change between 1800 and 2100

Key questions

What are the likely effects of the predicted sea level rises on coastal communities? How might the response of those communities depend on their level of development?

Impact on the atmosphere

Key terms

Enhanced greenhouse effect – The impact on the climate from the additional heat retained due to the increased amounts of carbon dioxide and other greenhouse gases that humans have released into the Earth's atmosphere since the industrial revolution.

Geo-sequestration – The technology of capturing greenhouse gas emissions from power stations and pumping them into underground reservoirs.

Radiative forcing – The difference between the incoming solar energy absorbed by the Earth and energy radiated back to space.

Soil organic carbon (SOC) – The organic constituents in the soil: tissues from dead plants and animals, products produced as these decompose and the soil microbial biomass.

Although the land and oceans will take up most of the extra CO_2, as much as 20 per cent may remain in the atmosphere for many thousands of years. This is significant because CO_2 is the most important gas for controlling the Earth's temperature. Carbon dioxide, methane and halocarbons are **greenhouse gases** that absorb a wide range of energy – including infrared energy (heat) emitted by the Earth – and then re-emit it. The re-emitted energy travels out in all directions but some returns to Earth where it heats the surface. Without greenhouse gases, the Earth would be a frozen −180°C. With too many greenhouse gases, the Earth would be like Venus, where the greenhouse atmosphere keeps temperatures around 400°C.

Clearly, the fact that there is a greenhouse effect at all is good for life on Earth. The problem that is facing us is that of an **enhanced greenhouse effect**. This is where the extra CO_2 and other greenhouse gases in the atmosphere are causing something called **radiative forcing**.

The concept of radiative forcing is that energy is constantly flowing into the atmosphere in the form of sunlight that always shines on half of the Earth's surface. Some of this sunlight (about 30 per cent) is reflected back to space and the rest is absorbed by the planet. Some of this absorbed energy is radiated back into the much colder surrounding space as infrared energy. If the balance between the incoming and the outgoing energy is anything other than zero there

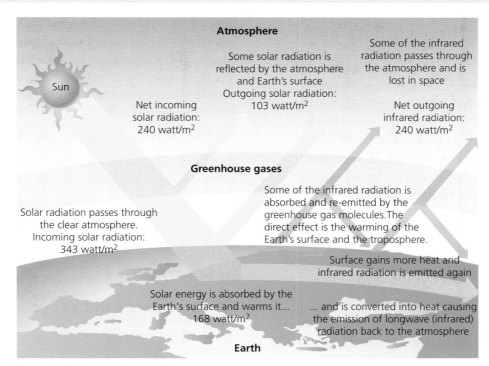

Figure 1.54 The greenhouse effect

has to be some warming (or cooling, if the number is negative) going on. The amount that the Earth's energy budget is out of balance is called the radiative forcing and is a measure of recent human activities. It is measured in watts/m² of the Earth's surface.

Studies have shown that prior to 1750 radiative forcing was negligible. Since then it has increased, not only because of increased greenhouse gas emissions but also changing albedos because of land use changes. Measurement of the actual amount of radiative forcing is difficult because of many complicating factors including natural changes in solar radiation and the effects of aerosols such as carbon particles from diesel exhausts (which lead to warming).

The current level of radiative forcing, according to the IPCC AR4, is 1.6 watts/m² (with a range of uncertainty from 0.6 to 2.4). This amounts to a total of about 800 terawatts – more than 50 times the world's average rate of energy consumption, which is currently about 15 terawatts.

If CO_2 levels continue to rise at projected rates, experts predict that the Earth will become much hotter, possibly hot enough to melt much of the existing ice cover. Figure 1.55 depicts projected surface temperature changes through 2060 as estimated by NASA's Global Climate Model.

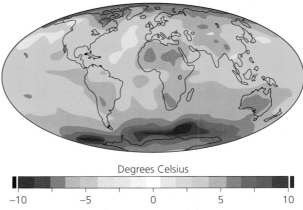

Figure 1.55 Predicted increases in surface air temperature 1960 to 2060 according to NASA's Global Climate Model

Key question

How might the way people live in various parts of the world contribute to rising levels of CO_2?

1.4 Water, carbon, climate and life on Earth

Carbon forms the key component for all known life on Earth. It bonds with other elements such as oxygen, hydrogen and nitrogen to form complex molecules. Water too makes life as we know it possible. Every drop cycles continuously through air, land and sea to be used by someone (or something) else in the cycle. Although there is a lot of water on Earth, only a fraction of one per cent supports all life on land. Climate change and growing populations are increasing the pressures on that reserve.

Water and carbon cycles and the atmosphere

As stated above, the increased emission of CO_2 is warming the atmosphere. This increased temperature results in higher evaporation rates and a wetter atmosphere, which leads to a positive feedback situation of further warming. Carbon dioxide causes about 20 per cent of the Earth's greenhouse effect, water vapour accounts for about 50 per cent and clouds 25 per cent. The rest is caused by small particles (aerosols) and minor greenhouse gases like methane.

When carbon dioxide concentrations rise, air temperatures go up. The oceans warm up and more water vapour evaporates into the atmosphere, which then amplifies greenhouse heating. Although CO_2 contributes less to the overall greenhouse effect than water vapour, scientists have found that it is CO_2 that sets the temperature. It controls the amount of water vapour in the atmosphere and therefore the size of the enhanced greenhouse effect.

This is summarised in Figure 1.56.

This rise in temperature is not all the warming that we will see based on current CO_2 concentrations. There is a time lag between the increase in CO_2 and increased warming because the ocean soaks up heat. This means that the Earth's temperature will increase at least another 0.6 °C because of carbon dioxide already in the atmosphere.

Climate change mitigation

This refers to efforts to reduce or prevent emission of greenhouse gases. The various means of mitigation are summarised in Figure 1.57.

Carbon capture and sequestration (CCS) technologies

Carbon capture and storage (CCS) is a technology that can capture up to 90 per cent of CO_2 emissions produced from the use of fossil fuels in electricity generation and industrial processes, preventing the carbon dioxide from entering the atmosphere.

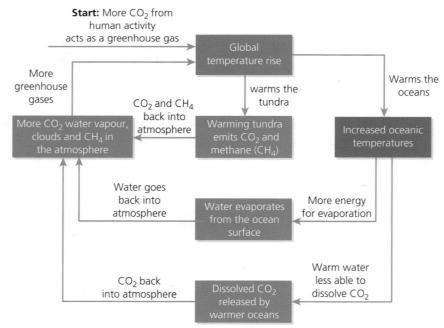

Figure 1.56 A system diagram for the impact of water and carbon on climate change showing positive feedback

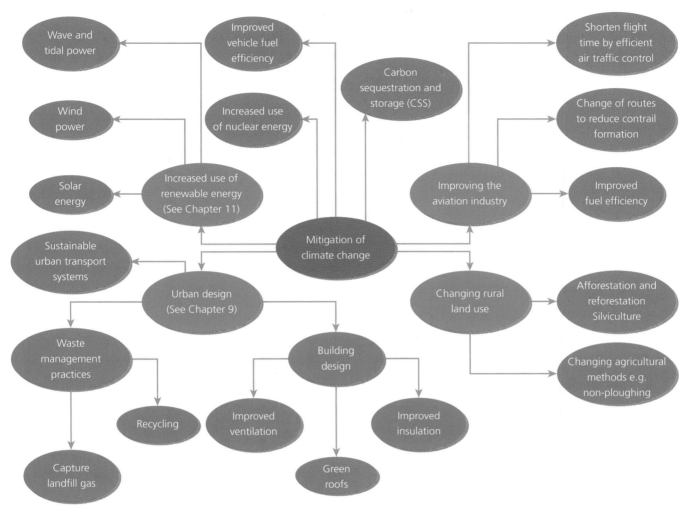

Figure 1.57 Management of climate change

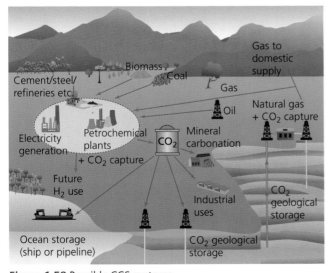

Figure 1.58 Possible CCS systems

The CCS chain consists of three parts (see Figure 1.58):

● **Capturing** the CO_2: Capture technologies allow the separation of CO_2 from gases produced in electricity generation and industrial processes by one of three methods: pre-combustion capture, post-combustion capture and oxy-fuel combustion.

● **Transporting** the CO_2 by pipeline or by ship to the storage location: Millions of tonnes of CO_2 are already transported annually for commercial purposes by road tanker, ship and pipelines.

● Securely **storing** the carbon dioxide emissions underground in depleted oil and gas fields, deep saline aquifer formations several kilometres below the surface or the deep ocean.

CCS systems could be used to extract a greater percentage of oil and gas out of existing reservoirs by the CO_2 being injected under such pressure as to force the oil or gas out. Although this would partly pay for the CCS technology, it would also enhance the original problem by producing more fossil fuel for burning. When CO_2 is stored in deep geological formations it is known as **geo-sequestration** (see Figure 1.59, page 42).

CO_2 is converted into a high pressure liquid-like form known as 'supercritical CO_2' which behaves like a runny liquid. This supercritical CO_2 is injected directly into sedimentary rocks. The rocks may be in old oil fields, gas fields, saline formations or thin coal seams. Various physical (for example, impermeable 'cap rock') and geochemical trapping mechanisms prevent the CO_2 from escaping to the surface.

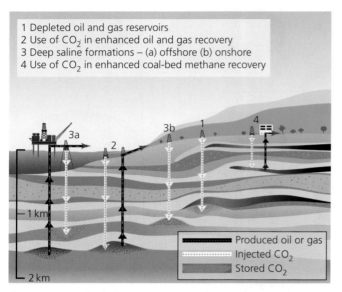

Figure 1.59 Methods for storing CO_2 in deep underground geological formations

Source: IPCC

Captured CO_2 could also be stored in the ocean by a variety of means, as shown in Figure 1.60. This method has the main disadvantage of the CO_2 causing acidification of the oceans and all the problems that arise from that.

An example of CCS in action is at the 110-megawatt coal power and CCS plant in Saskatchewan, called Boundary Dam, built by the provincial utility SaskPower. It is a coal-fired power station complex that has been retrofitted to capture 90 per cent of its CO_2 output (approximately 1 million tonnes per year).

The CO_2 will eventually be piped 66 km to the Weyburn Oil Unit and injected into an oil-bearing formation at 1,500 m depth. This will add pressure to the oil-bearing rock and so help push more oil out of the ground, a process called enhanced oil recovery (EOR). Until that is ready it will be injected into local salt formations. The capture process was started in October 2014 and CO_2 injection started in April 2015.

CCS imposes big costs and energy penalties: the Boundary Dam plant's CCS unit cost $800 million to build and consumes 21 per cent of the coal plant's power output in order to scrub out the carbon dioxide and compress it into a liquid for burial. It is hoped that this extra cost will be offset by the extra oil recovered from the Weyburn oil field.

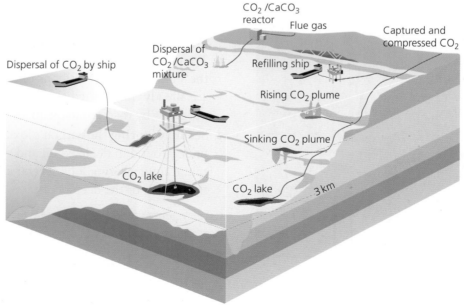

Figure 1.60 Methods of ocean storage

Source: IPCC

Changing rural land use

Carbon stores can be improved by ensuring that carbon inputs to the soil are greater than carbon losses from it. There are a variety of different strategies depending on land use, soil properties, climate and land area.

- **Grasslands:** These offer a global greenhouse gas mitigation potential of 810 million tonnes of CO_2 (in the period up to 2030), almost all of which would be sequestered in the soil. Soil carbon storage in grasslands can be improved by:
  - avoidance of **overstocking** of grazing animals
  - **adding manures and fertilisers** that have a direct impact on **soil organic carbon (SOC)** levels through the added organic material. There are also the indirect benefits of increasing plant productivity and stimulating soil biodiversity (for example with earthworms that help degrade and mix the organic material)
  - **revegetation**, especially using improved pasture species and legumes, can increase productivity, resulting in more plant litter and underground biomass, which can add to the SOC stock
  - **irrigation** and water management can improve plant productivity and the production of SOM.
- **Croplands:** Techniques for increasing SOC include the following:
  - **Mulching** can add organic matter. If crop residues are used, mulching also prevents carbon losses from the system.
  - **Reduced or no tillage** (ploughing and harrowing) avoids the accelerated decomposition of organic matter and depletion of soil carbon that can otherwise occur. Reduced tillage also prevents the break-up of soil aggregates that protect carbon.
  - Some **use of animal manure** or chemical fertilizers can increase plant productivity and thus SOC.
  - **Rotations of cash crops** with pasture or the use of cover crops and green manures have the potential to increase biomass returned to the soil.
  - Using **improved crop varieties** can increase productivity above and below ground, as well as increasing crop residues, thereby enhancing SOC.
- **Forested lands and tree crops:** Forests are able to reduce CO_2 emissions to the atmosphere by storing large stocks of carbon both above and below ground.
  - **Protection** of existing forests will preserve current soil carbon stocks.
  - **Reforesting** degraded lands and increasing tree density in degraded forests increase biomass density and therefore carbon density, above and below ground.
  - **Trees in croplands** (silviculture) and orchards can store carbon above and below ground. CO_2 emissions can be reduced if they are grown as a renewable source of fuel.

This is summarised in Figure 1.61

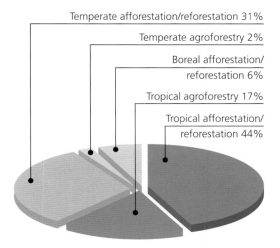

Temperate afforestation/reforestation 31%

Temperate agroforestry 2%

Boreal afforestation/reforestation 6%

Tropical agroforestry 17%

Tropical afforestation/reforestation 44%

Total carbon sequestration potential 38 Gt

Figure 1.61 Potential contribution of afforestation/reforestation and agroforestry activities to global carbon sequestration, 1995–2050
Source: IPCC

It is important to note that many of these mitigation schemes have different and unwanted side effects.

Improved aviation practices

In 2013, the global aviation industry carried 3 billion passengers, producing 705 million tonnes of CO_2. Although the industry has made major strides in reducing its production of CO_2 (for example, the Airbus A380 and the Boeing 787 both use less than three litres of fuel per 100 passenger km), the EU Directorate General for Climate Action predicts that by 2020 the global emissions of CO_2 will be 70 per cent more than in 2005 and could be 300 to 700 per cent more by 2050.

The ways that they could reduce their emissions are shown in Figure 1.62 (page 44). These must be treated with caution because many of them are still at the aspirational or theoretical stage.

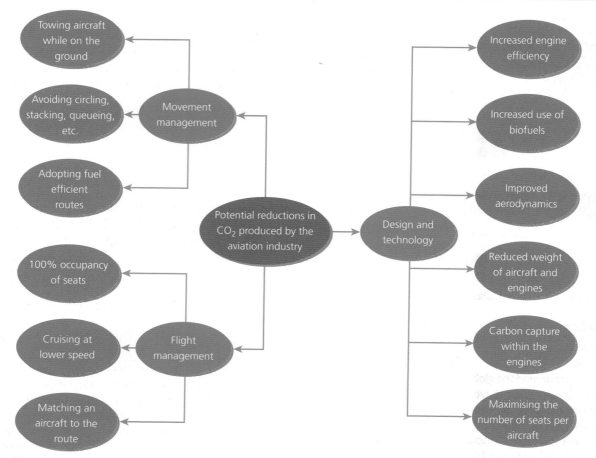

Figure 1.62 CO_2 mitigation within the aviation industry

Case study of a tropical rainforest setting: Water and carbon in the Amazon

The Amazon Basin is the world's largest rainforest and one of the most biodiverse. Its 300 billion trees and 15,000 species store one-fifth of all the carbon in the planet's biomass. Tropical forests have been present in South America for millions of years and were at one point spread over most of the continent. In the past the forest has shrunk back and then advanced again as ice ages came and went. Today's Amazon rainforest covers around 5.5 million km² and is spread across nine countries.

Figure 1.63 The location of the Amazon rainforest

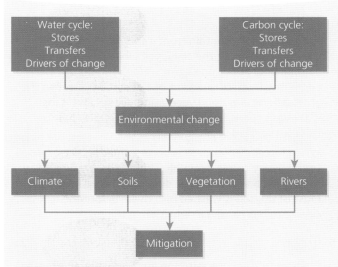

Figure 1.64 Key themes shown in this study

Carbon

The Amazon is estimated to store between 80 and 120 billion tons of carbon. Along with other tropical rainforests around the globe it forms a carbon sink of 1-3 GtC/year. Studies have shown that these forests have been increasing in above-ground biomass by 0.3–0.5 per cent per year and that the rising productivity of tropical forests is due to sequestering of increased CO_2 concentrations in the atmosphere (see Figure 1.4). This negative feedback seems to have offset rising atmospheric levels of CO_2.

Despite this, a study completed in 2015 has revealed that the Amazon forest is losing its capacity to absorb CO_2 from the atmosphere. From a peak of two billion tonnes of CO_2 each year in the 1990s, the net uptake by the forest has halved and now, for the first time, has been overtaken by fossil fuel emissions in Latin America.

It appears that for the Amazon basin, an increase in atmospheric CO_2 led to a growth spurt for the Amazon's trees; in time however, the growth stimulation feeds through the system, causing trees to live faster, and so die younger. This has led to a surge in the rate of trees dying across the Amazon.

Water

The average discharge of water into the Atlantic Ocean by the Amazon is approximately 175,000 m³/s, or around 15 per cent of the fresh water entering the oceans each day.

The Rio Negro, a tributary of the Amazon, is the second largest river in the world in terms of water flow, and is 100 m deep and 14 km wide near its mouth at Manaus, Brazil.

Average rainfall across the whole Amazon basin is approximately 2,300 mm annually. In some areas of the northwest portion of the Amazon basin, it can exceed 6,000 mm. Up to half of this rainfall may never reach the ground. It is intercepted by the forest canopy and re-evaporated into the atmosphere. Additional water evaporates from the ground and rivers or is released into the atmosphere by transpiration from plant leaves. Of the rainfall that is evapotranspired back into the atmosphere, about 48 per cent falls again as rain. Only about 30 per cent of the rainfall actually reaches the sea. The rest is caught up in this constant closed system loop.

Drivers of change

- Between 2000 and 2007, the Brazilian Amazon was deforested at a rate of 19,368 km² per year. During this time, an area of forest larger than Greece was destroyed.
- Brazil is the world's fourth largest climate polluter, with 75 per cent of their greenhouse gas emissions attributed to deforestation and land use change; 59 per cent of this is from loss of forest and burning in the Amazon.
- The removal of forest was done using slash and burn techniques which:
  - reduces the retention of humidity in the soil's top layer down to a depth of one metre
  - facilitates sudden evaporation of water previously retained in the forest canopy
  - increases albedo (reflectiveness) and temperature
  - reduces porosity of soil, causing faster rainfall drainage, erosion and silting of rivers and lakes.
- Any moisture that evaporates from deforested areas forms shallow cumulous clouds which usually do not produce rain.
- Forests emit salts and organic fibres along with water when they transpire. These act as condensation nuclei and assist in cloud and rain formation. Their loss inhibits the formation of cloud and reduces rainfall.
- If destroyed, the vast carbon store will be released into the atmosphere.
- There are a range of differences between tropical rainforest and the pasture land it is generally replaced with.
  - Forests absorb approximately 11 per cent more solar radiation.
  - The average temperature in the rainforest is approximately 24.1°C; in pastures it is 33°C.
  - The daily temperature variation of Amazon forest soils at 20 cm did not exceed 2.8°C, though under pastures it was 8°C.
  - The moisture content in the upper one metre of pasture soil is about 15 per cent less than under nearby forest.
  - Deeper forest roots can pump more soil moisture to the surface, producing 20–30 per cent more air humidity and consequently 5–20 per cent more precipitation than pastures.

Climate change

Studies investigating all tropical rainforest regions found a mean temperature increase of 0.26°C ± 0.05° every ten years since the mid-1970s and predict that by the year 2050, temperatures in the Amazon will increase by 2–3°C. There has also been evidence of more frequent and increased extremes in temperature.

Amazonia experienced falling amounts of rainfall between the 1920s and the 1970s but since then it appears there has been no significant change.

Vegetation change

There has been a massive net loss of forests in the Amazon Basin, about 3.6 million hectares per year between 2000 and 2010. Although most of this has been caused by deliberate deforestation, a significant amount has been as a result of climate change. Some species are limited by their tolerance to temperature change, drought and seasonality. Climate change can affect species sustainability by directly altering the conditions needed to grow and survive. Droughts and unusually high temperatures in the Amazon in recent years may also be playing a role in killing millions of trees although the tree mortality increases began well before an intense drought in 2005.

A 2009 study concluded that a 2°C temperature rise above pre-industrial levels would see 20–40 per cent of the Amazon die off within 100 years. A 3°C rise would see 75 per cent of the forest destroyed by drought over the following century, while a 4°C rise would kill 85 per cent.

Soil

Amazonian soils contain from 4 to 9 kg of carbon in the upper 50 cm of the soil layer, while pasturelands contain only about one kg/m². When forests are cleared and burned, 30–60 per cent of the carbon is lost to the atmosphere; unburned vegetation decays and is lost within ten years. The soil fungi and bacteria that used to recycle the dead vegetation die off.

When forest clearance first occurs, the soils are exposed to the heavy tropical rainfall. This rapidly washes away the topsoil and attacks the deep weathered layer below. Most of the soil is washed into rivers before the forest clearance has caused a reduction in the rainfall.

Rivers

Changes in total precipitation, extreme rainfall events and seasonality may:

- lead to an overall reduction in river discharge
- cause an increase in silt washed into the rivers, which could disrupt river transport routes
- lead to flash flooding
- destroy freshwater ecosystems; this could remove a source of protein and income to local inhabitants
- destroy water supply which fulfils the needs of Amazonian peoples.

Warming water temperatures may:

- kill off temperature dependent species
- change the biodiversity of the river system by introducing new species and killing others
- reduce water-dissolved oxygen concentrations, which could destroy eggs and larvae, which rely on dissolved oxygen for survival.

Mitigation

Figure 1.65 describes some of the strategies to reduce the effects of environmental change in Amazonia.

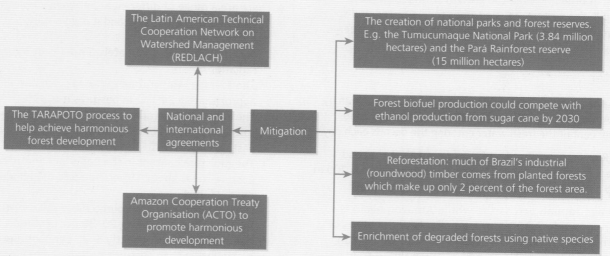

Figure 1.65 Strategies to reduce the effects of environmental change in Amazonia

Case study of a river catchment at a local scale: River Brock, Lancashire

The River Brock drains an entirely rural catchment. It is 17.8 km long, rising in the hills of the Forest of Bowland. The headwaters drain steep millstone grit hills in the north and carboniferous limestone in the south. There is peat on the high moors while the lower catchment is overlain by boulder clay. The flow is almost completely natural. Myerscough Agricultural College have a licence to abstract a maximum of 45.46 m³ daily with an annual maximum of 16,592.9 m³. At present this is being withheld.

The regime of the River Brock

This depends upon the pattern of rainfall, which can be highly variable. Table 1.1 indicates that rainfall for the Brock catchment was relatively low. Figure 1.67 shows a lower than average flow all year with a late winter maximum and low flow in the summer. 2012 was a particularly wet year in this part of the country as shown by Table 1.2. See figures on page 48.

Key questions

The source area of the River Brock is on west facing slopes. How will this influence the amount of rainfall (input) that enters the drainage basin?

The source area is peat moorland and rough grazing. How might this affect the transfer of water following precipitation?

How might the reintroduction of the Myerscough Agricultural College licence to abstract water affect:
- the discharge of the river
- the ability of the river to support a wide biodiversity?

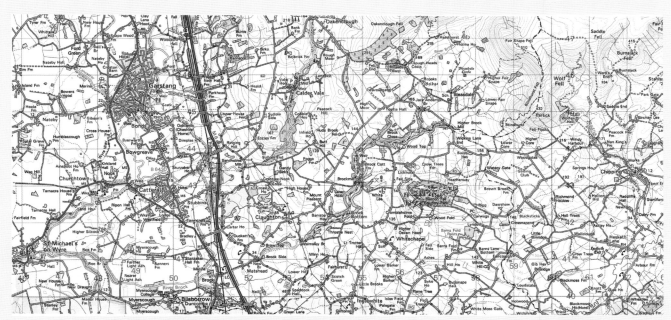

Figure 1.66 OS map extract showing the location of the River Brock (Map is reproduced at half of OS Landranger map scale of 1:50,000. Each grid square is 1 km²)

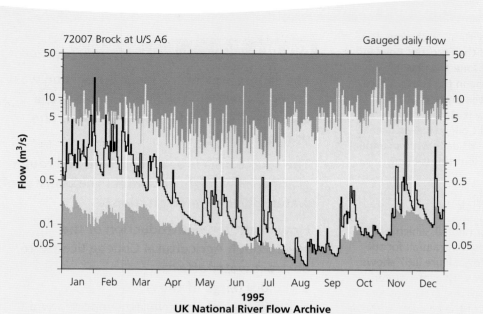

Figure 1.67 Flow data of the River Brock at Brock, 1995
Source: Centre for Ecology & Hydrology

Table 1.1 Monthly rainfall totals for Chipping, Lancashire, 1995

Month	Rainfall
Jan	45
Feb	107
Mar	50
Apr	97
May	77
Jun	63
July	92
Aug	110
Sep	67
Oct	147
Nov	154
Dec	68
Total	1077

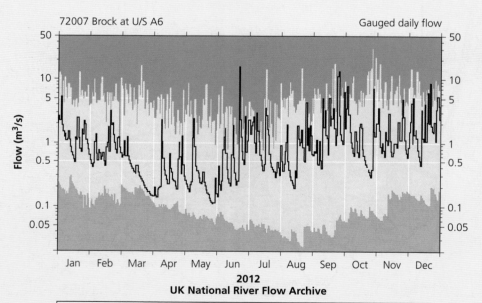

Figure 1.68 Flow data of the River Brock at Brock, 2012
Source: Centre for Ecology & Hydrology

Table 1.2 Monthly rainfall totals for Chipping, Lancashire, 2012

Month	Rainfall
Jan	188
Feb	124
Mar	31
Apr	98
May	79
Jun	213
July	166
Aug	159
Sep	313
Oct	195
Nov	190
Dec	225
Total	1981

Skills focus

You should be able to interpret the logarithmic scale used for discharge.

What statistical technique could you use to investigate the relationship between monthly rainfall and maximum (or minimum) discharge. Why use that technique?

Key questions

How does the annual flow for 2012 vary from that of 1995?

Account for the differences in the two patterns of flow.

How might the discharge of the river be affected by the following potential climatic changes?
- An increase in summer temperatures with a reduction in precipitation.
- An increase in the intensity of rainfall events in winter.

Downstream changes in the discharge

The River Brock is short enough to be studied almost from source to mouth. Measurements of discharge variations downstream could be compared to other drainage basin characteristics.

Flooding

The most serious flooding to affect the village of St Michael's on Wyre was in October 1980 when there was a combination of intense and prolonged rainfall. As a result comprehensive flood protection was put in place:
- In the area of St Michael's to the north of the River Wyre the Environment Agency constructed a flood storage basin at Catterall for a one-in-fifty-years flood. It has a capacity to store 1.7 million m³ of floodwater.
- Embankments were raised and strengthened in the lowland area close to the River Wyre.

- Warnings are issued for one-in-five-years floods and sandbags given out by the local council, which maintains a store of 2,000.
- In the area of St Michael's to the south of the River Wyre there is a 1 in 100 years probability of flooding with warnings and sandbag distribution.

The river Brock contributes to the flooding of the River Wyre.

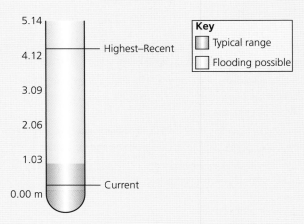

Figure 1.69 The discharge profile of the River Brock

Looking at Figure 1.69 we can see that the river level at or near the confluence with the River Wyre at 04.30 on the 26/05/2015 was 0.29 m. The typical river level range for this location is between 0.12 m and 0.93 m. It has reached a highest recorded level of 4.29 m on the 21/01/2008.

The steep sided nature of the Brock Valley means that flooding is not a problem until the river approaches the confluence with the River Wyre. Here the land is much flatter, the result of it being a former glacial outwash plain. Although the river is embanked before reaching St Michael's on Wyre, the agricultural nature of the land use makes it low priority for the Environment Agency.

Key question

The greatest discharge does not coincide with the worst floods. Why might this be the case?

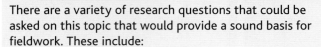

Review questions

1 How can a systems approach to the study of the water and carbon cycles explain how they influence climate?

2 What are the differences between changes to the carbon cycle bought about by natural variation and those brought about by human activity?

3 Summarise the attempts to mitigate the impacts of climate change. How successful have these attempts been?

4 For a drainage basin you have studied:
   ● what are the main hydrological characteristics of that drainage basin?
   ● how do those characteristics affect the shape of the storm hydrograph?

5 What aspects of your river catchment study could be used by other people to help them manage sustainable water supply and/or flooding?

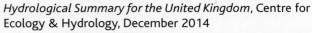

Fieldwork opportunities

There are a variety of research questions that could be asked on this topic that would provide a sound basis for fieldwork. These include:

● How do infiltration rate and infiltration capacity vary with slope, steepness, soil texture, geology and/or land use?

● How do rainfall and evaporation affect the level of water, velocity and discharge of a river?

● How do different sections of a river respond to a period of rain?

● How does the amount of interception vary with vegetation type and rainfall intensity?

● How do different land uses affect the amount of carbon stored in the vegetation and soils?

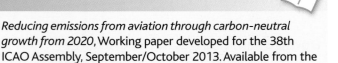

Further reading

Hydrological Summary for the United Kingdom, Centre for Ecology & Hydrology, December 2014

A typical snapshot of UK hydrology, this is one of a series of reports that could be used for research into changes in the water balance.

Management of the London Basin Chalk Aquifer: Status Report 2014, The Environment Agency.

This report aims to describe the current condition of groundwater levels and abstraction within the chalk and understand any changes seen in recent years. The report is designed to help decision makers manage groundwater levels to protect infrastructure and water resources.

Nina Morgan, *Carbon Capture and Storage: Analysing uncertainty,* Energy briefing paper, Communications Team, UK Energy Research Centre, UKERC Publication reference UKERC/BP/CCS/2012/001

This is a general overview of the state of the carbon capture industry in the UK. It looks at the economic argument for carbon capture as well as its possible future in the UK.

'The benefits of soil carbon', *The UN Environment Programme's Year Book 2012*, United Nations Environment Programme

This looks at the role played by soil carbon in regulating climate, water supplies and biodiversity, and maintaining the ecosystem services that we depend on.

Reducing emissions from aviation through carbon-neutral growth from 2020, Working paper developed for the 38th ICAO Assembly, September/October 2013. Available from the Air Transport Action group, www.atag.org.

A summary of the alternative approaches that the aviation industry could take to mitigate the effects of aviation on climate change.

Global Environment Outlook GEO5: Environment for the Future We Want, United Nations Environment Programme, first published in 2012, ISBN: 978-92-807-3177-4. Available online.

GEO-5 is designed to be the most comprehensive, impartial and in-depth assessment of its kind. It reflects the collective body of recent scientific knowledge, drawing on the work of leading experts, partner institutions and the vast body of research undertaken within and beyond the United Nations system.

Jeff Tollerson, 'Climate change: The case of the missing heat', *Nature*, Volume 505, 16 January 2014, pp. 276–8.

An article that shows that climate change is not constant. It considers how various factors can affect the steady growth of global temperatures.

Environmental consequences of ocean acidification: A threat to food security, Division of Early Warning Assessment, United Nations Environment Programme 2010

A detailed look at the causes and consequences of ocean acidification.

Question practice

A-level questions

1. Explain the concept of feedback in relation to the carbon cycle. (4 marks)

2. Figure 1.70 shows the changing amounts of carbon dioxide production by source. Using the figure, analyse the changes shown. (6 marks)

3. Study Figure 1.55 (page 39), which shows the predicted increases in surface air temperature between 1960 and 2060. Using the figure and your own knowledge, assess the impact of these temperature changes on life on Earth. (6 marks)

4. With reference to an area of rainforest you have studied, assess the extent to which the water and carbon cycles in that area may be related to attempts at global governance. (20 marks)

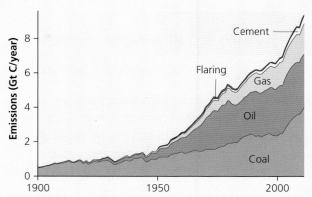

Figure 1.70 World carbon dioxide production by source

Source: US Department of Energy

AS level questions

1. The largest store of global water is:
   a) the atmosphere
   b) the biosphere
   c) the hydrosphere
   d) the lithosphere (1 mark)

2. Which of the following is a water store in a drainage basin?
   a) Evapo-transpiration
   b) Groundwater
   c) Infiltration
   d) Stemflow (1 mark)

3. Outline the potential impacts of water abstraction on the water cycle. (3 marks)

4. Figure 1.71 shows the projected atmospheric greenhouse gas concentrations as predicted by the United States Environmental Protection Agency. The **highest emissions** pathway is incomplete. Using Table 1.3 complete Figure 1.71 and analyse how the varying predictions may affect global climates. (6 marks)

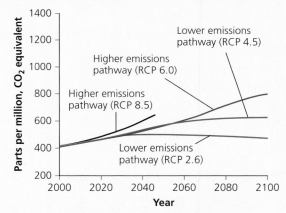

Figure 1.71 Projected atmospheric gas emissions

Source: US Environmental Protection Agency

5. Evaluate the view that changes in the global carbon cycle are a threat to Antarctica. (9 marks)

6. With reference to a river catchment you have studied, to what extent do different land surfaces within that drainage basin affect both the infiltration rates and the channel flow? (20 marks)

Table 1.3 Projected highest emissions pathway

Year	Projected atmospheric greenhouse gas emissions (ppm CO$_2$ equivalent)
2060	780
2080	1,030
2100	1,310

Chapter

2

Hot desert systems and landscapes

Drylands are found at all latitudes around the globe, but most hot desert environments are arid regions of the world located at mid and low latitudes on or around the tropics. These areas are arid, receiving little or no rainfall, with sparse or no vegetation. Due to their extreme climate hot desert environments are sparsely populated.

In this chapter you will study:
- the global distribution of mid- and low-latitude deserts and their margins
- the climate, soil and vegetation characteristics of hot deserts and their margins
- the processes leading to aridity in hot deserts and their margins
- the systems and processes leading to the development of hot desert landscapes
- desertification – its causes, consequences and alternative futures for local populations.

2.1 Introduction to hot desert environments and their margins

Deserts as natural systems

The introduction to Chapter 1: Water and carbon cycles outlined how many aspects of physical geography can be studied using a **systems** approach and you should read that section carefully before reading further here. This chapter explores the landscapes of hot desert environments and their margins and we will see that they are **open systems** with a range of **inputs**, **stores/components**, **flows/transfers** and **outputs** that combine to create one of the most distinctive types of landscapes on Earth. When there is a balance between the inputs and outputs then the system is said to be in a state of **dynamic equilibrium**. If one of the elements of the system changes, for example there is increased sediment blown into the desert but there is no corresponding change in the amount of sediment being removed from the desert, then some desert landforms may change and the equilibrium is upset. This is called **feedback**. Figures 2.1a and 2.1b illustrate the open system of a desert landscape and how desertification can be viewed as an example of a positive feedback system. They highlight many of the ideas that are explored in detail in the rest of the chapter.

Figure 2.1b Positive feedback in a hot desert environment - desertification

Key question

Can you think of any other examples of feedback in hot desert landscapes?

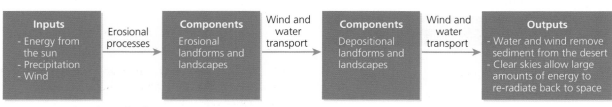

Figure 2.1a A hot desert landscape as an open system

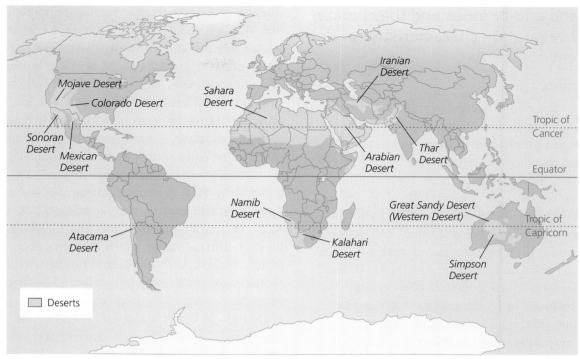

Figure 2.2 The location of major hot desert environments

Like all physical environments, the combination of erosional and depositional landforms that combine to produce characteristic hot desert landscapes are constantly changing elements (or components) of the geomorphological systems of which they are a part. As the system tends towards dynamic equilibrium, driven by the input of energy, the geomorphological processes present in hot deserts continually shape the desert landforms, creating their distinctive landscape features.

Global distribution

Deserts cover over a fifth of the surface of the Earth, however hot desert environments and their margins generally run in parallel belts north and south of the Equator in hot **arid** and **semi-arid** mid and low-latitude locations.

Almost all hot desert landscapes in the northern hemisphere are located towards the west of continents. The major hot desert areas of the northern hemisphere include:

● the Mojave, Sonoran, Great Basin and Chihuahuan deserts in North America
● the Sahara in North Africa and Somali-Ethiopian deserts in the Horn of Africa
● the Arabian, Iranian and Thar deserts stretching from the Red Sea through the Middle East to Pakistan and India in Asia.

In the southern hemisphere hot desert areas include:
● the Atacama desert in South America
● the Namib, Kalahari and Karoo deserts in southern Africa
● the Great Sandy (Western), Great Victoria and Simpson deserts, and others, in Australia.

Key question

What reasons can you think of to explain this distribution?

Key terms

Arid – The climate of an area that receives less than 250 mm of precipitation per year.

Aridity index – The ratio between precipitation (P) and potential evapotranspiration (PET).

Continentality – The impact of increasing distance from the coast on the climate of an area.

Desert – An arid environment receiving very low levels of rainfall.

Semi-arid – The climate of an area that receives between 250 and 500 mm of precipitation per year.

Water balance – The relationship between the annual precipitation received and the amount of water lost to potential evapotranspiration.

Water and aridity index and the climate, soil and vegetation characteristics of hot desert environments

Aridity index

A desert is most simply defined as an arid environment receiving very low levels of rainfall. Typically hot deserts are categorised as arid receiving less than 250 mm precipitation per year and semi-arid receiving 250–500 mm precipitation per year. However, this is very simplistic and the categorisation of deserts is varied and complex. Deserts are classified as hot and cold, sandy and rocky, foggy and sunny, and barren, and are found at both high and low extremes of altitude and in both coastal and continental interiors. Almost all definitions are linked to the lack of water, and the **water balance** which compares the mean annual **precipitation (P)** received with the mean annual **potential evapotranspiration (PET)**. PET is the amount of water that could be lost from the soil by plant transpiration and direct evaporation from the ground. The ratio of P and PET is known as the **aridity index.**

Hot deserts and their margins are considered to be areas described as **hyper-arid** to **arid**, and **semi-arid** on the aridity index as illustrated in Figure 2.3.

Classification	Aridity Index	Global land area (million km²)	Global land area (%)
Hyper-arid	AI <0.05	10.0	7.5
Arid	0.05<AI<0.20	16.2	12.1
Semi-arid	0.20<AI<0.50	23.7	17.7

Figure 2.3 The aridity index

Places with greater P than PET will have an aridity index of greater than 1.0, while those with PET greater than P will have an aridity index of less than 1.0. Of the hot deserts in Figure 2.2 (page 53), the Sahara has the highest aridity, followed by those in Arabia, East Africa, Australia and those of South Africa, with lower levels in the Thar and the deserts of North America.

Key question

What do you think the reasons are for this aridity?

Climate

Although hot deserts are warm throughout the year with very hot summer temperatures, they have climates characterised by extremes. Temperatures vary wildly both annually (between the hottest and coldest months) and **diurnally** (between daytime and night-time temperatures). With very low **humidity** levels in hot deserts, especially those in subtropical latitudes, cloud levels are extremely low. Clear skies allow significant amounts of **insolation** to reach the surface, almost twice that at more humid latitudes closer to the equator or at higher latitudes. Daytime temperatures can commonly rise to over 30°C, with highs of 45–55°C in many locations, and one claim for the highest temperature ever recorded, 57.7°C was at Al Aziziyah in Libya in 1922. The cloudless skies explain the high **diurnal range**, as temperatures drop very rapidly at night, with hot deserts losing twice as much heat at night as the more humid latitudes. Night-time temperatures can often drop below 0°C, with records of around –18°C not being rare. Therefore, the diurnal range can be over 30°C, with extreme ranges of over 40°C in places (Figure 2.4).

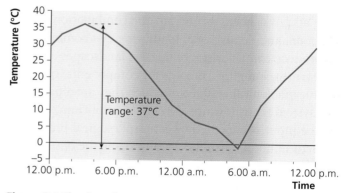

Figure 2.4 The diurnal temperature range in a hot desert

Semi-arid regions are also warm all year, with hot dry summers, average temperatures of around 21–27°C, temperatures rarely going above 38°C, and night-time temperatures being cool, around 10°C. As distance from the tropics increases so does the annual temperature range in hot deserts due to the increased seasonality of the climate. This means that desert margins have huge variations in temperatures, with those on the poleward side of arid areas, being significantly colder than those on the equatorial side.

Key question

Why do you think hot deserts have such large diurnal ranges?

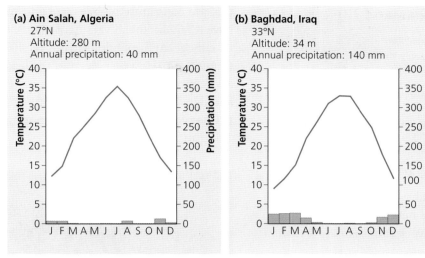

Figure 2.5 Climate graphs for two desert areas

The reality is that there is huge variety in the temperature patterns experienced in hot desert areas around the globe due to differences in their geographical situation. For example, those closer to the sea, like the Atacama, have a cooler, less extreme temperature range than those in continental interiors.

The amount of rainfall is usually extremely low, or occurs in rarer, short downpours following long dry periods (Figure 2.5). These cloudbursts can often deliver the total annual rainfall in a number of hours. The rate of evaporation often exceeds the rate of rainfall. Hot deserts are classified as receiving less than 250 mm of rainfall, but rates vary greatly around the world. Places like Iquique in the Atacama Desert in Chile receive almost no rain and are the driest, averaging less than 15 mm a year, with the central Sahara receiving a similar amount. American deserts receive slightly higher amounts at around 280 mm. Desert margins or semi-arid areas are often classified as receiving 250–500 mm a year, but like hot deserts they are generally dry and rainfall events are unpredictable and can occur as sudden downpours. Due to the slightly lower temperatures less moisture is lost due to **evapotranspiration**, and in areas with low night-time temperatures the condensation of dew can equal or exceed the amount of rainfall received in some areas.

Skills focus

You can easily find annual rainfall and temperature data for a hot desert area and could use this to construct and interpret a climate graph.

Soil

Around 17 per cent of the Earth's surface has desert soils. The extent of desert soils varies in different continents: Australia has about 44 per cent, Africa 37 per cent and Eurasia only 15 per cent.

When soils do develop in deserts they are often infertile with a thin **soil profile** and tend to be alkaline and quite saline. Rates of soil development are extremely slow in hot deserts due to:

- lack of moisture
- extremely high temperatures and high rates of evaporation
- sparse vegetation and limited organic material.

Desert soils are often characterised by a thick accumulation of basic mineral salts (often calcium or sodium compounds) at or near the surface. This occurs due to the process of **capillary movement** where any moisture in the soil or subsoil moves upwards through the tiny spaces between soil particles (capillaries). This capillary action is most effective when evaporation exceeds precipitation, hence its prevalence in arid environments.

The main zonal soil type for hot deserts and their margins is the soil order of **aridisols**. This is an order of soils including infertile alkaline and saline soils of desert areas characterised by accumulations of mineral salts at or near the surface (Figures 2.6, page 56). These aridisols fall into two main categories **sierozems** (in semi-arid areas) and **raw mineral soils** (in arid environments), although there are many other groups and sub-groups of soils also known as aridisols.

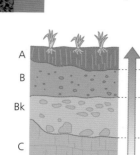

A horizon with prismatic structure, brown or grey

B horizon with clay accumulation

Bk has a thick zone of calcium carbonate accumulation

Salts are carried upwards in solution

Figure 2.6 Soil profile of a typical aridisol

Aridisols range from reddish-yellow to grey-brown in colour depending on the nature of the parent material and are generally very thin (usually less than 100 cm deep).

The **raw mineral soils** of arid areas have a coarse texture and are often rocky or gravelly due to **physical weathering**. **Chemical weathering** also occurs and where water exists in the subsoil capillary action leads to an accumulation of calcium and sodium salts near the surface. With a lack of moisture there is little downward movement of minerals (**leaching**), and arid soils tend to have limited variation between horizons. These soils tend to be slightly alkaline and are often unproductive rather than infertile, as minerals and nutrients are available but extremes of temperature and a lack of water are the main limiting factors for vegetation. If concentrations of salts are high enough hard salt crusts can form at the surface.

In arid to semi-arid areas receiving around 250 mm of rainfall **grey desert soils** or **sierozems** can form. The darker colour suggests the presence of some organic material as they often develop below 'desert-shrub' vegetation. These soils are often used for cultivation and with continued irrigation calcium-rich B horizons can develop beneath the thin A horizon, with rich accumulations of calcium carbonate or gypsum giving a lighter colour.

Key question

Why will soil profiles be different in more humid environments like the UK than in hot deserts?

Vegetation

Hot desert environments often look barren, desolate and lifeless, however other than the **sand sea** areas that are constantly mobile, arid and semi-arid areas have a remarkable diversity of plants. Plants are usually ground-hugging shrubs or short woody trees (Figure 2.7). However, even in the least hospitable, arid environments, vegetation cover remains sparse, and depending on the temperature and rainfall, **net primary productivity** values can range from near 0 to 120 g/m²/year.

Figure 2.7 Low-growing desert-shrub vegetation

The limited availability of water, extreme temperatures (in terms of heat and range) and the intense constant sunlight during the day, mean that plants have developed a range of physical and behavioural adaptations in order to:

● maximise use of, and limit the loss of moisture
● store moisture in their stem or leaves
● procure water with extensive and/or deep root systems
● respond rapidly to sporadic rainfall followed by rapid life cycles.

Figure 2.8 Saguaro cactus with adaptations for surviving in arid environments

Plants in desert areas have a number of features to help them survive.

- **Succulence**: Succulent plants, or **xerophytes**, contain compounds or cells in fleshy leaves, stems or roots where they can store water, for example, all cacti and non-cacti like aloes.
- Procuring water following brief rainstorms: The water does not penetrate deep into the soil or remain wet for very long so succulents have the ability to rapidly absorb huge amounts of water as plant roots can only take up water when the soil is more moist than the interior of the roots. The main adaptation to cope with these conditions is to have shallow (less than 10 cm) and very wide extensive root systems. These **phreatophytes** have a root system that lies just beneath the surface. The roots of the 12–18 m-high Saguaro cactus (Figure 2.8) can extend horizontally to the same distance as the height of the plant. The 60 cm-tall Cholla cacti lives in extremely arid conditions and has roots that extend up to nine metres. The Agaves are the exception to this adaptation as they have a limited root system, but have a leaf rosette that channels rain water directly to the base of the plant.
- Most succulents also have adaptations to conserve water, including:
  - thick waxy cuticles and closed stomates, making the leaf surface waterproof
  - small, spiky or waxy (or no) leaves to reduce surface area to limit transpiration.
- Many desert plants have adapted to protect their precious water supplies from thirsty desert animals. They do this by being spiny, bitter or toxic, or by simply living in inaccessible locations.
- Desert plants are also drought tolerant, or **ephemerals**, with adaptations like becoming dormant or losing their leaves during extreme dry spells, often appearing dead for months or even years at a time.
- A number of plants, including some ephemerals, will try to avoid drought altogether by having extremely short life cycles; coming into bloom rapidly after rainfall, then channelling all their life energy into producing seeds which can themselves then lay dormant until a future rainfall event.
- As evaporation rates are so high, desert plants also need to be **halophytes**, that is, tolerant to high levels of salinity, like the saltbush.

Fieldwork opportunities

The northern Sahara Desert is accessible close to the town of Erfoud to the east of Marrakesh in Morocco. During a field visit it would be quite simple to collect basic information about the characteristics of this hot desert environment without requiring large amounts of equipment. Field observations could be made on the following:
- climate
- soils
- vegetation.

The causes of aridity in hot desert environments

There are a number of factors that contribute to the arid nature of these regions:
- the general pattern of atmospheric circulation
- distance from oceans or **continentality**
- relief
- cold ocean currents.

Global atmospheric circulation

Figure 2.9 (page 58) illustrates how the latitudes where hot deserts tend to be located, between about 20–30° north and south of the equator, are generally dominated by high pressure throughout the year.

At the equator there is a net gain in energy as large amounts of incoming solar radiation, **insolation,** is received because the sun is directly overhead with a high **angle of incidence**. Air in contact with the Earth is heated and begins to rise, cooling, causing the water vapour within it to condense, form clouds and lead to precipitation. The rising air is replaced by air rushing in from the north and south creating an area of low pressure known as the **inter-tropical convergence zone (ITCZ)**. The rising air begins to cool and track polewards. At around 20–30° north and south this now cool denser air descends. As it does it warms and expands, meaning little cloud formation occurs, giving the clear skies responsible for the heat and aridity of these latitudes. These two circulation cells are known as the **Hadley cells**. Figure 2.9 also shows that the mid-latitude cells (**Ferrel cells**) have air descending at these mid-latitudes; the combination of the sinking air from both cells is what creates the high pressure at the surface.

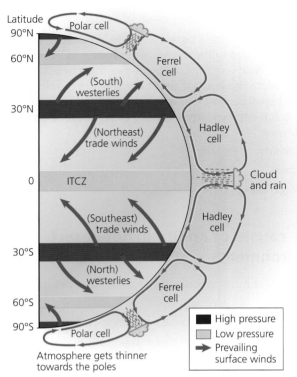

Figure 2.9 The tri-cellular model of atmospheric circulation

Distance from the sea

Continentality has an effect on both the rainfall and temperature received by maritime areas and continental interiors.

As Figure 2.10 illustrates, coastal locations have a more moderate climate with little annual variation in temperature and receive higher levels of rainfall. With distance from the sea temperatures are far more extreme, can exhibit higher extremes and are generally much drier as there is little moisture available for cloud formation. For example, parts of the Sahara in the centre of North Africa are over 2,000 km from the sea.

Relief

Around mid-latitude areas there are a number of extremely dry regions on the leeward side of mountain ranges that experience what is known as the **rain shadow** effect, for example Arabia to the west of the Himalayas and central Australia to the west of the Eastern Ranges. Figure 2.11 illustrates the effect of the southeast **trade winds** from the southern Atlantic meeting the Andes and the effect on the Atacama desert in the west of South America.

If moist air brought inland by prevailing winds meets a range of mountains it is forced to rise, leading to cooling, condensation, cloud formation and **relief rainfall** on the windward side. Once over the summit the air descends on the leeward side of the range. This sinking air warms leading to a drop in **relative humidity** and clear skies with little, or no, rainfall.

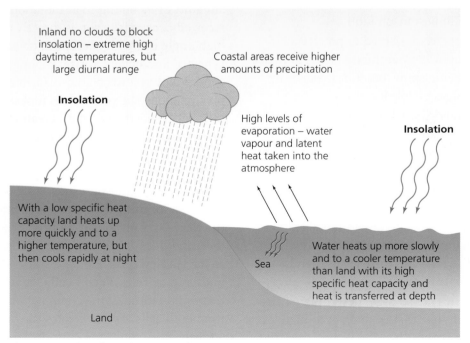

Figure 2.10 The effect of continentality

Figure 2.11 Rain shadow effect in South America

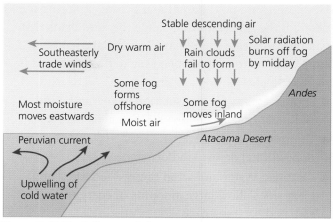

Figure 2.13 The influence of the Peruvian current

Cold ocean currents

The Atacama Desert in the sub-tropical latitudes of western South America illustrates the effect of cold ocean currents. There are a number of such currents that form part of the **global oceanic circulation** (Figure 2.12).

Any wind moving over these cold waters is cooled, which causes the relative humidity to increase and eventually moisture is condensed to create fog and mist offshore. As the land heats up quicker than the sea this may generate gentle onshore breezes which will take the fog and mist inland. However, the intense heating from the directly overhead sun soon burns this away. As the air was cool it could not hold much moisture so cloud formation is unlikely, and precipitation very rare. Some vegetation has even developed adaptations to take advantage of the moisture condensing on it as dew, such as in northern Chile where this may be the only moisture it receives for years at a time (Figure 2.13).

2.2 Systems and processes in hot desert environments

This book highlights how most aspects of physical geography can be thought of as natural systems with inputs, stores/components, flows/transfers and outputs; with geomorphological processes, driven by the input of energy, constantly changing the elements of these systems as they tend towards dynamic equilibrium.

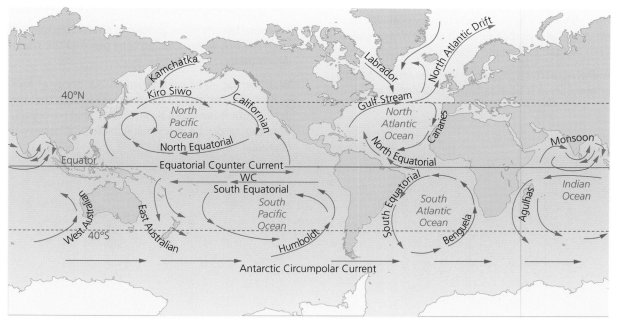

Figure 2.12 Oceanic circulation

Hot deserts are no different and before exploring the detail of the processes that shape the landforms that combine to create their characteristic landscapes, a simple summary of the open system in hot deserts is included below.

- Inputs: energy from insolation; water and wind; sediment (mostly sand-sized particles)
- Components: characteristic erosional and depositional desert landforms
- Outputs: Water and wind remove sediment and clear skies allow large amounts of energy to be re-radiated back to space.

You should also read the relevant material at the beginning of Chapter 1 under the heading 'Systems frameworks and their application' (page 2).

Key terms

Abrasion – Where material carried by moving wind or water hits exposed rock surfaces, thus wearing them away. Often referred to as a *sandblasting* or *sandpapering* effect.

Chemical weathering – The processes leading to the decomposition or breaking down of rocks due to chemical reactions. This most often requires the presence of water, and/or exposure to the air.

Deflation – Where wind removes dry, unconsolidated (loose) sand, silt and clay particles from the surface and transports them away.

Deposition – Occurs when the velocity of the wind decreases until it can no longer transport the grains it is carrying.

Endoreic streams – Where rivers occupy drainage basins that are closed and do not flow out to the sea or other rivers, but instead end inland in lakes or swamps.

Ephemeral streams – Streams that flow intermittently in hot desert areas following heavy thunderstorms.

Erosion – The wearing away of the Earth's surface by the mechanical action of processes of glaciers, wind, rivers, marine waves and wind.

Exfoliation – A process of mechanical weathering that results in the breaking, splitting or *peeling-off* of the outer rock layers. Also commonly known as 'onion skin weathering'.

Exogenous streams – Rivers that originate external to the desert in adjacent highlands and more humid environments, flow from outside of the desert and pass through it.

Insolation – The incoming solar radiation that reaches the Earth's surface.

Mass movement – The movement of material downhill under the influence of gravity, but may also be assisted by rainfall.

Saltation – A process where sand-sized particles are transported by bouncing and hopping along the surface.

Sediment – Any naturally occurring material that has been broken down by the processes of erosion and weathering and has then been transported and subsequently deposited by the action of ice, wind or water.

Surface creep – Where saltating particles return to the surface and hit larger particles that are too heavy to hop; they slowly creep (slide or roll) along the surface from a combination of the push of the saltating grain and the movement of the wind.

Suspension – Transportation by wind where the smallest particles, generally less than 0.2 millimetres, are held in the air.

Thermal fracture – The weathering of rock resulting from their rapid and repeated heating and cooling.

Transportation – The processes that move material from the site where erosion took place to the site of deposition.

Weathering – The breakdown and/or decay of rock at or near the Earth's surface creating regolith that remains *in situ* until it is moved by later erosional processes. Weathering can be mechanical, biological/organic or chemical.

Sources of energy in hot desert environments

Many of the landform features of hot deserts are created by the action of water and wind. To fully explain how these processes work it is important to understand where the energy comes from to drive the various **weathering** and **erosion** processes. **Insolation**, wind and run-off provide the energy to drive hot desert geomorphological systems. Together with sediment (mostly sand-sized particles) these are the main **inputs** in desert systems.

Insolation

Many processes that occur in hot deserts result from changes or differences in temperature. Therefore energy from the sun is a vital **input** in desert systems. We have already explored why hot deserts are arid, but why are they hot? In the low- to mid-latitude hot desert environments the sun is almost directly overhead during the day throughout the year – meaning there is 12 hours of daylight. Also, as the sun is high in the sky in these latitudes, the **angle of incidence** of the insolation is very high (Figure 2.14) meaning that any given amount of radiation is concentrated on, and so warms, a smaller area of earth than at higher latitudes.

In more moist environments the sun's energy is used to evaporate water on and under the surface, taking heat energy up into the atmosphere as **latent heat**, so keeping the ground cooler than it otherwise might be. In more arid areas the lack of moisture means the ground becomes excessively warm, making more energy available to heat the air in contact with the ground.

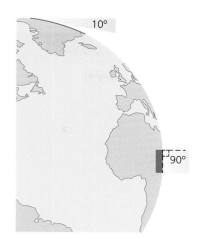

Figure 2.14 The height of the sun in the sky – angle of incidence

Winds

Wind is also an important input in hot deserts as it is a source of energy and driver of processes in desert systems as they tend towards dynamic equilibrium. With their mid-latitude location along the sub-tropical high-pressure belt many large hot desert areas are subject to localised winds blowing outwards towards the edge of the desert. These winds then act as an agent of erosion and transport. Examples of these local winds include:

- the *kharif* in the Somali-Chalbi desert in the Horn of Africa, blowing June to September
- the *irifi* easterly winds that blow in the Western Sahara around March
- famously, the *harmattan* in the Sahara blowing in winter months from November to March
- the *shamal* over the Arabian Gulf from late May to early July
- there are also more localised winds generated in the Sudanese Sahara known as *haboobs*.

Other hot desert areas have their own local winds. Although winds in desert areas are on average generally no stronger than elsewhere, it is their effects that are most notable, as we will see later in the chapter.

Table 2.1 Extremes of precipitation in deserts

Location	Mean annual precipitation (mm)	Highest precipitation in a single storm (mm)	Storm date
Chicama (Peru)	4.0	394.0	March 1925
Luderitz (Namibia)	16.7	102.0	16–22 April 2006
Masirah Island (Oman)	70.0	430.6	13 June 1977
Diego Aracena (Chile)	1.3	9.1	6 July 1992
El Djem (Tunisia)	275.0	319.0	25–27 Sept 1969
Ziyaratgah (Iran)	28.0	30.0	6 March 1997
Oodnadatta (Australia)	173.4	200.0	9 February 1976
El Cajoncito (Mexico)	220.7	237.5	29 Sept 1982

Source: Middleton, N. 2009, OUP: Deserts: A Very Short Introduction, p.17

Run-off

Rainfall in deserts is often described as 'spotty', in that it is often very localised and spatially and temporally unpredictable. Despite arid hot desert areas being defined as having less than 250 mm of rainfall annually, and some hyper-arid areas recording almost none, rare intense storms can produce huge amounts of rainfall in very localised areas (Table 2.1).

Together with the baked ground and limited vegetation cover, rapid **overland flow (run-off)** can be a significant agent of erosion and transportation of sediment.

Sediment sources, cells and budgets in hot desert environments

Hot desert landscapes are often dominated by loose **sediment**. There are a variety of origins, or **inputs**, of sediment into hot deserts. The sediment may simply be derived from the weathering of the underlying parent material. Secondly, sediment may be **fluvial** in origin; rivers may bring the sediment into deserts. If they are **ephemeral**, or dry up, then sediments may be left behind in dry river beds, or as rainfall is episodic and often heavy, desert rivers may flood and deposit material on the desert surface. Thirdly, other sediments originating beyond desert margins may be **aeolian** and be transported in and deposited by wind as **loess**.

Hot deserts have a range of landscapes and landforms depending on the combination of processes operating there. Areas dominated by erosion are a **source** of sediment and the system has a **net sediment loss**. In those dominated by the transportation and **deposition** of sediment, the system has a **net sediment gain**.

Large deserts can be significant sources of sediment themselves on a global scale. In fact satellite imagery often shows large clouds of *dust* being blown five thousand miles across the Atlantic to the Americas (Figure 2.15). Also the large loess deposits of central Asia and northern China, although not formed in hot deserts, were certainly transported in from adjacent deserts.

Figure 2.15 Saharan sediment being blown across the Atlantic

Geomorphological processes

As hot desert geomorphological systems develop over time and tend towards dynamic equilibrium a variety of processes take place. Driven by the input of energy, these are what change the characteristics of the components of the hot desert landscape.

Weathering

Weathering is the breaking down or decaying of material where it is (*in situ*). Unlike erosion it does not involve the transporting away of the broken down material, but instead produces material known as **regolith**. Weathering can be subdivided into a number of types: **mechanical weathering**, **chemical weathering** and **organic (or biological) weathering**.

Mass movement

Mass movement is the movement of material downhill under the influence of gravity, but may also be assisted by rainfall. However, because of the low levels of moisture and little vegetation there is limited soil in hot desert environments. Therefore where mass movement occurs it is dominated by **rock falls**, where small blocks of rock become detached from an exposed cliff face and fall freely to the base of the cliff, and **rock slides**, where there is a failure throughout the rock as a whole and the material collapses *en masse*, rather than as individual blocks. These create generally steeper slopes with the accumulation of coarse, angular material at their base.

Steep solid rock cliffs are quite common in hot deserts. If vertical jointing is a common feature of the rock forming these cliffs and this is underlain by rocks that are more easily eroded, then rock fall and cliff retreat will occur (Figure 2.16).

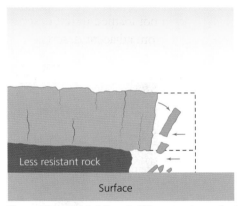

Figure 2.16 Rock fall and cliff retreat in vertically jointed rocks

Where the rock forming cliffs has more horizontal bedding planes then the cliff retreats in a more uniform manner due to the usual weathering and erosion processes. However, if the bedding planes of the geology dip then a ridge can form with a gentle slope forming parallel to the dip (a dip slope) and a steep cliff on the other side (Figure 2.17).

Figure 2.17 Mass movement in rocks with dipping bedding planes

Erosion

Erosion refers to processes that wear away the land surface by mechanical action. In general the agents of erosion include glaciers, rivers, wind and ocean waves and currents. Obviously wind and water are the only ones present in hot deserts. What distinguishes erosion from the processes of weathering is the removal of the *denuded* surface material by processes of wind and water transport (see below). The processes of erosion result in the **denudation**, or degradation of the surface of the Earth.

Transportation

Transportation refers to the processes that move material from the site where erosion took place to the site of deposition. In hot deserts the two agents that can perform transportation are moving water and wind. The material that is being transported is called the **load**. Generally water transports its load in the following ways:

- **Traction**: larger stones and boulders are rolled along the channel bed by the movement of water downstream. Traction occurs with higher levels of energy and **discharge**.

- **Saltation**: smaller stones and pebbles are bounced, or leap frog, along the channel bed. Discharge and energy levels need to be high enough to *pick up*, or thrust, small particles up into the water column, however the flow is not fast enough to keep the particle off the bed so it falls back to the bottom, having been moved a short distance downstream. Particles landing back on the bed may now dislodge other particles up into the moving water, thus repeating the bouncing action.

- **Suspension**: the smallest particles, generally silt and clay, are carried along in the moving water column by the flow of the water. Generally the turbulence of the moving water allows the river to not only pick up this load but also carry it downstream, only depositing the finest particles when the water is almost stationary.

- **Solution**: the transport of dissolved minerals within the mass of the moving water itself.

Deposition

As moving water and wind are not continuously present in hot desert environments depositional features are common. Water and wind both deposit – put down – their load when they experience a reduction in the amount of material they can carry, or a reduction in **capacity**. With a loss of energy both water and the wind will deposit their load. A common cause of deposition in hot deserts is the evaporation of water and streams simply dry up leaving their load behind.

Distinctively arid geomorphological processes

Weathering in hot deserts

Weathering processes are some of the main ways that typical desert landscapes are shaped. Most weathering in hot deserts is believed to be mechanical or physical, however it is due to a combination of factors and the reasons for the prevalence of weathering include:

- the regular heating and cooling of rocks
- the presence of moisture, even in the smallest quantities
- the presence of living organisms.

The role of different weathering processes is explored below, but it is clear that variations in the input of energy and water are important factors in controlling what changes take place.

Key question

Why do you think weathering plays such an important role in hot deserts?

Thermal fracture

A type of mechanical weathering, **thermal fracture**, results from the rapid heating and cooling of rocks in deserts, which is accentuated by the extremes of temperature experienced. With air temperatures during the day rapidly rising to in excess of 40°C the surface layers of rocks can get much hotter, up to 80°C in very exposed locations. At night, temperatures fall very rapidly, generally to below 10°C, but below 0°C in places is not uncommon. This huge **diurnal range,** and the extremes, creates a regular rhythm of heating and cooling that causes the exposed rock to also regularly expand during the day and contract at night. However, this is a rather simplistic understanding of how mechanical weathering operates on rocks in hot deserts and it is now believed that it is a combination of a range of weathering processes that cause rock to disintegrate (degrade) in a variety of ways.

Exfoliation

The mechanical weathering process of **exfoliation** results in the breaking, splitting or peeling off of the outer rock layers (Figure 2.18). This 'onion-skin weathering' was often believed to have simply resulted from pressure changes in the rock as only the outer layers were exposed to the cycles of heating and cooling. However, the reality is more complex. Exfoliation most often affects reasonably uniform coarse-grained crystalline igneous rocks. At depth the rock is under considerable pressure, and as erosion and weathering occur at the surface, removing surface material, the pressure on rocks lower down is decreased. This creates tension and to relieve this cracks form, running parallel to the surface. As the rock is exposed to the heating and cooling outlined above, salt-rich water is drawn to the surface by **capillary action**. These dissolved salts are deposited in the cracks and together with chemical weathering the cracks are enhanced and slabs of rock detach from the surface.

Figure 2.18 Examples of rocks affected by 'onion-skin' (or exfoliation) weathering

Chemical weathering

The rate of **chemical weathering** is slow in hot deserts because they have little soil and moisture levels are low. In more temperate and humid environments it is chemical weathering that helps in the development of soils, so where it is bedrock that is present at the surface in these arid environments different features are formed. Due to the lack of organic material any soil that does form in a desert tends to be a similar colour to the parent rock. Most of the chemical weathering that does occur in hot deserts results from the deposition of salts precipitated from rain water or brought to the surface by capillary action. Processes of chemical weathering in deserts that contribute to the mechanical breakdown of rocks include:

- **Crystal growth**: this weathering process is now seen as one of the main agents contributing to some

Figure 2.19a Hardpan

Figure 2.19b Desert varnish

of the mechanical processes outlined above. Most simply, as water that is present in spaces (pores, joints, bedding planes) evaporates, salts that were dissolved in the water are deposited. Over time larger and larger crystals develop. These salts have different thermal capacities to the surrounding rock so it is often believed that the differential heating and cooling, and so expanding and contracting, of these salts compared to the rock itself can assist in the physical breaking down of the rock.

- **Hydration**: this is another chemical process that is also linked to the mechanical breakdown of rocks in hot deserts. Some rocks are able to absorb any water that is available, even small amounts of morning dew. As they absorb the water the rock may physically swell causing the initiation of stresses in the rock, making it vulnerable to other mechanical breakdown. Where hydration causes salt minerals in certain rocks to alter chemically they can become weaker and thus more vulnerable to other kinds of chemical weathering. When water is added to the mineral anhydrite for example, gypsum is formed.

Other chemical weathering processes:

- **Hydrolysis**: this occurs where mildly acidic water reacts or combines with minerals in the rock to create clays and dissolvable salts; this itself degrades the rock, but both are likely to be *weaker* than the parent rock making it more susceptible to further degradation.
- **Oxidation**: this is the breakdown of rocks by oxygen and water and leads to a common feature of many hot deserts, the red-brown colour of many surface rocks, where minerals of iron have been oxidised.

As any available moisture is continually drawn to the surface and evaporated, many of these chemical weathering processes lead to accumulations of salts on or near to the surface. Many of these become *cemented* and create layers of **duricrust**, or **hardpans** (Figure 2.19a). The chemical nature of the underlying rock can lead to the formation of different kinds of **crusts**: in calcium-rich limestone areas gypsum is created which itself leads to the formation of **gypcrete**, while where lime is present **calcrete** is deposited.

Where iron and manganese oxides have been weathered from rocks at depth and are drawn to the surface in solution, the desert heat evaporates the water and a deep red stain, glaze or **desert varnish** (Figure 2.19b) is left on all exposed surfaces.

Block and granular disintegration

In rocks that are either heavily jointed, like granites, or have prominent bedding planes like limestone, the various mechanical processes outlined above can make masses of rock break down into large blocks (Figure 2.20, page 66). Many igneous rocks like granite have a uniform structure and form in large masses deep below the Earth's surface. When they form, regularly spaced fractures develop and as the weight of the rock above is reduced by erosion of the overlying rock and it is slowly brought towards the surface by uplift, these fractures open slightly becoming joints. It is then that even the smallest amounts of available water in the hot deserts can enter these cracks. Together with the constant heating and cooling, the water can now chemically weather the rock along these faults leading to blocks of rock literally breaking off (**block disintegration**). If temperatures do drop below freezing there is the possibility that **freeze thaw** weathering can occur, where during the day liquid water enters the joints, then freezes at night as temperatures drop below 0°C. As water freezes it expands by about

Figure 2.20 Weathering of well-jointed rock

9 per cent, thus exerting pressure on the surrounding rock. Repeated cycles of freezing and thawing can lead to blocks of rock being dislodged. (Traditionally this process was known as **congelifraction**).

Where rocks have a more granular structure they can be broken down into separate grains both by freeze thaw or differential thermal expansion and contraction caused by the huge temperature ranges in hot deserts. Even when only small amounts of moisture are available, possibly even from dew on cold desert mornings, it can find its way into pore spaces within the granular rock. If the temperature drops below freezing it expands, thus putting pressure on the surrounding grains of rock. With repeated freezing and thawing individual rock particles are dislodged. Where rocks have a granular structure different minerals will heat up and cool down at different rates during the diurnal heating and cooling cycle. This leads to individual particles of rock expanding and contracting at different rates leading to individual grains being broken off (**granular disintegration**).

The role of wind in hot desert environments

As one of the main inputs in hot desert systems, wind is important as a driver of change and contributes to a range of geomorphological processes occurring there.

Wind and **aeolian** processes are a very common feature of hot desert environments and their margins because:

- the cloudless skies and high angle of incidence means air at the surface is heated and rises, so cooler air moves in to replace it. It is this movement or air that results in winds
- many desert regions are relatively barren with few surface features to create friction to reduce wind speed, or provide shelter. Therefore, winds can blow unimpeded for considerable distances.

As wind has energy and is moving it has the ability to erode, transport and deposit sediment, and so has a significant impact on the landscape of hot desert regions.

Erosion

There are two processes of erosion by wind:

- **Deflation:** wind removes dry, unconsolidated sand, silt and clay particles from the surface and transports it away. Wind only removes the finer material creating a surface covered in a concentration of coarse and fine pebbles known as **reg** or **desert pavement** (Figure 2.21).
- **Abrasion:** this is often referred to as a sandblasting or sandpapering effect, where the material carried by the wind hits exposed rock surfaces and creates a range of erosional features. A number of factors affect the rate of abrasion: the direction, frequency and velocity of the wind, the **lithology** of the rocks and the size and nature of the particles carried by the wind.

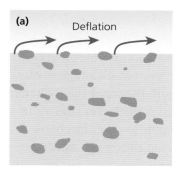

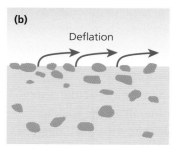

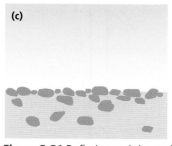

Figure 2.21 Deflation and the evolution of reg

Stage 1 Underlying basin structure within rocks is covered by a thick accumulation of sand

Stage 2 Strong winds remove some of the loose material to reveal part of the basin shape

Eddies

Figure 2.22 The formation of a deflation hollow

Wind can remove very significant amounts of material from the surface creating **deflation hollows** (Figure 2.22). Some of the largest deflation hollows are found in North Africa, where they can extend over hundreds of square kilometres. The Qattara Depression in Egypt is the deepest, reaching 134 m below sea level at its deepest point. There is some debate about the exact combination of processes that forms these huge desert depressions, but wind has undoubtedly contributed to removing the millions of tonnes of sand and other material that was once there. The depth of deflation hollows is controlled by the level of underlying groundwater.

Transportation

As wind is an almost ever-present feature of many hot desert environments and there is often large amounts of loose surface material, wind transport is extremely important in shaping the landscape. The transportation of sediment by wind is not only an important agent of change, in terms of processes of erosion and deposition, but represents a significant flow/transfer of material. The movement of particles by wind is dependent on a range of factors:

- the wind strength and direction
- the amount of turbulence
- the duration of the wind
- the relief and surface features
- the amount and nature of any vegetation.

Aeolian processes are able to transport material in three main ways. The nature, speed and effectiveness of this depends on the wind strength and the size of the particles (Figure 2.23).

- **Suspension:** this is where the smallest particles, generally less than 0.2 mm, are held in the air. The moving air is able to support the weight of these small particles and carry them indefinitely and appears as a layer of dust or haze close to the surface. If the winds are strong and persistent enough clouds of dust can be carried to significant altitudes and over hundreds or thousands of

kilometres, with 'red rain' of Saharan sand occasionally falling in the British Isles. With very high velocity winds vast amounts of silt-sized particles are lifted creating sand/dust storms.

- **Saltation:** this is a process where sand-sized particles are transported by bouncing and hopping along the surface. When there is turbulence particles are lifted by a gust of wind, usually only a few centimetres into the air (but sometimes as high as 2 m) then carried a short distance downwind before dropping back to the surface. Grains hitting the surface may hit other particles and cause them to jump into the air, be caught by the wind themselves and so continue the saltation. Particles are usually transported at one-half to one-third of the speed of the wind.

- **Surface creep:** this is where saltating particles return to the surface and hit larger particles that are too heavy to hop, they slowly creep (slide or roll) along the surface from a combination of the push of the saltating grain and the movement of the wind. It is believed that surface creep accounts for as much as 25 per cent of all grain movement in hot deserts.

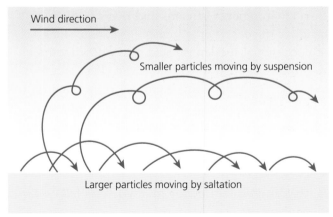

Figure 2.23 Wind transport

Deposition

Deposition occurs when the velocity of the wind decreases until it can no longer transport the grains it is carrying. Wind-deposited materials can often give

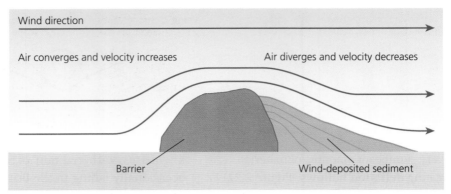

Wind direction

Air converges and velocity increases

Air diverges and velocity decreases

Barrier

Wind-deposited sediment

Figure 2.24 Wind deposition around obstacles

clues to the scale and direction of winds in the past. For example analysis of the huge wind-blown deposits of silt (loess) on the Loess Plateau in China, and similar but smaller accumulations in Europe and the Americas can help map the pattern of previous surface winds.

On a more local level, deposition may occur as the wind meets an obstacle and is slowed on the downwind side. This can lead to the formation of dunes (Figure 2.24).

Water in hot desert environments

As one of the main inputs in hot desert systems, water is an important driver of change and contributes to a range of erosional and depositional processes that shape the landscape. Water can also transport a large amount of sediment so is responsible for a significant flow/transfer of material.

Although rainfall and surface waters are rare in hot desert environments, streams and rivers originating outside of the desert or at higher elevations, along with occasional sudden isolated heavy downpours, can produce enough water to have an effect on the landscape. At higher altitudes in desert mountains and on high plateaux water acts mostly as an agent of erosion, while in flatter lowland areas depositional landforms are more common.

Sources of water

The origins of water flowing in hot deserts fall into the following categories:

● **Exogenous**: rivers originate external to the desert in adjacent highlands and more humid environments and flow from outside of the desert and pass through them. The volume of these rivers is large enough that they are perennial and maintain a flow all year, although they often do respond to seasonal rainfall

(or snow melt) patterns in the source area. These exogenous (or exotic) sources, including ground water and rivers, account for the greatest volume of water flowing in hot deserts. The Nile, flowing from its sources in the Ethiopian highlands and central Africa, flows north across the arid regions of North Africa to the Mediterranean at the Nile Delta. The Colorado River originates from the Rocky Mountains then flows across the arid regions of southwest USA to reach the Pacific Ocean (Figure 2.25). These rivers have been, and continue to be, of significant importance to the populations of the regions they flow across due to their historically reliable and constant flow.

Figure 2.25 The Colorado River

● **Endoreic**: this is where rivers occupy drainage basins that are closed and do not flow out to the sea or other rivers, but instead end inland in lakes or swamps. Examples include the River Jordan flowing into the Dead Sea and the Okavango that flows into Botswana and splits into a number of **distributaries** to form the Okavango Delta.

● **Ephemeral**: these are rivers that flow on the surface only periodically. They are a common source of water in hot desert areas and appear intermittently after heavy rainstorms. The suddenness, rapidity and intensity of the downpours can generate huge amounts of overland flow and very significant increases in channel discharge with extremely short **lag times**. This is the result of a number of factors:

- the scale of the precipitation event itself
- the lack of vegetation means that little **interception** occurs
- following extended periods of drought, and the removal of much loose surface material by aeolian processes, the desert surface is often baked hard, so the torrential nature of the rainfall exceeds its **infiltration** capacity.

The episodic role of water

The episodic nature of rainfall in hot desert regions is exemplified well in North Africa. Subtropical deserts like the Sahara remain arid most of the time due to the dominance of high-pressure conditions, giving average rainfall of 25 mm/year generally and only rising to 100 mm in the high plateaux of Ahaggar and Tibesti. In fact parts of western Algeria and southwest Egypt can go anywhere between two and five years without receiving more than 0.1 mm of rain in any 24-hour period. However, single rainfall events can deliver two to three times the annual average in short intense desert storms. Some west-facing slopes in parts of Algeria have recorded intensities of between 8.7 mm in 3 minutes and 46 mm in just over an hour.

Most of this variability occurs in the Sahara during the summer months as the West African Monsoon moves north towards the area and can on occasion allow areas of moist south-westerly air to push further north into the desert itself, giving short-lived areas of low-pressure. Disruption of the high-pressure cells along the sub-tropical high pressure belt can also lead to troughs, or *gaps*, in the high pressure at the surface, allowing more moist air to move north and create low pressure areas and rainfall on the southern edges of the Sahara. In desert areas where there are local water sources to provide moisture for the atmosphere, if conditions are favourable, the intense heating of the desert surface can lead to localised warming and rising of air and localised low pressure systems developing. With such intense heating and rapid rising and cooling of the air, large releases of latent heat during condensation can fuel uplift creating large but short-lived thunder storms.

During intense torrential downpours, both the impermeability of some desert surfaces and the sheer speed and amount of rainfall can lead to **sheet floods**. Initially the impact of the rain begins to dislodge and to move loose material across the surface (known as sheet wash or sheet erosion). As the volume of water increases, this **overland flow** intensifies and can rapidly develop into extensive but relatively shallow floods. This huge amount of overland flow can remove large volumes of material as sheet erosion, thus creating erosional features, but equally contributing to depositional features downstream. If this overland flow is funnelled into steep-sided narrow valleys, common in some hot desert areas (known as **wadis**, or **arroyos** in the Americas), **channel flash flooding** occurs. These can also move significant loads of sediment and can be very powerful due to the concentration of the water in such a narrow valley, creating a deep, fast-flowing torrent of water with huge potential for erosion and transport of sediment, and thus significant depositional features when the flow recedes.

Key question

What do you think the reasons are for such sudden thunderstorms in deserts?

2.3 Arid landscape development in contrasting parts of the world

Origins and development of landforms of mid- and low-latitude deserts

Aeolian landforms

The processes of wind erosion, transport and deposition outlined above can involve very significant amounts of material, so wind action (**aeolian processes**) play a significant role in shaping desert landscapes. Although similar processes happen in most desert environments the resulting landscapes will be unique in each location, reflecting the way erosional, transport and depositional processes interact with the local geology and climate.

Ventifacts

Ventifacts are exposed rocks lying on the desert surface that have been abraded or shaped by wind-blown sediment. This is usually sand, or finer silt and clay-sized particles and is often likened to a **sand-blasting** effect. Ventifacts can range in size from pebbles as small as a centimetre to huge boulders many metres in diameter. They are characterised by smooth and/or flattened sides (or facets) and sharp edges or ridges. There are many recognised 'faceted-types', with different numbers of flattened sides and sharp edges.

It is not completely understood how some ventifacts have multiple smoothed facets, but it is probable that either the rock changed position, thus presenting a different side to the prevailing wind, the prevailing wind changed direction over time, or there are areas with seasonal reversals of wind where the prevailing wind direction is different in summer and winter.

Yardangs

A **yardang** is a streamlined parallel ridge of rock often described as resembling the upturned keel of a boat. They are aligned in the direction of prevailing winds and neighbouring yardangs are separated by a wind-scoured groove or trough (Figure 2.26). The streamlined shape, with a rounded upwind face and long tapering ridge downstream, suggests that aeolian erosional processes are the dominant processes of formation. It is agreed that abrasion plays a key role in shaping and smoothing the ridge, and in eroding material from the trough. However, as many larger yardangs will have taken millions of years to take their current form it is thought agents such as deflation, sporadic fluvial incision, weathering and mass movement may also have played a role. Some yardangs are found in areas where the geology has horizontal rather than vertical bedding.

Yardangs found in contrasting arid environments vary in size quite considerably, with ridges ranging from a few centimetres high and several metres long, through all sizes to so-called 'mega-yardangs' measuring tens of metres high to many kilometres long. Several regions of the central Saharan desert have large numbers of these large features, including to the west of the Tibesti Mountains in northern Chad. These mega features, in some of the remotest deserts in the world, are so large and occur in such large concentrations that they are easily visible in remotely sensed Landsat satellite imagery.

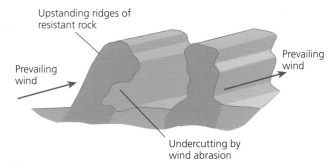

Figure 2.26 Yardangs

Zeugen

Zeugen is the term that has come to collectively refer to features such as rock pillars, rock pedestals and rock mushrooms, or yardangs that exhibit considerable undercutting. This range of names suggests a range of different characteristics, but all have been formed where less-resistant rock underlies a layer of more resistant geology at the surface (Figure 2.27a and Figure 2.27b). As wind transports huge amounts of sediment, mainly within two metres of the surface, abrasion is the main agent of erosion. Rock pedestals, mushrooms or pillars tend to be smaller features with dimensions of several metres, however where the features are elongated in the direction of the prevailing wind they can have similar dimensions to yardangs.

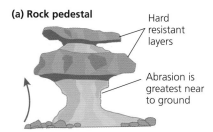

Figure 2.27a Rock pedestal

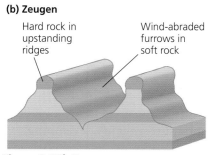

Figure 2.27b Zeugen

Barchans and sief dunes

Sand dunes are the main depositional landforms in hot desert environments and require certain conditions for formation to occur:

- a ready supply of sand
- prevailing winds strong enough to transport sediment, but not strong enough to remove more than is deposited
- steady winds from one dominant direction
- an obstacle, such as vegetation or rocks, to trap the sediment and encourage deposition to occur.

Where the supply of sand is great enough and a large extent of dunes develop, the landscape is referred to by the Arabic term **erg**. The specific size and shape of dunes varies widely depending on:

- the direction and strength of the prevailing winds
- the supply of sediment
- the morphology of the landscape where the dunes form
- the presence of any other winds.

Barchans are crescent-shaped sand-dunes with a gently sloping convex windward side with two *horns* or *wings* extending laterally which curve in a downwind direction (Figure 2.28). The leeward, downwind face (or slip face) of the dune is a steeper concave slope. Barchans only form where winds almost only come from one direction. As material is moved up the windward slope by saltation and creep and then slides, falls and flows down the leeward side, the whole dune moves downwind across the desert landscape. The leeward slope is kept steep by eddying of the wind as it descends down the

Figure 2.29 Dunes in the Namib Desert

dune. The height of barchans ranges from a few millimetres to giant dunes of between 500 m and 1,000 m. The wings advance more rapidly than the centre of the dune, as the rate of advance is inversely proportional to the height of the dune. Individual barchans do exist, but they often occur in groups or linear formations to form a constantly shifting 'sand-sea'.

Seif-dunes derive their name from an Arabic term *sword-dune*. These knife-edged ridges of sand or longitudinal dunes form long parallel ridges of sand separated by wind-scoured depressions. Groups, or chains, of seif dunes can extend over 100 km and be over 200 m high. The tops and sides of such dunes are often serrated due to local wind action and eddying.

Other contemporary classifications of dunes:

- Transverse dunes: resembling large-scale sand ripples, these are large ridges of sand with a steep downwind face and form in large groups or 'fields' of dunes. They occur in areas with an abundant,

(a) In plan

Prevailing wind

Gentle, slightly concave slope

Horn moves faster than centre of dune as there is less sand to move

X   Y

Maximum height: 30 m

Horn

Barchans migrate, moving forwards by up to 30 m yr^{-1}

Figure 2.28 Barchans and their formation

(b) In profile

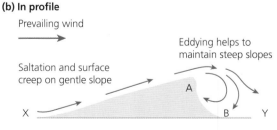

Prevailing wind

Saltation and surface creep on gentle slope

Eddying helps to maintain steep slopes

X   A   B   Y

A   Steep, upper slip slope of coarse grains with continual sand avalanches due to unconsolidated material (unlike a river, coarse grains are at the top)

B   Gentle, basal apron with sand ripples: the finer grains. as on a beach, give a gentler gradient than coarser grains

constant supply of sand, and can be the result of barchan dunes merging if their supply of sand increases significantly.

- Parabolic (or blowout) dunes: a horseshoe-shaped dune where the open end faces upwind. These are generally relatively stable features that have been colonised by vegetation, and occur where there is a constant supply of sand, often in coastal locations.
- Star dunes: these have variable slip face directions as they occur in areas with an abundant supply of sand but variable wind directions. These are often quite massive and permanent dome-like features extending over hundreds of kilometres and being hundreds of metres high, creating a **draa** landscape.

Skills focus

Construct an annotated sketch or diagram to show the characteristics and formation of an aeolian desert landform.

Landforms created by water

Even though most rainfall in hot desert environments is episodic and low intensity, there are occasional, sudden isolated heavy downpours which can have a significant effect on the development of desert landscapes. In mountainous upland areas and plateaux water most often acts as an agent of erosion, while in lowland regions depositional features dominate.

Wadis

Wadis are steep-sided, wide-bottomed, gorge-like valleys formed by fluvial erosion in arid and semi-arid regions. As the wadis are rarely filled with water the valley walls are steep and often covered with thick layers of weathered material, with a build-up

of sediment on the valley floor. These valleys are either permanently dry, due to climate change and the drying up of former rivers, or only occasionally occupied by ephemeral streams, which can flow as a torrent with very high levels of discharge following sudden storms. They can range from small channels a matter of metres in length to complex channel systems over 100 km long. When ephemeral channels do appear on the valley floor they are often **braided** due to the large amounts of previously deposited material that they have to find a pathway through. The stream pattern will change following each flash flood (Figure 2.30).

Bajadas and pediments

Where there is a distinct break in gradient as highland regions meet a more gently sloping lowland area or depression in hot desert environments, a **pediment** is found (Figure 2.31). These are gently sloping areas (less than 7°) of bare rock and debris. They are believed to be formed by a range of processes including fluvial erosion, alongside the deposition of material washed down from the uplands and deposited as the **capacity** and **competence** of the water is suddenly reduced by the abrupt change in slope angle.

If wadis, or valleys, containing ephemeral or perennial streams run down from uplands, depositional material may build up on these pediments due to the rapid loss of energy upon meeting the much gentler slopes. The deposited load forms **alluvial fans** as distributaries run out from wadis onto the pediment. The deposited material is **graded**, with the coarsest material left at the upstream end of the alluvial fan and the smallest material on the downstream of the fan. Alluvial fans can extend for a few metres to several kilometres from the mouth of the wadi.

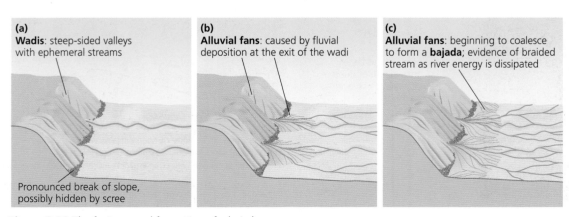

(a)
Wadis: steep-sided valleys with ephemeral streams

Pronounced break of slope, possibly hidden by scree

(b)
Alluvial fans: caused by fluvial deposition at the exit of the wadi

(c)
Alluvial fans: beginning to coalesce to form a **bajada**; evidence of braided stream as river energy is dissipated

Figure 2.30 The features and formation of a bajada

In areas where a linear mountain range has several parallel wadis in close proximity, the alluvial fans may coalesce forming a **bajada** covering, or just beyond, the pediment.

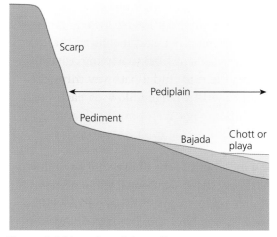

Figure 2.31 Location of the pediment

Playas

Permanent standing bodies of water or lakes are rare in hot desert regions because of the intermittent nature of the rainfall. Where ephemeral streams flow into inland basins or depressions **salt lakes**, **chotts** or **playas** form (Figure 2.32). Following the rare periods of rainfall the water quickly evaporates due to the intense heat, leaving a dry lakebed behind. These lakebeds consist of salts (evaporites) that were carried in by the streams and precipitated during the evaporation. Sodium chloride (NaCl) is the most common precipitate, giving the dry lakebed a colour resembling a beach (the possible origin of their name, as *playa* means beach in Spanish).

Figure 2.32 A salt lake in the Namib desert

Other salts common in these crusts include calcium sulphate (gypsum), sodium sulphate, magnesium sulphate, and potassium and magnesium chlorides. People have commercially exploited these rich salt deposits throughout history. Chott el Djerid in southern Tunisia occupies an intermontane basin, is over 100 km wide and is almost entirely bare salt flats.

Inselbergs and mesas and buttes

In regions where the bedrock is sedimentary with horizontal bedding planes with a resistant cap rock, fluvial erosion creates **mesas** and **buttes**. Derived from the Spanish for table, mesas are isolated flat-topped plateau or hill-like features, with steep slopes or cliffs on at least one side, often falling away to a wadi or canyon. Buttes closely resemble mesas in all ways except that they are much smaller. They are thought to be the remains of heavily dissected plateau where the water has eroded all but a thin isolated rock pillar. The lower slopes of both are covered by **scree** resulting from mechanical weathering and rock fall from the upper sections. Monument Valley National Park in Arizona famously is home to some of the most spectacular examples (Figure 2.33).

Figure 2.33 Mesas and buttes in Monument Valley National Park, USA

Inselbergs are more rounded steep-sided hills that rise abruptly from a lowland plain. They are characteristic of tropical savannah zones, but are found in a range of climatic zones and are generally composed of solid crystalline rocks such as granite. There has been considerable debate about the processes that formed them, but most agree that they are relic features of previous geomorphological processes. One theory suggests that the surrounding slopes retreated in parallel as pediments encroached into the remaining highlands in a process called **pediplanation**. Others argue they are the result of deep chemical weathering when the resistant rock was buried beneath the surface and has since been exposed by the subsequent removal of the surrounding regolith. Both mechanisms suggest that the desert environments which inselbergs are found in today were once more humid, supporting the theory that deserts experienced wetter periods (**pluvials**) in the past.

Fieldwork opportunities

The Grand Canyon National Park in southwestern USA is a fantastic location to study the landforms of a hot desert landscape. A fieldtrip would allow you to investigate a whole range of features including:

- open deserts
- plateaux
- mesas and buttes.

Studying the mighty Colorado River provides the opportunity to explore the role of water in a hot desert landscape, including the Grand Canyon itself.

Characteristic desert landscapes through time

Perhaps the one feature that initially comes to mind when thinking about **characteristic desert landscapes** is sand. However, hot desert landscapes include some of the most spectacular landscapes in the world, alongside others that are exceptionally monotonous. From seemingly endless flat plains to fields of massive dunes the size of mountain ranges, some hot deserts are also home to remarkable features that attract millions of tourists each year, including Uluru in the Red Centre of Australia and the Grand Canyon in southwestern USA. What separates a desert landscape from others is the lack of vegetation, leaving their geological fabric exposed and allowing the structure of their geomorphological features to be clearly visible.

No two hot desert landscapes are the same because of the unique interaction of processes, time and landforms in different locations. Each hot desert landscape is formed by the processes explored earlier in this chapter and is composed of unique assemblages of the features and landforms also discussed above. Some of the factors relating to time, process and landforms that lead to this fascinating diversity of hot desert landscapes include:

- the speed and nature of weathering. Rocks will be broken down quicker in areas with greater diurnal temperature ranges, where there are more rapid cycles of heating and cooling of the different minerals in different rocks
- the presence of moisture. The amount, and availability, of water in different places will lead to the formation of different landscapes, as it can speed up weathering processes, including the growth of salt crystals or frost shattering, especially if temperatures fluctuate rapidly. Where moisture is present hot deserts are more likely to have landscapes with biological and organic features, from lichens and bacteria, to grasses and cacti.

The speed at which both of the above shape the desert landscape is heavily dependent on time. Some deserts will have extreme temperature fluctuations on a daily basis, while others will have more moderate diurnal ranges; some will regularly have moisture present, while others will be completely arid for years.

We saw above how the landscape of a hot desert may be dominated by erosional or depositional landforms, and that those landforms could be either aeolian or fluvial. Again, the rate of landscape formation will vary as the presence of wind and flowing water may be constant, seasonal or sporadic depending on the climate.

It is also important to remember that water and wind not only have the power to shape the desert landscape of the present, but they may also have played a fundamental role in the geological history of the landscape. Some landscapes will have features that record a much wetter past, where great rivers may once have flowed through areas that are now almost completely arid, for example the presence of inselbergs, indicating that an area was once far wetter than it is today. Some desert landscapes may be dominated by features that are relatively recent, like alluvial fans at the mouth of wadis that may have developed over hundreds to a few thousands of years. Other desert landscapes are truly ancient, like the great 'shield deserts', including much of central Australia that is dominated by vast almost featureless plains, with rare outcrops of resistant rock, like Uluru. Wind and water have been slowly wearing away these landscapes since the ancient super-continents, like Gondwana, broke up between 100 and 200 million years ago. In truth no two deserts are exactly alike.

However, water action can lead to the formation of one of the most distinctive features of semi-arid regions, **badlands** (Figure 2.34). In areas with relatively impermeable less-resistant geology, run-off from heavy but sporadic rainfall moulds a distinctive landscape. Vegetation cover is never able to establish itself due to the aridity so there is little holding the regolith and bedrock together, so flowing water is able to create dramatic erosional and depositional landforms.

Badlands have a range of general features:

- wadis of all shapes and sizes with steep sides and debris-covered bottoms
- unstable slopes that regularly collapse as gullies erode headward into hillsides
- regular mass movement as slope failure and slumping are common
- as the many wadis and small gullies emerge onto lowland areas alluvial fans and bajadas are common
- as water carves out surface cracks pipes are formed which are eroded into larger caves as surface water flows into them
- as caves erode further over time, aided by weathering processes, natural arches form.

Due to their striking landscapes and often inaccessible locations, many badlands, especially in the USA, have proved popular with adventure tourists, as they remain undeveloped due to the lack of options for other viable economic activity. The unique landscape of some of the badlands of southern Tunisia near Matmata has been popular with film-makers for some time, with sequences for the original *Star Wars* film being filmed here, keeping the location popular with tourists to the present day.

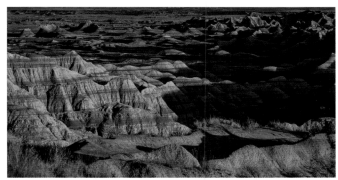

Figure 2.34 Badlands in South Dakota, USA

This section shows just how unique all hot desert landscapes are, with different factors giving rise to different combinations of landforms and features to give each desert its own characteristic landscape. Unlike badlands, not all assemblages of landforms form a named landscape type. Table 2.2 gives a summary of the dominant features of some of the world's most well-known hot deserts.

Table 2.2 Physical features of different desert landscapes

Desert	Characteristic landscape features
Arabian (Arabian Peninsula)	Almost entirely sandy, with some of the largest sand dune systems in the world
Australian Great Sandy, Simpson and Great Victoria Gibson and Sturt	Mostly sandy plains Mostly stony surfaces
Chihuahuan (Arizona, New Mexico, Texas and north central Mexico)	High flat plateau with some stony surfaces and sandy soil, broken by mountain ranges and distinctive mesas
Kalahari (southwestern Africa)	Extensive sand dunes interspersed with gravel plains
Mojave (Arizona, California, Nevada)	A varied landscape including sandy soils, gravel pavements and salt flats
Sahara Northern Africa	Vast ranges of dunes amongst mountains and rocky areas, and gravel plains and salt flats
Thar India and Pakistan	Mostly sand dunes with areas of gravel plains

Skills focus

It is important that you are able to describe and explain the features of characteristic hot desert landscapes. Use the library or the internet to research two contrasting hot deserts, then complete the following tasks.

Identify the different landforms in each desert.

Explain the processes that created these landforms.

Try to identify the age and origin of the features, for example, do they result from current processes or are they relic features that are being modified in the present day?

Draw and annotate a sketch of the landscape to describe the characteristic features.

2.4 Desertification

There have and continue to be a range of definitions of **desertification**, and it is generally accepted that it is not simply the 'advance of deserts', but it could include sand dunes encroaching on land. However, a UNESCO source in 2015 defines it as 'the persistent degradation of dryland ecosystems by human activities and by climate change'.

Key terms

Desertification – The persistent degradation of dryland ecosystems by human activities and by climate change.

The introduction to this chapter indicated that hot deserts are open systems and that the process of desertification is an example of positive feedback. It is also possible to think of desertification as an open system with inputs, components and outputs, although this is quite simplistic and the issue is developed in greater detail later in the chapter. This section goes on to explore the changing extent of hot deserts during the Holocene and the causes and impacts of desertification, alongside the possible future impacts of climate change on the landscapes and people in areas at risk of desertification in the twenty-first century.

The changing extent and distribution of hot deserts over the last 10,000 years

As the Earth has experienced colder and warmer periods during the **glacials** and **interglacials** of the Pleistocene the extent and distribution of desert environments has altered considerably. Records show the distribution of deserts has changed markedly since the last glacial

maximum about 18,000 years ago (Figure 2.35 – NB not all areas shown were hot deserts, but all were arid).

We clearly see that during the height of the last glacial maximum about 18,000 years ago, the extent of arid areas was huge, however almost all of these areas would not have been hot, as global temperatures were on average about 12°C lower than today. By about 8,000 years ago, during the 'Holocene Climate Optimum', temperatures were much warmer, with similar temperatures to today, but the extent of deserts is confined to very small areas, with present day deserts extending over considerably larger areas.

Key question

You may want to research the factors that control these long-term cycles of natural climate change.

Last glacial maximum (circa 18,000 YBP)

Key
☐ Extreme desert

Early Holocene (circa 8,000 YBP)

Key
☐ Extreme desert

Present potential vegetation

Key
☐ Extreme desert (<2% vegetation cover)

Figure 2.35 Changing extent of deserts over the last 18,000 years

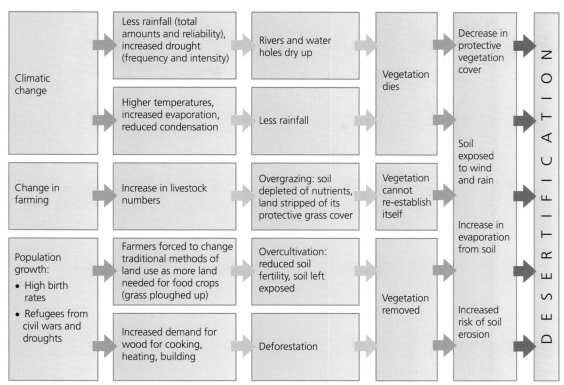

Figure 2.36 The causes of desertification in areas such as the Sahel

The causes of desertification

In the context of the definition and history given above, the causes of desertification are a combination of human factors intensified by physical processes and natural and human enhanced climate change. Concern about desertification is not a new phenomenon, but some would argue that it was the first global environmental issue to be acknowledged as occurring on a global scale.

Due to debate among some scientists about what is the most important cause of desertification: climatic factors or direct human action, there appears to be slow movement on agreeing the best way to tackle it. However, it is clear that whatever the causes of it, recent enhanced climate changes, together with direct human activities, are exacerbating desertification.

Climate change

The IPCC states that areas like the Sahel are experiencing:

- less and more unpredictable rainfall – in the last 25 years of the twentieth century the Sahel had the greatest loss of rainfall anywhere on the planet, and rates of decline were the fastest since instrumental measuring began
- higher temperatures – without available moisture local evapotranspiration is limited, and in desert areas this is often the only local water source for precipitation
- a reduced water supply as rivers dry up and the water table falls.

Human causes

Historically population levels in many hot deserts and their margins remained relatively low and stable and were well adapted to the environmental conditions. However, in recent decades people are having a direct effect on rates of desertification due to:

- population growth from natural increase (possibly as small local populations begin to move through stages of demographic transition) and migration as populations flee environmental problems or human conflicts in neighbouring areas
- population pressure on the land because of:
  - intensification of agriculture to feed larger local populations

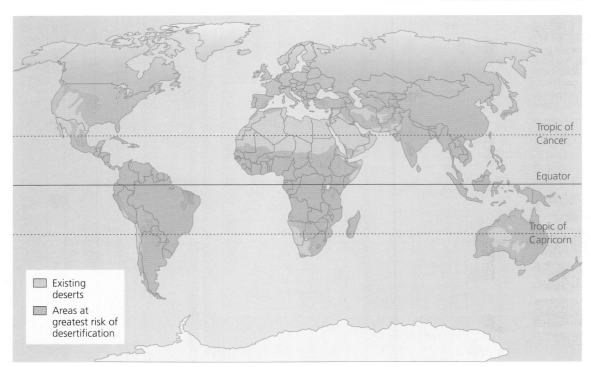

Figure 2.37 Areas at risk from desertification

- over-cultivation, overgrazing and deforestation, for a huge range of reasons in different locations. All remove vegetation which cannot then get re-established due to soil erosion following the exposure of the surface to wind and rain
- the increasing demand for fuelwood in some locations which cannot be underestimated.

All the above lead to **land degradation** and ultimately desertification.

Areas at risk of desertification

The areas of the Earth at greatest risk of desertification are shown in Figure 2.37.

According to the UN, approximately 25 per cent, or 3.6 billion hectares, of the Earth's surface is currently desertified. The UN also reports that around one billion people in some 100 countries are at risk of desertification, with around 12 million hectares of land being lost to desertification annually.

Impact of desertification on ecosystems, landscapes and populations

The impacts of desertification are extensive and can vary considerably from place to place, but general impacts on ecosystems, landscapes and populations are outlined below with some place-specific issues covered in the case study of the end of this chapter.

Climate change and possible futures for local populations in hot desert environments

Predicted climate change and its impacts

It is clear that the area of degraded land and deserts has increased over recent decades and is continuing to do so at present. It is also clear that the majority of climate scientists agree that during the last quarter of the twentieth century and start of the twenty-first the Earth has experienced significant climate change with unprecedented enhanced rates of global warming. Some sources suggest that the average temperature of the planet rose between 0.3 and 0.6°C during the last century, and that with every doubling of the concentrations of atmospheric CO_2 drylands could experience temperature increases of between 2 and 5°C.

Future desertification is not a simple problem to predict or solve. Desertification is not a problem with a single simple cause or equally with a single simple solution (Figure 2.38). Figure 2.39 (page 80) attempts to show how future climate change and desertification are both components of major feedback loops. The inner loops show how desertification and biodiversity loss are linked to climate change via soil erosion while the outer loop illustrates the link between climate change and biodiversity loss.

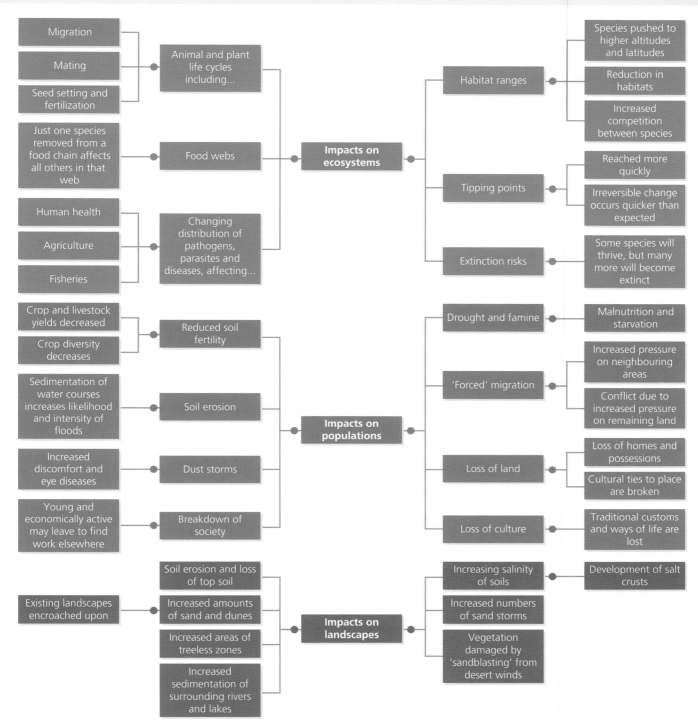

Figure 2.38 Impacts of desertification on ecosystems, landscapes and populations

The problems relating to future desertification are likely to affect many of those identified as the **least developed countries** by the UN, as many of these are located in, or have areas of hot deserts at their margins. The problems of very low levels of HDI (**Human Development Index**) will exacerbate and be exacerbated by the impacts of desertification, especially as many of these countries also have the most pressing population pressures. However, the pattern and rates of desertification are as varied as the rates of poverty and poverty-related problems. This suggests that other than global-level solutions to climate change, solutions to the problems posed by desertification are going to be found at **grassroots** levels by the local populations at risk.

Alternative possible futures for local populations

The Millennium Ecosystem Assessment commissioned by the UN in 2005 said that looking to the future is

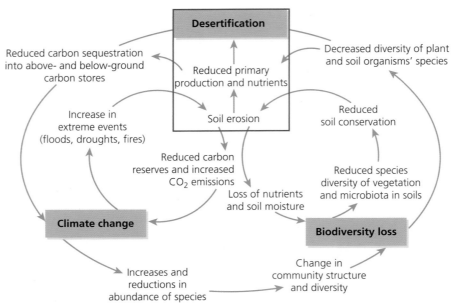

Figure 2.39 Links and feedback between desertification, global climate change and biodiversity loss

difficult and combatting desertification will depend on the development path the world follows. They suggested four scenarios, two of which see a world of increasing **globalisation** and two of which see a world of increasing **regionalisation**. In both scenarios they see opportunities for both **top-down** and **bottom-up** solutions including actions that are *reactive* to problems that already exist and actions that are *proactive* to try to limit future problems (Table 2.3).

Global officials may have mapped out four possible routes to dealing with desertification, but the consensus is that the area of desertified land will increase in the future. However, there are some key points on desertification and the future:

● Poverty and unsustainable landuse will continue to be the main causes of desertification.

● Relieving the human pressure on at-risk areas is strongly linked to poverty reduction.

● Population growth and increased demand for food will lead to an expansion of cultivated land and inevitable deforestation and land degradation.

● Climate change and desertification are linked, but the future impacts and mitigation of both are uncertain.

● Hot deserts and their margins are predicted to experience more extreme weather events such as droughts and floods.

● Proactive approaches are likely to be more effective in addressing economic problems and desertification.

● If local knowledge and local priorities are incorporated into future plans to help mitigate the vulnerability of local populations and ecosystems it is less likely that 'solutions' will be proposed that simply make the problem worse or move it elsewhere.

Table 2.3 Scenarios for addressing the problems of desertification

		Path of world development	
		Globalisation	**Regionalisation**
Approach to ecosystem management	**Reactive**	**Global orchestration** Suggests an increasingly globalised world where ecosystem management is reactive to current issues. 'Development' initiatives are top down and focus on equity, economic growth and public goods such as infrastructure and education.	**Order from strength** In a more regionalised world, countries, regions and communities are empowered to react to environmental problems as the 'development' focus is more bottom up but concentrates on security and economic growth. However, it does not look to issues that may arise in the future.
	Proactive	**TechnoGarden** In a globalised world, technology is shared to allow populations affected by future desertification to be proactive in prevention by using green technologies. The problems of those affected can be alleviated, thus alleviating future global-scale problems from developing.	**Adapting mosaic** A bottom-up proactive scenario where the emphasis is on empowering the development of local adaptations and learning.

- Local populations need to be helped to adapt to current and future pressures to reduce current and future vulnerability and improve their resilience.
- The extension of intensive irrigation systems will have a negative impact, increasing desertification.
- Current pressures on ecosystems due to climate change, overgrazing, over-cultivation and large-scale irrigation will get worse if they are not 'reacted' to.
- It is not inevitable that globalisation will lead to increased desertification.
- Globalisation could lead to increased co-operation and resource transfers to support those on the front line in areas affected.
- Organisations in a globalised world can be proactive in promoting the protection of local property rights and encourage all levels of governance to integrate environmental issues into their planning.
- In a fragmented regionalised world, the role of global organisations will be limited and there could be reduced interest in the exchange of technology and resources, and in issues beyond national borders.

Skills focus

In what ways do you think geographical information systems (GIS) and remotely sensed data could be useful for monitoring desertification?

Fieldwork opportunities

The Toubkal National Park and surrounding area, two hours drive from Marrakesh in Morocco, is accessible and provides the opportunity to study water management in the margins of a hot desert environment. Also, due to its location on the edge of the Saharan desert, it is possible to study the impacts and potential future impacts of desertification on the local populations and their way of life.

2.5 Quantitative and qualitative skills in hot desert environments

Quantitative and **qualitative** techniques in geography are defined and explored in Chapter 12, and you should read this section carefully.

It is very unlikely that you will be fortunate enough to visit and undertake fieldwork in a hot desert environment as part of this course, however it is important to reflect on how a range of quantitative and qualitative skills could be utilised in a study of hot desert landscapes. As most hot deserts are inaccessible for most of us, the case study on page 83 illustrates how a number of quantitative techniques could be employed in a sand dune landscape which, like hot deserts, is also dry and dominated by aeolian processes.

Key question

Why would you use quantitative methods over qualitative methods to assess the extent of desertification?

Quantitative data and hot desert environments in the twenty-first century

Satellite technology has proved to be one of the most powerful tools for collecting quantitative data about hot deserts, especially the ongoing process of desertification over the last 25 years. The **Landsat** collaboration between NASA and the United States Geological Survey (USGS) began in 1972, with Landsat 7 (1999) and 8 (2013) now providing some powerful evidence for the spread of desertification (Figure 2.40).

Figure 2.40 Landsat images show the shrinking of Lake Faguibine in Mali between 1974 and 2006

Landsat can provide images of the same geographical area over time. If the images are taken at the same point in the growing season they can indicate where significant changes to the landscape have taken place. Powerful geographical information systems (GIS) and geospatial mapping techniques can use high-tech software to apply different filters to the satellite imagery to identify changes to the desert surface, including vegetation cover, soil characteristics, water content/cover and surface temperatures, among others.

Increasingly scientists are working towards creating global databases of field observations, where researchers on the ground record changes in land use, which can be used alongside remotely sensed data to record the extent and impacts of desertification.

Qualitative data and hot desert environments in the twenty-first century

This chapter illustrated how understanding the causes of desertification is complex and a number of authors suggest that this provides opportunities to employ more qualitative research techniques. Some suggest that it is crucial to understand and explain the links between social and economic factors and local development pathways that produce and exacerbate land degradation and desertification. Quantitative empirical data can provide excellent data on the existence and extent of desertification but cannot explain the social and economic factors that drive local populations to pursue unsustainable activities that degrade the land.

One study on Lesvos in Greece promoted the use of in-depth interviews which, although prepared in advance, did not have a specific research hypothesis (Figure 2.41). Without using leading questions, researchers were able to collect a large range of information about the experiences and thoughts of the respondents (Iosifides and Politidis, 2005).

Figure 2.41 Soil degradation in Lesvos, Greece

The interview schedule used with local farmers in Greece focused on five themes:

- characteristics of the respondent and their farming practices
- views on rural development and development processes
- views on local socio-economic conditions
- community resilience (alternative employment and income opportunities and social relations)
- views of the local community on the problems of desertification and land degradation.

In-depth interviews generate huge amounts of information which then needs to be **coded** into different categories so that a detailed analysis can take place. The researchers suggest that it is impossible to formulate solutions to the problems of desertification without fully understanding the motivations and circumstances of the local populations of the areas affected, and that this can only be achieved using qualitative techniques.

Case study of a hot desert environment setting: Shifting dunes threaten settlements – measuring aeolian processes that shape sand dunes

Geographical context

In areas like the Sahel, the semi- to hyper-arid zone running in a belt east to west across Central Africa along the southern fringes of the Sahara desert, the notion of huge areas of sand dunes rapidly spilling out of the desert and swallowing up huge expanses of land is generally agreed to be a misconception. However, it is well documented that dunes have, at times, advanced towards the city of Nouakchott, the capital city of Mauritania on the western edge of the Sahara (Figure 2.42).

For areas concerned about encroaching sand dunes scientists would obviously need to take measurements to establish the size of the dunes and the rate at which the sand is advancing to establish the nature of the threat.

Figure 2.42 Dunes advance towards Nouakchott

Response

As geography students we could use a range of techniques to gather field data to measure and quantify some of the aeolian processes at work in shifting sand dunes. As most of us will not be able to visit sand dunes in mid- and low-latitudes the same techniques could be employed in coastal sand dunes closer to home where similar aeolian processes are at work. This case study includes examples of the types of fieldwork data and techniques that could be conducted in a local sand dune setting.

Investigating aeolian processes in a local sand dune setting

Fieldwork aim

To investigate how wind speed and direction affect the rate of sediment movement on a stretch of sand dunes.

Background

A fieldwork investigation was completed along a stretch of dunes along the Drigg Coast. Drigg is located on the Cumbrian coast in northwest England (Figure 2.43a and 2.43b, page 84).

Collecting field data

At regular 5 m intervals along the dune front, students could measure the rate at which the wind is removing/transporting sand. At each collection point a small plastic sheet is secured on the sand surface. Using an anemometer and weather vane the wind speed and wind direction are recorded. A known amount of sand is placed on the plastic sheet and a stop watched is used to record the length of time taken to remove the sand (Figure 2.44, page 84).

Using systematic sampling the wind direction and wind speed could be recorded at 5 m intervals along the dune front. A measured amount of sand is placed on the plastic sheet and the time taken for all the sand to be transported away recorded in the table below.

This primary data could then be analysed to assess areas of the dunes where sand is being transported most rapidly. Or, knowing the quantity of sand and the full length of the dunes, a calculation/estimation could be made of the amount of sand being moved along the

Distance along dunes (m)	0	5	10	15	20	25	30	40	45	50
Wind direction										
Wind speed (m/s)										
Time (s)										

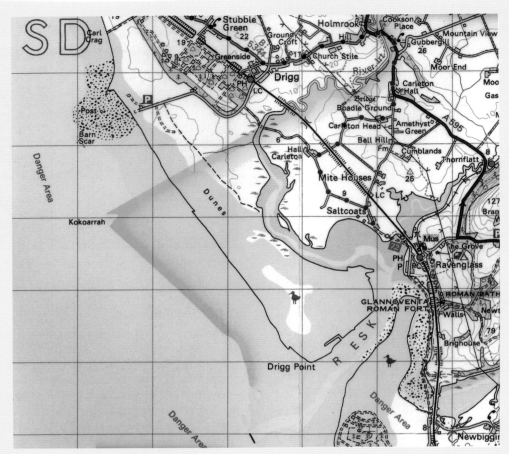

Figure 2.43a The Drigg Coast, Cumbria

Figure 2.43b Sand dunes on The Drigg Coast, Cumbria

Figure 2.44 Measuring the effect of wind on sand dunes

dunes in a given period, for example, in grams of sand per second per km.

Students could also investigate if the slope angle or profile of the dunes made any difference to the rate of sand transportation. So at each 5 m interval along the dunes a measuring tape, ranging poles and clinometer could be used to measure the profile of the dunes. The tape measure is placed at the high tide mark on the beach and taken perpendicular to the beach inland across the dunes in a straight line. Ranging poles are placed at 2 m intervals along the tape measure and the clinometer used to measure the angle between each pole (Figure 2.45).

Figure 2.45 Measuring the dune profile

Having recorded this data in a table like the one on page 83, students could then use it to construct cross sections through the dunes. A correlation could then be conducted to see if there is a relationship between the steepness of the dunes and the rate of sand transportation from the earlier investigation.

Skills focus

What techniques could you use to present and analyse this field data back in the classroom?

Case study at a local scale: Causes, impacts and implications of desertification for sustainable development

Aims of case study:

- to illustrate how population pressure can lead to desertification
- to analyse the relationships between people and landscape in areas affected by desertification
- to show how sustainable solutions can help people adapt to and mitigate the effects of desertification.

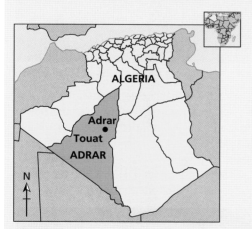

Figure 2.46 Location maps - Adrar, Touat region, Southern Algeria

Close to Adrar in the Touat region of southern Algeria (Figure 2.46) people have turned to a traditional approach to improve their **resilience** to the arid climate and **adapt** to and **mitigate** the impacts of the pressures of desertification on their precious water supply.

Geographical context

For centuries efficient and sustainable irrigation systems utilising the limited water supplies of the **oases** of this area of Northern Sahara have enabled local populations to survive in the hostile arid conditions. Small settlements traditionally developed around a small oasis, utilising ground water to sustain themselves and small-scale agriculture (Figure 2.47a). However, in recent decades water supplies, soil quality and the sustainability of these oases have come under threat due to:

- population growth – a natural increase of 1.5 per cent per annum and increased life expectancy (72 females, 70 males)
- intensification of agricultural production and over-cultivation – for example, moving towards monoculture of high yielding modern seed varieties (some of which are cash crops for export) and the use of chemical fertilisers and modern machinery (Figure 2.47b, page 86)
- most importantly the use of modern powerful water pumps removing huge quantities of water direct from groundwater stores. This is then used for jet irrigation where most is lost to high rates of evaporation while there is only an average of 50 mm of rainfall a year.

Figure 2.47a Small-scale farming near desert oasis in Touat region

Figure 2.47b More intensive farming in the Adrar region

This has created a number of issues:
- further intensification of agriculture to try and counter declining yields
- deeper and deeper wells have been dug to tap even deeper groundwater supplies
- more intensive use of fertilisers.

Response

Over the last two decades (prior to events following the Arab Spring in North Africa) there was a collaboration between a network of **non-governmental organisations (NGOs)** from the Mediterranean region and Algeria itself to develop a strategy to protect the oasis ecosystem of this region (including the Algerian association Touiza and The Mediterranean Co-operation of the MED Forum). Four oases were selected where activities took place, such as restoring palm trees to provide shade and help prevent soil erosion and campaigns to raise public awareness about preserving the oases. Ultimately this led to schemes to rehabilitate and preserve the traditional irrigation schemes known as **foggaras**.

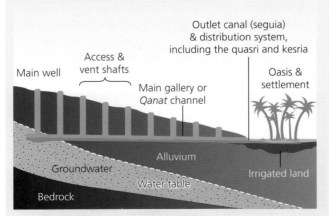

Figure 2.48 Diagram of a foggaras irrigation system

Foggaras are a series of man-made galleries and tunnels that harvest groundwater. They work where the oasis and village are located on the floor of a gentle-sided valley, or at the foot of a gentle slope. Wells are dug about 2.5 km *upstream* from the oasis and underground

pipes then run almost horizontally under the surface, at a very slight incline of about 1–2 mm/m. Water then flows gently downstream to the oasis and settlement. The galleries and access wells (located about every 20–30 m) are constructed out of locally quarried stone and palm trunks held together with a cementing mix of clay and straw (Figure 2.48).

The advantages of foggaras include:
- low gradient and gravity flow prevent erosion by fast-flowing water and do not require expensive diesel-driven pumps
- water flowing underground prevents evaporation and water loss
- locally sourcing building materials provides work and reduces the need for imported materials.

Once the water reaches the oasis it flows into an open-air canal called a *seguia*. Here it is collected in a small basin, or reservoir, called a quasri (Figure 2.49).

Figure 2.49 Water collected from a foggaras system in the quasri

With water being such a valuable commodity for the community they set up a 'water assembly' to decide how much water those who have water rights receive. A comb-like stone device (kesria) is used to divide the water from the quasri into various irrigation channels to be used freely by each of those with water rights (Figure 2.50). The 'water deciders' hold responsibility for distributing the water.

Figure 2.50 Water distributed via a kesria

Maintaining the foggaras system can be a dangerous and difficult job, so if galleries or wells do collapse more modern surveying techniques and equipment are used to help repair, or dig new wells, to direct water to the system. In oases where electricity is available, any spare/excess water downstream of the oasis can be gathered and pumped back to the main well at the head of the foggaras to maintain the supply and prevent the groundwater supply running low.

Local populations also benefit from:
- workshops on:
  - saving water
  - combating pollution and desertification more generally
  - preserving palm trees
  - rehabilitating existing but currently redundant foggaras
- being able to take their skills learnt on their own foggaras projects to other communities in the region and thus create a trickle-down effect.

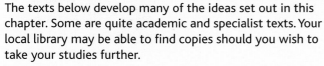

Review questions

1 Using a systems approach can you identify more examples of positive and negative feedback in hot desert systems?

2 What are the main reasons for the changing global distribution of hot desert landscapes over the last 10,000 years?

3 What are the different sources of water in hot desert environments?

4 For a local population in a hot desert environment you have studied:
- What are the predicted impacts of climate change?
- What are the possible future alternative outcomes?

5 What aspects of your coastal sand dune study could be used to help those preparing a Shoreline Management Plan to sustainably manage that stretch of coastline?

Further reading

The texts below develop many of the ideas set out in this chapter. Some are quite academic and specialist texts. Your local library may be able to find copies should you wish to take your studies further.

Middleton, N. (2009) *Deserts: A Very Short Introduction*. Oxford University Press

Hill, M. (2002) *Access to Geography: Arid and Semi-arid Environments*. Hodder Education

Harvey, A. (2012) *Introducing Geomorphology: A Guide to Landforms and Processes*. Dunedin

Press, F. & Siever, R. (2003) *Understanding Earth*. Freeman

Grotzinger, J. & Jordan, T. (2014) *Understanding Earth*. Freeman

Laity, J. (2008) *Deserts and Desert Environments*. Wiley-Blackwell

Thomas, D.S.G. (2011) *Arid Zone Geomorphology: Process, Form and Change in Drylands*. Wiley-Blackwell

Goudie, A.S. (2013) *Arid and Semi-arid Geomorphology*. Cambridge University Press

Granger, A. (2015) *The Threatening Desert: Controlling Desertification*. Routledge

Williams, M. (2014) *Climate Change in Deserts: Past, Present and Future*. Cambridge University Press

More information about foggaras may be found by completing an internet search for 'Algeria foggaras' or visiting www.unesco.org/mab/doc/ekocd/algeria.html

Iosifides, T. & Politidis, T. (2005), 'Conducting Qualitative Research on Desertification in Western Lesvos, Greece' In: *The Qualitative Report*, Volume 10, Number 1, March 2005, pp. 143-62. www.nova.edu/ssss/QR/QR10-1/iosifides.pdf

Question practice

A-level questions

1. Outline the role of wind on erosion processes in hot deserts. **(4 marks)**

Figure 2.51 Exposed soil profile in a hot desert soil in central Australia

2. Figure 2.51 shows an exposed soil profile in a desert soil. This exposed soil profile is an exposed section through the layers of a desert soil in central Australia. The deep orange, red and yellow colours are iron oxides and laterites formed under oxidising conditions.

   Using Figure 2.51 and your own knowledge analyse the relationship between climate, soil and vegetation to explain why the development of soils like those in Figure 2.51 is extremely slow.   **(6 marks)**

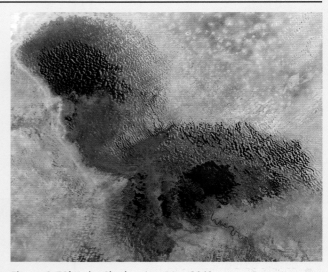

Figure 2.52b Lake Chad region, May 2013

3. Figures 2.52a and 2.52b show the region surrounding Lake Chad in 1972 and May 2013. Straddling the borders of three countries in West Africa, Lake Chad has been a source of fresh water for irrigation and domestic use for decades. Between 1963 and 2001 the lake shrank to nearly a twentieth of its original size, from approximately 25,000 km^2 to 1,350 km^2. Since the 1960s about half of the decrease in the size of the lake has been attributed to human use of water, with the rest attributed to changing patterns in climate.

   Using Figures 2.52a and 2.52b and your own knowledge, assess the possible impacts of climate change in this desert environment.   **(6 marks)**

4. Assess the relative importance of the roles of processes of erosion and processes of weathering in shaping hot desert landscapes.   **(20 marks)**

Figure 2.52a Lake Chad region, 1972

3

Coastal systems and landscapes

Coastal zones are dynamic environments with distinctive landscapes formed by the interaction of a range of wind, marine and terrestrial processes. Coastal environments are important to the human race. About half of the world's population live on coastal plains with over 50 per cent of the population living within 150 km of the sea.

In this chapter you will study:
● coasts as natural systems
● the systems and processes leading to the development of coastal landscapes
● the management of coastal landscapes, including coastal flooding and erosion
● sustainable approaches to managing coastal systems in the future.

3.1 Introduction to coastal systems and landscapes

Coasts as natural systems

The introduction to Chapter 1: Water and carbon cycles outlined how many aspects of physical geography can be studied using a **systems** approach and you should read that section carefully before reading further here. This chapter explores coastal environments, which are **open systems** with a range of **inputs, components/**

stores, flows/transfers and **outputs** that combine to create distinctive landscapes. When there is a balance between the inputs and outputs then the system is said to be in a state of **dynamic equilibrium**. If one of the elements of the system changes, for example there is increased deposition on a beach but there is no corresponding change in the amount of sediment removed from the beach, then the beach features may change and the equilibrium is upset. This is called **feedback**. Figure 3.1a illustrates the open system of coastal landscapes and Figure 3.1b (page 90) is an example of how a negative feedback mechanism on the coastal zone acts to stabilise coastal morphology and maintain a dynamic equilibrium. The figures also introduce many ideas that are explored in detail in the rest of the chapter.

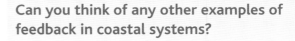

Key question

Can you think of any other examples of feedback in coastal systems?

Coastal landscapes, like all geomorphological environments, consist of a constantly changing assemblage of erosional and depositional landforms; they are the result of continuous change in the elements (or components) of the coastal system. As this system tends towards **dynamic equilibrium**, driven by the input of energy, the processes operating in coastal systems

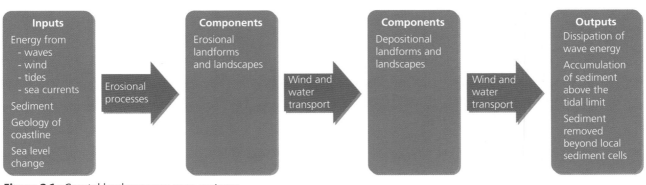

Figure 3.1a Coastal landscapes as open systems

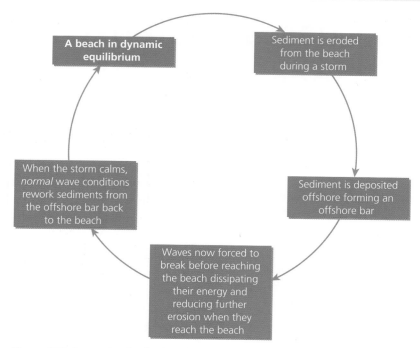

Figure 3.1b A negative feedback mechanism in a coastal environment

continually shape the coastal landforms, creating their distinctive landscape features. However, the coastline itself consists of a series of different zones where specific conditions prevail that depend on factors such as tides, wave action and the depth of the sea. Figure 3.2 illustrates how these zones relate to each other.

● **Backshore** is the area between the high water mark (HWM) and the landward limit of marine activity. Changes normally take place here only during storm activity.

● **Foreshore** is the area lying between the HWM and the low water mark (LWM). It is the most important zone for marine processes in times that are not influenced by storm activity.

● **Inshore** is the area between the LWM and the point where waves cease to have any influence on the land beneath them.

● **Offshore** is the area beyond the point where waves cease to impact upon the seabed and in which activity is limited to deposition of sediments.

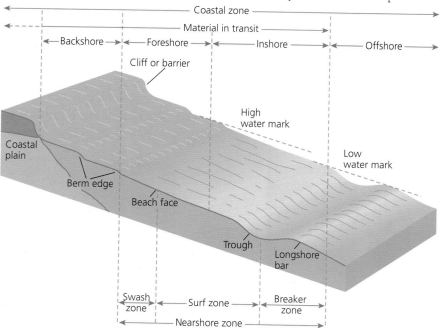

Figure 3.2 Coastal zones

- **Nearshore** is the area extending seaward from the HWM to the area where waves begin to break. It includes the following:
  - **Swash zone:** the area where a turbulent layer of water washes up the beach following the breaking of a wave.
  - **Surf zone:** the area between the point where waves break, forming a foamy, bubbly surface, and where the waves then move up the beach as **swash** in the swash zone.
  - **Breaker zone:** the area where waves approaching the coastline begin to break, usually where the water depth is 5 to 10 m.

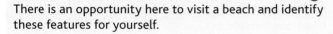

Fieldwork opportunities

There is an opportunity here to visit a beach and identify these features for yourself.

3.2 Systems and processes in coastal environments

This book highlights how most aspects of physical geography can be thought of as natural systems with inputs, stores/components, flows/transfers and outputs; with geomorphological processes, driven by the input of energy, constantly changing the elements of these systems as they tend towards dynamic equilibrium. Coastal environments are no different and before exploring the detail of the processes that shape the landforms that combine to create characteristic coastal landscapes, a summary of a simple coastal system is included below.

- Inputs:
  - energy from waves, winds, tides and sea currents
  - sediment
  - geology of the coastline
  - sea level change
- Components:
  - characteristic erosional and depositional coastal landforms
- Outputs:
  - dissipation of wave energy
  - accumulation of sediment above the tidal limit
  - sediment removed beyond local sediment cells.

You should also read the relevant material at the beginning of Chapter 1 (page 2) under the heading 'Systems frameworks and their application'.

Many of the characteristic features of coastal environments are created by the action of wind, waves, tides and sea currents. To fully explain how these processes work it is important to understand where the energy comes from to drive the various **weathering**, **erosion** and **mass movement** processes.

Key terms

Erosion – The wearing away of the Earth's surface by the mechanical action of processes of glaciers, wind, rivers, marine waves and wind.

Fetch – Refers to the distance of open water over which a wind blows uninterrupted by major land obstacles. The length of fetch helps to determine the magnitude (size) and energy of the waves reaching the coast.

Mass movement – The movement of material downhill under the influence of gravity, but may also be assisted by rainfall.

Weathering – The breakdown and/or decay of rock at or near the Earth's surface creating regolith that remains *in situ* until it is moved by later erosional processes. Weathering can be mechanical, biological/organic or chemical.

Sources of energy in coastal environments

The energy to drive the coastal system is provided by waves, wind, tides and currents.

Wind

Wind is a vital input into the coastal system as it is a primary source of energy for other processes, but is also an important agent of erosion and transport itself. Features of wind as an input into the coastal system:

- Spatial variations in energy result from variations in the strength and duration of the wind. Where wind speeds are persistently high and uninterrupted, wave energy is likely to be higher. Although local weather patterns may influence short-term changes in wind speed and direction, most coastlines will have a **prevailing wind direction**. That is, the wind will generally reach the coast from one direction. This is important as it is one factor that controls the direction that waves approach the coastline, and also the direction of the transport of material in the coastal zone.
- **Fetch** refers to the distance of open water over which a wind blows uninterrupted by major land obstacles. The length of fetch helps to determine the magnitude (size) and energy of the waves reaching the coast.
- Wind plays a vital role in wave formation. Waves are created by the transfer of energy from the wind blowing over the sea surface (referred to as the

'frictional drag' of the wind). The energy acquired by waves depends upon the strength of the wind, the length of time it is blowing and the fetch.

● Wind acts as an agent of erosion as it can firstly pick up and remove sediment from the coast (like sand from a beach) and use it to then erode other features. The most common type of wind erosion is **abrasion**, where the wind uses the material it carries to wear away landscape features. (These processes occur in hot deserts and are covered in detail in Chapter 2.) As wind is able to pick up and move material it is an important agent of moving sediment along the coast, inland from the shoreline and beyond the tidal zone (see below).

Key question

Looking at a map of the British Isles describe the areas where waves will approach the coast with the greatest fetch.

Waves

Once created and driven by the wind, waves are the primary agent of shaping the coast. Figure 3.3

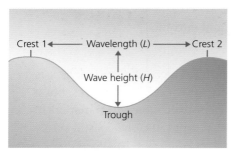

Figure 3.3 Wave terminology

illustrates some of the key features of waves. Wave characteristics include:

● **wave height:** this is the height difference between a wave crest and the neighbouring trough
● **wavelength** or **amplitude:** this is the distance between successive crests
● **wave frequency** or **wave period:** this is the time for one wave to travel the distance of one wavelength, or the time between one crest and the following crest passing a fixed point.

As waves approach shallow water, friction with the seabed increases and the base of the wave begins to slow down. This has the effect of increasing the height and steepness of the wave until the upper part plunges forward and the wave 'breaks' onto the shore (Figure 3.4). The rush of water up the beach is known

as **swash** and any water running back down the beach towards the sea is the **backwash**. Waves can be described as **constructive** or **destructive waves**.

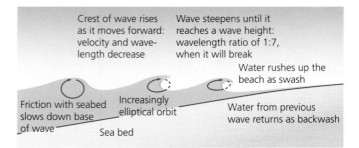

Figure 3.4 Wave movement

Source: Waugh, D. An Integrated Approach

Key terms

Backwash – The action of water receding back down the beach towards the sea.

Constructive waves – Waves with a low wave height, but with a long wavelength and low frequency of around 6–8/min. Their swash tends to be more powerful than their backwash and as a consequence beach material is built up.

Destructive waves – Waves with a high wave height with a steep form and high frequency (10–14/min). Their swash is generally stronger than their backwash, so more sediment is removed than is added.

Swash – The rush of water up the beach after a wave breaks.

Constructive waves

Constructive waves (Figure 3.5) tend to have low wave height, but with a long wavelength, often up to 100 m. They have a low frequency of around 6–8/min. As they approach the beach, the wave front steepens only slowly, giving a gentle spill onto the beach surface. Swash rapidly loses volume and energy as water percolates through the beach material. This tends to give a very weak backwash which has insufficient force to pull sediment off the beach or to impede swash from the next wave. As a consequence, material is slowly, but constantly, moved up the beach, leading to the formation of ridges (**berms**).

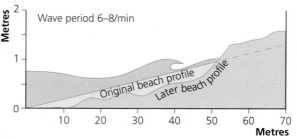

Figure 3.5 A constructive wave

Destructive waves

Destructive waves (Figure 3.6) have high wave height with a steep form and high frequency (10–14/min). As they approach the beach, they rapidly steepen and, when breaking, they plunge down. This creates a powerful backwash as there is little forward movement of water. It also inhibits the swash from the next wave. Very little material is moved up the beach, leaving the backwash to pull material back down the beach. Destructive waves are commonly associated with steeper beach profiles. The force of each wave may project some shingle well towards the rear of the beach where it forms a large ridge known as the **storm beach**.

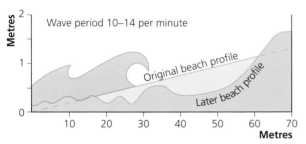

Wave period 10–14 per minute

Original beach profile

Later beach profile

Figure 3.6 A destructive wave

Most beaches are subject to an alternating cycle of constructive and destructive waves. (Figures 3.7a and 3.7b) Constructive waves build up the beach resulting in a steeper beach profile. This encourages waves to become more destructive (as destructive waves are associated with steeper profiles). With time, though, destructive waves move material back towards the sea, reducing the beach angle and encouraging more constructive waves. So the pattern repeats itself. This is another example of **negative feedback** that should maintain the state of **dynamic equilibrium**, but this is often impossible as other factors, such as wind strength and direction, are not constant.

Fieldwork opportunities

There is an opportunity here to visit a local beach and identify the kinds of waves present there.

Figure 3.7a A constructive wave

Figure 3.7b A destructive wave

Wave refraction

Another crucial factor in determining the effects of wave action is the topography of the coastline. When waves approach a coastline that is not a regular shape, they are **refracted** and become increasingly parallel to the coastline. Figure 3.8 shows a headland separating two bays and illustrates the process of **wave refraction**. As each wave approaches the coast, it tends to drag in the shallow water which meets the headland. This increases the wave height and wave steepness and shortens the wavelength. That part of the wave in deeper water moves forward faster, causing the wave to bend. The overall effect is that the wave energy becomes concentrated on the headland, causing greater erosion. The low-energy waves spill into the bay, resulting in beach deposition. As the waves pile against the headland, there may be a slight local rise in sea level that results in a **longshore current** from the headland, moving some of the eroded material towards the bays and contributing to the build-up of the beaches.

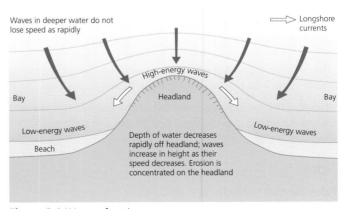

Waves in deeper water do not lose speed as rapidly

Longshore currents

High-energy waves

Bay

Headland

Bay

Low-energy waves

Low-energy waves

Beach

Depth of water decreases rapidly off headland; waves increase in height as their speed decreases. Erosion is concentrated on the headland

Figure 3.8 Wave refraction

Currents

The term **current** refers to the permanent or seasonal movement of surface water in the seas and oceans. There are three currents covered here:

- **Longshore currents** (sometimes known as **littoral drift**) occur as most waves do not hit the coastline 'head on' but approach at an angle to the shoreline. This generates a flow of water (current) running parallel to the shoreline. This not only moves water along the surf zone but also transports sediment parallel to the shoreline.

- **Rip currents** are strong currents moving away from the shoreline. They develop when seawater is *piled up* along the coastline by incoming waves. Initially the current may run parallel to the coast before flowing out through the breaker zone, possibly at a headland or where the coast changes direction. These can be extremely hazardous to swimmers and small boats.

- **Upwelling** is the movement of cold water from deep in the ocean towards the surface. The more dense cold water replaces the warmer surface water and creates nutrient rich cold ocean currents. These currents form part of the pattern of global ocean circulation currents.

Key question

What is the pattern of major ocean currents and what drives them?

Key terms

Longshore or littoral drift – Where waves approach the shore at an angle and swash and backwash then transport material along the coast in the direction of the prevailing wind and waves.

Wave refraction – When waves approach a coastline that is not a regular shape, they are **refracted** and become increasingly parallel to the coastline. The overall effect is that the wave energy becomes concentrated on the headland, causing greater erosion. The low-energy waves spill into the bay, resulting in beach deposition.

Tides – The periodic rise and fall of the level of the sea in response to the gravitational pull of the sun and moon.

Tides

Tides are the periodic rise and fall in the level of the sea. They are caused by the gravitational pull of the

sun and moon, although the moon has much the greatest influence because it is nearer. The moon pulls water towards it, creating a high tide, and there is a compensatory bulge on the opposite side of the Earth (Figure 3.9). In areas of the world between the two bulges, the tide is at its lowest.

As the moon orbits the Earth, the high tides follow it. Twice in a lunar month, when the moon, sun and Earth are in a straight line, the tide-raising force is strongest. This produces the highest monthly tidal range or **spring tide**. Also twice a month, the moon and sun are positioned at 90° (perpendicular) to each other in relation to the Earth. This alignment gives the lowest monthly tidal range, or **neap tides**. At this time the high and low tides are between 10 to 30 per cent lower than the average.

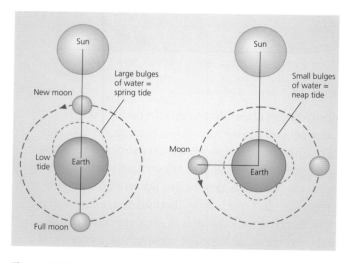

Figure 3.9 The causes of tides

The regular pattern of tides illustrated above is in fact significantly modified in individual locations by the morphology of the seabed, the proximity of land masses and the impact of the spinning forces of the Earth (**coriolis** force). Also, **tidal range** can be a significant factor in the development of a coastline. Tidal range is simply the difference in height of the sea water at high and low tide. This is not fixed due to the tide cycles outlined above. Tidal ranges determine the upper and lower limits of erosion and deposition and the amount of time each day that the littoral zone is exposed and open to **sub-aerial weathering**. Along the coasts of the Mediterranean Sea, tidal ranges are low. This restricts wave action to a narrow width in the coastal zone. In parts of the British Isles, however, tidal ranges are high. This gives a wide zone of wave attack, resulting in the formation of wide wave-cut platforms in many places. Table 3.1 shows a simple classification of tidal range.

Table 3.1 Classification of tidal range

Macrotidal	More than 4 m
Mesotidal	2 to 4 m
Microtidal	Less than 2 m

Tidal surges

Tidal, or **storm surges** are occasions when meteorological conditions give rise to strong winds which can produce much higher water levels than those at high tide. One area affected by this phenomenon is the North Sea and east coast of Britain. Depressions (intense low pressure weather systems) over the North Sea produce low pressure conditions that have the effect of raising sea levels. The sea level can rise by about one centimetre for every one millibar drop in pressure. Strong winds drive waves ahead of the storm, pushing the sea water towards the coastline. This has the effect of *piling-up* water against the coast. The shape of the North Sea means that often water is increasingly concentrated into a space that is decreasing in size (funnelling). High tides, especially those of a spring tide, intensify the effect.

Famously the North Sea was affected by a tidal surge in 1953. However, the storms and associated surges of December 2013 and January 2014 brought some places along the east coast of England higher water levels than 60 years earlier. Figure 3.10 shows the size and power of the waves on Cromer pier and promenade during 5–6 December 2013.

Figure 3.10 Tidal surge at Cromer, Norfolk, December 2013

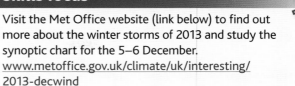

Skills focus

Visit the Met Office website (link below) to find out more about the winter storms of 2013 and study the synoptic chart for the 5–6 December.
www.metoffice.gov.uk/climate/uk/interesting/2013-decwind

Low energy and high energy coasts

Key terms

Coastal sediment budget – The balance between sediment being added to and removed from the coastal system, that system being defined within each individual sediment cell.

High energy coast – A coastline where strong, steady prevailing winds create high energy waves and the rate of erosion is greater than the rate of deposition.

Low energy coast – A coastline where wave energy is low and the rate of deposition often exceeds the rate of erosion of sediment.

Sediment cell – A distinct area of coastline separated from other areas by well-defined boundaries, such as headlands and stretches of deep water.

The processes outlined above combine to create **low energy coasts** and **high energy coasts** depending on local conditions. The following simple classification suggests the features of each.

Low energy coasts:
- Coastlines where wave energy is low.
- The rate of deposition often exceeds the rate of erosion of sediment.
- Typical landforms include beaches and spits.
- Examples include many estuaries, inlets and sheltered bays. The Baltic Sea is one of the best examples due to its sheltered waters and low tidal range.

High energy coasts:
- Coastlines where strong, steady prevailing winds create high energy waves.
- The rate of erosion is greater than the rate of deposition.
- Typical landforms include headlands, cliffs and wave-cut platforms.
- Examples of high energy coasts are the exposed Atlantic coasts of northern Europe and North America, including the north Cornish coast in south-west England.

Sediment sources, cells and budgets

Alongside energy, sediment is also an important input in coastal systems. Coastal sediment comes from a variety of sources, including:

- streams or rivers flowing into the sea
- estuaries
- cliff erosion
- offshore sand banks
- material from a biological origin – including shells, coral fragments and skeletons of marine organisms.

The source of sediment which leads to the build-up of certain depositional features around the British coast is in dispute. Research has suggested that sediment movements occur in distinct areas or **sediment cells**, within which inputs and outputs are balanced. Along the coastline of England and Wales, 11 of these sediment (or littoral) cells have been identified (Figure 3.11).

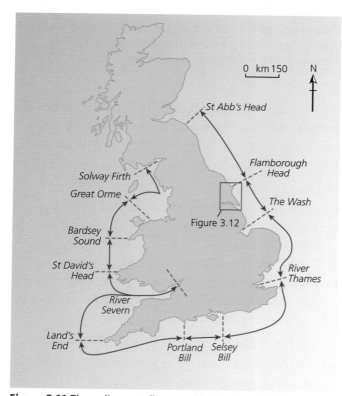

Figure 3.11 The sediment cells around England and Wales

These are distinct areas of coastline separated from other areas by well-defined boundaries, such as headlands and stretches of deep water. In theory,

sediment cells can be regarded as closed systems from which nothing is gained or lost. However, in reality it is easy for fine sediments to find their way around headlands and into neighbouring cells.

Sediment cells vary in size. The larger ones are divided into smaller sections (**sub cells**) to allow closer study and management. An example of a sub-cell is the one that operates between Flamborough Head and the Humber Estuary on the east coast of England (Figure 3.12).

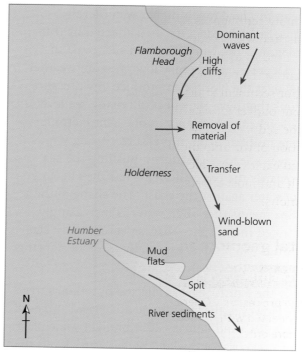

Figure 3.12 Detail of the Flamborough Head – Humber Estuary sub cell

Coastal sediment budgets

For a sandy beach or muddy shoreline to exist, sediment has had to come from somewhere and the combination of wave, current and tide action to be such that the material remains there. This introduces the concept of the **coastal sediment budget.** The coastal sediment budget is best thought of as being similar to a bank account, and is defined as the balance between sediment being added to and removed from the coastal system, that system being defined within each individual sediment cell. Figure 3.13 illustrates the concept of the coastal sediment budget.

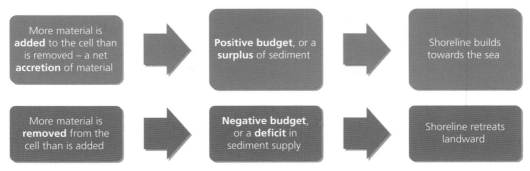

Figure 3.13 Coastal sediment budget

The processes of coastal erosion control the level of the deficit as they remove material from the shoreline, and if more sediment is removed than is added then the coastline will recede. The sediment budget can also be used to identify the sources that deliver sediment to the cell and to the sites where sediment is stored, or sediment sinks. Therefore, calculating the sediment budget for a cell requires the identification of all the sediment sources and sinks, and an estimation of the amount of sediment added and removed each year. Calculating budgets is extremely difficult and most efforts rely on complex models and estimations.

Coastal geomorphological processes

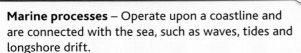

Key terms

Marine processes – Operate upon a coastline and are connected with the sea, such as waves, tides and longshore drift.

Sub-aerial processes – Includes processes that slowly (usually) break down the coastline, weaken the underlying rocks and allow sudden movements or erosion to happen more easily. Material is broken down *in situ*, remaining in or near its original position. These may affect the shape of the coastline, and include weathering, mass movement and run-off.

As coastal geomorphological systems develop over time and tend towards dynamic equilibrium a variety of processes take place. Driven by the input of energy these are what change the characteristics of the components of coastal landscapes.

Coastlines are affected by two sets of geomorphological processes:

● **Marine processes** are those that operate upon a coastline that are connected with the sea, such as waves, tides and longshore drift.

● **Sub-aerial processes** operate on the land but affect the shape of the coastline, such as weathering, mass movement and run-off.

Figure 3.14 introduces the main components of these geomorphological processes that shape the coastline.

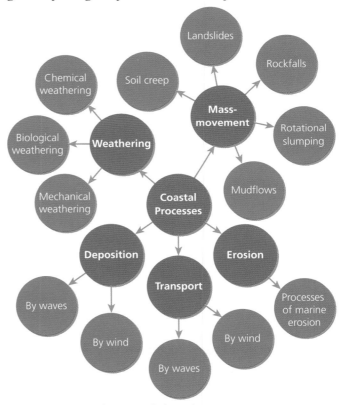

Figure 3.14 Coastal geomorphological processes

The interaction of these processes along the coastline is distinctive, creating the range of uniquely coastal landscapes.

Distinctively coastal processes

Processes of marine erosion

Waves often break on a coastline with considerable energy. It has been estimated that waves breaking against the foot of a cliff can generate energy of 25–30 tonnes m^{-2}. Coastlines are eroded in a number of ways:

- **Hydraulic action** refers to the impact on rocks of the sheer force of the water itself (without debris). This can exert enormous pressure upon a rock surface, thus weakening it. Such activity is sometimes referred to as **wave pounding**.

- **Wave quarrying:** A breaking wave traps air as it hits a cliff face. The force of water compresses this air into any gap in the rock face, creating enormous pressure within the fissure or joint. As the water pulls back, there is an explosive effect of the air under pressure being released. The overall effect of this over time is to weaken the cliff face. Storms may then remove large chunks of it. Some authors also use the term **cavitation** from the study of the effects of pressure changes in areas of rapid flow in rivers for this process.

- **Abrasion/corrasion:** The material the sea has picked up also wears away rock faces. Sand, shingle and boulders hurled against a cliff line will do enormous damage. This is also apparent on inter-tidal rock platforms where sediment is drawn back and forth, grinding away at the platform.

- **Attrition:** The rocks in the sea which carry out abrasion are slowly worn down into smaller and more rounded pieces.

- **Solution (corrosion):** Although this is a form of weathering rather than erosion, it is included here as it contributes to coastal erosion. The process of solution includes the dissolving of calcium-based rocks (for example, limestone). It is unlikely that seawater itself is the agent of this, as its pH is generally stable between 7.5 and 8.5, making it slightly alkaline. If the pH of seawater was generally below 7 then it would potentially kill much marine life. However, in localised areas where fresh water interacts with seawater, conditions for solution may occur, or carbon-based rocks at the coast may be broken down by water flowing from the land or rainwater which may be slightly acidic. Some geographers also include the effects of the evaporation of salts from water in the rocks to produce crystals. These expand when they form and put stress upon the rocks. Salt from seawater spray is capable of corroding several types of rock. This is most definitely a form of weathering and is expanded upon below.

The rate of coastal erosion is affected by a number of factors:

- **Wave steepness** and **breaking point:** Steeper waves are high energy waves and have greater erosive power than low energy waves. The point at which waves break is also important; waves that break at the foot of a cliff release more energy than those that break some distance from the shore.

- **Fetch:** How far the wave has travelled determines how much energy has been generated in it.

- **Sea depth:** A steeply-shelving seabed at the coast will create higher and steeper waves.

- **Coastal configuration:** Headlands attract wave energy through refraction.

- **Beach presence:** Beaches absorb wave energy and can therefore provide protection against marine erosion. Steep, narrow beaches easily dissipate the energy from flatter waves, while flatter, wide beaches spread out the incoming wave energy and are best at dissipating high and rapid energy inputs. Shingle beaches also deal with steep waves as energy is rapidly dissipated through friction and percolation.

- **Human activity:** People may remove protective materials from beaches (sand and shingle), which may lead to more erosion, or they may reduce erosion by the construction of sea defences. Sea defences in one place, however, may lead to increased rates of erosion elsewhere on the same coastline.

There is one more especially important factor that determines the nature of the erosional processes that take place on a particular coastline: geology.

Geology

Lithology refers to the characteristics of rocks, especially resistance to erosion and permeability. Very resistant rocks such as granite, and to a lesser extent chalk, tend to be eroded less than weaker materials such as clay. Some rocks, like limestone, are well-jointed, which means that the sea can penetrate along lines of weakness, making them more vulnerable to erosion. Variation in the rates at which rocks wear away is known as **differential erosion**.

The **structure** and variation of the rocks also affects erosion. When rocks lie parallel to the coast, they produce a very different type of coastline than when they lie at right angles (perpendicular) to the coast. Figure 3.15 shows two contrasting types of coastline that are found close to one another in Purbeck (southern England).

The southern part of the coast has the rocks running parallel to it – known as a **concordant** coastline. Here the resistant Portland limestone forms cliffs, and these

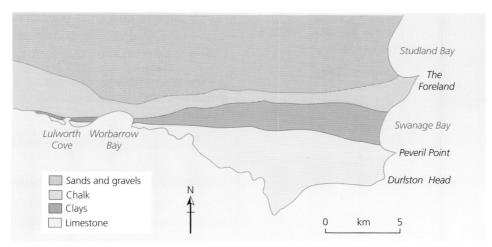

Figure 3.15 The Purbeck Coast

have protected the coast from erosion, only allowing the sea to break through in a few places (the large area of Worbarrow Bay and the small area of Lulworth Cove) to the clay behind.

To the east, the rocks run at right angles to the coast (known as a **discordant** coastline), allowing the sea to penetrate along the weaker clays and gravels and produce large bays (for example, Swanage Bay) flanked by outstanding headlands (The Foreland and Peveril Point).

The **dip** of the rocks is also a major factor. The steepest cliffs tend to form in rocks that have horizontal strata or which dip gently inland, whereas rocks that dip towards the coast tend to produce much more gently sloping features (Figure 3.16).

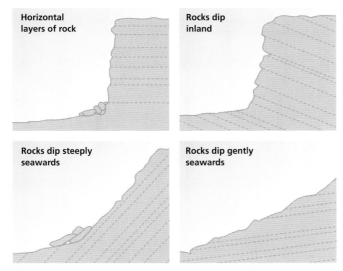

Figure 3.16 The influence of rock strata on coastlines

Key question

Why might some people suggest that the geology of the coastline is the most important factor in determining what coastal landforms are found there?

Processes of marine transportation

Wave and tidal energy that is not used for erosion or not lost through friction with the seabed can be used to transport material at the coast. The transportation of sediment by the sea is not only an important agent of change, in terms of processes of erosion and deposition, but represents a significant flow/transfer of material.

Figure 3.17 (page 100) illustrates that material is transported by seawater in a number of ways. These include:

- **Traction:** Large stones and boulders are rolled and slid along the seabed and beach by moving seawater. This happens in high energy environments.
- **Saltation:** Small stones bounce or leapfrog along the seabed and beach. This process is associated with relatively high energy conditions. Small particles may be thrust up from the seabed only to fall back to the bottom again. As these particles land they in turn dislodge other particles upwards, causing more such bouncing movements to take place.
- **Suspension:** Very small particles of sand and silt are carried along by the moving water. Such material is not only carried but is also picked up, mainly through the turbulence that exists in the water. Large amounts of suspended load, especially near estuaries, can cause a milky or murky appearance of the sea.

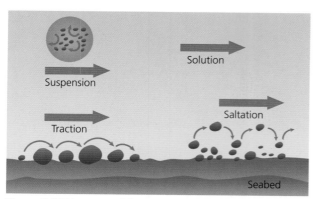

Figure 3.17 How material is transported by the sea

- **Solution:** Dissolved materials are transported within the mass of moving water.

As waves rarely approach the coastline head-on the processes of transport described above combine to move material along the coast.

Longshore (or littoral) drift occurs when waves approach the shore at an angle and material is pushed up the beach by the **swash** in the same direction as the wave approach. As the water runs back down the beach, the **backwash** drags material down the steepest gradient, which is generally perpendicularly back down the beach, where it is picked up by the following incoming wave. Over a period of time, sediment moves in this zigzag fashion along the coast (Figure 3.18). If the material is carried some distance it will become smaller, more rounded and better sorted.

Obstacles such as groynes (wooden breakwaters) and piers interfere with this drift and accumulation of sediment occurs on the windward side of the groynes, leading to entrapment of beach material. Deposition

of this material also takes place in sheltered locations, such as the head of a bay and where the coastline changes direction abruptly – here spits tend to develop.

Fieldwork opportunities

There is an opportunity here to visit a local beach and investigate how material is being transported along it.

Processes of marine and aeolian deposition

Marine deposition often takes place where the waves are low energy or where rapid coastal erosion provides an abundant supply of material. The sea lays down material when there is a reduction in energy resulting from a decrease in velocity or volume of water.

Situations when deposition occurs include:
- when sand and shingle accumulate faster than they are removed
- as waves slow following breaking
- as water pauses at the top of the swash before backwash begins
- when water percolates into the beach material as backwash takes it back down the beach.

Aeolian deposition

Aeolian processes refer to the entrainment, transport and deposition of sediment by wind. Wind plays an important role in shaping many coastlines and is an almost constant feature of most, not just because of the general pattern of prevailing winds that drives the waves. During the day, the wind on the coastal fringe is generally from the sea. Air moves in response to the small pressure differentials

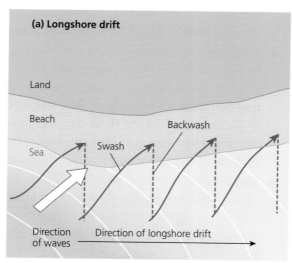

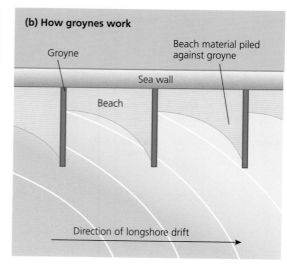

Figure 3.18 Longshore drift

set up by the warmer land and colder sea. When there is a large tidal range, large amounts of sand may be exposed at low tide thus providing a supply of sediment to be entrained (picked up) by the wind.

Sand-sized sediment is the most significant in terms of depositional features at the coast. Once entrained, sand is generally transported close to the ground and over relatively short distances. Sand is transported by wind in two ways, which are dependent on wind speed and how dry or moist the source of sand is.

- **Surface creep:** A process similar to traction, where wind rolls or slides sand grains along the surface.
- **Saltation:** Where the wind is strong enough to temporarily lift the grains into the airflow to heights of up to one metre for distances up to 20 to 30 m.

Wind action can shape and form a range of landforms at the coast, but the most obvious is a beach itself and perhaps the most distinctive are dunes, which are explored later in the chapter.

Sub-aerial processes

As mentioned above, as well as marine processes there are also **sub-aerial** (land-based) **processes** which shape the coastline. These come under the general headings of **weathering** and **mass movement**.

Sub-aerial weathering

Sub-aerial weathering includes processes that slowly (usually) **break down** the coastline, weaken the underlying rocks and allow sudden movements or erosion to happen more easily. Material is broken down *in situ*, remaining in or near its original position. Weathering processes are common at the coast due to the presence of air and water and cycles of wetting and drying, and can be categorised as:

- mechanical/physical weathering
- biological weathering
- chemical weathering.

Mechanical/physical weathering processes that occur at coasts depend on the nature of the climate. In latitudes where temperatures fluctuate above and below freezing, **freeze-thaw** action is common, especially as there is a ready supply of water. Water that enters cracks in the rocks freezes as temperatures remain below 0°C. As it freezes the water expands by almost 10 per cent, meaning the ice occupies more space and so exerts pressure on the surrounding rock. As the process repeats and continues, the crack

widens, and eventually pieces of rock break off. Where processes of erosion, weathering and mass movement remove overlying material, the rock beneath is said to experience **pressure release**. As the overlying mass is unloaded mechanisms within the rock cause it to develop weaknesses, or cracks and joints as it is allowed to expand. This makes the rock susceptible to other processes of erosion and weathering.

Biological weathering includes processes that lead to the breakdown of rocks by the action of vegetation and coastal organisms. Biological weathering is quite active on coastlines. Some marine organisms, such as the piddock (a shellfish), have specially adapted shells that enable them to drill into solid rock. They are particularly active in areas with chalk geology where they can produce a sponge-like rock pitted with holes. Seaweed attaches itself to rocks and the action of the sea can be enough to cause swaying seaweed to prise away loose rocks from the sea floor. Some organisms, algae for example, secrete chemicals capable of promoting solution. Some animals can also weaken cliffs as they burrow or dig into them, such as rabbits or some cliff-nesting birds.

Chemical weathering is common on coasts as it occurs where rocks are exposed to air and moisture so chemical processes can breakdown the rocks. **Solution** is the main chemical process and was included above as it combines with erosion to produce many distinctive features. Other processes include:

- **Oxidation** causes rocks to disintegrate when the oxygen dissolved in water reacts with some rock minerals, forming oxides and hydroxides. It especially affects ferrous, iron-rich rocks, and is evident by a brownish or yellowish staining of the rock surface.
- **Hydration** is included here as it makes rocks more susceptible to further chemical weathering, although it involves the physical addition of water to minerals in the rock. This causes the rock to expand, creating stress, which can itself cause the rock to disintegrate. The process weakens the rock and can create cracks, or widen joints allowing further chemical weathering to occur.
- **Hydrolysis** is where mildly acidic water reacts or combines with minerals in the rock to create clays and dissolvable salts; this itself degrades the rock, but both are likely to be *weaker* than the parent rock, thus making it more susceptible to further degradation.
- **Carbonation** occurs where carbon dioxide (CO_2) dissolved in rainwater makes a weak carbonic acid

(H_2CO_3). This reacts with the calcium carbonate ($CaCO_3$) in rocks like limestone and chalk to create calcium bicarbonate ($Ca(HCO_3)_2$) which then dissolves easily in water. Carbonation is more effective in locations with cooler temperatures as this increases the amount of carbon dioxide that is dissolved in the water.

- As well as naturally occurring CO_2 there are increasing levels of other gases associated with industry and the burning of fossil fuels in the atmosphere that also react with rainwater making it mildly acidic. The presence of sulphur dioxide and nitric oxides can create rainwater with weak sulphuric and nitric acids. This **acid rain** then reacts with various minerals in different rocks weakening or even dissolving them.

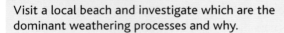

Fieldwork opportunities

Visit a local beach and investigate which are the dominant weathering processes and why.

Mass movement

Mass movement is common on coastlines, especially those that are steep. The nature of the mass movement experienced on a particular coastline is dependent on a number of factors:

- the level of cohesion within the sediment
- the height of the slope and slope angle
- grain size within the sediment
- temperature and level of saturation.

Mass movements are generally either rapid sudden failures of the slope or the effects of processes that develop over some time. Types of mass movement:

- **Landslides:** Occur on cliffs made from softer rocks or deposited material, which slip as a result of *failure* within it when lubricated, usually following heavy rainfall.
- **Rock falls:** These occur from cliffs undercut by the sea, or on slopes affected by mechanical weathering like frost action.
- **Mudflows:** Heavy rain can cause large quantities of fine material to flow downhill. Here the soil becomes saturated and if excess water cannot percolate deeper into the ground surface layers become very fluid and flow downhill. The nature of the flow is dependent on the level of saturation, type of sediment and slope angle. On relatively

gentle slopes the flows are often referred to as 'solifluction', creating lobe-like features towards the base of the slope.

- **Rotational slip, or slumping:** Where softer material overlies much more resistant materials, cliffs are subject to slumping. With excessive lubrication, whole sections of the cliff face may move downwards with a slide plane that is concave, producing a rotation movement. Slumps are a common feature of the British coast, particularly where glacial deposits form the coastal areas, for example, east Yorkshire and north Norfolk.

Figure 3.19 shows a typical rotational slump in an area where glacial deposits form cliffs on top of an impermeable layer.

- **Soil creep:** This occurs where there is a very slow, almost imperceptible, but continuous movement of individual soil particles downslope. There is some uncertainty about the exact causes of creep, but most geographers agree that the presence of soil moisture is important, together with a range of weathering and other processes.

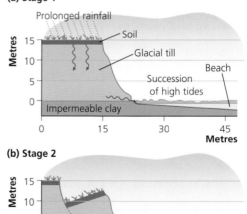

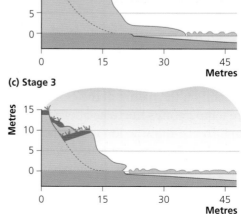

Figure 3.19 Rotational slumping
Source: AQA

Run-off

Run-off is another important process that operates on coastlines. It may take the form of a stream emerging in a bay, taking with it large quantities of load during times of flood, or it may be a stream cascading over a cliff excavating a V-shaped groove as it does so. It can also simply be water that flows over the surface to reach the coastline, the presence of which will also assist many of the mass movement processes above.

3.3 Coastal landscape development

Characteristic coastal landscapes

As the coastline is simply the zone where the land meets the sea, there is not one single *typical* landscape type that typifies all coasts. Around the world individual coastal locations will have a coastal landscape with features that reflect the interaction of a range of factors, including, among other things:

- coastal geology and lithology
- climate
- nature of tides and waves.

However, it is possible to place coastlines in the following simple classification. They can be either:

- concordant or discordant
- a cliffed coast, flat coast or graded shoreline
- an emergent or submergent coastline.

The characteristics of a particular coastal landscape will depend on whether it is:

- a high or low energy coast
- dominated by processes of erosion or deposition
- more or less intensely managed by people.

One thing is certain, and that is that coastal landscapes are not static and have many characteristic features that change over a range of timescales. Waves and tides continually shape and rework coastlines on a daily basis, yet as noted above the nature of these waves and tides is not static. Over longer periods of time processes of erosion, transportation and deposition create landscapes that we become familiar with, which over human timescales we classify as characteristic of a particular location. For example, the cliffs of parts of the Yorkshire coast

in Eastern England are seen as *characteristic* of that place, and the long sandy beaches and barrier reefs of Queensland, Australia are *characteristic* of that place. However, over even longer geological timescales coastal landscapes respond to more significant changes. For example, the Yorkshire cliffs above are composed of material deposited during recent glacial periods, before which that material would not have been there. Equally, the timescales followed by glacials and interglacials of ice ages create coastlines that have emerged from formerly high sea levels, like the raised beaches of the Atlantic coast of Ireland, or others that have become submerged as sea levels have risen, like the flooded glacial valleys forming the fjords of Norway and New Zealand.

This section explores a range of landforms that are part of characteristic coastal landscapes from around the world, including some from the British Isles. It is important that students understand the diversity of coastal landscapes that exist globally and, accordingly, study examples from a wide range of places.

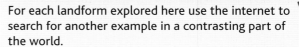

Skills focus

For each landform explored here use the internet to search for another example in a contrasting part of the world.

For each landform in this chapter, practise drawing an annotated sketch to describe its characteristics and explain its sequence of formation.

Origin and development of landforms of coastal erosion

The section above illustrated how the nature of the geology of the coastline can influence the landforms found at a coastline, and where this leads to differential rates of erosion one of the most recognisable combinations of landforms is created: headlands and bays.

Headlands and bays

Figure 3.21 (page 104) shows the impact of geology on a coastline. Areas with alternating more and less resistant rocks are a common feature of many coastlines. Initially erosional processes predominate in areas with less-resistant rock, forming bays, leaving the more resistant rock protruding out to sea as headlands. Because of refraction, the headlands then receive the highest-energy waves and are more vulnerable to the forces of erosion than are the bays. The bays experience low-energy waves that allow sediment to accumulate and form beaches. These then act to protect that part of the coastline.

Figure 3.20 Concordant coast of San Francisco Bay

When geology runs parallel to the coast, as in Purbeck, Dorset, or the San Francisco Bay area of the Californian coast, it is possible for marine processes to create headlands and bays. In San Francisco Bay (Figure 3.20), sea level rise over the last 10,000 years has slowly inundated a series of valleys running parallel to the coastline that correspond to the alignment of the complex geology of the Californian coast. Figure 3.21 illustrates how the differential erosion of alternating geology has also led to the formation of headlands and bays on the Purbeck coast.

Remember that it is important to think of landforms at places like Lulworth as being the result of systems operating at the coast, where the main inputs include:

- the geology and lithology of the coast, the angle of the dip of the coastline in front of the headland, the nature of the waves approaching the coast and the direction and strength of the prevailing wind.

The components (or processes) include:
- the differential rates of erosion of the different rocks
- wave refraction
- erosion of the headland
- deposition in the bay.

The outputs include:
- the characteristic features of the resulting landscape including:
  - the headland and bay
  - the erosional features of the headland
  - the depositional features in the bay.

Cliffs and wave-cut platforms

When high and steep waves break at the foot of a cliff their energy and erosive action is concentrated into a small area of the rock face. With erosion concentrated at its base the cliff begins to be undercut, forming a feature known as a **wave-cut notch**. Further erosion increases the stress on the cliff above and over time it will collapse; how long this takes will depend on the characteristics of the cliffs and waves.

The cliff line will begin to retreat and after successive collapses a gently sloping (less than 5°), relatively smooth, **wave-cut platform** is formed at the base of the cliff (Figure 3.22). If the platform is still in the tidal zone and continually exposed to cycles of marine erosion and sub-aerial processes, weaknesses in the rock surface may be exploited, and on closer inspection

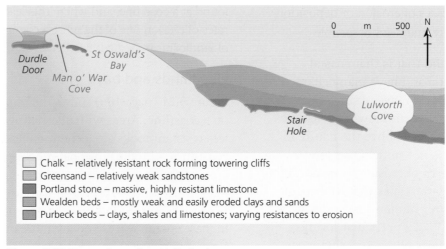

Figure 3.21 The geology of south Purbeck

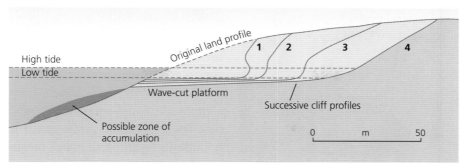

Figure 3.22 Formation of a wave-cut platform

what appears to be a smooth rock surface may in fact be quite rough and jagged, with features like rock pools and fissures.

The platform continues to grow and, as it does, the waves break further out to sea and have to travel across more platform before reaching the cliff line. This leads to a greater dissipation of wave energy, reducing the rate of erosion on the cliff and slowing down the rate of growth of the platform. There tends, therefore, to be a limit to how big the feature can grow and some experts have suggested that growth beyond 0.5 km is unusual.

> ## Key term
>
> **Wave-cut platform** – A gently sloping (less than 5°), relatively smooth, marine platform caused by abrasion at the base of the cliff.

Geos, caves, blowholes, arches, stacks and stumps

These features are all independently observable on coastlines around the world, but they also represent a sequence of events in the erosion of a cliff or headland. On any cliff line it is the weakest parts, such as cracks, fissures, joints and bedding planes that are attacked by the sea. Along a joint the sea will cut inland, widening the crack to form a narrow, steep-sided inlet known as a **geo**.

Figure 3.23 Erosional features of Cape Dyrhólaey, southern Iceland

In other circumstances the cliff is undercut and a **cave** is formed, usually from a combination of marine processes. If erosion continues vertically upwards, it is possible for the cave to be extended to the top of the cliff where a **blowhole** will form. Much more likely, if the cave is on the side of a headland, is that it will extend backwards to meet another on the other side, due to refraction around the headland. Eventually the conjoining of the caves will create a hole all the way through the headland, known as an **arch**.

As the cliff recedes and the wave-cut platform develops, the arch will eventually collapse due to gravity, following a combination of marine erosion from below aided by sub-aerial processes weakening the arch. This leaves an isolated portion of rock as a **stack** standing above the platform. In time, the sea will exploit the wave-cut notch at the base of the stack, leading eventually to its collapse. A small portion of the wave-cut platform may be left marking the former position of the stack. This is known as a **stump**.

There are many famous examples of coastlines around the world where these features stand out. Parus Rock, or Sail Rock, in the Black Sea and the Twelve Apostles in Victoria, Australia are well-known stacks. The south coast of Iceland has some excellent examples of many of these features including the caves, arches and stumps associated with the impressive headland at Dyrhólaey, the most southerly point of the island. (Figure 3.23)

On Purbeck, in Dorset, in the Portland stone (a highly resistant limestone) the sea has cut the famous arch of Durdle Door. Also on Purbeck the chalk escarpment culminates in The Foreland and its detached pieces that are known as Old Harry Rocks. Figure 3.24 (page 106) is a sketch of this area, where the sequence from headland to stack can be seen. Marine erosion and sub-aerial

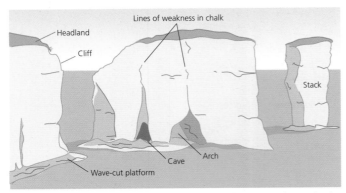

Figure 3.24 Coastal erosion features seen at Old Harry Rocks, Purbeck

processes will eventually reduce the upstanding parts of this area to a wave-cut platform.

Skills focus

This chapter has already highlighted the importance of thinking about coastal features using a systems approach. Look back at the section on headlands and bays. Now use a systems approach to help you understand the formation of the other landforms of coastal erosion. Make sure you are as specific as possible and include details from located examples you have studied. Landforms include:

- cliffs and wave-cut platforms
- geos, caves, blowholes, arches, stacks and stumps.

Origin and development of landforms of coastal deposition

Landforms of coastal deposition occur on coastlines where sand and shingle accumulate faster than they are removed. It often takes place where the waves are low energy or where rapid coastal erosion provides an abundant supply of material.

Beach characteristics

Beaches are found at the point where the land meets the sea and represent the accumulation of sediment deposited between low spring tides and the highest point reached by storm waves. They are mainly composed of sand and shingle. Sand tends to produce beaches with a more gentle gradient (usually below 5°) because its small particle size means the sand becomes compact when wet, and allows very little percolation. Most of the swash therefore returns as backwash, little energy is lost to friction, and material is carried down the beach. This leads to the development of **ridges** and **runnels** in the sand at the low-water mark. These run parallel to the shoreline and are broken by channels that drain the water off the beach (Figure 3.25).

Shingle may make up the whole or just the upper parts of the beach. The larger the size of the material, generally the steeper is the gradient of the beach (around 10–20°). This is because water rapidly percolates through shingle, so the backwash is somewhat limited in its ability to transport material back down the beach. This, together with the uneven surface, means that very little material is eroded from the beach.

At the back of the beach, strong swash at spring high tide level will create a **storm beach**. This is a ridge composed of the biggest boulders thrown by the largest waves, above the usual high tide mark. Below this will be a series of ridges marking the successively lower high tides as the cycle goes from spring to neap. These beach ridges are known as **berms** and are built up by constructive waves. **Cusps** are semi-circular-shaped depressions which form when waves break directly on to the beach and swash and backwash are strong. They usually occur at the junction of the shingle and sandy beaches. The sides of

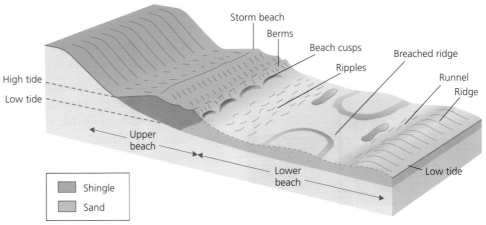

Figure 3.25 Beach features
Source: AQA.

the cusps channel incoming swash into the centre of the embayment and this produces a stronger backwash in the central area which drags material down the beach, deepening the cusp. Below this, **ripples** are developed on the sand by wave action or tidal currents.

The angle at which waves generally approach a coastline will determine the nature of the kinds of beach features that develop. Some authors use the terms **swash-** and **drift-aligned beaches** as a broad classification of beach types to reflect the nature of equilibrium between erosion and deposition a beach tends towards.

Swash-aligned beaches:
- are generally oriented parallel to the incoming wave crests
- experience minimal longshore drift
- can be found on irregular coastlines where longshore drift is impeded, and waves hit sections of the coast head-on.

Drift-aligned beaches:
- are generally oriented parallel to the direction of dominant longshore drift
- can have considerable amounts of sediment transported long distances along them
- initially develop where a section of coastline is fairly regular, or where the predominant wave direction is at an angle to the beach
- can extend out from the coastline if there is a sudden change in the direction of the coastline, for example upon reaching an estuary.

The idea of equilibrium is especially significant for drift-aligned beaches as without a continual supply of sediment the longshore drift would remove the sediment faster than it was deposited.

Fieldwork opportunities

There is an opportunity here to visit a local beach and investigate the characteristics of sediment in different locations on the beach.

Spits, tombolos, bars and barrier-beaches

A **spit** is an elongated, narrow ridge of land that has one end joined to the mainland and projects out into the sea or across an estuary, usually on a drift-aligned coast. Like other depositional features, it is composed of sand and/or shingle and the mixture is very much dependent upon the availability of material and the wave energy required to move it.

Figure 3.26 shows the formation of a spit. On the diagram, the prevailing winds and maximum fetch are from the southwest, so material will be carried from west to east along the coast by the process of longshore drift. Where the coastline changes to a more north–south orientation, there is a build-up of sand and shingle in the more sheltered water in the lee of the headland. As this material begins to project eastwards, storms build up more material above the high-water mark, giving a greater degree of permanence to the feature. Finer material is carried further eastward into the deeper water of the estuary, and as the water loses its capacity to transport it further is deposited, extending the ridge (spit) into the estuary.

Increasingly the end of the spit begins to curve round as wave refraction carries material round into the more sheltered water. The second most dominant wind direction and fetch may contribute to this, pushing the spit material back towards the mainland. The spit cannot grow all the way across the estuary as the material is carried seaward by the river and the deeper water at the centre inhibits growth.

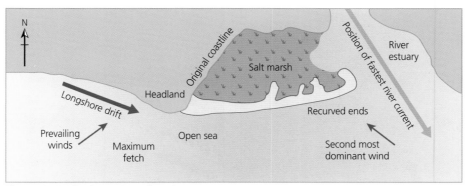

Figure 3.26 The formation of a spit

Figure 3.27a Spurn Head in East Yorkshire, illustrating some of the features of a simple spit

Figure 3.27b Sandy Hook Spit, New Jersey, USA

A simple classification of spits includes **simple** or **compound spits**. Figure 3.27a illustrates the characteristics of a simple spit. Simple spits:

- are either straight or recurved
- do not have minor spits, or recurved ridges, along their landward edge.

Figure 3.27b is an example of a compound spit. Compound spits:

- may have similar features to simple spits
- have a number of recurved ridges, or minor spits, along their landward side, possibly marking the position where they terminated in the past.

As spits mature, **sand dunes** can also develop as deposited sand dries out and is blown to the landward side of the spit, where it can accumulate and become stabilised by vegetation as species like marram grass get established. Also, as the spits increase in size, an increasingly large, more sheltered area develops between the land and the spit. Low-energy, gentle waves enter this area and deposit finer material such and silt and clay. These deposits build-up and are colonised by vegetation to become **salt marshes.**

Some examples of spits have been mentioned above but other well-known spits around the British coast are found at Borth (west Wales), Dawlish Warren (Devon), Orford Ness (East Anglia) and Blakeney Point (Norfolk). Other famous spits from around the world include Farewell Spit, New Zealand and Homer Spit in Alaska.

A spit that joins an island to the mainland is known as a **tombolo** (Figure 3.28a). The best example in Britain is Chesil Beach on the south coast of England. This links the Isle of Portland to the mainland and is about 30 km long. One of the most beautiful and often photographed tombolos from around the world has to be The Angel Road of Shodo Island, Japan.

If a spit develops across a bay where there is no strong flow of water from the landward side, it is possible for the sediment to reach across to the other side. In this case, the feature is known as a **bar** (Figure 3.28b).

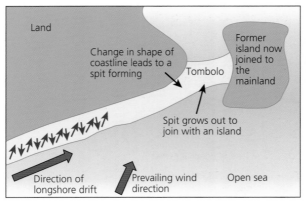

Figure 3.28a Formation of a tombolo

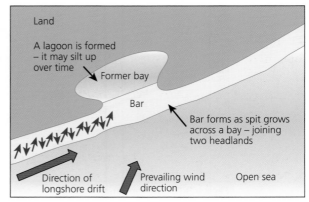

Figure 3.28b Formation of a bar

Some bars, however, may simply be the result of the onshore migration of material from offshore as sea levels rose following the last ice age. Like beaches, these bars may be completely or partially submerged by incoming tides. Slapton Ley, a bar found in Devon, is believed to have formed in this way. Some research suggests that it is a combination of this onshore migration and longshore drift that also formed Chesil Beach. The longest spit in the world, Arabat Spit in the Sea of Azov, is a bar, and joins towns in Ukraine to the north and Crimea to the south.

Where a ridge of beach material that remains semi-submerged accumulates seaward of the breaker zone then it is known as an **offshore bar**.

A **barrier beach** or **barrier island** is an elongated bank of deposited sand or shingle lying parallel to the coastline and not submerged by incoming tides. Where the bank is high enough to allow sand dunes to develop it is known as a barrier island. Often the sheltered area between the barrier beach and the land becomes a lagoon or coastal marsh, or in more tropical locations mangrove swamps may develop. One of the most famous barrier beaches is the Lido of Venice. They may be formed by the breaching of a spit or by constructive waves pushing a bar towards the land.

Skills focus

In the previous section you used a systems approach to help understand the formation of the various landforms of coastal erosion. Now do the same for the depositional landforms. These include:

- beaches, spits, tombolos, bars and barrier beaches.

Process, time, landforms and landscapes in coastal settings

It is important to view each unique coastal landscape as an assemblage of features and landforms that combine in that place to give it its own characteristic landscape. In order to understand why that landscape looks the way it does today, it is crucial to look beyond the processes that are happening in the present and think about how different processes may have shaped its features through time.

There is certainly not one combination of landforms that creates one distinct coastal landscape. Where

the sea and land interact there is a huge diversity of characteristic coastal landscapes including:

- bays
- estuaries
- beaches
- deltas
- dunes
- mud flats and salt marshes.

The list could go on and more landforms are explored below, but it is also important to recognise that with the large and ever-changing amounts of energy available from wind-driven waves, tides and currents, coastal landscapes are dynamic and constantly changing.

The characteristic coastal landscapes that we see today are a result of the factors that shaped them during the Holocene (roughly the time since the end of the last glacial period some 12,000 years ago). Therefore a range of factors has been important in producing the present coastal landscape features, and continues to do so today. These include:

- local tectonic processes
- sea level change – global and local
- climatic change – natural and that enhanced by human activity
- changing ocean currents and wave regimes
- natural disasters or events – including, for example, storms or tsunamis
- changing sources, types and amounts of sediment
- the changing nature of human activity.

All of the above, and more, have continually changed over the millennia, but all will have left their mark as features in the coastal landscapes of the present day, where contemporary coastal processes will inevitably continue to alter and modify them further.

Coastal sand dunes

Coastal sand dunes are accumulations of sand shaped into mounds by the wind. They represent a dynamic landform. Like other coastal landforms sand dunes can be studied using a systems approach. Important inputs include:

- a plentiful supply of sand
- strong onshore winds
- a large tidal range

- an obstacle to trap the sand
- vegetation growth to encourage further growth of the dune.

Sand is mostly moved inland by the process of saltation. Due to the differential heating of the land and the sea, localised differences in atmospheric pressure mean that during the day, the wind on the coastal fringe is generally from the sea. Air moves towards the area of slightly lower atmospheric pressure generated over the land as air is warmed by and rises over the warmer land, from slightly higher pressure over the colder sea. Where there is a large tidal range, large amounts of sand are exposed at low tide, and this further contributes to dune formation. The vegetation succession and development of the sand dune **psammosere** ecosystem is explored in detail in Chapter 5, but here the sequence of the development of sand dunes themselves (Figure 3.29) is as follows:

- Sand may become trapped by obstacles (seaweed, rock, driftwood, litter) at the back of the beach, possibly on the highest berm or storm beach.
- The first dunes to develop are known as **embryo dunes**. They are suitable for colonisation by grasses. These are able to grow upwards through the accumulating wind-blown sand, stabilising the surface. As a result low, hummocky dunes are formed. The long roots, and underground shoots (rhizomes) of marram grass also help to bind the

sand together. The presence of plants adds organic matter to these dunes, which aids water retention.

- Upward growth of embryo dunes raises the height to create dunes that are beyond the reach of all but the highest storm tides. These **foredunes (mobile dunes)** are initially yellow, because they contain little organic matter, but as vegetation cover increases, humus is added to the sand. As a result, the dunes look more grey in colour and in places can reach heights of 20 m.
- The dunes gradually become **fixed**. An organic layer develops as other types of vegetation colonise alongside the marram grass.
- In places **dune slacks** develop. These are depressions within the dunes where the water table is on or near the surface and conditions are often damp.
- Behind the yellow and grey dunes, the supply of sand is gradually cut off, giving smaller dune features. These areas may be referred to as **wasting dunes**, or are areas of **dune heath**.
- Within this system, it is possible to find **blowouts** where wind has been funnelled through areas and has removed the sand. Wildlife or human activity can often be a catalyst for the formation of blowouts.

Fieldwork opportunities

There is an opportunity here to visit a sand dune and investigate how their characteristics change along a transect from the beach inland.

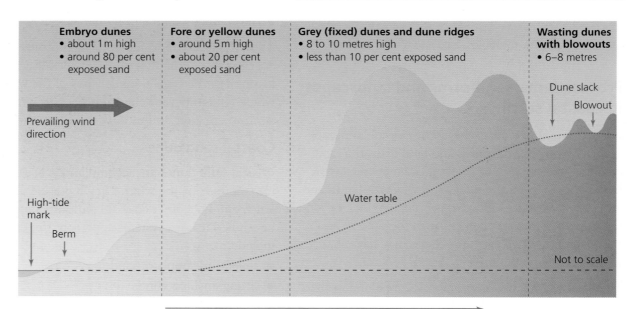

Figure 3.29 A typical sand dune transect

Landscapes associated with estuarine mudflats and salt marsh environments

Mudflats only develop on sheltered shorelines that are not exposed to powerful waves. Often they are located in estuaries where rivers meet the sea or on the landward side of a spit. They are low lying areas of the shore that are submerged at high tide and are composed of silt and clay. Mudflats develop in estuaries where the flow of fresh water out of the river is slow and the sea water flows into the river mouth with each high tide and out with each low tide. Where the saltwater flows gently into the estuary it brings large amounts of fine sediments; this meets the equally slow moving river which is also carrying its own load of fine silts and clays. As the two flows meet the fine particles settle out of suspension by the process of **flocculation**, where the individual clay particles aggregate together to form larger, heavier particles that can sink to the bed.

Figure 3.30 Mudflats of the Keer Estuary, in Morecambe Bay, northwest England

Key questions

Can you explain why mudflats only develop in the conditions described here?

Can you identify the main inputs into the development of a salt marsh landscape?

At low tide the inter-tidal area of mud is left exposed, with water only left in permanent channels and the generally smooth surface shows evidence of tidal action where the flowing water has carved into, or shaped the surface (Figure 3.30).

Some mudflats can be very extensive covering many tens of square kilometres, but they are not necessarily a permanent feature as they are very susceptible to changes in sea level, wave action or changes in discharge levels in the river or tidal flows. In the shelter of Morecambe Bay in northwest England, four river estuaries are home to the largest single area of continuous mud- and sandflats in the country. Mudflats are a common feature of coasts globally. Sizeable examples in the USA are the extensive mudflats in Cape Cod and Plymouth Bays off Massachusetts.

Over time mudflats can develop into **salt marshes**. Like the development of sand dunes, these ecosystems change slowly over time (Figure 3.31 and Figure 3.32, page 112). The vegetation succession that develops is known as

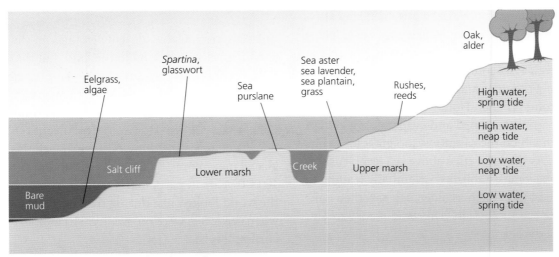

Figure 3.31 The structure of a salt marsh

Figure 3.32 Salt marsh in Morecambe Bay, northwest England

a **halosere** (tolerant of salty conditions) and follows these stages:

- Low-lying vegetation, such as eelgrass, may begin to grow on the mudflats. This slows the currents further and leads to further, more uneven, deposition.
- **Pioneers** begin to colonise the area. These plants are able to tolerate salt and periodic submergence by the sea. They are known as **halophytes** and examples include glasswort, sea blite and *Spartina*. *Spartina* has two root systems – a fine mat of surface roots to bind the mud and long, thick, deep roots that can secure up to two metres of deposited material. This enables the plant to trap more mud than other pioneers, and thus it has become the dominant vegetation on tidal flats in the British Isles.
- The pioneers gradually develop close vegetation over the mud and this allows colonisation by other plants such as sea aster, marsh grass and sea lavender. These form a dense mat of vegetation up to 15 cm high. The growth of vegetation has the effect of slowing the tidal currents even further and this, together with the vegetation's ability to trap particles, leads to more mud and silt accumulation. Dead organic matter also helps to build up the surface, which grows in height at anywhere between 1 and 25 mm per year.
- As mud levels rise, complex creek systems develop that channel the tides and these deepen as the marsh becomes higher. Hollows may form where seawater becomes trapped and evaporates, leaving salt-pans in which the salinity is too great for plants to survive. As the land rises above sea level, rushes and reeds become established, eventually leading to the growth of trees such as alder, ash and then oak to complete the succession. This land is now rarely covered by the sea.

Skills focus

It is important that you are able to describe and explain the features of characteristic coastal landscapes. After visiting coastal landscapes (or by using the library or internet to research contrasting coastal landscapes) complete the following tasks.

Identify the different landforms in each landscape.

Explain the processes that created the landforms.

Try to identify the age and origin of the features; for example, do they result from present-day processes or are they relic features that are being modified today?

Draw and annotate a sketch of the landscape to describe the characteristic features.

Sea level change

There are two ideas relating to sea level change that are explored in this chapter:

- **eustatic** change
- **isostatic** change.

Key terms

Eustatic change – A global change in sea level resulting from an actual fall or rise in the level of the sea itself.

Fjord – Former glacial valley drowned by rising sea levels.

Isostatic change – Local changes in sea level resulting from the land rising or falling relative to the sea.

Raised beaches – Areas of former wave-cut platforms and their beaches which are at a level higher than the present sea level.

Ria – Former river valley drowned by rising sea levels.

Tides are responsible for daily changes in the levels at which waves break on to the land, but the average position of sea level relative to the land has changed significantly through time. Many changes took place during the glacial and interglacial times of the Quaternary ice age. However, the current sea level reflects changes that have occurred since the last glacial maximum 18,000 years BP, and in the last 10,000 years in particular. Figure 3.33 illustrates what the world looked like during the last glacial maximum when sea levels were on average 110 metres below current levels.

Figure 3.33 Global sea levels 18,000 years ago

A typical sequence of sea level rise to reflect the advance and retreat of the ice would have run as follows:

- **Stage 1:** As the climate begins to get colder, marking the onset of a new glacial period, an increasing amount of precipitation falls as snow. Eventually, this snow turns into glacier ice. Snow and ice act as a store for water, so the hydrological cycle slows down – water cycled from the sea to the land (evaporation, condensation, then precipitation) does not return to the sea. As a consequence, sea level falls and this affects the whole planet. Such a worldwide phenomenon is known as a **eustatic** fall.

- **Stage 2:** The weight of ice causes the land surface to sink. This affects only some coastlines and then to a varying degree. Such a movement is said to be **isostatic** and it moderates the eustatic sea level fall in some areas.

- **Stage 3:** The climate begins to get warmer. Eventually the ice masses on the land begin to melt. This starts to replenish the main store and sea level rises worldwide (eustatic). In many areas this floods the lower parts of the land to produce **submergent** features such as flooded river valleys (**rias**) and flooded glacial valleys (**fjords**).

- **Stage 4:** As the ice is removed from some land areas they begin to move back up to their previous levels (**isostatic** readjustment). If the isostatic movement is faster than the eustatic, then **emergent** features are produced such as **raised beaches**. Isostatic recovery is complicated as it affects different places in different ways. In some parts of the world it is still taking place as the land continues to adjust to having masses of ice removed. Today, the southeast of the British Isles is sinking while the northwest is rising. This reflects the fact that the ice sheets were thickest in northern Scotland and that this was the last area in which the ice melted.

Figure 3.34 shows how the sea level rise from 18,000 to 10,000 years BP was beginning to make the coastline of Western Europe recognisable. However, there were still significant land bridges, especially the extensive grassy plain that stretched between the southern Baltic and eastern Britain (known as Doggerland).

Figure 3.34 The coastline of Western Europe 10,000 years ago

Skills focus

There is an opportunity here to research the British coastline to identify examples of emergent and submergent sections of coast.

Coastlines of submergence and emergence

As mean global temperatures continue to rise, there is an inevitable consequence for sea levels. As more standing ice melts, particularly in Antarctica and Greenland, fresh water will be released into the oceanic store. This could have serious implications for many islands in the Indian and Pacific Oceans and for low-lying coastal areas.

Submergent features

Rias are created by rising sea levels drowning river valleys. The floodplain of a river will vanish beneath the rising waters, but on the edges of uplands only the middle- and upper-course valleys will be filled with sea water, leaving the higher land dry and producing this feature. In Devon and Cornwall, for example, sea level rose and drowned the valleys of the rivers flowing off Dartmoor and the uplands of Cornwall. Good examples are the Fowey estuary in Cornwall and the Kingsbridge estuary in south Devon. Rias have a long section and cross profile typical of a river valley, and usually a dendritic system of drainage (Figure 3.35, page 114).

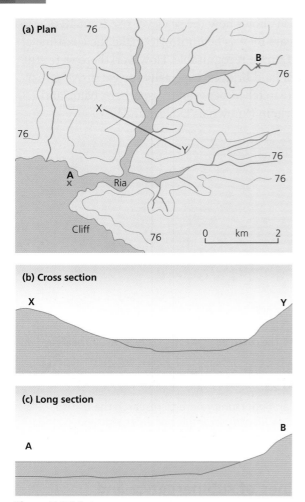

Figure 3.35 A ria

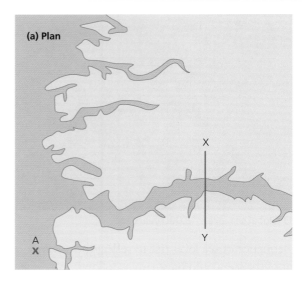

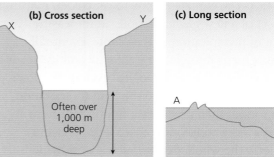

Figure 3.36 A fjord

Fjords are drowned glacial valleys typically found on the coasts of Norway, southwestern New Zealand, British Columbia in Canada, southern Chile and Greenland. The coast of western Scotland contains fjords which are not as well developed as those in the areas above because the ice was not as thick and did not last for the same length of time.

Fjords have steep valley sides (cliff-like in places) and are fairly straight and narrow (Figure 3.36). Like glacial valleys, they have a typical U-shaped cross section with hanging valleys on either side. Unlike rias, they are not deepest at the mouth, but generally consist of a glacial rock basin with a shallower section at the end, known as the threshold. They were formed when the sea drowned the lower part of glacial valleys that were cut to a much lower sea level. The threshold is thought to be due to reduced glacial erosion as the glacier came in contact with the sea and the ice became thinner. Good examples include Sogne Fjord in Norway, which is nearly 200 km long, and Milford Sound in New Zealand (Figure 3.37).

Figure 3.37 The entrance to Milford Sound fjord, New Zealand

Where the topography of the land runs parallel to the coastline and becomes flooded by sea level rise a **Dalmatian coast** is formed. Deriving their name from the Croatian coast in the Adriatic, the flooded valleys run parallel to the coast rather than at *right-angles* to it as in the case with fjords and rias. Here islands and peninsulas are aligned parallel to, but just offshore from, the mainland.

Skills focus

Use an atlas to find a map of the Adriatic Sea and draw and annotate a sketch map of the Dalmatian coastline of Croatia.

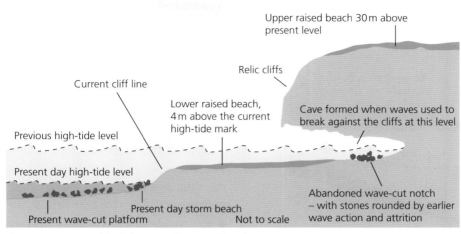

Figure 3.38 A simplified diagram of a raised beach

(Labels in figure:)

Upper raised beach 30 m above present level

Relic cliffs

Current cliff line

Lower raised beach, 4 m above the current high-tide mark

Cave formed when waves used to break against the cliffs at this level

Previous high-tide level

Present day high-tide level

Present wave-cut platform

Present day storm beach

Not to scale

Abandoned wave-cut notch – with stones rounded by earlier wave action and attrition

Emergent features

Raised beaches are areas of former wave-cut platforms and their beaches, which are at a level higher than the present sea level. Behind the beach it is not unusual to find old cliff lines with wave-cut notches, sea caves, arches and stacks. Raised beaches are common around the coasts of western Scotland where three levels have been recognised, at 8 m, 15 m and 30 m. Because of differential uplift these are only approximate heights.

On the west of the Isle of Arran there is a well-developed raised beach north of Drumadoon. This has a **relict cliff,** arches, stacks and caves, including the well-known King's Cave. This beach is around four or five metres above present sea level and is probably the equivalent of the lower raised beach in Figure 3.38. It was clearly produced when the sea was at that level, which initially suggests that the sea has fallen to its present level. However, we know that sea levels have *risen* considerably (eustatic) since the end of the last ice age, so the beach must have reached its raised position by isostatic rising of the land. The land locally must have risen faster than the eustatic rise in sea level to create this phenomenon.

Where a greater expanse of gently sloping formerly submerged land has been exposed by uplift or the lowering of sea levels, some geographers refer to this feature as a **marine platform**, or marine terrace. (Some see these terms, along with coastal terrace and perched coastline as interchangeable with 'raised beach'.) One distinction between a raised beach and a marine platform may be that in a marine platform what is now exposed is part of a gently sloping continental shelf, whose gentle gradient is now continued for some distance both offshore and inland.

Impacts of recent and predicted climate change on coasts

Sea level has been rising consistently over the last 10,000 years. In recent millennia it has risen quite slowly (about one or two millimetres per year), but the rate has increased recently to about four or five millimetres per year. The rate of rise in the future is uncertain with average predictions varying between 18 and 59 cm compared to 1990s levels by 2090.

Key question

Why do you think there is such uncertainty in the predicted sea level change by 2090?

Changes in sea level are the result of two processes:
- increases in the volume of the ocean
- subsidence of the coast.

There are two ways in which the volume of the ocean is increasing. First, and most obviously, as the Earth's climate warms, both naturally and due to current warming related to greenhouse gas emissions and a human-induced enhanced greenhouse effect, water currently stored on the surface as ice is released into the oceans as it melts. Second, as temperatures warm there will be thermal expansion of the oceans. Scientists have been improving their understanding of the impacts of both thermal expansion and the melting of ice caps and smaller valley glaciers, however, what is proving more difficult to predict is how the massive ice sheets will behave during future warming. This, alongside the complexity of the modelling involved, leads to quite a range in the extremes of the potential future sea level rise. Figure 3.39 shows how NASA's Earth Observatory project summarises this range.

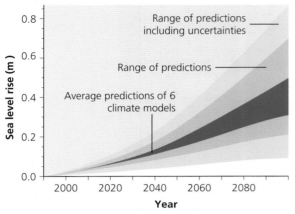

Figure 3.39 Range of possible predicted sea level increase

Rising sea levels could have many adverse effects, including increased coastal flooding and erosion, unless action is taken to mitigate it. In certain parts of the UK, notably the southeast, the rise in sea relative to the land may be greater than average, owing to localised subsidence. Since the last ice age, glacier load removal in the north and west of the UK has resulted in a gradual uplift. At the same time, gradual subsidence has occurred on the margins of the North Sea basin in the east and southeast. This has resulted in the loss of numerous villages from low-lying east coast areas since the compilation of the Domesday Book.

A number of areas around the British coastline are potentially at risk from rising sea levels. These include major conurbations, such as London, Hull and Middlesbrough, and high-grade agricultural land. Major road and rail links near the coast are also at risk, and several power stations are situated on low-lying ground.

As well as direct effects such as coastal erosion or the flooding of coastal areas, higher mean sea levels could have an impact on underground water resources. The zone where seawater mixes with fresh water in rivers is dynamic and a rise in sea levels can cause it to move upstream. A similar effect can occur between fresh water contained in rocks under the land and salt water in sea sediments, leading to intrusion of salt water beneath the land. This would adversely affect some points along the lower reaches of rivers where water is abstracted for domestic and irrigation purposes. These abstraction points would have to be moved upstream or become intermittent to avoid abstracting saline water.

3.4 Coastal management

Human intervention in coastal landscapes

Some coastal landscapes, including human and natural environments, are coming under increasing pressure from both natural processes and human activities. In response to this, a range of protection and management strategies have been put in place in many coastal areas. These solutions are often successful but, in some cases the solutions themselves cause other problems. Coastal management has two main aims:

- to provide defence against, and mitigate the impacts of flooding
- to provide protection against, and mitigate the impacts of coastal erosion.

Other aims of management include:

- stabilising beaches affected by longshore drift
- stabilising sand dune areas
- protecting fragile estuarine landscapes.

Key terms

Hard engineering – Making a physical change to the coastal landscape using resistant materials, like concrete, boulders, wood and metal.

Soft engineering – Using natural systems for coastal defence, such as beaches, dunes and salt marshes, which can absorb and adjust to wave and tide energy.

Management strategies can work either with or against natural processes. Working *with* nature means allowing the natural processes of erosion to occur (**managed retreat**) and not spending money on the defence of the coastal area. This is now applied to large stretches of coastline in the UK where there are few settlements. **Soft engineering** techniques such as beach nourishment are said to work with nature.

Working *against* nature usually occurs where there is significant capital investment – buildings and communications – in the coastal region that has to be protected. Protection involves constructing sea walls, revetments, groynes and other examples of **hard engineering**. The costs of such defences are justified by the potential expense of replacing sea-damaged buildings and infrastructure if they were not in place.

There are an ever-increasing range of approaches to coastal defence, some of which are outlined and illustrated in the next section.

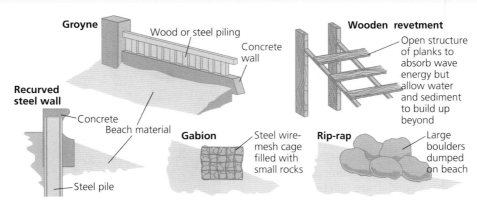

Figure 3.40 Examples of hard engineering solutions

Traditional approaches to coastal erosion and flood risk

Hard engineering

Hard engineering generally involves making a physical change to the coastal landscape using resistant materials, like concrete, boulders, wood and metal. Often each strategy is large scale and costs a significant amount of money. Each type of hard sea defence is built with a specific purpose (Figure 3.40):

- **Sea walls** (sometimes **recurved**) aim to dissipate wave energy. The recurved structure throws waves back out to sea, into the path of the next incoming wave, thus also reducing its wave impact. Sea walls also provide a physical barrier to flooding by raising the height of the coastline. Sea walls must have a continuous facing because any slight gap will be exploited by hydraulic action. They also need drain outlets so that any water that does get over them does not accumulate inland.

- **Rock armour** (rip-rap) consists of large boulders dumped in front of a cliff or sea wall to take the full force of the waves. The boulders are deliberately left angular in appearance to present a large surface area to the waves, and create gaps for water to filter through, again mitigating their impact on the coast. Usually the boulders are not secured in place so energy is taken out of the most powerful waves by rocking or slightly moving the massive rocks.

- **Gabions** operate on the same principle as rip-raps, but smaller boulders are contained within steel wire-mesh cages, each of which can be joined together to form larger structures, or walls.

- **Revetments** are concrete or wooden structures placed across a beach or coastline to take the full force of the wave energy, preventing further erosion of the coast.

- **Groynes** are wooden, stone or steel breakwaters built nearly at right angles to waves (usually 5 to 10° from the perpendicular to prevent scouring on the down-drift side of the groyne). They are built to control longshore drift by trapping sediment to create higher and wider beaches which will then also dissipate wave energy. The groynes themselves will also break up the waves as they hit the coast. Halting the bulk of longshore drift in an area may have serious effects down the coast where it will cut off the supply of beach material and could leave the coast there exposed to erosion.

- **Cliff fixing** is often done by driving iron bars into the cliff face, both to stabilise it and to absorb some wave power.

- **Offshore reefs** force the waves to break offshore, which reduces their impact on the base of cliffs. In some places redundant ships have been deliberately sunk parallel to the shore to both slow down approaching waves, and to act as a substructure for reef material to begin to colonise.

- **Barrages** are large structures built to prevent flooding on major estuaries and other large sea inlets. A barrage acts as a dam across an estuary and prevents incursion of seawater. Good examples of barrages are those that are part of the Delta Plan in the Netherlands and the Cardiff Bay barrage in Wales which was completed in 1999.

Hard engineering strategies are generally long lasting and effective over their planned lifespan; however they do have several disadvantages, including:

- structures can be expensive to build and to maintain (to repair a sea wall can cost over £5,000/m)
- defence in one place can have serious consequences for another area of the coast
- structures are sometimes an eyesore, spoiling the landscape and physically disrupting natural habitats.

The coastal town of Morecambe in Lancashire is used as an illustrative example of the use of hard engineering strategies.

Hard engineering: Heysham and Morecambe

Lancashire County Council, like all others on the UK coast, has four defence options:

- **Hold the line** – retain the existing coastline by maintaining current defences or building new ones where existing structures no longer provide sufficient protection.
- **Do nothing but monitor** – on some stretches of coastline it is not technically, economically or environmentally viable to undertake defence works. The value of the built environment here does not exceed the cost of installing coastal defences.
- **Retreat the line** – actively manage the rate and process by which the coast retreats.
- **Advance the line** – build new defences seaward of the existing line.

Figure 3.41 illustrates how some hard engineering techniques have been used along sections of the coast at Heysham and Morecambe on the Lancashire coast.

Background

The current defences along the 8.5 km stretch of coastline are the result of a comprehensive improvement scheme of the existing structures during a seven-phase programme between 1989 and 2007 costing £28 million. It includes a mix of traditional hard engineering strategies with more contemporary methods to improve the potential for sustainable management.

Strategy 1 – Rock armour/rip rap to enhance and protect the existing sea wall

Boulders of locally sourced limestone were placed along the majority of the existing promenade and sea walls from the western end of the promenade to about one kilometre east of the town centre.

Stones range from 0.77 to 7.00 tonnes, averaging 3.50 tonnes. In total 436,000 tonnes of rock armour was installed (Figure 3.42).

Figure 3.42 Existing sea wall and new rock armour

Strategy 2 – Breakwaters or rock groynes

Around ten breakwaters and rock groynes were built at intervals in front of the town. These included a number of fish-tail breakwaters (Figure 3.43). Just under a million tonnes of locally sourced limestone boulders were used.

Figure 3.43 Fish-tail breakwater (rock groyne)

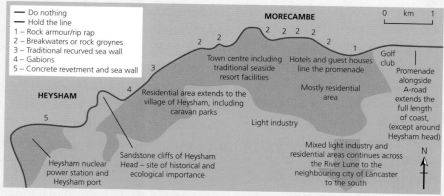

Figure 3.41 Heysham and Morecambe coastal defences

Strategy 3 – Traditional recurved sea wall

Much of the length of the promenade in front of both Heysham and Morecambe had existing traditional recurved concrete sea walls. In places these were repaired and reinforced (Figure 3.44) with a project of further improvements and repair begun in 2015.

Figure 3.44 Recurved concrete sea wall

Strategy 4 – Gabions

Over 500 cages, measuring 2 × 1 × 1 m, filled with small limestone boulders were used in various locations to reinforce the coast (Figure 3.45).

Figure 3.45 Gabions

Strategy 5 – Concrete revetment and sea wall

To the west of Heysham Head, extending to the port and power station, the existing sea wall and large concrete revetments were repaired and left in place (Figure 3.46).

Figure 3.46 Concrete revetment

Summary

Although a range of traditional hard engineering strategies were implemented, they were designed in ways that were sympathetic to Morecambe Bay's classification as a **Site of Special Scientific Interest (SSSI)** and **Special Area of Conservation (SAC)**, and aimed to support the sustainable development of the shoreline. For example, the local sourcing of limestone for the rock armour and rock groynes meant that it was of the same geological origin as the rocks of the coastline. As most of the quarries were less than 10 km away this also reduced the carbon footprint of its transportation. The location of the breakwaters and rock groynes is such that it has created a series of small bays to encourage beach formation, to act as a soft engineering strategy to reinforce the hard defences.

As well as encouraging beach formation between the breakwaters, 89,000 tonnes of sand and 19,000 tonnes of shingle were used for **beach nourishment** between Heysham and the eastern end of the promenade (another soft strategy).

Key question

Explain why Lancashire County Council has chosen to use the strategies outlined in Figure 3.41.

Soft engineering

Soft engineering uses natural systems for coastal defence, such as beaches, dunes and salt marshes, which can absorb and adjust to wave and tide energy. It involves manipulating and maintaining these systems, without changing their fundamental structures.

- **Beach nourishment:** This is the attempt to replace material that has been lost through longshore drift. It is not unknown for local councils to move material from one end of a beach to the other before the start of the tourist season.

- **Dune regeneration:** The fragile sand dune environment is easily disrupted by human activities. Most damage is caused by the removal of vegetation by either agriculture (overgrazing) or tourism (trampling the dunes). This can lead to blowouts during which large amounts of sand may be carried inland and deposited on valuable agricultural land. Management strategies to regenerate dunes include:
  - replanting vulnerable areas with plants such as marram grass and stabilising the surface with sacking or wire mesh
  - afforestation with quick-growing conifers
  - selective grazing
  - restricting access by fencing off areas
  - providing boardwalks for tourists
  - giving tourists information about potential damage.

- **Managed retreat:** This involves abandoning the current line of sea defences and then developing the exposed land in some way, perhaps with salt marshes, to reduce wave power. If old sea defences such as walls are abandoned, low-lying land will be flooded. This will be reclaimed naturally by marsh plants. The new area of marsh will act as a defence against rising sea levels. In this way the scale of hard sea defences can be reduced. There have been proposals in some areas to ban new developments on the coast. In California, for example, there are already requirements on some stretches of coastline that building must be a certain distance from the shore.

- **Land-use management:** Even if it is inevitable that a coastal area will be eroded or flood in the future, a local authority may be able to mitigate the impact.

This strategy involves addressing people's behaviour and educating the local community. Officials can help people plan for the future, and encourage land owners to think about how they can continue to use land that is at risk. For example, caravan parks on cliff tops can provide an income from the land, but be moved and re-sited quickly when the time comes. Giving land at risk of flooding over to grazing rather than growing crops means the sheep or cattle can be moved if storms are forecast. This strategy only works if the local population agree to having their use of land limited, and it cannot remedy damage that has already been done.

- **Do nothing:** In the first decade of the twenty-first century a school of thought has grown up that asks whether the coast *should* be protected. Tens of millions of pounds are spent annually in the UK on coastal protection and it might be cheaper to let nature take its course and pay compensation to those affected. The storms of December 2013 and January 2014 illustrated how trying to control the power of nature is often futile, with the traditional hard sea defences of places like Dawlish in Devon and Aberystwyth in west Wales proving ineffective against the powerful wind and waves. Some argue that the limited funds available for coastal protection should be targeted to places like this that have significant infrastructural or economic value for large numbers of people. This debate is not new; the House of Commons Select Committee on Agriculture suggested in 1998 that large tracts of land should be 'surrendered to the sea' as trying to protect them is a waste of money. Obviously those living in places deemed not worthy of protection may view the debate quite differently.

The Sefton Coast is used to illustrate how some soft engineering strategies have been used in an attempt to protect another stretch of the Merseyside and Lancashire coast.

Key question

Why do you think the authorities have chosen to adopt soft engineering strategies to protect the Sefton Coast stretch of coastline?

Coastal erosion and soft engineering at Formby Point on the Sefton Coast

The Sefton Coast, north of Liverpool, has the largest dune area in England, extending over 17 km, and ranging from 200 m to 4 km in width (Figure 3.47). The sand dune system around Formby Point experienced continual erosion during the twentieth century, losing 700 m from 1920 to 1970. Over the last 20 years the average rate of erosion has been around 4.5 m/year.

Table 3.2 Causes of coastal erosion at Formby Point

Physical (natural)	Human
Periodic storms combined with high tides	Dredging of beach material for the local foundry trade and glass industry
	Building of hard/fixed sea defences to the north (Birkdale to Southport) and south (Hightown to Crosby)
	Activities relating to the development of ports in Liverpool and Preston
	Spoil dumping to the north
	Human access (walking on the frontal dunes)
	Use of off-road vehicles (breaking up the dunes and destroying vegetation)
	Afforestation of a conifer plantation in the mid-twentieth century

The area around Formby Point and Ainsdale still attracts large numbers of visitors to the beaches, sand dunes and pine forests. The local populations of rare red squirrels and natterjack toads are also an attraction.

Some recent climate models suggest an average sea level rise of 0.3 m over the next 60 years and an increase in maximum wave height could lead to a significantly increased risk of erosion. In this scenario the dunes could play a vital role in mitigating flood risk for the local area. The immediate impacts of the erosion of Formby point are not constrained to the immediate area. As the coastline at the Point becomes straighter, material is being transported to the north and south where sand is deposited. In places like Crosby this can be problematic where sections of coastal footpaths and access roads are occasionally buried under considerable accumulations of sand.

The **Sefton Coast Management Scheme** was developed in the late 1980s and in the Formby and Ainsdale areas includes:

- 'planting' used Christmas trees on the seaward edge of the dunes to trap more sand and encourage dune regeneration
- fencing off areas of sand dunes to restrict pedestrian access

- wooden posts placed in front of the dunes to encourage dune regeneration
- building boardwalks to stop people trampling the dunes
- signage to direct people to the beach via routes avoiding the dunes
- ranger services to educate local children about the protection and conservation of the dunes
- banning off-road vehicles from the dune system
- controlling the extraction of sand for commercial purposes
- a debate about the impact the pine plantation has had on the natural dune landscape and whether removing the plantation and scrub cover would help to re-establish and maintain a broader spectrum of habitats.

All of these measures seek to protect the area for future use, in other words to be sustainable.

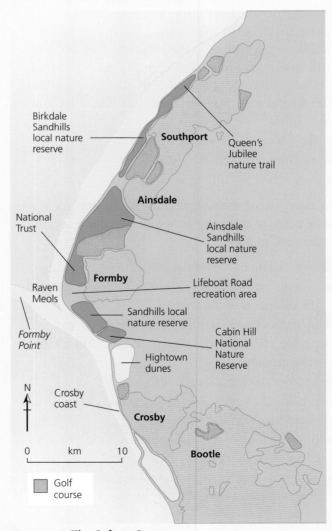

Figure 3.47 The Sefton Coast

Sustainable approaches to coastal flood and erosion management

Since the 1990s new approaches to the management and mitigation of coastal flood risk and coastal erosion have focused on sustainability. This section explores the policies used by the UK Government to make coastal management more sustainable around the British Isles, and the policies developed at a global level which seek to create a common and integrated approach around the world.

Shoreline management plans

Figure 3.11 (page 96) shows that the movement of coastal sediment around the British Isles mainly occurs within distinct cells, the boundaries of which rarely coincide with the administrative boundaries of the Local Authorities. To avoid a piecemeal approach to coastal management an integrated system of **Shoreline Management Plans (SMPs)** was introduced in 1995. There are 22 SMPs around the coast of England and Wales corresponding to the sediment cells and sub-cells discussed earlier. SMPs are designed to identify the most sustainable approach to managing the flood and coastal erosion risks to the coastline. They aim to plan for the:

- short term (0–20 years)
- medium term (20–50 years)
- long term (50–100 years).

As many of the short-term plans near completion, most have been, or are being reviewed to create a second generation of SMPs.

The key features of SMPs are that they:
- provide an assessment of the risks associated with the evolution of the coast
- provide a framework to address the risks to people and to the developed, historic and natural environment
- address risks in a sustainable way
- provide the policy agenda for coastal defence management planning
- promote long-term management policies for the twenty-second century
- aim to be technically sustainable, environmentally acceptable and economically viable

- ensure management plans comply with international and national nature conservation and biodiversity legislation
- incorporate a 'route map' to allow decision makers to make changes to the short- and medium-term plans to ensure long-term sustainability is maintained
- provide a foundation for future research and the development of new coastal management strategies in the future
- are 'live' working documents to be continually reviewed and updated, setting new targets for future management objectives.

Each SMP describes how each **management unit**, or stretch of coastline covered by the plan, is to be managed. Figure 3.48 is a reminder of the four management options available for each unit.

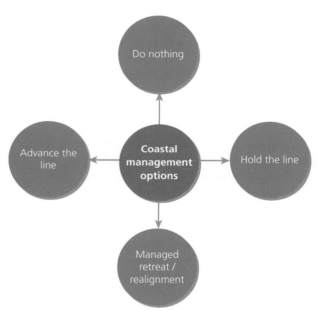

Figure 3.48 Options for coastal management

The areas covered by the 22 SMPs are:
- SMP 1 Scottish border to the River Tyne
- SMP 2 The Tyne to Flamborough Head
- SMP 3 Flamborough Head to Gibraltar Point
- SMP 4 Gibraltar Point to Hunstanton
- SMP 5 Hunstanton to Kelling Hard
- SMP 6 Kelling Hard to Lowestoft
- SMP 7 Lowestoft to Felixstowe
- SMP 8 Essex and South Suffolk

- SMP 9 River Medway and Swale Estuary
- SMP 10 Isle of Grain to South Foreland
- SMP 11 South Foreland to Beachy Head
- SMP 12 Beachy Head to Selsey Bill
- SMP 13 Selsey Bill to Hurst Spit
- SMP 14 Isle of Wight
- SMP 15 Hurst Spit to Durlston Head
- SMP 16 Durlston Head to Rame Head
- SMP 17 Rame Head to Hartland Point
- SMP 18 Hartland Point to Anchor Head
- SMP 19 Anchor Head to Lavernock Point
- SMP 20 Lavernock Point to St Ann's Head
- SMP 21 St Ann's Head to Great Ormes Head
- SMP 22 Great Ormes Head to Scotland

Skills focus

Visit the web link below and find the SMP for the coastline nearest to you and produce a summary of the key objectives for the management of that area. www.gov.uk/government/publications/shoreline-management-plans-smps

Integrated coastal zone management

The term '**integrated coastal zone management**' (ICZM) originated from the UN Earth Summit of Rio de Janeiro in 1992. The guidelines about how ICZM should operate were set out in the **Agenda 21** documents released following the summit. In subsequent decades different organisations with responsibility for the governance of coastal zones have developed their own policies and approaches for integrated coastal management. On the environment pages of the website of the European Commission, ICZM is described as follows:

> Integrated coastal management aims for the coordinated application of the different policies affecting the coastal zone and related to activities such as nature protection, aquaculture, fisheries, agriculture, industry, offshore wind energy, shipping, tourism, development of infrastructure and mitigation and adaptation to climate change. It will contribute to sustainable development of coastal zones by the application of an approach that respects the limits of natural resources and ecosystems, the so-called 'ecosystem-based approach'.

> Integrated coastal management covers the full cycle of information collection, planning, decision-making, management and monitoring of implementation. It is important to involve all stakeholders across the different sectors to ensure broad support for the implementation of management strategies.

Following Rio the European Commission developed its framework for ICZM to promote **integrated coastal management**. The background to this initiative includes:

- coastal zones are some of the most ecologically productive areas in the world
- the natural assets of coasts have for millennia made them popular for:
  - settlements
  - tourist destinations
  - business centres
  - ports
- around 200 million people live near Europe's coastline.

The Commission feels that this concentration of people and economic activity is putting great pressure on our coastal environment and creates excessive exploitation of natural resources, all of which is leading to:

- biodiversity loss
- habitat destruction
- pollution
- conflicts between stakeholders
- overcrowding in some locations.

Some argue that these impacts of human activity are not unique to the coastal zone. The commission suggests that the revised ICM initiative is required because:

- they are some of the areas most vulnerable to climate change and natural hazards
- coasts are especially at risk of:
  - flooding
  - erosion
  - sea level rise
  - extreme weather events

- the lives of people in many coastal communities are already changing due to the impacts of these issues.

For the well-being and economic prosperity of many coastal communities the condition of the natural environment is often crucial, therefore officials believe the long-term plans proposed by ICZM are vital to ensure sustainable development for generations to come.

In the past coastal management was often viewed as having a *sectoral approach*, where the local authorities made decisions for their stretch of coastline and other agencies and interest groups tried to manage their particular environment, cause or social group. In the view of the European Commission this led to decisions that undermined each other, an inefficient use of resources and failure to meet sustainability objectives. This is why ICZM is designed to integrate the interests of all stakeholders to avoid such problems. ICZM aims to co-ordinate policies that affect the coastal zone and the activities that take place there, including:

- nature conservation/protection
- aquaculture (farming marine organisms, for example, a lobster farm)
- fishing
- agriculture
- industry
- offshore wind energy
- shipping
- tourism
- infrastructure development
- developments to adapt to and mitigate climate change.

Key question

Can you think of any specific examples of situations where it would be important to co-ordinate the needs of different stakeholders?

ICZM aims to contribute to sustainable development by employing an '**ecosystem-based approach**' that operates within the limits of natural resources and ecosystems.

If conducted properly ICZM should operate in a cycle with each stage generating feedback to be addressed by the following stage (Figure 3.49).

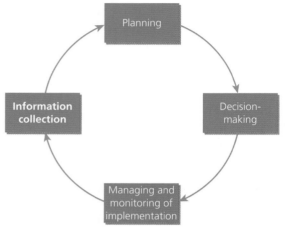

Figure 3.49 The cycle of integrated coastal zone management

In March 2013, the European Commission launched a Directive to establish a framework for **maritime spatial planning (MSP)**. This is to build on and work alongside ICZM by adopting some of the techniques already employed by land-use planners. It often involves mapping the activities of all stakeholders in a particular coastal zone, so that all those involved can visually assess the cumulative impacts of their actions on a particular ecosystem. This can then inform future plans. By utilising powerful GIS and geo-spatial mapping techniques various scenarios can be visually played-out to help agree upon the most sustainable way forward. MSP is hoped to play an integral role in improving the interaction between sea- and land-based activities, including:

- connecting offshore wind turbines to the electricity distribution network on land
- assessing the impacts of strategies to protect coastlines against erosion and flooding on activities like aquaculture and fisheries, and their impacts on marine ecosystems.

3.5 Quantitative and qualitative skills in coastal landscapes

Quantitative and **qualitative** techniques in geography are defined and explored in Chapter 12 and you should read this section carefully.

There are many opportunities to visit, and undertake fieldwork, in a coastal landscape as part of this course, and it is important to reflect on how a range of quantitative and qualitative skills could be utilised in a study of coastal systems and landscapes. The case study on page 126 illustrates how a number of quantitative techniques can be employed in a beach landscape.

Quantitative data and coastal systems and landscapes

Coastal landscapes provide a wealth of opportunities to employ a range of observational and measurement techniques to collect quantitative data. A good starting point is completing annotated field-sketches. Others include measuring:

- characteristics of waves, for example, identifying constructive and destructive waves
- the movement of material by longshore drift
- the size and shape of beach sediment
- the gradient of beaches and sand dunes – beach and dune profiles
- wind direction and wind speed
- the dimensions and characteristics of coastal landforms.

Other quantitative techniques could be used to identify, measure and map the types of sea defences used along a stretch of coast.

Quantitative investigations of coastal landscapes and systems can produce a large amount of useful data that readily lends itself to manipulation and analysis, with a range of data presentation and statistical methods, as explored in Chapter 12. Coastal fieldwork also produces quantitative data that is easily mapped using a range of geo-spatial mapping techniques to help give a clear visual representation of the findings of the research.

Qualitative data in coastal landscapes

This chapter illustrates how predicting the impacts of climate change on future sea levels is complex. Even more challenging will be assessing the increased challenges that this may pose for many vulnerable coastal populations around the world. Equally those tasked with the challenge of mitigating the worst impacts of any sea level rise, while improving the resilience of communities and helping them adapt to their changing circumstances will require a more holistic assessment of many aspects of their situations. Qualitative techniques such as **participant observation** and **in-depth interview** schedules would be one way of engaging with the views of the populations involved to inform the development of bottom-up solutions to future issues. The case study on page 130 exemplifies a region where such techniques could be employed – with marginal coastal communities of southern Bangladesh.

Perhaps on a more local level, questionnaires and interviews could be used with coastal communities in the British Isles to assess the views of various stakeholders about any proposed sea defences for a particular stretch of coast. The qualitative results of such surveys could then be analysed alongside more empirical data about factors such as financial cost and impacts on natural coastal processes to produce thorough cost–benefit or SWAT analyses of any proposed coastal management schemes, that include the views of all stakeholders.

Key question

Produce a list of stakeholders who would have an interest in a proposed coastal management scheme on the British coast.

Fieldwork opportunities

Here is an opportunity to compile a questionnaire that could be used with stakeholders to gather information about the views of the stakeholders involved in a coastal management scheme.

Case study of a coastal environment at a local scale: The coastal processes, landscape and sustainable management of Pevensey Bay, East Sussex

Geographical context

Pevensey Bay on the East Sussex coast, occupies a low-lying area of softer sediments between sandstone geology to the east at Hastings and chalk of the South Downs to the west (Figure 3.50). Over several thousands of years natural processes associated with sea level rise, and subsequent wave and longshore drift action, from west to east, have created a natural shingle barrier. This shingle bank extends for nine kilometres between Eastbourne and Bexhill, and although the shape of the coastline has continually changed through geomorphological processes over hundreds of years, and continues to be modified today, it has remained a natural defence against coastal flooding. The management of this coastline is challenging, as the modern-day settlements require that the coastline remains in a relative fixed position. However, as sea levels continue to rise, wave action and longshore drift continue to

move sediment in the littoral zone and storm frequency appears to increase, the shape of the coastline changes, and will continue to change, almost on a daily basis. So balancing the wishes of residents, economic costs and benefits and sustainability is an ongoing challenge.

The natural and man-made sea defences currently protect:

- the permanent flooding of a 50 km² area – including communities in Pevensey Bay, Normans Bay, Langley, Westham and parts of Pevensey
- 10,000 properties
- recreational and commercial sites
- A259 coast road and railway line from Hastings to Portsmouth
- two nature reserves and a SSSI wetland site
- numerous livestock and arable farms.

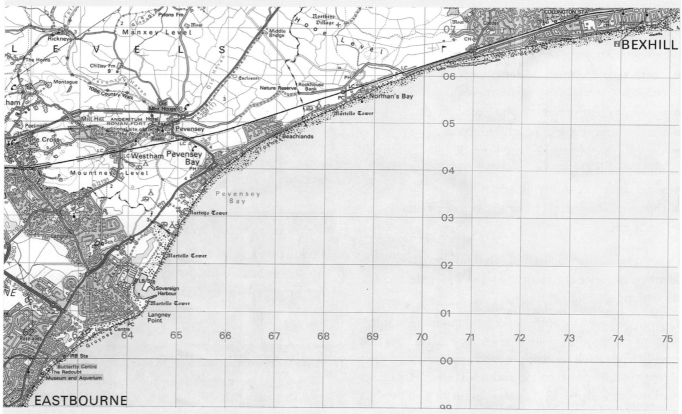

Figure 3.50 Ordnance survey map showing the location of Pevensey Bay, East Sussex (Map is reproduced at half of OS Landranger map scale of 1:50,000. Each grid square is 1 km²)

Coastal processes

Like all natural coastlines the shore at Pevensey Bay constantly tries to adapt to minimise the effects of the power of the sea. The prevailing winds from the southwest move beach material towards the northeast creating beaches that are more parallel to the waves. During major storm events the beaches become flatter, creating beach profiles that reduce wave energy and prevent waves reaching the top of the shingle bank, thus reducing the flood risk. In the past the shingle bank and beach have been maintained naturally, however research showed that the movement of sediment was increasingly irregular and there was a net loss of 25,000 m³ of beach every year. As flooding, rather than erosion, is the biggest risk here, soft man-made defences constructed of sand and shingle are seen as most effective in absorbing and breaking up the power of the waves.

Current sea defences

The management of the sea defences in Pevensey Bay is unique as they are organised and maintained by the country's first public–private partnership (PPP) coastal management scheme. The project began in 2000 after a realisation in 1997 that a 1-in-20-year storm event could breach the shingle beach. The contract was awarded to a consortium called Pevensey Coastal Defence Ltd (PCDL), was to run for 25 years and had an initial budget of £30 million.

The defences centre on the need to manage and improve the existing shingle beach. PCDL has therefore based its management on working with the natural environment of the local sediment sub-cell. The management is environmentally and aesthetically acceptable and is an example of development which maximises opportunities for future recreational and environmental projects.

The shingle beach stretches the length of Pevensey Bay. The crest height of the defence is six metres and the shingle bank extends seawards an average of 45 m. Traditionally this had been maintained with the use of 150 ageing wooden timber groynes. Since 2000 the project has removed each groyne as it fails, with the aim of retaining about 10, and focuses on a more sustainable soft management approach to work with the coast processes.

Key question

Why do you think the authorities have chosen not to replace the old wooden groynes at Pevensey Bay?

Table 3.3 Beach maintenance strategies in Pevensey Bay

Strategy	Notes
Recharge	The natural movement of sediment west to east along the coast is interrupted at Sovereign Harbour. Longshore drift still removes 25,000 m³ of sediment annually from the bay, transporting it to Bexhill. Lorries transport 5,000 m³ around the harbour each year, but the remaining 20,000 m³ of gravel and sand is dredged from the sea floor just off shore. This is the natural source of the original beach material so is a close match (Figure 3.51, page 128). In recent years, rather than using bulldozers to push it into place (causing noise and disruption), all 20,000 m³ have been left in situ and redistribution was left to the natural process of longshore drift. This provided more natural sorting of the beach material and saved considerable amounts of time and money.
Recycling	The rate of drift of material along the bay varies from place to place. Some areas have net accumulation, some net loss and others do not change. Dump trucks and bulldozers are used to redistribute sand and shingle from areas of gain. Recycling sediment is required following especially stormy weather, such as Winter 2013–14, and when conditions do not allow the dredger to operate close to shore. If relatively small amounts of sediment need recycling a 'box' or earth scraper is attached to the bulldozer to limit the use of larger plant machinery.
Bypassing	The southern arm of Sovereign Harbour interrupts the west to east movement of sediment causing it to accumulate. If this was allowed to happen it would eventually spill round into the harbour and silt-up the lock gates. During the winter months (so as not to disrupt the tourist season) trucks are used to transfer between 5,000 and 15,000 m³ of surplus shingle to the west (bypassing the harbour). This is either added directly to the beach or is stockpiled as an emergency source of beach material.
Reprofiling	Destructive waves in winter storms remove material from the crest of the shingle beach and deposit it lower down the beach. Over time wave action would naturally return the material back up the beach, but in the meantime areas of the coast may be left vulnerable. To prevent the shingle being transported away by longshore drift a bulldozer simply pushes the material back up the beach, re-instating the beach crest. During winter this and recycling can be necessary on an almost daily basis (Figure 3.52, page 128).
Groynes	Restoring the 150 groynes along the bay would have cost 40 per cent of the 25-year budget, about £12 million. Instead, as each groyne fails it is carefully removed, so that it does not become dangerous for beach users or break apart and float out to sea. The wood is reused to either repair the remaining groynes or made available to local residents, businesses and farmers to be recycled and reused. Around 10 groynes will be retained to maintain important discontinuities in the coastline. This will create a more open natural beach.
Beach surveys	Twice a month a quad bike with a GPS receiver surveys the beach. The data is used to produce three-dimensional models of the shingle bank and beach. The system works like a satellite navigation system in a car but is accurate to between 15 and 30 mm. These surveys help to maintain the sustainability of the project as replenishment, reprofiling and recycling can be targeted to only where needed.

Figure 3.51 Dredger recharges the beach at Pevensey Bay

Further information and updates on current progress managing Pevensey Bay coast can be found at www. pevensey-bay.co.uk.

Figure 3.52 Using a bulldozer to reprofile the beach after a storm

Key question

Why are the beach surveys considered such an integral part of the management strategy in Pevensey Bay?

Investigating coastal processes

Background

As geography students it would be possible to collect field data to understand processes like those occurring at Pevensey Bay. An investigation could be conducted to investigate the characteristics of a beach and the impacts of processes like wave action and longshore drift. This could include an analysis of:

- the shape of the beach
- the amount of beach material in different locations
- the nature of beach material.

Collecting field data

At regular intervals along the length of the beach (5 or 10 m depending on the length of the beach, time available and number of students) mark out a straight line **transect** from the shoreline to the back of the beach. Record the length of the transect and use a compass to note its orientation. Its position should be recorded on a base map. Using a tape measure, data is collected at regular intervals up the beach. Again, the length of the interval will depend on the length of the transect and time available. The beach may have many obvious changes in slope angle which may be used instead, especially as this

may give a more accurate profile when plotted back in class. Always start at the water's edge and work inland, as the tide may begin to come in.

Along the transect a beach profile could be measured and an analysis of the sediment completed. A summary of the method for each is outlined below.

Beach profiling and sediment characteristics

Two students are needed to complete each transect and could follow the method below:

- Student 1 stands at the seaward end of the transect with a ranging pole.
- Student 2 takes a second ranging pole further up the beach along the transect to the first break of slope.
- A tape measure is used to measure the distance between the ranging poles.
- Student 1 uses a clinometer to measure the angle between the tops of the two ranging poles to give the gradient of that section of the transect (Figure 3.53).

  (Also see Figure 2.45, page 85, for a similar technique on sand dunes.)
- This process is repeated for every break of slope to the top of the beach.

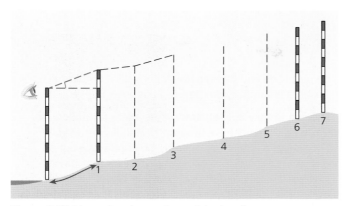

Figure 3.53 Measuring the gradient of the beach

In order to investigate the impacts of wave action and longshore drift on the sorting of sediment on the beach, data could be collected while measuring gradient along the transect. If a statistically significant number of transects are completed at 10 or more sites located systematically along the beach (for example, at 10 m intervals) an analysis of sediment change along the beach can be conducted. Collecting data along each transect will also allow an analysis of how waves and tides sort material up the beach.

- At each collection point a sample of at least 10 pebbles is selected.
- A quadrat could be placed adjacent to the ranging pole and the pebbles selected from within it.
- Using a ruler, or callipers, measure either the long-axis of each pebble or multiple axes to give an average size (Figure 3.54). (If the beach material is fine then a set of graduated sieves would be used to sort a sample of sediment from each quadrat to ascertain the average size of sediment at each site.)

- Students could also record pebble roundness, where each pebble is assessed against an exemplar Power's Roundness scale. (The scale ranges from very angular, sub-angular or rounded, to very rounded.)

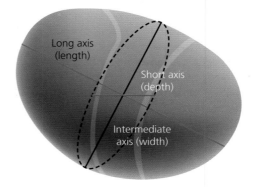

Figure 3.54 Measuring axes of pebbles

Findings

Back in class the results can be analysed to draw conclusions about the characteristics of the beach from waters-edge to the top of the beach. Also, the profile data can be used to calculate the cross-sectional area at each location, which could then give an assessment of the amount of beach material at that point. If time allows, measurements taken at the same location on different dates could give an assessment of how the amount of sediment at each location changes over time, to help draw conclusions about the effects of wave action and longshore drift. If data can only be collected on one visit, then an analysis of profile and sediment data for each transect along the beach could also allow conclusions to be drawn about the effects of longshore drift and wave action.

Table 3.4 Recording sheet for beach data

Date:		Weather conditions:											
Transect number:		Grid reference for start of transect:								Transect length:			
Beach profile	Break of slope on transect	Waters-edge	1	2	3	4	5	6	7	8	9	10	
	Distance from previous break	NA											
	Gradient												
Sediment	Average pebble size (cm)												
	Average pebble roundness												

Case study of a contrasting coastline beyond the United Kingdom: The Sundarbans Bangladesh

Background

The Sundarbans is a coastal zone occupying the world's largest delta, that extends over 10,000 km² of southern Bangladesh and India on the Bay of Bengal. The delta is formed from the sediment deposited by three of the world's great rivers, the Ganges, Brahmaputra and Meghna (Figure 3.55). The natural climax ecosystems of the Sundarbans are mangrove forests and swamps.

Figure 3.55 Map showing the Sundarbans Delta, Bay of Bengal

Coastal processes

Tidal action is the primary natural process that shapes this distinctive coastal landscape. A dense, well-developed, network of interconnecting river channels flows across the clay and silt deposits. Traditionally the location of the network of main channels remained relatively static due to the silts and clays being quite resistant to erosion. The larger channels are generally straight and up to two or more kilometres wide, flowing generally north to south due to the strong tidal currents. The extensive network of interconnecting smaller channels (or **khals**) drains the land with each powerful ebb tide.

The non-cohesive sediments like sand are washed out of the delta and deposited on banks, or chars at the river mouths, where the strong south-westerly monsoon winds then blow them into large ranges of sand dunes. With the protection of the dunes, finer silts washed into the bay are deposited, where wave action then adds and shapes further deposits of sand to form new islands. Vegetation establishes itself and eventually, if the natural succession can proceed, the dense mangrove forests,

home to the endangered Royal Bengal Tiger for which the area is famous, can develop (Figure 3.56).

This is a unique coastal landscape where for millennia these great rivers have brought rich sediment that, once shaped and maintained by tide and wave action, has created the diverse mangrove forests that have sustained local populations for generations. However, the equilibrium of the natural processes that exist here is very delicate and the increasing pressures that are placed on it may be jeopardising its very existence.

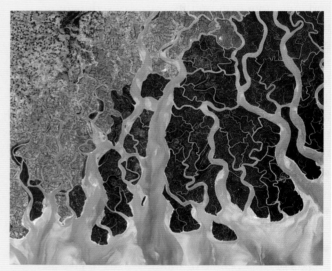

Figure 3.56 The Sundarbans of Bangladesh

Challenges and opportunities of the Sundarbans

The mangrove forests of the Sundarbans are an extremely important ecosystem for the local populations, providing many economic and environmental opportunities. If managed sustainably this coastal zone provides a wide range of goods and services (Figure 3.57).

Despite the wealth of goods and services outlined in Figure 3.57 many from outside view this area of coastal Bangladesh as uninhabitable. However, traditionally (and perhaps surprisingly) coastal erosion has not been a problem; the main challenges facing the local populations are outlined in Table 3.05.

GOODS

SERVICES

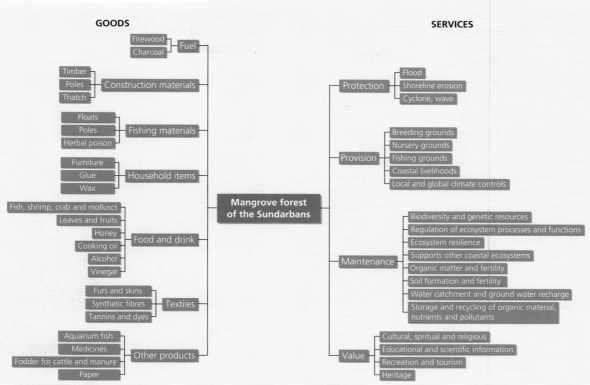

Figure 3.57 Goods and services of the mangrove forests of the Sundarbans

Table 3.5 Natural and human-induced challenges for the people of the Sundarbans

Natural	Human
Coastal flooding	Over-exploitation of coastal resources from vulnerable habitats
Cyclones	
High levels of salinity in the soil	Conversion of wetlands to intensive agriculture and settlements
Instability of the islands	Destructive fishing techniques
Accessibility and remoteness	Lack of awareness of the environmental and economic importance of the region
Human-eating tigers	Resource-use conflicts
	Lack of awareness of coastal issues by decision makers

The human response to challenges of the Sundarbans

Traditionally many local populations have lived successful productive lives in the Sundarbans of Bangladesh. Life is not always easy, but by living with their environment rather than against it they have coped with many of the natural challenges they face (Figure 3.58).

Figure 3.58 Chandpai Village, Sundarban National Park, Bangladesh

Resilience

The wealth of goods and services outlined in Figure 3.57 has allowed local populations to remain resilient to the challenges of this low-lying coastal landscape. In fact the mangrove forests provide a high level of resilience as they provide significant protection and shelter against:

● storm winds

● floods

● tsunamis

● coastal erosion.

It is said that a density of 30 trees per 0.01 hectares can reduce the destructive force of a tsunami by up to 90 per cent. In fact the coastal communities of the Sundarbans are said to be more resilient to natural disasters than other coastal communities elsewhere in Bangladesh.

The fertility of the soil and ecological diversity also provided a plentiful supply of a large range of nutritious foods. Despite the constant threats of flooding and storms the forests have an economic value, even if used for traditional activities such as fishing, gathering crustaceans, or timber and tannin production. Some estimate that just one hectare of mangrove forest has an annual economic value of over $12,000. This suggests that unlike many other local communities in other parts of Bangladesh, one of the poorest countries in the world, the mangroves could provide resilience against poverty, and opportunities for sustainable economic development in the future for the people living there.

Mitigation

Although vulnerable to many challenges in the coastal Sundarbans, it is said that its people moderate their risk in a number of ways. Some suggest that it is the strength of each community's resources that helps to mitigate the worst impacts of this challenging environment. Primarily, prior to more recent human pressures like deforestation, overfishing, intensive agriculture and settlement growth, the communities of the Sundarbans could utilise a number of **open access** natural resources. These included:

- *khas* land (government-owned lands that are supposed to be protected for use by local populations)
- wetlands and fisheries
- forests.

The ever-present threat of natural disasters has meant that in some areas of the Sundarbans there has been significant investment in the physical infrastructure (roads, telecommunications, schools, hospitals, cyclone shelters, flood protection and tube wells) which also mitigate the risks they face. Many communities of the Sundarbans are said to have good levels of social capital from the legal frameworks and services provided by a number of formal government and NGO organisations, alongside the traditional laws, tenets and social sanctions that communities have used for generations to manage their use of the region. As noted above there is some economic value to the resources of the area and together with financial resources like access to micro-credit, some Sundarbans communities have a greater economic safety net than other

vulnerable groups in Bangladesh. Figure 3.59 summarises the **livelihood assets** that help mitigate the challenges of the Sundarbans.

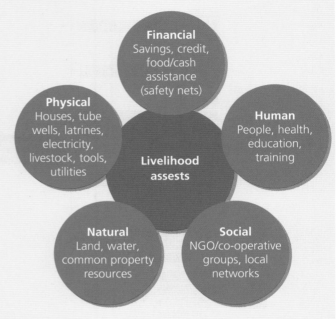

Figure 3.59 Livelihood assets of communities in the Sundarbans

In recent decades the level of resilience provided by these livelihood assets may be decreasing as poverty and marginalisation of some coastal communities increases due to:

- shrinking of the open access resources
- degradation of ecosystems
- corruption of both local and national political institutions
- conflicts over land ownership
- increasing deaths by tigers (if wives are widowed they can often face very limited opportunities in this male-dominated society, especially as many marry very young before completing their formal education).

Adapting to the Sundarbans

In the past it appears that people of the Sundarbans have adapted well to their coastal environment and in some places many have prospered on the natural bounty. However, it is almost inevitable that the impacts of climate change may pose the greatest risk to current and future generations in southern Bangladesh. Table 3.06 (page 133) summarises some of the future challenges facing the region alongside possible adaptations that could mitigate some of the risk and improve the community's resilience to climate change.

Table 3.6 Future challenges and adaptations in the Sundarbans

Challenges	Adaptations
Increased frequency and intensity of floods	Grassroots NGOs run education programmes to encourage farmers to return to more traditional ecologically-friendly methods
Flood waters remaining for longer	NGOs provide education and training about community level preparation for natural disasters – planning and preparation can significantly mitigate the impacts of natural disasters
Permanent embankments built by the commercial shrimping industry are encouraging the deposition of silt, thus raising water levels in the rivers	USAID trains communities to become resilient to future climate shocks. 30,000 people have received training on improving agricultural techniques
Rising temperatures contribute to increased salinity levels in the soil	Relaunch of the policy to build multi-purpose cyclone shelters equipped with communication equipment and megaphones, which also serve as primary school facilities (Figure 3.60). (Many were built in the 1970s, but have since become neglected)
Increased pesticide and fertiliser use is affecting water quality	New salt-tolerant rice varieties that can survive being submerged in sea water for over two weeks
Changes to seasonal patterns of rainfall	NGOs build latrines on higher ground and educate communities about water and sanitation, and water-borne diseases
	Installing storage tanks for rainwater in areas at most risk from inundation by salt water
	Distributing water-tight containers to store important belongings and papers during floods

Generations of people have been sustained by the coastal mangrove forests of the Sundarbans. However, population pressure and the unpredictable impacts of future climate change may prove too hard to adapt to for many.

Figure 3.60 Raised cyclone shelter nears completion on the edge of the Sundarbans

Opportunities for sustainable development in the future

The threats to the mangrove forests and peoples of the Sundarbans may appear bleak, however with timely intervention and careful management there are opportunities for the sustainable development of this unique coastal zone in the future:

- The region has many international and national designations already in place.
- There are opportunities for eco-tourism in the mangrove forest and wetland areas.
- There has been recent investment to improve communications in the region.

Further research

Find out more about the Sundarbans at:

www.eoearth.org/view/article/156336/

www.sundarbans.com.bd/

www.sundarbans.org/

www.usaid.gov/bangladesh/environment-and-global-climate-change

Fieldwork opportunities in coastal landscapes

Studies of coastal systems and landscapes provide many opportunities for collecting field data. Fieldwork could be used to research a range of human and physical geographical enquiries. There is potential to collect a broad range of quantitative and qualitative data from primary and secondary sources (as outlined above). Coastal fieldwork activities could fall into one of the three categories described below (with some examples):

- Investigating coastal processes:
  - the characteristics of waves and their impacts
  - the nature and impact of longshore drift
  - factors affecting rates of coastal erosion
  - factors affecting sub-aerial processes
- Investigating coastal landforms:
  - beach profiles
  - sediment analysis
  - features of headlands and bays
  - features of wave-cut platforms
  - evidence of sea level change
- Investigating coastal management:
  - the methods of coastal management used
  - impacts of coastal management
  - the success, or cost–benefit analysis of coastal management
  - the views of different stakeholders about coastal management.

Review questions

1 Using a systems approach can you identify more examples of positive and negative feedback in coastal systems?

2 What are the main sources of energy in coastal environments?

3 What are the reasons for eustatic, isostatic and tectonic sea level change?

4 For a local population in a coastal landscape you have studied:
   - What challenges does their environment pose for them?
   - What future challenges will they face and how might they adapt to them?

5 What aspects of your coastal fieldwork investigation could be used to help those preparing a Shoreline Management Plan to sustainably manage that stretch of coastline?

Further reading

The texts below develop many of the ideas set out in the chapter above. Some are quite academic and specialist texts but your local library may be able to find copies should you wish to take your studies further.

Harvey, A. (2012) *Introducing Geomorphology: A Guide to Landforms and Processes*. Dunedin Academic Press

Holden, J. (2012) *An Introduction to Physical Geography and the Environment, Third edition*. Pearson.

Masselink, G., Hughes, M. and Knight, J. (2011) *Introduction to Coastal Processes and Geomorphology, Second edition*. Routledge

Woodroffe, C. D. (2002) *Coasts: Form, Process and Evolution*. Cambridge University Press

Question practice

A-level questions

1. Explain a negative feedback mechanism in the coastal system. (4 marks)

2. Table 3.7 shows data collected on a beach in west Wales. The investigation aimed to determine the effects of longshore drift and other coastal processes on the size of shingle found along the beach.

   A 1-km stretch of beach was surveyed from north to south using a systematic sampling strategy. Twenty pebbles (collected the same distance from the sea) were measured at 50-metre intervals along the beach and the mean length of the pebbles is presented in the table below.

   This is the null hypothesis: There is no significant relationship between the size of sediment and the distance along the beach.

   Below is a partly completed Spearman's rank test where the formula to calculate the Spearman's rank correlation coefficient (R_s) is:

   $$R_s = 1 - \frac{6 \sum d^2}{n^3 - n}$$

Table 3.8 Critical values for R_s

n	0.05 (5%) significance level	0.01(1%) significance level
20	0.377	0.534
21	0.368	0.521
22	0.359	0.508

Complete Table 3.7 and interpret your Spearman's rank correlation coefficient using Table 3.8 (6 marks)

3. With reference to Figure 3.61 and your own knowledge, assess the role of deposition in the development of this area of the Morecambe Bay coastal landscape. (6 marks)

Table 3.7

Distance from southern end of beach (m)	Rank	Mean sediment length (cm)	Rank	Difference in rank (d)	d²
0	21	14.3	1	20	400
50	20	14.1	2	18	324
100	19	13.9	3	16	256
150	18	13.4	4	14	196
200	17	12.7	5	12	144
250	16	12.2	6	10	100
300	15	11.6	7	8	64
350	14	11.1	9	5	25
400	13	11.3	8	5	25
450	12	10.7			
500	11	10.4			
550	10	9.4			
600	9	9.4			
650	8	9.6	12	−4	16
700	7	8.9	15	−8	64
750	6	8.3	16	−10	100
800	5	7.7	17	−12	144
850	4	7.6	18	−14	196
900	3	7.2	20	−17	289
950	2	7.3	19	−17	289
1,000	1	6.9	21	−20	400
					Σ = 3068.5

4. 'Climate change in the twenty-first century will mean that the risks will outweigh the opportunities facing people living in a coastal area.

   With reference to a coastal area beyond the UK that you have studied, to what extent do you agree with this view? (20 marks)

Figure 3.61 Mudflat and saltmarsh landscapes at Keer Estuary, Morecambe Bay, Lancashire, in northwest England

AS level questions

1. Which process/activity can lead to eustatic sea level change?

   a) A localised change in sea level

   b) Melting of ice sheets on land leading to global change in sea level

   c) Rebound of land during interglacials

   d) Tectonic activity such as an earthquake   (1 mark)

2. Where is the backshore area of the coastal zone found?

   a) In between the high water mark and the landward limit of marine activity. Changes normally take place here only during storm activity.

   b) Lying between the high water mark and the low water mark. It is the most important zone for marine processes in times that are not influenced by storm activity.

   c) The area where a turbulent layer of water washes up the beach following the breaking of a wave.

   d) Where waves break on to the beach.   (1 mark)

3. Outline the role of waves in changing coastal landscapes.   (3 marks)

4. The following is a short article about the possible impacts of climate change in Morecambe Bay in northwest England. Using the extract and your own knowledge, assess possible impacts for Morecambe Bay of future sea level rise.   (6 marks)

Morecambe Bay

Located on England's west coast, Morecambe Bay Limestones National Character Area is a spectacular part of the UK.

The area includes the coastal and inland area lying to the south and west of the South Cumbria Low Fells and the northern and eastern margins of Morecambe Bay. Its character is defined by the distinctive coastal salt marshes, large areas of intertidal flats and the sweeping, constantly shifting, views across the waters of Morecambe Bay.

Projected sea level rise over the coming century, together with possible changes to storms, could reshape the coastline. Rising sea levels can be managed by either developing coastal defences, or adopting a 'managed retreat' approach. Installing defence mechanisms could affect access to whole sections of the coastline.

In the same way, practices used to manage river flooding can affect access along riversides, whichever approach is adopted. Hard defences are likely to change the character of the area in a significant way and floodplain restoration could result in periodic flooding.

On the other hand, climate change could result in new 'climate spaces' into which species and habitats can migrate, creating new opportunities for people to access and appreciate the local biodiversity.

December 2013

Source: https://climateandus.com

5. Assess the importance of wind and waves in the creation of **one** distinctive coastal landscape. (9 marks)

6. 'During the twenty-first century climate change will make the costs outweigh the benefits of using traditional hard engineering approaches to coastal flood and erosion risk.'

   To what extent do you agree with this view?

   (20 marks)

Glacial systems and landscapes

Glacial landscapes are dynamic environments in which the landscape continues to develop through contemporary processes but which mainly reflect former climatic conditions associated with the Pleistocene era. The operation and outcomes of fundamental geomorphological processes and their association with distinctive landscapes are really observable. In common with the other physical sections of this book, glaciers can be seen as natural systems. Glacial landscapes present the limited number of human inhabitants with unique economic, social and political opportunities and challenges.

Many of these areas are increasingly referred to as being fragile environments with important issues facing their biodiversity and sustainability.

In this chapter you will study:
● the nature and distribution of cold environments
● glaciers as natural systems and glacial systems including glacial budgets
● geomorphological processes and glacial landscape development
● human impacts on glaciated landscapes.

4.1 Introduction to glacial systems and landscapes

Glaciers as natural systems

The introduction to Chapter 1: Water and carbon cycles outlined how many aspects of physical geography can be studied using a **systems** approach and you should read that section carefully before reading further here. This chapter explores glacial systems and landscapes and we will see that glaciers are **open systems** with a range of **inputs, stores/components, flows/transfers and outputs**. When there is a balance between the inputs and outputs then the system is said to be in a state of **dynamic equilibrium**. If one of the elements of the system changes, for example there is increased ice accumulation but there is no corresponding change in the amount of meltwater, then the mass of the glacier may change and the equilibrium is upset. This is called **feedback**. Figures 4.1a and 4.1b (page 138) illustrate the open system of a glacier in a glacial landscape and are a simple example of a positive feedback system in a cold environment. They highlight many of the ideas that are explored in detail in the rest of the chapter.

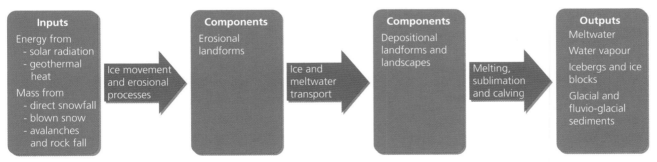

Figure 4.1a A glacier as an open system in a glacial landscape

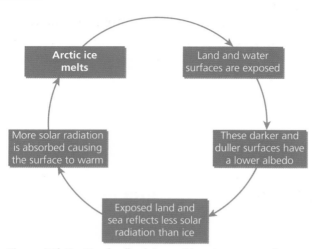

Figure 4.1b Positive feedback in a cold environment – the 'ice–albedo feedback'

Key question

Can you think of any other examples of feedback in glacial landscapes?

Like all physical environments, characteristic glacial landscapes are a combination of a range of constantly changing erosional and depositional landforms within a geomorphological system. Landscapes are shaped and reworked by the action of physical processes driven by the input of energy as the system tends towards dynamic equilibrium. There are a variety of characteristic landscapes in cold environments, including combinations of landforms that form glacial, glaciated, fluvio-glacial and periglacial landscapes, each with their own unique assemblages of erosional and depositional features.

4.2 The nature and distribution of cold environments

Physical characteristics of cold environments

The term 'cold environments' covers a broad range of landscapes, climates and ecosystems. Some receive large amounts of precipitation as snow while others receive the same low levels of precipitation as some hot deserts. By definition all are 'cold', but some of the coldest parts of Antarctica can be below −90°C with almost no areas of the whole continent ever achieving above 0°C, while some

have very seasonal temperature ranges and fluctuate around 0°C.

Latitude, altitude and climate characteristics

The distribution of cold environments is varied, and they are found in different locations with different climate characteristics dependent on their latitude and altitude. Cold environments can be grouped into the following classifications:

● **Polar:** Found at the Earth's northern and southern extremes, almost entirely within the Arctic and Antarctic circles, above approximately 66° north and south of the equator.

  – The polar Arctic region of the north largely consists of the Arctic Ocean that encircles the North Pole, and some islands and most northerly extremes of Asia, Europe and North America. The central polar area is continuously covered by a drifting ice pack. This sea ice averages 3 m in thickness and doubles in extent in winter to around 13 to 14 million km². The Arctic area has a very wide range of climates, but it is continuously cold with mean annual temperatures ranging from 4°C to −28°C. There are similarly wide-ranging levels of precipitation, with the overall average being only around 100 mm, with the Atlantic area of the region receiving significantly more.

  – The southern polar region includes the massive frozen wilderness of the **ice sheet** and surrounding ice shelves of Antarctica which cover 13 million km², and the periodically frozen expanse of the Southern Ocean. With such a large expanse of uninterrupted ocean this southern zone experiences strong westerly winds. Together with the coldness of the ocean and large landmass, this gives Antarctica a much colder climate than the Arctic. The continuously cold waters of the Southern Ocean mean that sea ice develops rapidly in the winter from about 2 million km² in February to around 16 million km² in September. The fluctuation in the extent of the sea ice in the Arctic is much less, due to the influence of the surrounding continents. Another recently observed difference between the poles is that the extent of the winter sea ice around Antarctica is increasing while that in the north is shrinking. The interior of Antarctica includes mountains and high plateaux. The combination of the atmospheric circulation patterns and the high altitude means that the

South Polar Plateau is the coldest location on Earth, with mean annual temperatures in places of around −55°C. The coastal areas of Antarctica are much milder with average annual temperatures of around −10°C. Average precipitation for Antarctica as a whole is around 200 mm/year. Rates range from three to five times this in coastal regions to very arid conditions on the Polar Plateau.

- **Alpine:** These are the most instantly recognisable 'glaciated' landscapes. Commonly they are areas of high relief, with peaks often rising well over 3,000 m from valley floors. The most striking landscapes are also associated with tectonically active mountain ranges, including the Himalaya, Karakoram and Tibet plateau in Asia, the Rockies, Cascades and Andes in the Americas, and the New Zealand Alps. It is the combination of tectonic uplift and rapid rates of erosion by water and ice that creates such spectacularly well-developed glaciated features. Older mountain ranges in more tectonically stable regions have lower relief and the features of their well-developed glaciated landscapes have often been *softened* and smoothed by subsequent weathering and erosion. Alpine features are found in areas that were once glacial, including the mountains of northern and western Britain, and those recently and currently occupied by glaciers including the European Alps and Norwegian mountains. Alpine landscapes can include small ice caps, mountain glaciers and tundra environments.
  - Alpine regions with active glaciers today occur at almost any latitude, but they are at high enough altitudes for snow and ice to remain throughout the year. Obviously temperatures drop below freezing for significant periods, often below −10°C, but in summer temperatures can rise well above freezing, with those in lower latitudes regularly above 20°C in summer months.
  - It is suggested that Alpine landscapes develop in a sequence following an *Ergodic* model (Flint, 1971), from a preglacial fluvial landscape which is then modified episodically during cycles of **glacials** and **interglacials** during **ice ages**. Recurring glacial erosion takes these landscapes through three stages to a mature alpine landscape with many of the features explored later in this chapter.
- **Glacial:** These are environments that are currently covered by ice sheets and **glaciers**. These are permanently cold enough for substantial masses of ice to remain on the landscape throughout the year. The two most significant areas are the ice sheets and ice shelves of Antarctica and Greenland. At present about 10 per cent of the land area of Earth is covered by ice, including glaciers, ice caps and ice sheets covering over 15 million km². Table 4.1 summarises the distribution of this ice.

Table 4.1 Land area covered by ice sheets and glaciers

Region	Estimated area (million km²)
Antarctica	13.5
Greenland	1.74
Rest of Arctic region	0.24
Asia	0.125
Alaska	0.041
Rest of North America	0.03
Andes	0.03
Scandinavia	0.003
Alps	0.004
Australasia	0.0012
Africa	0.0001
Total	15.7143

- **Periglacial** (including **tundra** regions): These are generally found in dry high-latitude areas that are not permanently covered by snow and ice, but are cold enough for periglacial processes to occur. Currently areas of northern Alaska and Canada, northern Scandinavia, Siberia and the islands of the Arctic Ocean such as Spitsbergen have active periglacial regions with tundra ecosystems. In these places the cold conditions create permanently frozen ground known as **permafrost**, which is often overlain by a more mobile unfrozen layer of soil called the **active layer**. Due to low levels of precipitation, cold temperatures and often thin poor soils, vegetation is sparse in periglacial environments. Some areas in lower latitudes but at higher altitudes do experience some periglacial processes.

The climate of cold environments

The main features of the **polar** climate are:
- mean monthly temperatures below freezing all year
- extreme cold in winter with monthly average temperature below −50°C
- precipitation below 150 mm/year, with all of it falling as snow (at the South Pole, precipitation is around 50 mm/year)

- strong winds blowing outwards from the centre of the continent. Winds of over 200 km/hr have been recorded. There is therefore a substantial wind-chill factor. The winds also whip up the powdery snow to create frequent blizzards and white-outs.

The main features of the **tundra** climate are:

- long and bitterly cold winters with temperatures averaging –20°C
- brief, mild summers with temperatures rarely above 5°C
- at least 8 months of the year when the temperature remains below 0°C
- small amounts of precipitation, less than 300 mm/year, most of which falls as snow
- frequent strong winds which increases the wind chill.

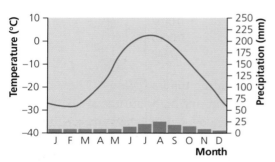

Figure 4.2a The climate of tundra regions: Barrow, Alaska

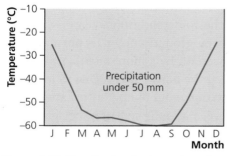

Figure 4.2b The climate of polar regions: Amundsen–Scott Station, South Pole

The causes of such climates are as follows.

- **The low level of insolation:** In summer the sun remains at a low level in the sky even though there is a period of continuous sunlight. Because the rays strike the surface obliquely, there is a wider surface area to heat. In addition, the longer passage through the atmosphere allows for increased absorption, scattering and reflection

of radiation. Less insolation therefore reaches the surface (see Figure 4.3). Continuous summer radiation ('the land of the midnight sun') makes the tundra days seem pleasantly warm, but this only lasts for a short period. In winter there is a corresponding period when there is no incoming solar radiation as the sun does not appear above the horizon.

- **The high albedo:** In areas of continuous snow cover, much of the incoming solar radiation is reflected off the ice/snow surface. This reduces the amount that can actually contribute to the warming of the atmosphere.
- **The high pressure systems** of polar regions means that frontal systems rarely penetrate these areas, giving low levels of precipitation.
- **The coolness of the air** is another factor in low precipitation. Very cold air can only hold low levels of water vapour. When precipitation occurs it is often little more than very light powdery snow.
- **Katabatic winds:** In Antarctica, masses of cold dense air flow down the valleys and off upland areas. Such movements are known as katabatic and are strong in Antarctica where there is a real difference between the interior and coastal areas. With few obstacles to hinder air movement, such winds can exceed 200 km/hr.

The vegetation of the Tundra

The main features of the vegetation are:

- a very low level of productivity
- very few plant species and therefore low biological diversity
- an absence of full grown trees (tundra means 'treeless plain' in Sami). There are, however, a number of dwarf varieties to be seen
- most flowering plants are perennials – they flower year after year. They have hardy seeds armoured by a thick seed case
- there are five types of plants, each occupying its own specialised niche: lichens, mosses, grasses, cushion plants and low shrubs (Figure 4.4). Lichens are the pioneer plants as they colonise bare areas. They have no roots and are able to absorb water and nutrients directly into their foliage. Together with mosses, they initiate soil formation.

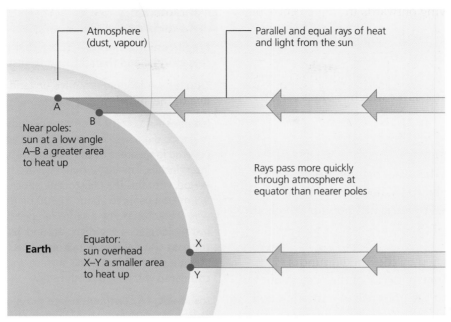

Figure 4.3 The effects of latitude on insolation

The tundra has many features that make plant growth difficult. These include the low levels of insolation and the very short growing season of not more than 50 to 60 days. Precipitation levels are also low with soil moisture frozen for much of the year. Soils are often waterlogged, as thawed water is unable to penetrate to depth because of the permafrost and with low temperatures, evaporation levels are low. Tundra plants also have to contend with strong desiccating winds and permafrost not far below the surface.

Tundra plants have therefore to overcome a number of restrictions imposed by the environment. To cope with such difficulties, plants have adopted some of the following features:

- ground hugging to avoid the cold, desiccating winds and to take advantage of any warmth close to the surface
- slow growth rates
- shallow roots because of the presence of permafrost
- the ability to photosynthesise at extremely low temperatures, even under a thin covering of snow
- reduced transpiration due to the small leaves and a thick cuticle
- low albedo of plant surfaces meaning that they can absorb more radiation than the surrounding surfaces
- perennial plants which can store food from year to year.

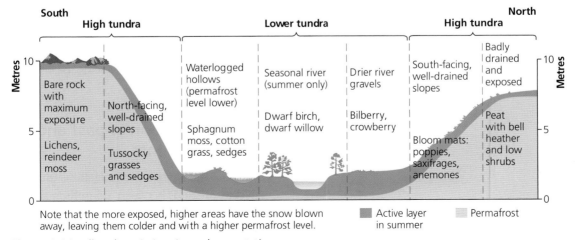

Note that the more exposed, higher areas have the snow blown away, leaving them colder and with a higher permafrost level.

Figure 4.4 Small-scale variations in tundra vegetation

Tundra soils

A distinctive soil type has developed underneath tundra areas. Its major features are:

- a lack of clearly differentiated soil horizons (layers)
- a thin surface organic layer which is often very acidic
- a uniform blue grey colour
- waterlogged in summer
- gleyed (because of the waterlogging, iron compounds are reduced to their ferrous form which is grey – as opposed to ferric compounds which are red-brown).

Surface litter is restricted by the limited tundra vegetation. The slow rate of decay of the vegetation due to the climate and the presence of only a few soil organisms also means there are limited amounts of organic matter in the soil. Soil organisms normally act as mixing agents and as there are few available to carry out this work, the soil horizons are poorly differentiated. The freezing of the soil in autumn leads to it being churned and the horizons become further distorted. Where the soil overlies bedrock, frost-heave raises fragments to the surface giving rise to typical periglacial features.

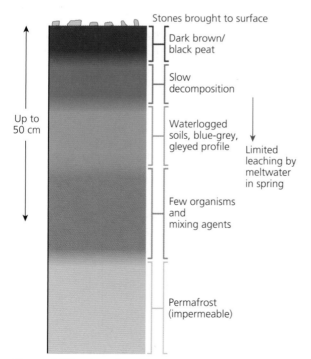

Up to 50 cm

- Stones brought to surface
- Dark brown/ black peat
- Slow decomposition
- Waterlogged soils, blue-grey, gleyed profile
- Limited leaching by meltwater in spring
- Few organisms and mixing agents
- Permafrost (impermeable)

Figure 4.5 Tundra soil profile

Key terms

Fragile environments – Those that are easily damaged or disturbed and then difficult to restore once destroyed.

Glacier – A large mass of ice on land moving downhill due to the influence of gravity.

Ice ages – A common term for periods when there were major cold phases known as glacials and ice sheets covered large areas of the world, interspersed with warmer interglacials. The last 2 million years, the Quaternary Period, contains the Pleistocene epoch lasting from about 1.8 million years to about 11,500 BP (before present). This was characterised by a series of episodes of glaciation, the last major glacial beginning around 120,000 BP reaching the **Last Glacial Maximum** about 18,000 BP. At that time about a third of the Earth's surface was covered by ice, compared with just a tenth today.

Ice sheet – A continental-size mass of ice, covering at least 50,000 km² that is dome-shaped with flow of ice outward from the centre. Ice sheets can be up to 2,000 m thick today; those in the Quaternary glaciation could have been twice that thickness.

Insolation – The radiant energy emitted by the sun; i.e. incoming solar radiation.

Periglacial – Processes and landforms associated with the fringe of, or area near to, an ice sheet or glacier.

Tundra – A Lapland term for the climate and vegetation type found across extensive areas of northern Siberia, Scandinavia, Canada and Alaska. Tundra lies betweenthe region of continual snow and ice of the Polar regions and the northern limit of tree growth (or the *taiga* environments) and at high altitudes above the tree line in the Alps, Rockies, Andes and Himalayas. Similar ecosystems are found around the ice-free fringes of Antarctica. It is characterised by a stony or marshy surface with mosses, lichens and other low-lying vegetation including shrubs and dwarf trees. Tundra has short summers but mean monthly temperatures do not exceed 10°C, which is just warm enough for snow and the surface layer of the permafrost (active layer) to melt. Currently tundra covers about 8 per cent of the land area of Earth, but during the glacials of the Quaternary glaciation tundra covered huge areas of central North America and Europe, including the British Isles.

Global distribution of past and present cold environments

Throughout the Pleistocene glaciations (during the Quaternary Period) between 1.6 million and about 12,000 YBP (years before present) glacials and interglacials have seen cold environments expand and contract across the surface of the Earth. There is still some debate about the number of cycles and extent of ice coverage in some locations during this epoch. Figure 4.6 shows the current extent of cold environments.

Figures 4.7a and 4.7b show some recent research that describes how the extent of the distribution of ice on Earth has changed over the last 18,000 years or so, since the **Last Glacial Maximum**.

Figure 4.7a Ice coverage from c. 18,000 years ago

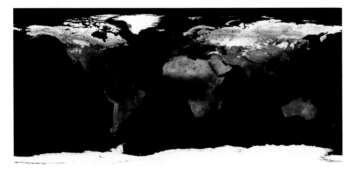

Figure 4.7b Ice coverage December 2014

Skills focus

Use the internet to investigate how scientists are using satellite technology and geo-spatial and digital imagery to map the ever-changing pattern of ice on Earth. Good places to start are:
www.bas.ac.uk/team/operational-teams/operational-support/mapping-and geographic-information/#project
http://visibleearth.nasa.gov/view.php?id=84418

4.3 Systems and processes in glaciers and glacial landscapes

Glaciers as natural systems

In the introduction to Chapter 1: Water and carbon cycles, you will have read that many aspects of physical geography, water and coastlines for example, can be studied using a systems approach. This is because certain aspects of physical geography are highly complex and geographers have attempted to simplify aspects of them so that relationships between components can be better understood. A system is an assemblage of interrelated parts that work together by way of some driving process. Before you read on in this section about glacial systems and how they relate to the development of the landscape, you should also read the relevant material at the beginning of Chapter 1 under the heading 'Systems frameworks and their application'.

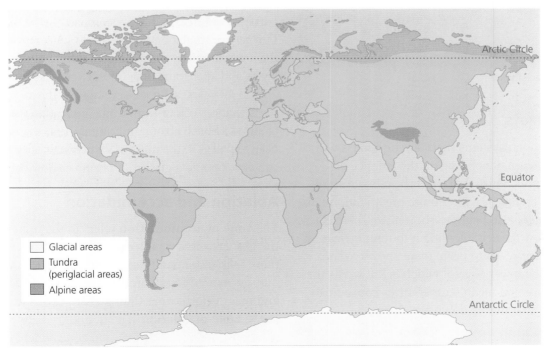

Glacial areas
Tundra (periglacial areas)
Alpine areas

Figure 4.6 Cold environments

Glacial systems

In order to understand how, when and where glaciers and ice sheets form, and what effect they then have on the landscape it is useful to view glaciers as **systems**. Like other physical systems glaciers have **inputs**, **stores**, **transfers** and **outputs**. Figure 4.8 shows a cross-section through an ice sheet to illustrate an idealised diagram of the **open system** of a glacier or ice sheet.

Snow and ice are the most important inputs to the glacier system. This comes from:

● direct snowfall

● blown snow

● avalanches from slopes surrounding the glacier.

Together these inputs are known as **accumulation**. These inputs are then transferred down-valley, under the influence of gravity, by glacier movement. Mass is lost from the system by either melting and evaporation, or the *calving* of ice blocks or icebergs. Together these outputs are known as **ablation**.

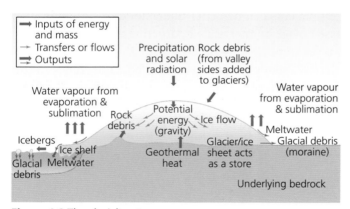

Figure 4.8 The glacial system

Glacial budgets

As glaciers are open systems their size and mass will increase or decrease depending on the balance of inputs and outputs. During glacials the climate starts to become colder and in winter more precipitation falls as snow. Summers also begin to shorten, so there is less time for the snow to melt allowing it to remain year on year. Initially, this leads to permanent snow cover in upland areas, the lower edge of which is known as the **snow line**. As annual mean temperatures continue to drop, the snow line moves down the slope.

Currently the snow line is at sea level in Greenland but at 6,000 m at the Equator. Britain's climate is currently

too warm for there to be a snow line, but it is estimated that if the Scottish mountains were about 250 m higher there would be. The aspect of a slope determines the altitude of the snow line, for example in the northern hemisphere it is higher on north-facing slopes and lower on south-facing slopes as they receive more insolation. In the European Alps the more habitable south-facing slope is known as the *adret* slope and the more inhospitable north-facing slope is called the *ubac* slope (Figure 4.9).

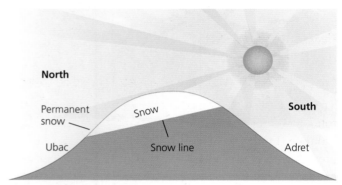

Figure 4.9 The snow line in the northern hemisphere

Snow initially falls as flakes with an open, feathery structure that traps air. As snow accumulates the lower layers are slowly compressed by the upper layers, turning this lower snow into a more compact form known as **firn** (usually after about two seasons of accumulation). The French term **névé** does refer to the layers of freshly accumulated snow but is often interchangeable with 'firn'. Meltwater also seeps into spaces between the snow particles and freezes, further compacting the mass. Further accumulation progressively squeezes air out of the lower layers by the additional weight of the upper layers and after some time (most experts suggest between 20 and 40 years) solid ice develops. As the ice forms the mass changes colour from white, indicating the presence of air, to a bluish colour, indicating the air has largely been expelled. If the mass of ice becomes large enough it can develop into a glacier and begin to flow downhill.

Ablation and accumulation

The **zone of accumulation** refers to the upper part of the glacier where inputs exceed outputs and therefore more mass is gained than is lost over a year. Where outputs exceed inputs in the lower part of the glacier, and mass is lost rather than gained, this is known as the **zone of ablation**. The boundary between the ablation zone and accumulation zone, where net gain and net loss are balanced, is known as the **equilibrium**

line. (This is not to be confused with the *firn line*, which is similar but different.) (Figure 4.10.)

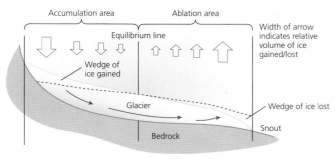

Figure 4.10 Glacier budget

Historical patterns of ice advance and retreat

The difference between the total accumulation and total ablation during one year is known as the **net balance** (Figure 4.11). **Temperate** glaciers in alpine areas have a negative balance in summer when ablation exceeds accumulation and a positive balance in winter when accumulation exceeds ablation. If the summer and winter budgets cancel each other out, the glacier appears to be stationary. In other words, the snout of the glacier remains in the same position while ice is still advancing down the valley from the zone of accumulation into the zone of ablation.

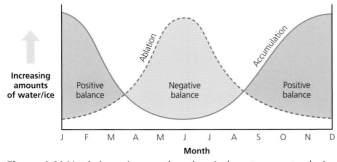

Figure 4.11 Net balance in a northern hemisphere temperate glacier

If accumulation begins to exceed ablation rates, then the snout moves down the valley. This is known as **glacial advance**. When ablation rates exceed accumulation the glacier begins to shrink in size and the position of the snout moves up the valley. This is called **glacial retreat**. It is important to note that despite the *position* of the snout moving back up the valley, the (retreating) ice continues to move downslope from the upper parts of the glacier.

During the glacials and interglacials of the Quaternary glaciers have advanced and retreated

Figure 4.12 Ice coverage in Europe 18,000 years BP

many times. Evidence for this is found in the landscape. At present glaciers cover about 6,500 km² in Europe, but 18,000 years BP ice coverage was much more extensive (Figure 4.12). Individual glaciers not only advance and retreat in response to long-term climate changes, but also gain and lose mass in response to more localised, or short-term changes in climate. Figure 4.13 illustrates the pattern of advance and retreat observed in one European glacier over a much shorter time period.

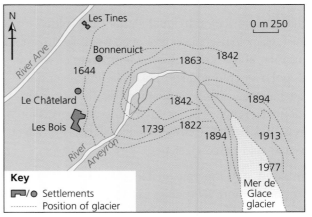

Figure 4.13 Changing snout position of the Mer de Glace, France

While most glaciers around the world are following the pattern of the Mer de Glace in Chamonix, where negative glacial budgets mean their snouts have retreated in the recent past, some glaciers in the Himalayas and Alaska in North America have positive annual net balances and their snouts are advancing, like Alaska's Hubbard Glacier (Figure 4.14).

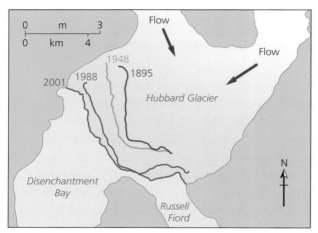

Figure 4.14 Alaska's advancing Hubbard Glacier

Warm- and cold-based glaciers

The characteristics of individual glaciers are unique. However, glaciologists have developed a simple three-fold classification of glaciers, including **temperate** or **warm-based**, **polar** or **cold** and sub-polar glaciers. Over the decades many authors have adapted this classification and adopted a range of other interchangeable terms for each type. For simplicity this chapter will concentrate on warm- and cold-based glaciers.

Warm-based, or temperate, glaciers (Figure 4.15) tend to be found in more temperate maritime locations. Good examples of this include the western coastal mountain ranges of North America, western Norway, southern Iceland, New Zealand and Patagonia (Chile). Here the alpine valley glaciers of the European Alps are included in the warm-based category. These tend to be smaller valley glaciers typically ranging from hundreds of metres to a few kilometres in width and hundreds of metres to tens of kilometres in length.

Figure 4.15 Franz-Joseph glacier in New Zealand

Cold-based or polar glaciers by definition are almost all found in areas of high latitude within the Arctic and Antarctic circles, in especially cold areas of Alaska, Canada and Antarctica (Figure 4.16). Here cold-based glaciers include some very cold valley glaciers but also refer to the much larger glaciers associated with the vast **ice caps** and **ice sheets** covering hundreds of km².

Figure 4.16 Commonwealth Glacier, Antarctica

The characteristics and development of temperate/warm-based glaciers

Temperate glaciers are found in places with high winter snowfall rates and spring and summer temperatures high enough to create rapid summer melt rates. The vast amounts of meltwater act as a lubricant and allow the glacier to be far more mobile than cold glaciers. The faster rates of glacial movement mean that warm-based glaciers are more likely to erode, transport and deposit material.

At the surface the thin layer (a few metres) of more recent snow and firn is subject to seasonal temperature fluctuations so melts rapidly at around 0°C in the summer melting. This surface layer insulates the layers of ice beneath it. With increasing depth in a glacier the ice is under ever increasing pressure from the surrounding ice, which has the effect of lowering the melting point of the ice. This is known as the **pressure melting point**. For example, at the base of a 2,000 m deep glacier the melting point of the ice is –1.27°C. All ice in a temperate glacier is at or near the *melting point* because of the warmer atmospheric temperatures, the weight of the ice above, and as temperature glaciers may be relatively thin a greater proportion of the ice is influenced by geothermal heat at the bed (Figure 4.17).

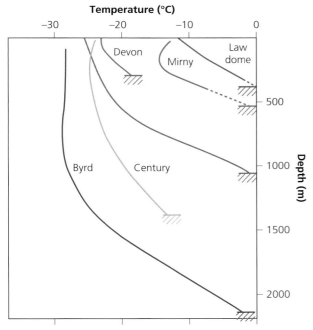

Figure 4.17 Temperature profiles of selected Greenland and Antarctic glaciers

The characteristics and development of polar/cold-based glaciers

Cold glaciers are generally found in places with low precipitation rates, even arid conditions, so receive little new snow each year and accumulation rates are low. There is little or no melting of ice in cold glaciers and so the ice can be very old, with the ice at the base of some Arctic and Antarctic ice sheets dating back around 100,000 years BP.

All ice, except the most upper surface layers that can be exposed to summer atmospheric temperatures of above 0°C, is below the *melting point*. There is little meltwater associated with these glaciers as atmospheric and sub-glacial (geothermal) heat sources are not great

enough to reach melting point. Other than limited melting of the surface in summer the majority of ice loss is due to *sublimation* and the calving of icebergs/blocks.

Movement in cold-based glaciers is much slower than in temperate glaciers as they are often frozen to their beds, thus most movement is due to **internal flow**. Much less erosion, transportation and deposition occurs.

While, as we will see later, many of the most striking features of the landscapes of many cold environments are the result of processes of erosion, transportation and deposition, it is important not to forget the role played by other geomorphological processes.

Geomorphological processes and glaciated landscape development

As glacial geomorphological systems develop over time and tend towards dynamic equilibrium a variety of processes take place. Driven by the input of energy these are what change the characteristics of the components of glacial and glaciated landscapes.

Weathering

Weathering is the breakdown and/or decay of rock at or near the Earth's surface creating regolith that remains *in situ* until it is moved by later erosional processes. Weathering can be mechanical, biological/organic or chemical. Although the low temperatures of most cold environments limit the impact of biological or organic weathering, mechanical and chemical weathering play a very important role in both shaping the environment and providing material for glacial erosion.

Frost action/freeze–thaw cycles

● **Frost action/freeze–thaw cycles** lead to **frost shattering**. This occurs in areas where temperatures rise above 0°C during the day but drop below freezing at night for a substantial part of the winter. Water that enters cracks in the rocks freezes overnight. As it freezes the water expands by just under 10 per cent, meaning the ice occupies more space and so exerts pressure on the surrounding rock. As the process repeats and continues, the crack widens, and eventually pieces of rock break off (Figure 4.18, page 148). On steep slopes this leads to the collection of material at the base, known as **scree**. In a glacial valley much of this material falls from the valley side onto the edges of the glacier (an input) and some finds its way to the base of the ice via numerous crevasses and *moulins* on the glacier's surface.

Figure 4.18 Evidence of frost shattering, Ecrins National Park, France

Nivation

- **Nivation** is a series of processes that operate underneath patches of snow in hollows, particularly on north- and east-facing slopes (in the northern hemisphere). Freeze–thaw action together with chemical weathering processes, operating under the snow, causes the underlying rock to disintegrate. As some of the snow melts in the spring, the weathered particles are 'flushed-out' of the hollow and moved downslope by the meltwater and **solifluction** (see below). The accumulation of more debris by repeated seasons of freezing and thawing, and the accumulation of more snow leads to the formation of **nivation hollows** which, when enlarged, can be the beginnings of a corrie (cirque) as outlined below.

Key question

Why might weathering processes have more impact in the mountains of Scotland than in Antarctica?

Ice movement

As ice accumulates to greater depths it begins to move due to the influence of gravity. As the ice mass moves downhill it does not all behave in the same way. It has great rigidity and strength, but under steady pressure it behaves as a plastic (mouldable) body. In contrast, when put under sudden compression or tension, it will break or shear apart. This gives two zones within the glacier (Figure 4.19):

- the upper zone where ice is brittle, breaking apart to form crevasses

- the lower zone which has steady pressure. Here meltwater resulting from pressure melting and from *friction* with the bedrock allows a more rapid, plastic flow.

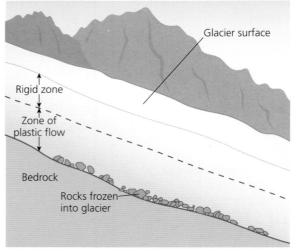

Figure 4.19 Zones within a glacier

Source: AQA

The technical details of ice movement are extremely complex, but several main types of movement can be identified:

- **Internal deformation** including **creep** that occurs when stress builds up within a glacier, allowing the ice to behave with plasticity and flow. This is common when obstacles are met. *Internal flow* within the glacier involves the ice crystals orientating themselves in the direction of the glacier's movement and sliding past each other. Internal deformation and creep are the main feature of the flow of cold, or polar, glaciers as, without the presence of meltwater, they tend to be frozen to their beds.

- **Rotational flow** that occurs within the corrie (cirque), the birthplace of many glaciers. Here ice moving downhill can pivot about a point, producing a rotational movement. This, combined with increased pressure within the rock hollow, leads to greater erosion and an over-deepening of the corrie floor.

- **Compressional flow** that occurs where there is a reduction in the gradient of the valley floor leading to deceleration and a thickening of the ice mass. At such points ice erosion is at its maximum (Figure 4.20).

- **Extensional flow** that occurs when the valley gradient becomes steeper. The ice accelerates and becomes thinner, leading to reduced erosion (Figure 4.20.)

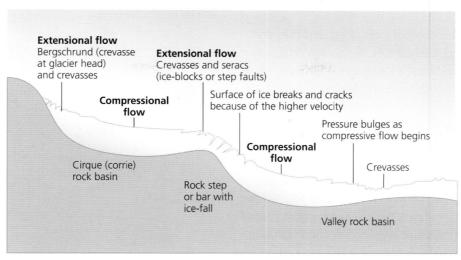

Figure 4.20 Extensional and compressional flow

- **Basal sliding (flow/slippage)** occurs because as the glacier moves over the bedrock there is *friction*. The lower levels of ice are also under a great deal of pressure and this, combined with the friction, results in some melting. The resulting meltwater acts as a lubricant, enabling the ice to flow more rapidly.

At certain times glaciers can move forward quite suddenly. These are known as **surges**.

- **Surges** occur when an excessive build-up of meltwater under the glacier leads to the ice moving rapidly forward, perhaps as much as 250–300 m in one day. Such surges represent a hazard to people living in the glacial valley below the snout. Following the increased sub-glacial volcanic activity in Iceland over recent years, glaciologists have been able to study these surges more closely.

Within a glacier itself there are different rates of movement. The sides and base of the glacier move at a slower rate than the centre surface ice (Figure 4.21, page 150). As a result, the ice cracks, producing crevasses on the surface. These also occur where extending flow speeds up the flow of the ice, where the valley widens or the glacier flows from a valley on to a plain like a *piedmont* glacier.

Glacial erosion

As warm-based glaciers are large bodies of ice and are moving downhill they have the ability to erode and shape the landscape. Glacial erosion tends to happen in upland regions and is carried out by two main processes:

- **Abrasion** occurs where the material the glacier is carrying rubs away at the valley floor and sides. It is often likened to the effect of sandpaper, as it is not the

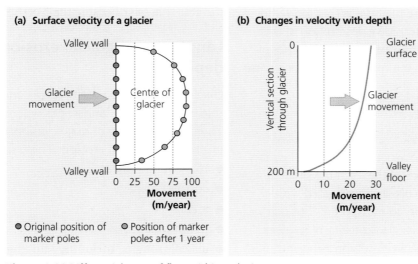

Figure 4.21 Differential rates of flow within a glacier

ice (the paper) that has the ability to actually wear away at the rock, but the material embedded in it (like the sand on the paper). The coarser, harder material may leave scratches on the rock known as **striations**, while the finer debris smooths and polishes rock surfaces. The debris involved is often worn down by the process into a very fine material known as **rock flour**. This is taken away from the glacier by sub-glacial meltwater streams, turning them a milky colour.

- **Plucking** involves the glacier freezing onto and into rock outcrops. As the ice moves forward it pulls away masses of rock. Plucking is mainly found at the base of the glacier where pressure and friction often result in the melting of the ice. There is a considerable amount of meltwater at the base of most glaciers, which is very close to freezing point, and often this will refreeze thus attaching the glacier to the bedrock. It is common in well-jointed rocks and those where the surface has been weakened by freeze–thaw action (frost shattering). Plucking leaves a very jagged landscape.

Glacial transport and deposition
Sub-glacial debris, en-glacial and supra-glacial moraine

As well as eroding rock over which it is flowing, a valley glacier is also capable of transporting large amounts of debris. Some of this may also be derived from rockfalls from the valley side. Material is transported on the surface of the glacier (**supra-glacial debris**) or buried within the ice (**en-glacial**). Material found at the base is known as **sub-glacial** and may include rock fragments that have fallen down crevasses or been washed down *moulins* as well as material eroded at the base. Figure 4.22 illustrates where material is transported by the glacier, including an alternative description of this material. Strictly speaking **moraine** applies to types of landform (explored below), but many textbooks now use the term in this way.

Glacial deposition

The huge amounts of material carried by a glacier will eventually be deposited. The majority of this will be the debris released by the melting of the ice at the snout. It is also possible for the ice to become overloaded with material, reducing its capacity to transport it. This can occur near to the snout, as the glacier melts, or in areas where the glacier changes between compressing and extending flow, thus reducing the mass of ice at that point.

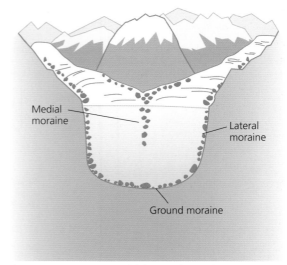

Figure 4.22 Transport of debris within a glacier

Material that is deposited directly by the ice is known as **till**; this has generally replaced the term **boulder clay** as this is quite specific in describing certain kinds of till.

The term till is used to describe an unsorted mixture of rocks, clay and sand that was mainly transported as supra-glacial or en-glacial debris and deposited when the ice melted. Individual stones tend to be angular to sub-angular, unlike river or beach material which is rounded and smooth. Till reflects the character of the rocks over which the ice has passed. The till deposited in south Lancashire, for example, includes rocks from the Lake District (for example, shap granite) and southern Scotland (for example, riebeckite from Ailsa Craig in the Firth of Clyde). In East Anglia the till contains pieces of granite from southern Norway. This indicates not only the passage of the ice but the fact that the sea level must have been considerably lower to allow ice to move over the area that later became the North Sea.

Landforms of glacial erosion and deposition

Glaciers are large powerful forces and play a significant role in shaping the landscapes where they are found. There are a number of major landforms that are mainly produced by glacial erosion. These include:

- roche moutonnée
- corries (cirques)
- arêtes
- pyramidal peaks
- glacial troughs, including hanging valleys and truncated spurs.

Fieldwork opportunities

There are several accessible areas of the United Kingdom where you can study closely all the above features of glacial erosion. The Lake District is one such area (see Case study 1); another is Snowdonia in North Wales. In Scotland, the Cairngorms show all these features. Further afield, the Chamonix area of France has alpine glaciers with deep glacial valleys and in Switzerland the Lauterbrunnen Valley is readily accessible. In Iceland, you can study glaciation as it is happening along with the volcanic activity of the island.

Due to the enormous amount of material eroded and transported by glaciers they also create a range of depositional landforms, including:

- drumlins
- erratics
- moraines (lateral, recessional, push and terminal/end moraines
- till plains.

Fieldwork opportunities

As well as the fringes of upland areas (the south Lake District, for example, has good examples of drumlins), many lowland areas of the United Kingdom show significant features of glacial deposition. There are moraines and fluvio-glacial deposits in East Anglia; the highest point of Norfolk, Beacon Hill (105 m), is a glacial deposit and is part of the moraines of the Cromer Ridge. Deposits are also found in the Vale of York, drumlins in the Ribble Valley (Lancashire/Yorkshire) and within many of the soils of lowland Britain (north of the Thames Valley) you will find stones which are small erratics which were brought south by the ice sheets and glaciers which once covered the country.

Erosional landforms

Corries and associated landforms

A **corrie** is an armchair-shaped rock hollow, with a steep back wall and an over-deepened basin with a rock lip. It often contains a small lake (**tarn**). In the British Isles, corries are mainly found on north-, northeast- and east-facing slopes where lower levels of insolation allowed increased accumulation of snow. The dimensions of corries will depend on the topography of the area and the size of the glaciers that formed them, but in the English Lake District, Red Tarn has a depth of 25 m, a back wall rising 300 m to the summit of Helvellyn and is around 300–400 metres across (see case study on page 176).

If several corries develop in a highland region, they will jointly produce other erosional features. An **arête** is formed when two corries lie back to back or alongside each other. If enlarged over many glaciations a narrow, steep-sided ridge between the two hollows is formed. An example is Striding Edge above Red Tarn on Helvellyn in the Lake District (see case study on page 176). If more than two corries develop on a mountain, the remaining central mass will survive as a **pyramidal peak**, which often takes on a very sharp appearance due to frost shattering (Figure 4.23). A classic example is the Matterhorn in the Alps bordering Switzerland and Italy.

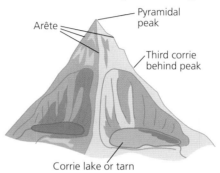

Figure 4.23 Arêtes and pyramidal peak

Corries result from the interaction of several processes (Figure 4.24, page 152). Glaciologists believe that the initial process is **nivation** that acts upon a shallow, pre-glacial hollow and enlarges it into an embryo corrie (this can be a slow process across several glacial periods within an ice age). As the hollow grows, the snow becomes deeper and is increasingly compressed to form firn and then ice. The back wall becomes steeper through the action of **plucking**. Due to the bowl shape of the hollow, the **rotational flow** of the ice, together with the debris supplied by plucking and frost shattering on the back wall, abrades the floor of the hollow, which over-deepens the corrie.

As the hollow becomes deeper, the thinner ice at its edge does not produce the same amount of downcutting and a rock lip develops on the threshold of the feature. Some thresholds are increased in height by the deposition of moraines formed when the glacier's snout was last in that position. During interglacials when the ice has melted, the corrie fills with meltwater and rainwater to form a small lake (tarn).

Glacial troughs and associated landforms

Glaciers flow down pre-existing river valleys as they move from upland areas. They straighten, widen and

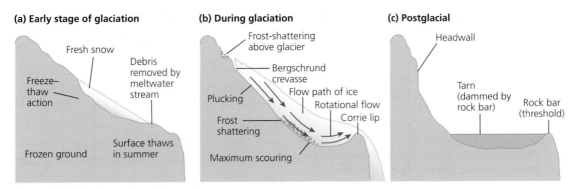

Figure 4.24 The formation of a corrie

deepen these valleys, changing the original V-shaped river feature into the U- or trough shape typical of glacial erosion. The action of ice, combined with huge amounts of meltwater and sub-glacial debris, has a far greater erosive power than that of the pre-glacial river that occupied the valley.

As both extending and compressing flow are present, the amount of erosion varies down the valley. Where compressing flow is present, the glacier will over-deepen parts of the valley floor, leading to the formation of **rock basins**. Scientists have suggested that this over-deepening is caused by increased erosion at the confluence of glaciers, areas of less-resistant geology or zones of well-jointed rocks.

Major features of glacial troughs:
- Generally straight with a wide base and steep sides – a U-shape.
- Stepped, long profile with alternating steps and rock basins (Figure 4.25).
- Some glacial valleys end abruptly at their heads in a steep wall, known as a **trough end**, above which lie a number of corries (Figure 4.26).
- Rock basins filled with **ribbon lakes**, for example, Wast Water in the Lake District.

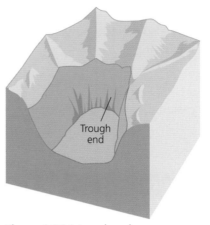

Figure 4.26 A trough end

- Over-deepening below the present sea level – this led to the formation of **fjords** when sea levels rose after the ice ages and submerged the lower parts of glacial valleys, for example, on the coasts of Norway and southwest New Zealand (Milford Sound).
- **Hanging valleys** on the side of the main valley (for example, the valley of Church Beck which flows down into Coniston Water in the Lake District). These are either pre-glacial tributary river valleys that were not glaciated, or smaller tributary glacial

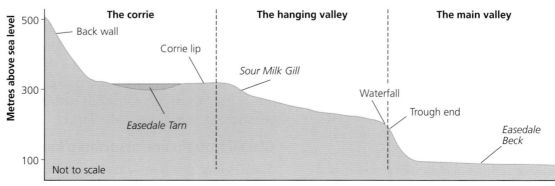

Figure 4.25 The long profile of a glacial valley: Easedale, Lake District

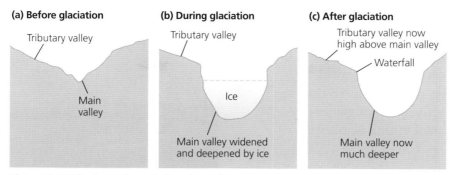

Figure 4.27 The formation of a hanging valley

valleys (Figure 4.27 and Figure 4.28). The glaciers in the tributary glacial valleys would have been smaller, containing less ice and so eroding less than in the main valley. Consequently the floor of the tributary valley was left higher than that of the main valley when the ice retreated.

Figure 4.28 Lady Bowen waterfall in Milford Sound, New Zealand, is a hanging valley

- Areas of land projecting from the river-valley side (spurs) have been removed by the glacier, producing **truncated spurs**.
- Small mounds of resistant rock on the valley floor are not always completely removed and as the glacier moves over these **roches moutonnées** are formed. These have an up-valley side (stoss slope) polished by abrasion and a down-valley side (lee slope) made jagged by plucking. These are relative small features, up to a few tens of metres long and less than 5 m high (Figure 4.29).
- After ice retreat, many glacial troughs were filled with shallow lakes which were later infilled, and post-glacial frost-shattering modifies their sides, developing screes altering the original glacial U-shape (for example, Great Landgdale, Lake District) (Figure 4.30).

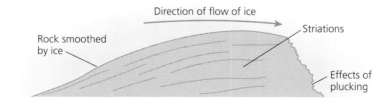

Figure 4.29 The formation of a roche moutonnée

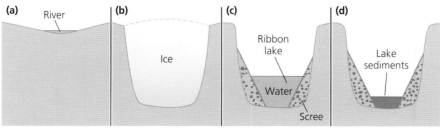

Figure 4.30 The formation of a glaciated valley

Depositional landforms

Glaciers can transport huge quantities of material, all of which is deposited following the melting of the ice. Two types of glacial deposit are recognised:

- **lodgement till**: sub-glacial material that was deposited by the actively moving glacier. A typical feature formed in this manner is a **drumlin**
- **ablation till**: produced at the snout when the ice melts. **Terminal (end)**, **push** and **recessional moraines** are typical features produced from ablation till.

Drumlins

The term drumlin is derived from the Gaelic word *druim*, meaning a rounded hill, and has been adopted to describe these streamlined elongated 'whaleback' hillocks (Figure 4.31). Main features of drumlins:

- They are smooth, oval-shaped small hills, often resembling half an egg.
- They can be up to 1.5 km in length (although most are much smaller) and up to 50–60 m in height.
- They have a steeper 'upstream' end and a more gently sloping 'downstream' end. They are elongated in the direction of ice advance.
- They are often found in groups know as **swarms** and, given their shape, this is sometimes referred to as a 'basket of eggs' topography (Figure 4.32).
- They are formed from unsorted till, and most are believed to have little internal structure.
- They are found on lowland plains such as the central lowlands of Scotland. There is a well-known swarm at Hellifield in the Ribble Valley. Many are found at the lower end of glacial valleys.

Figure 4.32 Risebrigg Hill in North Yorkshire is a drumlin, lying within a swarm of drumlins known as the Hills of Alslack

There has been considerable debate about the origin of drumlins for over 100 years and there are a few popular theories on their mode of formation:

- The first suggests that the ice moulds and shapes a preexisting landscape of glacial drift (ground moraine) deposited in a previous glacial advance.
- Secondly, it is thought that the moving glacier deposits till around a nucleus formed by some unevenness in the floor, such as protruding rock or mass of frozen till. This material is then moulded as the ice continues to move over it.
- However, one of the most widely held views is that drumlins result from the ice becoming overloaded with debris. This reduces its capacity to carry material and deposition occurs at the base of the ice. Again, this material is shaped by further glacier advance.

Erratics

Sometimes it is possible to find a large block of rock that has been moved from one area and deposited in another which has a very different geology. These are known as **erratics** (Figure 4.33).

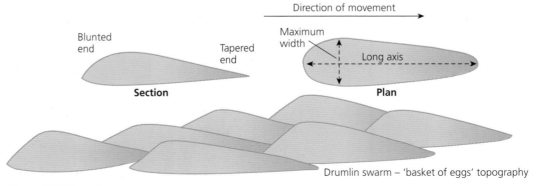

Direction of movement

Blunted end

Tapered end

Section

Maximum width

Long axis

Plan

Drumlin swarm – 'basket of eggs' topography

Figure 4.31 Drumlins

Figure 4.33 Norber erratics in the Yorkshire Dales

Moraines

Moraines are lines, or a series of mounds of material, mainly running across glacial valleys. The main type is the terminal or end moraine that it found at the point where the snout of the glacier reached its furthest extent 'downstream'. Features of **terminal moraines**:

- They are a ridge, or mounds, of material stretching across a glacial valley.
- They are elongated at right angles to the direction of ice advance.
- They are often steep-sided, particularly the ice-contact side, and can reach up to 50–60 m high.
- They are often crescent-shaped, moulded to the form of the snout.
- They are formed from unsorted ablation material (till).

Terminal moraines are formed when ice melts during a period of snout standstill, when it has reached its furthest point down the valley, and the material it has been carrying is deposited. This is why they contain a range of unsorted material, from clay to boulders.

As the glacier retreats, it is possible for a series of moraines to be formed along the length of the valley, marking points where the retreat paused for some time. These are known as **recessional moraines** (Figure 4.34).

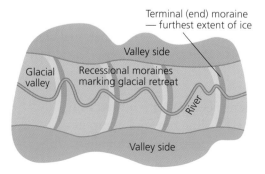

Figure 4.34 Moraines

If the climate cools for some time, leading to glacier advance, a previously deposited moraine may be shunted forwards up into a mound known as a **push moraine**. Such features are recognised by the orientation of the individual pieces of rock, which may have been pushed upwards from their original horizontal position.

Till plains

Often found behind a terminal moraine, or towards the margins of former ice sheets, in low-lying areas, **till plains** are wide areas of generally flat relief created by a till sheet. The underlying bedrock may itself be quite uneven, but it is hidden by a thick covering of glacial till, sands and gravel. The depth of the till sheet can vary greatly, being thicker where depressions exist in the underlying surface. The material tends to be quite compacted giving a poorly drained surface with bogs, lakes and slow-moving meandering streams. There are large till plains south of the Great Lakes in North America, marking the margins of the earlier Laurentide ice sheets.

Characteristic glaciated landscapes

The preceding sections illustrate the significance of the role of glaciers in shaping the landscape. Regions where glaciers have returned many times during the glacials of the Quaternary Ice Age, such as the European Alps, are often said to have a classic glaciated landscape. There are many valleys in the Alps of Austria, Italy, France and Switzerland that exhibit many of the features explored above. A good example is the Lauterbrunnen Valley close to Bern in Switzerland (Figure 4.35).

Figure 4.35 The glaciated features of an alpine valley, the Lauterbrunnen Valley, Switzerland

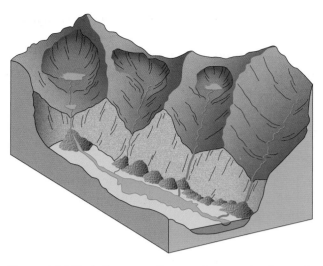

Figure 4.36 Block diagram to illustrate features of a glaciated alpine valley

Key question

What features of a glaciated valley can you identify in Figure 4.35 (page 155) and Figure 4.36?

Skills focus

Using Figures 4.35 (page 155) and 4.36 make a simple pencil sketch of a glaciated valley and annotate the features of a glaciated valley outlined in this chapter.

Fluvio-glacial processes

When glaciers melt huge quantities of meltwater are produced which has the capacity to transport much debris. Meltwater streams often flow under considerable pressure, either sub-glacially or in confined channels. This gives them high velocity and turbulent flow. Meltwater channels can therefore pick up and transport a larger amount of material than normal rivers of a similar size. When the meltwater discharge decreases, the resultant loss of energy causes the material to be deposited. As with all material deposited by flowing water, the heavier particles are dropped first, resulting in sorting of the material. Deposits may also be found in layers (stratified) as a result of seasonal variations in the meltwater flow. Therefore, the landscapes of regions 'downstream' of glacial or formerly glaciated areas are often dominated by **fluvio-glacial** landforms.

Fluvioglacial landforms
Erosion

Meltwater channels are some of the main features of fluvio-glacial erosion, however understanding their origin can be complex. Some key features of meltwater channels are described below.

● They are classified according to where they flow in relation to the glacier during the glacial period (Figure 4.37).

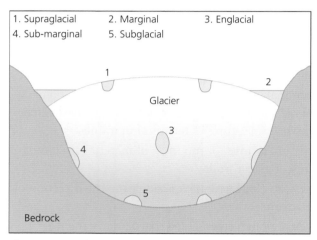

Figure 4.37 Meltwater channel types

● Sub-glacial streams are able to flow uphill due to hydrostatic pressure.
● Only meltwater flowing in channels in contact with the bedrock floor leaves any erosional trace in the post-glacial landscape.
● It is believed that beneath current glaciers two types of channel occur:
  – those cut downwards into bedrock by headward erosion, known as Nye channels
  – those cut upwards into the ice itself by meltwater, known as Röthlisberger channels. These only leave any postglacial trace if the channel becomes choked with sediment (see below).
● It is also believed that with high discharge and its turbulent flow, large meltwater channels can create sub-glacial valleys that are often deep and riddled with potholes.

It is only in recent decades with advances in remotely operated technology and computer-generated digital imagery that glaciologists are beginning to be able to actually see what is happening in en-glacial and sub-glacial streams.

Pro-glacial lakes and overflow channels are also features of fluvio-glacial erosion, with some well-known examples in the British Isles. During deglaciation, lakes develop on the edges of the ice, some occupying large areas. Overflows from these lakes which cross the lowest points of watersheds will create new valleys. When the ice damming these meltwater lakes totally melts, many of the new valleys are left dry, as drainage patterns revert to the pre-glacial stage. In certain cases, however, the postglacial drainage adopts them, giving rise to new drainage patterns.

Large meltwater lakes of this kind occurred in the English Midlands (Lake Harrison), the Vale of Pickering in North Yorkshire (Lake Pickering) and the Welsh borders (Lake Lapworth) at the end of the last glaciation. The River Thames is thought to have followed a much more northerly course before the Quaternary glaciation – its modern course was formed when ice filled the northern part of its basin and forced it to erode a different route.

The River Severn is also believed to have been diverted during the last glaciation. Figure 4.38 shows the stages of this process.

- **Stage 1 Pre-glacial**: The River Severn flowed northwards to enter the Irish Sea in what is now the estuary of the River Dee. The present Lower Severn was a shorter river flowing from the Welsh borderlands to the Bristol Channel.
- **Stage 2 The last ice age:** Ice coming down from the north blocked the River Severn valley to the north. The water from the blocked river formed a huge pro-glacial lake known as Lake Lapworth. The lake eventually overflowed the watershed to the south to join the

original Lower Severn. In the process it cut through a solid rock area, creating the gorge at Ironbridge.

- **Stage 3 Deglaciation** and the post-glacial period: As the ice retreated to the north the way should have been left open for the two rivers to return to the pre-glacial situation. The route north, however, was blocked with glacial deposits, and as the Ironbridge Gorge had been cut very deep (lower than the exit to the north), the new drainage adopted this rather than its former route. The River Severn now flows from central Wales to the Bristol Channel.

Deposition

Landscapes shaped by fluvio-glacial deposition are mainly composed of the following features: **eskers**, **kames** and **outwash plains**, and pro-glacial lakes (Figure 4.39, page 158). Lakes on the outwash plain often have layered deposits in them called **varves**.

Eskers

Eskers have the following main features:
- They are long ridges of material running in the direction of ice advance.
- They have a sinuous (winding) form, 5–20 m high.
- They consist of sorted coarse material, usually coarse sands and gravel.
- They are often stratified (layered).

Eskers are believed to be deposits made by sub-glacial streams. The channel of the stream will be restricted by ice walls, so there is considerable hydrostatic pressure which enables a large load to be carried and also allows the stream to flow

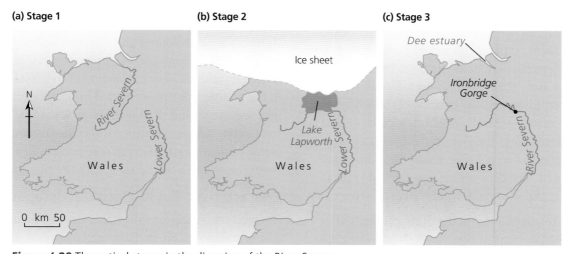

(a) Stage 1   **(b) Stage 2**   **(c) Stage 3**

Figure 4.38 Theoretical stages in the diversion of the River Severn

Source: AQA

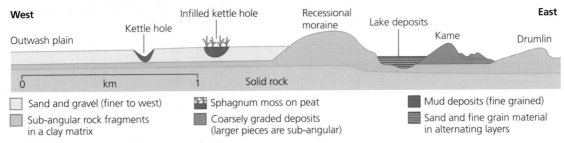

West — East

Outwash plain · Kettle hole · Infilled kettle hole · Recessional moraine · Lake deposits · Kame · Drumlin

0 — km — 1 — Solid rock

☐ Sand and gravel (finer to west)
☐ Sub-angular rock fragments in a clay matrix
☐ Sphagnum moss on peat
☐ Coarsely graded deposits (larger pieces are sub-angular)
☐ Mud deposits (fine grained)
☐ Sand and fine grain material in alternating layers

Figure 4.39 Features of a landscape modified by fluvio-glaciation and lowland glaciation

uphill for short distances. This accounts for the fact that some eskers run up gentle gradients. The load builds up the bed of the channel above the surrounding land, or, as noted above, close to the snout of the glacier the hydrostatic pressure may drop leading to an overburdening of the channel and deposition and the formation of a cast of the sub-glacial tunnel. A ridge is left when the glacier retreats during deglaciation. In some areas, the ridge of an esker is combined with mounds of material, possibly kames (see below). Such a feature is known as a **beaded esker**. Eskers are common in formerly glaciated areas of Europe and North America.

Kames

Kames are mounds of fluvio-glacial material (sorted, and often stratified, coarse sands and gravel). They are deltaic deposits left when meltwater flows into a lake dammed up in front of the glacial snout by recessional moraine deposits. When the ice retreats further, the delta kame often collapses. Kame terraces are frequently found along the side of a glacial valley and are the deposits of meltwater streams flowing between the ice and the valley side.

Outwash plains (sandur)

Outwash plains are found in front of the glacier's snout and are deposited by the meltwater streams issuing from the ice. They consist of material that was brought down by the glacier and then picked up, sorted and dropped by running water beyond the position of the ice front. The coarsest material travels the shortest distance and is therefore found near to the glacier; the fine material, such as clay, is carried some distance across the plain before being deposited. The deposits are also layered vertically, which reflects the seasonal flow of meltwater streams.

Meltwater streams that cross the outwash plain are **braided**. This happens as the channels become choked with coarse material because of marked seasonal variations in discharge. On the outwash plain there

is often a series of small depressions filled with lakes or marshes. These are known as **kettle holes**. It is believed that they are formed when blocks of ice, washed onto the plain, melt and leave a gap in the sediments. Such holes then fill with water to form small lakes. Aquatic plants become established in the lakes and this leads over time to the development of a marshy area and then peat.

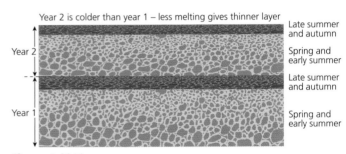

Year 2 is colder than year 1 – less melting gives thinner layer

Year 2 — Late summer and autumn / Spring and early summer
Year 1 — Late summer and autumn / Spring and early summer

Figure 4.40 Varves

Lakes on the fringe of the ice are filled with deposits that show a distinct layering. A layer of silt lying on top of a layer of sand represents one year's deposition in the lake and is known as a varve (Figure 4.40). The coarser, lighter-coloured layer is the spring and summer deposition when meltwater is at its peak and the meltwater streams are carrying maximum load. The thinner, darker-coloured and finer silt settles during autumn and through the winter as stream discharge decreases and the very fine sediment in the lake settles to the bottom. Varves are a good indicator of the age of lake sediments and of past climates as the thickness of each varve indicates warmer and colder periods. Also, any organic material, such as seeds and pollen, caught-up in these layers can be used for paleoclimatic research.

Characteristic fluvio-glacial landscapes

The preceding sections illustrate the significance of the role of meltwater on shaping the landscape. Many areas in southern Iceland on the edges of the warm-based glaciers and ice sheets have fluvio-glacial landscapes composed of a range of fluvio-glacial features and landforms. The most recognisable of these are the expansive sandur or outwash plains.

Figure 4.41 The fluvio-glacial landscape of the Skeidarár Sandur in southern Iceland

The Skeidarár Sandur in southern Iceland (Figure 4.41) is a vast outwash plain from the Vatnajökull icecap. It covers 1,300 km² and exhibits the classic features of fluvio-glacial landscapes including:

- deep accumulations of black gravel, sand and silt originating from the local volcanic rock
- criss-crossed by constantly shifting braided streams; evidence of abandoned channels as the seasonally changing discharge levels affect rates of erosion and deposition
- very sparse vegetation and little evidence of human habitation. This results from the constant risk of floods or glacial burst due to volcanic eruptions in this area. The last eruption under this glacier was in 1996
- close to the glacier snout are ridges or steps that may result from kames or kettle features.

Key question

Why might it be difficult to establish settlements in these kinds of landscape?

Periglacial features

Here the term **periglacial** refers to landscapes that, although not actually glaciated, are exposed to very cold conditions with intense frost action and the development of permanently frozen ground or permafrost. Some geographers apply the term more narrowly to only apply to areas affected by freeze–thaw cycles in close proximity to ice sheets.

At present, areas such as the tundra landscapes of northern Russia, Alaska and northern Canada, alongside high mountainous areas such as the European Alps, experience a periglacial climate. In the past, however, as ice sheets and glaciers spread, many areas which are currently temperate were subject to such conditions.

The climate of periglacial regions is marked by persistently low temperatures, with an annual average of between –15°C and –1°C, with average annual rainfall (excluding snowfall) ranging between about 120 mm and 1,400 mm. Summers are short and temperatures can sometimes reach above 15°C, but remain well below freezing in winter, with some areas occasionally reaching –50°C.

Permafrost

Permafrost generally refers to permanently frozen ground. In areas where temperatures below the ground surface remain below 0°C continuously for more than two years permafrost will occur. Currently, it is estimated that permafrost covers 20–25 per cent of the Earth's land surface. If water is present in the soil a significant amount will freeze, *cementing* the mineral and organic particles together. The ground below this frozen layer that remains unfrozen is known as **talik**. Figure 4.42 outlines how temperature changes with depth in permafrost.

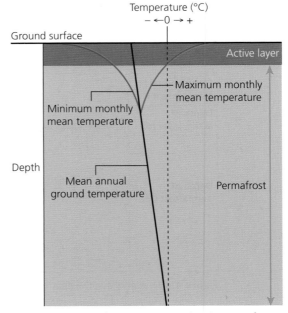

Figure 4.42 Typical temperature regime in permafrost

When summer temperatures rise above freezing, the surface layer thaws from the surface downwards to form an **active layer**. The depth of this layer varies considerably and depends upon local conditions, but may extend to 4 m. As the ice in this layer melts, large volumes of water are released.

There are three categories of permafrost (Figure 4.43):

- **Continuous permafrost** is found in the coldest regions, reaching deep into the surface layers. In Siberia today, it is estimated that the permafrost can reach down over 1,500 m. In the very coldest areas, there is hardly any melting of the uppermost layer.

- **Discontinuous permafrost** occurs in regions that are slightly warmer, where the ground is not frozen to such great depths. On average the frozen area will extend 20–30 m below the ground surface, although it can reach 45 m. There are also gaps in the permafrost under rivers, lakes and near the sea.

- **Sporadic permafrost** is found where mean annual temperatures are around or just below freezing point. In these places, permafrost occurs only in isolated spots where the local climate is cold enough to prevent complete thawing of the soil during the summer.

Periglacial mass movement processes

There are a number of processes of mass movement that occur in periglacial landscapes:

- **Solifluction** occurs when summer temperatures rise enough to melt the huge amounts of water held as ice in the upper layers of the permafrost. Due to the impermeable layer of still frozen ground below, this water cannot drain away, and as there is little evaporation due to the low temperatures, this surface layer becomes very wet. Excessive lubrication from this water reduces the friction between the soil particles. On slopes as gentle as 2°, this saturated layer becomes quite mobile and soil begins to move downslope. The downslope movement of saturated soils can happen in many environments, but when related to the freezing and thawing of the active layer it is also referred to as **gelifluction**.

- Scientists also have a term that refers to any flows of earth within the still frozen permafrost. This is **congelifluction**.

- **Frost creep** refers to the gradual downslope movement of individual soil particles due to the alternating freeze–thaw cycles within the active layer. This relates to **frost heave** (see below).

- **Rock falls** result from **freeze–thaw action (frost shattering)**, and can involve the movement of large amounts of material. This process has already been described on page 147 because of its role in providing large amounts of the erosive material in glaciers. **Scree** slopes (accumulations of **talus**) at the foot of slopes resulting from frost shattering are a common feature of periglacial landscapes. On relatively flat areas, extensive spreads of angular boulders are left, which are known as **blockfield** or **Felsenmeer** (sea of rocks) (Figure 4.44).

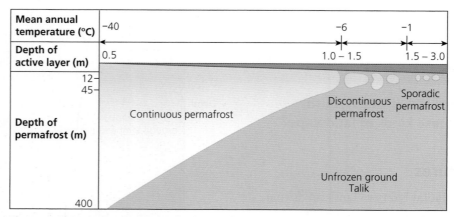

Figure 4.43 Variations in the depth of permafrost

Figure 4.44 Felsenmeer on Mount Washington, USA

Periglacial landforms

There is a clear link between the landforms of periglacial landscapes and the processes that form them. The main landforms and the processes that form them are explored below, and illustrated in Figure 4.45.

Patterned ground

The repeated cycles of freezing and thawing of the active layer can produce a range of landforms collectively referred to as **patterned ground**. The formation of specific examples of patterned ground can be quite complex, but the main process involved is known as **frost heave**.

As the active layer starts to freeze, ice crystals begin to develop. They increase the volume of the soil and cause an upward expansion of the soil surface. Frost heave is most significant in the fine-grained material and as it is uneven it forms small domes on the surface.

Within the fine-grained material there are stones which, because of their lower specific heat capacity, heat up and cool down faster than the surrounding finer material. Cold penetrating from the surface passes through the stones faster than through the surrounding material. This means that the soil immediately beneath a stone is likely to freeze and expand before other material, pushing the stone upwards until it reaches the surface. Depending on the specific local conditions a range of features can form, summarised in Table 4.2 (page 162).

Ice wedges

Ice wedges are another feature that creates a type of **patterned ground**. These are relatively narrow cracks or fissures in the upper layers of the ground filled with ice (sometimes extending below the level of the permafrost). Due to the presence of air bubbles in the ice the wedge has a milky appearance. They begin as small cracks, less than 5 cm across, but over hundreds of years can reach over 10 m wide and extend many metres into the ground. There are some similarities in how they begin to form to cracked surface features in other areas like dried mud flats. However, in periglacial landscapes these result from **ground contraction** due to extreme cold, generally in areas of continuous permafrost.

During the winter the freezing of the soil in the active layer causes the soil to contract. In the following summer the initial small crack is filled with meltwater (called an **ice vein** when it freezes). This process occurs repeatedly through the cycle of winter and summer widening and deepening the crack to eventually form the ice wedge.

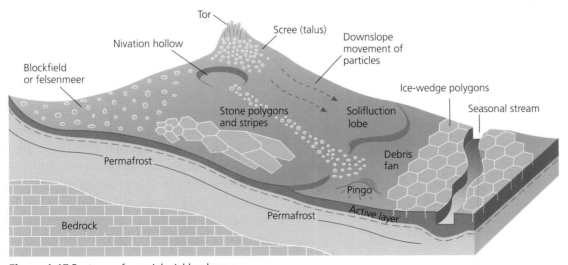

Figure 4.45 Features of a periglacial landscape

Table 4.2 Characteristics of types of patterned ground

Category	Features summary	Formation summary
Circles	Circular-shaped feature	Usually form on flat ground
	Non-sorted circles – without a stone border	Processes vary for individual types of circle
	Sorted circles – with a stone border	Formed by seasonal frost heave
	Usually about 0.5 to 3 m diameter	
	Slightly domed with polygonal surface cracks	
	Occur singularly or in groups	
	Within the circle the soil is usually fine grained silt and clay	
	If the soil is coarser with gravels they are called stony earth circles	
Nets	Similar in appearance to circles and polygons, though neither fully circular or polygonal in shape	Similar to circles and polygons
	May be an intermediate form between circles and polygons on very gently sloping ground	
	Creates a **hummocky** terrain	
Polygons	Multi-sided raised mounds of sorted material	Only seen on flat ground
	Approximately 10 m in diameter	Freezing causes domes to form
	Finer particles make up the central mound surrounded by a border of larger stones	The seasonal cycles of frost heave move stones to the surface
	The larger the border stones the larger the diameter of the polygon	Gravity acts on the larger stones making them roll downslope to form the border – thus giving these their alternate name *sorted polygons*
	Not to be confused with ice-wedge polygons (Figure 4.47)	It is also believed that ground contraction also plays a role in the sorting and shaping of the polygons (see below)
Steps	Step-like features formed on slopes	Created by frost-action of the permafrost
	More gently sloping than the slope on which they are formed	Different types result from the processes that formed the parent feature
	They are either terraces with stones or vegetation forming a banked edge (depending on their origin)	
	They are derived from either circles, nets or polygons	
	Similar scale to their parent feature – up to 10–20 m wide	
Stripes	Lines of stones running downslope	Similar origins to steps
	Some are parallel lines of sorted material with stone borders	Formed from the processes of frost heave
	Some are unsorted with vegetation borders	Due to the slope, upon reaching the surface stones form linear features
	Found on gradients over 6°	
	Believed to have developed from circles, nets or polygons	

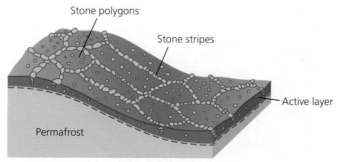

Figure 4.46 Stone polygons and stripes

The way the ground contracts creates a pattern of cracks on the surface, which when viewed from above, have a similar shape to stone polygons (Figure 4.46). These are therefore known as **ice-wedge polygons** (Figure 4.47). These are often larger than stone polygons and are generally 15–30 m in diameter, but can reach 100 m across. These are also generally only found on flat or very gently sloping surfaces. If the climate warms enough for all the ice to melt it may be replaced by fined grained sediment that infills the void forming an **ice-wedge cast**.

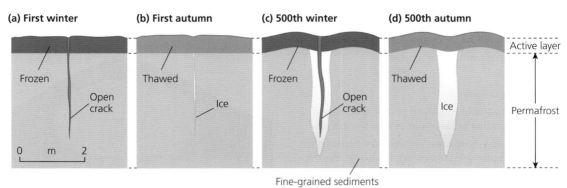

(a) First winter **(b) First autumn** **(c) 500th winter** **(d) 500th autumn**

Frozen | Thawed | Frozen | Thawed | Active layer

Open crack | Ice | Open crack | Ice | Permafrost

0 m 2

Fine-grained sediments

Figure 4.47 The formation of ice wedges

Pingos

Pingo is an Inuit term for a dome-shaped, ice-cored mound of earth in periglacial landscapes (Figure 4.48). Pingos range in size and age. They can be a few metres in height and diameter, but the oldest pingos, found on the MacKenzie Delta, are believed to be about 1,000 years old and are over 50 m high and over 300 m in diameter (some have been recorded over 600 m across). Some researchers suggest that they grow by 2 to 3 cm a year.

The ice core results from a slow *injection* of pressurised water into the dome. There are two types of pingo due to the different origins of this water. The specific

Figure 4.48 A pingo in northern Canada

processes underlying the formation of both types are complex. Table 4.3 provides a summary of the key features of each.

Table 4.3 Key features of closed and open system pingos

Closed system	
Famously found on the MacKenzie Delta area, of the Northern Territories, Canada, so known as **MacKenzie Delta type**	**Stage 1** Unfrozen lake
Generally found in areas of continuous permafrost	
Develop beneath lakebeds	Permafrost
The growth of the ice core is *hydrostatic*	Unfrozen sediments (talik)
Deep lakes (over 2 m) may remain unfrozen in winter	**Stage 2** Sediments begin to freeze
The permafrost layer at the lakebed is insulated from the cold and thaws	Lake drains
An area of unfrozen waterlogged ground is now sandwiched between the lake and underlying permafrost	Permafrost Liquid water begins to gather beneath
The lake may begin to drain, the lakebed is no longer insulated, so the waterlogged bed begins to freeze	
Due to localised differences in pressure between the lake, freezing lakebed and underlying permafrost, the newly freezing water *gathers* together to form an ice lens that expands, pushing the lakebed sediments above it up into the classic dome shape.	**Stage 3** Raised dome with cracked surface	Ice core
The ice lens continues to grow as long as there is still unfrozen ground in the lakebed as a source of pressurised water to add to the ice core (Figure 4.49)	Permafrost Liquid water lens due to hydrostatic pressure

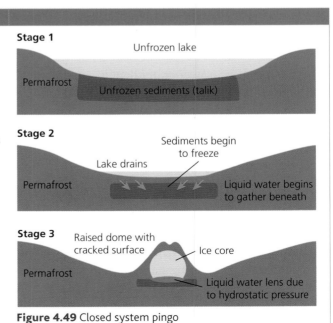

Figure 4.49 Closed system pingo

Open system

Common in Greenland and Alaska, so known as **East Greenland type**

Generally found in areas of discontinuous permafrost

Found in valley bottoms

The growth of the ice core is *hydraulic*

Water is able to seep into the upper layers of the ground and flows from higher surrounding areas under artesian pressure

Water accumulates in flat low-lying areas between the upper layers of permafrost or soil and frozen ground beneath the water, then freezes

The freezing ice core expands thus doming the overlying layers into the classic pingo shape

They can grow as pressurised water continues to flow in from its surroundings (Figure 4.50)

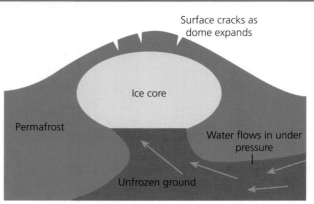

Figure 4.50 Open system pingo

As the pingo grows, increasing cracks and instability of the steep sides may expose the ice inside. As the ice melts the pingo collapses. When the core has completely melted all that remains is a doughnut-shaped mound possibly containing a small lake in the centre (Figure 4.51). Some present-day temperate regions have fossil pingos evidencing their colder past climates.

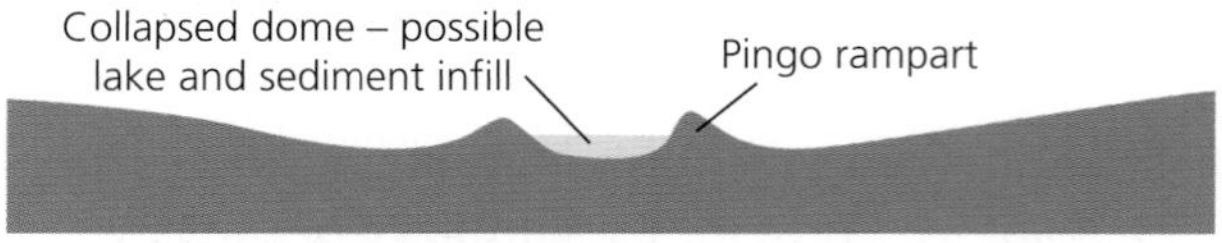

than the slope it occupies. Some geographers believe they are associated with the processes of frost-action and creep as outlined above. However, others believe they may well be formed by the trampling of animals traversing steep slopes! Many researchers have not found any in landscapes inaccessible to animals!

Figure 4.51 Fossil (collapsed) pingo

Solifluction (gelifluction) lobes and sheets

Where solifluction (or gelifluction) (as outlined above) occurs on slopes of around 1–20° solifluction sheets are formed. These have a smooth surface and can extend over 100 m across a slope.

Where slopes have a steeper gradient of 10–20° then solifluction lobes will form. These have a tongue-like appearance that extends downslope and can be up to 50 m in width and 5 m high. Depending on the local subsurface conditions and slope processes solifluction terraces can also form (Figure 4.52).

Terracettes

Terracettes are small periglacial features. They are often referred to as 'sheep tracks' as they have the appearance of small narrow pathways running horizontally along the side of a slope. They are narrow steps, with a *tread* surface usually only tens of centimetres wide that runs generally parallel to the contours of the slope. They are bounded by a low steep border (*riser*) that is steeper

Figure 4.52 Solifluction lobes and solifluction terraces

Thermokarst

Thermokarst refers to depressions in the ground surface resulting from ice melting within the permafrost. This melting causes localised subsidence giving a ground surface similar to that found in areas of limestone geology known as *karst*, where similar hollows, depressions and collapsed structures result from the erosion and weathering of the limestone bedrock. As thermokarst only results from temperature changes and not erosion and weathering, features occur where human activity inadvertently warms the surface layers of the permafrost (see Figure 4.60, page 173).

Characteristic periglacial landscapes

The preceding section illustrates a range of landforms and processes typical of a periglacial landscape. Owing to the open and sparsely vegetated nature of the landscapes, rates of erosion by water and wind can also be high, continually remodelling and shaping the contemporary landscape. In regions where there are high levels of seasonal meltwater the drainage will be dominated by **braided channels** (see Figure 4.41, page 159).

Unobstructed winds blowing across periglacial landscapes can reach high velocities. These winds can shape the landscape in the way explored in Chapter 2, with both erosional features like ventifacts and depositional features formed from fine wind-blown sediments (loess).

Process, time, landforms and landscapes in glaciated settings

It is important to view each unique glaciated and periglacial landscape as an assemblage of features and landforms that combine in that place to give it its own characteristic landscape. In order to fully understand why that landscape looks as it does today it is crucial to look beyond the processes that are happening in the present and think about how different processes may have shaped its features through time.

This is especially important because the rate of development of different landscapes in different cold environments varies considerably. Some landscapes are shaped rapidly by the almost constant action of processes of erosion, weathering and deposition. For example, where warm-based glaciers are moving rapidly they will be eroding and depositing material far quicker than in a landscape being shaped by a cold-based glacier. It is also the case that many weathering processes operate at a far faster rate in more temperate cold environments. For example, cycles of freezing and thawing may happen dozens of times during a Scottish winter, on an almost daily basis, while in some of the coldest parts of the polar regions the freezing and thawing cycle may operate over a timescale of millennia. Therefore, time has an extremely important role to play in controlling the pace of the processes that shape cold environments, even at the most local level.

To fully understand the features of a characteristic glaciated landscape it must be viewed in its entirety as an assemblage of a range of landforms. So the characteristic glaciated landscape of the European Alps is not just the glacial trough, but includes arêtes, hanging valleys, ribbon lakes, striations and truncated spurs amongst many others. It also is important to remember that the landscape is not simply the result of the current processes we can witness happening today, or imagine having happened in the *recent* past, but is also the result of cycles of processes that have occurred over a very much longer timescale. This **land systems** approach to studying landscapes gives a much broader geographical context to the study of glacial landscapes.

Ice ages have come and gone on Earth, and we are currently in the Quaternary period, which spans the last 2.58 million years. In that time there have been many glacial and interglacial fluctuations, initially operating on a 41,000-year cycle and on about a 100,000-year cycle more recently. In fact there have been eight glacials in the last 740,000 years. During each of these, ice advanced from the polar regions and, in the northern hemisphere, each glacial advance slowly altered the landscape shaped by the previous ice advance. So the glaciated landscape we see today, in the current interglacial, is also the result of cycles of processes through millennia.

As periglacial landscapes often border current glacial landscapes then it is true that they too are ever-changing landscapes. In the past, during glacials, landscapes that are currently periglacial may have been glacial and covered by ice, so the characteristic landscape we see today is not just the result of recent periglacial processes, but a combination of short-term and long-term cycles of glacial and periglacial activity.

Skills focus

It is important that you are able to describe and explain the features of characteristic glaciated and periglacial landscapes. After visiting a glaciated landscape (or by using the library or internet to research contrasting glaciated or periglacial landscapes) complete the following tasks.

Identify the different landforms in each landscape.

Explain the processes that created the landforms.

Try to identify the age and origin of the features; for example, do they result from present-day processes or are they relic features that are being modified today?

Draw and annotate a sketch of the landscape to describe the characteristic features.

4.4 Human occupation and development

Despite their inhospitable climates and difficult terrain, cold environments have been home to limited human populations throughout the Holocene since the Last Glacial Maximum. However, as global population continues to expand and economic development seeks to exploit untapped natural resources, some may say that the human impact on cold environments is increasing.

This section initially identifies and explores:

- human impacts on fragile, cold environments
- cold environments as human habitats
- varying human impacts on cold environments over time.

The section then concludes with case studies of these relationships in specific landscapes.

Human impacts on glaciated landscapes

It is possible to identify four main categories of ways in which humans have and do impact cold environments.

A fragile environment

Scientists are increasingly concerned about the concept of **environmental fragility**. The term **fragile environment** was defined earlier. In all natural environments there is a sensitive balance between non-living (climate, geology, soils) and living (fauna and flora) components. Some natural environments appear to be quite robust and easily respond to and cope with major natural events like forest fires and volcanic eruptions. Others are very susceptible to disturbance by people, easily damaged by human activity and then slow to recover (if at all). It is said that many of the cold environments explored in this chapter would fall into this category and be classed as **fragile environments**.

Specific examples of the ways in which people affect cold environments are explored below (Figure 4.53), but reasons for their fragility and limited ability to recover include:

- short summers, long winters, cold temperatures and limited rainfall limits plant growth
- the slow rate of plant growth means that any disruption to the ecosystem takes a long time to be corrected
- some estimate that it could take over 50 years for an area of tundra, for example, to return to its former state after interference. (It could take 50 years for tyre tracks in the permafrost to be completely revegetated!)

Human activity, biodiversity and sustainability

Despite their hostile conditions people are present in almost all cold environments; there are few places that people have not exploited. Table 4.4 gives a summary of a range of cold environments and examples of some of the activities undertaken in them.

Table 4.4 Human activities in cold environments

Cold environment	Human activities (Past and present)
Antarctica and Southern Ocean	Sealing and whaling; mining and oil extraction; tourism; scientific exploration and research; chemical, sewage and waste dumping; maritime transport
Arctic basin	Sealing and whaling; mining; oil and gas exploration; tourism; maritime transport; small-scale hunting and fur trapping; forestry; caribou herding
The Rockies, Andes, Himalayas and European Alps	Tourism; forestry; agriculture (arable and pastoral); energy production (HEP); transport routes (using accessible glaciated valleys as routeways)
Siberia	Hunting, fishing and fur trapping; mining; forestry; oil, gas and coal extraction; fresh water supply; military bases

Table 4.4 illustrates the limited range of human activities that are viable in many cold environments. Presently it may not be that the range of human activities is increasing in these areas that makes them susceptible to damage, but that the scale of some of the activities may be increasing and so too their impacts.

One of the reasons for the limited range of human activities in these landscapes is their inaccessibility and remoteness, but also the limited range of living and non-living natural resources. The limited biodiversity

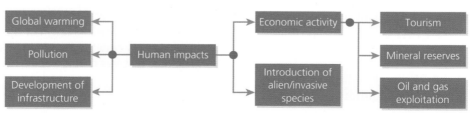

Figure 4.53 Categories of human impacts in cold environments

of most cold environments also affects their ability to be used **sustainably**. Some suggest that their fragility makes it almost impossible for humans to use these environments in a way that meets present needs without compromising them for future generations.

For example, the low productivity and limited species diversity mean that the plants are very specialised and any disruption causes difficulty when it comes to regeneration. In such circumstances, species have great difficulty in adapting to a changed environment.

Wide fluctuations occur in the amount of energy held in each trophic level of different food chains because population numbers change rapidly. For example, variations in the numbers of lemmings and arctic hares, both of which are liable to short-term and long-term fluctuations, have consequences for the populations of their predators, such as arctic foxes and snowy owls.

Disruption to the functioning of the biome has long-term implications. This is why there has been so much concern over the proposed exploitation of resources such as the oil reserves of north Alaska that fall within the Arctic National Wildlife Refuge.

Recent and prospective impact of climate change

Climate change is a major issue in twenty-first century Geography and is explored widely elsewhere. The majority of climate scientists agree that the Earth's climate is changing, with recent decades having a pattern of accelerated warming of average global temperatures. The Earth's climate constantly changes naturally, as evidenced by the cycles of glacials and inter-glacials discussed above. However, many argue that there has been an enhanced

rate of change recently that can only be explained with the inclusion of the impacts of human activity into the analysis. While the causes of climate change are not straightforward, the proposed impacts also vary widely across the globe.

Causes of climate change – natural vs. human

Many theories have been put forward to explain climate change. Table 4.5 summarises some suggested causes.

Table 4.5 Suggested causes of climate change

Natural	Changes in solar activity (the solar constant and sunspot activity)
	Variations in the eccentricity of the Earth's orbit and axial tilt (affecting the amount of solar radiation reaching the surface)
	Meteorite impact
	Volcanic activity (increasing dust and altering the composition of gases in the atmosphere)
	Plate movement (redistribution of land masses, affecting amounts of absorption and reflection of insolation)
	Changes in oceanic circulation
Human	Changes in atmospheric composition, especially the build-up of carbon dioxide and other greenhouse gases from agriculture, industry, transport, energy production and waste (the enhanced greenhouse effect)
	Deforestation – trees act as a major carbon store; if they are felled and burnt, or allowed to rot, then the carbon returns to the atmosphere

Although there are fewer human factors listed, it is this that is seen as being the most likely explanation for the recent changes in climate. The thin layer of gases that form the Earth's atmosphere is crucial for maintaining life on Earth and without them the planet would be 30°C colder. This is known as the **greenhouse effect** (Figure 4.54).

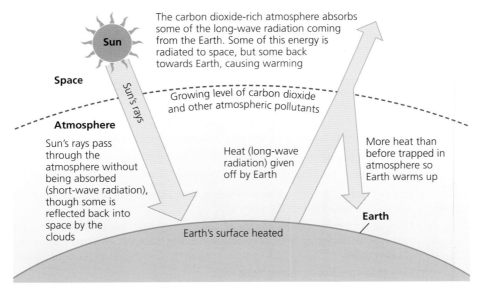

Figure 4.54 The greenhouse effect

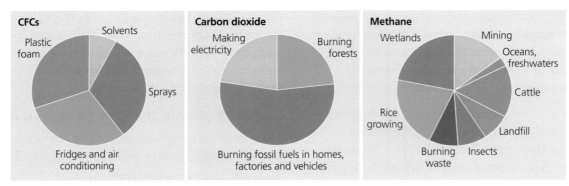

Figure 4.55 Origin of greenhouse gases

The gases that are responsible for this greenhouse effect and trap this heat include carbon dioxide, chlorofluorocarbons (CFCs), methane, nitrous oxide and ozone. If the natural balance of atmospheric gases remains stable, and the amount of solar radiation is unchanged, then the Earth's temperature remains the same. However, Figure 4.55 suggests that there is a clear link between human activity, economic development and the balance of these gases in the atmosphere, suggesting that human activity is responsible for the enhanced greenhouse effect and thus current climate change.

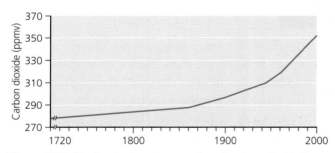

Figure 4.56 Changes in the concentration of carbon dioxide in the atmosphere, 1720–2000

The concentration of carbon dioxide in the atmosphere increased by about 15 per cent during the twentieth century and the current rate of increase is about 2 per cent per year (Figure 4.56). Together with increases in other greenhouse gases like methane and nitrous oxides, this has upset the natural balance and led to an increased rate of global warming in recent decades.

Recent and prospective impact of climate change in cold environments

Most geographers and climate scientists agree that the human enhanced greenhouse effect is already changing Earth's climate and already having impacts. Many of these issues are not new and have been discussed for two to three decades. In fact some impacts that were predicted 20 or 30 years ago are now occurring, with scientists making new predictions for the coming decades. Table 4.6 summarises some of these impacts.

Table 4.6 Impacts of climate change in cold environments

Type	Impacts	Evidence/details
Observable	Shrinking glaciers	Globally 90 per cent of glaciers are losing ice mass
		A gradual decrease in ice mass has been recorded since the 1850s, with a short period of gain around 1970, but in the decades since the rate of loss is unprecedented
		Millions of people rely on glacial meltwater as their freshwater supply.
		Glaciers respond to local climate, so some glaciers in some regions do advance at certain times
	Reduced ice on rivers and lakes	In the northern USA most lakes are freezing later (about 1 day per decade) and thawing earlier (about 1.5 days per decade) than 150 years ago
		Many isolated communities, and the oil and gas industries, rely on frozen rivers and lakes to transport vital supplies, including drilling equipment, building materials, etc., that are too big and bulky to fly in at other times
		Early thaws are also increasing flood risks

Type	Impacts	Evidence/details
Observable	Shifting plant and animal ranges	Many predict a poleward shift in major high-latitude northern tree belts including the taiga and boreal conifer forests. These major changes will take time and so changes in treeline will be slow
		However, studies in the 1990s showed that the northern margins of individual species have changed markedly. For example, of 35 species of European butterflies 63 per cent had ranges that shifted north (up to 240 km in the 20th century), while only 3 per cent shifted south (Parmesan et al, 1999)
		Bird watchers in Britain are spotting changing patterns in the migration of certain bird species that over-winter here
Predicted and now occurring	Loss of sea ice	A small area of the Arctic Ocean has ice cover all year. It is at its smallest in September. Studies have predicted it is shrinking, and with the use of remotely sensed images showed that 2012 was the smallest on record. Arctic ice is also becoming thinner making it more vulnerable to enhanced rates of melting in the future
	Accelerated sea-level rise	Studies show that the rate of sea level rise for most of the twentieth century was about 1.7 mm per year. Satellite monitoring then showed that mean global sea levels rose by around 3 mm per year between 1993 and 2009. In 2013 the IPCC report said the current average rate is around 3.2 mm per year. (Thermal expansion and the addition of meltwater contribute to the increase)
	Longer, dryer summers	By the early 2000s satellite data confirmed that longer Arctic summers were thinning the ice dramatically. Some suggest a 40 per cent thinning since the 1960s. Polar bears rely on sea ice from which to hunt seals and without it they are finding it increasingly hard to find enough food. In some places in Alaska and northern Canada polar bears have been straying into towns searching for food
Predicted future change	Highest expected temperature change to be in the highest northern latitudes	Average global temperatures could rise between 1.4 and 5.8°C by 2100
		Latitudes between 40°N and 70°N could rise between 5 and 8°C by 2100, due to the loss of sea ice and snow cover reducing albedo rates (less insolation will be reflected by the surface)
	Continued contraction of area covered by snow and ice	The UN predicts mid-northern latitudes in Europe and North America may see the largest rates of decline in snow and ice cover. One estimate predicts that loss of snow and glaciers in the mountains of Asia alone will affect the lives of 40 per cent of the world's population Areas of Canada and Siberia may receive increased snowfall due to more humid climates
	Permafrost to thaw to increasing depths	A 6°C temperature increase in the Arctic by 2100 would lead to a 30 to 85 per cent loss of near-surface permafrost. This could permanently change local ecosystems and hydrology and increase the risk of wild fires and soil erosion. Local infrastructure such as buildings and roads could be affected causing economic and social problems for local communities. The impact could be global as models suggest that melting permafrost could release huge amounts of carbon dioxide and methane, thus accelerating atmospheric warming. This is a positive feedback known as the **permafrost carbon feedback**
	Further decrease in extent of sea ice	The UN predict that the Arctic may be free of sea ice by 2100, and that Antarctic sea ice may decline at similar rates, but not thin as much. Changes in sea ice could have major impacts on ocean circulation and thus the major climate patterns of the world, which is proving extremely difficult to model and predict. Increasing ice-free periods in the Arctic could have major impacts on world shipping, in addition to disrupting the lives of local populations
	Shrinking of Greenland ice sheet leading to sea level rise	The ice sheet is expected to almost completely melt by 2100 contributing to projected sea level rise by 2100 of between 18 and 59 cm above 1990s levels. Other factors like thermal expansion will be important and modelling the future is complex
	Possible growth of Antarctic ice sheet due to increased snowfall	Warmer temperatures could increase the amount of precipitation. Other theories suggest that the extent of the ice sheet may be increasing due to acceleration of glacial movement to the sea
	Possible significant impacts of small temperature increases on fragile tundra ecosystems	The IPCC suggest even a 1–2°C rise in global temperatures above 1990 levels jeopardises unique and already threatened ecosystems
		If temperatures rise by more than 1.5–2.5°C, 20–30 per cent of species could be at risk of extinction; if the rise is more than 3.5°C models predict extinctions of 40–70 per cent of known species
	Increased issues of invasive species in warmer tundra environments	This could increase extinction rates of indigenous species that are outcompeted for resources. Local populations' traditional ways of life may be affected

Cold environments as human habitats
The distribution of population within the cold environments

Despite the inhospitable characteristics of cold environments, many people live and work in these places. The proportion of indigenous and non-indigenous populations alters as changes in development affect the challenges and opportunities for people in these regions. Over 4 million people live and work within the Arctic Circle (Figure 4.57).

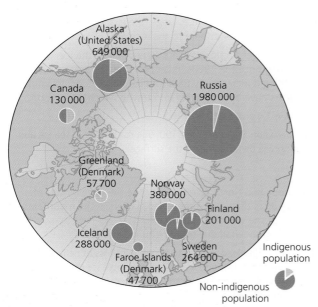

Figure 4.57 Population distribution in the Arctic

Key question

Why do you think the non-indigenous population is higher than the indigenous population in almost all parts of the Arctic?

Opportunities and risks presented by cold environments

The physical environment impacts human activities. The harsh conditions of cold environments present challenges for human occupation and development. These conditions include:

- very low temperatures
- short summers and therefore short growing seasons
- low precipitation

- thin, stony, poorly developed soils
- permafrost
- surface thaw in summer leading to waterlogging and possible flooding
- snow lying for long periods
- blizzards.

Many indigenous people have adapted their way of life to cope with the climate and make the most of the limited resources. The traditional economic activity of indigenous populations in many tundra regions included hunting and fishing, caribou and reindeer herding, fur trading and even hunter gathering. These communities often lived at low population densities, only took resources from the natural environment that they needed, and had social structures and cultural systems that actively promoted the protection of their environment. In some cases the natural environment is revered in a religious or spiritual way. This way of life therefore is said to have been **sustainable**. A case study of one local population is developed later in this chapter.

The peoples living in these cold environments face challenges posed by the physical environment itself, but life is made increasingly difficult by other, more variable and human issues. Such obstacles to life include:

- a shortage of labour and skills due to low population density
- increasing unpredictability of climate events – late winter freezes, or early spring melts
- remoteness and inaccessibility
- lack of permanent jobs and limited educational opportunities after secondary level
- absence of modern conveniences common in urban areas
- remoteness can lead to a feeling of economic or political neglect from national government
- impacts from outside on natural resources that are relied upon, including climate change, commercial fishing, etc.

Human adjustments to life in cold environments

Although challenging, cold environments hold many opportunities for people, so over millennia people have adapted to surviving in them.

Traditional adaptations in the Arctic tundra

The **Inuit**, indigenous people of areas of North America, adapted all aspects of their lives to survive in the harsh tundra landscape, including their:

- lifestyle
- shelters
- food
- transport.

They lived in small communities, with small family size and low birth rates. Spread over a large geographic area at very low population density they put little pressure on the scarce resources. They were often nomadic to limit the impact on resources in each location, but also to follow migrating animals. Therefore, in the summer they built portable tents out of driftwood and seal skins, while in winter they needed more substantial shelter so would build snowhouses, or igloo, or in more settled communities log-and-sod houses. These log-and-sod dwellings would be built partially in holes in the ground, so the surrounding earth provided more protection from the elements.

Inuit would make the most of the short summer by hunting, gathering and preserving as much food as they would need to see them through the winter. They would often dry fish or seal meat so that it would last through the year. Many indigenous communities of the north have adapted to be able to safely eat raw meat, which again limits the need for extra resources. Experience and tradition tells them how much food to hunt, and they will not take more than is needed. All parts of any prey are used: for clothing, utensils or even building materials.

In the winter many Inuit communities use dog sleds to exploit the frozen landscape, and in the summer use single person kayaks or larger open boats called umiaks. All of the above are used for transport and hunting. In many cold environments being mobile is a key to survival.

Key question

Why do you think people would choose to live in such harsh physical environments?

Contemporary adaptations in the Arctic tundra

Establishing permanent settlements and developing activities such as oil and gas extraction has required major technological advances, primarily because the permafrost creates a unique set of problems for construction work and engineering. Many innovations have come from efforts to overcome the problems posed by the permafrost. New methods of construction have been employed to protect the permafrost and prevent the subsidence of buildings and infrastructure. Although they are successful, these methods are more expensive than conventional construction, adding to the costs of living in the region, and continual maintenance is often necessary. Some of the methods are described below.

- Smaller buildings, such as houses, can be elevated above the ground on piles driven into the permafrost. The gap below the building allows air to circulate and remove heat that would otherwise be conducted into the ground (Figure 4.58).

Figure 4.58 Buildings on stilts and insulated ducting for water and power in Prudhoe Bay, Alaska

- Larger structures can be built upon aggregate pads, which are layers of coarse sand and gravel, 1–2 m thick. This substitutes the insulating effect of vegetation and reduces the transfer of heat from building to ground. These pads can also be placed beneath roads, railways and landing strips.
- In large settlements **utilidors** (Figure 4.59, page 172) have been built. A utilidor is an insulated box, elevated above the ground that carries water supplies, heating pipes and sewers between buildings. Pipes cannot be buried underground

because of the damage that would be caused by freezing and thawing in the upper soil layers.

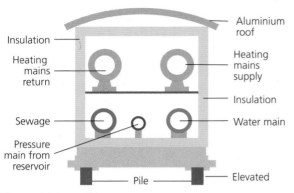

Figure 4.59 A utilidor

Typical forms of occupation and economic activity in different cold environments

Different cold environments present different challenges and opportunities, so people have learnt to occupy and exploit them in different ways. Table 4.7 summarises these activities.

Table 4.7 Economic activity in cold environments

Cold environment	Characteristics and economic activity
Alaska	With abundant natural resources, the remote wild landscapes have seen the development of oil, mining, fishing, timber and tourism
Northwest Territories, Canada	Its size and remoteness has limited much economic development, but expanding sectors include diamond mining, oil and gas, the fur trade and tourism
European Alps	Being easily accessible with steep mountain slopes, high snowfall rates and high living standards makes them popular for tourists wanting winter sports. Equally, the beautiful scenery and hot sunny weather is popular with hikers, climbers and sightseeing tourists in the summer months. The glacial troughs and narrow valleys are good locations for the production of renewable energy in the form of HEP and wind turbines, and serve as essential routeways for transport
Siberia	Its size and remoteness makes it inaccessible, however its rich ore and mineral wealth has led to the development of mining of iron, silver, copper and gold, alongside coal, oil and gas extraction. The vast forests provide for a large timber industry

Cold environment	Characteristics and economic activity
Iceland	Being an island, Iceland has traditionally relied on the fishing and whaling industries. Today it also exploits its geothermal energy to produce vast amounts of clean electricity; cheap flights with budget airlines have opened the island up to sightseeing and adventure tourists
Antarctica	The extreme climate, protected status, lack of pollution and human population, make it suitable for adventure tourism and scientific research
Southern Ocean	The cold nutrient-rich waters have always been used for fishing. In the late 1990s hundreds of thousands of tonnes of krill and Patagonian toothfish were harvested, before international agreements came in to try to reduce the impact. With only limited visitor numbers allowed onto Antarctica, cruise ships have become increasingly common in the waters surrounding the continent

Changing impact of human activity on fragile cold environments over time

In North America the number of Inuit was always small in terms of the vast area in which they lived, so very little pressure was put on the environment, which traditionally remained relatively undisturbed.

In the north of Europe, the Sami followed the seasonal movement of reindeer and as long as they lived at low population densities they knew their environment could sustain them, provided they managed its use sustainably.

From the seventeenth century onwards, the resources of tundra areas began to be exploited by outsiders. The major forms of economic activity outlined above, like sealing, whaling, trapping for fur and mining, particularly for gold, began to grow. Mining led to the establishment of permanent settlements whereas other activities tended to be temporary or seasonal. All activities began to bring larger numbers of people than had lived there previously. In the twentieth century, tundra regions were exploited on a much larger scale, with dramatic impacts on the lifestyle of indigenous people as well as on the physical environment.

In the tundra many of the negative human impacts occur when people try to establish permanent settlements and infrastructure. Problems are caused when vegetation is cleared from the ground surface. This reduces insulation and results, in summer, in the deepening of the active layer. Even minor disturbances, such as vehicle tracks, can greatly increase melting, because the vegetation is very slow to re-establish itself.

Buildings speed up this process by spreading heat into the ground. The thawing of ground ice leads to the development of unnatural **thermokarst**, a landscape of topographic depressions characterised by extensive areas of irregular, hummocky ground interspersed with waterlogged hollows. The damage caused by this form of ground subsidence can be seen in tilted and fractured older buildings and in damage to roads, railways and airfield runways (Figure 4.60).

Figure 4.60 Thermokarst under a car park, Fairbanks, Alaska

Apart from the production of thermokarst, human impacts on the physical environment include:
- hunting: over exploitation
- transport: risks of spillages, especially in places like Alaska where oil and gas extraction dominate; road vehicles damaging the ground
- tourism: vegetation removal, litter and waste do not easily degrade. There have been well-publicised campaigns about this issue even on the slopes of Mount Everest where huge amounts of garbage and human waste are left by climbers every season
- general air pollution
- research: even in the Antarctic, where the number of people allowed on the landmass is limited, their presence, including transport and dealing with waste, will undoubtedly have some impact.

All of these factors occur in local areas, but they are adding to the global effects of climate change, melting the snow and ice of these environments.

Potential for sustainable development in cold environments

Some suggest that it is their lack of development that gives many cold environments the potential for sustainable development in the future. Much of the Arctic tundra and Antarctica typify the common perception of wild and natural places. Their remoteness and the extreme physical processes keep them inaccessible to mass tourism and the excesses of economic development. Conservationists believe wilderness areas have intrinsic value and possess outstanding qualities that are worth conserving for the future. Areas such as these have an aesthetic value for people seeking spiritual refreshment and contemplation. Scientifically, they are important because:
- there is a need to maintain the gene pool of wild organisms to ensure that genetic variety is maintained
- animal communities can be studied in their natural environment
- wilderness is a natural laboratory for the scientific study of ecosystems
- there is a need for pure natural systems to be used as a yardstick against which managed or mismanaged systems can be compared.

There are good reasons for conserving wilderness regions, but they also often contain a range of exploitable resources. Pressure to develop these resources comes from national and transnational groups that require both energy sources and raw materials to support industrial growth. Balancing developmental pressures against the need to conserve the essential values of wilderness is the increasingly difficult task of management. Sustainable development has an important role to play here, but there is disagreement about how it may be successfully applied in many wilderness environments.

In 1964, the Wilderness Act in the USA designated a number of wilderness areas. The largest number of designated areas in any state is in Alaska, which instituted its own wilderness legislation, the National Interest Lands Conservation Act, in 1980.

How cold environments are managed at present

Antarctica as a **Global Commons** is explored in detail in Chapter 7 and you should study that section on page 322. Figure 4.61 summarises some of the key aspects of the management of Antarctica.

The whole of Antarctica is now designated as a 'reserve devoted to peace and science' by the protocols of the Antarctic Treaties. However, it is not just the natural environment that is protected; other protocols protect the cultural and historical importance of the continent. Some sites have been designated as Historic Sites and Monuments (HSMs). Visit the website of the British Antarctic Survey to find out more about HSMs including the cross at Cape Evans on Ross Island,

erected in 1913 as a memorial to Captain Scott's party that perished in 1912 (Figure 4.62).

Figure 4.62 Memorial cross at Cape Evans for Captain Scott's expedition

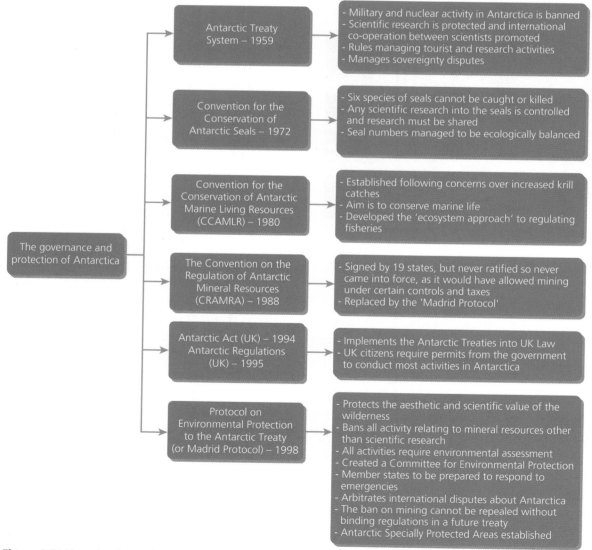

The governance and protection of Antarctica		
Antarctic Treaty System – 1959	- Military and nuclear activity in Antarctica is banned - Scientific research is protected and international co-operation between scientists promoted - Rules managing tourist and research activities - Manages sovereignty disputes
Convention for the Conservation of Antarctic Seals – 1972	- Six species of seals cannot be caught or killed - Any scientific research into the seals is controlled and research must be shared - Seal numbers managed to be ecologically balanced
Convention for the Conservation of Antarctic Marine Living Resources (CCAMLR) – 1980	- Established following concerns over increased krill catches - Aim is to conserve marine life - Developed the 'ecosystem approach' to regulating fisheries
The Convention on the Regulation of Antarctic Mineral Resources (CRAMRA) – 1988	- Signed by 19 states, but never ratified so never came into force, as it would have allowed mining under certain controls and taxes - Replaced by the 'Madrid Protocol'
Antarctic Act (UK) – 1994 Antarctic Regulations (UK) – 1995	- Implements the Antarctic Treaties into UK Law - UK citizens require permits from the government to conduct most activities in Antarctica
Protocol on Environmental Protection to the Antarctic Treaty (or Madrid Protocol) – 1998	- Protects the aesthetic and scientific value of the wilderness - Bans all activity relating to mineral resources other than scientific research - All activities require environmental assessment - Created a Committee for Environmental Protection - Member states to be prepared to respond to emergencies - Arbitrates international disputes about Antarctica - The ban on mining cannot be repealed without binding regulations in a future treaty - Antarctic Specially Protected Areas established

Figure 4.61 Managing Antarctica

Managing cold environments: alternative futures

Currently the Antarctic Treaties and Environmental and Madrid Protocols comprehensively protect Antarctica. However, there is a growing debate over these questions: Should current treaties be adapted or removed to encourage development in Antarctica? Or should the Treaty be maintained in its current form?

Arguments about possible future development in Antarctica:

- Energy and mineral resources elsewhere are becoming depleted:
  - Antarctica is believed to have large reserves of coal and oil, and precious metals such as gold and silver.
  - Can we envisage Antarctica having quarries, offshore oil rigs and oil wells on land?
  - Mining is banned until at least 2048, and any prospecting or extraction in and around Antarctica would be prohibitively expensive due to its remoteness and climate – so do conservationists need to be concerned?
- Tourism numbers visiting Antarctica remain high, with over 27,000 in 2008–09 (numbers have fallen since with controls on cruise ships in the early 2010s).
  - If tourism were to be encouraged as an economic activity could we envisage Antarctica having a commercial airport and port, with tourist hotels and shops?
  - The growth of adventure tourism relies on the affluent clientele being able to buy into a unique experience. If more tourists were allowed to visit current sites operators may seek to expand operations into new sites that have not been visited by people so far.
- In some parts of the world commercial fishing is depleting fish stocks to near extinction levels.
  - The waters of the southern ocean still have plentiful fish stocks.

  - Even though fishing is heavily regulated, fishing levels around the continent are already potentially undermining conservation efforts.
  - If restrictions were lifted, would levels of overfishing quickly reach the same as in other areas, or would the remoteness and hostile conditions naturally limit commercial fishing activity?
- **Bioprospecting** is a growing area of scientific research, with many keen to search for genetic and biochemical resources from Antarctica's flora and fauna.
  - Antarctica's biodiversity is unique, with many companies already expressing an interest in research.
  - The continent's animal and plant life is of interest, because their unique adaptations to deal with the harsh conditions may be useful for future valuable commercial products, like medicines or synthetic materials. There is already great interest in chemicals found in some species that act like natural antifreeze, which could have many commercial uses.
  - Bioprospecting is happening in Antarctica as it is often classed as scientific research, and if profits from resulting products could be targeted towards conservation and protection of the wilderness should bioprospecting be promoted?

This debate is too extensive to develop further here. Many geographers may instinctively fall on the side of continued protection, and maintaining and strengthening of the current protocols and treaties, but a full discussion of **sustainable development** in Antarctica would need to at least consider some of the questions above. Other cold environments will have their own unique issues raised by discussions of their future sustainable development.

Case study of a glaciated environment at a local scale: The Helvellyn area of the English Lake District

The present form of the Lake District owes much to the work of the Pleistocene glaciation, modifying and deepening the pre-glacial drainage system. Many of the lakes (Ullswater and Windermere, for example), radiate like the spokes of a wheel outward from the centre of ice accumulation. These ribbon lakes occupy glacially deepened valleys. Many of the smaller lakes, known as tarns, occupy the basins of corries which are found at higher altitudes and were in the areas most favourable for snow accumulation and thus became the source of glaciers. Such glaciers over-deepened the original hollows and when the ice finally melted, they were filled with meltwater to form small lakes. Over 150 glacial corries have been identified within the Lake District.

The Helvellyn Range runs approximately north-south, including the peak of Helvellyn itself (950 m altitude), which is part of a ridge of high ground over 600 m extending for about 11 km. Most of the corries of the area are found on the east side of the ridge, with many facing northeast, this being the orientation most conducive to snow accumulation (Figure 4.63).

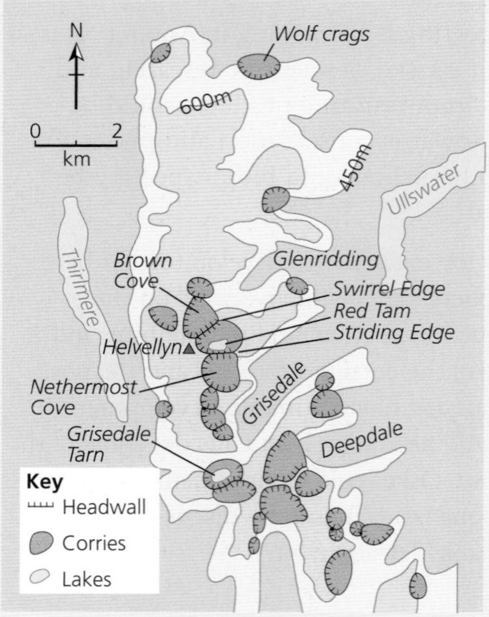

During phases of ice build-up, erosion within these corries intensified as local glaciers expanded within them. As many of the corries were close together, this erosion meant that only small, knife-edged ridges were left separating the hollows. Several of such aretes have formed within the area, particularly Swirral Edge and the much photographed Striding Edge (Figure 4.64).

The valleys of Grisedale and Glenridding provide excellent examples of glacial troughs. The head of Grisedale contained the largest glacier in the range during the last phases of the Pleistocene glaciation. Fed by ice from at least four corries, the glacier extended down the valley to an altitude of about 215 m. Above this altitude in the valley today, there are abundant moraines marking the position of the glacial snout. In the Grisedale valley, it is believed that the ice in these latter stages was not as extensive and moraines from earlier glacial periods are not as well formed as they have been subject to more post-glacial erosion. This valley opens out beyond the Helvellyn range and the deep rock basin cut earlier in the Pleistocene is now filled by a ribbon lake, Ullswater. A number of bands of hard rock running across the valley floor proved more resistant to glacial erosion and as a result the whole rock basin has an irregular profile. At one of the points between the resultant smaller individual rock basins, there is a roche moutonnée forming an island in the lake.

Figure 4.63 Corries, arêtes, glacial troughs and lakes in the Helvellyn area

Figure 4.64 Striding Edge, Helvellyn

Case study of a contrasting glaciated landscape from beyond the UK: The Athabasca Glacier

The Columbia Icefield is the largest field of its kind in the Rocky Mountains of North America, covering an area of 325 km² between the Banff and Jasper National Parks, although some modern calculations have suggested that this may have shrunk to under 230 km². The icefield has a surface elevation of about 3,000 m and the depth of ice reaches a maximum of over 360 m. It feeds eight major glaciers, perhaps the most accessible of which is the Athabasca Glacier (Figure 4.65 and Figure 4.66).

The Athabasca Glacier, like many others, has been in retreat for a number of years, receding at an average rate of between one and three metres per year. In recent times, however, there are indications that the rate of melting has quickened and studies have shown that on its top surface it could be losing at least 5 m/year.

Since 1750, the glacier has retreated nearly two kilometres from its maximum 'little Ice Age' extent. As it has done so, a large amount of glacial deposition has formed in advance of its snout. This has made the area an excellent one for the study of moraines.

Push moraines: Immediately in front of the snout there are seasonal push moraines. These small ridges are spaced between 2 and 20 m apart, and are typically between 0.7 and 2 m high. They are produced by the seasonal shift of the snout, in which there is retreat and dumping of debris in summer, followed by forward movement of the snout as the glacier gains mass in winter. This causes the debris to be bulldozed up into a ridge that is then

left behind as the glacier resumes its retreat in the next summer. As long as the glacier's snout continues to retreat a greater distance in summer than it advances in winter, it leaves behind a series of these moraines that can be used

Figure 4.66 The Athabasca glacier

to track glacial retreat year by year.

Recessional moraines: Further beyond the snout there are recessional moraines marking places where the snout had remained relatively stationary at various periods of time during its retreat. These moraines are larger than the seasonal push moraines, ranging between three and six metres in height. The furthest of these has been dated to 1870 and there are also moraines marking a pause in retreat in 1938, 1950 and 1960 (Figure 4.65). Knowing the ages of these moraines has enabled the study of how plants colonise fresh glacial debris.

Lateral moraine: Along the southeast side of the glacier is a 1.5 km long lateral moraine standing 124 m above the valley floor. Its large size results from a high input of rock debris from the slopes above owing to the slopes being composed mainly of well-jointed limestone which is weathered easily by freeze–thaw action. This accumulation of debris along the side of the glacier is preventing the ice beneath from melting as rapidly as the rest of the glacier. Over time the lateral moraine will lower as the ice beneath it slowly melts.

Pro-glacial lake: Between the snout of the glacier and the last recessional moraine, meltwater has built up to form Lake Sunwapta. As its outlet through the moraine is progressively lowered by meltwater erosion, the lake will eventually disappear.

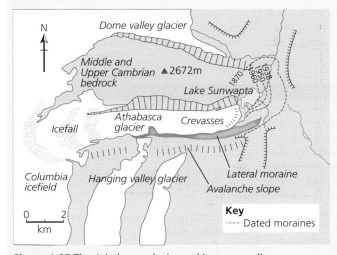

Figure 4.65 The Athabasca glacier and its surroundings

Case study of a contrasting glaciated landscape from beyond the UK: The Sápmi region of tundra, northern Europe

Geographical context

The Sápmi region, also known as Lapland, is the area stretching across the far north of Norway, Sweden, Finland and the Kola Peninsula of Russia, that is the cultural home of the Sami indigenous people (Figure 4.67).

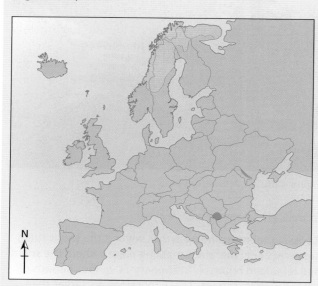

Figure 4.67 The Sápmi region or northern Europe

Approximately 2 million people live in the region with a mix of ethnic origins, including the Sami, Russians, Norwegians, Swedes and Fins. Individuals from each of these national groups would also class themselves as ethnically Sami.

A challenging place to live

Sami people have lived successfully in this cold environment for centuries, but this does not mean that life in this tundra region of the Arctic north is not challenging.

Traditional adaptations of the Sami people of Sápmi

People have bred and herded reindeer in Sápmi for over 1,000 years (Figure 4.68). Due to the diversity of environments in different parts of Sápmi different forms of reindeer breeding developed in different areas.

Figure 4.68 A Sami reindeer herder at a round-up in northern Norwegian Sápmi

Until the mid-twentieth century, many family groups remained fully nomadic, moving their herds to the shelter of the coniferous forests in the winter and the

Table 4.8 Challenges of the Sápmi region for the Sami people

Social	Economic	Environmental
The Sami are now a minority population in the region and their culture and way of life has been oppressed for a long time. In recent decades there have been significant efforts to protect the rights of, and promote the cultures and livelihoods of the indigenous population	For those who have chosen to follow a more modern settled life, the Sápmi region as a whole is rich in mineral deposits, has ice-free ports for transport and is also seeing an increase in tourism and military installations, which have brought wealth to the area	The Sami traditionally live mostly inside the Arctic Circle, which has short milder summers, but long dark winters
Many young Sami are attracted to the more settled comforts and education and employment opportunities of a modern urban life so are leaving their traditional communities	However, all of the above add to the challenges already experienced by those wishing to maintain traditional economies based on reindeer herding, fishing, hunting and small-scale agriculture	The relief is generally a low plateau, but with many lakes, rivers and marshes, making long distance travel quite difficult
Cultural traditions are threatened as communities become ageing populations with fewer people for traditions to be passed on to	Only 10 per cent of the Sami manage to remain solely as nomadic reindeer herders, as activities like commercial logging threaten their lands	Vegetation is generally sparse, but the south is very densely forested
		Rivers have been dammed for hydroelectricity production; the frozen lakes and ice can no longer be used to walk the reindeer to new grazing lands, so lorries and boats have to be used

rich pastures of the high mountains in the summer. This continual movement limited the impact on either environment. Tame reindeer were used for transport and to lead the herd to new grazing each day. The Sami survived on the reindeer themselves, consuming their meat and milk. Every part of the reindeer was processed into food, clothes or utensils, including skin, fur, sinews, bone and antlers.

With strong links between family groups and culture and traditions going back hundreds of years, knowledge about interacting with nature and understanding when to move grazing areas under changing climatic conditions was passed from generation to generation.

They also understood the importance of limiting their impact on, and protecting the natural resources around them, so many Sami communities had very low fertility levels. By spacing births a number of years apart it meant the number of young children that could not walk with the animals was at a minimum, and most waited until children could move independently with the herd before having more children. This naturally kept population numbers low thus limiting the impact on the environment and resources. The traditional Sami lifestyle could be viewed as sustainable, as it provided for their needs without negatively affecting their environment. However, during the second half of the twentieth century there have been many challenges to this traditional way of life, with some questioning the viability of sustainable development in the area in the future.

Possible sustainable development of the Sápmi region

Many view the growth of settlements and modern industry in Sápmi as unsustainable. If the traditional way of life of the Sami is seen as a sustainable use of this region then a number of issues will need to be overcome.

There are three conditions needed for the Sami and their reindeer to thrive:

- Access to old forests with good supplies of ground and tree-dwelling lichens.
  - Most Sami do not own their land.
  - Many of their grazing rights are informal and based on historical customs.
  - Commercial forestry is clearing large areas.
  - The right of the Sami to freely graze their reindeer in many areas of Sápmi is protected, but increasing numbers of landowners are making legal challenges to this.

- A family needs over 500 reindeer to live off without any other income.
  - As other landuses limit the amount of grazing land available this can put pressure on the size of the herd.
- Reindeer need to be able to move freely to find fresh sources of food.
  - Modern infrastructure, like roads and railways and HEP reservoirs, along with settlement growth, blocks traditional migratory routes.

However, many international organisations and the four national governments of the region increasingly seek to protect indigenous peoples and promote their traditions and ways of life. This seems like a difficult balancing act, with so many different Sami communities spread across areas of four separate nations who all also see the needs of others who wish to exploit the natural wealth of this region.

The Tåssåsen Sámi and their environment

An example of a Sami community is the Tåssåsen Sámi community, Östersund, Jämtland County, Sweden. The Tåssåsen Sami community lives in central parts of the Swedish mountain range west of the town of Östersund in Jämtland County and, with other Southern Sami groups, like those near Sveg to the south, has called this part of northern Europe home since prehistoric times (Figure 4.69).

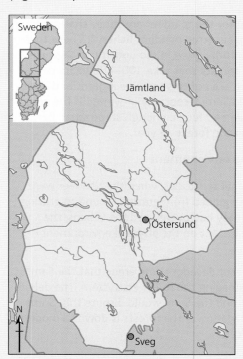

Figure 4.69 Map showing the location of Jämtland County, Sweden

By the early 2000s this was one of the few remaining communities of Southern Sami living a traditional way of life.

Background

- Grazing grounds covered 111,000 hectares – with winter grazing including an area 100 km by 30 km of old forest.
- For Sami they live quite far south, so occupy mainly mountain grazing.
- Narrow valleys are used to direct the herd and bring them together to be counted and inspected.
- The reindeer herd numbered 5,500.
- The community numbered 50 individuals in families split into 15 grazing groups.
- Their culture and traditions were very strong and unlike other communities the young people had stayed and kept the population young and viable.

In the summer the Tåssåsen guide their reindeer up onto the higher mountain slopes to graze the fresh ground-growing lichens to supplement the lichens growing on the trees. As the winter approaches the herds are guided back down to the lower forested slopes. Here the trees provide shelter and the snow cover on the forest floor is much lighter and easier to dig through to reach the lichens beneath, than that outside of the forest.

The best lichen for the reindeer are those that grow hanging from trees (tree pendent lichens). These tree pendent lichens only grow in old undisturbed spruce, pine and birch forests that take 100–200 years to mature. A small area of this remains in Jämtland, used by the Tåssåsen Sami, especially in winter. If the area of forest is large and fertile enough some small herds could remain in the forest all year.

Impact on the environment

In their traditional territories, when the reindeer were able to be freely guided from grazing ground to grazing ground before they had overgrazed the lichens, the Sami had little impact on the physical environment of Jämtland.

However, in recent decades some argue that the Sami's reindeer are damaging the forest ecosystem. The debate is complicated, and the Sami would dispute this. Some claim any degradation of the forests is down to modern forestry practices.

Figure 4.70 Tåssåsen reindeer herd in winter

The Tåssåsen Sami and the future

Like all indigenous people living in the cold environments of the north, climate change is a potential threat to the Tåssåsen Sami. As outlined above, this area of the northern hemisphere is projected to have the highest rates of warming. Possible impacts:

- Lichen growth is dependent on winter snow and ice conditions. Warmer winters with less snow would reduce the amount of lichen for grazing.
- Lichen occupy a very specific ecological niche. Even slight warming will increase competition with other plant species, again reducing the amount for grazing.
- Later onset of, and warmer winters will mean fewer rivers and lakes freeze, thus limiting the movement of herds between grazing grounds. Some areas could become overgrazed.
- Reindeer and the Sami are in tune with the changing seasons. Some groups may be slow to adjust to rapid changes in seasonal patterns.
- Summer warming may increase insect populations. The reindeer could lose energy and time on dealing with insect harassment rather than foraging. Many Sami have already made the decision to end their nomadic way of life and farm reindeer in a more settled commercial way. However, this ends hundreds of years of culture and tradition. Also, it is a more capital intensive way of farming requiring the input of more capital for fixed pasture, fences and feed, so is more expensive and requires fewer people. Those that adopt these new approaches quickly have to diversify into other modern commercial activities to make the reindeer profitable and support all members of the community.

It may be that the predicted effects of climate change mean the traditional grazing lands cannot support the reindeer herds in the future anyway. However, there is a more immediate and tangible threat to the traditional way of life of the Tåssåsen reindeer herders of Jämtland.

Much of the forest area that they have used traditionally is now owned privately, largely by commercial logging businesses. Since the 1990s there have been well-publicised court cases brought by the commercial forestry industry to attempt to ban the reindeer herders from their lands. Community leaders like Olof Johannsson say this could be the end of a culture dating back over 1,000 years.

Further research

You could use the following websites to research the Sápmi region and the Sami in more detail:

http://samenland.nl/lap_sami_si.html

www.oloft.com/casestudy.html

www.sacredland.org/index.html@p=91.html

http://news.bbc.co.uk/1/hi/world/europe/629818.stm

Further reading

The texts below develop many of the ideas set out in this chapter. Some are specialist texts and quite academic and your local library may be able to find copies should you wish to take your studies further.

Woodward, J. (2014) *The Ice Age: A Very Short Introduction*. Oxford University Press

Benn, D. & Evans, D. (1998) *Glaciers and Glaciation*. 1st Edition. Arnold

Benn, D. & Evans, D. (2013) *Glaciers and Glaciation*. 2nd Edition. Routledge

Raw, M. and Knight, J. (2007) *Glaciation and Periglaciation Advanced Topic Master*. Philip Allan

Redfern, D. (2010) *Climate Change: Contemporary Case Studies*. Philip Allan Updates

Harvey, A. (2012) *Introducing Geomorphology: A Guide to Landforms and Processes*. Dunedin Academic Press

Anderson, D. (2004) *Glacial and Periglacial Environments, Access to Geography Series*. Hodder & Stoughton

The following websites are also good sources of additional information on some of the topics covered.

To discover more about the characteristics of the range of types of patterned ground and more detail on the theories about the processes that formed them, and to see a range of photographic examples visit: http://permafrosttunnel.crrel.usace.army.mil/permafrost/patterned_ground.html

Issues of managing Antarctica are far-reaching and complex. Suggested websites include: www.bas.ac.uk/about/antarctica/environmental-protection/ www.discoveringantarctica.org.uk/alevel_5_4.html www.ats.aq/e/ep.htm

Review questions

1 In Tundra regions, what is the relationship between:
   a) the climate and the vegetation?
   b) the climate and the soil?

2 What do you understand by the term 'fragile environment'?

3 Explain how glacial ice is formed.

4 Why do glaciers advance and retreat?

5 Contrast the processes of glacial abrasion and glacial plucking.

6 Describe the different ways in which material can be transported by a glacier.

7 Select a feature that has been formed largely through glacial erosion. Describe it and with the aid of diagrams explain its formation.

8 Compare and contrast lodgement till and ablation till.

9 Describe the distribution of periglacial environments.

10 With reference to an alpine or tundra region you have studied, discuss how people have responded to the social, economic and environmental challenges of living in that cold environment.

Question practice

A-level questions

1. Explain the development of cold-based glaciers. (4 marks)

2. Figures 4.71 and 4.72 show the annual minimum extent of Arctic sea ice in summer 1979 and summer 2015. Figure 4.73 shows the change in minimum volume of Arctic sea ice between 1979 and 2013. Figure 4.74 shows the change in the extent of Arctic sea ice each September between 1981 and 2015 compared to the average for that period.

   Using Figures 4.71– 74 analyse the trends in Arctic sea ice. (6 marks)

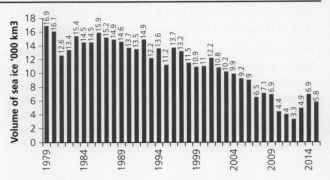

Figure 4.73 Minimum volume of Arctic sea ice between 1979 and 2013

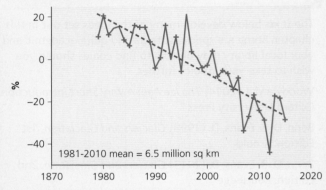

Figure 4.74 Changing extent of Arctic Sea ice each September between 1979 and 2015 compared to 1981–2010 average

Figure 4.71 Annual minimum extent of Arctic sea ice reached 21 September 1979
Source: NASA

3. Using Figure 4.75 and your own knowledge, assess the role of water in the development of landscapes in cold environments like that shown in Figure 4.75. (6 marks)

Figure 4.75

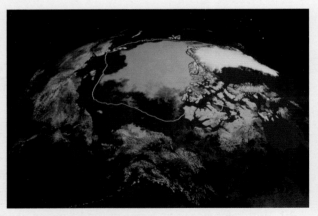

Figure 4.72 Annual minimum extent of Arctic sea ice reached 11 September 2015
Note: the gold line represents the average minimum extent of sea ice for the 36-year period
Source: NASA

4. Assess the potential challenges and opportunities for future development in a glaciated landscape you have studied beyond the UK. (20 marks)

AS level questions

1. What is meant by the term 'a fragile environment'?

    a) One that has experienced significant negative impacts from human activity

    b) One that is easily damaged and disturbed but is easy to restore once destroyed

    c) One that is easily damaged or disturbed and then difficult to restore once destroyed

    d) One that is in pristine natural condition and will not change in the future   (1 mark)

2. Which of the following are all types of ice movement?

    a) Rotational flow, solifluction and nivation

    b) Internal deformation/creep, basal sliding and rotation flow

    c) Compressional flow, extending flow and frost action

    d) Abrasion, plucking and creep   (1 mark)

3. Distinguish between warm- and cold-based glaciers.   (3 marks)

4. Figures 4.76a–4.76c locate the snout of the Vowell Glacier in British Columbia.

    With reference to Figures 4.76a–4.76c, interpret the evidence that this glacier is changing.   (6 marks)

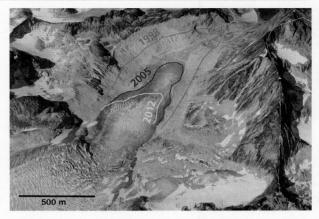

Figure 4.76b Google Earth image of Volwell Glacier British Columbia, 2005

Figure 4.76c Google Earth image of Volwell Glacier British Columbia, 2012

5. Analyse the links between climate, process and landform which lead to the formation of characteristic glaciated landscapes.   (9 marks)

6. 'The twenty-first century will bring increased challenges to people living in periglacial landscapes and they will have to develop new strategies to overcome these challenges.'

    To what extent do you agree with this view?   (20 marks)

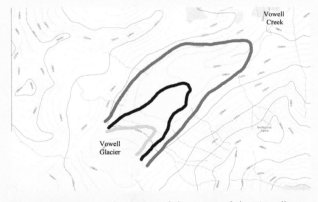

Figure 4.76a Mapped position of the snout of the Vowell Glacier in British Columbia

In this chapter you will focus on the atmosphere and the lithosphere (the earth's crust and the section immediately below it – the upper mantle), which intermittently but regularly present hazards to human populations, often in a dramatic and sometimes catastrophic fashion. By exploring the origin and nature of these hazards and the various ways in which people respond to them, you will be able to engage with many dimensions of the relationships between people and the environment in which they live. The study of natural hazards will also give you the opportunity to exercise and develop your observational skills, measurement and geospatial mapping skills, together with data manipulation, interpretation and presentation of statistics.

You should be aware that in this chapter we have presented you with the details of a number of hazard events, which will widen your knowledge and understanding of the hazards themselves. You are expected to have studied, in terms of impacts and human responses, one recent volcanic event, one recent seismic event and one recent wildfire event. With recent tropical storms, you are expected to have studied two events 'in contrasting areas of the world'.

In this chapter you will study:
- the concept of hazard in a geographical context
- volcanic hazards
- seismic hazards
- storm hazards
- wildfires

5.1 The concept of a hazard in a geographical context

The concept of a natural hazard and its potential impact

A **natural hazard** is a perceived event that threatens both life and property. Natural hazards often result in disasters that cause some loss of life and/or damage to the built environment and create severe disruption to human activities. Natural or environmental hazards include volcanic activity, seismic events (earthquakes) and tropical storms (hurricanes/typhoons/cyclones). All of these can therefore cause disruption to human systems, including death and injury, property and communication system damage and the disruption of economic activities.

These forms of hazard thus pose a risk to human populations. That risk is increased because we build shanty towns on unstable tropical slopes, urbanise volcanic zones, live in areas with active faults and on coasts susceptible to hurricanes and tsunamis. The problem is also exacerbated by the failure to recognise a potential hazard and act accordingly.

Natural hazards, and their effects on people, tend to have the following common characteristics:
- their origins are clear and the effects that they produce are distinctive, such as earthquakes causing buildings to collapse
- most natural hazards only allow a short warning time before the event (some hardly at all)
- exposure to the risk is involuntary, although this applies to the populations of less well developed countries. In developed areas, most of the people who occupy hazardous areas are often well aware of the risks, which they choose to minimise or even ignore
- most losses to life and damage to property occur shortly after the event although the effects of natural hazards can be felt in communities long after that time (disease, disruption to communications and economic activities)

- the scale and intensity of the event requires an emergency response.

Key terms

Adaptation – In the context of hazards, adaptation is the attempts by people or communities to live with hazard events. By adjusting their living conditions, people are able to reduce their levels of vulnerability. For example, they may avoid building on sites that are vulnerable to storm surges but stay within the same area.

Fatalism – A view of a hazard event that suggests that people cannot influence or shape the outcome, therefore nothing can be done to mitigate against it. People with such an attitude put in place limited or no preventative measures. In some parts of the world, the outcome of a hazard event can be said to be 'God's will'.

Natural hazards – Events which are perceived to be a threat to people, the built environment and the natural environment. They occur in the physical environments of the atmosphere, lithosphere and the hydrosphere.

Perception – This is the way in which an individual or a group views the threat of a hazard event. This will ultimately determine the course of action taken by individuals or the response they expect from governments and other organisations.

Risk and vulnerability

Risk is the exposure of people to a hazardous event presenting a potential threat to themselves, their possessions and the built environment in which they live. People though, consciously put themselves at risk from natural hazards and the question has to be why do they do it?

Possible reasons include the following:

- **Hazard events are unpredictable:** We cannot predict the frequency, magnitude or scale of a natural hazard event.
- **Lack of alternatives:** Due to social, political, economic and cultural factors, people cannot simply uproot themselves from one place and move to another, giving up their homes, land and employment.
- **Changing the level of risk**: Places that were once relatively safe may have become, through time, far more of a risk. Deforestation, for example, could result in more flooding from torrential rain associated with tropical storms and there could also be a greater risk from landslides.
- **Cost/benefit:** There are many hazardous areas that offer advantages that in people's minds outweigh the risk that they are taking. Californian cities, for example, have a high risk from earthquakes, but people see the many advantages of living there as greater than the potential risk.
- **Perception:** see below.

Vulnerability to physical hazards means the potential for loss. Since losses vary geographically, over time and among different social groups, vulnerability therefore also varies over time and space. Researchers at the University of South Carolina have examined all the variables which link risk and vulnerability and have come up with the model shown in Figure 5.1.

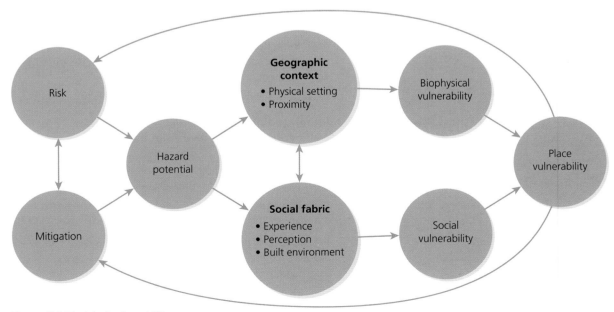

Figure 5.1 Model of vulnerability

This raises the question, is the risk the same for all people in an area? In other words, are some people more vulnerable than others? A similar sized natural hazard event can have widely varying impacts in different parts of the world. People's wealth and the level of technology that they can apply do affect the degree to which the hazard event will impact upon them. Richer people and countries can protect themselves by building sea defences, constructing earthquake-resistant buildings, providing better emergency services, etc. They can also be better prepared by being made more aware of the risk through education. The people of cities in poorer countries are more vulnerable. As such urban areas have grown, more and more people have been forced to live in hazardous areas such as very steep hillsides that are prone to landslides, and in the lowest lying parts where they are at risk from tropical storms and tsunamis.

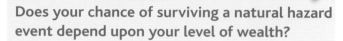

Key question

Does your chance of surviving a natural hazard event depend upon your level of wealth?

The perception of natural hazards

People react to the threat of hazards in different ways because of the way in which individuals receive and process information. Perception is influenced by many factors including:

- socio-economic status
- level of education
- occupation/employment status
- religion, cultural/ethnic background
- family and marital status
- past experience
- values, personality and expectations.

Perception of a hazard will ultimately determine the course of action taken by individuals in order to modify the event or the responses they expect from governments and other organisations.

There is often a great difference in the perception of a hazard between peoples of differing levels of economic development. In wealthier areas there is a sense that the better that you are prepared, the more able you will be to withstand the impact of the hazard and perhaps even prevent the disaster from taking place. This is usually based upon government and community action, and is backed by capital that will fund technologically-based solutions. The sense of helplessness in the face of natural hazards tends to increase with the level of poverty and the deprivation of the people. Even in wealthier countries there are groups of disadvantaged people who tend to look upon natural hazards as part of their way of life, as they are seen as unavoidable, just as the bulk of people in poorer countries see the impacts of these events as being part of the conditions of poverty.

People may perceive natural hazards in the following ways:

- **Fatalism (acceptance):** Such hazards are natural events that are part of living in an area. Some communities would go as far as to say that they are 'God's will'. Action is therefore usually direct and concerned with safety. Losses are accepted as inevitable and people remain where they are.
- **Adaptation:** People see that they can prepare for, and therefore survive the event(s) by prediction, prevention, and/or protection, depending upon the economic and technological circumstances of the area in question.
- **Fear:** The perception of the hazard is such that people feel so vulnerable to an event that they are no longer able to face living in the area and move away to regions perceived to be unaffected by the hazard.

Management of natural hazards

Key terms

Community preparedness/risk sharing – This involves prearranged measures that aim to reduce the loss of life and property damage through public education and awareness programmes, evacuation procedures, the provision of emergency medical, food and shelter supplies and the taking out of insurance.

Frequency – The distribution of a hazard through time.

Integrated risk management – The process of considering the social, economic and political factors involved in risk analysis; determining the acceptability of damage/disruption; deciding on the actions to be taken to minimise damage/disruption.

Magnitude – The assessment of the size of the impact of a hazard event.

Prediction – The ability to give warnings so that action can be taken to reduce the impact of hazard events. Improved monitoring, information and communications technology have meant that predicting hazards and issuing warnings have become more important in recent years.

Key terms

Primary effects – The effects of a hazard event that result directly from that event. For a volcanic eruption these could include lava and pyroclastic flows. In an earthquake, ground shaking and rupturing are primary effects.

Resilience – The sustained ability of individuals or communities to be able to utilise available resources to respond to, withstand and recover from the effects of natural hazard events. Communities that are resilient are able to minimise the effects of the event, enabling them to return to normal life as soon as possible.

Secondary effects – These are the effects that result from the primary impact of the hazard event. In volcanic eruptions these include flooding (from melting ice caps and glaciers) and lahars. In an earthquake, tsunamis and fires (from ruptured gas pipes) are secondary effects.

People respond to natural hazards and the threats that they can pose by seeking ways to reduce the risk. Responses can come from individuals, the local community with people working together, and from national governments and international agencies.

Community **resilience** is the sustained ability of a community to utilise available resources to respond to, withstand and recover from the effects of natural hazards. Communities that are resilient are able to minimise the effects of a hazard, making the return to normal life as effortless as possible.

A key feature of the modern approach is that hazards are best combated by efficient management. Modern management techniques, with their gathering of information, careful analysis and deliberate planning, aim to make the most efficient use of the money available to confront natural hazards. A process known as **integrated risk management** is often used which incorporates identification of the hazard, analysis of the risks, establishing priorities, treating the risk and implementing a risk reduction plan, developing public awareness and a communication strategy, and monitoring and reviewing the whole process. The governments of many countries use such schemes. An example developed by the New Zealand government is shown in Figure 5.2.

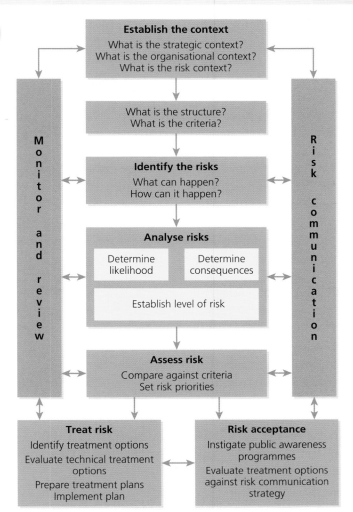

Figure 5.2 The process of risk management

People and organisations therefore try to manage natural hazards in the following ways:

● **Prediction:** It may be possible to give warnings that will enable action to be taken. The key to this is improved **monitoring** in order to give predictions which means that warnings can be issued. The National Hurricane Centre in Florida is a good example of an agency demonstrating how prediction can depend upon monitoring, through the use of information from satellites and land-, sea- and air-based recordings.

● **Prevention:** For natural hazards this is probably unrealistic although there have been ideas and even schemes such as seeding clouds in potential tropical storms in order to cause more precipitation, which in theory would result in a weakening of the system as it approached land.

● **Protection:** The aim is to protect people, their possessions and the built environment from the impact of the event. This usually involves modifications to the built environment such as improved sea walls and earthquake-proof buildings. One way in which governments can act, and people react, is to try to change attitudes and behaviour to natural hazards which will reduce people's vulnerability. **Community preparedness (or risk sharing)** involves prearranged measures that aim to reduce the loss of life and property damage through public education and awareness programmes, evacuation procedures and provision of emergency medical and food supplies and shelters. There can also be attempts to modify losses through insurance (richer areas) and international aid (in poorer regions).

All attempts at management must be evaluated in terms of their success. Successful schemes include the use of dynamite to divert lava flows on Mt Etna and pouring sea water on lava flows in Iceland. On the other hand, the Japanese felt that they were well prepared for earthquakes and yet in 1995 the city of Kobe suffered the Great Hanshin earthquake, which destroyed over 100,000 buildings (with three times that number damaged) and a death toll of over 6,000 with 35,000 injuries. (For more details see the later local case study, page 224.)

All this can be put together as the disaster/risk management cycle. This illustrates the ongoing process by which governments, businesses and society plan for and reduce the impact of disasters, react during and immediately following an event, and take steps to recover after an event has occurred. Appropriate actions at all points in the cycle lead to greater preparedness, better warnings and reduced

vulnerability or the prevention of hazard events during the next cycle. The complete cycle includes the shaping of public policies and plans that either modify the causes of the hazard events or mitigate their effects on people, property and infrastructure.

One of the main goals of disaster management, and one of its strongest links with development, is the promotion of sustainable livelihoods and their protection and recovery during such events. Where this goal is achieved, people have a greater capacity to deal with disasters and their recovery is more rapid and long lasting.

The Federal Emergency Management Agency (FEMA) was created in the USA in 1978. The agency's primary purpose is to co-ordinate the response to a disaster that has occurred in the United States and that has overwhelmed the resources of local and state authorities. They have created major analysis programmes for floods, hurricanes and earthquakes. Their operations are carried out very much along the lines shown by the disaster/risk management cycle model shown in Figure 5.3.

Disaster/response curve: To show that hazard events can have varying impacts over time, in 1991 Park devised his impact/response model (Figure 5.4). This model shows an early stage, before the disaster strikes, where the quality of life is normal for the area. Here people try their best to prevent such events and prepare in case they should happen. When the event happens, the quality of life suddenly drops with people taking immediate action to preserve life and, if possible, the built environment.

The next stage Park called **relief**, where medical attention, rescue services and overall care are delivered. This can last from a few hours to several days if the event has been very damaging. From this point the quality of life of the people of the area starts to slowly increase.

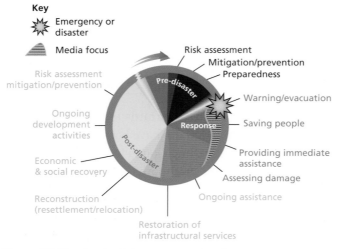

Figure 5.3 The disaster/risk management cycle

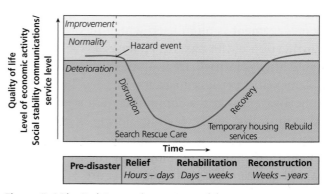

Figure 5.4 The Park impact/response model

Next comes **rehabilitation**, where people try to return the state of things to normal by providing food, water and shelter for those most affected. This period can last anything from a few days to weeks.

Finally comes **reconstruction**, where the infrastructure and property are reconstructed and crops regrown. At this time people use the experience of the event to try to learn how to better respond to the next one. This period can take from weeks to several years.

Park also showed how different events can have different impacts. This is shown in Figure 5.5 by the speed of the drop in the quality of life, the duration of the decline and the speed and nature of recovery. The difference in the three lines could be related to the type of hazard, the degree of preparedness or the speed of the relief effort, and the nature of recovery and rebuilding.

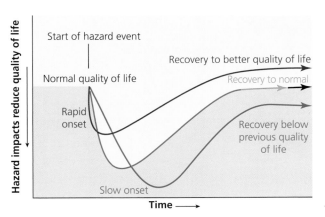

Figure 5.5 Variations within the Park model

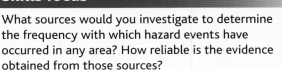

Key question

Select one of the case studies of a hazard event to be found later in this chapter and describe how the Park model can show the impact through time of that event.

Distribution, frequency and magnitude

Distribution refers to the spatial coverage of the hazard. This can refer to the area affected by a single event, some of which can have a very localised effect, while others have a much wider effect, such as tsunamis which can cross large oceans. Volcanic eruptions have been known to have a global effect with the spread of dust and the consequent short-term climatic change. Distribution of a hazard can also refer to the areas where the particular hazard is likely to

occur. Earthquakes and volcanoes, for example, are generally associated with tectonic plate boundaries whereas tropical cyclones usually occur between 5° and 25° north and south of the equator.

Frequency refers to the distribution of the hazard through time whereas **magnitude** assesses the size of the impact. At its most basic level, the frequency-magnitude principle leads us to expect many small insignificant events and, in the long term, increasingly fewer events as magnitude rises.

Skills focus

What sources would you investigate to determine the frequency with which hazard events have occurred in any area? How reliable is the evidence obtained from those sources?

The effects of natural hazard events

The effects are the impact that an event has upon both the physical and human environments. Some commentators differentiate between primary and secondary effects when considering hazard impacts. For an earthquake, for example, the primary effect would be the ground shaking and cracking followed by the secondary effects of soil liquefaction, landslides, tsunamis and the effects on people and the built environment such as collapsing buildings, fires, flooding and the knock-on effects which could be with the population for a long time. Communication systems could be out of order, the ability to produce food crops may take some time to be restored and the economy of a region may be so damaged that the legacy of the hazard event will be around for years.

5.2 The theory of plate tectonics

Before the development of plate tectonic theory, Earth scientists divided the interior of the Earth into three layers: the crust, the mantle and the core.

The **core** is made up of dense rocks containing iron and nickel alloys and is divided into a solid inner core and a molten outer one, with a temperature of over 5000 °C. This heat is produced mainly as the result of two processes: **primordial** heat left over from the Earth's formation and **radiogenic** heat produced by the radioactive decay of isotopes, particularly Uranium-238, Thorium-232 and Potassium-40.

The **mantle** is made up of molten and semi-molten rocks containing lighter elements such as silicon and oxygen.

The **crust** is even lighter because of the elements that are present, the most abundant being silicon, oxygen, aluminium, potassium and sodium. The crust varies in thickness – beneath the oceans it is only 6–10 km thick but below the continents this increases to 30–40 km and under the highest mountain ranges this can be as high as 70 km.

The theory of plate tectonics has retained this division, but new research has suggested that the crust and upper mantle should be divided into the lithosphere and asthenosphere.

The **lithosphere** consists of the crust and the rigid upper section of the mantle and is approximately 80–90 km thick. This is the section of the Earth that is divided into seven very large plates and a number of smaller ones (Figure 5.6). Plates are divided into two categories, oceanic and continental, depending on the type of material from which they are made (see Table 5.1).

The **asthenosphere** lies beneath this layer and is semi-molten on which the plates float and move.

Table 5.1 Differences between continental and oceanic crust

	Continental crust	**Oceanic crust**
Thickness	30–70 km	6–10 km
Age	Over 1,500 million years	Less than 200 million years
Density	2.6 (lighter)	3.0 (heavier)
Composition	Mainly granite; silicon, aluminium, oxygen (SIAL)	Mainly basalt; silicon, magnesium, oxygen (SIMA)

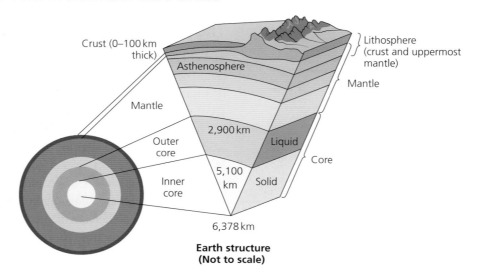

Earth structure (Not to scale)

Figure 5.6 The structure of the earth

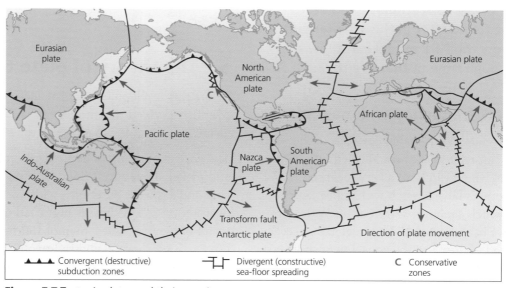

Figure 5.7 Tectonic plates and their margins

The theory

Plate tectonic theory revolutionised the study of Earth science. As soon as maps of the Atlantic Ocean were produced, people began to notice that the continents either side seemed to fit together remarkably well – the bulge of South America fitting into the indent below West Africa (Figure 5.7). Francis Bacon had noticed this fit as early as the seventeenth century but it did not attract any serious attention as no one thought that the continents could move.

In 1912, a German, Alfred Wegener, published his theory that a single continent existed about 300 million years ago. He named this super-continent Pangaea and maintained that it had later split into the two continents of Laurasia in the north and Gondwanaland in the south. Today's continents were formed from further splitting of these two masses. Wegener published this theory of continental drift and claimed that it was supported by several pieces of evidence that these areas were once joined.

Geological evidence for the theory includes:
- the above-mentioned fit of South America and west Africa
- evidence of a late-Carboniferous glaciation (290 million years ago), deposits from which are found in South America, Antarctica and India. The formation of these deposits cannot be explained by their present position; they must have been formed together and then moved. There are also striations on rocks in Brazil and West Africa which point to a similar situation
- rock sequences in northern Scotland closely agree with those found in eastern Canada, indicating that they were laid down under the same conditions in one location.

Biological evidence for the theory includes:
- fossil brachiopods found in Indian limestone are comparable with similar fossils in Australia
- fossil remains of the reptile *Mesosaurus* are found in both South America and southern Africa. It is unlikely that the same reptile could have developed in both areas or that it could have migrated across the Atlantic
- the fossilised remains of a plant which existed when coal was being formed have been located only in India and Antarctica.

Development of the theory

Wegener's theories were unable to explain how continental movement could have taken place and his ideas gained little ground. From the 1940s onwards, however, evidence began to accumulate to show that Wegener could have been correct.

The mid-Atlantic ridge was discovered and studied along with a similar feature in the Pacific Ocean. Examination of the ocean crust either side of the mid-Atlantic ridge suggested that **sea-floor spreading** was occurring. The evidence for this is the alternating polarity of the rocks that form the oceanic crust. Iron particles in lava erupted on the ocean floor are aligned with the Earth's magnetic field. As the lavas solidify, these particles provide a permanent record of the Earth's polarity at the time of the eruption (palaeomagnetism). However, the Earth's polarity reverses at regular intervals (approximately every 400,000 years). The result is a series of magnetic 'stripes' with rocks aligned alternately towards the north and south poles (Figure 5.8). This striped pattern, which is mirrored exactly on either side of a mid-oceanic ridge, suggests that the oceanic crust is slowly spreading away from this boundary. Moreover, the oceanic crust gets older with distance from the mid-oceanic ridge.

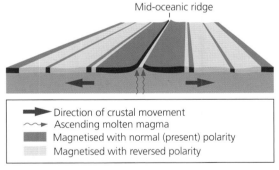

Figure 5.8 Magnetic 'stripes' on the Atlantic Ocean floor

Sea-floor spreading implies that the earth must be getting bigger. As this is not the case, then plates must be being destroyed somewhere to accommodate the increase in their size at mid-oceanic ridges. Evidence of this was found with the discovery of huge oceanic trenches where large areas of ocean floor were being pulled downward in a process known as **subduction.**

Hot spots around the core of the earth generate thermal convection currents within the asthenosphere, which cause magma to rise towards the crust and then spread

before cooling and sinking (Figure 5.9). This circulation of magma is the vehicle upon which the crustal plates move. The crust can be thought of as 'floating' on the denser material of the asthenosphere. This is a continuous process, with new crust being formed along the line of constructive boundaries between plates (plates moving away from each other – divergent) and older crust being destroyed at destructive boundaries (plates moving towards each other – convergent). Where two crustal plates slide past each other and the movement of the plates is parallel to the plate margin, there is no creation or destruction of crust. At these conservative margins, there is no subduction and therefore no volcanic activity. (See information on the San Andreas fault zone (California) in the seismic hazards section.)

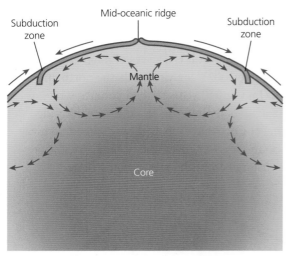

Figure 5.9 Convection currents and plate movements

Landforms associated with plate movements

The movement of the plates are therefore responsible for the production of the following landforms.

- **Ocean ridges:** The longest continuous uplifted features on the surface of the planet, they are formed when plates move apart in oceanic areas. The space between the plates is filled with basaltic lava upwelling from below to form a ridge. Volcanic activity also occurs along these ridges, forming submarine volcanoes which sometimes rise above sea level, such as Surtsey to the south of Iceland.
- **Rift valleys:** Such valleys form when plates move apart on continental areas. In East Africa, for example, the brittle crust fractures as sections of it move apart. Areas of crust drop down between parallel faults to form the valley. An area between two parallel rift valleys forms an upstanding block, known

as a **horst.** The line of the African rift is thought to be an emergent plate boundary, the beginning of the formation of a new ocean as eastern Africa splits away from the remainder of the continent.

- **Deep sea trenches:** Where oceanic and continental plates meet, the denser oceanic plate is forced underneath the lighter continental one (subduction). The downwarping of the oceanic plate forms a very deep part of the ocean known as a trench. Off the western coast of South America, the Nazca plate is subducted under the South American plate forming the Peru-Chile trench. A similar process happens when two oceanic plates move towards each other. On the western side of the Pacific Ocean, the Pacific plate is subducted beneath the smaller Philippine plate forming the very deep Marianas trench.
- **Island arcs:** During subduction, the descending plate encounters hotter surroundings, and this coupled with the heat generated from friction, begins to melt the plate. As this material is less dense than the surrounding asthenosphere, it begins to rise towards the surface as plutons of magma. Eventually these reach the surface and form complex, composite and explosive volcanoes. If the eruptions take place offshore, a line of volcanic islands forms. These are known as island arcs. The Marianas Islands and Guam form a very good example, running parallel to the Marianas trench.
- **Young fold mountains:** The plates forming continental crust have a much lower density than the underlying layers, so there is not much subduction when such plates meet. As such plates move towards each other, their edges and the sediments between them are forced up into fold mountains. As there is little subduction, there is no volcanic activity. Material is also forced downwards to form deep mountain roots. At present, the Indo-Australian plate is moving northwards into the Eurasian plate. The previous intervening ocean, known as the Sea of Tethys, has had its sediments forced upward in large overfolds to form the Himalayas, a process which is continuing today. Sediments that have accumulated on the continental shelf, along the edge of a plate, can also be uplifted as the plate edges buckle during the subduction of a denser oceanic plate. The Andes, running down the area where the Nazca plate is being subducted beneath the lighter South American plate, are a good example. Molten magma rising from depth also gives rise to a number of explosive volcanoes in the Andes.
- **Volcanoes.**

Alternative ideas on plate movements

For a long time, the generally accepted view of plate motion was that it was brought about by convection currents. It is now believed that forces behind plate motion are not as simple as would be explained entirely by convection currents.

Forces that act on plates can also be generated at their boundaries. These forces push from the ridge, drag the plates down at the trenches, or act along the sides of plates at conservative boundaries.

At constructive boundaries, the upwelling of hot material at ocean ridges generates a buoyancy effect that produces the ocean ridge which stands some 2–3 km above the ocean floor. Here, oceanic plates experience a force that acts away from the ridge, known as **ridge push,** which is the result of gravity acting down the slope of the ridge. The occurrence of shallow earthquakes, resulting from the repeated tearing apart of the newly-formed crust, indicates that there is also some frictional resistance to this force. Some experts do not like the term 'ridge push' and would prefer to name the process **gravitational sliding.**

The situation at destructive plate boundaries is more complex. A major component is the downward gravitational force acting on the cold and dense descending plate as it sinks into the mantle. This gravity-generated force pulls the whole oceanic plate down as a result of the negative buoyancy of the plate. This force is known as **slab pull.** Due to the pushing of the subducting plate against the overriding plate, there is frictional resistance that gives rise to both shallow and deep earthquakes in subduction zones.

Each plate, however, moves at its own rate, which suggests that the relative importance of the driving and retarding forces must vary from plate to plate. It therefore seems unlikely that any single agent is the sole driving mechanism of plate motion; plates are therefore controlled by a combination of forces.

5.3 Volcanic hazards

Key terms

Lahars – These are formed by volcanic ash mixing with water and flowing downhill. Essentially they are volcanic mudflows. In the Philippines, if a typhoon occurs after a volcanic eruption, then lahars can result.

Lava – Molten rock (magma) flowing onto the surface. Acid lava solidifies very quickly, but basic lava (basaltic) tends to flow some distance before solidifying (for example, on the Hawaiian Islands).

Lithosphere – The layer of the Earth which consists of the crust and the upper section of the mantle. It is this layer which is split into a number of tectonic plates.

Pyroclastic flows – Also known as **nuées ardentes**, formed from a mixture of hot gas (over 800°C) and tephra. After ejection from the volcano they can flow down the sides of a mountain at speeds of over 700 km per hour. Some volcanologists apply the term nuées ardentes when the cloud is formed only from hot gas.

Tephra – The solid matter ejected by a volcano into the air. It ranges from volcanic bombs (large) to ash (fine).

Nature and distribution

Most volcanic activity is associated with plate tectonic processes and is mainly located along plate margins.

Volcanic activity is therefore found at the following sites:

● Along **oceanic ridges** where plates are moving apart and magma is forcing its way to the surface, cooling and forming new crust. As the plates move further apart, this new crust is carried away from the ridge (**sea-floor spreading**). The best example is the mid-Atlantic ridge where Iceland represents a large area formed by volcanic activity (Figure 5.10). Volcanoes formed here have fairly gentle sides because of the low viscosity of the basaltic lava. Eruptions are frequent but relatively gentle (effusive). (See material on Eyjafjallajökull, Iceland.)

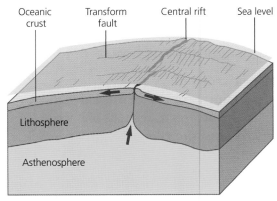

Figure 5.10 Cross section of the mid-Atlantic ridge

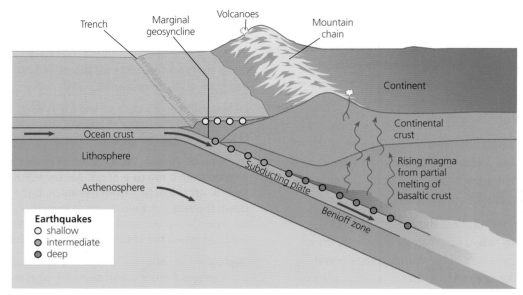

Figure 5.11 Cross section of oceanic/continental plate convergence at a destructive margin

● On or near **subduction zones**: The line of volcanoes, or 'ring of fire' that surrounds the Pacific Ocean is associated with plate subduction. The deeper the oceanic plate descends, the hotter the surroundings become. This, together with the heat generated from friction, begins to melt the oceanic plate into magma in a part of the subduction zone known as the Benioff zone. As it is less dense than the surrounding material, this molten magma begins to rise as plutons of magma. Eventually, these reach the surface and form volcanoes (Figure 5.11). The andesitic lava, which has a viscous nature (flows less easily), creates complex, composite and explosive volcanoes (compared with the basaltic emissions on constructive margins). If the eruptions take place offshore, a line of volcanic islands known as an **island arc** can appear, for example, the West Indies. Island arcs can also occur when oceanic plates meet and one is forced under the other, with the consequent subduction processes taking place. On the western side of the Pacific Ocean, the Pacific plate is being dragged beneath the smaller Philippines plate. A line of volcanic islands including Guam and the Marianas has been formed from magma upwelling from the Benioff zone.

● Associated with **rift valleys**: At constructive margins in continental areas such as east Africa, the brittle crust fractures as sections of it move apart. Areas of crust drop down between parallel faults to form rift valleys (Figure 5.12). The crust here is much thinner than in neighbouring areas, suggesting that tension in the crust is causing the plate to thin as it starts to split. Through this

thinning crust, magma forces its way to the surface to form volcanoes and there are a number of active volcanoes situated along the rift in east Africa. (See the material on Mt Nyiragongo.)

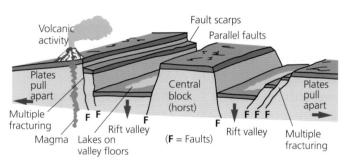

Figure 5.12 Cross section of a rift valley

● Over **hot spots**: In certain places, a concentration of radioactive elements below the crust causes a hot spot to develop. From this, a plume of magma rises to eat into the plate above. When this lava breaks through to the surface, active volcanoes form above the spot. The basaltic lava flows slowly and forms huge but (relative to the area covered) flattish volcanoes, sometimes referred to as 'shield' volcanoes. Figure 5.13 shows the example of the Hawaiian Islands in the Pacific Ocean. The hot spot is stationary, so as the Pacific plate moves over it, a line of volcanoes is created. The one above the hot spot is active and the remainder form a chain of islands with extinct volcanoes. The oldest volcanoes have put so much pressure on the crust that subsidence has occurred. This, together with marine erosion, has reduced these old volcanoes to seamounts below the level of the ocean. Figure 5.13 shows the line of the Hawaiian islands and their ages. From this evidence it is clear that the Pacific plate is moving northwest (see also Figure 5.14).

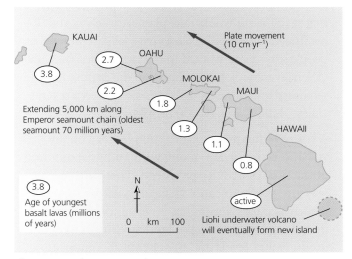

Figure 5.13 The Hawaiian hot spot

Skills focus

There is no volcanic activity taking place in the British Isles at present but this was not always the case. Consult the geological record to find evidence of past volcanic activity and map this on an outline of the British Isles.

The magnitude and frequency of events

Volcanic eruptions show an enormous variation. On the one hand they can involve the tranquil effusion of a sluggish lava; on the other they can take the form of huge explosions that eject gas and dust and can blot out the sun for many years, bringing on global climatic change. The main method of measurement of **magnitude** has been the **volcanic explosivity index (VEI)**, a logarithmic scale running from 0 to 8. Quiet lava-producing eruptions score 0–1 on the index whereas the eruption of Mt St Helens (USA) (1980) scored 5 and Pinatubo (Philippines) (1990) was rated at 6 on the scale. Colossal eruptions on the scale of 7–8 occur very infrequently. Volcanologists estimate that the last eruption at 7 on the scale was Tambora (Indonesia) in 1815 and to find the last eruption which was graded 8, we have to go back around 73,000 years to Toba (Indonesia). Critics of the VEI point out that it does not take into account gas emissions or the atmospheric/climatic impact of eruptions.

To determine the **frequency** of eruption of any volcano, its previous history of activity can be interpreted by volcanologists using the deposits associated with the volcano itself and those within the wider region it can effect.

The impacts of volcanic activity

A volcanic event can produce a variety of effects, the impact of which can range from the area immediately around the volcano to the whole planet. The impact presented by a volcanic eruption can be categorised into primary and secondary effects. **Primary effects** are brought about by material ejected from the volcano and described below:

● **Tephra:** Solid material of varying grain size ranging from **volcanic bombs** to **ash**, all ejected into the atmosphere.

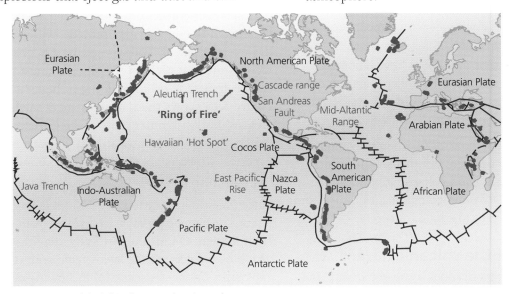

Figure 5.14 Global distribution of active volcanoes

- **Pyroclastic flows (also known as nuées ardentes):** Very hot (over 800°C), gas charged, high-velocity flows made up of a mixture of gas and tephra. These usually hug the ground and flow down the sides of the volcano with speeds of up to 700 km per hour. The Roman city of Pompeii (Italy) was destroyed in 79 AD by such flows from Mt Vesuvius.
- **Lava flows.**
- **Volcanic gases:** These include carbon dioxide, carbon monoxide, hydrogen sulphide, sulphur dioxide and chlorine. In 1986, carbon dioxide emissions from the lake in the crater of Nyos (Cameroon) killed 1,700 people.

Secondary effects:

- **Lahars (volcanic mud flows):** Melted snow and ice as a result of the eruption combined with volcanic ash forms mud flows that can move down the course of river valleys at high speeds. In 1985, a lahar destroyed the Colombian town of Armero after an eruption of the volcano Nevado del Ruiz. Only a quarter of the 28,700 population survived.
- **Flooding:** When an eruption melts glaciers and ice caps, serious flooding can result. This happened in Iceland in 1996 when the Grimsvotn volcano erupted.
- **Volcanic landslides.**
- **Tsunamis:** Sea waves generated by violent volcanic eruptions such as those formed after the eruption of Krakatoa (Indonesia) in 1883. Tsunamis from this eruption are estimated to have killed 36,000 people. (For details on tsunamis see the section on seismic hazards, pages 203–204)
- **Acid rain:** Volcanoes emit gases which include sulphur. When this combines with atmospheric moisture, acid rain results.
- **Climatic change:** The ejection of huge amounts of volcanic debris into the atmosphere can reduce global temperatures and is believed to have been an agent in past climatic change.

Volcanic events become hazardous when they impact upon people and the built environment, killing and injuring people, burying and collapsing buildings, destroying infrastructure and bringing agricultural and other economic activities to a halt.

Management of the volcanic hazard

- **Prediction:** Locating volcanoes is straightforward, but it is very difficult to predict when activity will take place. The already mentioned Colombian volcano, Nevado del Ruiz, came to life in late 1984 with small-scale activity but volcanologists, although they knew the danger a major eruption would pose to the surrounding area, were unable to predict when that major event would take place. As the volcano continued with small-scale activity for several months, people stayed and worked in the area. When the final violent event took place, almost all the population had remained and devastating lahars swept down several valleys, killing over 20,000 people.

A study of the previous eruption history of any volcano is important, along with an understanding of the type of activity produced. There are several ways in which volcanologists are seeking to give a fairly accurate timing for an eruption. These include the monitoring of land swelling, changes in groundwater levels and the chemical composition of groundwater and gas emissions. It is also possible to monitor seismic activity looking for the shock waves that result from magma moving towards the surface, expanding cracks and breaking through other areas of rock.

- **Protection:** In this case, protection usually means preparing for the event. Monitoring a volcano, as shown above, will possibly identify a time when the area under threat should be evacuated. The governments of several countries with volcanoes, such as New Zealand, have made risk assessments and from them produced a series of alert levels in order to warn the public of the threat.

Geological studies of the nature and extent of deposits from former eruptions and associated lahars and floods may also provide evidence for hazard assessment. Following this, it is possible to identify areas at greatest risk and land use planning can be applied to avoid building in high risk areas. Figure 5.15 shows the risk assessment made for the area around Mt Rainier (part of the Cascade Range in the USA), one of the most studied volcanoes in North America, which is not surprising as 3.5 million people live and work in proximity to it.

Once the lava has started to flow and is fairly viscous, it may be possible to divert it from the built environment. On Mt Etna in Sicily, digging trenches, dropping blocks into the lava stream and using explosives has been successful in slowing down the flow and, in some cases, diverting it. In 1973, the inhabitants of Haeimaey (Iceland) were able to divert a lava flow by pouring seawater on the front so it would quickly solidify. In parts of the Hawaiian Islands, barriers have been built across valleys to protect settlements from lava flows and lahars.

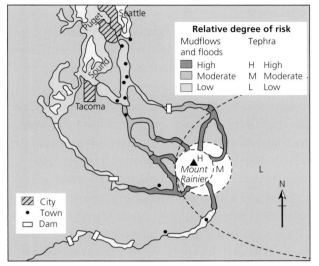

Figure 5.15 Risk assessment of Mt Rainier, Cascade Range, USA

Many devastated areas in poorer countries require aid for considerable periods of time as volcanic events can be prolonged and very damaging to the local economy. Such aid is needed for monitoring, evacuation, emergency shelters and food, long-term resettlement of the population and the restoration of the area's economic base and infrastructure.

Responses can also be seen as both short and long term. Immediate responses are those which take place just before or after the volcanic event such as warnings, evacuation and attempts to stop lava flows. In the long term, responses include the monitoring of a volcano, research into finding new methods of prediction and even building barriers in anticipation of the direction of lava flows and lahars.

Volcanic events

The eruption of Mt Nyiragongo, Congo, January 2002

Mount Nyiragongo lies in the Virunga Mountains in the Democratic Republic of the Congo (Figure 5.16) and it is associated with the African Rift Valley.

The main crater of the volcano is 250 m deep, two kilometres wide and often contains a lava lake. Since records of the area began in the nineteenth

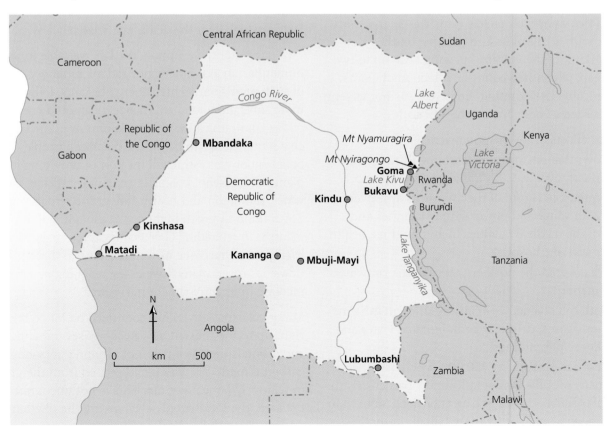

Figure 5.16 The location of Mt Nyiragongo

century, the volcano has erupted over 30 times, and together with its neighbouring volcano, Mount Nyamuragira, it is responsible for around 40 per cent of Africa's recorded volcanic eruptions. Lava erupted by Nyiragongo is very fluid and has been known to move downhill at speeds of over 90 kilometres per hour.

- **Risk and vulnerability:** The Goma area is part of a fertile agricultural region. The combination of altitude (1,500 m), extended growing season and the fertile volcanic soils has encouraged large numbers of people to settle there, despite the presence of the volcano. Since 1882, there have been 34 eruptions, although each period of activity does not always involve lava flows. Because the lava is slow flowing, the volcano does not pose a huge threat to people, but the flows can overwhelm agricultural areas and large swathes of the built environment. In recent times, though, people have also become very vulnerable to the emissions of poisonous carbon dioxide gas seeping from the ground around the volcano.

Although there had been some seismic activity in the area, the eruption in January 2002 was unexpected, although warnings of the lava flows enabled most people to flee from their effects. This was a large eruption, with a 13 km fissure opening on the southern flank of the volcano spewing lava, up to two metres deep, which flowed in the direction of Goma and Lake Kivu. The eruption, however, was not violent, being recorded as only one on the VEI scale.

The major effects of the eruption were:
- lava flows that destroyed at least one third of Goma, a town with over 200,000 inhabitants (Figure 5.17)
- the commercial centre of the town was destroyed along with water and power supply facilities and many of the medical facilities including three health centres and one hospital
- the lava covered the northern third of the runway at Goma airport
- the death toll reached 147 (largely through inhaling poisonous gases, drinking contaminated water and the collapse of buildings)
- it was estimated that over 350,000 people fled the area, many over the border to Rwanda, particularly the town of Gisenyi. This caused an enormous problem in providing food and shelter in this small country

- sulphurous lava entered Lake Kivu and poisoned the water which was a major source of drinking water in the area
- there was a fear that the rise in temperature of Lake Kivu, due to the lava, could allow toxic gases (carbon dioxide and methane) to be released from the lake bed
- several earthquakes accompanied the eruption, one measuring over 5 on the Richter scale. These tremors were strong enough to cause structural damage to some buildings in the area
- thousands of people required medical attention, first from the effects of smoke and fumes from the lava, which caused eye irritation and respiratory problems, and secondly from such complaints as dysentery linked to the drinking of contaminated water
- there was a vast amount of looting from abandoned homes and commercial properties. (Many people were killed when a petrol store, which they were attempting to loot, exploded. It is believed that the looters were hoping to sell the petrol in order to buy food.)

The initial response to the eruption was that the authorities were able to issue a 'Red Alert' for Goma and the surrounding area, which enabled a full evacuation to take place. This prompt response was one of the factors which kept the death rate relatively low.

Two days after the eruption, the UN was able to ferry in humanitarian aid. Emergency rations initially consisted of such provisions as high energy biscuits, followed by more substantial food aid (maize, beans and cooking oil) as communications began to improve. The UN also set up camps to house the displaced people. The organisation has estimated the cost of providing food, blankets, household utensils, temporary shelter, clean water, sanitation and health care to the refugees at $15 million. A much higher cost, though, will be incurred in rebuilding Goma's infrastructure, homes and livelihoods. One of the main problems was that the lava flows destroyed many businesses, resulting in a massive increase in unemployment in the area.

Minor volcanic activity has continued on Mt Nyiragongo right up to the present day. Recently several children and animals died as a result of inhaling poisonous carbon dioxide gas which had seeped from the ground in relatively large amounts. This is not uncommon in the area; the local people even giving the phenomenon a name: 'Mazuku'.

Key question

Study Figure 5.1 (page 185), the model of vulnerability. Using the information contained in this study of a volcanic event, replace the generalities of the model with the real details of the Nyiragongo eruption. What does the completed figure tell you about the extent of the risk to people and the environment (and their vulnerability) in living in the area around Mt Nyiragongo?

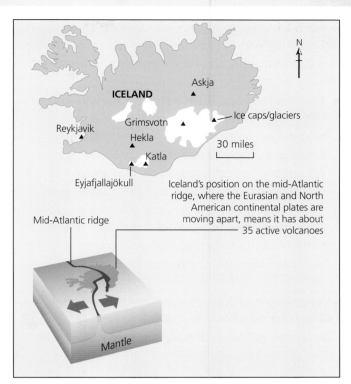

Figure 5.18 Iceland's major volcanoes and its position on the mid-Atlantic ridge

Figure 5.17 The impact of the eruption of Mt Nyiragongo, Goma, Congo in 2002

The eruption of Eyjafjallajökull (Iceland), April 2010

Iceland's position on the mid-Atlantic ridge means that it has a number of active volcanoes, the major ones being shown on Figure 5.18. In the past, they have proved devastating. Hekla, it has been estimated, has erupted around 20 times since AD 874, pouring out a total of 8 km³ of lava from a line of fissures that stretches over 5 km across the mountain. Many of these volcanoes are covered by ice caps feeding glaciers, which means that when they erupt much ice is melted resulting in flooding and huge ash clouds also develop. As the magma hits the ice it suddenly cools, forming a glass-like material which instantly disintegrates. Explosions of gas within the main vent pulverise the fragments into very tiny particles. As the gas and steam blown from the vent carry this ash into the atmosphere it can rise to great heights where it is caught in high-altitude winds (jet stream). As this ash is carried away from the volcano it disperses and becomes invisible but it can be sucked into the engines of aircraft so endangering their operation.

In early 2010, seismic activity was detected in the area of Eyjafjallajökull. This gave geophysicists evidence that magma was pouring from underneath the crust into the volcano's magma chamber. Minor eruptions followed and then the volcano went quiet for a short period. On 14th April, however, an explosive eruption occurred (4 on the VEI scale) in the top crater, which had the following effects:

- huge amounts of ice were melted causing floods to rush down the nearby rivers, requiring nearly 1,000 people to be evacuated
- as the eruption continued, large quantities of ash poured from the volcano into the higher levels of the atmosphere (Figure 5.19, page 200). Once picked up by the jet stream, this ash cloud was blown towards Europe. Many European countries were forced to shut down their air space; the largest such shut down since World War II. It has been estimated that this cost the airlines £130 million per day for the six days that the airspace was closed. This affected several million people and had knock-on travel effects across the globe.

By the end of 2010, Eyjafjallajökull was considered to be dormant, but volcanologists switched their attention to its neighbour, Katla, as its last eruption in 1918

produced five times the amount of ash as the eruption of Eyjafjallajökull. There have been signs that magma is rising beneath Katla and a major eruption may not be very far away.

● **Risk and vulnerability:** There is very little risk to people's lives from this volcano as so few of them live in this part of Iceland, although some had to be evacuated after the most recent eruption. The real threat comes via the ash cloud produced by such volcanoes. This means that people are vulnerable at some distance away from the volcano, not in terms of their lives, but in their ability to travel, as flights would be severely restricted due to the ash cloud.

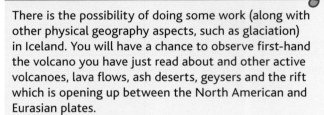

Fieldwork opportunities

There is the possibility of doing some work (along with other physical geography aspects, such as glaciation) in Iceland. You will have a chance to observe first-hand the volcano you have just read about and other active volcanoes, lava flows, ash deserts, geysers and the rift which is opening up between the North American and Eurasian plates.

Figure 5.19 Ash cloud from the eruption of Eyjafjallajökull

Other volcanic events worth studying

Etna (Sicily, Italy) 1991–99

Montserrat (West Indies) 1995–97

Merapi (Indonesia) 2010

Mt Ontake (Japan) 2014

5.4 Seismic hazards

Key terms

Earthquake – As the crust of the Earth is constantly moving, there tends to be a slow build up of stress within the rocks. When this pressure is released, parts of the surface experience, for a short period, an intense shaking motion. This is an earthquake.

Retrofitting – In earthquake-prone areas buildings and other structures can be fitted with devices such as shock absorbers and cross-bracing to make them more earthquake proof.

Tsunami – Giant sea waves generated by shallow-focus underwater earthquakes, violent volcanic eruptions, underwater debris slides and landslides into the sea.

Nature and distribution

As the crust of the Earth is mobile, there tends to be a slow build up of stress within the rocks. When this pressure is suddenly released, parts of the surface experience an intense shaking motion that lasts for just a few seconds. This is an **earthquake**. The point at which this pressure release occurs within the crust is known as the **focus**, and the point immediately above that on the earth's surface is called the **epicentre** (Figure 5.20). The depth of the focus is significant in the effects on the surface and three broad categories of earthquake are recognised:

● **shallow focus (0–70 km deep):** these tend to cause the greatest damage and account for 75 per cent of all the earthquake energy released

● **intermediate focus (70–300 km deep)**

● **deep focus (300–700 km deep)**

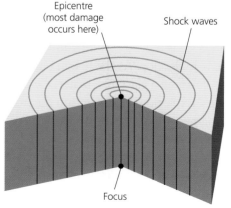

Figure 5.20 The focus and epicentre of an earthquake

The vast majority of earthquakes occur along plate boundaries (Figure 5.21), the most powerful being associated with destructive margins. At conservative margins, the boundary is marked by a fault, movement along which produces the earthquake. Perhaps the most famous of these is the San Andreas Fault in California, which represents the boundary between the North American and Pacific plates. In reality, the San Andreas system consists of a broad complex zone in which there are a number of fractures of the crust (Figure 5.22).

Some earthquakes occur away from plate boundaries and are associated with the reactivation of old fault lines. In September 2002, an earthquake (4.8 Richter) occurred in the UK Midlands. The epicentre was at Dudley, west of Birmingham, and was believed to have been caused by movements on an old fault line called the Malvern lineament.

It has been suggested that human activity could be the cause of some minor earthquakes, through building large reservoirs which puts pressure on the underlying rocks, or subsidence of deep mine

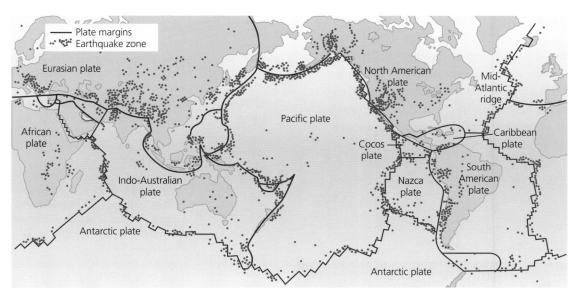

Figure 5.21 Global distribution of earthquakes

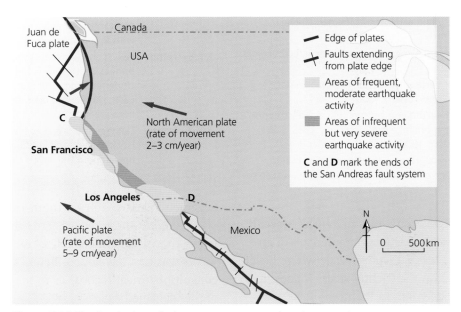

Figure 5.22 The San Andreas fault system: a conservative plate margin

workings. In recent times, people have become increasingly worried about the process of fracking (hydraulic fracturing of rock in order to release gas). In Oklahoma (USA), for example, some areas of the state have experienced a tremendous increase in minor earthquakes since fracking began in 2009.

Magnitude and frequency

The **magnitude** of earthquakes is measured in a variety of ways. The **Richter scale** (Table 5.2) is a logarithmic scale – an event measured at 7 on the scale has an amplitude ten times greater than one measured 6 on the scale. The energy release is proportional to the magnitude, so that for each unit increase on the scale, the energy released increases by approximately 30 times. Some geologists have been unhappy with the fundamentals of the Richter scale for some time and now it is more usual to use the **moment magnitude scale (MMS)** which identifies energy release. Its scale is the same as that of Richter, that is, 1–10. The largest earthquake ever recorded was estimated to be the Valdivia (Chile) event in 1960 (9.5 MMS).

Table 5.2 The Richter scale

Number (Logarithmic)	Effects
1–3	Normally only detected by seismographs, not felt
4	Faint tremor causing little damage
5	Widely felt, some structural damage near epicentre
6	Distinct shaking, less well-constructed buildings collapse.
7	Major earthquake causing serious damage (e.g. Kobe 1995, Turkey 1999)
8	Great earthquake causing massive destruction and loss of life (e.g. Mexico City 1985, San Francisco 1906)
9–10	Very rare great earthquake causing major damage over a large region and ground seen to shake

Another scale which has come into use is the **Mercalli scale** (Table 5.3) which measures the intensity of an event and its impact. It is a 12-point scale where Level 1 is approximate to 2 on the Richter scale. It goes up to Level XII (approximately 8.5 on Richter).

Table 5.3 The Mercalli scale

I. Not felt	Not felt except by a very few under especially favourable conditions
II. Weak	Felt only by a few persons at rest, especially on upper floors of buildings
III. Weak	Felt quite noticeably by persons indoors, especially on upper floors of buildings. Many people do not recognise it as an earthquake. Standing motor cars may rock slightly. Vibrations similar to the passing of a truck. Duration estimated
IV. Light	Felt indoors by many, outdoors by few during the day. At night, some awakened. Dishes, windows, doors disturbed; walls make cracking sound. Sensation like heavy truck striking building. Standing motor cars rocked noticeably
V. Moderate	Felt by nearly everyone; many awakened. Some dishes, windows broken. Unstable objects overturned. Pendulum clocks may stop
VI. Strong	Felt by all, many frightened. Some heavy furniture moved; a few instances of fallen plaster. Damage slight
VII. Very Strong	Damage negligible in buildings of good design and construction; slight to moderate in well-built ordinary structures; considerable damage in poorly built or badly designed structures; some chimneys broken
VIII. Severe	Damage slight in specially designed structures; considerable damage in ordinary substantial buildings, with partial collapse. Damage great in poorly built structures. Fall of chimneys, factory stacks, columns, monuments, walls. Heavy furniture overturned
IX. Violent	Damage considerable in specially designed structures; well-designed frame structures thrown out of plumb. Damage great in substantial buildings, with partial collapse. Buildings shifted off foundations
X. Extreme	Some well-built wooden structures destroyed; most masonry and frame structures destroyed with foundations. Rails bent
XII. Extreme	Few, if any (masonry) structures remain standing. Bridges destroyed. Broad fissures in ground. Underground pipelines completely out of service. Earth slumps and land slips in soft ground. Rails bent greatly
XIII. Extreme	Damage total. Waves seen on ground surfaces. Lines of sight and level distorted. Objects thrown upward into the air

Source: USGS

Seismic records enable earthquake frequency to be observed, but these records only date back to 1848 when an instrument capable of recording seismic waves was first developed. Before that we have to consult historical records as to the date and the effects of the identified event.

The effects of seismic events

The initial, or **primary** effect (impact), of an earthquake will be **ground shaking**, the severity of which will depend upon the magnitude of the earthquake, its depth, the distance from the epicentre and the local geological conditions. In the Mexico City earthquake of 1985, for example, the seismic waves that devastated the city were amplified several times by the ancient lake sediments upon which the city is built.

Another primary effect is **ground rupture**, which is the visible breaking and displacement of the earth's surface, probably along the line of the fault. Ground rupture poses a major risk for large engineered structures such as dams, bridges and nuclear power stations.

Secondary effects are as follows:

- **soil liquefaction:** when violently shaken, soils with a high water content lose their mechanical strength and start to behave like a fluid
- **landslides/avalanches:** slope failure as a result of ground shaking
- **tsunamis:** giant sea waves generated by shallow-focus underwater earthquakes involving movements of the sea bed, or landslides into the sea (see below)
- **fires:** resulting from broken gas pipes and collapsed electricity transmission systems
- **effects on people and the built environment:** collapsing buildings; destruction of road systems and other forms of communication; destruction of service provision such as water, electricity and gas; flooding; disease; food shortages; disruption to the local economy, either subsistence or commercial. Some of the effects on the human environment are short term; others occur over a longer period and will depend to a large extent on the ability of the area to recover.

Tsunamis

Tsunamis are giant sea waves (tsunami means 'harbour wave' in Japanese) generated by shallow-focus underwater earthquakes (the most common cause), volcanic eruptions, underwater debris slides and large landslides into the sea. Tsunamis have a very long wavelength (sometimes over 100 km) and a low wave height (under one metre) in the open ocean, and they travel quickly at speeds of over 700 km per hour (some tsunamis take less than a day to cross the Pacific Ocean) but, when reaching shallow water bordering land, increase rapidly in height.

Quite often, the first warning given to coastal populations is the wave trough in front of the tsunami which results in a reduction in sea level, known as a drawdown. Behind this comes the tsunami itself, which can reach heights in excess of 25 m. The event usually consists of a number of waves, the largest not necessarily being the first. When a tsunami reaches land, its effects will depend upon:

- the height of the waves and the distance they have travelled
- the length of the event (at source)
- the extent to which warnings could be given
- coastal physical geography, both offshore and on the coastal area
- coastal land use and population density.

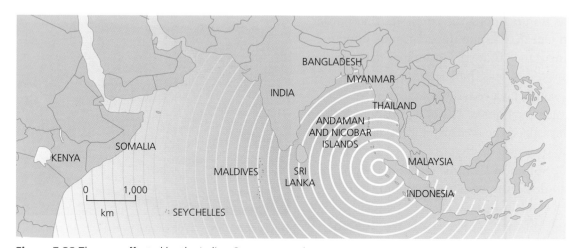

Figure 5.23 The area affected by the Indian Ocean tsunami

The wave will wash boats and wooden coastal structures inland, and the backwash may carry them out to sea. People are drowned or injured by the tsunami as both the water itself and the debris that it contains are hazards. Buildings, roads, bridges, harbour structures, trees and even soil are washed away, most tsunamis having an effect at least 500–600 m inland, depending upon the coastal geography. Tsunamis generated by the explosion of the volcano Krakatoa in 1883 have been estimated to have drowned over 35,000 people and produced waves that travelled around the world, the highest being over 40 m.

Around 90 per cent of all tsunamis are generated within the Pacific Basin, associated with the tectonic activity taking place around its edges. Most are generated at convergent plate boundaries where subduction is taking place, particularly on the western side of the Pacific and the bordering eastern side of the Indian Ocean (25 per cent of all events). Some recent events which have taken place in that area are described below:

● December 2004, **east Indian Ocean off Sumatra, 'the Boxing Day' tsunami:** Generated by a powerful magnitude 9.0 earthquake some 25 km below the Indian Ocean floor off northwest Sumatra (one of the biggest earthquakes ever recorded). The tsunamis which spread across the Indian Ocean killed an estimated 300,000 people in bordering countries and devastated hundreds of communities (Figure 5.23, page 203). Extensive damage was caused to coastal communications, particularly bridges and railway lines and damage to local economies left people unable to feed themselves. A positive result was that a warning system was set up among the countries that border the Indian Ocean. Despite the enormous human cost, the insurance industry estimated that the disaster could cost less than $5 billion.

● July 2006, **south Java coast:** Generated by a 7.7 magnitude earthquake 112 miles offshore, a tsunami devastated the area around Pangandaram resulting in over 600 deaths.

● April 2007, **Solomon Islands:** Generated by a 8.0 magnitude earthquake, the tsunami swept across the islands killing at least 15 people.

● March 2011, off the Pacific coast of the **Tohoku region of Japan** (Figure 5.24) Earthquake occurred 70 km offshore and of a magnitude of 9.0: the most powerful earthquake ever recorded in Japan (it was estimated that it moved the main island of Japan, Honshu, 2.4 m east). Tsunamis generated reached over 40 m in height and in some places penetrated over 10 km inland. These tsunamis devastated part of Japan's Pacific coast, causing nearly 16,000 deaths with over 6,000 people injured and 2,500 missing. It was estimated that over 300,000 people were also displaced from their homes. The built environment suffered with 127,000 buildings destroyed, 277,000 'half collapsed' and 750,000 partially damaged. In addition, the Fukushima nuclear power plant was seriously affected resulting in the evacuation of people living within 20 km of the facility. The assessed cost of the damage by insurers was over $30 billion while it has also been estimated that the economic cost to Japan was $235 billion. Tsunami waves ran right across the Pacific to North and South America; in Chile the tsunami created waves 2 m high.

The geological record indicates that huge tsunamis have affected areas such as the Mediterranean Basin (for example, the Santorini eruption around 1450 BC) and the North Sea area (for example, the Storegga slide, around 7250 BP) resulting from huge submarine debris slides off Norway which produced tsunamis of over 6 m in height in Scotland and other areas bordering the North Sea. It is believed that these tsunamis continued across the Atlantic to hit the coastlines of Spitsbergen, Iceland and Greenland.

Figure 5.24 Devastation caused by the Tohoku tsunami, Japan 2011

Managing the seismic hazard

- **Prediction:** The prediction of earthquakes is very difficult. Regions at risk can be identified through plate tectonics, but attempts to predict a few hours before the event are questionable. Such attempts are based upon monitoring groundwater levels, release of radon gas and unusual animal behaviour. Fault lines such as the San Andreas can be monitored and the local magnetic fields measured. Areas can also be mapped on the basis of geological information and studies made into ground stability in order to predict the impact of an earthquake and to produce a hazard zone map that can be acted upon by local and even national planners.

  Close studies of fault lines can sometimes indicate the point along the fault where the next earthquake could be due. A study of the pattern of the events along the San Andreas fault between 1969 and 1988 indicated the existence of a **'seismic gap'** in the area of Loma Prieta (that is, the area had not had any real seismic activity for the past twenty years). This area suffered an earthquake in October 1989 that measured 6.9 MMS and was the worst to hit the San Francisco region since 1906. In total, 63 people died as a result of the event with over 3,700 seriously injured. Because of the seismic survey, this event was not entirely unexpected, but like all earthquakes, it was not possible to predict it precisely. Such a system, however, would not work for events such as the one at Northridge, as this took place on an unknown fault line.

- **Prevention:** Trying to prevent an earthquake is thought by almost all to be impossible. This, however, has not stopped studies into the feasibility of schemes to keep plates sliding past each other, rather than 'sticking' and then releasing, which is the main cause of earthquakes. Suggestions so far as to lubricating this movement have focused on water

  and oil. Some people have even gone as far as to suggest nuclear explosions at depth!

- **Protection:** Since earthquakes strike suddenly, violently and without warning, it is essential that everyone from civil authorities to individuals are prepared. In the USA, the Federal Emergency Management Agency's (FEMA) programme has the following objectives:
  - to promote understanding of earthquakes and their effects
  - to work to better identify earthquake risk
  - to improve earthquake-resistant design and construction techniques
  - to encourage the use of earthquake-safe policies and planning practices.

Protection therefore means preparing for the event by modifying the human and built environments in order to decrease vulnerability. It also includes attempts to modify the loss by insurance and aid. Some of the ways are described below:

- **Hazard-resistant structures:** Buildings can be designed to be aseismic, in that they can be earthquake resistant. There are three main ways in which this can be achieved:
  - By putting a large concrete weight on top of the building which will move, with the aid of a computer programme, in the opposite direction to the force of the earthquake in order to counteract stress.
  - Putting large rubber shock absorbers in the foundations which will allow some movement of the building.
  - By adding cross-bracing to the structure to hold it together better when it shakes.

  Older buildings and structures, such as elevated motorways, can be **retrofitted** with such devices to make them more earthquake proof. A comparison between the 1989 Loma Prieta earthquake (6.9 MMS) and the 1988 event in Armenia (6.8 MMS) shows the effects of different types of buildings. The greater earthquake-proof buildings of California resulted in 63 deaths, whereas in Armenia, over 25,000 people died, many inside buildings that collapsed as a result of soft foundations and no earthquake-proofing features. In the town of Leninakan, for example, over 90 per cent of more modern 9–12 storey pre-cast concrete frame buildings were destroyed.

- **Education:** For many areas, this is the main way that loss of life can be minimised. Instructions are issued by the authorities in how to prepare for such events by securing homes, appliances and heavy furniture, and assembling 'earthquake kits'. Children have earthquake drills at school as do people in offices and factories. Government offices and many companies in Japan observe Disaster Prevention Day (1 September) which marks the anniversary of the Tokyo (Kwanto Plain) earthquake in 1923. Following the Loma Prieta event (1989), the American Red Cross issued a list of supplies that people should keep at hand in case of an earthquake. These include water (at least a three-day supply for all persons in the house, and pets!); a whole range of foodstuffs (particularly canned and high energy foods); clothing and bedding; first aid kit; and tools and supplies (to include radio, torch batteries, can opener, matches, toilet paper, small fire extinguisher, pliers and aluminium foil).

- **Fire prevention:** 'Smart meters' have been developed that can cut off the gas if an earthquake of sufficient magnitude occurs. In Tokyo, the gas company has a network that transmits seismic information to a computer which then informs employees where to switch off major pipelines, so reducing the number of fires.

- **Emergency services:** These need careful organisation and planning. Heavy lifting gear needs to be available and many people should be given first aid training, as it could be some time after the event that trained medical personnel arrive. Much of the preparation in California involves the establishment of computer programs that will identify which areas the emergency services should be sent to first.

- **Land-use planning:** The most hazardous areas in the event of an earthquake can be identified and then regulated in terms of land use. Certain types of buildings should be put in areas of low risk, such as schools and hospitals. It is also important to have sufficient open space, as this forms a safe area away from fires and aftershock damage to buildings.

- **Insurance:** In richer areas, people are urged to take out insurance to cover their losses, the only problem being that for individuals, this is very expensive. In the Kobe earthquake in Japan in 1995, for example, only seven per cent of the people were covered by earthquake insurance.

- **Aid**: Most aid to poorer countries has generally been to help in the few days after the event, providing medical services, tents, water purification equipment, search and rescue equipment, etc. Aid over the longer term is much more problematical; it is something which is needed for the reconstruction of the built environment and redevelopment of the economy.

- **Tsunami protection:** Tsunamis cannot be entirely predicted, even if the magnitude and location of an earthquake is known. Certain automated systems can be installed to give warnings, the best of which uses bottom pressure sensors, attached to buoys, which constantly measure the pressure of the overlying water column. Regions with a high tsunami risk use warning systems (such as a klaxon) to warn the population before the wave reaches land. The Pacific Warning System is based on Hawaii. It monitors earthquake activity and issues warnings to countries around the Pacific edge if tsunamis are likely. Some countries have built prevention walls up to 12 m in height. These have not proved very effective, as large tsunamis are likely to overwhelm them.

Seismic events

Haiti (West Indies), January 2010

In January 2010 an earthquake of magnitude 7.0 MMS and depth of 13 km, struck the Caribbean country of Haiti, the poorest country in the western hemisphere. The epicentre was located 25 km west of the capital, Port-au-Prince (Figure 5.25). Following the initial event, the area recorded at least 50 aftershocks measuring 4.5 or greater.

- **Risk and vulnerability:** The risk from physical hazards is high in Haiti. It lies at the junction of the North American and Caribbean plates and between two fault zones: the Septentrional Zone to the north and the Enriquillo-Plantain system that runs directly beneath Port-au-Prince. The 2010 earthquake was the seventh major event recorded since observations began in the 1550s. Earthquakes also occur in the neighbouring country of the Dominican Republic; a tsunami generated by an earthquake there in 1946 killed 1,800 people in Haiti and injured many others. People are also at risk from tropical cyclones, with frequent flooding and widespread damage. Several storms in 2008 killed over 800 people.

People in Haiti are also very vulnerable when hazard events occur. There has been a long history of national debt, extreme poverty and poor housing conditions that tend to exacerbate the death toll as the effects often last long after the earthquake (often several years).

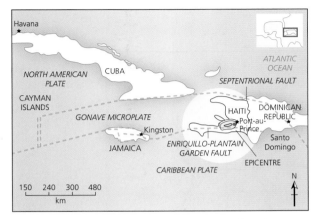

Figure 5.25 The location of the 2010 Haiti earthquake

Major effects of the event:

● The estimated death toll varied from 100,000 to 316,000. The Haitian government was accused of inflating the higher figure in order to gain more aid from the international community. More reliable estimates from outside agencies put the figure around 160,000, with at least 300,000 people injured.

● An estimated 2.3 million people were made homeless.

● The earthquake had a devastating effect on the built environment (Figure 5.26, page 208). It was estimated that 250,000 residences and 30,000 commercial properties were either destroyed or needed to be demolished. One of the major reasons for this was that construction standards were low in Haiti and the country had no building codes. Many government and public buildings were damaged or destroyed including the Palace of Justice, the National Assembly, the Supreme Court, Port-au-Prince Cathedral, the main art museum and the National Palace. The main prison was also destroyed, leading to the escape of around 4,000 inmates.

● Vital infrastructure necessary to respond to the event was severely damaged or destroyed. This included major roads (blocked by falling debris or cracked), the control tower at the airport and the harbour.

● All the hospitals in Port-au-Prince were destroyed and many medical facilities in other parts of the country were seriously damaged.

● There was also considerable damage to the communications infrastructure such as the public telephone service and also to cell phone masts.

● The education system collapsed with 1,300 schools destroyed and the university buildings seriously damaged.

● There was a breakdown in law and order in parts of the country with sporadic violence and looting reported.

● The collapse of water supplies, a lack of basic sanitation and thousands of unburied bodies led to a spread of disease. It has been estimated that over 8,000 people died in a cholera outbreak.

A major relief programme was directed towards the disaster, but rescuers, relief/aid agencies and national governments faced extreme difficulties in dealing with the aftermath of the earthquake. The country, for example, had little heavy lifting equipment, so it was difficult to look for survivors in fallen buildings. One immediate response from the US government was to deploy 3,500 soldiers to the country. In the UK, a large search and rescue team with dogs was sent to look for survivors. One difficult task was the removal of dead bodies, the number of which was so enormous that they had to be buried in large communal graves.

All of this meant that recovery for Haiti was an extremely slow process and even three to four years after the event, the effects are still obvious within the country. By 2014, over 170,000 people were still in 'displacement camps' (from a 1.5 million peak), 23 per cent of children were not at primary school, 70 per cent lacked access to electricity and 600,000 were said to be still 'food insecure'. On the positive side, 50 per cent of the debris has been removed, new building codes have been established, part of the country's debt has been written off and most agencies are resolved to make sure that the recovery is sustainable. Overall, it has been estimated that the final cost of the event could be around $8 billion.

Key question

Study Figure 5.1 (page 185), the model of vulnerability. Using the information contained in this study of an earthquake, replace the generalities of the model with the real details of the Haiti event. What does the completed figure tell you about the extent of the risk to people and the environment (and their vulnerability) in living in Port-au-Prince?

Figure 5.26 Devastation caused by the Haiti earthquake, January 2010

Christchurch (South Island, New Zealand), February 2011

In September 2010, the region of Canterbury suffered an earthquake of magnitude 7.1 MMS, the epicentre of which was 40 km west of Christchurch (Figure 5.27). The earthquake of 2011 was of a lower magnitude, 6.3 MMS, but had much greater and devastating effects. Many seismologists regard this event as an aftershock from that of 2010. The epicentre was only 10 km southeast from the centre of Christchurch, which in part accounts for the impact of the event on structures already weakened by the 2010 earthquake. Following on from February 2011, a number of strong aftershocks were recorded (over 5.5 MMS) in the next few months, which further damaged buildings, particularly in the city centre and the eastern suburbs. One estimate put the depth of the earthquake at only 5 km, which would also account for the amount of damage that it caused.

- **Risk and vulnerability:** New Zealand is an earthquake-prone country as it lies on the 'Pacific Ring of Fire', which is geologically active. The country experiences over 20,000 earthquakes every year, only 200 of which can actually be felt. Christchurch does not lie in New Zealand's major earthquake zone, although there were earthquakes in the region in 1869, 1888 and 1901, the effects of which were mainly damage to property (including the spire of Christchurch Cathedral). As the epicentres were outside of Christchurch, the risk to people was minimal as most of the area, apart from the city, was sparsely populated.

The inhabitants of Christchurch did not see themselves as particularly vulnerable. Apart from the central business district, many of the buildings in the city are of single-storey construction and the housing density is low. Also, with its geological position in mind, there are very stringent building regulations in place.

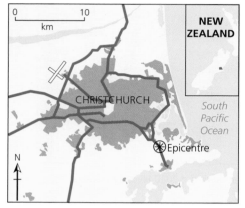

Figure 5.27 Location of the Christchurch earthquake

Effects of the earthquake:

- The death toll was 185, of whom 115 were in the Canterbury Television building which collapsed and then caught fire.
- Nearly 7,000 people were treated for injuries.
- The city centre of Christchurch was devastated. In the centre, at least 1,000 buildings collapsed or were demolished afterwards as they were dangerous. All of this meant that the city centre was cordoned off for some considerable time after the earthquake, with shops and other services having to relocate to outside of the central district.
- The spire of Christchurch Cathedral was brought down.
- Around 10,000 houses in the suburbs had to be demolished. It was also estimated that over 100,000 dwellings had been damaged.
- Harbour facilities at Lyttleton were extensively damaged.
- There was widespread soil liquefaction across the suburbs. Areas affected could not be built upon after the earthquake. Water was released during liquefaction leading to flooding in some suburban areas.

- At the Tasman glacier, some 200 km from the epicentre, ice was dislodged and fell into Lake Tasman causing a tsunami in the lake over 3 m high.
- By 2013, the New Zealand government estimated that the rebuild cost would be NZ$ 40 billion.

While damage occurred to many older buildings, particularly those with unreinforced masonry and built before stringent building codes were introduced, many high rises built within the past thirty years performed well. Some economists have estimated that it might take the New Zealand economy at least 50 years to recover from this event (Figure 5.28).

Key question

The earthquakes in Haiti and at Christchurch were of a similar magnitude but the death toll in each area was completely different. Why was this?

Figure 5.28 Devastation caused by the Christchurch earthquake, New Zealand, 2011

Other seismic events worth studying

Northridge (Los Angeles, USA) 1994

Gujarat (India) 2001

Indian Ocean (off Sumatra) 2004

L'Aquila (Italy) 2009

Tohoku (Japan) 2011

Bohol (Philippines) 2013

Kathmandu (Nepal) 2015

5.5 Tropical cyclone (typhoons, hurricanes) hazards

Key terms

Mitigation – Mitigation strategy is designed to reduce or eliminate risks to people and property from natural hazards. Money spent prior to a hazardous event to reduce the impact of it can result in substantial savings in life and property following the event.

Storm surge – A rapid rise in sea level in which water is piled up against a coastline to a level far in excess of the normal conditions at high tide. Usually produced during the passage of a tropical storm when wind-driven waves pile up water against a coastline combined with the ocean heaving upwards as a result of much lower air pressure.

Nature, distribution and magnitude

Tropical revolving storms (hurricanes, cyclones, typhoons) are intense low-pressure weather systems that develop in the tropics. These violent storms usually measure some 200–700 km in diameter. They begin with an area of low pressure, resulting from surface heating, into which warm air is drawn in a spiralling manner. Such small-scale disturbances can enlarge into tropical depressions with rotating wind systems and these may continue to grow into a much more intense and rapidly rotating system – the tropical revolving storm. It is not entirely clear why tropical storms are triggered into becoming so intense, but we do know that several conditions need to be present:

- An oceanic location with sea temperatures above 27°C – this provides a continuous source of heat in order to maintain rising air currents.
- An ocean depth of at least 70 m – this moisture provides latent heat; rising air causes the moisture to be released by condensation and the continuation of this drives the system.
- A location at least 5° north or south of the Equator in order that the Coriolis force can bring about the maximum rotation of the air (the Coriolis force is weak at the Equator and will stop a circular air flow from developing).
- Low level convergence of air in the lower atmospheric circulation system – winds have to come together near the centre of the low-pressure zone.
- Rapid outflow of air in the upper atmospheric circulation – this pushes away the warm air which has risen close to the centre of the storm.

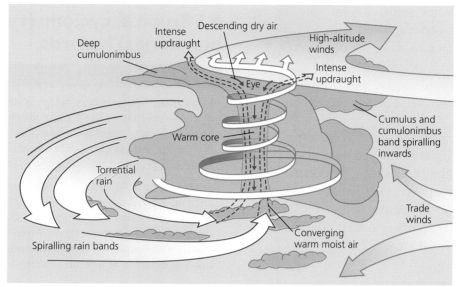

Figure 5.29 The structure of a tropical revolving storm

The tropical revolving exists while there is a supply of latent heat and moisture to provide energy and low frictional drag on the ocean surface. Once the system reaches maturity, a central eye develops. This is an area 10–15 km in diameter in which there are calm conditions, clear skies, higher temperatures and descending air. Wind speeds of more than 300 km/hour have been observed around the eye. Figure 5.29 shows the structure of a typical mature tropical revolving storm. Once the system reaches land or colder waters polewards, it will decline as the source of heat and moisture is removed.

- **Distribution:** Such storms occur between 5° and 20° degrees north and south of the Equator. Once generated, they tend to move westwards and are at their most destructive:
  - in the Caribbean Sea/Gulf of Mexico area where they are known as **hurricanes** (11 per cent of all tropical revolving storms)
  - on the western side of Central America (eastern Pacific) (17 per cent)
  - in the Arabian Sea/Bay of Bengal area where they are known as **cyclones** (8 per cent)
  - off southeast Asia where they are known as **typhoons** (this is the major area for such storms with one-third of all storms every year occurring here)
  - off Madagascar (southeast Africa) (11 per cent)

- off north-western and north-eastern Australia (20 per cent). Off north-western Australia they are given the local name, **willy-willies**.

- **Magnitude:** Tropical revolving storms are measured on the **Saffir-Simpson scale**, a five-point scale based upon central pressure, wind speed, storm surge and damage potential. A scale 5 event, for example, has:
  - central pressure at 920 mb or below
  - wind speeds at 250 km/hr (69 m/sec) or greater
  - storm surge at 5.5 m or greater
  - damage potential that refers to 'complete roof failure of many buildings with major damage to lower floors of all structures lower than three metres above sea level. Evacuation of all residential buildings on low ground within 16–24 km of coast is likely.

The average lifespan of a tropical storm is 7–14 days. Every year, around 80–100 such storms develop around the world, with approximately 80 per cent of these going on to become tropical revolving storms.

The impacts of tropical revolving storms

The vulnerability of people to storm events depends upon a range of factors, both physical and human:
- the intensity of the storm (scale 1–5)
- speed of movement, that is, the length of time over the area

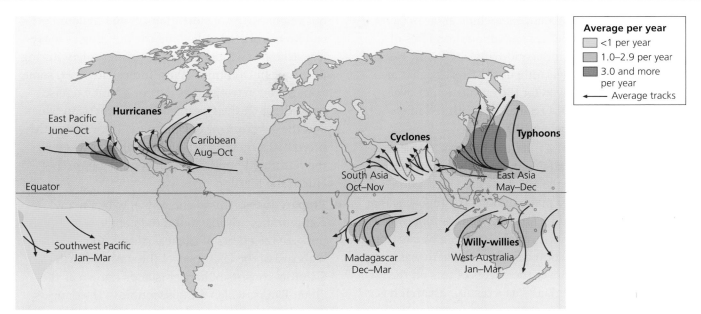

Figure 5.30 Global distribution and seasons of tropical revolving storms

- distance from the sea
- physical geography of the coastal area – width of coastal plain or size of delta, and location of any mountain ranges relative to the coast
- the preparations made by a community
- warnings and community response.

The major impacts that a tropical revolving storm can have on an area are described below:

- **Winds:** They can often exceed 150 km/hr (over 250 km/hour in a scale 5 event). Such high winds can cause structural damage to buildings (even collapse) and roads, bridges, etc. They can bring down electricity transmission lines and devastate agricultural areas. The huge amounts of debris that are flung about are a serious threat to peoples' lives.

- **Heavy rainfall:** It is not unknown for rainfall to exceed 200 to 300 mm, bringing about severe flooding, landslides and mudslides. If there is high relief near the coastal area, rainfall could increase to over 500 mm/day.

- **Storm surges:** High sea levels result when the wind-driven waves pile up and the ocean heaves upwards as a result of the lower atmospheric pressure. These storm surges can have a devastating effect on low-lying coastal areas such as river deltas where the flooding can extend a long way inland. Storm surges cause the majority of deaths in such events and agricultural areas can suffer for a long time

as soil is contaminated by sea water (see details of Hurricane Katrina (2005) on page 213).

Managing the tropical revolving storm hazard

- **Prediction:** The prediction of tropical cyclones depends upon the state of monitoring and warning systems. Weather bureaux, such as the National Hurricane Centre in Florida (USA), are able to access data from geostationary satellites and from both land- and sea-based recording centres. The USA also maintains a round-the-clock surveillance of tropical storms that have the potential to become hurricanes by the use of weather aircraft. Such information is compared with computer models so that a path can be predicted and people warned to **evacuate** an area. It is essential that such warnings are correct, as there is a high economic cost associated with evacuation, and as with all false alarms, people might become complacent and refuse future advice. It has been estimated in the USA that to evacuate coastal areas, the cost is roughly one million dollars per kilometre of coastline due to losses in business and tourism and the provision of protection. As cyclones tend to follow an erratic path, it is not always possible to give more than 12–18 hours warning, which means that in most poorer areas where communications are poor, this is insufficient for a proper evacuation. Some areas, however, have established warning systems that do give the people a chance to take

some precautions. Examples include areas of Central America (such as Belize) and along the shores of the Bay of Bengal (India and Bangladesh). In 1997, a tropical cyclone warning in the Cox's Bazaar area of Bangladesh allowed for the evacuation of over 300,000 people. As a result, the death toll in the disaster was kept below 100.

- **Prevention:** Like other natural hazards, tropical cyclones cannot really be prevented, but there has been research into the effect of cloud seeding in order to cause more precipitation. The theory is that if the cyclone could be forced to release more water over the sea, then this would result in a weakening of the system as it approached land. There was some concern expressed over the effects of this on the global energy system, and as a result, research has not continued.

- **Protection:** As with all natural hazards, this refers to being prepared. Predicting the landfall of a tropical cyclone will enable evacuation to take place, together with the emergency services being put on full alert. If evacuation does take place, then protection units, such as the National Guard in the USA, have to be called in to prevent homes and commercial properties from being looted. People are also made aware of how to strengthen their homes and commercial properties in order to withstand the high winds. **Cyclone/hurricane drills** can be practised along the lines of those that are carried out in earthquake-prone areas. In Florida, for example, there is 'Project Safeside', a hurricane awareness programme that is composed of precautionary drills for use in schools and the Emergency Operations centre of the state. Where storm surges are a problem, **land-use planning** can identify the areas at greatest risk and certain types of development can be limited in such areas. **Sea walls, breakwaters** and **flood barriers** can be built and houses put on stilts. The sea wall that was built to protect Galveston (Texas) from further flooding, after the storm surge of 1900 that killed over 6,000 people, was so expensive that it is unlikely to be repeated elsewhere. In the USA, local authorities are required to address this problem by limiting expenditure on developments in high-risk areas and directing population away from them, as well as having sound plans to reduce evacuation times and for post-disaster redevelopment. In poorer areas, however, the need for land usually

outweighs such considerations. Some structures can be **retrofitted** (adjusting the building to make it resistant to winds) to ensure greater safety during an event. In Dominica (West Indies), in 1994, some homes were retrofitted as part of a joint exercise carried out by the Organization of American States and the government of Dominica. The value of the project was shown in the following year when all the retrofitted buildings withstood the impact of Hurricane Marilyn.

- **Preparedness:** This is one way in which many communities are able to use their abilities in order to mitigate the effects of tropical storms. On the eastern seaboard of the USA, part of the state of North Carolina consists of barrier islands known as the Outer Banks, which are susceptible to the dangers brought by hurricanes. Local communities have got together to put into a practice the Outer Banks **Mitigation** Plan. Its major aims are the saving of lives and money, property and natural resource protection, reducing future vulnerability, speedy recovery from the effects of hurricanes and post-disaster funding. Also in the plan is a great emphasis on making as much information as possible available to the people of the Outer Banks area.

As with other natural phenomena, people in richer areas are urged to take out insurance, whereas in poorer countries it is important that following a cyclone event aid is available, both in the short and long term, as damage to the economic base of the affected area is likely to last for a number of years.

People's ability to resist natural hazards, such as cyclones, very much depends upon a range of political and economic factors. Poorer areas suffer more because land-use planning, warning systems, defences, infrastructure and emergency services are inadequate and this usually results in a higher death toll. Richer countries will have some planning systems in place, sophisticated warning arrangements, better defences and infrastructure and emergency services that are much more comprehensive and better prepared. Countries such as the USA therefore suffer much lower death tolls from individual hurricanes but they have to bear a much greater monetary loss because of the damage to the built environment. It is true to say, however, that the loss of a house in a richer country will probably be covered by insurance, but the simpler dwelling of the inhabitant of a poor country may very well be a greater loss, taking into account the time invested in it and the years of irreplaceable and uninsured savings that it represents. The cyclone that hit Bangladesh in 1991 was responsible for an estimated 131,000 deaths and a monetary cost of $1.7 billion, whereas a hurricane of similar strength that hit Florida in 1992 (Hurricane Andrew) killed only 60 people but caused damage estimated at $20 billion.

Tropical storm events

Hurricane Katrina (southern USA), August 2005

In 2005, Hurricane Katrina was the eleventh named tropical storm, fourth hurricane and first Category 5 hurricane in what was to become one of the most active hurricane seasons ever recorded in the Atlantic area. The storm first developed on 23 August over the Bahamas and was upgraded and named Katrina the next day.

On 28 August, Katrina was upgraded to Category 4 and it became clear that it was heading for the coasts of Mississippi and Louisiana (Figure 5.31). It continued to intensify that day, rapidly becoming a Category 5 hurricane with sustained windspeeds of 280 km/hr and gusts of up to 345 km/hr and a central pressure of 902 mb, making it the fifth most intense Atlantic basin hurricane on record.

Advance warnings were in force by 26 August and the possibility of 'unprecedented cataclysm' was already being considered. President Bush declared a state of emergency on 27 August in Louisiana, Alabama and Mississippi, two days before the hurricane was expected to make landfall. Risk assessments conducted in preparation for such an event had been published; for example the *National Geographic* magazine had run an article less than 12 months before the storm occurred. When it happened, however, the authorities found it difficult to respond to the sheer scale of the disaster. At a news conference on 28 August, shortly after Katrina had been upgraded to a Category 5 storm, the mayor of New Orleans ordered that the city be evacuated.

- **Risk and vulnerability:** As New Orleans lies on the Gulf Coast of the USA, it is in an area which can expect severe tropical storms at certain times of the year. Since 1924 there have been 33 Atlantic hurricanes that have reached Category 5 and Katrina was the only one to hit New Orleans. The city is also at risk because the original flood control measures had resulted in the shrinkage of soils, which meant that over 50 per cent of the land was now below sea level.

This made much of the population vulnerable to a severe storm, of which Katrina at Category 5 was certainly a good example. Much of the levee system that had been built to protect the city had not been constructed to modern standards

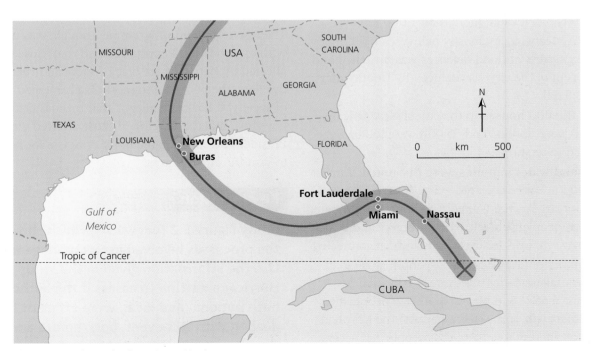

Figure 5.31 The path of Hurricane Katrina

and the replacement work was only 60–90 per cent complete. The situation was severe enough for the magazine *Scientific American* to state that 'New Orleans is a disaster waiting to happen.' As a consequence, levee failure was inevitable given the severity of the storm surge. In the aftermath it was alleged that the population had been made more vulnerable through a lack of leadership, an evacuation plan being drawn up by the state authorities only 19 hours before Katrina's landfall.

Most of the population managed to leave in private cars and on school buses, but some 150,000 people remained, mostly by choice. As a last resort, refugees who had been 'left behind' were encouraged to seek protection in the massive Louisiana Superdome. Basic supplies of food and water were delivered to support 150,000 for three days. When Katrina hit Louisiana on the morning of 29 August, it was accompanied by a massive storm surge, up to 10 m high in places. Although it had weakened to a Category 3 storm as it reached the border of Louisiana and Mississippi, the storm was so intense that its powerful winds and storm surge smashed the entire Mississippi Gulf Coast as it passed through.

Main effects of hurricane Katrina:

- More than 1 million people became **refugees**, displaced from their homes. One month after the storm, refugees from Katrina were registered in all 50 states. Although 75 per cent of evacuees had stayed within 400 km of their homes, tens of thousands had located more than 1,000 km away.
- In New Orleans, a city mostly below sea level, the complex system of flood defences was breached and 80 per cent of the city was deluged by **flood water** (Figure 5.32).
- Of the 180,000 **houses** in the city of New Orleans, 110,000 were flooded and 55,000 were too badly damaged to repair.
- **Power and water supplies were disrupted.** Almost 233,000 km² were declared a disaster zone, an area almost as large as the United Kingdom. An estimated 5 million people were left without power and it took almost two months for everybody to be reconnected.
- The number of **fatalities**, 1,242 people, was far below initial estimates. Of these, 1,035 were in Louisiana and 228 in Mississippi. Few people died in the aftermath of the storm because the USA had the infrastructure to ensure that people had access to clean water, medical care and food supplies.

- Many **oil installations (rigs and refineries)** in the Gulf area were damaged. An immediate effect of this was that the price of oil shot up, affecting people worldwide. For the first time, the price of petrol in the United Kingdom reached £1 per litre.
- The famous French Quarter of New Orleans was severely damaged, thus **reducing tourism revenues.**
- **Looters** ransacked the abandoned homes and shops in New Orleans. This resulted in the deployment of the National Guard, which was given orders to treat looters ruthlessly.
- The **financial cost** of the storm broke all records in the USA, with damage estimated to be in the region of $200 billion.
- Many **businesses** were affected by storm damage. Most were adequately insured, which led to some major **insurance companies** issuing profit warnings to their shareholders in light of the large number of claims which they had to meet.

Government aid was rapidly assigned to help recovery. The US Senate authorised a bill assigning $10.5 billion in aid in the first week of September. On 7 September another $51.8 billion was allocated from Federal funds. Other countries also responded to the disaster, even Afghanistan! The public donated $1.8 billion to the American Red Cross alone; more than the amounts raised for 9/11 and the 2004 Indian Ocean tsunami appeal.

The rescue and aid programme, though, was not without its critics. Reports claimed that those most affected by Katrina were black American urban dwellers, the poorest and most disadvantaged members of society. It has been alleged that the authorities would have responded differently if those affected were white. President Bush also came in for criticism for being slow to visit the affected area.

Key question

Study Figure 5.2 (page 187), which shows the processes involved in risk management. Use the information given in the study of Hurricane Katrina to assess if the authorities, both national and local, were efficient in dealing with this event, both before and after the storm had affected New Orleans.

Figure 5.32 Destruction caused by Hurricane Katrina, New Orleans, 2005

Typhoon Haiyan (Philippines), November 2013

Typhoon Haiyan (also known as Yolanda) was one of the strongest tropical cyclones ever recorded and the deadliest Philippine typhoon in modern history. It originated from an area of low pressure in the Federated States of Micronesia (western Pacific Ocean) on 2 November. Tracking westward it gradually developed into a tropical storm by 4 November (Figure 5.33). It then began a period of rapid intensification that brought it to typhoon intensity by 5 November. By 6 September, the system was assessed as a Category 5 or super-typhoon with wind speeds already reaching over 250 km/hr, with gusts reaching over 300 km/hr.

On 7 November Haiyan reached the Philippines where it caused great destruction in the area around Tacloban, which was struck by the northern eyewall, the most powerful part of the storm. By this time, central pressure had dropped to 895 mb and the highest wind speeds reached 315 km/hr (calculated as sustained for one minute). This meant that Haiyan was the most powerful storm ever to strike land. After crossing the Philippines, Haiyan continued westward towards southeast Asia, eventually reaching the north of Vietnam on 10 November as a severe tropical storm. It eventually dissipated as a tropical depression over southern China on 11 November.

- **Risk and vulnerability:** Tacloban lies in the centre of the Philippines where tropical storms cross from the Pacific Ocean towards southeast Asia, which makes the region the most affected in the world by this type of hazard.

In recent years many Filipinos have moved in large numbers from countryside areas to cities such as Tacloban looking for opportunities to work and make money. This has led to high densities of population and often overcrowded residential areas, which makes the people more vulnerable when events such as Haiyan occur. Many of the people have built poorly constructed houses on the flat land nearest the sea and such locations make people there vulnerable to storm surges and flooding. Others have built houses on steep hillsides, which again makes them vulnerable to the high rainfall of tropical

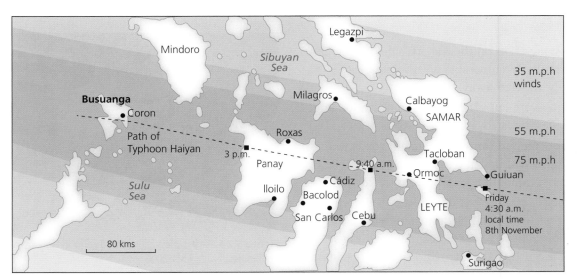

Figure 5.33 The track of Typhoon Haiyan (wider area and central Philippines)

cyclones which, on slopes where vegetation has been removed, will cause landslides that can devastate whole communities. Poor people are especially vulnerable to the aftermath of such events. Unable to provide for themselves, they look to regional and national government agencies who often do not have the resources to cope in such emergencies. This makes people, such as those in Tacloban, dependent on foreign government help and on NGOs, such as charitable organisations.

Apart from the severe winds, Haiyan also caused:

- very heavy rainfall. In one area of the Philippines, nearly 300 mm was recorded, much of which fell in under 12 hours
- huge storm surges, measuring up to six metres, which were responsible for most of the damage and loss of life.

Major effects of the storm:

- There were over 6,000 **fatalities**, with over 1,000 people declared missing. Bodies were being discovered well into 2014. Some estimates have put the death toll as high as 10,000.
- There was widespread **damage to buildings and infrastructure** (Figure 5.34). The American army officer in charge of relief flew over Tacloban and observed 'I do not believe there is a single structure that is not destroyed or severely damaged in some way.' Some journalists described the devastation as 'off the scale' and 'apocalyptic'.
- It was estimated that nearly 2 million people were made **homeless** and more than 6 million people **displaced**, with at least 20,000 fleeing to the capital, Manila.
- The low-lying areas on the eastern side of Tacloban were particularly vulnerable with some areas completely **washed away.**
- Large areas were **cut off**, with communication and power systems destroyed.
- With **flooding and landslides**, major roads were blocked (also with fallen trees) which made relief work almost impossible except from the air.
- Tacloban city's **airport** was damaged by the storm surge (the terminal building destroyed by a 5.2 m storm surge) but facilities had to be repaired quickly as it became the centre of the international relief operation.

- Widespread **looting** was reported.
- **Local government** in many areas collapsed as so many local officials were killed during the storm.
- The **total damage** has been estimated at $2.9 billion.

After the event it was feared that there would be substantial outbreaks of diseases such as cholera, leading to an even higher death toll. Prompt action by the WHO and other relief agencies ensured that outbreaks, such as that of dysentery, were kept isolated and to a minimum.

Immediately after the storm, the United Nations stated that 'access remains a key challenge as some areas are still cut off from relief operations. Unknown numbers of survivors do not have basic necessities such as food, water and medicines and remain inaccessible for relief operations, as roads, airports and bridges were destroyed or covered in wreckage.' Many countries contributed to the relief effort through cash donations or practical help. The UK government deployed two navy ships to the Tacloban area supplying over 200,000 tonnes of aid, while the Americans sent numerous ships and aircraft and over 13,000 service personnel.

Key question

Both New Orleans (Katrina) and Tacloban (Haiyan) were hit by very powerful tropical storms but the estimated cost of the damage to each area was completely different. Why was this?

Figure 5.34 Devastation caused by Typhoon Haiyan in Tacloban, Philippines, November 2013

Other tropical storms worth studying

The Bangladesh cyclone (BOB 01)(Bay of Bengal) 1991

Hurricane Andrew (USA) 1992

Hurricane Mitch (Central America) 1998

Cyclone Nargis (Myanmar) 2008

Hurricane Sandy (USA) 2012

5.6 Fires in nature
The nature of wildfires

Wildfire is a natural process in many ecosystems and can be a necessary and even beneficial process in some of them. It is also true to say that fire can be a very destructive process in the natural world, one which has both natural and human causes.

Fires can be major events, occurring on a large scale and causing widespread destruction and killing much wildlife. Surface fires sweep rapidly over the ground, consuming plant litter, grasses and herbs, and scorching trees, and it is possible for ground temperatures to rise to over 1000 °C. At higher levels there are crown fires that spread through the canopy of trees.

The nature of the fire will depend upon the types of plants involved, strength of the winds, topography of the area in question and the behaviour of the fire itself. Once vegetation has dried out, the nature of the fire will depend largely upon the wind, as the largest fires occur in dry windy weather with low humidity. Wind drives the fire forwards and burning embers that ignite more vegetation are more easily spread in windy weather. The key factors therefore in any fire are the climate and the nature of the plants involved.

Key terms

Retardants – Chemicals sprayed on to fires in order to slow them down. They are composed of nitrates, ammonia, phosphates, sulphates and thickening agents.

Pyrophytic vegetation – Pyrophytes are plants adapted to tolerate fire. Methods of survival include thick bark, tissue with a high moisture content and underground storage structures.

Causes of wildfires and their spread

For a natural fire to occur and to spread, two things are needed:

- **An ignition source:** in the case of natural fires, lightning is by far and away the main cause. Climate will affect the frequency of electrical storms, particularly one in which there is hardly any rainfall. Increasingly, fires are the result of human intervention, particularly those which occur in and around settlements. Such fires are started by falling power lines, carelessly discarded cigarettes, children playing with matches, camp fires and agricultural fires (controlled burning) which get out of hand. In some cases, there is evidence that devastating fires of this kind have been deliberately started by arsonists. In the USA and Australia, greater access to wild areas by tourists has increased the danger of fires through some of the ways listed above.
- **Fuel:** the fuel has to be of sufficient quantity and dry enough to burn. Climate affects the frequency and duration of droughts during which the vegetation and litter has an opportunity to accumulate and dry out. Climate also affects the type of vegetation that will grow in an area and the rate at which litter can be produced.

Distribution

Wildfires are essentially a rural hazard and can occur in most environments, although with the continued expansion of human habitation, wildfires now occur within the boundaries of even substantial settlements in California, Australia and the countries of southern Europe. Areas that are most susceptible are those with a combination of dry vegetation and lightning strikes. Clearly, areas with a dry season are most likely to be affected as are those regions of the world with a semi-arid climate and susceptible to drought. Such regions of the world include:

- parts of Australia
- USA and Canada – including California and Florida (Figure 5.35, page 218)
- southern Europe – including southern France, Italy, Greece, Turkey and Mediterranean islands such as Sicily, Cyprus, Corsica and Sardinia.

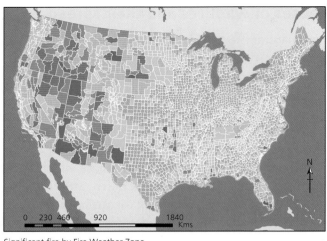

Significant fire by Fire Weather Zone

■ No significant fires ■ 1-53 ■ 54-177 ■ 178-382 ■ 383-677 ■ 678+

A significant fire is defined as a fire which burned at least 10 acres.

Figure 5.35 Wildfires in the USA (1980-2014)

Fires and natural ecosystems are often closely linked, particularly Mediterranean climate regions and the savannah grasslands. Fire can clear vegetation and aid new seed germination, stimulate the growth of certain plants and rid an area of insects and some parasites. Some species are **pyrophytic** in that they can withstand fire through some mechanism such as resistant bark, for example the baobab tree. In Australia, plants such as banksia need fire for their woody fruit to open and thus regenerate.

Fire has not been a real hazard of tropical rainforest areas due to the humid climate, but recent burning for forest clearance has often got out of hand, resulting in widespread fires that burned out of control for long periods. Both the Amazon Basin (Brazil) and south-east Asia (Indonesia) have suffered from such fires in recent years.

A good example of an area where wildfires are a continuing problem is the Los Angeles Basin in California. Wildfires that occur here often get a great deal of media coverage (in some cases globally) as it is likely that a wildfire event could threaten the homes of famous film and TV personalities. Wildfires are common in the Los Angeles area and pose a serious threat for the following reasons:

● much of the area is covered by drought-resistant chaparral, which is a scrub vegetation, as the area is too dry for much tree growth. This vegetation, after the summer drought, can be tinder dry.

● a dry wind called Santa Ana, that descends from the local mountains, increases the dryness of the vegetation to the point where a spark, lightning or a carelessly discarded cigarette can cause a major fire. This wind also allows for the easy spread of fire, which makes it very difficult to control.

● much of the area, outside the centre of Los Angeles, consists of low-density building where the natural vegetation has been allowed to remain between properties, which exposes a large number of them to the fire risk.

Key question

Some natural environments are more likely than others to experience fire as a hazard. Identify such environments and describe their global situation.

The effects of wildfires

The direct effects of a wildfire event are:

● **Loss of crops, timber and livestock:** Forest fires can have a huge impact in timber producing areas, with the loss of trees that will take many years to replace. In the USA, it has been estimated that over $10 million per day is spent fighting such fires.

● **Loss of life:** Although many fires are events from which people can get out of the way, some fires move so fast that people can be trapped, although this is not usual. In the Victoria (Australia) bushfires of 2009, 173 people lost their lives (see page 220).

● **Loss of property:** At one time, only a few rural communities were at risk from fires, but with urban expansion, the fringes are now susceptible, for example Sydney (Australia) and Los Angeles (see above). The cost of damage and of fighting the fires can run into hundreds of millions of dollars. After such events, large numbers of people can be left homeless.

● **Release of toxic gases and particulates:** The Southeast Asia 'haze' is fire-related large-scale air pollution that occurs regularly and is still present in parts of the region. It is largely caused by illegal agricultural fires due to slash-and-burn practices in Indonesia. The resulting pollution covers a number of countries in the region including Singapore, Malaysia, Thailand and Vietnam. This introduces a political dimension, as the governments of these countries have demanded that the Indonesian government take action against companies responsible for illegal forest fires. The size of the haze produced by the Indonesian fires of 1997-98 can be seen in Figure 5.36. These fires were described by one commentator as 'probably amongst the two or three, if not the largest forest fires, in the last two centuries of recorded history'.

- **Loss of wildlife:** In Indonesia, the fires of 1997 destroyed the Wein River orangutan sanctuary on Kalimantan.
- **Damage to soil structure and nutrient content:** With the intense heat generated at ground level, wildfires can destroy many soil nutrients and lead to an alteration in the soil's structure.

Secondary effects can be:

- **Evacuation:** Many people will flee from the area of the fire. Such people will not be allowed back into the affected area, often for a long period, if not forever. Emergency shelters/accommodation will have to be found along with food, etc.
- **Increased flood risk:** In certain environments, where rain comes in heavy bursts, the loss of so much vegetation, and the consequent decrease in interception, can lead to increased flooding.

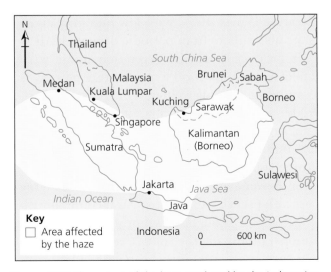

Figure 5.36 The extent of the haze produced by the Indonesian fires of 1997–98

Dealing with the fire hazard

There are several ways of dealing with this hazard, such as planning to mitigate the effect of an event, coping with the wildfires as they happen and addressing the effects.

Preparedness, mitigation and prevention

There are several ways in which the hazard can be managed before the event occurs:

- **Managing the vegetation:** This is done by controlled burning to get rid of much of the litter and by creating firebreaks in the vegetation in advance rather than during the event.
- **Managing the built environment:** This is done by increasing the gap between houses and vegetation and by incorporating more fire-resistant methods in construction (using more stone and brick rather than wood; fitting spark arresters to chimneys).
- **Modelling:** This involves studying the ways in which fires behave with computer simulations in order to comprehend and predict fire behaviour. Figure 5.37 (page 220) shows a fire propagation model.
- **Education:** In areas susceptible to wildfires, it is important to make people aware with regard to home safety and how to avoid starting fires.
- **Warning systems:** These can be put in place by establishing lookout towers and even air patrols. In tourist areas, notice boards at strategic locations could carry warnings of the fire hazard.
- **Community action:** In Victoria (Australia) in February 1983, there were a series of disastrous fires that claimed 47 lives, destroyed over 2000 homes and cost around AU $200 million. As a result, a community education programme was established known as 'Community Fireguard', whose purpose was to assist people in developing their own fire survival strategies. To relay the message, the authorities avoided the 'top down' method they had been using, instead focusing on identifying the most vulnerable areas in fire-prone communities and then generating local interest to make residents aware of how they could be responsible for their own safety. From this, the residents developed their own fire survival techniques such as local warning systems, ensuring that buffer zones were maintained, conducting brush and street cleanups, running equipment training sessions and preparing emergency plans.

• **Being well insured**: in wealthier countries, residents are urged to take out insurance against fire damage, although this can be very expensive in fire-prone areas.

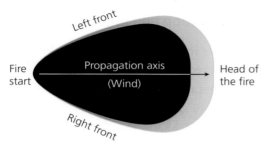

Figure 5.37 Fire propagation model

Dealing with the event as it happens

This involves fighting the fires as they happen, which can be a very dangerous operation and it is not unknown for firefighters to suffer fatalities when they are involved in this action. Spraying with water and chemicals both from the ground and in the air is costly. Firefighters also work on the ground beating out the flames and bulldozing large breaks in the vegetation, preventing the fires jumping in certain directions. **Retardants**, to slow the pace of the wildfire, are often sprayed on fires in areas which are fairly inaccessible or when human safety and structures are endangered.

Addressing the effects

Once the fire has been extinguished, it is important that, in the short term, communities and governments first try to repair the damage caused by the fire. After this, long-term action will mean that attention will be directed towards making sure that areas are well prepared for the next fire. Action will be as follows:

• **replanting trees (1)**: particularly in commercial forestry areas in order to improve the economy of areas devastated by fire.

• **replanting trees (2)**: after vegetation has been removed by fire, there is a great danger of flash flooding and mudslides on, what are now, unprotected slopes. Planting trees should stabilise slopes and should also lead to an improvement in water quality.

• **preparedness**: this means making sure that people are better prepared for the next fire by having emergency supplies ready, along with survival kits, etc., and by setting up community groups such as that in Victoria (Australia).

Wildfire event

Victoria (Australia), February 2009

In early February 2009, a series of bushfires swept across the Australian state of Victoria (Figure 5.38). These were named the 'Black Saturday bushfires' and were Australia's all-time worst bushfire disaster, resulting in the highest loss of life from fires in the country's history.

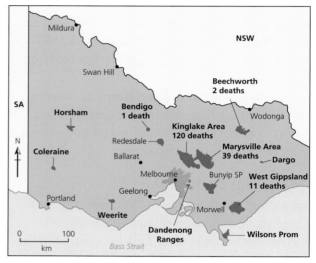

Figure 5.38 Wildfires in Victoria, Australia, February 2009

Risk and vulnerability

Victoria is covered by eucalyptus forests which have an oil-rich foliage which easily burns. The summer climate often features periods where the temperature is over 40 °C with low humidity and the state also suffers from periodic droughts. All this can produce tinder-dry vegetation, and it will only take a spark to set off a major fire. The area also can be susceptible to high winds (over 100 km/h) blowing out of Australia's interior, which will not only produce drier conditions but can fan flames and drive them over considerable distances. Lightning strikes are also a feature of the summer weather. Power cables also cross the state, bringing this essential service to isolated rural communities, but these can be brought down by adverse weather conditions, producing the spark that sets off a fire. In rural areas, there has been depopulation in recent years which has led to a decline in rural services which makes it more difficult to fight the fires when they occur.

Background to the fires

In late January 2009, an exceptional heatwave affected the south-eastern parts of Australia. Temperatures were regularly above 40 °C, some of the hottest days in Victoria's history. The configuration of the pressure systems also meant that hot tropical air was directed towards the area. On the morning of 7 February, hot, north-westerly winds of over 100 km/h hit the state, also causing very low humidity (under 10 per cent).

Fires started in a number of areas (Figure 5.38) from various sources. Fire investigators in Victoria have identified a number of these including lightning strikes, sparks from falling power cables, sparks from power tools and machinery and, in some cases, the fires being deliberately started (arson).

The fires had the following effects:

- **Fatalities:** 173 deaths were attributable to these fires including one firefighter
- **Property loss:** it was estimated that around 3,500 properties were destroyed including over 2,000 houses and 1,500 farm buildings
- **Displaced people:** it was estimated that over 7,000 people were displaced from their homes
- **Forest loss:** an area of around 450,000 ha was burnt (over 1 million acres)
- **Agricultural losses:** this included the loss of livestock (around 12,000 head, including sheep, cattle and horses); stored fodder, grain, hay and silage; loss of standing crops, pasture, fruit trees, vines and olives; plantation timber; loss of thousands of kilometres of boundary fences
- **Parks damaged:** large areas of national parks and other protected areas were destroyed, along with some of the wildlife they contained. The RSPCA estimated that over 1 million animals perished during the fires
- **Electricity supply:** disrupted to over 60,000 residents

- **Looting:** a number of cases of people looting empty properties were reported
- **Cost:** it has been estimated that the fires cost A$4.4 billion of which the largest component was insurance. The total cost of insurance claims has been estimated at A$1.2 billion.

In October 2009, it was announced that a new fire hazard system would replace the one existing before the fires. Under this new system, every day during the fire season the Bureau of Meteorology (BOM) would forecast an outlook for the *Fire Danger Index (FDI)* by considering the predicted weather including temperature, relative humidity, wind speed and dryness of vegetation. On the highest risk days, residents would be advised to leave the potentially affected areas.

In response to the event, new building regulations for bushfire-prone areas were fast tracked by the government. As part of an ongoing debate, the government was also urged by experts to ban housing in the highest fire risk areas, which they claimed were 'some of the most dangerous in the world'. The government also came under criticism for allowing rebuilding in Marysville, which had been 90 per cent destroyed during the event.

Key question

Study Figure 5.1 (page 185), the model of vulnerability. Using the information contained in this description of a fire, replace the generalities of the model with the real details of the event in Victoria. What does the completed figure tell you about the extent of the risk to people and their environment (and their vulnerability) in living in parts of Victoria?

Case study of a multi-hazard environment beyond the UK: The Philippines

The Philippines is a country off the mainland of southeast Asia consisting of an archipelago of over 7,000 islands (Figure 5.39). It is also home to over 100 million people. By virtue of its geographical position it is highly prone to natural disasters. Its position on the western rim of the 'Pacific Ring of Fire' brings both **earthquakes** and **volcanic activity**, with secondary impacts such as **lahars**. It is also situated in an area where tropical disturbances (**typhoons**) cross from the Pacific towards southeast Asia making the Philippines the most exposed country in the world to this particular hazard. Tropical storms bring a great deal of **flooding** and the threat of **landslides**, which again add to the hazardous nature of life in these islands. All of this hinders the attempts by government to reduce the incidence of poverty and to reduce the number of people and assets vulnerable to these hazards. The cost to the Philippine economy has been estimated at 0.5 per cent of national GDP and indirect and secondary impacts further increase this cost. In addition to the significant economic cost, there are substantial social and environmental impacts. In 2013, for example, the country was subject to one severe earthquake (Bohol), touched or crossed by seven typhoons, including Haiyan (see above) and eight tropical storms. At the same time, the major volcanoes were still rumbling away with intermittent small-scale emissions of lava, steam and gas.

Key question

Is it possible in a multi-hazard environment that some of the hazards may be linked?

Major threats to the population of the Philippines are described below:

Earthquakes: These are a fairly common occurrence in the Philippines. In the twenty-first century, between 2000 and 2013, there were 17 earthquakes with a magnitude of at least 5.6, which have caused numerous deaths and damaged a large number of buildings. The most devastating in recent years have been:

- **1976 off Mindanao (Moro Gulf):** this earthquake (magnitude 7.9) caused tsunamis which inundated the western shore of Mindanao killing up to 8,000 people

Figure 5.39 Map of the Philippines

- **1990 Luzon:** (magnitude 7.8) killed over 1,600 people and caused widespread ground rupturing and soil liquefaction
- **2013 Bohol:** a magnitude 7.2 earthquake which killed over 200, injured 800, and caused damage to tens of thousands of buildings.

Volcanic activity: The Philippines has a number of active volcanoes on its islands (opinions vary, but 23 would seem to be a reasonable figure). Among these are three volcanoes which have a history of violent eruptions:

- **Mayon:** This is the most active, having erupted nearly 50 times in the last 400 years, with the most destructive event coming in 1814. The last eruption was in 2014 (Figure 5.40). Mayon is noted for its lahars, with volcanic ash mixing with heavy rainfall from tropical storms to produce rivers of mud.
- **Pinatubo:** Erupted violently in 1991, being the second largest eruption on the planet in the twentieth century. Pinatubo can be deadly as large numbers of people live in the vicinity of the volcano (500,000 people within 40 km). Early warnings and evacuations probably saved at least 5,000 people in 1991 (Figure 5.41).

The explosion was so violent that it hurled gases, ash and steam into the upper atmosphere where they continued to affect global temperatures for at least two years (average global temperature down 0.5°C).

This was followed by Typhoon Yunya, with very heavy rainfall combining with volcanic ash to produce deadly lahars. Final death toll was around 850.

- **Taal:** One of the most active volcanoes in the country, having 33 eruptions noted on the historical record. All of these eruptions have come from a volcano that sits in the middle of a crater lake formed when a huge explosion in prehistoric times formed the caldera of Taal. In the historical record it is thought that Taal has been responsible for 5,000–6,000 deaths as it lies close to populated areas.

Tropical disturbances: Numerous storms cross the country every year bringing with them the risk of severe flooding, landslides and lahars if combined with a volcanic eruption. Such disturbances account for the highest numbers of deaths from natural hazards in the Philippines, most fatalities being the result of storm surges. For an account of the passage of a typhoon, see the previous section on Typhoon Haiyan (November 2013).

In a multi-hazard environment such as the Philippines, people are often at a greater risk than elsewhere. Not only are they vulnerable to the main hazards of earthquakes, volcanic activity and typhoons, but these hazards can be linked to others, in some cases being the direct cause of those hazards. These hazards – tsunamis, landslides/mudslides, flooding, fires and lahars – are sometimes seen simply as secondary effects of the main hazard, but they are often recognised in their own right as a hazard. Earthquakes can lead on to tsunamis, fires and landslides and in a human context, the hazard of crime (looting). Heavy rainfall associated with typhoons can cause serious flooding and landslides. The combination of volcanic activity (ash) and the same heavy rainfall can initiate lahars which can devastate whole communities.

Figure 5.41 The Pinatubo Volcano eruption in the Philippines, 1991

In the past two decades, the Philippines has experienced over 300 natural disasters and Filipino officials insist that the typhoon hazard is becoming more severe, in part because of climatic change. Over the past decades, Filipinos have flocked to risky low-lying areas, havens for cheap, crammed housing where there is a clear link between poverty and vulnerability to natural disasters. Rapid urbanisation has exacerbated the problem, with its tightly packed, flimsily constructed housing and the environmental degradation such as deforestation that has only added to the problem (steep, unprotected slopes leading to rapid run-off, flooding and the potential for landslides).

The Philippines institutional arrangements and disaster management systems have tended to rely on a response or reactive approach, when they should be, in contrast, taking a more effective proactive approach in which disasters can be mitigated, even avoided, by appropriate land-use planning, construction and other preventative measures which avoid the creation of disaster-prone conditions. There has been widespread emphasis on post-disaster relief and short-term preparedness (forecasting/evacuation), rather than on mitigation and post-disaster support for economic recovery. The current system tends to be more of a centralised top-down administrative system than one which is community based and gives rise to local initiatives. What is required is initiatives which emphasise the bottom-up approach that will help the most vulnerable communities cope with the hazards when they occur.

Some programmes have been started along these lines. The Philippines National Red Cross (PNRC), in conjunction with foreign Red Cross organisations, has started programmes at community level. The aim of these is to reduce the impacts of natural disasters by encouraging people to collaborate in protecting their

Figure 5.40 Lava cascading down the slopes of the Mayon Volcano in Manila, 2009

lives and the resources upon which they depend. Such plans address the following areas:

- co-operation and partnership with governmental bodies in order to gain financial support for mitigation measures and to ensure that the programmes have long-term sustainability
- training local volunteers in disaster management
- identifying risk through land-use mapping and determining which mitigation measures might be possible
- initiating mitigation measures which could be physical (sea walls, dykes), health related (clean water supplies) or planning tools (land-use plans, evacuation plans)
- dissemination of information to the whole community.

Many commentators have pointed to the great resilience of the Filipino people in the face of the problems in living in a multi-hazard environment. Others have simply described the attitude of the people as fatalistic; they accept that these events are part of living in such an area (even 'God's will') and that losses are inevitable from time to time.

The central government, however, is trying to do something about the situation. The United Nations claims that the Philippines already has some of the best risk-reduction laws in the world, but unfortunately most of these are still on paper. With a country of over 7,000 islands, too much of the responsibility for reducing disaster risk falls to local government where money is not always wisely spent. Recent government legislation, however, calls for 70 per cent of disaster spending to be used on long-term plans with only 30 per cent going on emergency aid. The main problem of living in a multi-hazard environment remains, though, as one government official put it, 'resources are stretched; even before we could recover from one disaster, here is the next one.'

Key question

What attitudes have the people of the Philippines taken towards living in such a multi-hazard environment?

Case study at a local scale of a place in a hazardous setting: Kobe, Japan

On 17 January 1995, an earthquake of magnitude 6.8 devastated the Japanese city of Kobe (Figure 5.42). The Great Hanshin Earthquake, as it is known, had an epicentre only 20 km from the central parts of the city and was the costliest urban disaster up to that point in time, economic damage being put at $100 billion (or 2.5 per cent of Japan's GDP). Over 6,000 people were killed in the earthquake, with over 35,000 injured and nearly a quarter of a million people made homeless. Those people whose dwellings had been seriously damaged or destroyed faced financial ruin too, as only three per cent of the city's buildings were insured. Kobe was also struck by a typhoon six months after the event.

Risk and vulnerability: The Japanese felt that they were well prepared for earthquakes given their research into the hazard and the wealth of the country which had enabled them to spend money on several forms of risk management. Some commentators, however, believe that there was an illusion of preparedness among the people and the city authorities and that they were caught unaware by the severity of the event. There were still too many older, traditional houses in the city. Many of these had heavy tiles on the roof (to withstand typhoons) but these tiles killed and injured many people when the supports collapsed. Many houses also had not been retrofitted, along with much of the transport infrastructure.

Figure 5.42 The location of Kobe

It was also suggested that as Kobe had not experienced a serious earthquake for 400 years, the people and the authorities had not maintained sufficient emergency supplies.

Many commentators have accused the Japanese government of poorly managing the aftermath of the event. They point out that they made people more vulnerable to the event by being slow to react to the scale of it, by not encouraging enough community volunteers and by refusing offers of help from foreign governments.

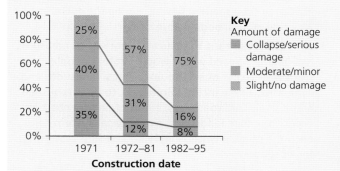

Data from building survey in Central Kobe

Figure 5.43 Building damage in central Kobe as a result of the Great Hanshin Earthquake

Skills focus

Figure 5.43 is an example of a compound bar graph. Some of these graphs are often difficult to read but we have made it easier by joining the segments bar to bar and even giving the percentage figures for each year shown. Joining the segments, without giving the percentage (or other) figures, will enable you to see if any particular component is increasing or decreasing. Students often produce bar graphs that are too complicated, with too many bars or segments. This loses the greatest assets of the method – simplicity and clarity of presentation.

The scale of the disaster and its complexity posed tremendous economic and logistical problems for the people of Kobe. There were also enormous emotional problems to overcome with the trauma of a shattered community in which people were struggling to readjust. Many in the city, though, saw this as an opportunity to put in place sustainable development plans that took disaster threats into account. The total bill for this reconstruction has been estimated at $120 billion, but Kobe seems to have coped well and many experts have commented on how impressive its efforts have been (Figure 5.44 and Figure 5.45).

Within a short time after the disaster, the city had achieved the following:

- Electric power, gas, water, telecommunications, and major road and rail links were restored within months of the event.
- Within several months, 48,000 housing units had been supplied to take care of the homeless who made up 20 per cent of the city's population after the event.

- 70 per cent of the port operations were restored within one year.
- Almost all of the debris resulting from the event was removed within a short time, with 60 per cent of it successfully re-used in landfill sites.
- 15 months after the event, manufacturing output of the Kobe region was up to 96 per cent of what it had been before the earthquake.

Figure 5.44 Damage caused by the Great Hanshin Earthquake, Kobe, Japan, January 1995

Figure 5.45 Reconstruction after the Great Hanshin Earthquake, Kobe, Japan

The Japanese government issued a document after the earthquake, setting out its policy towards such disaster-stricken areas. This was entitled '**Disaster Management in Japan**' and it summarised its aims in the following statement:

The objective of providing the recovery and reconstruction of a disaster-stricken area is to aid victims to return to normal life, restore facilities with the intention of preventing disasters in the future, and implementing fundamental development plans that focus on safety in the community. In view of the decline in social activities

in a community following a disaster, it is important that recovery and reconstruction measures are conducted as swiftly and smoothly as possible.

Apart from the reconstruction, it has been regarded as very important that the city as a whole should learn lessons from the events of January 1995. Learning from the Kobe event has enabled the government to put into place measures which have meant a more rapid and effective response to later earthquakes in the country. Since the event, there have been a vast number of conferences and research studies that have analysed the experience from various professional points of view. From these, the city council considers the following to be the most important lessons to be learned from the earthquake and its aftermath:

- **Improved seismic resistance of existing buildings:** Large numbers of houses collapsed and many of those remaining do not meet modern seismic standards. The city council has indicated that all older buildings will be retrofitted.
- **Improve the fire fighting capacity:** As a vast number of buildings were damaged in the fires that followed the earthquake, the city council is determined to explore all possible sources for extinguishing future fires, including rivers and the sea.
- **Protecting lifelines:** Backup systems must be put in place wherever possible, despite the cost. With telecommunications, for example, they have constructed a duplicate fibre optic system.
- **Community participation:** The city council now understands that in order to reduce future risks, the community needs to be better informed through **awareness and education programmes.** Areas of concern revealed by the disaster included the stockpiling of resources in readiness for an event, the ability of ordinary people to fight fires and the availability of basic tools for search and rescue.

- **NGOs:** The city council is also encouraging a range of NGOs, as they seemed flexible during the event in a way that official bodies were not.
- **Disaster-resistant measures:** A variety of measures has been proposed which would protect people from the effects of earthquakes, fires, landslides and typhoons. These include firebreaks to be established along rivers and roadways, a mountain greenbelt in order to reduce landslides, a new canal system to ensure a reliable water supply and backup systems for hospitals.

The mayor of Kobe sent out this message from his city to other communities that face the same hazard problem:

Wherever you live, city or village, you have to keep in mind that you must protect yourself. You have to prepare to protect yourself, but you cannot fight alone. You must have a system to fight disasters, where you must co-operate with relations and your colleagues. Each community has to have its own plan to respond in case of emergency. Your community may not be enough, though, so then you can work with other communities, the government and with other countries. But first, you must protect yourself, for yourself, wherever you are in the world.

Key question

To what extent are the measures now put in place in Kobe sustainable?

Further reading

Bankoff, G. (2015) *Culture of Disaster: Society and Natural Hazard in the Phillipines* (Routledge)

Bishop, V. (2001) *Hazards and Responses* (Collins)

Bryant E. (2004) *Natural Hazards* (Cambridge University Press)

Chaffey J., Frampton S., McNaught A., and Hardwick J. (2000) *Natural Hazards; Causes, Consequences and Management* (Hodder & Stoughton)

Nagle G. (1999) *Focus on Geographical Hazards* (Nelson Thornes)

Ross S. (1998) *Natural Hazards* (Nelson Thornes)

Rothery D. (2010) *Volcanoes, Earthquakes and Tsunamis* (Teach Yourself Series)

Skinner M. (2003) *Hazards* (part of the *Access to Geography* series) (Hodder & Stoughton)

Warn S. & Holmes D. (2008) *AS/A2 Geography Contemporary Case Studies: Natural Hazards and Disasters* (Philip Allan)

Eyjafjallajökull (Iceland) and Mount Ontake (Japan) – Resources on both are available from the Geographical Association (GA)

Review questions

1 Explain why volcanic and earthquake events should not be called 'acts of God'.

2 Why are poor people more vulnerable to a natural hazard event?

3 Why are urban areas, particularly large ones, considered to be very vulnerable to natural hazard events?

4 Why do certain areas experience both volcanic and seismic events?

5 In the context of natural hazards, what do you understand by 'fatalism'?

6 Using examples, explain why, in different countries, human responses to tropical storms are not the same.

7 What are the factors that contribute to the development of tropical storm systems?

8 How have people sought to manage the volcanic hazard through prediction?

9 What determines the impact of a tsunami?

10 'Resources are stretched, even before we could recover from one disaster, here is the next one.' Does this really sum up the attitude of people living in a multi-hazard environment?

Question practice

A-level questions

1. In the context of natural hazards, what is meant by the term 'primary impact'? (3 marks)

2. What are secondary impacts arising out of seismic events?

   a) Secondary impacts are those which take a long time to occur after the event and usually their impact is small.

   b) Secondary impacts arise in the aftermath of a seismic event and are a consequence of primary impacts.

   c) Secondary impacts come about as a result of aftershocks, such as landslides.

   d) After the initial ground-shaking in a seismic event, the secondary impact is the rupturing of the ground surface. (1 mark)

3. In this chapter, which covers natural hazards, consult Figure 5.15 Risk assessment of Mt Rainier, Cascade Range, USA (page 197). Using this figure, assess the scale of any possible future eruption and its potential impact. (9 marks)

4. How far is it possible to mitigate against the hazards associated with earthquakes? (20 marks)

AS questions

1. Describe the characteristics of **two** of the hazards associated with tropical storms. (4 marks)

2. Which part of the structure of the Earth is known as the lithosphere?

   a) Part of the upper mantle of the Earth on which the plates move

   b) The crust of the Earth which is divided into plates

   c) The layer of the Earth which consists of the crust and the upper section of the mantle

   d) The combined mass of water found on, under, and over the surface of the Earth (1 mark)

3. In this chapter, which covers natural hazards, consult Figure 5.15 Risk assessment of Mt Rainier, Cascade Range, USA (page 197). Describe the distribution of the risks shown by this assessment. (6 marks)

4. Where do most earthquakes occur and why are those locations earthquake-prone? (6 marks)

5. Compare the damage caused by two tropical storms that you have studied. (6 marks)

6. With reference to **one** seismic event that you have studied, evaluate the management of and the responses made to that event. (20 marks)

Chapter 6

Ecosystems under stress

This section focuses on the biosphere and in particular the nature and functioning of ecosystems and their relationships to the nature and intensity of various human activities. You have to study the impact of population growth and economic development on ecosystems at various scales; this affords you the opportunity to engage with fundamental contemporary people–environment issues including those relating to biodiversity and sustainability.

In this chapter you will study:
- ecosystems and processes, including ecosystems in the British Isles over time and local ecosystems
- ecosystems and sustainability
- marine ecosystems
- biomes.

Key terms

Biodiversity – The variation of life forms within a given ecosystem, biome, or for the entire Earth. Biodiversity is often used as a measure of the health of biological systems.

Biomass – The mass of living biological organisms in a given area or ecosystem at a given time. Biomass can refer to species biomass, which is the mass of one or more species, or to community biomass, which is the mass of all species in the community.

Biome – A major habitat category, based on distinct plant assemblages which depend on particular temperature and rainfall patterns, for example, tundra, temperate forest and rainforest.

Climatic climax – A biological community of plants and animals which, through the process of ecological succession, has reached a state of dynamic equilibrium with its climate and soils.

Ecosystem – A system in which organisms interact with each other and the environment.

Fauna – The animals of a particular region, habitat or geological period.

Flora – The plants of a particular region, habitat or geological period.

Food chain – An arrangement of the organisms of an ecological community according to the order in which they eat each other, with each organism using the next lower organism in the food chain as a source of energy.

Food web – A scheme of feeding relationships, resembling a web, which unites the member species of a biological community.

Net primary production (NPP) – The rate at which an ecosystem accumulates energy or biomass, excluding the energy it uses for the process of respiration. This typically corresponds to the rate of photosynthesis minus respiration by the photosynthesisers.

Nutrient cycling – The movement of nutrients in the ecosystem between the three major stores of soil, biomass and litter.

Plagioclimax – The plant community that exists when human interference prevents the climatic climax vegetation being reached.

Seral stage – An individual stage within a sere, for example, colonisation or stabilisation.

Sere – The entire sequence of stages in a plant succession. Different seres are named after the starting point of the succession, for example, lithosere, hydrosere, psammosere and halosere.

Sub-climax – The development of an ecological community to a stage short of the expected climax because of some factor, such as repeated fires in a forest, which arrests the normal succession.

Succession – the series of changes in an ecological community that occur over time.

Trophic level – An organism's position in the food chain. Level 1 is formed of autotrophs which produce their own food. Level 2 consume level 1 and level 3 consume level 2, etc.

6.1 Ecosystems and processes

An **ecosystem** is a community of living and non-living things that work together (Figure 6.1). Ecosystems have no particular size; they can be as large as a rainforest or a lake or as small as a tree or a puddle. The water, water temperature, plants, animals, air, light and soil all work together consuming and emitting energy as they feed, reproduce and die. Because the flow of energy in an ecosystem is uni-directional, they can be considered **open systems**. All the parts of an ecosystem work together in a balanced way.

Nature of ecosystems

Structure

Each ecosystem has two main components:
- abiotic
- biotic.

Abiotic components

Abiotic factors are all of the non-living things in an ecosystem. They influence the structure, distribution, behaviour and inter-relationships of all the organisms in that ecosystem. Abiotic factors vary among different ecosystems. For example, abiotic factors found in aquatic ecosystems may be things like water depth, pH, sunlight, turbidity (amount of water cloudiness),

salinity (salt concentration), available nutrients (nitrogen, phosphorous, etc.) and dissolved oxygen. Abiotic variables found in terrestrial ecosystems include things like rain, wind, temperature, altitude, soil, pollution, nutrients, pH, types of soil and sunlight.

- **Climatic factors:** These include precipitation, temperature, sunlight, wind, humidity, etc. Precipitation varies by latitude, continentality, topographical position (for example, rain shadow) and altitude. Some places have year-round precipitation while others have seasonal drought or very little precipitation at all. Temperature varies by latitude: locations near the equator are warmer than locations near the poles or the temperate zones. Continental areas have a wider range of temperatures than coastal (maritime) locations. Humidity influences the amount of water and moisture in the air and soil, which, in turn, affects rainfall. Wind can bring precipitation or can have a drying effect depending on the humidity of the air.

- **Topography** is the layout of the land in terms of elevation, gradient and aspect. For example, south-facing slopes in the northern hemisphere will receive more sunlight (energy) than those facing north.

- **Altitude** is the height above sea level. The characteristic life forms in mountains depend on elevation because of the changes in climate with altitude.

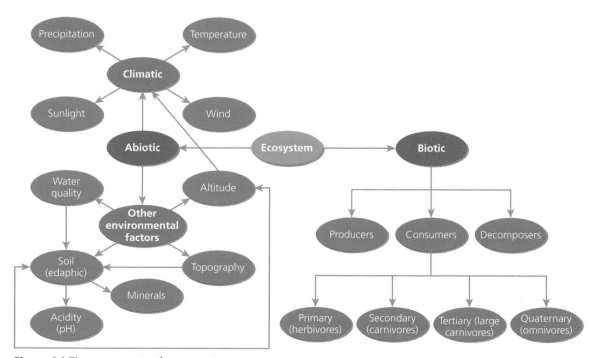

Figure 6.1 The components of an ecosystem

- Temperatures drop with altitude (approximately 1°C/100 m). Winters can be severe.
- Mountains force air masses to rise, cooling the air and eventually causing precipitation. In the lee of the mountains there is a rain shadow.
- Wind speed increases with height. Stronger winds cause physical damage; they discourage tree growth above certain elevations (the tree line). Only specifically adapted ground-hugging plants, such as grasses, sedges, mosses, shrubs and lichens can survive the extreme cold and winds of alpine zones.

This dependency causes life zones to form: bands of similar ecosystems at similar altitude as shown in Figure 6.2.

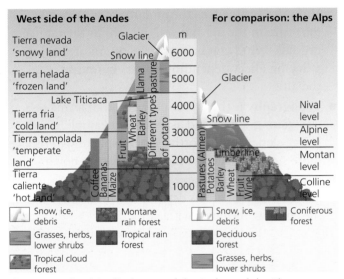

Figure 6.2 The altitudinal zones of the Andes and the Alps

- **Soils** are rich ecosystems in themselves. They are composed of both living and non-living matter. The study of plant–soil interactions is a rapidly growing area in terrestrial ecology. Recent research combining above-ground and below-ground approaches has shown that vegetation–soil interactions are the driving force for many ecosystem processes and their responses to global change.

Biotic components

The biotic components are the living organisms including plants, animals and micro-organisms (bacteria and fungi) that are present in an ecosystem.

These biotic components can be classified into three groups:

- **Producers:** Green plants have chlorophyll that enables them to convert solar energy into chemical energy by the process of photosynthesis. As the green plants manufacture their own food they are known as **autotrophs**.

The chemical energy stored by the producers is utilised partly by the producers for their own growth and survival with the remaining stored in the plant parts for their future use.

- **Consumers:** Animals lack chlorophyll and are unable to synthesise their own food. Therefore they depend on the producers for their food. They are known as heterotrophs.

There are four types of consumer:

- Primary consumers: These are the animals that feed on the producers. They are called **herbivores.** Examples are rabbit, deer, goat and cattle.
- Secondary consumers: The animals that feed on the herbivores and are called the **primary carnivores.** Examples are cats, foxes and snakes.
- Tertiary consumers: These are the large carnivores which feed on the secondary consumers. Examples are lions and wolves.
- Quaternary consumers or omnivores: These are the largest carnivores, which feed on the tertiary consumers and are not eaten by any other animal. Examples are lions and tigers.

- **Decomposers:** Bacteria and fungi belong to this category. They break down the dead organic materials of producers (plants) and consumers (animals) for their food and release into the environment the simple inorganic and organic substances produced as by-products of their metabolisms.

These simple substances are re-used by the producers, resulting in a cyclic exchange of materials between the biotic community and the abiotic environment of the ecosystem.

Energy flows

An ecosystem is an **open system**. The inputs and outputs are shown in Figure 6.3 (page 231).

Organisms require energy for growth, maintenance, reproduction, locomotion, etc. Hence, for all organisms, there must be both a source and a corresponding loss of energy.

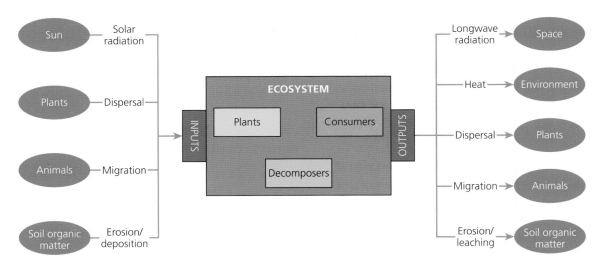

Figure 6.3 Inputs and outputs of a typical ecosystem

All the energy used by living things comes ultimately from the **sun**. Energy enters living systems as a result of **photosynthesis** by plants and some bacteria. Often, less than two per cent of the incoming sunlight is captured; of this more than half of the energy captured by plants is used in respiration and is lost as heat (Figure 6.4).

There are two types of organisms that have direct access to the energy in plant tissues:

- Herbivores feed on the plant while it is alive.
- Decomposers feed on the plant after it is dead.

Much of the energy in herbivore **biomass** is taken by:

- **carnivores**: these meat eaters survive mainly by eating herbivores
- decomposers.

In most ecosystems, the majority of the energy goes to the decomposers. For example, in an area of grassland only ten per cent of the energy in plants is taken by grazing animals (such as antelope). Herbivores then use almost all of their energy intake on respiration and maintaining their bodies (**maintenance**); the rest goes to herbivore **biomass** (the flesh and blood of the animal).

Almost all of the energy taken in by carnivores goes to maintenance. The decomposers receive most of the plant energy and use up over half of that in maintenance. The rest may be locked up in **soil organic material** or taken by organisms that feed on decomposers. Ultimately, all of the energy originally captured by plants is transformed and lost as heat; energy is not recycled.

Biomass and net primary productivity

Biomass is a term in ecology for the mass of living organisms in a given ecosystem. Biomass can refer to:

- the mass of all living species in a given ecosystem (community biomass)
- the mass of one species within an ecosystem
- the mass of decaying material. This includes leaf litter and decaying organic material on a forest floor. It can be a substantial component of, for example, a forest ecosystem.

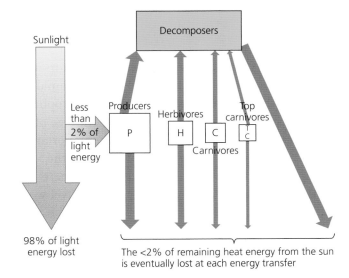

Figure 6.4 Generalised energy flow and heat loss through an ecosystem

Phytoplankton, along with some larger rooted plants such as kelp, make up most of the plant biomass of marine aquatic systems. In the animal kingdom, small aquatic fauna such as krill comprise a disproportionate amount of the animal biomass.

In forest ecosystems trees and shrubs may comprise a considerable amount of the plant biomass along with the forest floor detritus. Faunal statistics show that arthropods (such as insects) comprise a very high percentage of animal biomass, with large mammals making up a relatively small percentage. In the case of organisms other than animals and plants, fungi dominate the biomass.

Plants characteristically comprise the greatest part of the biomass of terrestrial ecosystems. It is important to note that the contribution of individual species within an ecosystem to the community biomass may change considerably from season to season and even from year to year.

Key question

Why might there be seasonal changes in biomass components?

Biomass can be expressed as the average mass per unit area (for example, tonnes/ha or g/m^2) or simply as the total mass in the community. It can also be expressed in units of energy (for example, joules/m^2).

The product of photosynthesis is a carbohydrate, such as glucose (a sugar), and oxygen which is released into the atmosphere. The **primary productivity** of a community is the amount of biomass produced through photosynthesis per unit area and time by plants. Primary productivity is usually expressed in units of energy (for example, joules/m^2/day) or in units of dry organic matter (for example, kg/m^2/year). Globally, primary production amounts to 243 billion tonnes of dry plant biomass per year. The total energy fixed by plants in a community through photosynthesis is referred to as **gross primary productivity** (GPP). A proportion of the energy of gross primary productivity is used by plants in a process called **respiration**. This provides a plant with the energy needed for various plant physiological and morphological activities.

Subtracting respiration from gross primary production gives **net primary productivity (NPP)**, which represents the rate of production of biomass that is available for consumption by **heterotrophic** organisms (bacteria, fungi and animals).

Food chains, trophic levels and food webs

As has been outlined above, the source of all food is the activity of producers or autotrophs; mainly **photosynthesis** by plants. The producers feed the primary consumers (herbivores) who in turn feed the secondary consumers (carnivores). These can be a source of food for tertiary consumers.

Such a path of food consumption is called a **food chain**. This chain follows a direct, linear pathway from one species to another (Figure 6.5). Examples of simple food chains include:

● Phytoplankton → zooplankton → fish → seal → killer whale
● Grass → grasshopper → frog → heron

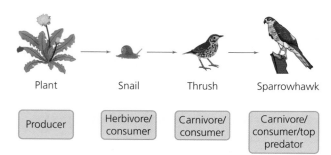

| Producer | Herbivore/consumer | Carnivore/consumer | Carnivore/consumer/top predator |

Plant — Snail — Thrush — Sparrowhawk

Figure 6.5 A typical food chain for a suburban garden in the UK

Each level of consumption in a food chain is called a **trophic level**.

The trophic level of an organism is the position it occupies in a food chain. The number of steps an organism is from the start of the chain is a measure of its trophic level. Food chains start at trophic level 1 with primary producers such as plants. They then move to herbivores at level 2, predators at level 3 and typically finish with carnivores or apex predators at level 4 or 5. Figure 6.6 is called a number pyramid and shows the number of organisms in each trophic level. It does not take into consideration the size of the organisms and it over-emphasises the importance of small organisms. In a pyramid of numbers the higher the trophic level the fewer organisms there are. If the biomass of the organisms at each level were taken into account rather than numbers then the mass of larger (but fewer) organisms (for example, a tree) would be better represented. This is called a **pyramid of biomass**.

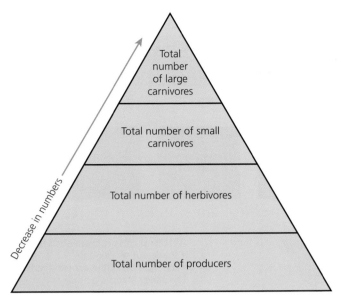

Figure 6.6 A number pyramid showing trophic levels

Most food chains are interconnected. Animals typically consume a varied diet and, in turn, serve as food for a variety of other creatures that prey on them. These interconnections create **food webs**.

The food web in Figure 6.7 only contains 25 species. Most food webs are much more complicated than this, numbering in the hundreds to thousands of species.

Skills focus

Describe the food web shown in Figure 6.7

Succession

Ecological succession is the gradual process by which ecosystems change and develop over time. Nothing remains the same and habitats are constantly changing. There are two main types of succession:

- **Primary succession** is the series of community changes which occurs on an entirely new habitat which has never been colonised before. For example, a recent lava flow.
- **Secondary succession** is the series of community changes which takes place on a previously colonised but disturbed or damaged habitat. For example, after felling trees in a woodland, land clearance or a fire.

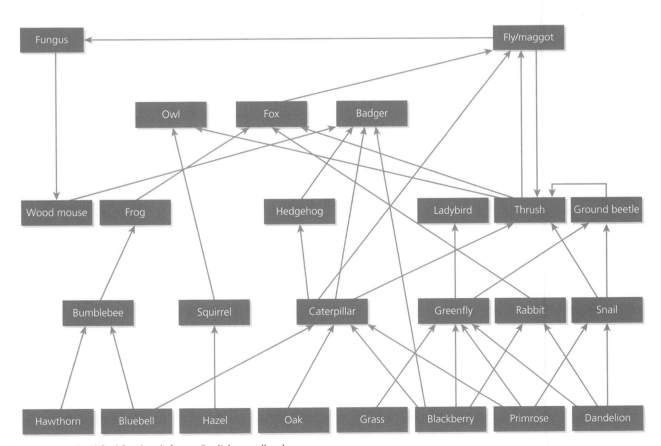

Figure 6.7 A simplified food web for an English woodland

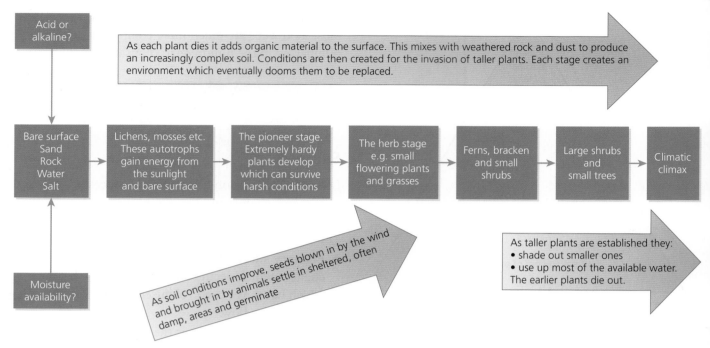

Figure 6.8 A generalised model of succession

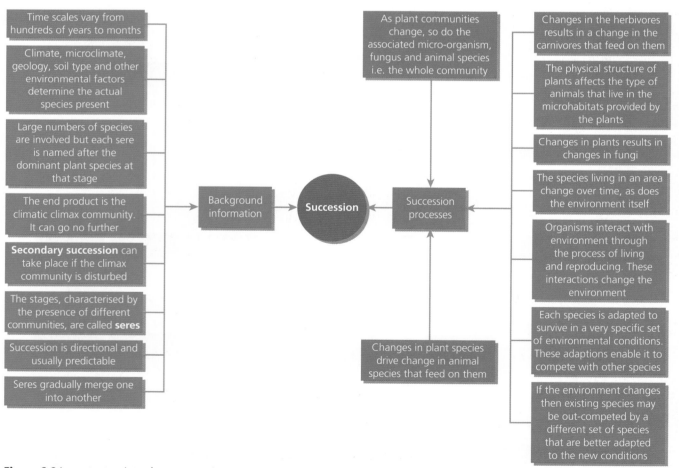

Figure 6.9 Important points about succession

Succession is the term used to signify the changes in the composition of a community of plants and animals over time (Figure 6.8 and Figure 6.9). These changes result in an increase in the complexity of the structure and species composition (**species diversity**) of such a community. Plants invade an area when the conditions are suitable and then die off when the succession leads to unfavourable local conditions. This means that there is a change in the dominant plant species with time.

Successions can start on a variety of surfaces and their names reflect this:

- **Lithosere**: A succession that starts on a newly exposed bare rocky surface, for example, newly erupted volcanic lava or rock exposed as an ice sheet retreats.
- **Psammosere**: A succession that starts on bare sand, for example, coastal sand dunes.
- **Hydrosere**: A succession that starts in fresh water, for example, a pond.

- **Halosere**: A succession that starts in salt water conditions, for example, a salt marsh.

Primary succession or **prisere** develops by the gradual colonisation of a lifeless abiotic surface. A **succession** goes through several stages; the entire sequence of stages is called a **sere** and the stages are referred to as **seral stages.** As succession occurs each new stage:

- is structurally more complex than the one it replaces
- contains more biomass
- has more species
- has greater net primary productivity
- has greater flows of nutrients and energy.

The species living in a particular place gradually change over time but so does the physical and chemical environment within that area.

An example of a lithosere is shown in Table 6.1.

Table 6.1 A lithosere in the UK

Bare rock surface	This is initially colonised by bacteria and single-celled photosynthesisers that are able to survive on few nutrients and get most of their energy from the sun. The surface conditions are often dry and the soil little more than particles of weathered rock
Seral stage 1 Colonisation	The first plant species to colonise an area are called pioneers. These are lichens that are adapted to the severe (dry, windy, soil-free) conditions (Figure 6.10). They begin to break up the rock to form a thin layer of proto-soil. As they die they add dead organic matter to weathered rock and windblown dust. This creates a simple soil which improves water retention. Mosses are then able to develop
Seral stage 2 Establishment	As the soil develops further, ferns and small herbaceous plants and grasses begin to grow. **Species diversity** increases. There are more invertebrates living in the soil, which increases the organic content. This enables the soil to hold more water.
Seral stage 3 Competition	Larger plants begin to establish themselves. These include shrubs and small trees. They use up a lot of available water and shade the ground. Some of the earlier colonisers are unable to compete and die out. Herbivores become established and predators begin to move in
Seral stage 4 Stabilisation	Fewer new species colonise. Complex food webs develop. This stage is dominated by larger, fast-growing trees such as birch and rowan. Top predators are found at this stage
Seral climax	This is the final seral stage. It represents the maximum possible development that a community can reach under the prevailing climatic (temperature, light and rainfall) conditions. This is called a **climatic climax** community. The total number of larger plant species falls as a few large species dominate the area. In the case of southern England, these are broad-leaved deciduous trees. For example, oak and ash

Figure 6.10 Lichen colonising bare rock in sub-arctic Canada

Key questions

Under what circumstances would a lithosere develop in the UK today?

How might succession on an area of urban dereliction vary from that of a lithosere?

Fieldwork opportunities

Succession studies provide an opportunity for fieldwork. This could be carried out on a piece of wasteland in an urban area.

Climatic climax, sub-climax and plagioclimax

A convenient concept to describe the end product of succession is the idea of a **climatic climax** community (Figure 6.11). This is the final stage in succession. It is an association of plants and animals which has attained a state of equilibrium within a given area.

Many ecologists believe however that the time required to achieve this state is unrealistically long. In most cases, external disturbances and environmental change occur so frequently that the realisation of a climax community is unlikely and therefore it has come to be regarded as a less useful concept. Nevertheless many authors and nature-enthusiasts continue to use the term 'climax' to refer to what might otherwise be called mature or old-growth communities.

The British Isles is in a temperate deciduous forest **biome**, so broad-leaved deciduous forest, dominated by oak and ash is the *normal* **climatic climax community for much of England and Wales**. At a more local scale however, the situation is far more complex. Heather moorland, coniferous forest, marshland and alpine (montane) communities are widely distributed, occurring wherever local **soil**, drainage or relief factors restrict the growth of deciduous forest.

It should also be noted that human activity over thousands of years has cleared almost all of the UK's **primary** forest, replacing it with agriculture, plantations and other **secondary growth**.

At any point in the development of an ecosystem, succession can be halted by an arresting factor. This can occur suddenly or gradually and can have a permanent or temporary effect. The result is a **sub-climax** community where the climatic climax has not been reached. Table 6.2 summarises these arresting factors.

Table 6.2 Arresting factors in ecological succession

Arresting factor	Possible cause(s)
Topoclimax	A change in topography. This could be a landslide, a volcanic eruption (lava or ash) or deposition of mud following a river flood
Hydroclimax	A change in drainage. This could be caused by a raised water table following increased precipitation
Biotic climax	The introduction of an alien species to an area. For example, rabbits are the most significant known factor in species loss in Australia
Plagioclimax	An environment maintained by management. This could be forest clearance in Amazonia for ranching, for example. It includes all agriculture, sports fields, parks, gardens, etc.

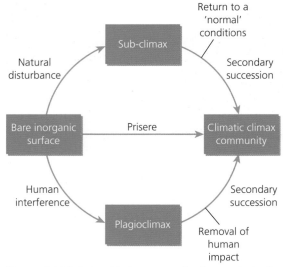

Figure 6.11 Relationship between sub-climax, plagioclimax and secondary succession

Plagioclimax communities are determined by planned or unplanned human activity (Figure 6.12). The process of succession has either been halted before reaching climatic climax or it has been deflected to a different climax. This could be caused by:

- cutting down existing vegetation
- burning (for example, to remove forest)
- planting trees/crops
- grazing (for example, goats in the Mediterranean region)
- draining wetland.

The plagioclimax will remain as long as the human activity continues. If this activity stops then the succession will revert to its original path unless there has been a permanent change to other environmental conditions (Figure 6.13).

Figure 6.12 A golf course: an example of a plagioclimax

Figure 6.13 An abandoned golf course: an example of secondary succession.

Key question

Why might Kinder Scout (Figure 6.32, page 252) and its blanket bog be considered a plagioclimax?

Mineral nutrient cycling

Nutrients are the chemical elements and compounds needed for organisms to grow and function. Energy flow and **nutrient cycling** are interdependent. The rate of nutrient cycling may affect the rate that energy can be trapped. For example, plants cannot make new cells (grow) if essential nutrients are absent.

Nutrients are stored in three compartments within an ecosystem:

- Soil: A mixture of weathered rock, air, water and decomposed organic matter on the surface of the Earth.
- Litter: The amount of dead organic matter on top of soil.
- Biomass: The total of plant and animal life in an ecosystem. These nutrients are cycled from one store to another as shown in Figure 6.14.

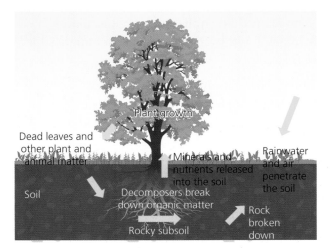

Figure 6.14 A generalised mineral nutrient cycle

All nutrient cycles can be summed up in a Gersmehl diagram (see Figure 6.15). In this model the size of the circles is in proportion to the amount of nutrients they store and the thickness of the lines is proportional to the amount of nutrient flow.

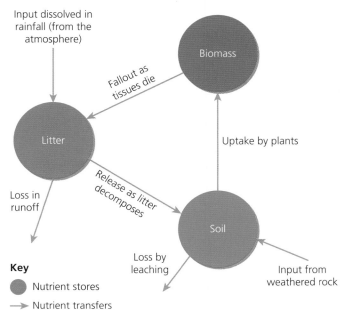

Key
- Nutrient stores
→ Nutrient transfers

Figure 6.15 A model of the mineral nutrient cycle

Key question

How will the sizes of the three nutrient stores and the nutrient transfers shown in Figure 6.15 vary between winter and summer for an area of deciduous woodland in southern Britain? Draw Gersmehl diagrams to show your answer.

Terrestrial ecosystems

The term **ecosystem** (a contraction of ecological system) is generally understood to be the entire assemblage of organisms (plant, animal and other living beings) living together in a certain space with their environment, functioning as a loose unit. Together, these components and their interactions with, and relationships to, each other form a dynamic and complex new whole, functioning as an 'ecological unit', with additional characteristics that cannot be found in the individual components. Nor could any organism live completely on its own without involving any other species of organism. A terrestrial ecosystem is one found only on landforms, that is, all ecosystems that are not aquatic or marine. They range in scale from sand grains to continents, though the latter tend to be called **biomes.**

Example of a terrestrial ecosystem in the UK: The chalk downlands of the South Downs

The South Downs is a chalk ridge that extends from near Winchester to the high sea cliffs of Beachy Head 100 km to the east. The ridge is an escarpment that reaches heights of over 200 m. The most dramatic feature of the South Downs is the steep, mostly northerly-facing scarp slope (Figure 6.16) that is broken only by the gaps created by the rivers, Meon, Arun, Adur, Ouse and Cuckmere. The gently sloping south-facing dip slope supports well-drained, easily worked soils.

The South Downs is comprised of tilted layers of relatively soft chalk containing bands of flint nodules. Both the chalk and the flint are the remains of animals and plants in the Cretaceous sea over 100 million years ago. The soils derived from the chalk of the South Downs are mostly thin, well-drained and poor in several minerals and nutrients.

The South Downs supports significant areas of unimproved calcareous grassland, with extensive areas occurring along the north-facing escarpment, and more widely scattered fragments on the south-facing, shallow sloping, predominantly arable dip slope.

Figure 6.16 Looking west along the north-facing scarp slope of the South Downs at Fulking

Abiotic factors

Climate

This area of southern England combines the highest average daytime temperatures found in the British Isles with the highest sunshine averages on the British mainland. Rainfall is below the UK average with 950 mm/yr. The warm, dry dip slope and cooler, damper scarp provide a range of conditions for wildlife. Species at the northern extent of their distribution occur in the South Downs.

Topography

Steep slopes on the scarp prevented ploughing and today support most of the relict ancient woodland and chalk grassland. Gently sloping, south-facing dip slopes have hot, dry conditions which favour growing cereal crops. Very little semi-natural habitat remains on the dip slope. The gaps in the ridge created by southward flowing rivers are narrow, steep sided and flat floored.

Geology, soils and drainage

The underlying chalk gives rise to thin, infertile, well-drained soils. These lack many minerals, especially potassium and phosphates, though they are of course rich in calcium. This soil type, known as **rendzina**, only allows slow rates of plant growth. It is largely responsible for the diversity of small, low-growing herbs able to co-exist in downland turf.

Pockets of deeper soils are present on the plateau, dip slope and valley bottoms. These more fertile areas are either cultivated or support woodland. In a few places a thin covering of wind-blown soil called 'loess' remains over the chalk. This is thought to derive from dry conditions during the last Ice Age. It produces a very unusual plant community known as 'chalk heath'. The soils of the river floodplains that cut through the chalk downs derive from water-borne silt and are highly fertile.

The poorly drained areas of the floodplains were traditionally used as grazing marsh, but with increased drainage has come conversion to arable use.

Biotic factors

There are many different **habitats** (that is, places where living things live) and these are summarised in Figure 6.17.

Two of these habitats are described below:

- **Chalk heath** is an extremely rare habitat found on the South Downs in a few, scattered locations. It occurs where deposits of acidic soil created by windblown deposits overlie the chalk. Characterised by acid-loving species such as heathers growing alongside typical chalk-loving plants, chalk heath was probably much more widespread in the past. Its vulnerability to disturbance of the soil and scrub encroachment has

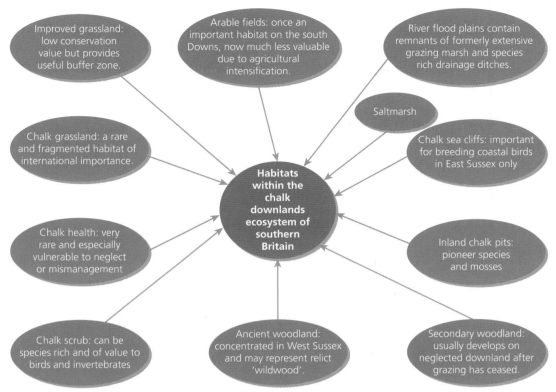

Figure 6.17 The range of habitats found in the chalk downlands of southern Britain

allowed this habitat to dwindle in size to present day minute levels. Lullington Heath National Nature Reserve has what is regarded as the best example of chalk heath in Britain, but there are other chalk heath sites within the South Downs.

- **Unimproved chalk grassland** is typically species rich (with up to 56 species in one square metre) and is maintained by sheep or rabbit grazing. Once covering extensive areas of the South Downs, mainly in East Sussex, this habitat is now scarce, fragmented and under threat from changing patterns of land use. Legal protection for the remaining areas of unimproved chalk grassland is increasing by designating some locations as **Sites of Special Scientific Interest (SSSI)** and others to Annex 1 of the EC Habitats Directive. The best examples are now under consideration as **Special Areas of Conservation**.

Factors influencing change in the downland area

- Many of the habitats within the South Downs National Park have become highly **fragmented** in recent years. In particular chalk grassland, chalk heath and unimproved grazing marsh mostly occur as isolated patches within agricultural and/ or unmanaged land. The implications for wildlife of fragmentation are numerous and include increased risk of local extinctions. On small, isolated sites species may not be able to sustain themselves without appropriate management. Chalk grassland fragments, for example, are vulnerable to scrub invasion.

- Increased edge to area ratio. When a habitat is next to intensively worked farmland, the edge effects of spray drift (pesticide or fertiliser) and run-off can soon cause an erosion of the high value habitat from the edges inwards.

- Species movement between small, isolated patches of available habitat can be severely restricted, especially for many of the characteristic chalk grassland plants which do not produce wind-dispersed seeds. The same applies to the many invertebrates which will not willingly cross open arable fields, like molluscs and some butterflies.

- The myxomatosis epidemic of the mid 1950s onwards showed that rabbit grazing is vital for the maintenance of many chalk habitats. Although some fluctuation still occurs in rabbit populations, they have largely recovered to the extent that they are once again a major source of grazing pressure on the South Downs.

- Leisure use of the South Downs is increasing with activities such as hang gliding, mountain biking and four-wheel drive rallies, all of which can disturb rare species and damage the grass cover, leading to erosion of the thin soil.

Key question

The South Downs have recently been designated as a national park. How does this help in protecting the downland ecosystem?

Global changes influencing ecosystems

Climate change

> The natural environment underpins all aspects of our lives. It will be affected by climate change, and yet we will be increasingly reliant on it to help us manage the impacts that a changing climate will bring.
>
> Defra, 2010f

Ecosystems are highly vulnerable to climate change; this can aggravate existing stress factors such as pollution, land conversion and invasive non-native species. While some species could benefit from climate change, far more are set to lose out, according to UK government estimates.

The UK may see changing patterns of wildlife and plants as species try to adapt by moving northwards, or have to compete with new non-native species. Habitats may come under increasing pressure – from salt marshes threatened by sea-level rise to beech woodland susceptible to summer droughts. Species could also experience reduced food supply if earlier breeding periods are at odds with the food available at the time.

Figure 6.18 and Figure 6.19 (page 241) summarise the UK government view of what the impacts could be.

On a global scale atmospheric warming is likely to be the greatest cause of species extinctions for millions

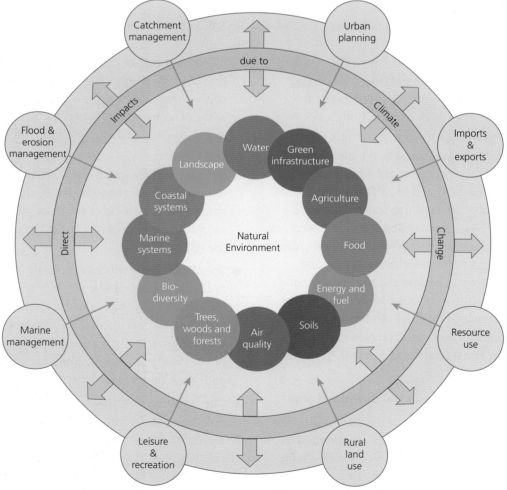

Figure 6.18 An illustration of the constituent parts of the natural environment and the direct and indirect impacts of climate change

Source: Adapted from Defra, 2010

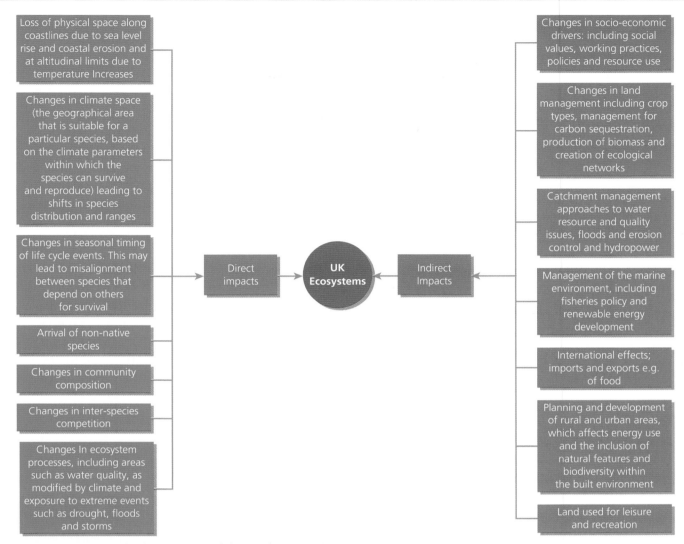

Loss of physical space along coastlines due to sea level rise and coastal erosion and at altitudinal limits due to temperature Increases

Changes in climate space (the geographical area that is suitable for a particular species, based on the climate parameters within which the species can survive and reproduce) leading to shifts in species distribution and ranges

Changes in seasonal timing of life cycle events. This may lead to misalignment between species that depend on others for survival

Arrival of non-native species

Changes in community composition

Changes in inter-species competition

Changes In ecosystem processes, including areas such as water quality, as modified by climate and exposure to extreme events such as drought, floods and storms

Changes in socio-economic drivers: including social values, working practices, policies and resource use

Changes in land management including crop types, management for carbon sequestration, production of biomass and creation of ecological networks

Catchment management approaches to water resource and quality issues, floods and erosion control and hydropower

Management of the marine environment, including fisheries policy and renewable energy development

International effects; imports and exports e.g. of food

Planning and development of rural and urban areas, which affects energy use and the inclusion of natural features and biodiversity within the built environment

Land used for leisure and recreation

Direct impacts

UK Ecosystems

Indirect Impacts

Figure 6.19 Direct and indirect impacts of climate change on UK ecosystems

of years. The **Intergovernmental Panel on Climate Change (IPCC)** says a 1.6°C average rise may put 20–30 per cent of species at risk. If the planet warms by more than 3°C, most ecosystems will struggle.

Climate change, along with habitat destruction and pollution, is one of the important factors that can contribute to species extinction. The IPCC also estimates that 20–30 per cent of the plant and animal species studied so far are at risk of extinction if temperatures reach levels predicted to occur by 2100. Projected rates of species extinctions are 10 times greater than recently observed global average rates and 10,000 times greater than rates observed in the distant past (as recorded in fossils). Examples of species that are particularly climate sensitive and could be at risk of significant losses include animals that are adapted to

mountain environments, such as the pika (Figure 6.20) and animals that are dependent on sea ice habitats, such as polar bears.

Figure 6.20 A pika in the Canadian Rockies

Key question

What other species may be at risk from climate change?

Human exploitation of the global environment

From 1960 to 2015 the global population increased by 150 per cent, going from 3 billion in 1960 to nearly 7.5 billion in 2015. Low-income countries have accounted for most population growth in the past 55 years, but this is beginning to change. Some high-income countries such as the USA are still experiencing high rates of population growth, while some developing countries such as China, Thailand, North Korea and South Korea have very low rates.

Global economic activity increased nearly fivefold between 1980 and 2010, despite the downturn in 2006. Average income per person almost doubled during this period. As per capita income grows consumption changes, with wide-ranging potential for effects on ecosystems. The share of additional income spent on food declines and the consumption of industrial goods and services rises. The composition of people's diets changes, with less consumption of starchy staples (rice, wheat, potatoes) and more of fat, meat, fish, fruits and vegetables.

The impact of science and technology on ecosystems is most evident in the case of food production. Much of the increase in agricultural output over the past 40 years has come from an increase in yields per hectare rather than an expansion of the area under cultivation. For instance, wheat yields rose over 200 per cent, rice yields rose over 100 per cent and maize yields rose nearly 160 per cent in the 40 years leading to 2015. At the same time there were some unintended effects, such as **eutrophication** (see Figure 6.23, page 244) of freshwater systems and hypoxia in coastal marine ecosystems, both of which result from excessive application of inorganic fertilisers. Advances in fishing technologies have contributed significantly to the depletion of marine fish stocks.

For terrestrial ecosystems, the most important direct drivers of change in ecosystems have been:

- land-use change, mainly conversion of natural vegetation to growing crops. Cropped areas currently cover approximately 30 per cent of the Earth's surface

- the application of new technologies which have contributed significantly to the exploitation of farmland and timber stocks

- deforestation and forest degradation, which affect 8.5 per cent of the world's remaining forests, nearly half of which are in South America. Deforestation and forest degradation have been more extensive in the tropics over the past few decades than in the rest of the world, although data on boreal forests is especially limited and the extent of the loss in this region is less well known. Approximately 10 per cent of the drylands and hyper-arid zones of the world are considered degraded, with the majority of these areas in Asia.

Freshwater ecosystems have been affected by water abstraction, invasive species and pollution, particularly high levels of fertiliser. The introduction of non-native invasive species is one of the major causes of species extinction in freshwater systems. For example, after the American signal crayfish was introduced to Britain, it reduced the now-endangered native white-clawed crayfish population by more than 50 per cent.

Local ecosystem

Ponds

Figure 6.21 A pond in rural Lancashire

A pond is an example of a distinctive local ecosystem (Figure 6.21). In the UK **Biodiversity Action Plan (BAP)**, ponds are defined as small, permanent or seasonal water bodies that are up to two hectares in size (less than three football pitches). They are depressions where water collects, and are often quiet environments with a shallow depth. This shallowness allows the sun to penetrate to the bottom, which in turn allows plants to grow. Pond plants (**flora**) either grow entirely underwater or partially on the surface. Some plants such as reeds and bulrushes will also grow along the pond's edge. Ponds in the UK support a large variety of animal (**fauna**) and plant life, such as birds, small fish, amphibians (frogs, toads and newts), insects, protozoa, algae, fungi and lily pads.

A pond's ecosystem, like all others, consists of abiotic environmental factors and biotic communities of organisms. Abiotic environmental factors of a pond's ecosystem include:

- temperature
- water flow
- salinity
- the percentage of dissolved oxygen.

A water body's salinity may determine the different species present. For instance, marine organisms tolerate salinity, while freshwater organisms will not thrive when exposed to salt. In fact, freshwater ecosystems often have plant species present which will absorb salts that are dangerous for freshwater organisms.

A pond ecosystem consists of four **habitats**:

- **The shore**: The nature of the shore determines the type of organisms found there. Rocky shorelines may not allow plants to grow, while muddy or sandy shorelines attract grasses, algae, earthworms, snails, protozoa, insects, small fish and micro-organisms.
- **The surface film**: The pond's surface is inhabited by pond skaters and free-floating organisms.

- **Open water**: An open-water habitat permits sizeable fish (carp, pike), plankton, phytoplankton and zooplankton to grow. Phytoplankton includes a large variety of algae, while zooplankton refers to insect larvae, etc. Fish feed on plankton or tiny organisms. Deep water provides space for many creatures, such as the great diving beetle, a voracious predator that feeds on smaller invertebrates, tadpoles and even small fish hiding among the pond weed. Snails graze on the algae that stick to the leaves of larger plants that grow in deeper ponds.
- **The pond bottom**: This habitat varies depending upon the pond's depth. Shallow ponds (less than 30 cm) with sandy bottoms provide a nesting environment for earthworms, snails, and insects. Deeper ponds (over 30 cm) often have muddy bottoms, which allow various micro-organisms, such as flatworms and dragonfly nymphs, to reproduce and survive.

A pond's **food chain** (see Figure 6.22) has three basic **trophic levels**:

- The first trophic level represents the producers and autotrophs, such as phytoplankton and plants. Producers prepare their own food with the energy emitted from the sun through a process known as photosynthesis.
- The second trophic level consists of herbivores, such as insects, crustaceans and invertebrates that inhabit the pond and consume the plants.

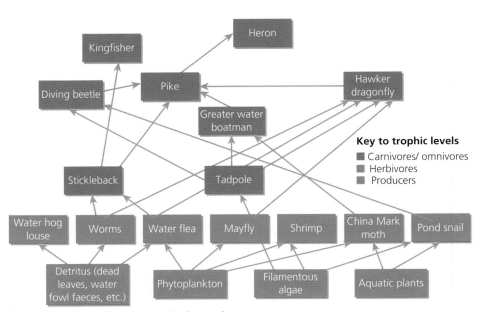

Figure 6.22 Generalised food web of a pond

- The third trophic level comprises of carnivores, such as various sizes of fish, which feed on both the plants and herbivores of the first and second trophic levels.

Decomposers located at the bottom of the food chain help decompose dead organic matter, which further breaks down into carbon dioxide and essential nutrients, such as nitrogen, phosphorus and magnesium. These nutrients supply the necessary life force for the autotrophs, etc., to produce food for the second trophic level organisms; this results in the perpetual flow of energy in the pond's ecosystem.

The importance of ponds

Ponds support many species of plants and animals – more than 100 UK BAP priority species are associated with them. They are particularly good for insects, such as damselflies and dragonflies.

Ponds provide important homes for amphibians including the protected great crested newt and common toad. They are also home to water voles, grass snakes and Daubenton's bats. Waterbirds such as swans, moorhens and tufted ducks rely on ponds for feeding and nesting, while waders like lapwing, redshank and snipe probe the muddy margins for invertebrates. Ponds provide stepping stones between isolated patches of habitat, linking up the countryside and allowing species to move about freely.

Recent research shows that 80 per cent of wildlife ponds in the UK are in a 'poor' or 'very poor' state and we have lost almost 500,000 in the last 100 years. This loss has a major effect on wildlife as they provide many species with a suitable breeding and feeding habitat. Ponds have been, and continue to be, lost to urban development and land-use change, agricultural drainage and in-filling, fragmentation and through poor management.

Fresh waters are adversely affected by a number of human pressures, including:

- excessive groundwater and surface water abstraction, which has caused many spring-fed ponds to dry up
- pollution, especially eutrophication. This is explained in Figure 6.23
- the effects of inappropriate land use or poor management, for example, afforestation, land drainage and overgrazing
- introduction of invasive plant and animal species, for example, the floating pennywort. This is a North American plant which was introduced to the UK

in the 1980s by the aquatic nursery trade. It roots in shallow margins of slow-flowing water bodies, particularly in ditches, slow-flowing dykes and lakes. It causes de-oxygenation of the water, which in turn affects fish and invertebrate populations and causes a choking of ponds and drainage systems

- urban, industrial, and agricultural development within catchments
- inappropriate development of recreation and navigation
- inter-basin water transfer schemes and the construction of dams and reservoirs.

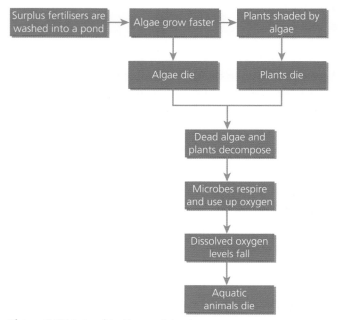

Figure 6.23 Eutrophication explained

Succession in a pond: a hydrosere

In time, an area of open freshwater such as a pond will silt up and dry out. If left alone for long enough (over 200 years) it will eventually become woodland. During this process, a range of different habitats, such as swamp and marsh, will succeed each other (Figure 6.24).

Stage 1: Open water

Deep freshwater will not support rooted, submerged plants because there is not enough light for photosynthesis at the bottom of the pond. There will be micro-organisms and plankton floating in the water.

Stage 2: Rooted plants

Over time, sediments will be transported into the pond by streams or rainwater. Large amounts of sediment can be deposited in this way. The water depth will gradually decrease, allowing rooted, submerged

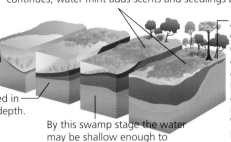

Deep freshwater will not support rooted submeged plants as there is insufficient light. There will be floating blue and green algae.

Swamp plants that grow in partly submerged conditions gradually die out as the pond edge rises above water level. Yellow iris continues, water mint adds scents and seedlings begin to grow.

Alder/willow carr. The ground is no longer completely saturated and anaerobic. Willows and alder trees that like moist ground dominate and they shade out the marsh undergrowth. Woodland floor plants like ferns and sedges appear.

Sediments are deposited in the pond reducing its depth. There will be rooted submerged plants like pondweed or rooted plants like lilies with floating leaves.

By this swamp stage the water may be shallow enough to allow emergent plants like yellow iris, branched bur-reed and greater reedmace.

Figure 6.24 A hydrosere

plants, such as starwort and pondweed to grow. Waterlilies, which are rooted, but with floating leaves, may also become established. The added plant life is accompanied by a variety of invertebrates and fish. The large numbers of different species present give rise to complex food webs.

The vegetation traps and holds more and more of the incoming sediment so that the water becomes shallower. Decomposing dead plant and animal matter provides food for detritivores and increases the nutrients in the water. This promotes plant growth.

Stage 3: Swamp stage

By this stage, the water may be too shallow to support fully submerged plants. Instead, emergent plants, such as yellow iris grow partly in and partly out of the water. These plants tend to have tall, flexible spear-shaped leaves. This allows them to cope with large differences in water level, always having some portion of the leaves above the water for photosynthesis.

The water in swamp areas teems with invertebrate life and provides an ideal place for amphibians to breed.

Stage 4: Marsh stage

Swamp plants which are adapted to grow in partially submerged conditions will gradually die out as the marsh floor progressively rises above the water level.

Some plants such as yellow iris, which grow equally well in swamp or marsh conditions, will continue to grow, while other marsh plants such as water mint become established. Tree seedlings, such as willow, which favour wet soil conditions, will become well established and begin to grow up.

Willow has a very high transpiration rate, transferring large quantities of water from the sediment into the

atmosphere. Together with the silt-trapping effect of the marsh plants, this greatly increases the rate at which the marsh dries out.

Stage 5: Carr and woodland stages:

The soil is still wet, but no longer completely waterlogged and air is able to enter along root channels. Willow and alder dominate the ground and shade out the lower-lying marsh vegetation. This is replaced by a variety of woodland floor plants including sedges, rushes, ferns and small flowering herbs which are adapted to low light levels and which flourish in wet conditions. Willows are able to support a wide range of invertebrates (more than 450 different species). This means that there is plenty of food for insectivorous birds.

In drier areas, increasingly aerobic decomposition will accelerate nutrient recycling, increasing the organic and nutrient content of the soil. Tree transpiration will continue to dry out the soil to the point where climax tree species such as oak, beech or ash can become established.

Pond conservation

One national example of pond conservation is the Million Ponds Project. This is a 50-year initiative set up by various charities and led and co-ordinated by Freshwater Habitats Trust. It aims to create an extensive network of new ponds across the UK, to reverse a century of pond loss and ensure that once again the UK has over one million countryside ponds.

Its first phase (2008–2012) aimed to create networks of clean water ponds across England and Wales. It also sought to change attitudes, so that pond creation is embedded as a routine activity in land management practices. The team worked with landowners and managers to create over 1,000 ponds for around half (49) of the 105 rare and declining pond species that are a national priority for conservation action under the UK government's Biodiversity Action Plan (BAP).

To achieve its aims, the project:

- funded new ponds and pond complexes for threatened freshwater plants and animals
- provided technical and on-the-ground support for pond creation
- engaged with and trained a wide range of stakeholders
- published best practice information in the extensive Pond Creation Toolkit, including the BAP species map
- raised the profile of ponds with policy makers and the media.

One critical element of the project is that these new ponds must have clean water. This is important because most countryside ponds are now badly damaged by pollution, and evidence shows that pond wildlife is declining across the UK.

Phase II of the Million Ponds Project is now underway in England and Wales, running from 2012 to 2020, and includes:

- identifying Important Areas for Ponds (IAPs) nationally – these are core areas in which to protect existing ponds, and from which stepping stones can be created into the wider countryside
- developing targeted landscape-scale pond creation projects in areas like the New Forest and in more intensively managed landscapes, such as catchments in Oxfordshire and Leicestershire
- investigating the potential to further embed pond creation into national policy initiatives, such as conservation credits, which could potentially create many thousands of clean water ponds in the future
- working with academic partners to develop applied research projects.

In 2014 and 2015 the project aimed to expand the Million Ponds Project to Northern Ireland and Scotland. Discussions with Natural England indicate that clean water pond creation will be an important part of the England Biodiversity Strategy objective to ensure that 90 per cent of Priority Habitats are in good condition by 2020 – with thousands of new high quality ponds created each year.

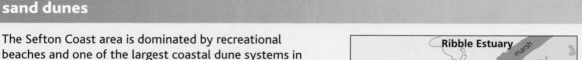

Case study of a specified ecosystem at a local scale: The Sefton Coast sand dunes

The Sefton Coast area is dominated by recreational beaches and one of the largest coastal dune systems in Great Britain, which extends 17 km along shore, 4 km inland, covers an area of 2,100 ha and shows dune heights reaching up to 30 m (Figure 6.25).

The sand dune area is well protected. It is a designated SSSI (Special Site of Scientific Interest), SAC (Special Area of Conservation), SPA (Special Protection Area) and under the RAMSAR convention on wetlands. There are three NNRs (National Nature Reserves) and three LNRs (Local Nature Reserves) within the borough of Sefton. Key species of this area include the sand lizard, natterjack toad, great crested newt, red squirrel, dune helleborine and pendulous flowered helleborine. There are 460 species of flowering plants recorded, including 33 locally or regionally rare species.

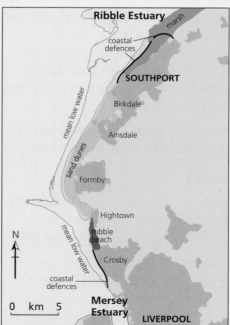

Figure 6.25 The location of the Sefton Coast sand dunes

Figure 6.26 Formby Point dune system

Succession on sand dunes

Sand dunes form above the highest high-tide mark behind a beach (Figure 6.26). As one moves from the bare sand of the beach to the dune heath and forest inland the location is older and vegetation and animal life have had longer to colonise and adapt to the environmental conditions. This sand dune succession is known as a **psammosere**. Each stage of a psammosere is known by its location on the coastline (Figure 6.27).

- **The strand line**: This is bare sand and it is inhospitable for plant growth. It is:

 - dry and salty, lacking in plant nutrients
 - has a high pH from fragments of sea shells
 - unstable.

There are a few highly specialised plants that have evolved strategies to grow in such a location. These are the **primary colonisers** of the psammosere and are restricted in their distribution to this very stressful type of habitat. Examples are:
 - sea rocket
 - prickly saltwort.

These annual plants may form miniature sand dunes as sand accumulates around the plant body and the plant is able to grow upwards a little through the accumulating sand. Alternatively, they may be washed away by the waves brought in on the next high tide.

- **Embryo dunes**: Sand accumulation which persists above the high-tide line is colonised by the first perennial plants in dune succession. These are specialised grasses such as:
 - sand couch
 - lyme grass.

Both of these grasses are able to grow upwards through accumulating wind-blown sand and as a result low, hummocky dunes are formed.

- **Mobile dunes**: Upward growth of the embryo dunes allows the surface to be raised so it is out of the reach

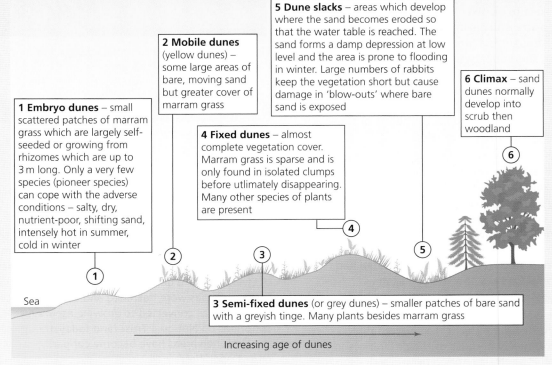

1 Embryo dunes – small scattered patches of marram grass which are largely self-seeded or growing from rhizomes which are up to 3 m long. Only a very few species (pioneer species) can cope with the adverse conditions – salty, dry, nutrient-poor, shifting sand, intensely hot in summer, cold in winter

2 Mobile dunes (yellow dunes) – some large areas of bare, moving sand but greater cover of marram grass

3 Semi-fixed dunes (or grey dunes) – smaller patches of bare sand with a greyish tinge. Many plants besides marram grass

4 Fixed dunes – almost complete vegetation cover. Marram grass is sparse and is only found in isolated clumps before utlimately disappearing. Many other species of plants are present

5 Dune slacks – areas which develop where the sand becomes eroded so that the water table is reached. The sand forms a damp depression at low level and the area is prone to flooding in winter. Large numbers of rabbits keep the vegetation short but cause damage in 'blow-outs' where bare sand is exposed

6 Climax – sand dunes normally develop into scrub then woodland

Sea

Increasing age of dunes

Figure 6.27 Plant succession on the Sefton Coast sand dunes

of all but the highest storm tides. This, together with the washing by fresh rainwater, results in a slightly less salty and unstable substrate. Under these conditions marram grass is able to colonise (see Figure 6.27, page 247). Marram grass is the major dune building grass. It is tall and robust (but flexible in the wind) and very effective at trapping sand by reducing the windspeed at the surface. It is able to grow upwards through accumulating sand at rates of up to one metre a year. The environment is still extremely stressful; marram is the dominant and only species present.

On the landward side of the frontal dunes the surface is more sheltered from the onshore winds and the effects of sea spray. More (still highly adapted) plants are able to colonise and the species diversity starts to increase. Good examples of these plants are:

- sea holly
- sea spurge.

As the conditions for plant growth improve with increasing distance from the sea, the species diversity continues to increase. A much smaller and very fine-leaved grass starts to fill in the bare spaces between the clumps of marram. This is:

- red fescue.

If conditions remain stable at the sand surface red fescue and mosses will continue to cover the bare sand between the patches of marram.

- **Fixed dunes:** When the vegetation has developed so that it forms a more or less complete cover, the dunes are said to be 'fixed'. These fixed dunes still represent a stressful environment. The pH is still very high, drought is a problem and nutrients are still in very short supply. In addition to these abiotic factors, the dunes may be affected by grazing or trampling. A thin, brown, organic layer has, however, started to form at the surface of the soil. If the dunes are grazed, for example, by rabbits or sheep, a **fixed dune grassland** will develop.

On the Sefton Coast the rabbit population was severely affected by the introduction of a disease called myxomatosis in the 1950s. The absence of grazing pressure caused by the loss of rabbits allowed a different type of vegetation, large coarse grasses and a woody plant called creeping willow, to develop. In recent years grazing by domestic stock has been re-introduced to the Ainsdale area and rabbit populations are now healthy again.

- **Dune slacks:** There are a number of wet and sometimes very large depressions in the dunes. These are called dune slacks. The water table fluctuates annually because of the differences in rainfall and evapotranspiration between the summer

and winter months. Because of this, many dune slacks are flooded during the winter period. The wet sand is colonised by plants and a succession occurs. This succession on wet sand, however, is very different to that on the dry dunes. A range of wetland plants are important and early vegetation can be extremely species-rich with plants such as sedges and horsetails together with a number of mosses. These slacks are the breeding ground for the natterjack toad (Figure 6.28).

- **Scrub and woodland:** Since the rabbit population was almost destroyed in the 1950s tall woody plants that invaded naturally have become widely established on the dunes. These include:
  - birch
  - hawthorn.

In addition, pines from the plantations seed into the open dunes.

Figure 6.28 A protected damp slack designed to encourage natterjack toads.

Skills focus

How might you survey the succession in a location such as Formby?

The human impact on the dunes of the Sefton Coast

From the middle of the nineteeth century, after the Southport to Liverpool Railway was built, vast amounts of 'night soil' (human manure) from the backstreets of Liverpool were dumped in the dunes in Freshfield, Formby. Dunes were flattened and about 80 ha of duneland was converted to growing asparagus. This continues on National Trust land today, though most of the flattened areas have become car parks, caravan sites or just open fields.

Uncontrolled public access in the 1960s and 1970s caused severe dune erosion at Formby Point. Dogs and people cause disturbance to ground-nesting birds. Off-road vehicles damaged large areas of sand dunes in the 1970s and 1980s until controlled by the introduction of countryside wardens.

Over 25 per cent of the Sefton Coast dunes are managed as golf courses. Management techniques such as drainage, irrigation, mowing, fertilising and reseeding has damaged habitats. However, much golf management is sympathetic to the conservation of the dune–links landscape and its characteristic species and habitat types. Approximately 70 per cent of each golf course area is out of play and direct golf management.

Management

In the early twentieth century the area was planted with pine trees. The aim was to stabilise the dunes and turn the 'wasteland' into a more productive estate providing a timber crop, woodland products and improved opportunities for agriculture and game. The first pines were planted around 1900 and areas continued to be planted until the 1960s. Today, these woodlands support an important population of the red squirrel, now rare in England. The slacks were generally not planted, although drainage ditches were put in to lower the water table. This drainage has led to a loss of wet slacks, an important habitat for protected species such as the natterjack toad.

Once the woodland canopy was formed, trees shaded out the area beneath and so no other plants were able to grow. Huge areas of natural dune landscape with specialised sand dune plants and animals were lost in this way. The trees also cause the water levels in the soil to drop; pine needles fall on the surface and alter the structure and chemistry of the soil. Together these create very different conditions from the original dunes. The needles are very acid and there is a very large accumulation of plant litter on the soil surface. In these pine woodlands, the pine trees are often the only plant species present (Figure 6.29).

In one part of the area, the Ainsdale National Nature Reserve, the seaward plantations have been removed. This has allowed the recolonisation by specialised plants such as yellow bartsia, and animals such as the sand

lizard and natterjack toad. Monitoring of the project shows that this is already a success. The landward plantations have been allowed to stay because of their importance to the red squirrel.

Figure 6.29 Pine woodland on the dunes at Formby

Conservation of fixed dunes

The fixed dunes and associated species suffer from a number of factors threatening to reduce their nature conservation value. The spread of scrub and rank vegetation, in particular, leads to associated problems of soil development and the drying out of slacks. In order to manage this there has been:

- **scrub cutting and clearance:** this allows light and warmth to reach the surface so colonisers can begin to grow
- **mowing** to control the height and density of creeping willow in dune slacks where it was rapidly overwhelming the short botanically diverse dune turf
- **turf-stripping and excavation:** some of the slacks have been scraped to provide breeding pools for the natterjack toad.
- **grazing:** domestic grazing was re-introduced onto the Ainsdale dunes by English Nature. It has controlled target species such as creeping willow; there has been an increase in species diversity with a corresponding return to a mix of vegetation with bare sand patches.

6.2 Ecosystems and sustainability

Biodiversity

> The preservation of biodiversity is not just a job for governments. International and non-governmental organisations, the private sector and each and every individual have a role to play in changing entrenched outlooks and ending destructive patterns of behaviour.
>
> *Kofi Annan*, Secretary General of the United Nations

Biodiversity, which is short for **biological diversity**, is the term used to describe the whole variety of life on Earth. It encompasses the diversity of all living things, from human beings to micro-organisms, the diversity of all the habitats in which they live and the genetic diversity of individuals within a species.

All of these are part of the global ecosystem – the Earth's entire collection of living things and the environment in which they live. Ecosystems form the basis for all forms of life support, including provision of food, clean air and water, climate and disease regulation and many others. Biodiversity underpins the health of ecosystems.

Measuring biodiversity

From the very broad definition given above it is clear that no single measure of biodiversity will be adequate. Biodiversity cannot be captured in a single number because measurements have two components:

- the number of entities, for example, the number of individuals, the number of species, the number of different habitats, etc.
- the degree of difference (dissimilarity) between those entities.

Measuring biodiversity is commonly used to provide evidence for the need for conservation or for planning more generally. Often **indicator species** are used as a way of measuring biodiversity. For example, stoneflies are found in clean, fast-moving streams with a gravel or stone bottom. The presence of stonefly nymphs indicates highly oxygenated (healthy) water. Mussels are used to measure the health of coastal environments worldwide.

This method can be very useful but it introduces an aspect of how we value different components of biodiversity. For example, we are more likely to use the abundance of birds or butterflies on a farm as a measure of biodiversity than the richness of microbes in the soil.

Species richness is one of the most commonly used measures of biodiversity and it refers to the number of different species; it tells us how many different species there are but does not convey information on the frequency of individuals of each of those species.

The advantages of this:

- Species richness has proven to be measurable in practice. It is doable.
- A lot of information already exists on patterns in species richness, and this has been made available in scientific literature. This allows comparisons over time.
- Species richness acts as a proxy measure for many other kinds of variation in biodiversity. In general, greater numbers of species tend to indicate more genetic diversity (in the form of a greater diversity of genes through to populations), more species diversity and greater ecological diversity.
- It is commonly seen as the unit of practical management and is most easily applied to legislation and political debate. For a wide range of people, variation in biodiversity is pictured as variation in species richness.

The disadvantages of this:

- There is a lack of agreement as to what constitutes a species. For example, it is not an adequate method of defining single cell organisms.
- Some assemblages of organisms may be very closely related (for example, two species of rat) while others are diverse (for example, a species of rat and a species of snail).

A more recently developed measure of biodiversity is the **Living Planet Index (LPI)**. The LPI measures trends in thousands of vertebrate species populations and provides information on changes in the abundance of the world's vertebrates. It can quickly convey information on which habitats or ecosystems have species that are declining most rapidly. This information can be used to define the impact humans are having on the planet and for guiding actions to address biodiversity loss.

Trends in biodiversity

Global trends in biodiversity

Through geological time, global biodiversity has steadily increased from when life as we know it started about 550 million years ago (mya). In that time there have been five episodes of what scientists call **mass extinctions** (for example, the end of the dinosaurs about 65 mya). Some modern scientists believe that we are now in another period of mass extinction, this time caused by man.

Certainly, population sizes of vertebrate species measured by the LPI have more than halved in recent years. It shows a decline of 52 per cent between 1970 and 2010. This means that the number of mammals, birds, reptiles, amphibians and fish across the globe is, on average, about half the size it was 45 years ago. This is a much bigger decrease than had been reported previously.

Biodiversity is declining in both temperate and tropical regions, but the decline is greater in the tropics. The tropical LPI showed a 56 per cent reduction in 3,811 populations of 1,638 species from 1970 to 2010. The 6,569 populations of 1,606 species in the temperate LPI declined by 36 per cent over the same period. Latin America shows the most dramatic decline – a fall of 83 per cent.

Exploitation through hunting and fishing and habitat degradation and loss are the main causes of decline (Figure 6.30). Climate change is the next most common primary threat, and is likely to put more pressure on populations in the future.

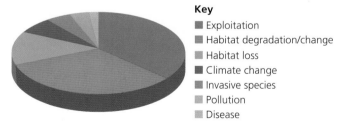

Key
- Exploitation
- Habitat degradation/change
- Habitat loss
- Climate change
- Invasive species
- Pollution
- Disease

Figure 6.30 Primary threats to LPI populations according to WWF

Terrestrial species declined by 39 per cent between 1970 and 2010, a trend that shows no sign of slowing down. The loss of habitat to make way for human land use – particularly for agriculture, urban development and energy production – continues to be a major threat, made worse by hunting.

The LPI for freshwater species shows an average decline of 76 per cent. The main threats to freshwater species are habitat loss and fragmentation, pollution and invasive species. Changes to water levels and freshwater system connectivity – for example through irrigation and hydropower dams – have a major impact on freshwater habitats.

Marine species declined 39 per cent between 1970 and 2010. The period from 1970 through to the mid-1980s experienced the steepest decline, after which there was some stability, before another recent period of decline. The steepest declines can be seen in the tropics and the Southern Ocean – species in decline include marine turtles, many sharks and large migratory seabirds like the wandering albatross.

Many species have completely disappeared from areas dominated by human influences (Figure 6.31, page 252). Even in protected areas, native species are often out-competed or consumed by organisms introduced from elsewhere. Extinction is a natural process, but it is occurring at an unnaturally rapid rate as a consequence of human activities. Already we have caused the extinction of 5–20 per cent of the species in many groups of organisms, and current rates of extinction are estimated to be 100–1,000 times greater than pre-human rates.

Key question

Why should we be concerned over this loss of species?

Example of a local area subject to biodiversity change

One of the most extensive, semi-natural habitats in the UK is **blanket bog.** It ranges from Dartmoor in the south to Shetland in the north. It is a globally restricted peatland habitat confined to cool, wet, typically oceanic climates. Blanket bogs comprise of generally upland **peat** bogs that are fed by rainfall alone. Peat is an accumulation of dead organic material, mainly plant remains, that is unable to break down because of the wet, acidic and cold conditions in which it is found. These peat bogs take thousands of years to form and support mosses, insects, rare birds and even carnivorous plants. Peat depth is very variable, with an average of 0.5–3 m being fairly typical.

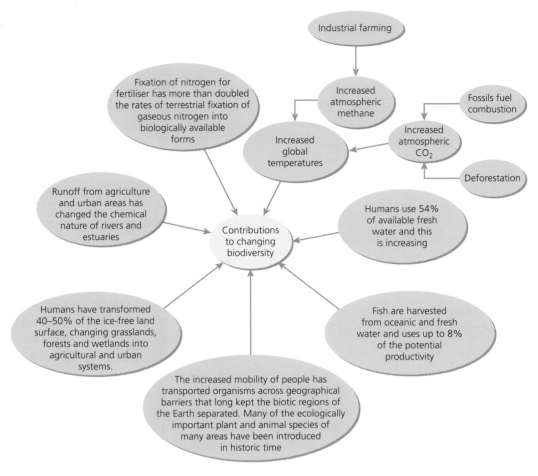

Figure 6.31 Contributions to changing biodiversity

Because peat is essentially semi-decomposed plant matter, it represents a huge carbon store: more carbon is stored in UK peat than in a combination of all the forests in Britain and France combined! When these habitats are healthy, and peat is actively forming, they are building up carbon, so they have an important role to play in helping to combat climate change.

One area of blanket bog is **Kinder Scout, Derbyshire**. Kinder Scout is an area of moorland and National Nature Reserve in the Dark Peak of the Derbyshire Peak District in England. Part of the moor, at 636 m above sea level, is the highest point in the Peak District. The 7,000-year-old blanket peat bog sits on top of a plateau (Figure 6.32 and Figure 6.33) formed from horizontal layers of an impermeable sandstone called Millstone Grit. The land is owned and managed by the National Trust.

Figure 6.32 The plateau of Kinder Scout

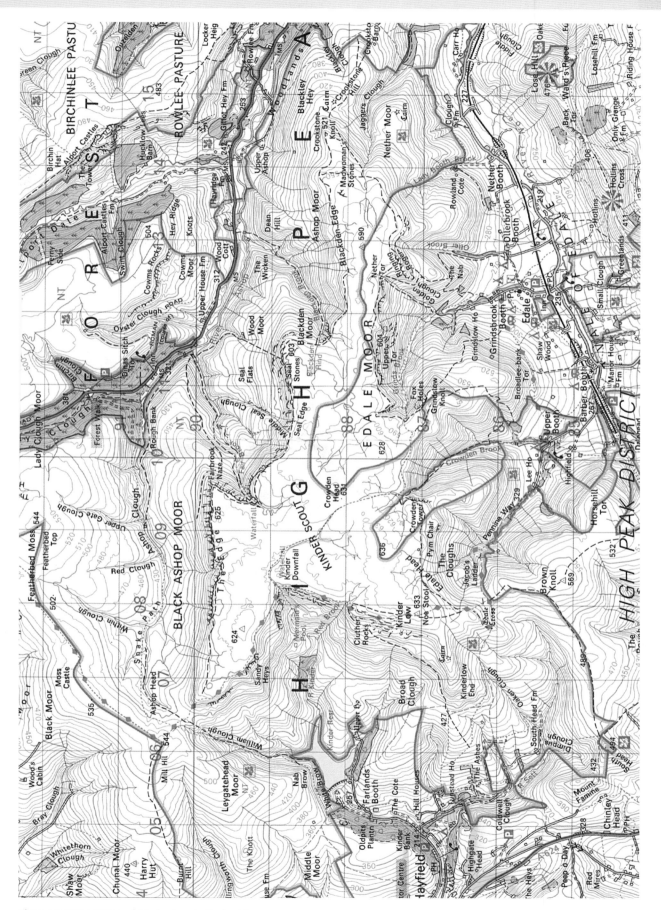

Figure 6.33 Location of Kinder Scout, Derbyshire

Kinder Scout is also one of the most damaged areas of moorland in the UK and its future is in jeopardy as a result of catastrophic wild fires, a long history of overgrazing by sheep, air pollution and the routes that thousands of visitors have taken.

In spring 2011 the National Trust started a five-year project to restore the peat bogs of Kinder Scout. They started by fencing much of the moorland off from sheep. The fencing should be in place for 15 years, at which point the Trust and other bodies will take stock of the impact.

They planted nearly half a million pods of cotton grass in 2011 (they are dropped by helicopter but then are planted by hand); these help to bind the loose peat together and combat erosion. This is part of a longer-term reseeding programme. Following on from that, heather brash was spread across large areas of Kinder to encourage heather to re-seed. Heather is cut and harvested from nearby local moors in winter and is spread over the worst eroding areas of blanket bog. It has four purposes:

- acting as a skin, thus reducing the effects of erosion
- providing a microclimate for seeds to grow in as they are protected from the harsh weather

Table 6.3 Causes and impacts of biodiversity change on blanket bog on Kinder Scout

Cause	Consequence	Impacts and potential impacts
Climate change		
Increased mean temperatures	Longer growing season	Bracken has become invasive at higher altitudes
		Potentially increased nitrogen deposition leading to changes in plant species
		Mire vegetation (that is, plant types that thrive in wet conditions) may become less dominant
Hotter summers	Increased evapotranspiration and lowering of water table	Changes in species composition. Increase in the release of dissolved organic carbon leading to declining water quality
Drier summers	Drought	Shift in the dominance of species
	Drier ground conditions	Possible wind erosion of dried peat
	Wildfire	Possible increased agricultural potential
		Changes in red grouse populations
Wetter winters	Increased overland flow	Loss in peat stability; increased slides
Storm events	Increased rainfall intensity	Gullying erosion
Other influences		
A long period of over-grazing	Overgrazing	Loss of biodiversity
Access to walkers from nearby industrial towns and cities since the 1940s	Footpath erosion. Damage to surface vegetation that holds the peat in place	Large areas of peat lost
		Exposed dried-out peat subject to wind and gullying erosion
Increased wildfires	Loss of peat	Diminution of species numbers
Draining of peat bog	Lowering of water table	Sphagnum mosses have been replaced by other species such as heather and moor grass
		Species of moorland birds that nest and feed in the heather, such as golden plover and curlew, as well as the mountain hare are threatened
Industrial pollution		Reduction in number of species

- providing seeds – heather is cut in winter when the seeds are ripe and thus they are transported onto the moors
- providing a fungi that helps moorland plants to thrive.

Large numbers of erosion gullies have been blocked each winter, involving the creation of small *dams* made of stone, timber or plastic piling, to reduce peat erosion, help keep the peat wet and encourage the new vegetation to thrive.

Skills focus

Using Figure 6.32 (page 252) and Figure 6.33 (page 253), describe the landscape of Kinder Scout and the surrounding area.

Describe the pattern of drainage for the Kinder Scout area. How has that helped Kinder Scout to develop into a blanket bog?

Key question

What are the costs and benefits of the type of conservation scheme that has been carried out on Kinder Scout?

Human population and ecosystems

An ecosystem is a dynamic complex of plant, animal, and micro-organism communities and their non-living environment interacting as a functional unit. Humans are an integral part of ecosystems.

Ecosystem services are the benefits that people obtain from ecosystems. The **Millennium Ecosystem Assessment (MEA)** analysed 24 ecosystem services and found that 15 were being degraded or used unsustainably. The decline in services affects the world's disadvantaged people most strongly, impedes sustainable development globally and, in developing countries, represents a considerable barrier to achieving the UN's Millennium Development Goals of reducing poverty and hunger.

The MEA grouped ecosystem services into four categories:

- **provisioning services,** such as the supply of food and water
- **regulating services,** which help to stabilise ecosystem processes such as climate and water storage and purification

- **supporting services**, including soil formation and nutrient cycling; and
- **cultural services**, such as recreational, spiritual, religious and other non-material benefits.

The MEA also considers human well-being to consist of five main components:

- the basic material needs for a good life
- health
- good social relations
- security
- freedom of choice and action.

Figure 6.34 (page 256) shows one attempt to highlight the importance of ecosystems and biodiversity to human populations. Human well-being is the result of many factors, many are directly or indirectly linked to biodiversity and ecosystem services while others are independent of these.

Many people have benefited over the last century from the conversion of natural ecosystems to human-dominated ecosystems and the exploitation of biodiversity. At the same time, however, these losses in biodiversity and changes in ecosystem services have caused some people to experience declining well-being, with poverty in some social groups being made worse.

Although biodiversity and ecosystem services experience change due to natural causes, there are also five indirect drivers of change:

- demographic
- economic
- sociopolitical
- cultural and religious
- scientific and technological.

Current change is driven by these **anthropogenic** factors, in particular growing consumption of ecosystem services (as well as the growing use of fossil fuels), which results from growing populations and growing per capita consumption. Figure 6.35 (page 256) is an illustration of this. Global economic activity increased rapidly from 1950 onwards and despite occasional economic downturns, is expected to continue to rise. At the same time the global population reached 6 billion in 2000 and is projected to reach 8.1–9.6 billion by 2050.

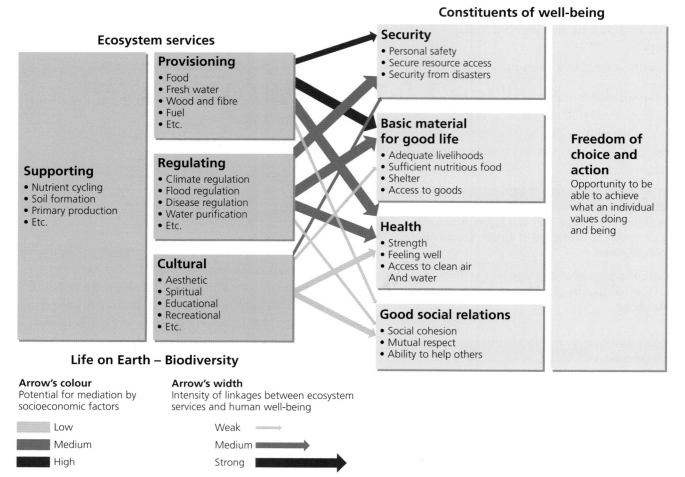

Figure 6.34 Linkages among biodiversity, ecosystem services and human well-being

Source: UN Millennium Ecosystem Assessment

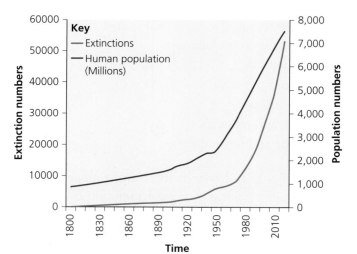

Figure 6.35 Species extinction and human population

Source: US Geological Survey

The UK ecosystem services

In 2011, a major UK government-sponsored report, the National Ecosystem Assessment (NEA) attempted to put a value on the UK's ecosystems. The conclusion was that the parks, lakes, forests and wildlife are worth billions of pounds to the economy. It concluded that the health benefits of merely living close to a green space are worth up to £300 per person per year.

> The natural world is vital to our existence, providing us with essentials such as food, water and clean air – but also cultural and health benefits not always fully appreciated because we get them for free.
>
> *Caroline Spelman* Environment Secretary

The economic benefits of nature are seen most clearly in food production, which depends on organisms such as soil microbes, earthworms and pollinating insects. If their health declines – as is currently happening in the UK with bees – either farmers produce less food, or have to spend more to produce the same amount.

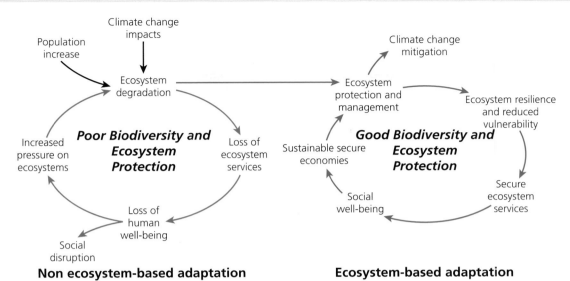

Figure 6.36 Beating the vicious cycle of poverty, ecosystem degradation and climate change

The report also concluded that:

> Humans rely on the way ecosystems services control our climate – pollution, water quality, pollination – and we're finding out that many of these regulating services are degrading.

> About 30 per cent of the key ecosystem services that the UK rely on are degrading and about 20 per cent are getting better. Our air quality, however, has improved a lot.

Ecosystem development and sustainability

Figure 6.36 shows two different scenarios that could result from human population growth and climate change. One way, the positive feedback loop, shows how poor ecosystem management leads to decline in human well-being and eventual further decline in ecosystems. Good ecosystem management, on the other hand, causes there to be a negative feedback loop which leads to climate mitigation and social well-being.

Key question

What type of activities might be considered 'good ecosystem management'?

6.3 Marine ecosystems

The marine ecosystem is the largest ecosystem on Earth because oceans and seas account for almost 70 per cent of its surface. It can be divided into smaller ecosystems, such as rocky shores, salt marshes, submarine canyons, hydrothermal vents, deep basins, seamounts and coral reefs.

Coral reef ecosystems

There are two broad groups of corals:

Deep-sea cold-water coral reefs

These are a recent discovery on Europe's margin. They are found in the deeper, nutrient-rich high latitude waters where ocean currents accelerate and prevent sedimentation. Colonial stone corals such as Lophelia pertusa and Madrepora oculata occur along much of the northwest European continental margin and in some Scandinavian fjords and have recently been discovered in the Mediterranean. They grow slowly (5–25 mm/year), but over time they form extensive reefs. The largest reef yet discovered, off the coast of Norway's Røst Island, is 40 km long and 2–3 km wide.

Tropical coral reefs

Tropical coral reefs are found between 30° north and south of the equator, in areas where surface water temperatures do not drop below 16 °C (Figure 6.37, page 258). The total area of the world's tropical coral reefs is around 284,300 km². These reefs are able to grow upwards at rates of 1–100 cm/year. They can form huge structures over incredibly long periods of time, making them the largest and oldest living systems on earth. For example, Australia's 2,000 km-long Great Barrier Reef was formed over the course of five million years.

A coral reef is composed of calcium carbonate, or limestone. This is absorbed from the water by colonies of coral polyps and coralline algae. Most of the underlying foundation of the reef is dead, made up of

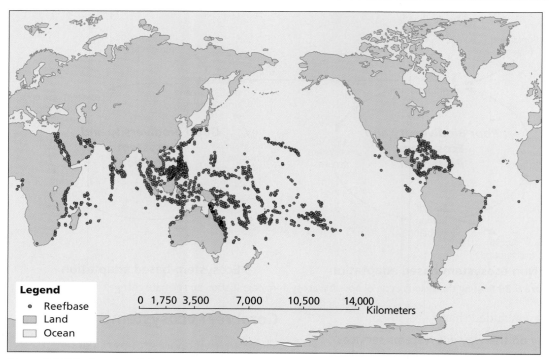

Figure 6.37 The location of the major warm water stony coral reefs of the world

layer upon layer of coral skeletons. The living reef is built over the top of this by tiny coral polyps adding new limestone to the massive base structure. Tropical coral polyps have small, symbiotic algae, or zooxanthellae, growing inside them. The algae get shelter and food (in the form of nutrients from captured plankton) from the polyp, while the polyp gets some food from the algae via photosynthesis. This photosynthesis means that the algae (and so the coral) need sunlight to live, explaining why tropical corals only grow where the sea is shallow, clear of sediment and away from river mouths.

The algae also give corals their colour. If they become stressed, such as if the water temperature becomes too high, they leave the polyp. This exposes the white, calcium carbonate skeletons of the coral and is called coral bleaching.

Coral reefs are divided into three main types:

- Fringing reefs grow directly from a shore with no area of open water between the beach and the inshore edge of the reef. This is the most common type of reef. Because there is no intervening lagoon to act as a buffer between the land and the reef, they are particularly sensitive to pollution and sedimentation from rivers.
- Barrier reefs are extensive linear reef complexes that develop parallel to the shore, separated from it by a lagoon (a wide band of sea water of varying depth).

- Atolls are roughly circular reef systems surrounding a central lagoon.

An example of a tropical coral reef system, the Jamaican coral reef

Jamaica, the third largest Caribbean island is 230 km long by 80 km wide with 891 km of coastline and a coral reef area of 1,240 km². Well-developed fringing reefs occur along most of the north and east coasts, while patchy fringing reefs occur on the broader shelf of the south coast (Figure 6.38). In addition to the reefs surrounding mainland Jamaica, there are reefs and corals on the neighbouring banks and shoals within Jamaica's Exclusive Economic Zone including Morant and Pedro Cays.

There have been significant changes to Jamaican coral reefs over the past 30 years due to the combined effects of human and physical influences.

Physical influences

- It is thought that **coral bleaching** is caused by unusually warm sea surface temperatures. Coral bleaching events worldwide have been attributed to sea surface temperatures (SSTs) rising and staying as little as 1 °C higher than the usual average monthly maximum SST during the hottest months of the year. In Jamaica, significant coral bleaching

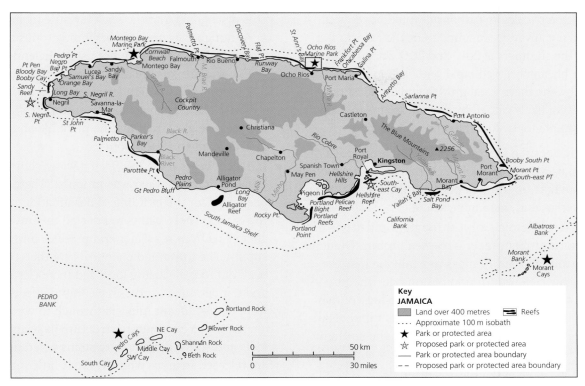

Figure 6.38 The coral reefs of Jamaica

and death has occurred when SSTs were at 29.3 °C or higher for July, the hottest month, in which the average SST is 27.5 °C. When a coral bleaches, it loses its zooxanthellae and will die within a matter of weeks unless the zooxanthellae can be replaced. In 2005, sea surface temperatures were high and on average 34 per cent of Jamaican corals bleached. In places this rose to 95 per cent. Fortunately, less than half (13 per cent) suffered subsequent mortality.

- Coral reefs develop best where the salinity of the sea water is between 34 and 37 parts per 1,000. Where rivers flow into the sea there are gaps in the reef because of the lowering of the salinity locally. This can be seen at the mouths of the South Negril River, the Rio Cobre and the Rio Bueno.
- Studies regarding links between the acidification of sea water and coral development have been inconclusive. Although scientists think that coral's ability to form an

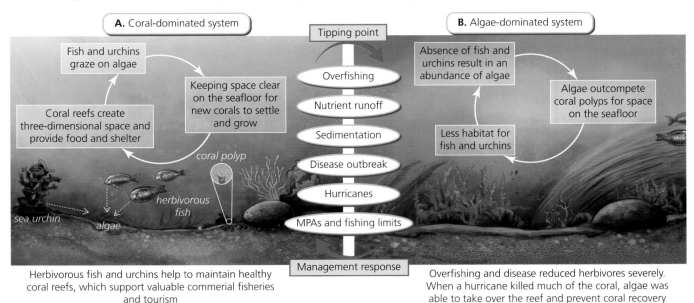

Herbivorous fish and urchins help to maintain healthy coral reefs, which support valuable commerial fisheries and tourism

Overfishing and disease reduced herbivores severely. When a hurricane killed much of the coral, algae was able to take over the reef and prevent coral recovery

Figure 6.39 Changes in the coral reefs of Discovery Bay, Jamaica

exoskeleton of calcium carbonate would be reduced, this has not always been found to be the case. In fact during the summers of 2010 and 2011 scientists in the Caribbean observed rapid acidification events driven by enhanced coral growth and calcification, which were likely caused by an increase in food supply thanks to phytoplankton blooms.

Not only does the formation of $CaCO_3$ produce CO_2, well-fed corals respire more CO_2 than their symbiotic algae can take up through photosynthesis – making the water around them more acidic.

- A disease epidemic swept through populations of an important reef grazer – the sea urchin, Diadema antillarum, and led to mass mortality across the Caribbean. Sea urchin populations declined by 99 per cent on Jamaica's coral reefs. The lack of fish and the urchins meant that there was rapid algal growth (an algal bloom). These limit the sunlight available and cause lack of oxygen in water. When oxygen levels decline, marine animals, coral reefs, seagrass beds and other vital habitats suffer and may die.

Coral cover declined from over 50 per cent in the 1970s to less than 5 per cent by the early 1990s. During the same period, macroalgae increased to a peak of 92 per cent cover on the reef. This is illustrated in Figure 6.39 (page 259).

- Major hurricanes such as Hurricane Allen (1980) and Hurricane Katrina (2005) have damaged Jamaica's reefs. The violence of the waves fragment and kill branching coral in shallow water. Even the soft corals are damaged by objects being tossed around in the waves. Sometimes the scars that resulted were rapidly overgrown by fast growing opportunistic algae.

Human influences

- Pollution comes in many forms including:
  - Onshore development such as bauxite mining is particularly important for the economy of Jamaica. This activity also releases large quantities of particulate matter (sediment) that can be deposited in marine areas.
  - Coastal developments such as tourist hotels in Jamaica are creating problems for the satisfactory treatment of sewage. Coastal hotels each have their own disposal systems but much waste

is discharged into the sea after little treatment. Although they are now beginning to install treatment plants, many are not used efficiently.
  - Excess nutrients are discharged to the marine environment through fertilisers from agriculture. These nutrients may lead to eutrophication (an excessive growth of marine plant life and decay). This leads to algal blooms.

- In recent years there has been a water shortage in Jamaica; in 2012 water was rationed. There have been calls for the building of desalination plants on the island. These plants remove all the unwanted chemicals from sea water and produce a fresh water supply. The unwanted chemicals are then flushed back out to sea causing localised increases in salinity and temperature. Studies have found effects ranging from no significant impacts through to widespread alterations to community structure in seagrass, coral reefs and soft-sediment ecosystems when discharges are released in areas where there is little mixing of sea water.

- The fish population on Jamaican reefs has been steadily declining as a result of over-fishing. One study indicated that fish biomass had already been reduced by up to 80 per cent by the late 1970s.

The future prospects for coral reefs

In the Caribbean the coral reefs have collapsed and been replaced by beds of algae. The proportion of the reef covered by live coral has fallen from 50 per cent in the 1970s to just 8 per cent in 2013, changing the fish communities dramatically. In fact a third of reef-building corals are in danger of extinction and reefs the world over are in serious decline.

To lose the reefs would be to lose the planet's most diverse ecosystems. People would also suffer. More than 450 million people live close to coral reefs and rely on them as sources of tourism revenue and protein and as buffers that dampen the energy of incoming storms.

Climate change is the biggest threat. The warming atmosphere heats up the oceans causing increased coral bleaching events. Further increases in dissolved CO_2 will make the oceans more acidic, depleting the carbonate ions that the corals need.

In 2012, ecologist Roger Bradbury provided a bleak outcome in an opinion piece for *The New York Times*, saying, 'There is no hope of saving the global coral reef ecosystem.' Opponents of this viewpoint state that not all corals, or all reefs, are the same. Some, like those in American Samoa, have genetic advantages that allow them to thrive in shallower warmer waters and some can recruit strains of algae that tolerate higher temperatures. Others grow in waters that are naturally acidic, where carbon dioxide seeps from the ocean floor. Corals can acclimatise to changing conditions, and there is some evidence that reefs which bleach extensively for one year are better able to handle warmer waters a decade later. Reefs can also change at the community level, shifting from sensitive species like the elkhorns to sturdy, robust ones like big boulder corals.

Humans can help by setting up marine protected areas – underwater national parks – where fishing is forbidden. Not only do they allow local reefs a chance to recover, but they can seed nearby areas with coral larvae. This will slow down the decline but if predictions of ocean warming in the next 100 years are correct then coral reefs will die.

In Jamaica, marine protected areas account for 180,000 ha or approximately 15 per cent of the country's coastal waters (and 1.1 per cent of Jamaica's total marine area).

6.4 Biomes

Concept of a biome

Climate, topography and soil determine the changing character of plant and animal life, as well as ecosystem functioning, over the surface of the earth. Although no two locations contain exactly the same assemblage of species, we can group biological communities and ecosystems into categories based on climate and dominant plant form, which give them their overall character. These categories are referred to as biomes.

A biome, therefore, is a large geographical area of distinctive plant and animal groups which are in the climatic climax stage of succession; a global-scale ecosystem. Each biome gets its name from the dominant vegetation and climate found there. There are 11 major terrestrial biomes plus several marine and aquatic ones. Each biome consists of many ecosystems whose communities have adapted to the differences in climate and the environment within that biome.

Ecosystems belonging to the same biome but in different parts of the world develop a similar vegetation structure and ecosystem functioning (including productivity and

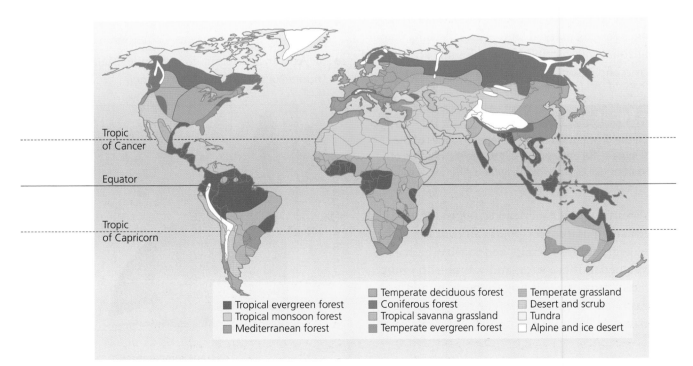

Figure 6.40 Vegetation zones of the world

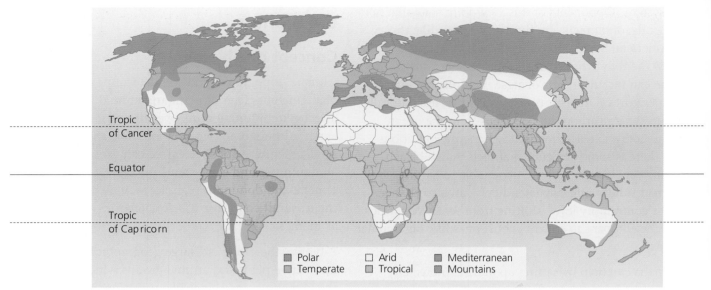

Figure 6.41 Climate zones of the world

nutrient cycling). An example of this is the way in which the woodland/scrubland ecosystems characteristics of Mediterranean climates (cool, wet winters and hot, dry summers) look similar and function in the same way whether in southern California, southern France, Chile, South Africa or Australia.

The major world biomes are based upon the underlying properties of temperature and soil moisture regimes; these are largely determined by latitude, climate, topography and the native vegetation that is adapted to these local conditions. Comparing maps of the world's vegetation (Figure 6.40, page 261) and climate (Figure 6.41) zones reveals marked similarities. A third map showing zonal soil distribution at the same scale would also bear comparison.

Skills focus

Describe the distribution of the vegetation zones of the world (Figure 6.40) and compare it to the distribution of the climate zones of the world (Figure 6.41).

Robert Whittaker, an American ecologist, plotted rainfall vs. temperature for points all over the globe on a single graph. He then looked at what biomes had developed at those sites, and was able to group the different biomes according to mean annual temperature and precipitation. The results are summarised in Figure 6.42.

Contrasting biomes–Biome 1
Tropical evergreen forest

These are mostly distributed between the latitudes 10° north and 10° south of the equator. They are found in:

● the Amazon river basin of South America, with more than half of it in Brazil; this area holds about one-third of the world's remaining equatorial forests

● the equatorial part of Africa including large parts of the Republic of Congo, the Central African Republic and Gabon

● the Guinea coast of Africa in isolated pockets stretching from Liberia to Cameroon

● SE Asia and Oceania, for example, Malaysia, Indonesia and Papua New Guinea.

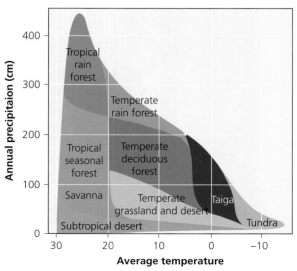

Figure 6.42 Whittaker biome diagram

Main characteristics of the tropical evergreen forest

The climate

The equatorial climate as shown in Figure 6.43 has no seasons. It is hot and wet all year round. The average temperatures rarely go below 24°C or rise above 28°C. There is a low diurnal range of temperature; during the daytime it might reach above 30°C, but the nights remain very warm. The high temperatures are because the sun is high in the sky all year round. They are, though, moderated by the presence of cloud.

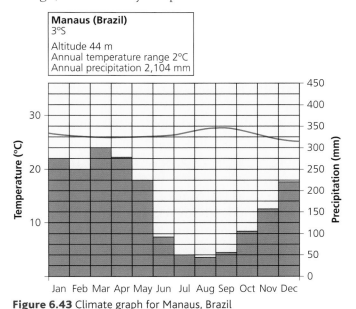

Figure 6.43 Climate graph for Manaus, Brazil

Tropical forests are to be found where the trade winds from the northeast and the southeast come together in the inter-tropical convergence zone (ITCZ, a low-pressure zone). At these low latitudes, intense solar energy from the overhead sun induces convection and rising air that loses its moisture through frequent rainstorms. Rainforests are subject to heavy rainfall, often over 2,000 mm/year. In these equatorial regions, rainfall may be all year round without apparent 'wet' or 'dry' seasons, although forests on the northern and southern fringes of the region do have a short dry season. Even in seasonal forests, the period between rains is usually not long enough for the leaf litter to dry out completely. During the parts of the year when less rain falls, the constant cloud cover is enough to keep the humidity high and prevent plants from drying out. The rainforest has been likened to a natural greenhouse.

The pattern of rainfall in the rainforest varies during the day. In the morning, skies are generally clear, though there may be a low mist. Evapotranspiration is rapid as the sun beats down on the humid forest and the low-pressure conditions allow this air to be rapidly uplifted. As the air rises it cools and water vapour condenses into clouds. These clouds continue to build until they become towering, grey, cumulonimbus clouds in the early afternoon. In the middle of the afternoon, heavy rain, often with thunder and lightning, returns the previously uplifted moisture back to ground level. The cycle begins again and the day ends as it started with clear skies.

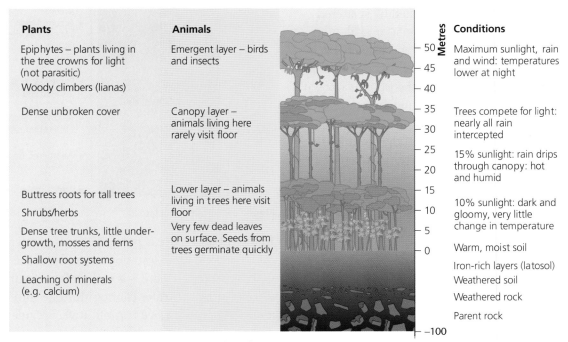

Figure 6.44 The layered structure of tropical rainforest

Day and night are more or less the same length at the equator. Dawn arrives at around 6 a.m. and night falls quickly at 6 p.m. There is little twilight. Twelve hours of sunlight every day allows photosynthesis to take place all year.

On the forest floor there is little breeze, though this can increase 40 or 50 m up. Right on the equator there is a distinct lack of wind because this is where the trade winds converge.

The vegetation

The stable climate, with evenly spread rainfall and warmth, allows most rainforest trees to be evergreen, constantly shedding and replacing leaves all year round. The moisture in the rainforest from rainfall, constant cloud cover and transpiration creates intense local humidity. Each canopy tree transpires some 760 litres of water annually; this translates to roughly 190,000 litres of water transpired into the atmosphere for every hectare of canopy trees. Large rainforests (and their humidity) contribute to the formation of rain clouds and generate as much as 75 per cent of their own rain.

Primary rain forests are very productive. The net primary productivity is 2,200 g/m²/year. This means that for every square metre of land, the forest produces 2,220 g of living matter. They are characterised by a unique vegetative structure consisting of several layers (Figure 6.44, page 263). The **canopy** refers to the dense ceiling of leaves and tree branches formed by closely spaced crowns of forest trees. The canopy is 20–40 m above the forest floor and is penetrated by scattered **emergent** trees that can reach above 50 m. Below the canopy ceiling are multiple leaf and branch levels known collectively as the **understorey.** The lowest part of the understorey, 1.5–6 m above the floor, is known as the **shrub layer,** made up of shrubby plants and tree saplings. The canopy screens light from the forest floor so that less than 5 per cent of the incident sunlight reaches it; this makes the ground layer quite open. When a tree dies naturally it brings down others as it falls, creating a small clearing in the forest. Saplings grow quickly taking advantage of the light. The fallen tree is rapidly decomposed, assisted by detrivores, and by the hot, humid conditions.

Fungi are very common at all levels of the rainforest. They assist the detrivores in decomposing the litter.

It is estimated that 90 per cent of the animal and bird species that exist in the equatorial forest biome reside in the canopy. Since the tropical rainforests are estimated to hold 50 per cent of the planet's species, the canopy of rainforests worldwide may hold 45 per cent of life on Earth.

The soil

Rainforest soils are called **latosols** (Figure 6.45). These can be more than 40 m deep. The constant hot wet climate of the rainforest provides perfect conditions for chemical weathering of the bedrock and there is a constant supply of minerals from the parent rock to the soil. **Ferrallitisation** is the name for the process by which the bedrock is broken down by chemical weathering into clay minerals and sesquioxides (hydrated oxides of iron and aluminium).

The red colour of the soil is partly the result of the presence of iron and aluminium minerals. As there is a moisture surplus in the equatorial climate (because rainfall exceeds evapotranspiration), there is downward movement of water through the soil. Silica minerals are washed out of the A horizon and transported downwards by this water in a process known as **leaching** or **eluviation**. Iron and aluminium compounds are less soluble and are left behind. The iron compounds give the soil its rich red colour.

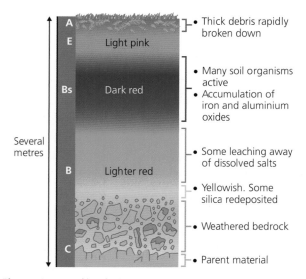

Figure 6.45 Profile of a latosol

In the rainforest areas precipitation is greater than potential evapotranspiration for much of the year. During this time there is a **soil moisture surplus**; the soil is saturated. Water cannot infiltrate into the soil and so leads to surface run-off. Rivers regularly flood during the wettest months.

Nutrient cycling

In the rainforest, nutrient cycling is very fast (Figure 6.46). The hot, damp conditions on the forest floor allow for the rapid decomposition of dead plant material, which in turn provides plentiful nutrients that are easily absorbed by plant roots. However, as these nutrients are in high demand from the rainforest's many fast-growing plants, they do not remain in the soil for long and stay close to the surface of the soil. If vegetation is removed, the soils quickly become infertile and vulnerable to erosion.

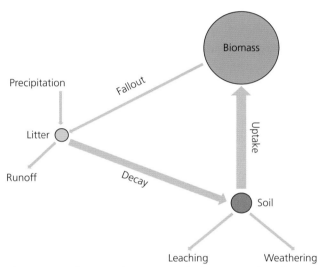

Figure 6.46 The mineral nutrient cycle for a tropical rainforest

Ecological responses

The vegetation has developed and adapted to the physical conditions of the rainforest in a number of ways:

- Leaves have drip tips and waxy surfaces. Others have jagged edges and holes in the leaves. All these allow the excess rainfall to be easily shed. This discourages growth of bacteria and fungi on the leaf surface.
- The trees grow rapidly upward towards the light. This results in tall slender trunks with few branches. All the leaves are concentrated in the umbrella-like crowns where they can absorb sunlight and photosynthesise. Because some of the trees are so tall, the tops can be subject to strong winds and so flexible trunks and buttresses help hold them up because the roots themselves are so shallow. Bark is thin because there is no need for protection against low temperatures.
- Most of the minerals needed by the trees are close to the surface of the soil and so the roots do not penetrate deep into the soil but spread outward, just below the forest floor.

- Climbing vines such as lianas are found throughout the rainforest. They have thick, woody stems and come in various lengths (up to 1,000 m) and varying shapes. They begin life on the forest floor but depend on trees for support as they climb upwards towards the sunlight. They attach themselves to trees with sucker roots, or tendrils wind themselves round a tree's trunk. When they reach the top of the canopy they often spread to other trees or wrap themselves around other lianas.
- Because the forest floor is so dark there are many epiphytic plants that grow on branches, trunks and even the leaves of the trees. One example, the bromeliads, has thick, waxy leaves that form a bowl shape in the centre for catching rainwater that is absorbed through hairs on its leaves. They can hold several litres of water and are miniature ecosystems in themselves, providing homes for several creatures including frogs and their tadpoles, salamanders, snails, beetles and mosquito larvae.
- Epiphytic orchids have aerial roots that cling to the host plant, absorb minerals and absorb water from the atmosphere.

The rainforest is home to more than half of the world's animals and they too have developed and adapted to the environment. Each layer of the rainforest has its own assemblage of animals adapted to each particular environment. Large predators such as the jaguar and omnivores like wild pigs live on the forest floor while monkeys, snakes, bats, birds and some amphibians live in the canopy. Some methods of adaptation include:

- camouflage, for example, the jaguar
- bright colours, for example, the poison arrow frog
- excellent climbing ability, for example, the gibbon.

Human activity and its impact on the tropical evergreen forests

The main impact of human activity is forest clearance. This problem is particularly acute in the Brazilian rainforest. Between May 2000 and August 2005, Brazil lost more than 132,000 km² of forest – an area larger than Greece. The state of Rondônia in western Brazil, once home to 208,000 km² of forest, has become one of the most deforested parts of the Amazon. Figure 6.47 (page 266) shows the area in satellite photographs taken 15 years apart.

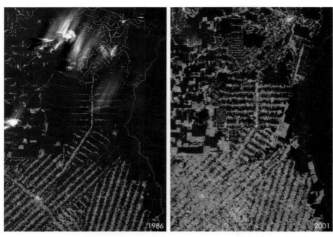

Figure 6.47 Satellite images of the same location in 1986 and 2001 show deforestation in the state of Rondonia, Brazil

In Brazil, the INPE, or National Institute of Space Research produces deforestation figures annually. These estimates are derived from 100 to 220 images taken during the dry season in the Amazon by the Landsat satellite. The loss between 2001 and 2014 is shown in Table 6.4.

Table 6.4 Estimated rainforest loss in Amazonia

Year	Estimated remaining forest cover in Amazonia (km²)	Annual forest loss (km²)	Percentage of 1970 cover remaining	Total forest loss since 1970 (km²)
2001	3,505,932	18,165	85.5	594,068
2002	3,484,538	21,394	85.0	615,462
2003	3,459,291	25,247	84.4	640,709
2004	3,432,968	27,423	83.7	668,132
2005	3,413,022	18,846	83.2	686,978
2006	3,398,913	14,109	82.9	701,087
2007	3,387,381	11,532	82.6	712,619
2008	3,375,413	11,968	82.3	724,587
2009	3,367,949	7,464	82.2	732,051
2010	3,360,949	7,000	82.0	739,051
2011	3,354,711	6,238	81.8	745,289
2012	3,350,140	4,571	81.7	749,860
2013	3,344,297	5,843	81.6	755,703
2014	3,339,446	4,848	81.4	760,551

Skills focus

What type of graph would be most suitable to display the information in Table 6.4?

Construct your chosen graph using the figures in Table 6.4. Describe the trends shown by the graph.

Brazilian deforestation is strongly correlated to the economic health of the country: the decline in deforestation from 1988 to 1991 matched the economic slowdown during the same period, while the fast rate of deforestation from 1993 to 1998 paralleled Brazil's period of rapid economic growth.

Reasons for forest clearance

- In many tropical countries, the majority of deforestation results from the actions of poor subsistence cultivators. **Shifted cultivators** is a term used to describe these small-scale farmers and their families who find themselves squeezed out of traditional farmlands in areas far outside the forests. They migrate to the last unoccupied lands available. They generally have little understanding of the ecological workings of tropical forests, and so they practise unsustainable forms of exploitation. The best estimate of numbers is somewhere between 200 million and 600 million and satellite surveys indicate that they account for roughly two-thirds of deforestation. They represent one of the biggest human migrations ever in such a short space of time.

- Virtually all forest clearing is done by fire. Though these fires are intended to burn only limited areas, they frequently escape agricultural plots and pastures and char pristine rainforest, especially in dry years like 2005.

- A large portion of deforestation can also be attributed to commercial exploitation of forest resources. Timber, such as mahogany from Brazil and teak from Burma (Myanmar) has been removed in an unsustainable and illegal manner.

- There are a small number of large landowners that clear vast sections of the Amazon rainforest for cattle pasture to feed the world's increasing demand for cheap beef. Large tracts of forest are cleared and sometimes planted with grasses.

- Favourable taxation policies, combined with government subsidised agriculture and colonisation programs, have encouraged the destruction of the Amazon rainforest. Low taxes overvalue agriculture and pasture and make it profitable to convert natural forest for these purposes when it otherwise would not be so.

- Brazil is one of the world's top producers and exporters of sugar cane, soya beans, oranges and other products. In the nine states of the Brazilian Amazon, the area under intensive mechanised agriculture grew by more than 3.6 million ha from 2001 to 2004. The greatest increase in area planted to soya bean was in Mato Grosso, the Brazilian state with the highest deforestation rate (40 per cent of new deforestation).

- Soya has become one of the most important contributors to deforestation in the Brazilian Amazon. Brazilian scientists have developed a new strain of soya bean that flourishes in the rainforest climate which, combined with high prices, has served as an impetus to expanding soya cultivation.
- Soya farms cause some forest clearing directly, but they have a much greater impact on deforestation by occupying already cleared land, thereby pushing ranchers and subsistence farmers ever deeper into the forest. Soya farming also provides a key economic and political impetus for new highways and infrastructure projects, which accelerate deforestation by others.

Impacts of deforestation

Impacts can be physical, economic, social and cultural. They mainly occur on a local scale, but some impacts can be global.

- As habitats shrink, plant species become endangered and the food chain within the forest is disrupted. Some animal species, for example tigers and orang-utans, are threatened by extinction.
- The vegetation protects the latosol soils from the regular heavy tropical downpours. Once the trees are removed the topsoil is open to erosion and to leaching of nutrients and minerals. Run-off causes sediment to block river channels and increases flooding.
- During the dry season, areas downstream of deforestation can be prone to long droughts which interrupt river navigation, wreak havoc on crops and disrupt industry.
- The microclimate of the forest is disturbed by deforestation – the daily water cycle of rapid evapotranspiration followed by afternoon precipitation cannot occur and there is less cloud cover, so a greater temperature range occurs. Rainforest is also the source of some of the rainwater that falls on agricultural areas outside the forest. Forest removal could lead to drought and agricultural failure.
- Burning associated with forest clearance leads to local air pollution. In 1997, Kuala Lumpur experienced smog which covered 3,000 km². This was the result of smoke from forest fires in Indonesia combined with car exhaust fumes.
- The forests' ability to absorb the carbon dioxide is reduced. At the same time, there is an increased presence of carbon dioxide released from the burning trees. This leads to increased atmospheric carbon dioxide and climate change.
- Deforestation can have economic benefits in terms of income from mining, farming and exports of hardwood. However, the culture of indigenous people is destroyed and they may be forced to move from their land.

Development issues

Population change

There are about 50,000,000 **indigenous** people living in and depending on the world's rainforests. The 'slash and burn' or 'shifting' agriculture practised by these indigenous groups is sometimes blamed for the destruction of large areas of tropical forest, but in fact they have developed the only demonstrably sustainable agriculture for tropical forests.

Many of the native societies of the rainforest have already been destroyed. In Brazil alone, 87 tribes were wiped out by Western diseases or genocide in the first half of the twentieth century. In the Indonesian province of Irian Jaya, the military is estimated to have killed up to 150,000 tribal people under the guise of counter-insurgency measures. In the Congo Basin the Baka, BaAka and Bakola pygmies were formerly hunter-gatherers but are now becoming increasingly settled, both through their own choice and because of government policies.

At the same time as the indigenous population is in decline, the overall population of the rainforests is increasing. In Amazonia, Manaus has grown to have a population of more than 2 million. It is the most populous city of the Amazon rainforest. In fact the Brazilian census of 2000 indicated that the population of the Amazon region was already about 70 per cent urbanised and increasing rapidly, much faster than the rest of Brazil. Also, whereas Brazil has mostly experienced steadily declining rates of population increase, the Amazon region continues to see recurrent spikes and contractions in growth rates, reflecting waves of migration to different parts of the region. This growth is mirrored in the Peruvian Amazon area.

Economic development

The national economies of the nine countries that have territory within the Amazon basin have all been boosted by the activities that have led to forest clearance. Mining, forestry, oil exploration and agriculture have brought economic benefits in the form of local employment and, where legal, tax revenue. Some of this money has been re-invested in infrastructure development (mainly roads) that have accelerated development and led to more forest clearance.

In Brazil, around the city of Manaus, there is a growing electronics industry, a result of government policies of opening up the area. Many major multi-national technology manufacturers, like LG, Samsung and Philips, have a presence there and their business is swelling the population further. The increase in workers in the city led to the decision to build a bridge over the Amazon to open up the south bank of the river to development (Figure 6.48).

Figure 6.48 The Rio Negro Bridge over the Amazon

Opened in late 2011, the Manaus-Iranduba Bridge over the Rio Negro branch of the Amazon River was the first to cross the world's greatest river system. It gives more commuters access to dormitory towns where developers are already building more housing. Environmentalists fear that the bridge, combined with new gas pipelines, roads and rising populations, could open up the rainforest to further destruction.

Research has shown that forest clearance may eventually lead to a collapse in agriculture because of changes to the regional climate (less rainfall, higher temperatures). In Brazil, based on existing trends of deforestation, loss of carbon sequestration and related feedbacks on rainfall, temperature and biomass, the researchers project a 34 per cent fall in pasture productivity and a 28 per cent decline in soya bean yields by 2050.

Agricultural change

The traditional economy of the rain forest is a mix between **hunting/fishing/gathering** and **shifting cultivation**. Numbers of people were so small and the area so large that these activities remained sustainable. Small areas of forest were cleared and farmed for two to four years. As the soil became less fertile new areas were cleared and the old ones abandoned. Soil had time to recover its fertility before the area was revisited many years later and once again cleared.

Population pressure from areas outside the rainforest has increased the demand for land and food. This has resulted in two broad categories of farmer in the rainforest: the **shifted cultivators** and **large corporations**.

On land that is only marginally useful for agriculture, poor land farmers use slash-and-burn techniques to clear it, but on a much larger scale than traditional practices. Instead of burning only one to four hectares, they burn hundreds to thousands of hectares after felling a tract of forest and leaving it to dry. This burning releases the minerals locked up in vegetation and produces a layer of nutrient-rich material above the otherwise poor soil. The cleared area is quickly planted and supports vigorous growth for a few years, after which the nutrient stock is depleted and large amounts of fertiliser are required to maintain production. This fertiliser can then be washed into rivers, affecting fish and other aquatic life. The land is abandoned when the soil is so deprived of nutrients that even fertiliser cannot restore it. It reverts to scrub. Drought-resistant grasses may move in or cattle ranchers may plant imported African grasses for cattle grazing. The land is now only marginally productive and only a limited number of cattle can subsist in the area.

Where the land is suitable for agriculture, large, single, cash crops like rice, citrus fruits, oil palms, coffee, coca, opium, tea, soya, cacao, rubber and bananas are cultivated. Some of these crops are better adapted to such conditions and last longer on cleared forest lands. Problems with this type of monoculture (single crop plantations) in the tropics, besides the loss of forest are:

- planting of a single crop makes the crop highly vulnerable to disease and pests. In natural rainforest, widespread infestations are rare because individuals of a given species are widely dispersed
- the planting of monocultures can be economically risky because of price fluctuations common in international commodities markets
- a single cold spell or drought can devastate a substantial segment of the agricultural economy.

Agricultural companies are clearing more rainforest than ever before, especially in the Amazon, where large tracts of rainforest are being converted into soya bean plantations. In Asia, especially Malaysia and Indonesia, large areas of rainforest are being cleared for oil palm plantations to produce palm oil, which is used widely in processed food, cosmetics and soap (Figure 6.49). Today palm oil is found in some 50 per cent of packaged snack foods, a proportion that is growing because palm oil is the cheapest type of vegetable oil.

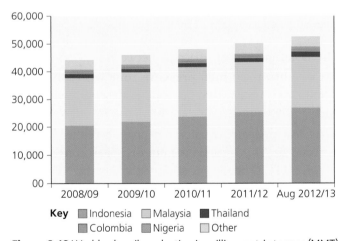

Figure 6.49 World palm oil production in million metric tonnes (MMT)

Extensive cattle ranching has caused deforestation in virtually every Amazon country; it accounts for up to 80 per cent of current deforestation. Brazil has 88 per cent of the Amazon herd, followed by Peru and Bolivia. Grazing densities are low as is productivity. There have been various attempts to slow down the rapid growth in ranching but poor communications, government inefficiencies and corruption have reduced their effect.

Implications for biodiversity

The destruction of natural habitats deprives the animals of the rainforest of the basis for their existence, causing an irreversible loss of biological diversity. One example, the orangutan, is particularly vulnerable because they are dependent on large contiguous forest areas. Fragmentation occurs when a once-continuous habitat becomes divided into separate fragments, resulting in limited breeding opportunities, reduced food resources and increased human conflicts for wildlife. In their search for food, orangutans often get lost in the palm oil plantations, where they are regarded as pests. They have been shot or clubbed to death by the palm oil plantation workers. According to the UN, there is a risk that no wild orangutans will remain outside of protected areas by 2020.

Sustainability

One example of how rainforests can be managed sustainably is taking place in Brazil, in parts of Amazonia such as Paragominas, where much local forest is either destroyed or damaged. This is because the usable timber would be ripped out of a stretch of forest and the rest would then be burned. When farming was tried on this cleared land, it was frequently found to be unprofitable. As a result, there are now about 165,000 km² of abandoned land in Brazilian Amazonia.

Changing attitudes has meant that timber firms are coming to see unharmed woodland as an asset that, properly managed, can yield a good income forever. Studies have shown that sustainable management of forests, also known as **reduced-impact logging (RIL)**, can be more profitable than the conventional methods of timber extraction. One such study, conducted near Paragominas, found that RIL was 12 per cent cheaper than conventional logging.

In RIL schemes, the area to be exploited is divided into up to 30 blocks, one of which has timber extracted each year, before being left alone for 29 years. This is enough for the forest to regenerate successfully, because in addition to rotation, the schemes take care to leave the oldest specimens of the exploited species standing. As well as providing cover from the tropical sun, the spreading branches of these tall trees re-seed the block with new specimens. RIL reduces the damage further by plotting the position of each block's valuable trees on a computer, which then works out the shortest set of access roads that needs to be carved out to remove the felled trees. Lumberjacks are also taught ways of felling trees that avoid damaging those around them.

Biome 2

The savanna grassland biome

It is to be found in (Figure 6.50):

- Northern-central Australia, straddling the tropic of Capricorn
- The Cerrado grassland of southern Brazil and the Llanos of Colombia and Venezuela
- Parts of India and Pakistan straddling the tropic of Cancer
- Sub-Saharan Africa, both north and south of the equator. In East Africa, in the upland areas, these grasslands are found right on the equator. It is the African Savanna that most people associate with this biome and which will be used as an example for the rest of the chapter.

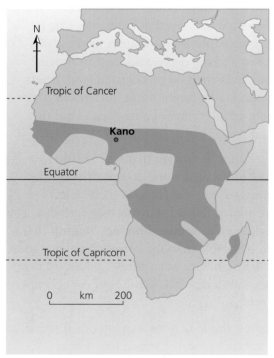

Figure 6.50 The savanna regions of Africa

Main characteristics of the savanna grassland biome

The climate

The typical climate is a tropical wet and dry climate that shows seasonal variations in wind direction, precipitation and temperature. It is transitional between the equatorial rainforests, where rain can be expected all year, and the hot deserts, which have minimal precipitation. Variations occur with increasing distance from the equator. Ninety per cent

of the annual precipitation falls within the wet season (Figure 6.51) and although the sun is higher in the sky, the rain clouds reflect insolation back out of the atmosphere and reduce the temperatures. During the dry season, with lack of cloud, heating can be intense and temperatures high.

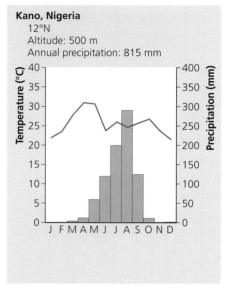

Figure 6.51 Climate statistics for Kano, Nigeria

The savanna areas of Africa lie between:

- the **subtropical high pressure zone** where the falling air is very dry. (This has a desert climate)

and

- the **low pressure** equatorial region where there is rising air. This is called the **inter-tropical convergence zone (ITCZ)** because it is located where the **NE trade winds** meet (converge with) the **SE trade winds.**

The low pressure ITCZ migrates throughout the year. This migration is, in part, due to the migration of the overhead sun. The higher the sun is in the sky, the greater the heating effect.

In June, the sun is overhead at the tropic of Cancer (23½ °N). The ITCZ then migrates northwards. It can reach as far north as 20 °N but often does not. By December, the overhead sun has moved to the tropic of Capricorn (23½ °S). The ITCZ follows the overhead sun in East Africa, but because of the shape of the continent and the fact that the land heats up more than the sea, the ITCZ remains at between 5 °N and 10 °N in West Africa.

The ITCZ draws in both hot dry air from the sub-tropical desert and hot moist air from the warm Atlantic Ocean.

In January, this moist air is drawn on to the coastal area of West Africa. As it reaches the ITCZ, it rises and heavy **convectional rainfall** occurs. To the north of the area, air is drawn south-eastwards from the sub-tropical high pressure area of the Sahara desert. This is dry and dusty. It is a regular occurrence and is called the **harmattan**. This evaporates any moisture and is responsible for the dry season and drought of the inland part of West Africa.

As the ITCZ migrates northwards, this moist air is drawn further north, causing the belt of rain to move with the ITCZ. By July, the ITCZ reaches its northern limit.

The result of this is that there are variations in precipitation and temperature such as:

- on the equatorial rainforest margins more than 1,000 mm of rain falls per year; this is the rainy season and can last up to 11 months
- on the desert/semi-arid margins there is less than 500 mm of rainfall per year with only 1 or 2 months of rainy season. As one moves further away from the equator the reliability of the rainfall decreases
- on the equatorial rainforest margin, temperatures range from 22 °C in the wet season to 28 °C in the dry season
- on the desert margins the temperatures range from 18 °C in the wet season to 34 °C in the dry season.

Skills focus

Contrast the savanna climate as illustrated by Figure 6.51 with that of Manaus (Figure 6.43, page 263).

The soil moisture budget

The vegetation of the savanna depends on the amount of water available (the moisture budget) for plant growth. This is illustrated by the use of a soil moisture diagram.

Figure 6.52 shows that precipitation is greater than potential evapotranspiration between July and September, whereas the reverse is true between October and June. There are four distinct periods that can be seen in the soil moisture budget for this area:

- **Soil moisture** recharge occurs through July and early August and this is when precipitation first becomes greater than evapotranspiration. Rainwater begins to fill the empty pores in the soil. When they are full the soil is said to have reached its field capacity.

- **Soil moisture surplus** occurs in late August and September. When the soil becomes saturated rainwater has difficulty infiltrating the ground. This causes surface runoff and explains the high river levels and flooding.
- **Soil moisture utilisation** occurs from October, as evapotranspiration begins to exceed precipitation. There is more water evaporating from the ground surface and being transpired by plants than is falling as rain. Water is also drawn up the soil by capillary action and this leads to further evapotranspiration.
- **Soil moisture deficit** occurs by December when the soil moisture is used up and there is a water deficit. Plants can only survive by being drought resistant or through irrigation. This period lasts until precipitation again becomes greater than evapotranspiration in early July and soil moisture recharge can begin.

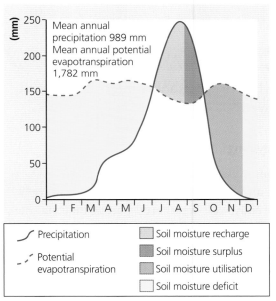

Figure 6.52 Soil moisture for Navrongo, northern Ghana (11 °N)
Source: AQA

This area in northern Ghana therefore shows the following characteristics:

- a lengthy period of moisture deficit
- a short period of moisture surplus
- total annual potential evapotranspiration greater than total annual precipitation.

The soil

The rapid downward movement of water in the soils during the rainy season washes away (leaches) the silica from the soil leaving a layer rich in iron and aluminium oxides that is characterised by a red colour. When the thin topsoil is removed, the exposed layer of iron and aluminium can harden to form a cement-like crust called a **laterite** (Figure 6.53).

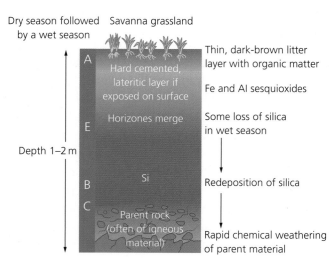

Figure 6.53 Profile of a ferruginous soil (laterite)

Key question

Under what circumstances might the topsoil be removed?

Nutrient cycling

Ferruginous soils are characterised by their infertility, although there are exceptions. The very old and already well-weathered parent material has, in many areas, been leached of nutrients by the very high rainfall of the wet season. Those areas of greatest wet season rain tend also to be the most infertile.

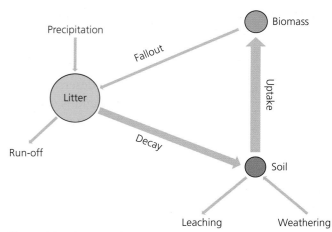

Figure 6.54 The mineral nutrient cycle for savanna grassland

During the dry season, plants die back and litter builds up which, if not destroyed by fire, forms a thin layer of humus. This low level of nutrients is further reduced by aerobic bacteria which are particularly active in areas with a long wet season and where the temperature remains above 25 °C for substantial periods of time. The activity of these bacteria under such conditions occurs at a rate above that of plant growth, thus breaking down plant matter faster than it is produced (Figure 6.54).

During the wet season leaching removes nutrients from the soil and that can result in a hard crust forming that hinders plant growth. Biomass varies seasonally, with elephant grasses growing up to 2 metres high in the wet season.

Ecological responses

Adaptations by vegetation

Figure 6.55 shows that the vegetation of the wetter areas consists of tall coarse grasses (elephant grass) with many deciduous trees. It is known as the tree savanna. In drier areas towards the desert margins, shorter tussock grass becomes dominant, with bare soil

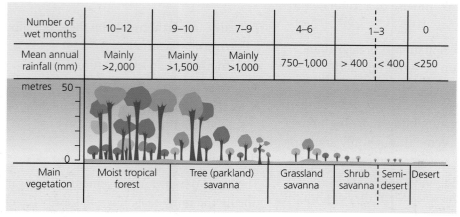

Number of wet months	10–12	9–10	7–9	4–6	1–3	0
Mean annual rainfall (mm)	Mainly >2,000	Mainly >1,500	Mainly >1,000	750–1,000	> 400	< 400	<250

| Main vegetation | Moist tropical forest | Tree (parkland) savanna | Grassland savanna | Shrub savanna | Semi-desert | Desert |

Figure 6.55 Section south to north through West Africa showing variations in vegetation

between the tufts of grass. These areas are known as the grassland and shrub savannas.

Trees in these areas tend to be deciduous, losing their leaves in the dry season, although some evergreens are also present. These have hard leathery leaves to reduce transpiration losses. Other plants are microphyllous (small-leaved), for the same reason.

In the **tree savanna**, a parkland exists. Here isolated trees have low umbrella-shaped crowns that shade root areas and reduce soil moisture evaporation. The trees show xerophytic characteristics (adaptations to dry surroundings), with dense cell fluids, hard waxy leaves, thorns and protected stomata, which all reduce water loss. The two main trees of the savanna are:

● Acacia (Figure 6.56), which has a crown structure often flattened by the trade winds. It loses its leaves in the dry season. It has long tap roots that are able to get water from deep underground. There are also roots close to the surface that are able to make use of dew from the area shaded by the tree's crown.

Figure 6.56 Acacia in the savanna of Amboseli National Park, Kenya

● Baobab ('upside-down tree'), which has a thick spongy trunk, long tap roots and bears leaves for only a few weeks. Like the acacia, the baobab is pyrophytic – it can withstand fire, mainly because of its insulating bark.

In the **grassland savanna** the grasses between the trees are shorter and sparser. They are perennial, dying back during the dry season and regrowing from root nodules when it rains. The grasses are tussocky, enabling them to retain some moisture. The naturally created straw dies down and protects the roots. They also have adaptations that discourage animals from grazing on them; some grasses are too sharp or bitter tasting for some animals, but not for others. This

means that all animals have something to eat. Many grasses grow from the bottom up so that growth tissue does not get damaged by grazers.

In the **shrub** or **scrub savanna** there are many acacia trees, thorn bushes and short tufted grasses. Many species generate short stems from a single stock, with deep, branched roots and dormant seeds that compete for water. In some plants, even the stems may be capable of photosynthesis, so plants can have fewer leaves and retain more water. Some grasses are feathery and wiry and turn their blades away from the strong sun to reduce water loss.

The animal life (fauna)

Most of the animals on the savanna have long legs (for example, wildebeest) or wings (for example, swifts) that enable them to go on long migrations. Many burrow underground (for example, the meerkat) to avoid the heat or raise their young. The savanna is a perfect place for birds of prey like hawks and buzzards. The wide, open plain provides them with a clear view of their prey, hot air updrafts keep them soaring, and there is the occasional tree to rest on or nest in. Animals do not sweat to lose body heat, so they lose it through panting or through large areas of exposed skin, or ears like those of the elephant.

The savanna has a large range of highly specialized plants and animals. They all depend on each other to keep the environment in balance. There are over 40 different species of hoofed mammals that live on the savannas of Africa. Up to 16 different species of browsers (those who eat leaves off trees) and grazers can coexist in one area. They do this by having their own food preferences, browsing/grazing at different heights, having different times of day or year to use a given area and having different places to go during the dry season.

These different herbivores provide a wide range of food for carnivores, like lions, leopards, cheetahs, jackals and hyenas. Each species has its own preference, making it possible to live side by side and not be in competition for food.

Other fauna include locusts, which can destroy large areas of grassland with devastating speed, and termites, which aerate the soil and break down up to 30 kg of cellulose per hectare each year. In some areas up to 600 termite hills per hectare can be found; this can have a significant effect on the upper layers of the soil.

Human activity and its impact on the savanna in Africa

Most of the African savanna vegetation types are used for grazing, mainly by cattle, goats, sheep or game. In some areas of southern Africa, crops and subtropical fruit are cultivated. Urban growth is limited because the hot, moist climate and diseases (such as sleeping sickness and malaria) hinder dense human settlement.

Human activity has two main effects on the vegetation:

- Grass is burnt off to ensure better growth of young grass next season for grazing. This regular burning makes it difficult for young trees and bushes to become established. Their place is taken by herbaceous plants and by the few indigenous woody plants that can survive fire, like acacia and baobab, which are therefore common in the savanna.

- Woody plants are killed by cattle eating their foliage. Thorny, animal-repellent trees and shrubs, such as acacia, therefore become numerous.

There is a belief among some biogeographers that humans have had a much greater influence than climate on the development of savanna vegetation. Some even suggest that grassland may therefore not be the climatic climax community but a plagioclimax.

For thousands of years, herders of the East African savannas have penned their cattle overnight in brush-walled corrals, called bomas. These remain in use for about a year, resulting in tons of manure that fertilizes these small areas. After abandonment, a lush carpet of grass springs up. These fertile 'glades', some as large as a football field, remain distinct from the surrounding area for over 100 years. Researchers in Kenya found that trees close to the edges of glades grew faster and were generally larger than trees elsewhere in the savanna. They also found more animal life there. The conclusion was that for many years after the pastoralists move on they leave fertile footprints across the landscape that significantly alter the dynamics of the entire ecosystem.

Conservation of the savanna is variable. There are huge areas of National Parks (e.g. the Kruger National Park in South Africa) as well as game reserves such as the Central Kalahari Game Reserve. Much of the area is used for game-farming and can thus be considered effectively preserved, provided that sustainable stocking levels are maintained. The importance of tourism and big game hunting in the conservation of the area must not be underestimated.

Development issues

Population change

Population densities for the African Savanna areas have traditionally been low. The nomadic way of life does not support large numbers of people and permanent settlers are presented with a variety of challenges from endemic disease, unreliable water supply and poor soils.

Recently, as human population carries on growing, the risk of human–wildlife conflict increases. This is particularly true of parts of the African savanna where new settlements have expanded into traditional wildlife habitats. One conflict is between settlers and the growing number of elephants. Many poor farmers have taken drastic action against elephants to protect their crops. This includes shooting, poisoning and other physical attacks. They also turn a blind eye to poaching. This has led to the slaughter of up to 40,000 elephants a year.

A more humane form of management has recently been developed in Kenya. Wire fencing is erected around agricultural plots. Honey bee nests are built on the fences. As the elephant brushes against the wire it disturbs the bees and they become angry. The sound of the bees frightens the elephants away.

This method also allows the farmer to harvest the honey and make an income from what had formerly been a problem.

Population growth outside the savanna areas of Africa is putting pressure on the continent's food resources. By 2050, the continent's population is expected to double and per capita GDP income will triple. This combination of people and prosperity means that Africa's food needs will increase by 400 per cent. It is estimated that satisfying this increased demand will require at least 140 million hectares of new farmland. This is made worse by growing efforts to turn Africa into a source of food and fuel for the rest of the world.

Scientists believe that growing crops like maize and soya in the savanna region would not be carbon efficient. Indeed, converting Africa's wetter savannas will have high environmental costs, which is unsurprising if we look at savannas elsewhere, such as Cerrado and Gran Chaco; these regions have recently undergone rapid agricultural transformation, resulting in large carbon and biodiversity losses.

Economic development

Although there are nodes of development based on natural resources such as metal ores and diamonds, Africa's savanna areas are dominated by agriculture and game conservation. Throughout the twentieth century game parks and reserves have been created. These in turn have attracted tourist money and created prosperity and jobs for local people. Examples of these reserves include the Kruger National Park of South Africa, The Hwange National Park in Zimbabwe and the Serengeti National Park in Tanzania.

Implications for biodiversity and sustainability

Two opposing forces are at work in much of the African savanna. The need to grow food puts pressure on the land and reduces biodiversity. The attraction of tourist income creates the need for reserves and protection programmes to maintain biodiversity.

Key question

What are the arguments for and against conservation in the savanna regions of Africa?

Case study of a specified region experiencing ecological change: The Sundarbans

Figure 6.57 Location of the Sundarbans

Geographical setting

The Sundarbans is a region in the Bay of Bengal composed of vegetated low lying islands with elevations ranging from 0.9–2.1 m above mean sea level, interspersed with a maze of tidal waterways (Figure 6.57). The region is spread across West Bengal in India and southern Bangladesh and is famous for its unique mangrove forests. The Sundarbans are part of the delta formed by the confluence of the Ganges, Padma, Brahmaputra and Meghna rivers. They have been designated by the Worldwide Fund for Nature (WWF) as one of 867 **ecoregions**. An ecoregion is a contiguous area characterised by well-defined similarity in flora and fauna as well as geomorphology, climate and soils. Ecoregions do not follow any political boundaries or landscape alterations by humans. The Sundarbans have also been recognised by UNESCO as a World Heritage Site.

The Sundarbans forest is about 10,000 km² across India and Bangladesh, of which 60 per cent lies in Bangladesh. It is home to many rare and globally threatened wildlife species such as the estuarine crocodile, the royal Bengal tiger, the water monitor lizard, the Gangetic dolphin and the olive Ridley turtle.

In June to September the monsoon regularly brings heavy rains and frequent devastating cyclones that cause widespread destruction. Annual rainfall can exceed 3,500 mm and daytime temperatures can exceed 48°C during these monsoon months.

The Sundarbans have a wide range of flora (334 plant species) and fauna (693 species).

The Sundarbans delta is an ecologically fragile and climatically vulnerable region that is home to over 4.5 million people.

Threats

The Sundarbans are subject to threats from both human activity and natural occurrences. These are summarised in Figure 6.58, page 276.

Threats to the Sundarbans have resulted in:

- a decline in standing volume of the two main commercial mangrove species
- depletion of its biodiversity. Five mammal species – the Javan rhinoceros, water buffalo, swamp deer, gaur and probably hog deer – have become locally extinct since the beginning of the twenty-first century.

Issues and/or impacts arising from the environment and/or changes to that environment

- Between 2001 and 2008, the area under agriculture in the Indian Sundarbans had reduced from 2,149 km² to 1,691 km².
- Because of rising sea levels and shrinking forest, humans and tigers are fighting for space. Farmers are forced into the forest to hunt for honey, fish or collect crabs, putting them at risk of tiger attack.

- As a result of over-exploitation of aquatic species in the last 15 years, coastal fishing has seen a decline in catch-per-unit effort – from 150–200 kg per haul to 58–65 kg per haul.

Responses to the issues outlined above

- The entire Sundarbans is reserved forest, established under the Indian Forest Act, 1878, when both India and Bangladesh were part of the British Empire. Since then, a number of wildlife refuge areas have been created on both sides of the border. These sanctuaries are supposed to prohibit the entry of people, cultivation of land, damage or destruction of vegetation, hunting or capturing of wild animals, introduction of exotic species, straying of domestic animals, causing of fires and pollution of water. Much of this is not working as population pressure grows.

- A large part of this ecoregion has been designated as a World Heritage Site both in India and Bangladesh. This is because of the wide variety of species present and the fact that the area is an example of delta formation and the subsequent colonisation of the newly formed deltaic islands and associated mangrove communities.

- Temporary fishing camps are established every season by local people in order to ensure a fresh supply of fish to the local markets. Men also go out into the forest to hunt, collect honey or cut firewood. Over the years, thousands of people have been killed by storms, eaten by tigers and crocodiles, or died of disease. The widows of those killed by tigers are known as 'Tiger widows', often being forced from their homes because they are unable to support their families.

- It is thought that roughly 60 per cent of the male workforce in the Indian Sundarbans is moving away from the region. They have moved to work in construction and mining jobs in Mauritius or in the Indian states of Kerala, Gujarat, Karnataka and Maharashtra.

- Many others, unable to make ends meet, are opting to work in the brick kilns, which have replaced the water bodies that earlier dotted the delta.

- Cyclones alone account for about 45 per cent of all deaths in the Sundarbans. Since some islands are far more vulnerable to extreme climate conditions plans have been put forward to move people from such areas to more stable zones and urban areas outside the Sundarbans. This would significantly decrease the population at risk and help land left behind to be converted into protected areas.

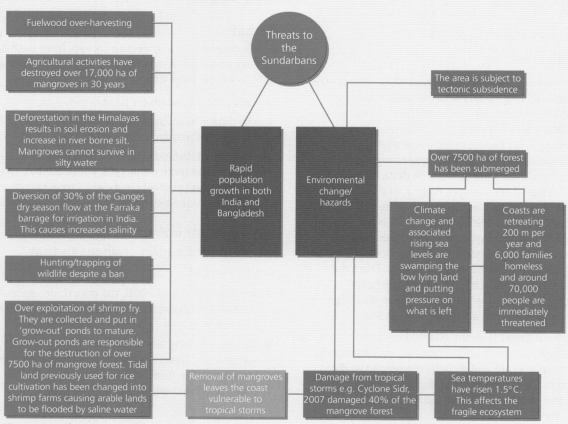

Figure 6.58 Threats to the Sundarbans

Fieldwork opportunities

1 Ponds offer a chance for fieldwork. Here are two questions that could be followed up:

- How does vegetation change from the edge of the water inland?
- How do soil characteristics/microclimate change with distance from the water?

2 Sand dunes offer a chance for fieldwork. Three questions that could be studied are:

- How does vegetation change from the seashore inland?
- How do soil characteristics/microclimate change with distance from the seashore?
- Compare two areas of the dune system, for example, ungrazed/grazed sections of the fixed dunes.

Review questions

1 How can a systems approach to the study of ecosystems explain the role organisms play in the environment?

2 Explain why climate change appears to be the biggest threat to many ecosystems.

3 What are the similarities and differences between the ways different aboriginal people have adapted to changing circumstances within their biome?

4 How do ecosystem services affect you?

5 Why is there a wide range of forecasts for the future of ecosystems? What does this tell us about our fundamental understanding of how the planet works?

Further reading

Living Planet Report, WWF

The *Living Planet Report* is the world's leading, science-based analysis of the health of our planet and the impact of human activity. WWF believes that humanity can make better choices that translate into clear benefits for ecology, society and the economy today and in the long term.

The Social Dimension of Ecosystem-based Adaptation, No 12, 2013

UNEP Policy Series on Ecosystem Management, Issue No. 7

The United Nations Environment Programme produces a series of policy reports. These are part of a toolkit to encourage debate on various policy issues related to ecosystem management. The titles above are examples. These are available from the UNEP website.

The UK National Ecosystem Assessment

The UK National Ecosystem Assessment (UK NEA) was the first analysis of the UK's natural environment in terms of the benefits it provides to society and continuing economic prosperity. Part of the Living With Environmental Change (LWEC) initiative, the UK NEA commenced in mid-2009 and reported in June 2011. It was an inclusive process involving government, academic, NGO and private sector institutions. Only available from the UK National Ecosystem Assessment website. http://uknea.unep-wcmc.org

Biodiversity and Climate Change – a Summary of Impacts in the UK, Joint Nature Conservation Committee.

This gives an overview of how biodiversity of different ecosystems in the UK is being affected by climate change.

The UK Biodiversity Action Plan (BAP)

This is also available from the Joint Nature Conservation Committee at jncc.defra.gov.uk

Assessing biodiversity in Europe – the 2010 report, EEA Report No 5/2010, European Environment Agency

This considers the status and trends of pan-European biodiversity, and the implications of these trends for biodiversity management policy and practice.

Projects abroad: Conservation management plan, Kenya

An example of conservation in the African savanna. www.projects-abroad.org/_downloads/us/conservation-management-plan/kenya-management-plan.pdf

Question practice

A-level questions

1. Which of the following statements correctly describes characteristics of a lithosere?

   a) It is a plant succession that begins life on a sandy surface.

   b) As soil gets thicker it is less able to retain water causing small plants to die out.

   c) Small plant species are able to survive under larger ones by sharing the sunlight and available water.

   d) The earliest colonisers are lichens, which are able to use the energy of the sun to create their own food and water.   (1 mark)

2. Which statement most accurately describes a typical savanna climate?

   a) It is equatorial: hot all year round with a wet season and a dry season.

   b) It is subtropical, with hot dry summers and a distinctive wet season in the mild winters.

   c) It is tropical, with a dry season and a wet season that coincides with the passage of the overhead sun.

   d) It is tropical, with mild dry winters and hot wet summers.   (1 mark)

3. Figure 6.7 (page 233) shows a simplified food web for an English woodland. With reference to the figure, analyse the relationships between different species within the food web.   (6 marks)

4. With reference to a local ecosystem you have studied, evaluate the success of conservation strategies in managing that ecosystem.   (9 marks)

5. Analyse how development issues such as population change, economic development and agricultural change affect the sustainability of warm water coral reefs.   (9 marks)

6. To what extent has a region you have studied that is experiencing ecological change been affected by **either** changing demographic and cultural characteristics **or** economic change and social inequalities.   (20 marks)

Part 2

Human Geography

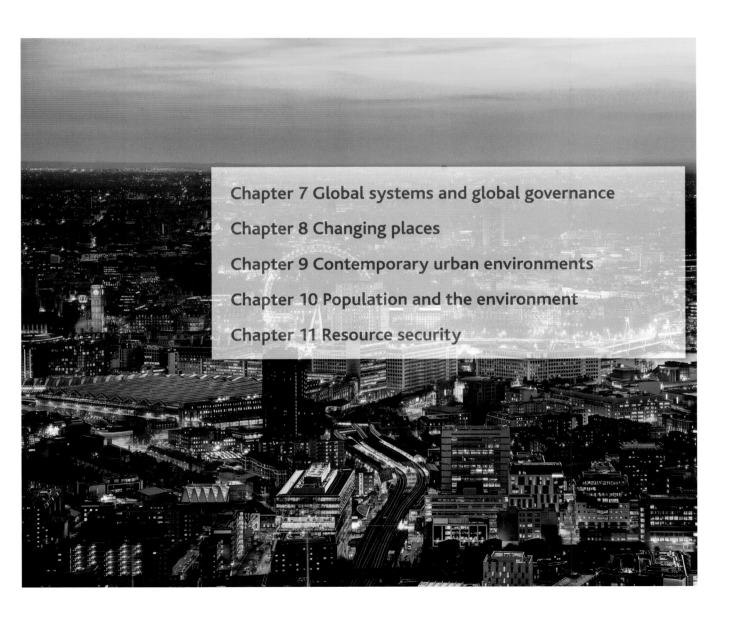

Chapter 7

Global systems and global governance

Our world is more connected now than it ever has been. The global economy and society have altered significantly in recent years as a result of the process of globalisation. Arguably, this has been positive for overall economic development but has also produced some negative consequences in terms of socio-political conflicts and environmental sustainability. There are few subjects either as controversial or as in need of better awareness than attempts to manage and govern human affairs on a global scale.

As a geographer, being able to recognise and explain links between economic, political and social change, as well as engaging with contemporary world issues, is important for understanding both international trade and the workings of the global community.

In this chapter you will study:
- globalisation
- global systems
- international trade and access to markets
- global governance
- Antarctica and the Southern Ocean
- the protection of Antarctica.

In the late 1960s, Canadian philosopher Marshall McLuhan predicted the advent of a 'global village' – a 'flat world' where free rein is given to economic and information flows. This would be reflected by the increasingly international manner in which organisations would operate:
- by thinking globally (not within national boundaries)
- by acting globally (being present in many countries)
- by making 'planet-wide' decisions.

The notion of the global village is perhaps exaggerated, even with regard to the movement of goods. Many obstacles to trade remain and in some sectors globalisation has barely begun. The 2008 global financial crisis has slowed the process of globalisation down, albeit temporarily. Recovery from this crisis has also been hit by setbacks such as international conflicts. Progress has been made with international trade and access to markets but the world is still not 'flat'.

Key question

How far has the world moved towards being a 'global village'?

International trade gives rise to a **'world economy'** in which prices, supply and demand are affected by global events. Trade has grown enormously since the Second World War; in manufactured goods alone it has grown from around US$100 billion in 1956 to US$19 trillion in 2013. Globalisation, however, is not just about an increase in trade.

7.1 Globalisation

It has been said that arguing against globalisation is like arguing against the laws of gravity.

Kofi Annan, former Secretary-General of the United Nations

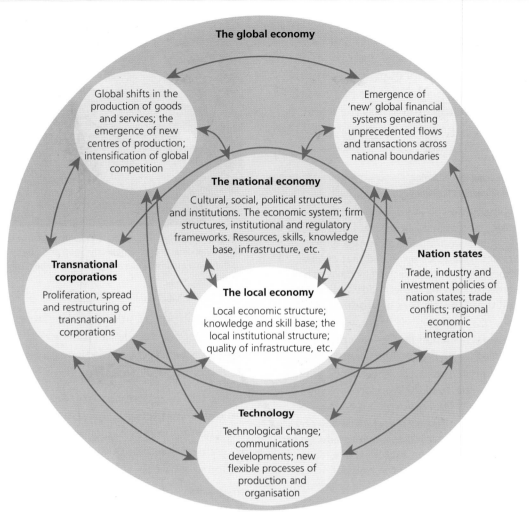

Figure 7.1 The global economy as a 'system' of interconnected elements and scales

Dimensions of globalisation

The term 'globalisation' was used to describe an unprecedented integration of world economies in the 1990s under the combined influence of the information technology revolution and the opening up of former Communist Bloc countries to the market economy. The focus of globalisation has been primarily on economic relationships such as **international trade**, **foreign direct investment** and international **capital flows**.

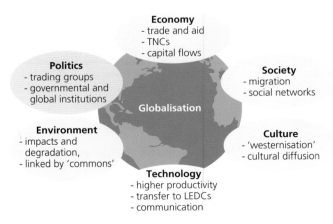

Figure 7.2 Factors and dimensions of the globalisation process – aspects of these are transferred more easily through global networks

Globalisation has since been expanded to encompass a wider range of dimensions including media, culture and social, political, environmental and even biological factors, for example the control of pandemics or the impacts of climate change. The process involves the spread of ideas and information across the world, but in doing so it creates a growing uniformity shared by very different places (Figure 7.3).

Figure 7.3 Globalisation can lead to **uniformity** through recognition of international brands. The Body Shop and Starbucks are familiar to UK high streets but these are located in Beirut, Lebanon

Factors of production

The roots of globalisation lie in international trade and the increasing accessibility of markets, which open up to the wider global community. This means that goods and services have to be produced that can be traded. A number of productive resources have to be combined in order to provide these goods and services; economists call these the '**factors of production**'. They are as follows:

- **Land** – All natural resources provided by the earth including minerals, soils, water, forests, etc.
- **Labour** – The human resource available in any economy. The quantity and quality of the workforce are key considerations to any producer of goods and services.
- **Capital** – In economic terms this refers to any physical resource that can be regarded as a man-made aid for production, such as buildings, factories, machinery, etc. '**Capital flows**' can involve the transfer of these physical resources from one place to another. However, in reality it usually refers to the flow of investment finance used to provide this capital.
- **Enterprise** – This is a very particular form of human capital describing those who take the risk of establishing businesses and organising the production of goods or provision of services.

The increased international flows of capital, labour and enterprise have been both a cause and an ongoing consequence of the globalisation process.

Flows of capital

For the purpose of understanding '**international capital flows**', capital includes all money that moves between countries which is used for investment (for example in land and physical capital), trade or production.

In the late twentieth century, **deregulation of world financial markets** meant that the activities of financial institutions such as banks, insurance companies and investment companies were no longer confined within national boundaries.

Figure 7.4 shows the capital flows experienced in the global economy system. It distinguishes between a 'core' area and a 'periphery' based on Frank and Wallerstein's core-periphery model of a world system. This assumed that global power is concentrated in the hands of a relatively small block of developed nations which they called the 'core'. Periphery countries were seen as those that are less developed and have been exploited and have suffered from a lack of investment, **leakages** and out-migration.

Table 7.1 The main dimensions of globalisation

Form of globalisation	Process caused by:	Characterised by:
Economic	increase in free trade growth of transnational corporations faster, cheaper transport global marketing	long distance flows of goods, capital and services as well as information and market exchanges
Cultural/social	migration global communication networks impact of western culture through media, sport, leisure and celebrity	spread of ideas, information and images
Political	growth of Western democracies and their influence on poorer countries decline of centralised (communist) economies (though communist political control is still strong in China and Russia)	the diffusion of government policy and development of market economies in former communist states

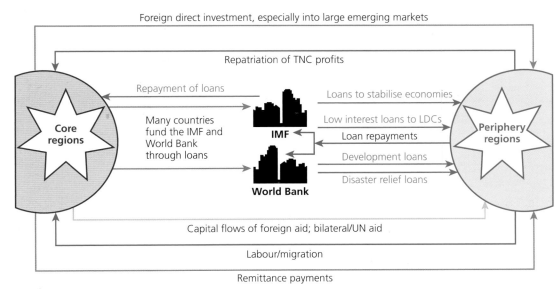

Foreign direct investment, especially into large emerging markets

Repatriation of TNC profits

Repayment of loans

Core regions

Many countries fund the IMF and World Bank through loans

IMF

Loans to stabilise economies

Low interest loans to LDCs

Loan repayments

Development loans

Disaster relief loans

World Bank

Periphery regions

Capital flows of foreign aid; bilateral/UN aid

Labour/migration

Remittance payments

Figure 7.4 Flows of finance and capital between regions in the global economy

The concept is largely outdated in the world of today because rapid growth of large medium-income economies such as the **BRIC** and, more recently, **MINT** countries means there is now a continuum of development, so many more regions of the world might be envisaged as 'core'.

Key terms

BRIC – An acronym used to identify a group of four countries – Brazil, Russia, India and China – whose economies have advanced rapidly since the 1990s.

Diaspora – A large group of people with a similar heritage or homeland who have moved and settled in places all over the world.

Leakages (economic) – Refers to a loss of income from an economic system. It most usually refers to the profits sent back to their base country by transnational corporations – also known as profit repatriation.

MINT – An acronym referring to the more recently emerging economies of Mexico, Indonesia, Nigeria and Turkey.

Figure 7.4 shows the salient flows of money between richer areas and emerging markets in the 'periphery'. These flows assist emerging economies to develop and can reduce disparities between the core and the periphery.

● **Foreign direct investment (FDI):** This is investment made mainly by TNCs (occasionally by governments) based in one country, into the physical capital or assets of foreign enterprises. The investing company may make its overseas investment in a number of ways. For example by setting up a subsidiary company, by acquiring shares or through merger or joint venture.

● **Repatriation of profits:** TNCs investing in overseas production will normally take any profit made from that investment back to their home-country headquarters. This is sometimes called an economic **leakage**. The majority of these flows return to companies based in richer countries.

● **Aid:** This is an important source of financial support for poor countries. It can take many forms and can be provided through the UN (multilateral) from contributions made by a number of richer countries (sometimes known as **Official Development Assistance – ODA**). It can also be provided bilaterally from one government to another, usually with mutual co-operation conditions applied. (Aid can also be supplied in the form of technology and expertise via **non-government organisations (NGOs)** or as food and relief at times of disaster.)

● **Migration:** The majority of out-migration of labour takes place from poorer to richer countries. This will exacerbate disparities as the less developed nations lose their most skilled and talented labour, who will pay taxes and spend much of their earnings in their destination country (although they habitually send **remittances** back to their country of origin).

● **Remittance payments:** These are transfers of money made by foreign workers to family in their home country. Figure 7.5 (page 284) shows how remittances have become the second most important source of income in developing countries, above international aid. India receives more remittance payments from their **diaspora** than any other country.

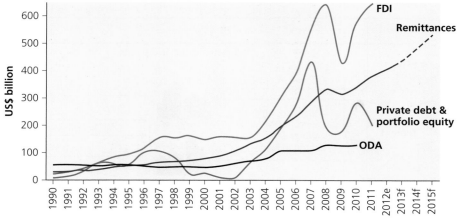

Figure 7.5 Remittances are becoming an increasingly important source of income for developing countries. Unlike FDI, remittances did not dip as significantly during the global financial crisis of 2008–09

Source: World Development Indicators and World Bank estimates

Flows of labour

Labour markets are not as free flowing as financial markets in the process of globalisation. People move less easily around the world than money because of restrictions on immigration. However, in recent years there has been a phenomenal rise in the numbers of migrants crossing international borders, mainly to seek better employment opportunities (Figure 7.7). Much of the movement has been from developing countries in South Asia, Africa and Latin America to the richer areas of North America and Europe. Another major destination for the movement of labour has

been around the oil-rich Gulf States of Kuwait, Qatar, Bahrain, Saudi Arabia and the United Arab Emirates, where the construction boom has provided plentiful employment opportunities.

Key facts about the movement of labour:

- Despite increases in cross-border movements, most migrants move over short distances within the same region or between neighbouring regions (especially in sub-Saharan Africa).
- North America, Europe and the Gulf countries in western Asia attract migrants from furthest afield.

The remittance dilemma in Somalia, 2012

Some of the least developed countries rely more heavily on remittances than other sources of capital. According to the World Bank, 40 per cent of Somalians rely on remittances to meet their basic needs.

In Somalia, they play an important role in supporting economic development and people's livelihoods, accounting for 50 per cent of GNI and 80 per cent of all investment in the country.

In 2012, concern that some of the money was falling into the hands of terrorist groups led to many US and UK banks and money transfer agencies withdrawing the service. The effects in Somalia were devastating and protests were organised by human rights groups supporting Somalians.

The concerns of US and UK financial institutions are based on conditions in Somalia, including:

- its informal economy, with little government regulation
- the lack of anti-money-laundering laws

- the lack of due diligence, SARs (suspicious activity reports) for money transmitters
- the fact that some remittances may end up in the hands of terrorist group Al-Shabaab, which receives donations from domestic and foreign sympathizers.

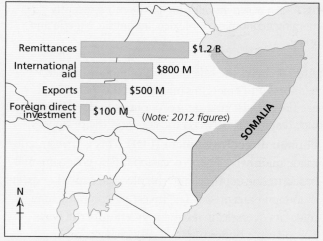

Figure 7.6 The remittance dilemma for Somalians

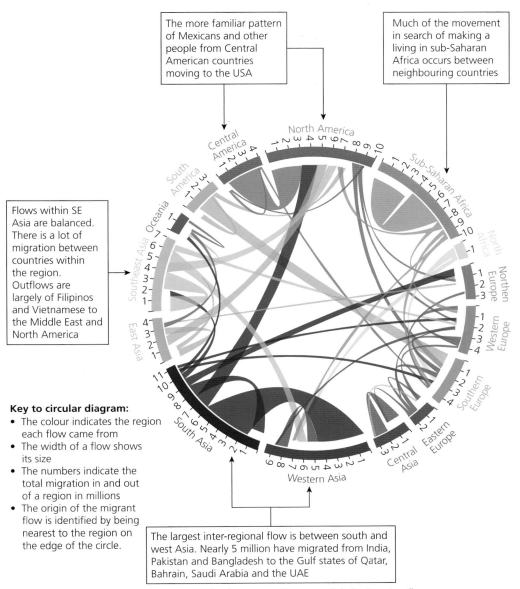

The more familiar pattern of Mexicans and other people from Central American countries moving to the USA

Much of the movement in search of making a living in sub-Saharan Africa occurs between neighbouring countries

Flows within SE Asia are balanced. There is a lot of migration between countries within the region. Outflows are largely of Filipinos and Vietnamese to the Middle East and North America

Key to circular diagram:
- The colour indicates the region each flow came from
- The width of a flow shows its size
- The numbers indicate the total migration in and out of a region in millions
- The origin of the migrant flow is identified by being nearest to the region on the edge of the circle.

The largest inter-regional flow is between south and west Asia. Nearly 5 million have migrated from India, Pakistan and Bangladesh to the Gulf states of Qatar, Bahrain, Saudi Arabia and the UAE

Figure 7.7 The global flow of people. Circular diagram to illustrate global migration flows between 196 countries from 2005–2010

- The bulk of economic migrants moving between continents are not the poorest but are those with some education and financial means.
- The largest regional flow of labour in the world is in Asia. Between 2005 and 2010 around 5 million workers moved from south Asia to west Asia.
- Over the past 25 years more people have migrated from Asia to North America and Europe but at the same time both the Gulf states and the Tiger economies of south east Asia have become attractive destinations for those seeking employment.

Figure 7.8 shows how flows of labour and remittances from source to the migrants' home reflect the key patterns described above.

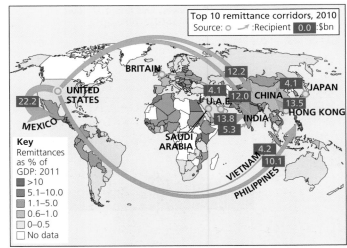

Figure 7.8 'Rivers of gold' – the top 10 remittance corridors in the world

Flow of products

The international movement of products is facilitated, especially for developing countries, by the reduction in costs of trade, which includes **transaction, tariffs**, and **transport** and **time** costs.

- **Transaction costs** have been reduced by the improvements in flows of data and the ease with which capital can be transferred to pay for transactions.
- **Transport and time** costs have been reduced by the process of **containerisation** which has enabled more complex and long distance flows of products, as does air transport which can speed delivery and reduce costs of more valuable or perishable cargo.
- The most obvious regulatory barriers to trade are tariffs, which with the encouragement of the **World Trade Organization (WTO)** have generally been reduced in global trade.

Key terms

Containerisation – A system of standardised transport that uses large standard-size steel containers to transport goods. The containers can be transferred between ships, trains and lorries, enabling cheaper, efficient transport.

Protectionism – A deliberate policy by government to impose restrictions on trade in goods and services with other countries – usually done with the intention of protecting home-based industries from foreign competition.

Tariffs – A tax or duty placed on imported goods with the intention of making them more expensive to consumers so that they do not sell at a lower price than home-based goods – a strategy of protectionism.

Flow of services

Services are economic activities that are traded without the production of material goods, for example, financial or insurance services. They can also be sub-divided into:

- **high level services** – services to businesses such as finance, investment and advertising
- **low level services** – services to consumers such as banking, travel and tourism, customer call centres or communication services.

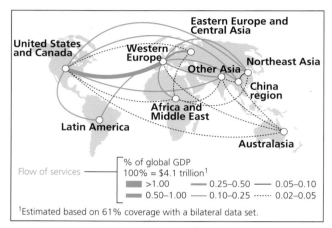

Figure 7.9 Global flow of services 2011

Source: UN Comtrade; IMF Balance of Payments; World Development Indicators, World Bank; Mckinsey Global Institute analysis

Services such as banking, insurance and advertising depend on communication and the transfer of information. They are **footloose** and can locate anywhere and advancing technology means they can still serve the needs of customers worldwide.

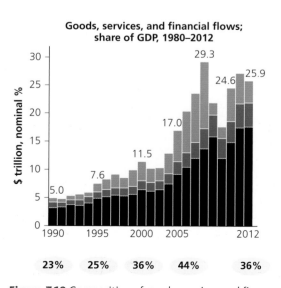

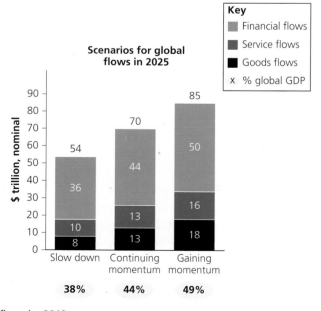

Figure 7.10 Composition of goods, services and finance flows by 2012

High-level services have increasingly been concentrated in cities in the more developed world, such as London, New York and Tokyo, which are the major centres of global industrial and financial control. Other cities have grown in importance such as Frankfurt, home to the European Central Bank of the EU, Toronto and Zurich. With the empowerment of East Asian economies, Hong Kong, Singapore, Seoul and Shanghai have also become major global financial centres.

A growing number of transnational service **conglomerates** have emerged, seeking to extend their influence on a global scale, such as HSBC Holdings in banking and financial services, Omnicom (US) in advertising and TUI Group (Germany) in travel and tourism.

One trend has been the decentralisation of low level services from the developed to the developing world. Call centre operations, for example, have moved from the UK to India where labour costs are generally 10–20 per cent lower than in the UK.

Flow of information

Information flows are governed by the movement of people through migration and by the speed of data and communication transfers. Both are responsible for the transfer of cultural ideas, language, industrial technology, design and business management support.

Digitisation and satellite technology have transformed these flows of information, which are now supported by:

- improvements to global telephone networks, making communication cheaper and easier
- mobile telecommunication technology
- email and the internet, which enable large amounts of information to be exchanged instantly across the globe
- live media coverage available on a global scale because of satellite technology.

The importance of information flows should not be underestimated because of their contribution to the expansion of **knowledge-intensive goods and services**. Such goods and services include those which have an intensive research and development (R&D) component and use highly skilled and educated labour. They include high-tech products such as semiconductors, pharmaceuticals, computer technology and business services such as international law, accounting and engineering. These industries need the exchange of ideas and flow of expertise to flourish.

Figure 7.11 Coca-Cola in Africa: the unmistakeable Coca-Cola brand even reaches poorer parts of Senegal. Coca-Cola pay for paintwork and often for refrigeration units for vendors

Key terms

Conglomerates – A collection of different companies or organisations which may be involved in different business activities but all report to one parent company – most transnational corporations are conglomerates.

Economies of scale – The cost advantages that result from the larger size, output or scale of an operation as savings are made by spreading the costs or by rationalising operations.

Global marketing

Marketing is the process of promoting, advertising and selling products or services. When a company becomes a global marketeer, it views the world as one single market and creates products that fit the various regional marketplaces. It will usually develop a recognisable '**brand**' and employ one marketing strategy to advertise the product to customers all over the world. The ultimate goal is to sell the same product, the same way, everywhere. Having one marketing campaign on a global scale like this generates **economies of scale** for the organisation, which reduces their costs.

Coca-Cola is an example of a company with a single product; only minor elements are tweaked for different markets (Figure 7.11). The company uses the same formulas (one with sugar, the other with corn syrup) for all its markets. The bottle design is recognisable in every country but the size of bottles and cans conforms to each country's standard sizing.

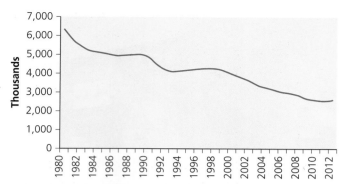

Figure 7.12 Loss of manufacturing jobs in the UK since 1980

Patterns of production, distribution and consumption

Globalisation created a new international division of labour with two main recognisable groups:

- the highly skilled, highly paid, decision-making, research and managerial occupations which, on a global scale, are largely concentrated in more developed countries
- the unskilled, poorly paid assembly occupations, which tend to be located in developing countries that have lower labour costs.

This simple division has undergone radical changes in the last 40 years. Many countries that were classified as less economically developed have become **newly industrialised countries** (NICs). These countries have developed their own industrial and commercial bases and their own TNCs which have spread their wings globally. This started with the four Asian '**tiger**' economies of Hong Kong, Singapore, South Korea and Taiwan, which were followed in succession by the **BRIC** economies, the most rapidly growing and prominent of these being China, and more recently the so-called **MINT** economies. Development has come to regions surrounding these new 'cores', particularly to south east Asia and Latin America.

Production

In 1954, around 95 per cent of manufacturing was concentrated in the industrialised economies of Western Europe, North America and Japan. The products were largely consumed in the country of origin.

Over time, decentralisation has occurred, largely as a result of foreign direct investment by **transnational corporations (TNCs)** into those developing countries able to take on manufacturing tasks at a competitive price. Lower land and labour costs, as well as incentives offered by governments in developing countries, have encouraged many TNCs to relocate the **production** side of their business abroad. This filtering down of manufacturing industry from developed countries to lower wage economies is known as **global shift**.

The **transfer of technology** made by TNCs has enabled countries in the developing world to increase their productivity, without raising their wages to the same levels as developed countries.

One of the consequences of global shift has been **deindustrialisation** in the richer countries and a loss of jobs in the manufacturing sector (Figure 7.12). For example, employment in manufacturing in the UK fell by around 50 per cent in the 30 years from 1983 to 2013, though decline has now steadied and productivity is rising again. More than 50 per cent of all manufacturing jobs are located in the developing world and over 60 per cent of exports from those countries to the developed world are of manufactured goods.

Global shift is not the only factor that caused the decline in manufacturing in developed countries. Outmoded production methods, products at the end of their life cycles and poor management have all contributed to the decline in manufacturing in these regions. Reversal of this trend has been prompted by foreign TNCs investing in deindustrialised regions.

Manufacturing transfers around the world with great ease and not only because of lower costs. Other factors affecting the location choice of entrepreneurs of some of the largest manufacturing companies include:

- the availability of a **skilled and educated workforce**
- the opportunity to **build new plant** with the latest and **most productive technology**
- **government incentives** in the form of tax breaks or enterprise zones to entice companies to invest and relocate
- **access to large markets** without tariff barriers, enabled through trade agreements.

Ebbs and flows in car production

Ebbs and flows in manufacturing are exemplified by the motor vehicle industry, the most global manufacturing industry. For example, since 1994 the US vehicle giants Ford and General Motors have relocated much of their component assembly to maquiladora plants in Mexico to take advantage of the cheaper production costs and the non-tariff barriers that are part of NAFTA (North American Free Trade Agreement). At the same time, the US car industry has been revitalised by investment in new plant and technology by foreign vehicle manufacturers such as Toyota.

The UK lost most of its domestic car industry as part of deindustrialisation from the 1980s onwards. However, as a result of investment by the Japanese companies Honda, Nissan and Toyota at factories in Swindon, Sunderland and Derby respectively, the UK is now one of the most productive car manufacturers in Europe (Figure 7.13). These plants give the Japanese TNCs access to the large and lucrative EU market, underlining the importance of free trade agreements as one of the factors in determining the location of production.

UK vehicle manufacture has been further boosted by investment by TNCs from BRIC economies, especially by Indian conglomerate Tata who bought Jaguar Land Rover (JLR) from Ford in 2008.

Figure 7.13 Aerial view of the Nissan factory at Washington, near Sunderland, the most productive car plant in Europe

Key term

Maquiladora – A manufacturing operation (plant or factory) located in free trade zones in Mexico. They import materials for assembly and then export the final product without any trade barriers.

Distribution and consumption

Product consumption still lies predominantly in the richer countries of the developed world. Products being manufactured in emerging NIC economies are largely exported and sold to countries in Europe, North America and Japan. For example, Dyson, a UK-based vacuum cleaner manufacturer, moved the manufacture and assembly of its products to Malaysia but still sells the bulk of its vacuum cleaners in the UK and other parts of Europe.

However, the pattern is changing. As emerging NICs develop, their populations are becoming more affluent and starting to demand similar consumer products to those being exported from their own countries. Different patterns for distribution and consumption are likely to appear in the future with a definite shift from west to east as the centre of gravity of economic activity (consumption and production). Forecasts suggest that:

- consumption will drive trade patterns more than production location decisions and so the fastest growing trade route will be between India and China
- as Asia becomes more competitive a growing share of the region's exports will be to other countries in Asia
- Western companies specialising in finance have enormous potential to benefit from the expansion of trade in financial services in the Asia-Pacific region.

Factors in globalisation

A number of influential factors have combined to increase the breadth and depth of links between nations and trading groups over the past 30 years. Figure 7.14 (page 290) identifies most of the key factors which work together to drive the process of globalisation further into the twenty-first century.

Links between countries have grown significantly since the development of digital computer technology and particularly since the advent of the internet, which has enabled speedy and 24/7 global communication. The emergence of English as the accepted global language of business has also eroded barriers.

There are few borders or boundaries for the constant flow of data, so ours is a distinctly digital age with around 7 billion mobile phone subscriptions globally and nearly 3 billion internet users.

A number of other technologies, systems and relationships have evolved to support globalisation and international trade.

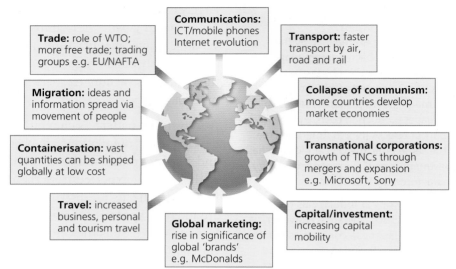

Figure 7.14 Factors that have combined to accelerate the pace of globalisation since the 1990s

Government support

One of the economic objectives of most governments is to increase exports from their country. Governments at national level will have trade departments whose function will be to ease and facilitate exports. For example, the UK government have the **UK Trade and Investment (UKTI)** department. They offer support and advice on all aspects of trade to encourage businesses, especially first-time exporters, to trade their goods overseas.

In some larger developing countries with less developed infrastructure, such as Pakistan, exporters are encouraged by government to use **dry ports**, located inland and nearer to their business. Dry ports save the exporter time and transport costs as all shipment arrangements and

customs documentation are completed locally before the goods are shipped to a seaport such as Karachi.

Financial

In the past trade was hindered by problems in exchanging finance for goods and exchange rate concerns. Deregulation of financial markets allowed arrangements for the removal or relaxing by governments of barriers to movements of finance. Communications technology has also removed these concerns and made international trading easier and faster, even for smaller enterprises. High-speed electronic trading systems and global exchange connectivity mean that financial transactions between importers and exporters can be completed quickly and securely.

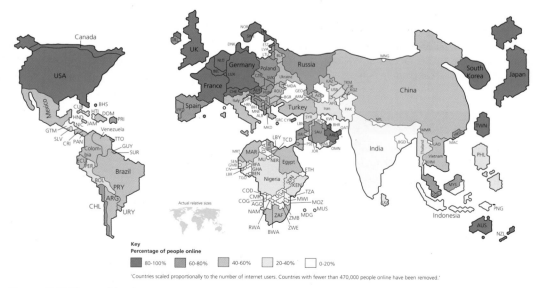

Figure 7.15 The world online. A cartographic representation of numbers with internet access globally and proportion within each nation

Figure 7.16 Containers stacked at Southampton docks can be transferred from ships to rail or road transport

Transport

Products and commodities can be shipped more quickly and in larger quantities as a result of integrating the following technologies, which have all contributed to easing international trade:

- increased size of aircraft; integrated air traffic networks
- growth of low-cost airlines and air freight companies
- the use of standardised containers – by sea, rail, road and air (Figure 7.16)
- handling and distribution efficiencies
- computerised logistics systems
- high speed rail networks.

Security

The trading community faces a number of security issues including supply chain security, crime and anti-terrorism, food and bio-security, fiscal and anti-smuggling. A number of initiatives have been introduced to alleviate threats, such as the World Customs Organization (WCO) and more regionalised measures such as an EU initiative to introduce a 'secure operator' quality label, awarded to operators meeting EU minimum standards of security.

In response to terrorist incidents of recent years, governments feel that they cannot leave the trade 'stable door' open. These measures have tightened up security for the benefit of business and have in some ways facilitated trade. Arguably though, their piecemeal introduction and lack of cohesion have had the opposite effect, by increasing costs and creating delays for the shipment of goods.

Management and information systems

The evolution of improved transport and communication systems has led to a production revolution which has transformed how companies manufacture products and distribute them worldwide. New processes of high volume production enable substantial economies of scale (cost reductions) on a global level. To benefit from these cost advantages, globally minded companies have invested in:

- large production and assembly plants capable of exploiting technology's economies of scale (for example, robotics in the automotive industry)
- global marketing and distribution networks to ensure that sales keep pace with the increase in production
- globally capable management.

These investments in international production, distribution and management are increasingly organised within **global value chains**, where the different stages of the production process are located across different countries. The **global production network** of any international organisation therefore covers the spatial inter-relationships necessary to support worldwide flows of information, raw materials, components, sub-assembled parts and finished products.

Corporations in different industries and economic sectors organise their global production networks in different ways in order to gain and maintain a competitive edge. For example, the fast fashion industry is reliant on sufficiently fast transport from a cluster of suppliers (mainly in Asia) to enjoy the short lead times necessary to be present in geographically disparate markets. Computer manufacturers such as Apple (see textbox on page 309) and Dell both co-ordinate a production network that span the Americas, Europe and Asia, and are reliant on outside suppliers for various components and peripherals. On the other hand, the automotive industry relies on components or sub-assembled parts arriving simultaneously for final assembly in large-scale production plants. For example, Renault still builds most of its vehicles in France but each vehicle has thousands of components supplied from all over the world, such as the batteries for the electric-powered Renault Zoe, which are supplied by a Korean manufacturer.

The management of such global production networks demands the remote management of production and distribution lines, which has been enabled by information technology and the internet giving businesses:

- virtually free telecommunications and video conferencing
- integrated ICT management systems; these ICT systems are usually supplied by third-party service providers (for example, GT Nexus, Wang and Unisys) and facilitate greater visibility of shipment control at each stage of passage.

Similarly, 'just-in-time' (JIT) technology and lean production management mean greater efficiency in the supply chain for manufacturers, ensuring that the correct supply of components arrives when they are needed and costs are cut by reducing the quantities of goods and materials held in stock. The principle underlying JIT is that production is 'pulled through' by specific customer orders, rather than 'pushed through' to build up stock. JIT therefore involves:

- producing and delivering finished goods 'just in time' to be sold
- sub-assembled goods 'just in time' to be assembled into finished goods
- parts 'just in time' to produce sub-assembled goods
- materials 'just in time' to be made into parts.

Inevitably, these systems designed to enable remote management and increase cost efficiency have led to:

- a spatial separation between higher-order business activities (such as research and development, design and engineering, marketing and advertising) which are located at corporation headquarters and strategic hubs around the world and lower-order activities (such as production and assembly) which are based either at low production-cost locations or in proximity to large markets for the finished goods
- global corporations increasingly focusing on a few key strategic activities and extensively outsourcing non-strategic activities
- rapid growth of the logistics and distribution 'solutions' industry, increased competitiveness between service providers within this industry and the rise of logistics as a profession.

Trade agreements

Since the 1950s, trade agreements have been formed by countries joining together to form trade blocs in order to stimulate trade between themselves and to gain economic benefits from co-operation. There are various forms that trade groupings might take, including free trade areas, **customs unions**, **common markets** and economic or monetary unions such as the European Union (EU). The UK is a member of the EU economically and politically, but not a member of its monetary union, known as the 'Eurozone', where there is a common currency – the Euro (€). Figure 7.17 shows a number of different regional trading blocs on a global scale. These are not exclusive; many other regional agreements exist and some overlap as many states are members of more than one agreement.

These unions or groups of countries usually allow free trade between group members in a free trade area; trade barriers are eliminated among the participating states. However, common external tariffs or trade restrictions will exist around the group member states.

Not all trade agreements are regionally based. For example the **Organization of Petroleum Exporting Countries (OPEC)** is made up of members mainly from the Middle East but also from South America and Africa. Its focus is on the trade of oil globally, which is the single most important traded **commodity**.

Others groupings are more loose knit and contain members with an interest in co-operation and development of trade, but with no formal trade agreement.

There are a number of advantages to nations which group together as trading entities:

On a global scale:

- to improve global peace and security and reduce conflict
- to increase global trade and co-operation on trade issues
- to help members develop their economies and standard of living.

Regionally (within each group):

- to compete on a global level with other trading entities
- to have a bigger representation in world affairs
- to allow freedom of movement of trade
- to allow people seeking work to move between countries more easily

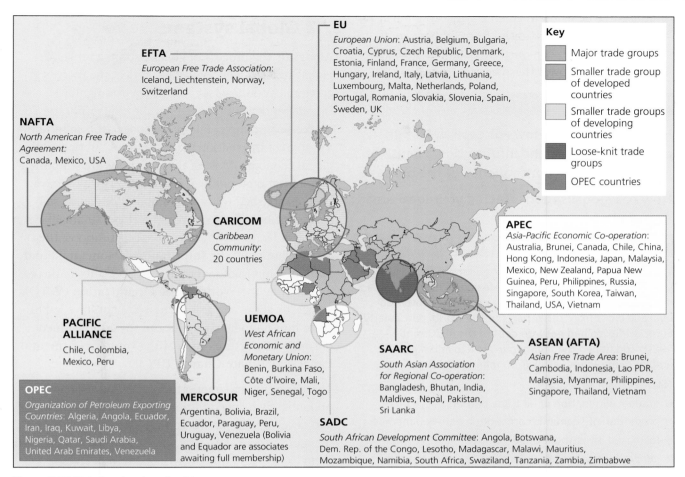

EFTA
European Free Trade Association:
Iceland, Liechtenstein, Norway,
Switzerland

EU
European Union: Austria, Belgium, Bulgaria,
Croatia, Cyprus, Czech Republic, Denmark,
Estonia, Finland, France, Germany, Greece,
Hungary, Ireland, Italy, Latvia, Lithuania,
Luxembourg, Malta, Netherlands, Poland,
Portugal, Romania, Slovakia, Slovenia, Spain,
Sweden, UK

NAFTA
*North American Free Trade
Agreement*:
Canada, Mexico, USA

CARICOM
*Caribbean
Community*:
20 countries

Key
Major trade groups
Smaller trade group
of developed
countries
Smaller trade groups
of developing
countries
Loose-knit trade
groups
OPEC countries

APEC
Asia-Pacific Economic Co-operation:
Australia, Brunei, Canada, Chile, China,
Hong Kong, Indonesia, Japan, Malaysia,
Mexico, New Zealand, Papua New
Guinea, Peru, Philippines, Russia,
Singapore, South Korea, Taiwan,
Thailand, USA, Vietnam

**PACIFIC
ALLIANCE**
Chile, Colombia,
Mexico, Peru

UEMOA
*West African
Economic and
Monetary Union*:
Benin, Burkina Faso,
Côte d'Ivoire, Mali,
Niger, Senegal, Togo

SAARC
*South Asian Association
for Regional Co-operation*:
Bangladesh, Bhutan, India,
Maldives, Nepal, Pakistan,
Sri Lanka

ASEAN (AFTA)
Asian Free Trade Area: Brunei,
Cambodia, Indonesia, Lao PDR,
Malaysia, Myanmar, Philippines,
Singapore, Thailand, Vietnam

OPEC
*Organization of Petroleum Exporting
Countries*: Algeria, Angola, Ecuador,
Iran, Iraq, Kuwait, Libya,
Nigeria, Qatar, Saudi Arabia,
United Arab Emirates, Venezuela

MERCOSUR
Argentina, Bolivia, Brazil,
Ecuador, Paraguay, Peru,
Uruguay, Venezuela (Bolivia
and Equador are associates
awaiting full membership)

SADC
South African Development Committee: Angola, Botswana,
Dem. Rep. of the Congo, Lesotho, Madagascar, Malawi, Mauritius,
Mozambique, Namibia, South Africa, Swaziland, Tanzania, Zambia, Zimbabwe

Figure 7.17 Global regional trading blocs

- to negotiate trade advantages as a group with other groups
- the possibility of developing a common currency to prevent currency fluctuations and simplify transactions
- to support particular sectors of a national economy (for example, agriculture within the EU)
- to share technological advances
- for remote regions or declining industrial regions to receive support from the larger organisation (for example, the EU Regional Fund helps regions such as southern Italy and western Ireland)
- to raise standards in education and healthcare across the region
- to spread democracy, human rights and possible political and legal integration (for example, the EU).

As trade agreements may lead to more economic, social and political integration, such as in the EU, a number of possible disadvantages can arise, including:

- some loss of sovereignty – decisions are centralised by what some see as an undemocratic bureaucracy

- some loss of financial controls to a central authority such as a bank (for example, European Central Bank, which oversees monetary policy in the Eurozone)
- pressure to adopt central legislation (for example, in Europe, Bosman ruling on soccer transfers, food standards and labelling)
- certain economic sectors are damaged by having to share resources (for example, the UK sharing its traditional fishing grounds with other EU nations such as Spain and France).

Trade deals can be assessed by looking at how successful they are in reducing barriers such as tariffs. In 2011 the rich **Organisation for Economic Co-operation and Development (OECD) group** studied 55 regional trade agreements to discover if barriers to agricultural produce were lowered.

- Deals between rich and emerging economies had lifted the number of goods that are traded duty-free from 68 per cent to 87 per cent over the previous ten years.

- In duty-free deals between emerging economies the proportion had advanced from 28 per cent to 92 per cent which demonstrates that regional trade agreements can lower these barriers.

Key questions

How much are the flows of capital, labour, products, services and information a cause of globalisation or a consequence of the process?

How has globalisation affected your local area?

North American Free Trade Agreement

The North American Free Trade Agreement (NAFTA) was signed by the USA, Canada and Mexico in 1994. The main driver for this agreement was the challenge presented by trade blocs from other parts of the world, particularly Europe. Mexico had got into debt in the 1970s and 1980s and hoped that economic growth and higher employment would result from joining NAFTA.

NAFTA's main aims are:

- gradual elimination of all trade barriers
- promotion of economic competition between members
- increased investment opportunities
- generally improved co-operation between the three member states.

Supporters of NAFTA have pointed out that:

- trade between member countries tripled between 1993 and 2007
- manufacturing grew in the USA, with increased employment
- Mexico receives increased foreign investment (as foreign TNCs establish plants in the country in order to gain access to Mexico's NAFTA trading partners' markets)
- Mexican workers receive higher wages and increased sales from agriculture reduces migration.

Opponents of NAFTA have pointed out that:

- some Canadian companies have closed because of competition from lower-cost US firms
- some US firms have moved to Mexico and American jobs have been lost
- food surpluses from the USA and Canada could be dumped in Mexico, reducing food prices and affecting the agricultural economy there
- the growth of US-owned, labour-intensive, export-orientated companies (maquiladoras) on the Mexican border keeps wage rates down
- Mexico could be exploited because of its natural resources and less stringent pollution laws.

7.2 Global systems

Alongside the process of globalisation, a range of global systems have evolved to reflect the increased economic, political, social and environmental interdependence that exists in the contemporary world. For example, Figure 7.4 (page 283) shows how a system of international capital flows 'lubricates' the global economy. Most of these systems are supported by international political organisations which have been established to provide stability or consensus across nations. For example, the global financial system is largely facilitated by two global institutions: the **World Bank** and the **International Monetary Fund**. Similarly, international trade and access to markets is overseen by the **World Trade Organization (WTO)**. Concerns about the global environment have led to more international summits and the establishment of support structures such as the **Intergovernmental Panel on Climate Change (IPCC)**.

These systems and supporting organisations have undoubtedly helped to improve co-operation, stability and development. However, they are often still led by the more powerful nations with vested interests, causing them to steer and influence the pattern of change for their own advantage. This can lead to increased inequality, conflict and injustice as less developed nations are limited in how they are able to respond.

The role of international financial institutions

It is difficult to view Figure 7.4 without noticing the pivotal role of the twin intergovernmental finance institutions, the International Monetary Fund (IMF) and the World Bank. Their role is fundamental in supporting the structure of the world's economic and financial order by regulating and acting as intermediaries in the flow of international capital. These institutions were established together at the end of the Second World War, in an attempt to steady the global economy and provide financial stability.

The IMF acts as a regulator of financial flows and stabiliser of the system. The World Bank is the provider of support for less developed countries and aims to reduce poverty. Both have been criticised for their

Table 7.2 Differences between the International Monetary Fund and the World Bank

International Monetary Fund	World Bank
Oversees the global financial system	Promotes economic development in developing countries
Offers financial and technical assistance to its members	Provides long-term investment loans for development projects with the aim of reducing poverty
Only provides loans if it will prevent a global economic crisis – the international 'lender of last resort'	Via the **International Development Association (IDA)**, provides special interest-free loans to countries with very low per capita incomes (less than US$865 per year)
Provides loans to help members tackle balance of payments problems and stabilise their economies	Encourages start-up private enterprises in developing countries
Draws its financial resources from the quota subscriptions of member countries	Acquires financial resources by borrowing on the international bond market
Has a total staff of 2,300 from 185 member countries and always elects a European managing director	Is a larger organisation with 7,000 staff from 185 countries and always has an American president

approach to their respective tasks. As part of their conditions for financial assistance, the IMF has been known to impose severe cuts on education and welfare spending by governments in developing countries. Similarly, the World Bank has been criticised because the conditions attached to loans have not always had the effect of reducing poverty. The Bank was also responsible for funding major '**top down**' projects, such as large multi-purpose dams to provide hydro-electric power (HEP) in less developed countries, which did not help to reduce poverty. Since the 1990s the Bank claims to support more '**bottom up**' sustainable development projects.

Key terms

Bottom up – When local people are consulted and supported in making decisions to undertake projects or developments that meet one or more of their specific needs.

Top down – When the decision to undertake projects or developments is made by a central authority such as government with little or no consultation with the local people whom it will affect.

The World Trade Organization (WTO)

The WTO deals with the global rules of trade between nations. It is the global institution responsible for facilitating international trade and its role and main aims are as follows:

- to supervise and liberalise trade by reducing barriers
- to act as an arbitrator sorting out trade problems between member governments

- to negotiate to reach agreements that become legal ground rules for international commerce
- to provide stability by giving trading nations confidence that there will be no sudden policy changes.

The WTO, which came into being in 1995, is the successor to the General Agreement on Tariffs and Trade (GATT), which was established in the wake of World War II. The WTO currently has over 160 members, over three quarters of which are developing or least-developed countries. The WTO is run by its members and all decisions are taken by consensus.

The WTO holds series or 'rounds' of talks on particular issues:

- 1986–1994 – the '**Uruguay Round**' made much progress on reducing barriers for trading manufactured industrial goods; this progress is ongoing.
- The most recent round of talks started in 2001 in Qatar's capital and is known as the **Doha Development Agenda**. Its focus has been on reforming trade in **agricultural produce**, especially between advanced and developing economies.

It was hoped that an agreement could finally be reached on the Doha Round in Geneva in 2008.

Hopes for the Doha talks in Geneva (2008) were that:

- tariffs could be reduced by 30 per cent
- a reduction could be made in subsidies paid to produce farm products

by richer developed countries adhering to these reductions, it would mean:
  - a benefit in trade for developing countries
  - reduced prices of food for consumers in MEDCs
  - fairer prices for farmers in emerging economies.

Problems

- The USA, EU and Japan insisted that, in return, the larger trading nations of developing economies, notably Brazil, China and India, would open their markets to Western manufactured goods.
- Emerging nations insisted on larger cuts in farm subsidies and tariffs paid to protect farmers in the USA and EU.

The Doha trade talks eventually collapsed in Geneva because of disagreement, mainly between the USA, China and India who would not compromise on the size of their tariffs. Also the USA would not agree to allow India and China to use 'safeguard clauses', which allowed developing nations to impose emergency quotas on imports. Since then further efforts to reach an agreement have stalled.

Successes

- A year later in Geneva, a separate agreement was reached between the EU and Latin American countries to end a long-standing trade dispute about the trade in bananas. This **bilateral agreement** gave a little more hope that the Doha round might be re-invigorated.
- Further success was achieved in 2013 in **Bali**, when the WTO achieved its first **multilateral** trade agreement in nearly twenty years with the **Bali Package**. The package was an agreement between all 159 WTO members on '**trade facilitation**' – primarily measures to speed up the movement of traded goods and to reduce costs by removing red tape in customs procedures.

Success at Bali gave impetus and hope that multilateral agreements are still achievable. However, agreements on trade in agricultural produce still prove to be a sticking point between richer nations (especially the USA and EU) and emerging economies, with the former unwilling to compromise on tariffs they impose on imports nor on the subsidies they offer to domestic producers.

Key terms

Bilateral agreement – An agreement on trade (or aid) that is negotiated between two countries or two groups of countries.

Common markets – A group formed by countries in geographical proximity in which trade barriers for goods and services are eliminated. (This may eventually apply to removing any labour market restrictions, as in the EU.)

Customs unions – A trade bloc which allows free trade with no barriers between its member states but imposes a common external tariff to trading countries outside the bloc (for example, the European Union).

Multilateral agreement – An agreement negotiated between more than two countries or groups of countries at the same time.

Key question

Why do there need to be systems in place to support world trade?

Issues associated with interdependence
Unequal flows in global systems
Effect of globalisation on the international labour market

Globalisation enables workers to move more freely around the world. The main flows are, predictably, from developing countries towards wealthier regions.

Table 7.3 Positive and negative effects of international movement of labour

Positive effects of labour movement	Negative effects of labour movement
Reduced unemployment where there is a lack of work – opportunities to seek work elsewhere	Countries find it difficult to retain their best talent – attracted away by higher wages
Reduces geographical inequality between workers (for example, eastern Europeans working in UK)	Loss of skilled workers causes a training gap
Addresses important skill and labour shortages (for example, the UK has recruited nurses from the Far East)	Outsourcing of production from high-wage to low-wage economies causes unemployment in more developed countries
Some workers return to their country of origin with new skills and new ideas	With greater movement of labour there is a greater risk of disease pandemics

Key term

Outsourcing – A cost saving strategy used by companies who arrange for goods or services to be produced or provided by other companies, usually at a location where costs are lower.

Outsourcing

The practice of **outsourcing** is largely one directional, that is, taking manufacturing or service jobs from high wage economies in Europe and North America and having them undertaken by a sub-contracting organisation in a lower wage economy such as China. Outsourcing provides jobs and investment in one country but often takes them away from another country. The consequences of outsourcing for the original country are usually negative:

- **Loss of jobs:** This has a knock-on effect in communities, especially where one large employer has outsourced. Unemployment means there is less spending in the local economy, so service workers, such as shopworkers, lose their jobs and services will close down. This is known as the **de-multiplier** effect.
- **De-industrialisation** of the economy: The closure of manufacturing companies because of outsourcing eventually leads to the closure of local suppliers. Areas go into decline with derelict factories, etc.
- **Structural unemployment:** The skill set of local workers is no longer compatible as the jobs they trained for have now moved abroad. They are often ill-equipped for the new types of work that enter their local economy. High investment from government is required to retrain workers and it may take a generation before a new workforce, with the education and skills required for the local economy, emerges.

Inequality issues

Globalisation should increase prosperity for all and make the planet more equal in terms of income distribution. There are two measures of inequality to consider:

- the difference *between* richer countries and low-income countries and whether the difference between the two is increasing or decreasing
- the inequality in incomes that exists *within* each country and how this is being affected by globalisation.

Indicators suggest that globalisation is reducing global inequality through the transfer of capital and income from richer to poorer economies. Paradoxically, it may be increasing inequality within countries as richer members of societies cope better with the changes in jobs and technology.

Inequalities between countries

As communication and transport increase the integration of economies, developing countries are closing the gap with their rich-world counterparts. The development continuum, with the exception of some of the least developed countries, has become more condensed. The fastest growing economies continue to be in Asia and although countries in sub-Saharan Africa have a large gap to make up in living standards, their economies are now growing more quickly than most developed economies.

Table 7.4 (page 298) shows the considerable differences in average income between twelve developed and developing economies both in 1985 and in 2014. The figures, however, support the idea that developing countries have increased their average incomes more rapidly than richer countries, and also tend to have faster economic growth rates (Figure 7.18).

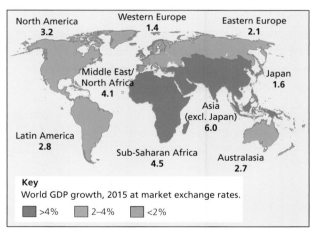

Figure 7.18 Global differential rates of GDP growth by continent

Skills focus

Consider methods of presenting the data in Table 7.4 (page 298) graphically, in particular to show relationships between the size of GDP per capita and growth rates.

This data could also provide the basis for a statistical analysis of relationships using a Spearman's rank correlation test.

Table 7.4 Trends in GDP per capita and economic growth between more and less developed economies from 1985 to 2014

Country	GDP per capita (US$) in 1985	GDP per capita (US$) in 2014	Increase in GDP by times	Percentage economic growth rate, 2014
Brazil	1,637	11,612	x7.1	1.8%
Canada	13,991	50,271	x3.6	2.3%
China	293	7,594	x25.9	7.0%
India	303	1,631	x5.4	6.5%
Japan	11,466	36,194	x3.2	1.8%
Kenya	312	1,338	x4.2	5.7%
Mexico	2,369	10,361	x4.4	4.0%
Nigeria	344	3,185	x9.2	5.8%
Peru	825	6,594	x8.0	4.9%
Portugal	2,706	22,081	x8.1	0.8%
UK	10,611	45,603	x4.3	2.5%
USA	18,269	54,630	x3.0	3.2%

Source: World Bank, World development Indicators and The Economist

Inequalities within countries

The measure usually used to indicate levels of inequality of income distribution within a country is the **Gini Index**, named after Italian statistician Corrado Gini. It is based on the idea of a Lorenz curve as shown in Figure 7.19. The 45° line represents an equal distribution of income so the further the curve is away from the line, the more unequal the distribution. The Gini index aggregates the inequalities in people's incomes into a single measure, with a coefficient score of between zero and one. An index score of one means a country's entire income goes to one person; a score of zero means that income is equally divided among the population.

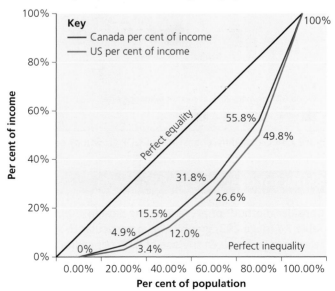

Figure 7.19 Lorenz curve and Gini index for Canada and the USA. The Lorenz curves plotted here show that Canada has a more equitable distribution of wealth than the USA

The level of inequality varies widely around the world. Developing economies are usually more unequal than rich ones, for example, South Africa has a score of 0.60.

Skills focus

The Lorenz curve is used in Figure 7.19 to demonstrate the degree of unevenness (in this case comparing income inequality between two countries) from a completely even distribution (represented by the 45° line). Consider how the Lorenz curve graphic might be used to compare and represent other measures of uneven distribution within physical or human geography. It might be used to present such comparisons in your fieldwork investigation.

Within many developing economies, the Gini index shows that inequality has worsened. Sub-Saharan Africa saw its Gini index rise by 9 per cent between 1993 and 2008 and China's score has increased by 34 per cent in the last twenty years. The majority of people live in countries where income disparities are larger than they were a generation ago. The main exception to trend is in Latin America, previously the most unequal continent, where Gini coefficients have generally fallen over the past ten years.

This general increase in inequality doesn't just apply to developing regions. Many countries, including Britain, Canada and even egalitarian Sweden, have seen a rise in the share of national income taken by the top one per cent. Globally the numbers of very wealthy people have increased significantly and this concentration of wealth at the very top is part of a much broader rise in disparities.

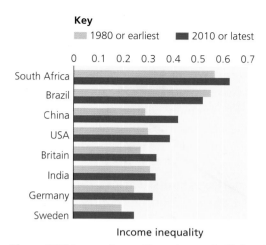

Figure 7.20 Income inequality – changes in Gini coefficient over time (0 = perfect equality; 1 = perfect inequality)

There seem to be two possible explanations for this paradox:

● **Outsourcing:** Contrary to popular belief, TNCs based in developing countries often employ skilled workers and pay higher wages. A report from the OECD found that average wages paid by foreign transnationals are 40 per cent higher than wages paid by local firms. By contrast, unskilled workers, particularly those in rural areas, tend not to have such opportunities. The result is that inequality increases.

● **Investment:** Those with some money to begin with will gain from investment and benefit more from growth, whereas those with no money stay rooted in poverty. Only with further development will equality increase.

Geopolitical issues

One of the arguments for globalisation is that it will lead to greater political stability. A sense of the world as one community makes it easier for governments to work together on common goals.

In his 1999 book, *The Lexus and the Olive Tree,* Thomas Friedman (a US writer on globalisation) outlined the idea of the 'Golden Arches Theory of Conflict Prevention', which holds that *'No two countries that both had McDonald's had fought a war against each other since each got its McDonald's.'* The theory has been disproved on a number of occasions since then. However, he makes a legitimate point about economic integration decreasing the likelihood of armed conflict between countries.

There are many causes of conflict though, and shortages of food, water and energy resources brought about by global growth may lead to conflict between nations in some regions. In addition, globalisation has not brought equal prosperity to all. Some groups may feel a sense of injustice or ideological opposition to the richer nations driving the process. This may give rise to civil conflict within countries and rapid global communications can hasten the spread of such conflicts. For example, 'The Arab Spring', sparked in Tunisia in 2010, quickly spread across North Africa and the Middle East.

Trade can also be used as a 'weapon' in conflict, either between nations or between trade blocs. In 2006, the UN Security Council imposed trade sanctions against Iran because of its refusal to suspend its uranium enrichment nuclear programme. This severely harmed

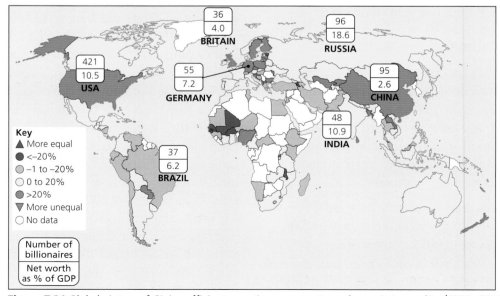

Figure 7.21 Global picture of Gini coefficient over time – percentage change in inequality (1980–2010)

Iran's integration into the world economy until it was recently resolved. Russia's supposed backing of separatists in the Ukraine has brought sanctions against them from the EU and the USA. Analysts argue that this ongoing political battle between Russia and the West is a feature of increased '**nationalism**' that is actually slowing down globalisation.

Key question

Why has globalisation caused more equality between average incomes globally but more inequality in income distribution within countries?

7.3 International trade and access to markets

The notion of **comparative advantage** is why most nations trade. The theory behind comparative advantage suggests that countries should specialise in providing goods and services that they excel at producing. They trade these for the things they are not as good at producing. It is easier to make these trading exchanges if there are fewer barriers. Theoretically production should increase in each country, and globally, because each country is concentrating on what it does best.

For much of the twentieth century trade remained relatively limited because of regulations, **protectionism** and high transportation costs.

Barriers to trade and protectionism

A barrier to trade is a government-imposed restraint on the flow of international goods or services. The most common barrier to trade is a **tariff** – a tax on imports. Although free trade has eroded the practice of protectionism, it is still a strategy used by governments in both developed and developing countries to protect their established or emerging industries.

There are a number of other barriers to trade, which are used as either protectionist strategies or sometimes as a means to affect diplomatic relations.

- **Import licence:** This is a licence issued by a national government authorising the importation of goods from a specific source.

- **Import quotas:** These set a physical limit on the quantity of goods that can be imported into the country.
- **Subsidies:** These are grants or allowances usually awarded to domestic producers to reduce their costs and make them more competitive against imported goods. (Export subsidies are also used by governments to encourage domestic producers to sell their goods abroad.)
- **Voluntary export restraints:** This is a diplomatic strategy offered by the exporting country to appease the importing country and deter it from imposing trade barriers.
- **Embargoes:** These involve the partial or complete prohibition of commerce and trade with a particular country. They are usually put into practice for political rather than commercial reasons.
- **Trade restrictions:** Other import restrictions may be based on **technical** or **regulatory** obstacles such as the quality standards of goods being imported, or how they are produced. For example, the EU attempts to put restrictions on the import of goods knowingly produced using child labour.

Trends in the volume and pattern of international trade

The emergence of **free market** ideas at the end of the twentieth century saw a push to remove existing barriers to international trade. With an increase in the mobility of factors of production, especially capital, **regional trade agreements (RTAs)** emerged and the global trade framework was strengthened by the General Agreement on Tariffs and Trade (GATT) (established in 1947). In 1995, GATT was replaced by the World Trade Organization (WTO) which continued the work of gradually lowering the barriers to international trade, with **free trade** as its aim. Reaching agreements has not been easy but average trade tariffs have shrunk to a tenth of their level when GATT first began operating. World trade has increased at a faster rate than global economic growth and the flow of most products has eased. In recent years, however, particularly since the global financial crisis of 2008–2009, global agreements have become increasingly difficult to reach. International trade flows are still expanding but have not yet regained the pace of the 1990s and 2000s. During major recessions international trade inevitably stalls and this can cause a setback to the economic aspect of globalisation.

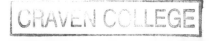

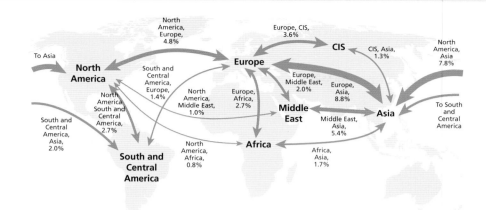

Figure 7.22 Global patterns of inter-regional trade, 2011 (World trade includes intra-EU trade; trade within regions and with unspecified destinations accounts for 54 percent of world trade in 2011)
Source: WTO

The pattern of international trade has changed as a result of the greater integration of economies. Geographically it is still dominated by a few large economic blocs, mainly North America, Europe and East Asia (Japan, South Korea and Taiwan) (Figure 7.22). The USA, Germany and Japan alone account for about 25 per cent of global trade and when the other **G7 countries** are included (UK, Canada, Italy and France) together they account for nearly 50 per cent of global trade. This dominance is now being challenged by emerging economies, which account for a growing share; China accounts for the largest growth. These changes are reflected in the balance of trans-oceanic trade. Trans-Pacific trade is now growing much faster than Trans-Atlantic trade.

In 2013, the total value of world trade in merchandised goods was US$18.8 trillion and in services it was US$4.6 trillion. Despite some recent slowing of growth in trade, forecasts suggest the following by 2020:

● World trade in goods will total around US$35 trillion and services will be around US$6 trillion.

● Intra-regional trade within Europe will be worth more than US$7 trillion – still more than Asian intra-regional trade (US$5 trillion) despite the more rapid growth of the latter.

● The Asia-Pacific region will experience the fastest growth in global trade.

● There will be some growth in sub-Saharan African countries as they develop and possibly become points for the assembly of finished goods.

● Europe will be the most important market for sub-Sahara's exports.

● The machinery and transport sector, which includes consumer electrical products, will make the largest contribution to trade.

Trends in investment

Foreign direct investment (FDI) is an important source of funding for development in all countries, especially for the least developed economies wanting to catch up. Whether they are countries just beginning to modernise, or those of the rich **OECD** group, it is difficult to 'jumpstart' economic growth without an injection of capital.

Each year more than US$1 trillion in FDI flows into countries around the world, but the distribution is far from equal. In 2011, the **UN Conference on Trade and Development** (UNCTAD) listed the top ten recipients of FDI as shown below.

1. USA:		$258 billion
2. China:		$220 billion
3. Belgium:		$102 billion
4. Hong Kong (China):		$90 billion
5. Brazil:		$72 billion
6. Australia:		$66 billion
7. Singapore:		$64 billion
8. Russia:		$53 billion
9. France:		$45 billion
10. Canada:		$40 billion

Some of these countries have **natural resources** that attract foreign investment but another pull is the size of their populations, or more precisely the size of their **markets**.

UNCTAD reports that countries with the greatest share of FDI as a proportion of their national income fall into **two main groups**:

1. Countries known for **natural resource development** – including Mongolia, Liberia and Congo Republic. These have attracted investment from large mining corporations such as Arcelor Mittal, a Luxembourg-based TNC.

2. Countries known for **financial business services** – including Singapore, Hong Kong SAR (China) and Luxembourg.

In summary, there are three main attractions that pull in investment:

● plentiful natural resources
● large and accessible consumer markets
● financial services.

Fair trade and ethical investment

Fair trade

Fair trade is a social movement whose goal is to help producers in developing countries achieve better trading conditions and to promote sustainability. The movement focuses mainly on agricultural-based products (or their direct derivatives) and includes coffee, tea, cocoa, fruit, sugar, honey, bananas, cotton, flowers, wine and chocolate. Goods may also include handicrafts or valuable minerals such as gold. These are traditionally the products which are exported from developing countries to richer nations.

Fair trade supporters argue that those producing the commodities do not get an equitable deal from the organisations they supply their produce to. These may be transnational corporations, food processing companies or buyers from supermarkets in developed countries. Buyers are able to force down the prices of individual suppliers because these suppliers:

● have little market influence
● are extremely reliant on the income from their goods.

Members of the fair trade movement advocate the payment of higher prices to producers, as well as helping them to achieve improved social and environmental standards. International fair trade organisations organise producers into **co-operatives** to combine their produce. This gives them more influence in governing market conditions and the power to negotiate better deals with buyers or to

supply direct. The goods are labelled as 'fair trade' goods with the International Fair Trade Certification mark, so that 'ethical consumers' can recognise that they are buying goods for which the producer received a fair price.

There are also '**alternative trading organisations**' such as Cafedirect, Oxfam Trading and Traidcraft which also focus on the application of fair trade. The pattern is slightly different in that they are retailers who buy directly from suppliers at a fair trade price.

It is important not to confuse fair trade with free trade, especially as the two ideas can be conflicting. The terms are used interchangeably but they mean completely different things. Table 7.5 summarises the key differences between the two.

Table 7.5 Fairtrade is NOT free trade

	Free Trade	Fair Trade
Main goal	To increase nations' economic growth	To empower marginalised people and improve the quality of their lives
Focuses on	Trade policies between countries	Commerce among individuals and businesses
Primarily benefits	Multinational corporations, powerful business interests	Vulnerable farmers, artisans and workers in less industrialised countries
Critics say	Punishing to marginalised people and environment, sacrifices long term	Interferes with free market, inefficient, too small scale for impact
Major actions	Countries lower tariffs, quotas, labour and environmental standards	Businesses offer producers favourable financing, long-term relationships, minimum prices and higher labour and environmental standards
Producer compensation determined by	Market and government policies	Living wage and community improvement costs
Supply chain	Includes many parties between producer and consumer	Includes fewer parties, more direct trade
Key advocate organisations	World Trade Organization, World Bank, International Monetary Fund	Fairtrade Labelling Organizations International, World Fair Trade Organization

Source: Fair Trade Resource Network

Ethical investment

Ethical investment is a form of ethical consumerism where investors make a deliberate choice to invest capital based on the activities of the firm or organisation they are investing in. The investor uses personal principles and beliefs as the main filter to make investment decisions. For example, some choose to eliminate certain industries entirely from their investment plans, such as tobacco, firearms or polluting energy companies. Ethical investment is usually seen as a socially responsible choice and most investors make their decisions based on environmental or social concerns.

Trading relationships and patterns

The main trading entities are the USA, which is part of the North American Free Trade Association (NAFTA) and the European Union (EU). Becoming increasingly important are the groupings of emerging nations such as the Association of South East Asian Nations (ASEAN), which also form the Asian Free Trade Area. China is a member of Asia-Pacific Economic Co-operation (APEC) but is a major economic force and trading entity in its own right.

Conflict and co-operation

Tensions have arisen between trading entities as they all want to ensure the best deals for their citizens, their workers and for the businesses based within their group (including TNCs). The failure to achieve an outcome from the WTO's Doha Round is largely because of disagreements between the EU and USA together arguing with emerging economies such as China and India. Equally, the banana trade escalated into a trade war between the EU and the USA, with the latter backing their home-based TNCs who accused the EU countries of trading unfairly.

Larger trade deals are on the horizon, such as the proposed **Trans-Pacific Partnership (TPP)** involving free trade between the USA and countries in Asia and South America. Another deal known as the **Transatlantic Trade and Investment Partnership (TTIP)** between the USA and the EU also seems set to go ahead.

The USA has traditionally been a 'protectionist' economy, lagging behind in entering any formal trade agreements (NAFTA was not formed until 1994), but it is now attempting to negotiate trade agreements that will place it at the 'core' of international trade. The development of both the TTP and TTIP agreements is arguably a strategic move by the USA to make deals with both the West and the East and may be a response to the emerging threat of China as an economic superpower and driver of trade.

Trans-Pacific Partnership (TPP)

The Trans-Pacific Partnership Agreement is a free trade agreement currently being negotiated by twelve countries: USA, Australia, Brunei, Canada, Chile, Japan, Malaysia, Mexico, New Zealand, Peru, Singapore and Vietnam.

It has been criticised, because the negotiations lack transparency, but will be one of the most important trading agreements globally as this region has seen the most rapid recent growth in trading. The topics being negotiated by the group extend beyond traditional trade matters and are thought to set binding rules on a whole range of subjects including investment, patents and copyrights, financial regulation and labour and environmental standards, as well as trade in industrial goods and agriculture.

Transatlantic Trade and Investment Partnership (TTIP)

This is a free trade agreement currently being negotiated by the EU and the USA. It has similarly been criticised for its 'covert' nature.

As a bilateral trade agreement, TTIP aims to reduce the regulatory barriers to trade for big business, including food safety law, environmental legislation, banking regulations and the sovereign powers of individual nations.

There are many opponents to the proposed agreement, particularly social and environmental NGOs such as Friends of the Earth. Their major concern is that it will give more power to large TNCs above that of democratically elected governments. Opponents believe TTIP will undermine democracy and social provision in both regions. They claim that such an agreement will threaten public services (such as the NHS) as well as consumer protection, data protection and the environment. It is even claimed that a feature of the agreement will be something called Investor State Dispute Settlement (ISDS) which will give large corporations the ability to sue EU governments (and thus taxpayers) if their profits are affected by any change in government policy.

The role of China

China is a member of the **G20 group**, an international forum for the governments and central banks of 20 major economies. It includes the G7 countries and the EU as a single member. It was established in 1999 to give a voice to the major developing economies (including the BRIC countries) who felt that the WTO was not fully serving their interests.

The '**Group of 77 and China**' is another forum which was established in 1964 when China was seen as a less developed economy. The aims of this group are to:

- represent the interests of the world's poorest countries
- help development in these countries
- reduce poverty and disease and improve human rights.

The importance of China's role within this group is:

- China has a common interest, with the poorest 77, in reducing poverty
- China hopes to increase its influence in world affairs by being part of this group
- the group needs a powerful ally to effect international co-operation and change.

China's phenomenal economic growth rate in the last 25 years has been based on the expansion of manufacturing of consumer goods for richer 'Western' countries, particularly those in Europe and North America.

China's influence as a large NIC has not been confined to industrial expansion within its own borders. The increased confidence of being an economic superpower has enabled the Chinese to spread their wealth and influence by investing in other parts of the world, including in Africa (Figure 7.23). Many Western-based TNCs are often too wary of investing in African countries, especially those which have a recent history of civil war. Chinese entrepreneurs, however, have seen the lack of development in many African countries as an opportunity for investing in resource development and increased trade.

China's role in Africa is varied: one clear objective is to extract a range of primary resources including metals to support industrial expansion in China, for example, investment in old copper mines in Zambia and Botswana. However, there is evidence that China's investment in Africa is helping some of the poorer countries to develop infrastructure, as well as healthcare and education.

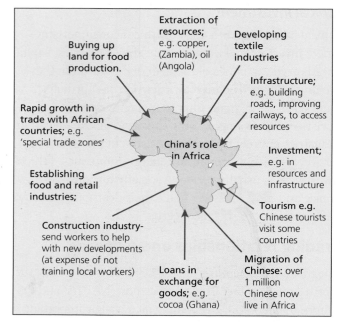

Figure 7.23 Chinese interests in Africa

Latin America

As another 'emerging' region Latin America's relations with other trading entities will be important in the future. At the moment they have two distinct trading blocs – Mercosur and the Pacific Alliance, each with different approaches to how they develop their trading links (Figure 7.24).

Mercosur: Formed in 1991 and comprises of Brazil, Argentina, Uruguay, Paraguay and Venezuela (Bolivia and Ecuador have applied to join). It is a traditional customs union and operates very similarly to the EU. It is sometimes called the 'Common Market of the South' and allows the free movement of labour between member states. The nature of its produce allows trade globally but it tends to view the EU and North America as its main markets.

Pacific Alliance: Formed more recently in 2011 and comprises of Chile, Peru, Colombia and Mexico (Costa Rica, Panama and Guatemala have applied to join). Its key differences to Mercosur are that it has been more open to making bilateral agreements with other nations and trading entities and has a tendency to see the Asia Pacific and the USA as its main market. The countries in this group will be part of the TPP agreement.

Although much younger with a slightly smaller overall population, the Pacific Alliance group has been growing more quickly both in terms of

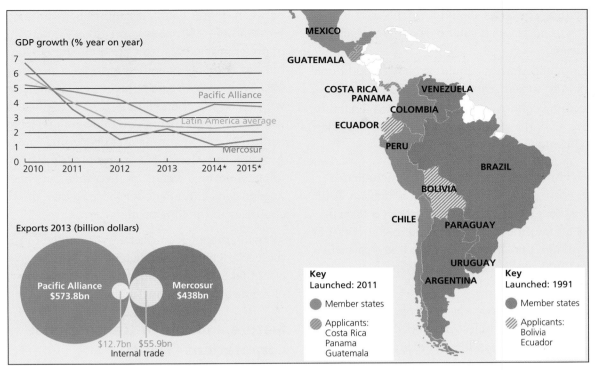

Figure 7.24 Mercosur and the Pacific Alliance in Latin America

individual economies and volume of trade. There is the possibility in the future that the two groups could merge to form one Latin American union, which would be a major player globally.

Differential access to markets

In one form or another, **special and differential treatment (SDT)** agreements have been a defining feature of the multilateral trading system for most of the post-war period.

During the latter part of the twentieth century, there were a growing number of newly independent nations following de-colonisation in Africa, Asia and the Caribbean. It was recognised that there was a danger that protectionist policies would exclude many developing countries from trading freely with more developed nations (they would inevitably encounter huge tariffs or quota barriers). In order to prevent this, the United Nations Conference on Trade and Development (UNCTAD) assisted developing countries in placing these issues in the GATT (and later WTO) rounds of talks from the 1970s onwards.

The category of the least developed countries (LDCs) was created by the UN General Assembly in 1971 with a view to providing the least developed among the developing countries with special support measures to overcome their specific disadvantages. In addition to official development assistance (ODA or multilateral aid) these support measures also included preferential access to developed markets in trade agreements. The provision of SDTs was to enable some of the world's poorest countries to:

- tackle the structural handicaps that characterise LDCs, which are:
  - a low level of income
  - a concentration on export of primary goods (commodities)
  - vulnerability to export price volatility
- engage in world trade on a more advantageous level by:
  - providing incentives for export diversification
  - allowing more stable export revenues
- promoting faster income growth and development.

The following arguments were advanced for less developed countries to gain SDT agreements:

1. Special and differential treatment is an acquired political right.
2. Developing countries should enjoy privileged access to the markets of their trading partners, particularly the developed countries.

3. Developing countries should have the right to restrict imports to a greater degree than developed countries.

4. Developing countries should be allowed additional freedom to subsidise exports.

5. Developing countries should be allowed flexibility in respect of the application of certain WTO rules, or to postpone the application of rules.

Examples of trade preference schemes include the EU's Everything but Arms (EBA) Agreement in 2001 to accept access for all products from LDCs (except arms and ammunition) on a duty-reduced, quota-free basis. Similarly, the US extends duty-free treatment to nearly 2,000 products from qualifying African countries.

Trade has played an important role in promoting economic development in less developed countries and differential access has allowed many countries to diversify their economies, achieve economic take-off and lift some of their population out of poverty. In particular it has helped many countries to develop service and tourism industries, and in some cases manufacturing, though overall this has been less successful. SDTs are not without their problems:

- Not all countries listed as LDCs are members of the WTO and the time taken to accede after application can be lengthy (on average 8 to 10 years).
- The WTO Doha Round recognised that SDT measures need to be made more effective and operational.
- Some of the lack of effectiveness of the SDTs is due to the fact that LDCs are not fully aware of them and, when aware, do not make productive use of them.
- Some measures have not been tailored to the conditions prevailing in most LDCs.
- There is concern among some richer nations that non-reciprocal and preferential trading agreements given to less developed countries will result in cheap imports flooding markets and undermine their own industrial base (this has already happened with the growth of newly industrialised countries).

The agreements are difficult to apply in a fair or standardised manner and the lack of reciprocity in the agreements has deterred some developed countries from participation. In many cases, they have been replaced by unilateral trade agreements between poorer countries and richer trading partners. For example, Mexico was previously a beneficiary of preferential access to Canadian and US markets but has arguably benefited more as a member of NAFTA having to accept reciprocal arrangements with its co-members. Similarly, sub-regional free trade areas and customs unions among developing countries are expanding and deepening in Asia, Latin America and Africa. These groupings greatly enhance the negotiating leverage of their members in trade negotiations and may offer more to least developed countries in the long term.

More international co-operation is needed to address these shortcomings of differential access because it is clear that SDT agreements have a role in removing some of the inequalities in international trade.

Nature and role of transnational corporations (TNCs)

Transnational corporations (TNCs) are companies that operate in at least two countries, with a headquarters based in one country but with business operations usually in a number of others. TNCs take many different forms and are based in different economic sectors. They no longer only originate from the more developed region; emerging economies also have TNCs which are major global companies.

TNCs may operate in more than one country for a number of reasons:

- **to escape trade tariffs** – for example, Nissan's decision to produce cars in Sunderland was largely to gain barrier-free access to the lucrative EU market
- **to find the lowest cost location for their production** – for example, Hewlett-Packard in Malaysia
- **to reach foreign markets more effectively** – for example, McDonald's
- **to exploit mineral or other resources available in foreign countries** – for example, BP in Azerbaijan.

The TNC of the twenty-first century has been called a 'globally integrated enterprise' because it:

- is prepared to **locate different functions of the business anywhere** – based on getting the right cost, skill and environment
- pursues a goal to **integrate production** and deliver value worldwide – state borders mean less and less to corporate practice.

There are certain characteristics thought to be common to TNC organisations, including:

- maximising global economies of scale by organising production to reduce costs
- sourcing raw materials or components at the lowest cost
- controlling key supplies
- control of processing at each stage of production
- branding of products/services so they are easily recognisable
- outsourcing of production.

Spatial organisation

Transnational corporations have become increasingly flexible in the global location of their assets (Figure 7.25).

Traditionally, the company headquarters are based in a major city in the home country. Most TNCs have subsidiary headquarters in each continent, or in countries where their main operations are based.

In order to maintain their position competitively, TNCs engage in research and development (R&D) activities. These also tend to be based in the country of origin and will often locate near to centres of higher education, to take advantage of a graduate labour market, or to make use of university research facilities.

Production

Production operations of TNCs involved in the **primary sector** will be based wherever there are unexploited resources. This tends to be in developing economies as reserves in more developed countries have largely been depleted. However, a combination of rising world prices and new technologies can make access to new reserves of raw materials viable in the home country, for example, the recent development of hydraulic fracturing, known as 'fracking', has revitalised oil and gas industry bases in North America.

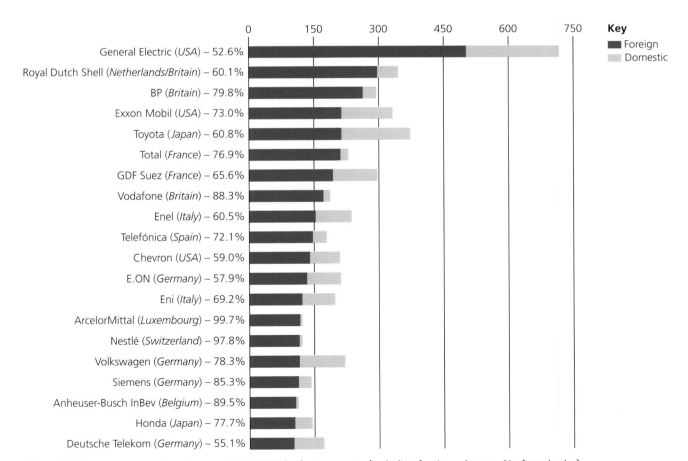

Figure 7.25 Largest transnational companies in 2011 by foreign assets (including foreign sales as a % of total sales)

Figure 7.26 South Korean company Samsung has located LED TV production in Kaluga in western Russia. Russian industry gains from the transfer of Samsung's technology, skills and capital investment

For TNCs in the **secondary sector**, production operations have largely been located in the manufacturing regions of developing countries, especially in south east and south Asia. These areas attract TNCs because:

- labour costs are lower
- there is investment in education, which makes it easier to train workers
- the work ethic means workers are willing to work long hours, with relatively few holidays, in a non-unionised labour environment
- there may be government incentives such as tax-free breaks, enterprise zones with low business rates or less restrictive environmental regulations.

The spatial organisation of production has allowed for TNCs from emerging economies to locate production where it gives access to large markets in developed regions. Kia Motors (South Korea) has factories in Slovakia and the USA to gain access to EU and NAFTA markets.

Service-based TNCs in the tertiary sector are more footloose and will locate operations either where there are relatively low labour costs balanced with good education or in proximity to their markets. Language is another important consideration here: lower level services, such as call centres, are outsourced to India, for example, because of the high proportion of well-educated English-speaking workers, who offer a lower cost alternative.

Key terms

Agglomeration – When companies in similar industries locate near to each other because of the benefits gained by sharing ideas and resources – called 'agglomeration economies'.

Multiplier (effect) – A situation where an initial injection of investment or capital into an economy (at any scale) in turn creates additional income by, for example, increasing employment, wages, spending and tax revenues.

Linkages, trading and marketing patterns

One of the features of TNCs is the ability to expand and gain more control of their industry and markets. They do this by integrating different parts of the business through investment, takeovers and mergers.

There are two types of integration:

- **Vertical integration:** An arrangement in which the **supply chain** of a company is owned entirely by that company, from raw material through to the finished product. This gives the TNC control over its supplies and stocks and reduces costs because of economies of scale. A good example is BP in the oil and gas industry. BP has negotiated exploration and production rights in over 40 oil and gas fields worldwide. It owns or jointly owns (with other energy companies) oil pipelines and has its own shipping fleet. The company owns refineries, usually based in the countries where end products are sold, and in the UK has over 1,100 retail service stations.

- **Horizontal integration:** A strategy where a company **diversifies** its operations by expansion, merger or takeover to give a broader capability at the same stage of production. This can be either complementary or competitive to its existing business. For example, Kraft Foods takeover of Cadbury in 2010 gave them a more diverse base in the grocery and confectionery market. Following a series of de-mergers, the newly formed Kraft Foods Group merged with Heinz in 2015 for the same purpose.

Benefits and costs of TNCs

Table 7.6 Outline of how TNC operations affect different participants positively and negatively

	For the host country	For the TNC	For country of origin (TNC base)
Benefits	Generates jobs and income Brings new technology Gives workers new skills Has a **multiplier effect**	Lower costs because of cheaper land and lower wages (fewer unions) Greater access to new resources and markets Fewer controls such as environmental legislation	Cheaper goods Can specialise in financial services and R&D occupations
Problems	Poor working conditions Exploitation of resources Negative impacts on environment and local culture Economic leakages/repatriation of profits	Ethical issues such as the image of environmental damage or 'sweatshops' can be detrimental to their reputation Social and environmental conscience	Loss of manufacturing jobs De-industrialisation Structural unemployment

Key question

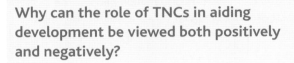

Why can the role of TNCs in aiding development be viewed both positively and negatively?

Apple Inc.: a transnational corporation

Apple Inc. – producer and retailer of computer technology and mobile electronic devices

Apple Inc. is a US transnational electronic technology corporation with its company headquarters based in Cupertino, northern California (Figure 7.27).

Apple produces many of the familiar high-tech products that are marketed under its universal 'i' brand. These include iMacs, iPhones, iPods and iPads and in 2015 it launched a new 'smartwatch' product 'Apple Watch' and a new version of Apple TV

Figure 7.27 Apple Inc. Head Office Campus in Cupertino, California. A new Apple headquarters (Campus 2) will open in the city in 2016.

The company started business in 1976, in the early days of personal computer manufacture. In 1982 it took over the smaller MacIntosh organisation and achieved more success by launching a new brand of desktop MacIntosh, later AppleMac computers. Apple computers earned a growing reputation for quality and they attracted a growing niche market of brand-loyal customers. Since 2000, it has experienced phenomenal growth as an organisation because of its development of mobile and Wi-Fi devices. Apple is now the world's:

- second-largest IT company by revenue (after Samsung Electronics)
- third-largest mobile phone manufacturer
- largest music retailer (through its iTunes store)
- number one global brand by value (US$145 billion).

It has 98,000 full-time employees and over 450 retail stores in sixteen countries. In 2014 it was the eleventh largest TNC globally with total assets of US$207 billion.

Apple's success has been due to a number of factors:

- stylish and well-designed products
- slick marketing and branding – generating a growing number of customers with brand loyalty
- innovative products

- focus on highly mobile devices, which fits their market's needs
- selling, via the internet, ancillary products such as music and apps.

Their market is predominantly in richer, developed countries. In 2011, 44 per cent of product sales were in the USA. Its average market profile consists of young (average age 31); wealthy (earning 26 per cent more than national average) and educated (58 per cent graduates) who are prepared to pay a premium price for their products. Over the past five years, however, its most rapid growth has been outside the USA, especially in the Asia-Pacific region, largely due to the global demand for the iPhone.

Spatial organisation

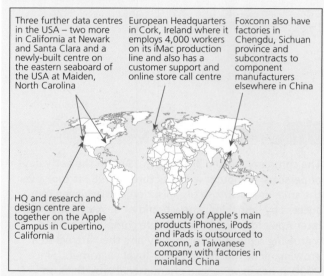

Three further data centres in the USA – two more in California at Newark and Santa Clara and a newly-built centre on the eastern seaboard of the USA at Maiden, North Carolina

European Headquarters in Cork, Ireland where it employs 4,000 workers on its iMac production line and also has a customer support and online store call centre

Foxconn also have factories in Chengdu, Sichuan province and subcontracts to component manufacturers elsewhere in China

HQ and research and design centre are together on the Apple Campus in Cupertino, California

Assembly of Apple's main products iPhones, iPods and iPads is outsourced to Foxconn, a Taiwanese company with factories in mainland China

Figure 7.28 Location of Apple's main global operations

Apple is a truly global company with a distinctive geography. Their main products are designed in Silicon Valley, California, made in mainland China by Foxconn, a Taiwan-based company, and sold all over the world, especially to consumers in developed countries (Figure 7.28).

Most of Apple's employees are based in the developed world, mainly in the USA. This reflects the high-tech and knowledge-intensive nature of its business.

- Its headquarters and research and design centre are together on the Apple Campus in Cupertino, California.
- It has three further data centres in the USA – two more in California at Newark and Santa Clara and a newly-built centre on the eastern seaboard of the US at Maiden, North Carolina.

The concentration of R&D and data centre locations in Silicon Valley, California is a good example of **agglomeration**, which is typical of high-tech industries where information exchange and access to well-qualified and expert staff is crucial.

- Apple it has its European headquarters in Cork, Ireland, where it employs 4,000 workers on its iMac production line and also has a customer support and online store call centre.
- Assembly of Apple's main products, iPhones, iPods and iPads, is **outsourced** to Foxconn (Figure 7.29). Its main production base is in Foxconn City, Shenzen in Guangdong province, north of Hong Kong. Foxconn also have factories in Chengdu, Sichuan province and in turn subcontracts to component manufacturers elsewhere in China.
- Of its 453 retail stores, 110 are in Europe and the Middle East and 25 are in China, where it has increased its market by 600 million people. The Chinese stores have been very successful and Apple has plans to open more there.

Production

Apple's mainstream products are produced in China for a number of reasons:

- a large source of highly skilled, hard-working but low paid workers – good for China as it provides some investment into the country and generates jobs
- a number of, mainly Taiwanese, companies competed for the Apple manufacturing contract, which forced down production costs
- Shenzen was the location of China's first and most successful **Special Economic Zone** (SEZ) offering a number of incentives to attract foreign companies.

Figure 7.29 Workers on the iPhone production line at the Foxconn complex

Foxconn City is a business park and has a number of factories belonging to different manufacturers who assemble high-tech products for well-known brand names including Sony, Hewlett-Packard and Dell, thus it is dubbed 'China's Silicon Valley'. The park is self-contained with high security and most workers live on site where there are dormitories, shops and cafes. There are over 400,000 workers employed in Foxconn City. Wages paid on the site are around US$150 to US$200 per month which is above the average minimum wage and higher than in other parts of Shenzen and China.

Impacts on countries in which it operates

Since its rapid growth, Apple has been the subject of a number of controversial claims about its business operations and the impacts they have in the countries in which they operate. These demonstrate the risks of locating overseas and of outsourcing production to foreign companies and reflect general criticisms levelled at TNCs.

Ireland

Apple's European HQ is based at Hollyhill, on the north side of Cork. It is the only fully Apple-owned manufacturing facility in the world. Along with a number of other foreign blue-chip firms, Apple was lured to locate in Ireland by the government's 12.5 per cent corporation tax, the second lowest in the EU. Generally Apple's impact in Ireland and in Cork in particular has been positive:

- It employs 4,000 workers directly on its iMac production line and call centre, Cork's largest private employer.
- Indications are that Apple's presence in Cork has generated up to a further 2,500 jobs for workers employed as part of the supply chain or in ancillary work.
- The company's presence in Cork has attracted other high-tech firms to the area.
- It has also attracted a highly skilled workforce and provided an inspiration for local education, research and development.
- The company has expanded and contributed to infrastructural improvements in the city.
- Together with other companies locating in the south and west of Ireland (for example, Galway), Apple has enhanced the Republic's reputation for hosting high-tech TNCs which contributed to the 'Celtic Tiger' economy of the 1990s and early 2000s.

Negative aspects of Apple's involvement in Ireland:

- Many of the more highly skilled workers at Hollyhill are foreign nationals (mainly from the EU), so Apple is accused of not creating sufficient work for local people. The counter-arguments to this are that this has helped Cork become a more vibrant cosmopolitan city and that at least 60 per cent of the workers are Irish (though the majority of these are production line workers).

Tax practices

Apple has been accused of corporate tax avoidance. As a TNC operating in a number of countries, Apple can use subsidiary firms in other countries to declare profits and pay a lower rate of tax. In 2014, both the US Senate and the EU commission investigated Apple's tax liabilities. Both claimed that Apple had received favourable treatment on tax from the Irish government, a claim that was strenuously denied. It also caused tension between the USA and the EU, as each claimed that Apple owed them tax on their profits. These tensions expose a flaw in an increasingly globalised world. Taxation systems need to be updated to accommodate large TNCs such as Apple, which manufactures most of its products in China but derives the majority of its profit from 'intellectual property', marketing, patenting and branding, which can be funnelled through a maze of subsidiaries.

Apple is far from alone in claiming that profits from large chunks of its intellectual property are generated outside the USA.

China

Labour practices

- **Working conditions:** In 2006 it was reported that 200,000 workers who lived and worked in the Shenzen factory were regularly working more than 60 hours a week for around $100 a month, half of which was taken up by living expenses. Media reports used the term 'sweatshop conditions' and reported enforced overtime – none of which enhanced Apple's reputation.
- **Health and safety:** In 2010 fifty workers at Lianjian Technology (a company subcontracted by Foxconn) in Suzhou, Jiangsu province, were poisoned by a toxic chemical used to clean iPad screens and decided to sue Apple. In order to reduce costs adequate ventilation had not been installed. Most cases were settled out of court.

- **Student and child labour:** Foxconn's use of students and children is part of its objective of maintaining a low-cost and flexible labour force. Employees under 18 are subjected to the same working conditions as adults. Provincial authorities supported the policy by allowing them to be graded as interns or trainees; university students were forced to work as a condition of graduating.

- **Suicides:** Of all the controversies, suicides reported in 2009–10 probably brought most damage to Apple's reputation as a business. In 2009, a Foxconn factory worker committed suicide after coming under pressure following the disappearance of a prototype model of an iPhone4. By the end of 2010, a total of 14 suicides had occurred, largely as a consequence of the severe working conditions.

These practices and their consequences have put enormous strain on the business relationship between Apple and Foxconn, but it is a very difficult and costly relationship for Apple to extricate itself from. In response, Apple now has a *Supplier Code of Conduct* and it audits supplier factories regularly.

Environmental issues

Criticisms have largely come from Greenpeace. As a global environmental campaigning group, they carry a lot of influence and produce a *Green Electronics Guide* which ranks companies on their environmental performance. Apple was criticised on four main counts:

- its reliance on non-renewable resources to supply electricity to its data centres

- use of toxic chemicals such as PVC and brominated flame retardants in their manufacturing processes

- factories in China were discharging pollutants and toxic metals into local water supplies, threatening public health

- the lack of recyclability of many Apple products.

Apple has responded by launching its *Green my Apple* campaign to improve its green credentials. It now use 75 per cent renewable energy, investing in large solar farms to power its data centres. It made the decision to remove PVC plastics and brominate chemicals from its products and has started to promote recycling of its products. As a result it has moved up the Green Electronics ranking from eleventh in 2006 to sixth by 2012.

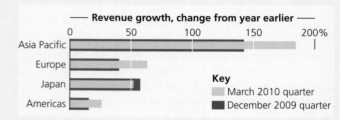

Figure 7.30 Growth of Apple's products internationally

Conclusion

Globalisation of Apple is not solely a positive or negative process. Apple is providing employment to many and helping China to grow and develop but at what cost to the workforce? Further investment and development in South America and Africa may follow in time.

Apple's response to criticisms about labour practices and environmental issues demonstrates the concern shown by large TNCs regarding ethical purchasing and a willingness to respond. This in itself is a positive feature of globalisation.

World trade in bananas

The banana trade raises a variety of environmental, economic, social and political problems. Bananas are one of the world's favourite fruits and globally one of the most commonly eaten.

Of all agricultural products, bananas are the fourth most important food product within least developed countries, being a staple food for around 400 million people. Of all fruits it is the greatest in terms of volume produced and is one of the five most consumed fruits on the planet. It contains large quantities of energy (90 calories per 100 grams) and one banana provides more than an adult's daily potassium requirement.

Globally, bananas are the fifth most traded agricultural commodity and in 2013 a record 16.5 million tonnes were exported, primarily from Latin America and the Caribbean.

The banana industry

Bananas are grown predominantly in hot, rainy lowlands of tropical regions. In a number of countries, such as India, Brazil and much of Africa, large quantities are produced but are mostly consumed domestically. India is the largest producer of bananas globally and exports to the Middle East and other parts of Asia. Similarly the Philippines are a large producer and export to Japan and other parts of East Asia. The main commercial producing-regions for export, however, are geographically concentrated in Central America and the Caribbean. Some of the countries in these regions are very dependent on banana exports.

Banana varieties are susceptible to diseases and almost all the bananas are treated with chemicals throughout the production cycle. Commercial plantations operated by large TNCs apply around 30 kg of active ingredients per hectare, per year. These include fungicides, insecticides and herbicides. In addition, fertilisers are applied regularly and after harvesting the fruit is washed with a disinfectant. With the exception of cotton, the banana industry has the largest agrochemical input into the environment.

Banana plantations also cost the environment in terms of deforestation (as land is cleared), waste (for every one tonne of bananas produced there are two tonnes of waste), soil fertility (because of contaminants) and loss of biodiversity (especially aquatic life as pollutants run into water courses).

The banana trade

World trade is dominated by two different groups of producers: the ACP group (Africa, Caribbean and Pacific) and the so-called 'dollar producers' of Central American republics (primarily Ecuador and Colombia) controlled by large US TNCs.

Trade of bananas follows the traditional pattern of developing regions exporting a low-value primary product to more developed countries. Exports are dominated by Latin America and the Caribbean countries, which produced 13 million tonnes (nearly 80 per cent) of bananas for the export market. The leading producers are Ecuador, Costa Rica, Colombia, Guatemala, Peru and Honduras. Smaller countries in Central America are now exporting at a faster rate, though Ecuador remains the main exporting country. In Asia, which produces 17 per cent of the export market, the main country producing commercially is the Philippines, while in Africa exports are smaller (0.80 million tonnes in 2013). The two main producers are Côte d'Ivoire and Cameroon.

The largest importers are the EU and the USA: in 2013, each consumed approximately 27 per cent of the total exported (about 4.5 million tonnes for each region).

As with almost all commodities produced in developing regions but consumed in richer countries, around 90 per cent of the price paid by the end consumer stays in the richer 'north' and never reaches the producer, who has most of the risks of producing a perishable fruit. The largest slice is taken by retailers and bananas are one of the biggest profit-makers in supermarkets.

Banana exporters – main banana exporting regions in the world

Banana importers – main areas of the world importing bananas

Banana consumers – these regions eat most bananas on a total consumption basis (tonnes). (Most are major growing regions eating their own produce and not exporting, for example, India.)

Figure 7.31 Pattern of main exporters, importers and consumers of bananas

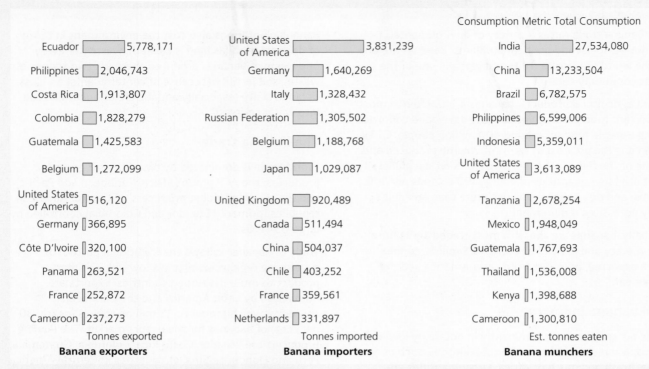

Consumption Metric Total Consumption

Banana exporters		Banana importers		Banana munchers	
Ecuador	5,778,171	United States of America	3,831,239	India	27,534,080
Philippines	2,046,743	Germany	1,640,269	China	13,233,504
Costa Rica	1,913,807	Italy	1,328,432	Brazil	6,782,575
Colombia	1,828,279	Russian Federation	1,305,502	Philippines	6,599,006
Guatemala	1,425,583	Belgium	1,188,768	Indonesia	5,359,011
Belgium	1,272,099	Japan	1,029,087	United States of America	3,613,089
United States of America	516,120	United Kingdom	920,489	Tanzania	2,678,254
Germany	366,895	Canada	511,494	Mexico	1,948,049
Côte D'Ivoire	320,100	China	504,037	Guatemala	1,767,693
Panama	263,521	Chile	403,252	Thailand	1,536,008
France	252,872	France	359,561	Kenya	1,398,688
Cameroon	237,273	Netherlands	331,897	Cameroon	1,300,810
Tonnes exported		Tonnes imported		Est. tonnes eaten	

Figure 7.32 Main exporters, importers and consumers of bananas

Price when leaving exporting country

- Loading the ship 1.5%
- Rail and lorry transport to harbour 1.5%
- Harvesting and packing 7.5%
- Plantation maintenance 4.5%

Price in retail shop

- Retailer 42%
- Ripening facility and wholesaler 15%

Price when entering importing country

- Import duties and importer 7%
- Unloading in Europe 5%
- Overseas shipment and insurance 11.5%
- Taxes and exporter 4.5%

Figure 7.33 Contribution to prices at each stage of banana production, export and sale

In the past, 80 per cent of the banana trade was dominated by just four large transnational companies **Chiquita, Dole, Del Monte** (all US-based TNCs) and **Fyffes** (based in Ireland). The other important producer is **Noboa**, which is a national corporation based in Ecuador. These companies are integrated vertically up the chain. They own or contract out plantations to other producers; they have their own sea transport and ripening facilities, and their own distribution networks in consuming countries. This chain allows them significant economies of scale gains so they can sell bananas in the USA and EU markets at a very low price (Figure 7.33). They repatriate profits to their countries of origin.

Most bananas for export are grown on large **monoculture** plantations in Latin America and increasingly in Africa. The remainder of banana production not controlled by TNCs, is produced on smaller-scale family farms, particularly in the Caribbean.

As recently as 2002, the big five companies controlled nearly 60 per cent of the market but their share has now fallen by volume to 45 per cent. They are still major stakeholders in the business and have responsibility for and influence over labour standards on the plantations that they own or source from.

Organisation of the banana trade has changed in recent years (Figure 7.34). The big companies have freed themselves of direct ownership of plantations, in favour of guaranteed supply contracts with medium- and large-scale producers. An increasing number of national growing companies based in Ecuador, Costa Rica and Colombia sell their produce either to the banana TNCs (as distributors) or directly to retailers in the developed countries, for example, Wal-Mart and Tesco. There has been a shift in power and retailers in the grocery sector in importing countries are increasingly dominating the supply chain. As grocery market share becomes

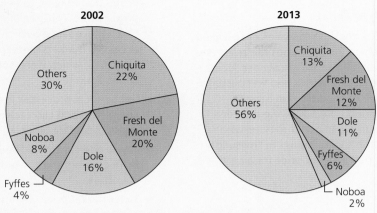

Figure 7.34 Changing market share of the largest banana traders

Source: Banana Link

concentrated in the hands of fewer retailers, suppliers have little option but to accept conditions such as low prices, discounts and delayed payments or otherwise risk being taken from the supplier list.

Issues

Trade wars

Bananas were the subject of one of the longest trade disputes in history, lasting 20 years from 1992 until the 2009 Geneva Banana Agreement was reached, coming into effect in 2012.

- The dispute started in 1975 when EU countries negotiated a trade agreement with former European colonies. The agreement was the **Lomé Convention** and was made with 71 African, Caribbean and Pacific countries (ACP countries), many of whom were banana producers.

- These countries were given special and differential treatment (SDT) with preferential tariff-free import quotas to supply EU markets. The idea was to enable these former European colonies to develop independently without having to use overseas aid.

- The agreement was extended to a list of banana suppliers to the EU including Cameroon, Dominican Republic, Belize, Ivory Coast, Jamaica, Ghana, Surinam and the Windward Isles.

- The effect of the deal was to **protect** the mainly smaller, family-run farms in the Caribbean and Africa from competition with the large Latin American producers, whose bananas were produced more cheaply on mechanised plantations.

- At the time, the US transnationals which controlled the Latin American crop were supplying around 75 per cent of the EU market, while only 7 per cent came from Caribbean suppliers.

- Despite this, in 1992, the TNCs filed a complaint to the WTO that the EU practice was unfair trade.

In 1997 the WTO ruled against the EU and the Lomé Convention and ordered the EU to cease the discrimination.

- The dispute was not resolved as the EU proposals did not satisfy the larger producers. This led to a trade war between the USA and the EU as the US government **retaliated** under pressure from TNCs (mainly Chiquita) and imposed WTO-approved sanctions on a range of EU products.

- A compromise was eventually reached in Geneva in 2009 between the EU and 11 Latin American countries. The EU agreed to gradually reduce tariffs on Latin American bananas; the agreement was ratified in 2012.

There are still concerns from the ACP producers that they are not able to compete. Of the Caribbean countries, only the Dominican Republic (which had a larger number of medium-scale plantations), Belize and the Windward Isles are competing successfully with the larger producers but around 80 per cent of bananas entering the EU now come from these sources. Their focus on producing organic and fair trade bananas is meeting the needs of a growing market in richer EU countries.

Race to the bottom

Because of the low prices paid to suppliers by supermarkets, many of the larger companies are relocating their plantations, increasingly to West Africa, as companies search for lower labour costs and weaker legislation. This is called pursuing a 'race to the bottom' in terms of social and environmental standards. Employers usually sub-contract labour so plantation work is increasingly casual. The work involves long shifts in unbearable heat and many workers fail to earn enough to cover their basic needs.

Fair trade and organic bananas

There has been a steady growth in the sales of so-called 'sustainable' bananas, which includes both fair trade and organic produce. This will help smaller-scale producers in the Caribbean and in parts of Africa and will partially counter the deterioration of conditions in banana production. There is undoubtedly a growing market segment of ethical consumers in richer nations who are becoming aware of the shortcomings in the supply chain and are willing to pay a higher price for a certified product.

Summary

The world trade in bananas demonstrates a number of relevant points about trade, especially in primary commodities:

- Mass production in developing countries will have negative environmental consequences.
- Transnational corporations have a large element of control of markets and can influence political decisions.
- WTO will support free trade against protectionist activities or agreements at all costs, even if the protection may be to help development (the lobbying power of US TNCs calling for free trade seems to have taken precedence over the 'special and differential access to markets' agreements for least developed countries, which were not very effective in this situation).
- Geopolitical processes mean that trade disputes can spread and can escalate to trade wars between regional trading blocs.
- Power and control of food production has shifted away from growers and towards retailers in high-income countries.
- Supermarket price wars may ultimately decide where and how food is produced.
- More ethical, sustainable consumer markets are growing but relatively slowly and only in places that can afford products bought at a higher price.

Key question

How does the global trade in bananas reflect injustice in free trade arrangements?

Geographical consequences of global systems and the impacts of globalisation

> Globalisation will make our societies more creative and prosperous, but also more vulnerable.
>
> *Lord Robertson* Former UK Defence Secretary and Secretary-General of NATO

Key question

Will globalisation bring prosperity to all as promised, or is it mainly driven to serve greed and increase the wealth of those already rich?

The following have been the main beneficiaries from the process of globalisation:

- **Newly industrialising countries:** Some medium-income nations have developed rapidly as a result of inward investment and have emerged to become major economic powers competing with the richer developed regions of the world.

- **Transnational corporations:** Large companies have grown in a number of different industrial and service sectors and although they are mainly based in the developed countries, TNCs from emerging economies have also become global powers.
- **International organisations:** Organisations such as the IMF, the World Bank and the WTO have all contributed to the globalisation of economies and have consolidated their position and control in world affairs.
- **Regional trading blocs:** More trade agreements have evolved which benefit their members, though some would argue that it is at the expense of the nation state.

Economic consequences of international trade

Developing countries in Asia, Africa and Latin America are now some of the fastest growing economies globally, such as the **MINT** countries – Mexico (economy growing at 4 per cent annually); Indonesia (5.9 per cent); Nigeria (5.8 per cent) and Turkey (4 per cent). In 2015, the top 10 fastest growing economies were all in Asia or Africa.

Table 7.7 Benefits and costs of free trade

Benefits	Costs
Lower prices for consumers Greater choice Access to larger, wealthy markets for TNCs Greater economies of scale through increased specialisation Greater foreign competition may weaken domestic monopolies (for example, UK supermarkets) Competition leads to greater innovation Access to cheaper raw materials for TNCs	The **injustice** of free trade not giving sufficient 'protection' to emerging industries in developing economies so they cannot compete with developed countries in the free market More developed economies are still protected by tariffs on agricultural imports The unjust exploitation of workers and poor working conditions Diseconomies of scale as a result of difficulties co-ordinating subsidiary companies

Risks of economic interdependence

One of the most significant disadvantages of globalisation is the increased risk associated with the interdependence of economies. A negative economic **shock** in one country can quickly spread to other countries. This is particularly the case if it affects a country's banking system. For example, the financial crisis in East Asia in the 1990s was triggered by the collapse of a few Japanese banks. Similarly the 2008–09 global financial crisis was initiated by the collapse of lending in the US sub-prime housing market.

Social and cultural impacts

- Globalisation has allowed for greater sharing of ideas, lifestyles and traditions. It means that people have greater access to foreign culture such as film, music, food, clothing and other goods and services that were not previously available domestically.

- The over-standardisation of many goods and services has led to increased cultural homogeny. This has damaged individual traditions and means less diversity. Access to foreign (mainly Western) cultures is causing the distinction between countries to fade, diminishing the world with one 'corporate' identity.

- In a bid to offset criticism of the standardisation of products on a global scale, a strategy called **glocalisation** has been adopted by some TNCs. This involves thinking globally but acting locally to reduce threats to cultural dilution. Products and services are likely to be more successful when they are customised for the local market. McDonald's is a commonly cited example of this strategy as the restaurant's menu is often customised to suit local tastes (Figure 7.36, page 318).

Key term

Glocalisation – A term used to describe products or services that are distributed globally but which are fashioned to appeal to the consumers in a local market.

- Increased awareness of global news and events – which can be positive by informing people of environmental issues such as climate change, but can be used negatively, for example, for propaganda purposes by terrorist organisations.

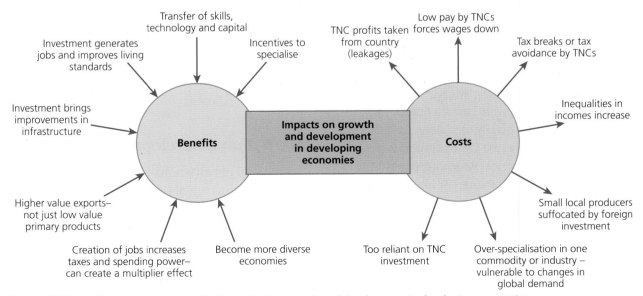

Figure 7.35 Benefits and costs of globalisation affecting growth and development in developing economies

Figure 7.36 Glocalisation – global brands going local. McDonald's Mac Burrito in Mexico City

Environmental impacts

With greater movement and use of resources, the world's population is putting ever-increasing pressure on the Earth's finite resources. Environmentalists argue that globalisation is concerned primarily with economic costs and largely ignores environmental costs. Problems include:

- more transportation, which increases greenhouse gas emissions
- depletion of non-renewable resources
- TNCs outsourcing production to countries where environmental standards are less strict
- weak or non-existent controls allowing pollution of the air, land, rivers and seas
- more waste from packaging
- IMF-enforced spending cuts reducing many nations' spending on the environment
- greater movement giving a higher risk of invasive species being introduced.

7.4 Global governance

Global governance refers to the ways in which global affairs affecting the whole world are managed. In international relations generally, decisions are made by individual state governments. Co-operation is usually negotiated between countries, which agree to abide by similar rules given in signed treaties or international laws.

As the world becomes more interdependent, global economic governance is increasingly important for achieving sustainable development across all nations. This is especially true for governance of the **global commons** where **environmental sustainability** is at the forefront of our concerns, not least our utilisation of the continent of Antarctica.

Regulating global systems

In this increasingly economically integrated world, some argue that governments are losing their influence and that corporations now have more power to control resources, capital and labour.

Key terms

Environmental sustainability – A state in which the demands placed on the environment can be met without reducing the quality of the environment for the future.

Global commons – Resource domains or areas that lie outside the political reach of any one nation state.

Global governance – A movement of political integration aimed at negotiating responses to problems that affect more than one state or region.

Non-government organisations (NGOs) – Any non-profit, voluntary citizens' group with a common interest, which is organised on a local, national or international level. Sometimes referred to as a 'civil society organisation' (CSO).

United Nations – An international organisation founded in 1945 made up of 193 member states whose aim is to promote international peace and co-operation.

In response to the acceleration of interdependence between countries the term '**global governance**' is used to name the process of designating laws, rules or regulations intended to manage global systems, for example the laws governing international trade. In recent years, global governance has focused on a number of international issues:

- reducing environmental problems
- trade and investment inequities
- reduction of poverty
- human rights violations
- civil conflict
- financial instability.

International development agencies

A number of organisations take responsibility for promoting growth, stability and the development of all regions sustainably, both economically and environmentally. Many of the organisations operating internationally are connected to the **United Nations (UN)** such as the **United Nations Development Programme (UNDP)**.

Figure 7.37 The UN Global Goals for Sustainable Development 2015

Despite the existence of the UN, there is no 'world government' with the political authority to exercise jurisdiction over executive, judicial, legislative or military affairs for the whole planet. The UN has a mainly advisory role; its purpose is to foster co-operation between state governments, rather than exerting authority over them. Similarly, international organisations such as the World Trade Organization (WTO) and the World Health Organization (WHO) exist to take responsibility for managing specific aspects of an increasingly inter-related world, such as regulating the rules of trade or dealing with international public health issues.

There are also a number of different **non-governmental organisations (NGOs)** with varying objectives but generally either to ensure justice and equality for people across the world or to campaign for environmental sustainability (Figure 7.38). The international strength and support for some of these organisations means that they have the authority to challenge the excesses of negative TNC impacts and of governments supporting them.

The United Nations Development Programme

The UNDP's aims are the eradication of poverty and the reduction of inequalities and exclusion. They operate in over 170 countries. The main work of UNDP has been to lead the drive in meeting the **Millennium Development Goals (MDGs).** These eight anti-poverty targets were set in September 2000 and the world committed to achieve them by 2015. Members of the UN are now in the process of defining **Sustainable Development Goals (SDGs)** (Figure 7.37) for the next 15 years. Their aim is to strengthen post-2015 frameworks for:

● development
● disaster risk reduction
● climate change
● global sustainable development.

Protection	Prevention	Promotion	Transformation
providing relief to victims of disaster and assisting the poor	reducing people's vulnerability, through income diversification and savings	increasing people's chances and opportunities	redressing social, political and economic exclusion or oppression
'Give A Man A Fish'	'Teach A Man To Fish'	'Organise a Fishermans' Co-Op'	'Protect Fishing & Fishing Rights'

Figure 7.38 Different roles of NGOs

The new agenda is to finish the work of the MDGs, leaving no countries behind. It is due to be adopted by UN member states at the Sustainable Development Summit in New York in September 2015. For progress on the Millennium Development Goals and to clarify the new SDGs, visit the UNDP website at www.undp.org/.

World Trade Organization (WTO)

As a response to globalisation the WTO's focus is on the issue of trade and investment inequities. Injustice has been partially alleviated by some of the processes of globalisation but has undoubtedly been accentuated in other ways. The WTO's commitment to liberalising trade means they are responsible, through negotiations, for combating exploitative practices used by some countries and TNCs, and for removing the protectionist policies adopted by some governments and trading blocs.

World Summit on Sustainable Development (WSSD)

The **Earth Summit** held in Rio de Janeiro in 1992, was important in setting out **Agenda 21** (see textbox), a voluntary action plan agreed by many governments to develop strategies for long-term **sustainable development**. Ten years after Rio, the World Summit on Sustainable Development (WSSD) in Johannesburg brought together thousands of participants, including heads of state, national delegates and leaders from NGOs and businesses. The WSSD focused the world's attention and direct action towards meeting difficult challenges. These included improving people's lives and conserving our natural resources in a world that is growing in population, with ever-increasing demands for food, water, shelter, sanitation, energy, health services and economic security. Further UN conferences were held to involve government leaders at the highest level, including Rio +20 in 2012. These summits were held:

- to reinvigorate the global commitment to sustainable development
- to assess progress on sustainable development goals and targets
- to advance international co-operation on sustainable development.

According to the WSSD Plan of Implementation, 'good governance, within each country and at the international level, is essential for sustainable development.'

Key term

Sustainable development – Development which recognises that the needs of the present have to be met but doing this without affecting the needs of future generations.

Agenda 21

At the **Earth Summit** in Rio di Janeiro in 1992, the international community adopted Agenda 21, a global 'blueprint' for sustainable development. Agenda 21 action plans are intended to be cascaded down through national governments and local authorities, to apply strategies that will encourage more sustainable individual lifestyles and behaviour. The fact that most parts of the UK now have various recycling options as part of their waste collection is a tangible outcome stemming from Agenda 21.

It is seen as a **'top down'** approach which is trying to encourage a **'bottom up'** response, as the ideas initiated in the Agenda are designed to filter down from the UN to national government level and then down to regional and local government decision-making in order to influence the actions of individuals. For example, each local authority in the UK has a Local Agenda 21 (LA21) strategy in line with the UN objectives and actions.

As a global-level agreement, Agenda 21 perhaps demonstrates the difficulty in imposing a strategy for all countries to follow. It has provoked much controversy, not least in the USA where the agreement has seen considerable opposition from some groups. These opponents see Agenda 21 as:

- an attack on personal liberties – for example, car usage, family size
- having a focus on population control (religious groups in particular are against this) – Agenda 21 encourages population sustainability in its ethos but there is no requirement to enforce population policy
- an attack on the idea of 'private property'

Those opponents also view the 'precautionary principle' (which, as part of Agenda 21, suggests that any development should proceed with caution and weigh up environmental impacts) as being 'guilty (of environmental harm) until proven innocent'.

On the other hand, there are many committed environmentalists in the USA who see these points as an overreaction to an action plan they believe is necessary to reduce the ecological footprint of US citizens.

United Nations Environment Programme

The body responsible for supporting a coherent structure of international environmental governance is the **United Nations Environment Programme (UNEP)**.

UNEP has been the leading global environmental authority since 1972 and has grown in stature alongside increasing concern about human impacts on the global environment. UNEP's mission statement is *'to provide leadership and encourage partnership in caring for the environment by inspiring, informing and enabling nations and peoples to improve their quality of life without compromising that of future generations.'* This is a clear reference to the original sustainable development definition given to us by the pre-Rio Brundtland Commission in 1987.

UNEP's work is now part of the UN system-wide preparations for the post-2015 UN Development Agenda – the Sustainable Development Goals.

World summits on climate change

Atmosphere, including global temperature, climate change and ozone depletion, is one of the global commons. In so far as human activity can influence these processes, it can be considered to be outside the political reach of any one nation state and subject to international governance. The United Nations Framework Convention on Climate Change (UNFCCC) is the body responsible for overseeing negotiations on reducing greenhouse gas emissions between nations.

At the UN Paris Climate Summit in December 2015, a historic, legally-binding climate deal was struck by world leaders, which will come into force in 2020. After nearly twenty years of disagreement (mainly between the USA and rapidly industrialising economies such as China and India) since the Kyoto Protocol of 1997, all 187 countries at the summit will combine pledges and work together to combat this important global threat. The main aim is to **hold global temperature rises to a maximum of 1.5°C** in order to avert the worst effects of global warming and climate change. The key features of the deal include:

- that all countries will voluntarily cut emissions
- a long-term aim to reduce the net emissions to zero in the second half of the twenty-first century

- that richer developed countries, and some wealthy developing countries, will pledge $100 billion each year to help developing countries adapt to climate change
- a review mechanism to increase pledges every five years (if they are insufficient) to keep warming below 2°C
- a loss-and-damage mechanism for addressing losses that vulnerable countries face from climate change (such as rising sea levels and increasing storm intensity).

Interactions at all scales

Success in global governance and in regulating the extremes resulting from unchecked globalisation can best be achieved if there is clear communication and understanding of strategies at all scales from global to local (or vice versa). This clarity is often provided by non-governmental organisations which operate across boundaries and at all levels. The 'Agenda 21' action plan has equally attempted to engage governments and individuals at all levels.

Non-governmental organisations (NGOs)

As part of the globalisation process, NGOs have expanded their scope from local and national settings and have increasingly become 'international' organisations.

In the twenty-first century NGOs have emerged as a global force to:

- democratise decision-making (for example, persuading governments to consider 'bottom-up' approaches)
- protect human rights
- provide essential services to the most needy.

There is a distinction to be made between **operational** and **advocacy** (campaigning) NGOs. Both are funded by charitable donations but some also receive money from governments and others from businesses (though this may compromise their independence):

- **Operational NGOs:** Those providing frontline support services to the needy (for example, Oxfam); tend to raise money for each project they undertake.
- **Advocacy NGOs:** Those who focus on campaigns to raise awareness to gain support for a cause (for example, Friends of the Earth); derive money from donations and, in some cases, from membership subscriptions.

In reality, NGOs work increasingly in partnership with other stakeholders like governments and international organisations.

NGOs are becoming more and more important in supporting development. They are often the only co-ordinated organisations to provide a voice for the poor of the world. With greater communication and awareness of issues greater than individual countries, many NGOs such as Amnesty International, Médecins sans Frontières and Greenpeace have emerged. They draw together people focusing on global issues and do not tie themselves to the interests of any government.

Key question

What range of issues might global governance attempt to tackle?

7.5 Global commons

The 'global commons' refers to resource domains or areas that lie outside of the political reach of any one nation state. It is a term used to describe supra-national 'spaces' in which common shared resources can be found. International law recognises four global commons:

- the high seas
- the atmosphere
- Antarctica
- outer space.

Some commentators would argue that **cyberspace** has also emerged as a new domain which meets the definition of a 'global common' by being a resource that is shared by all but is not controlled by any single nation. This includes the development of the worldwide web, one of the main drivers of globalisation.

Principle of common heritage

International law is guided here by the '**principle of the common heritage of mankind**'. This principle establishes that some localities belong to all humanity and that the resources there are available for everyone's use and benefit. It includes taking into account future generations and the needs of developing countries.

Historically, access to these resources (with the exception of fishing and whaling on the high seas) has been difficult.

More recently however:

- advances in science and technology have given easier access to a range of resources
- greater scarcity of resources, especially minerals, fuels and food has put increasing pressure on the global commons to provide resources for a needy and developing world.

These changes mean that the concept of common heritage is being put under increasing pressure.

'The tragedy of the commons'

This concept explains why shared 'common access resources' of any type are likely to be overexploited. If individuals act independently and according to their self-interest, this will be contrary to the interest of the whole group because the shared resource will become depleted.

This is exactly what has happened on a global scale to fish stocks, which have been overfished leading to depletion, marine pollution and loss of sustainability.

It is clear to see the need for international law and clear rules of global governance to protect the commons from over-exploitation.

Legal issues and institutional frameworks

Each of the global commons is covered by a number of international laws or treaties of one kind or another:

- the high seas by the UN Convention on the Law of the Sea (UNCLOS)
- the atmosphere by the United Nations Framework Convention on Climate Change (UNFCCC), the Montreal Protocol on substances that deplete the ozone layer and the Kyoto Protocol which sets limits on carbon emissions into the atmosphere
- Antarctica by the **Antarctic Treaty Systems** (ATS)
- outer space by the 1979 Moon Treaty which governs exploration and exploitation of its resources and the Treaty on Principles Governing the Activities of States in the Exploration and Use of Outer Space.

Key question

What are the global commons and how can they be protected?

7.6 Antarctica and the Southern Ocean

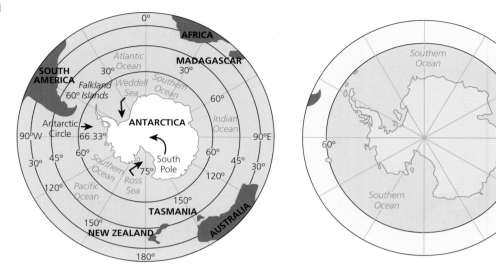

Figure 7.39 Map of Antarctica and location of the Antarctic and Southern Ocean

Location and geography

Most of Antarctica is south of the Antarctic Circle, with the exception of parts of the East Antarctic coastline and the Antarctic Peninsula which extends northwards from West Antarctica to about 63°S. Much of the continent's coastline is fringed by ice shelves. The largest of these are the Ross Ice Shelf in the Ross Sea and the Ronne Ice Shelf in the Weddell Sea. Each of these ice shelves covers an area greater than the British Isles.

In the spring of 2000, the International Hydrographic Organization delimited a fifth world ocean – the **Southern Ocean**, also known as the Antarctic Ocean, from the southern portions of the Atlantic, Indian and Pacific Oceans. It comprises the southernmost waters of the world, taken as being south of 60°S latitude. It has the unique distinction of being a large circumpolar body of water totally encircling the continent of Antarctica and encompasses 360 degrees of longitude.

The Southern Ocean extends from the coast of Antarctica north to 60 degrees south latitude, which coincides with the Antarctic Treaty Limit. At approximately this latitude, cold, northward flowing waters from the Antarctic meet with warmer sub-Antarctic waters. This is known as the **Antarctic Convergence** zone.

Key term

Antarctic Convergence – A curve continuously encircling Antarctica where cold northward flowing Antarctic waters meet the relatively warmer waters of the sub-Antarctic

The Antarctic Convergence is a zone approximately 32 to 48 km wide, varying somewhat in latitude seasonally and in different longitudes, extending across the Atlantic, Pacific, and Indian Oceans between the 48th and 61st parallels of south latitude. Here the cold, northward flowing Antarctic waters predominantly sink beneath sub-Antarctic waters. This forms associated areas of mixing and upwelling currents which are very high in marine productivity, especially Antarctic **krill.**

The Antarctic Convergence is a natural boundary rather than an artificial one like a line of latitude. It separates:

- two distinct hydrological regions
- areas of distinct climate
- areas of distinctive wild life.

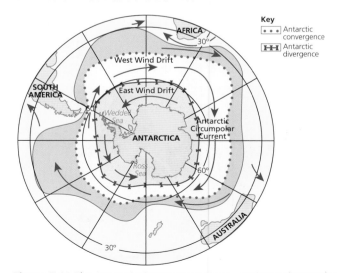

Figure 7.40 The Antarctic Convergence Zone – Circumpolar Current and Divergence Zone of easterly and westerly flowing currents

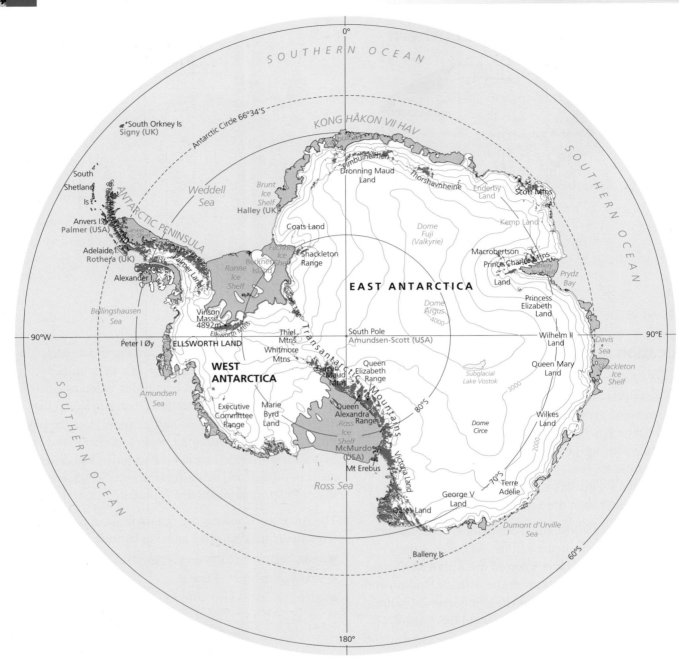

Figure 7.41 Physical map of Antarctica

Figure 7.42a Nunataks – Una Peaks at Cape Renard in Graham Land on the Antarctic Peninsula

Figure 7.42b Dry valley – Wright Valley, one of the McMurdo Dry Valleys near the Ross Sea, with wind carved rock in the foreground

Figure 7.40 (page 323) shows the Convergence Zone at around 60 degrees latitude south; this gives rise to the largest surface current in the world called the **Antarctic Circumpolar Current** which flows around Antarctica. This current effectively blocks warmer waters travelling southwards. The Circumpolar Current flows in an eastward direction being driven by westerly winds, and is otherwise known as the West Wind Drift. A lesser current, closer to Antarctica, flows in the opposite direction westwards; it is known as the East Wind Drift and is driven by easterly polar winds. The East Wind Drift features prominently in the Ross and Weddell Seas. The area where these opposite flowing currents meet is known as the zone of **Antarctic Divergence**.

Physical geography and climate of Antarctica

Antarctica is a unique continent as its mountainous and rocky terrain is almost entirely (97 per cent) covered by glacial ice (Figure 7.41). The continent is unevenly divided into East (Greater) Antarctica and West (Lesser) Antarctica by the Transantarctic Mountains. The Transantarctic range contains many peaks above 4,000 m and it also divides the West and East Antarctic ice sheets. The East Antarctic Ice Sheet is both larger and thicker than the West Antarctic Ice Sheet, as well as being much older. The Ellsworth Mountains are found in West Antarctica and contain the continent's highest peak – Mount Vinson (4,892 m).

With an average height of over 2,300 m above sea level, Antarctica has the distinction of being the highest continent. While it does have high mountains, it is the thickness of the ice sheets which give it the highest average elevation of all the continents.

Climate

In addition to being the highest continent, Antarctica is also distinctive for being the coldest, windiest and driest. Climatic features include:

- average temperature is −49°C, though it can be as low as −89°C
- mean annual wind speed of 50 miles per hour (mph) resulting from the convergent katabatic winds; gales can reach 200 mph
- mean annual precipitation, especially in the interior, is low, at less than 50 mm per annum
- it can best be described as a **polar desert.**

The thick ice sheets are not the result of heavy snowfall but of the accumulation of small inputs of snow and frost (which far exceed ablation rates) over a very long period of time.

Very little of Antarctica is free from ice and any ice-free areas owe their existence to specific local-scale factors. For example, high mountain peaks called **nunataks** are small areas of rock emerging above the ice sheets, like islands in a sea of ice (Figure 7.42a). High winds and steep slopes prevent snow and ice from accumulating on parts of these mountains.

Dry valleys are another intriguing type of landscape found in Antarctica (Figure 7.42b). These are found in high altitude areas of extreme aridity.

Although conditions are harsh on land in Antarctica, the marine conditions support a diverse ecosystem. Surrounding sea temperatures are warmer than the land and the upwelling of cooler water from the ocean depths brings nutrients, which in turn supports **phytoplankton** (microscopic algae that are the base of the marine food web). Blooms of phytoplankton provide food for krill, which many species depend on as a food source.

Some coastal areas of Antarctica, particularly along the Peninsula, have micro-climatic and topographic conditions which cause enough melting during summer months to allow some land to remain free of glaciers. The areas on the western Peninsula are likely to see even more climatic differences because of climate change.

Key question

Why does Antarctica present a unique environment that needs to be protected?

Threats to Antarctica

The discovery of the islands of the Southern Ocean in the eighteenth century led to the start of exploitation of the area. A number of economic activities have taken place in the region. The first, and initially most damaging, of these was the culling of seals for their fur.

Sealing began in the eighteenth century on and around the island of South Georgia. By 1800, the fur seals of South Georgia were wiped out and interest then centred on the South Shetland Islands. Within three years, over 300,000 seals had been killed and the population had been virtually eradicated. This was exploitation at its worst, with no thought given to future development.

Further threats to Antarctica followed during the nineteenth and twentieth centuries and many still exist.

Fishing and whaling

Whaling began in the nineteenth century. The main targets were blue and right whales; the main products, oil and whalebone (baleen). As the whale population of the North Atlantic became reduced by massive exploitation, the whalers turned their attention to the Southern Ocean. Whalers sailed from several countries in the northern hemisphere, particularly Norway, the USA and the UK.

Whaling was a highly profitable business and whaling stations were established on South Georgia and the South Shetlands. In 1904, the Norwegians developed Grytviken on South Georgia, which at its height employed over 300 people. The range of products increased to include meat meal, bonemeal, meat extract and, in later years, frozen whale meat. Grytviken was abandoned in 1965 because whale stocks were becoming seriously depleted and whaling was no longer commercially viable. Whale populations dramatically declined due to overkilling and many species became endangered.

The establishment of the **International Whaling Commission** (IWC) in 1946 eventually led to an end to most whaling in 1985. Most, but not all, whaling nations agreed to halt the slaughter, as stocks of many species were running dangerously low.

In 1994 the IWC established the Southern Ocean Whale Sanctuary, an area of 50 million square kilometres surrounding Antarctica where they banned all types of commercial whaling. Only Japan opposed the agreement; they, along with Norway and Iceland, form the main pro-whaling lobby. The status of the Southern Ocean Sanctuary is reviewed every 10 years. Commercial whaling is prohibited, though Japan has continued to hunt whales inside the Sanctuary in accordance with a provision in the IWC charter permitting whaling for the purposes of scientific research. Japan's position on whaling has caused international tensions over the issue.

Fishing has now replaced whaling in the area. In the 1960s, Russian ships began to exploit the Southern Ocean for a number of fish species, including the Antarctic rock cod. Concerns have been expressed recently over the number of fish being taken, particularly fishing for krill by Russia and Japan.

Krill

Krill are tiny shrimp-like crustaceans which swim in massive numbers and underpin the entire food web of the Southern Ocean. Virtually everything in the Antarctic region is dependent on krill – seabirds, penguins, seals, whales and other fish, such as the toothfish, which are another major fisheries target. A loss in the krill biomass will affect most other species.

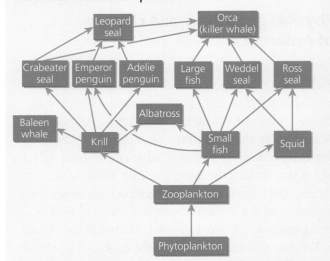

Figure 7.43 Antarctic and Southern Ocean food web reliant on the shrimp-like krill

Krill are being caught primarily as a healthy protein food in East Asia and as a health food supplement in the form of **krill oil** all over the world. They are an excellent source of protein and krill oil is extremely rich in omega-3 fatty acids. Cartons of krill oil capsules can command very high prices (as much as £40 for 100 capsules for red krill oil) in health food retail outlets. They are also valuable for **astaxanthin**, a carotenoid chemical compound which is taken by humans as an antioxidant health supplement.

The sudden recent demand for krill products has caused a dramatic increase in krill fishing in the Southern Ocean. Unsustainable fishing practice may cause a krill population crash, which would have a catastrophic impact on the Antarctic ecosystem.

Commercial fishing is becoming a significant threat to the Southern Ocean and Antarctica. There is a high risk of overfishing of target species, by-catch and direct destruction of marine habitats by ships and fishing gear. Ships also dump waste into the ocean, contaminating the water and killing organisms.

Climate change

The impact of climate change on Antarctica is a widely debated topic as many people believe that global warming is melting the polar ice caps, sea ice and land ice. Others take the view that global warming is not affecting the Antarctic at all as the land ice is growing. Only some of the land ice on the eastern side of Antarctica is growing but in actuality, it is the eastern **sea ice** that is increasing dramatically. The western side however is losing a significant amount of land ice and is not gaining any sea ice. Climate change is affecting different parts of Antarctica in different ways.

Western Antarctic Ice Sheet and Antarctic Peninsula

The Antarctic Peninsula is particularly sensitive to small rises in the annual average temperature; this area has experienced air temperature increases of nearly 3°C in the region in the last 50 years, which is a much faster rate than the average for global warming, according to the **Intergovernmental Panel on Climate Change (IPCC)**. The temperature of Western Antarctica has risen and, as a result, the West Antarctic Ice Sheet has thinned significantly.

This warming is not restricted to the land and Southern Ocean temperatures to the west of the Antarctic Peninsula have also increased by over 1°C since 1955. It has also been established that the Antarctic Circumpolar Current is warming more rapidly than the global ocean as a whole. This warming has had the following effects on the physical and living environment of Antarctica:

- distribution of penguin colonies has changed
- melting of snow and ice cover has increased colonisation by plants
- decline in the abundance of Antarctic krill
- glaciers and ice shelves fringing the Peninsula have retreated; some have collapsed entirely
- the ice shelves will not add to sea level change but as they break up it increases the flow rate of glaciers behind them – it is the melting of these glaciers that will cause sea levels to rise.

The temperature of Antarctica as a whole is predicted to rise by a small amount over the next 50 years. Any increase in the rate of ice melting is expected to be at least partly offset by increased snowfall as a result of the warming.

East Antarctica and sea ice expansion

The temperature of Eastern Antarctica has risen by a much smaller amount and unlike the Peninsula there is no significant loss of ice of any kind.

The paradox is that while the world is warming, sea ice is expanding on the eastern side of Antarctica towards the Indian and Pacific Oceans. In 2013, the extent of Antarctic sea ice was at an all-time record extent of 20 million square kilometres, larger than the continental land mass itself. Sea ice is a thin layer between one to two metres thick. It is formed when water is cooled sufficiently by the surrounding atmosphere. Once it is formed it can be blown by the wind or taken by currents, ultimately expanding and bonding with other floating sheets of ice.

There are four reasons why climate change may contribute to Antarctic sea ice expansion:

- Increasing westerly winds around the Southern Ocean, caused by climate change and ozone depletion are driving the seas northwards.
- More rain and snow resulting from climate change are layering the Southern Ocean with a cooler denser layer on top.
- These storms are also freshening the local water (it becomes less salty) thus raising the temperature needed for sea ice to form.
- Increased melting of continental land ice creates more floating icebergs which contribute to sea ice formation.

Ocean acidification

Carbon dioxide (CO_2) enters the atmosphere by the burning of fossil fuels and other emissions. This creates carbonic acid which makes the slightly alkaline ocean become a little less alkaline.

Polar and sub-polar marine ecosystems are projected to become so low in carbonate ions within this century that waters may actually become corrosive to unprotected shells and skeletons of organisms currently living there. Loss of these organisms will disrupt food webs.

The search for mineral resources

There has never been any commercial mining in Antarctica as mining is completely banned by the

Antarctic Treaty. There are no current or known future plans by any of the Antarctic Treaty nations to reverse this decision. Nevertheless, the future demand for resources will undoubtedly put pressure on the vast mineral reserves that are to be found on the continent (Figure 7.44).

Antarctica is known to have mineral deposits, though any sizeable deposits that are easy to reach are rare and currently not economically viable to mine. One of the main problems is the vast covering of moving ice streams and glaciers.

In the late 1970s and early 1980s some members of the Treaty were secretly trying to formulate a new minerals convention which would have allowed exploration and possible future exploitation of mineral and gas reserves. The Mineral Convention was adopted in 1988 but never came into force because it was not ratified by all members.

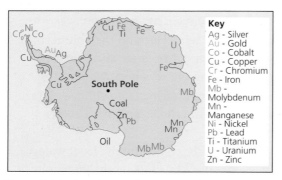

Figure 7.44 Mineral map of Antarctica showing known deposits of minerals

Tourism and scientific research

People have only inhabited Antarctica for about 100 years, but during that period they have had a large influence on the purity of the region. The continent is not populated, except by scientists at a small number of permanent research stations. Scientists and tourists are the main human threats to Antarctica.

Scientific researchers working either in the interior near the South Pole or in coastal areas are well briefed about the need for care of the Antarctic environment, but any activity is bound to create some degree of disturbance to the environment and wildlife. Impacts are caused by vehicle exhausts, construction of buildings and related facilities such as fuel storage,

runways and the disposal of waste such as rope, fuel drums and plastics, all of which endangers living organisms.

Figure 7.45 Penguin 'posing' for tourist, Antarctica

Antarctic tourism is of three types:

- camping trips for naturalists, photographers and journalists
- ship-board visits, largely by cruise ships but also by converted Russian ice breakers
- over-flights – these have restarted after an interval of nearly 20 years following the crash of an Air New Zealand DC10 on Mount Erebus, in which all passengers died.

Tourists go to Antarctica to see the glacial landscapes and the wildlife, particularly seals, whales and penguins (Figure 7.45). They also go for the remoteness and isolation and the chance to test themselves in adverse weather conditions. Tourists may be interested in historic sites, such as McMurdo Sound with its huts dating from the Scott and Shackleton expeditions. Tourism is concentrated in the short southern summer period, from mid-November to March.

Ship-borne tourism in Antarctica takes the form of 'expeditions'. The first tourist ship arrived in 1958; since then tourism has increased, catering for a niche market of adventure travellers and natural history lovers. Most of the ships are comparatively small, with an average capacity of between 50 and 100 people. Tourists are carefully briefed on the code of conduct in terms of behaviour ashore, adherence to health and safety requirements and rules about wildlife observation.

Most cruises follow the **Lindblad Plan** which manages itineraries in a way to ensure that any impacts are negligible. When visiting any one of around 200 possible sites, the group is divided into boatloads of around 20, each led by an expert guide. Each site may be visited only every two or three days to minimise impacts. Captains of cruise ships are required to observe this, and rules laid down by the **International Association of Antarctic Tour Operators (IAATO)**, when taking parties ashore.

Research on the impacts of tourism is undertaken by the Scott Polar Research Institute. Published findings show that the Antarctic environment has been little affected:

- Antarctic tourism is a well-run industry with a sound record for environmental concern.
- Guidelines are widely accepted by operators and tourists alike.
- Damage to vegetation (especially the fragile moss mat) is due to natural causes, such as breeding seals. Tourists are usually scrupulous in not walking on areas of fragile vegetation.
- No litter is attributed to tourists; they are more concerned about the waste they see around the scientific research stations.
- Virtually no stress is caused to penguins by tourists visiting their breeding colonies. However, tern colonies seem to suffer from disturbance.
- Seals are largely indifferent to the presence of humans.
- Out of 200 landing sites surveyed, only 5 per cent showed any wear and tear. These need to be rested, but at present there is no mechanism to implement this type of management.

Despite these encouraging signs, there are some concerns:

- The Antarctic ecosystem is extremely fragile – disturbances leave their imprint for a long time (footprints on moss can remain for decades).
- The summer tourist season coincides with peak wildlife breeding periods.
- The land-based installations and wildlife are clustered in the few ice-free locations on the continent.
- There is the possibility of land-based tourism being developed.
- The demand for fresh water is difficult to meet.

- Visitor pressure is felt on cultural heritage sites such as old whaling and sealing stations and early exploration bases.
- There is evidence that over-flying by light planes and helicopters causes some stress to breeding colonies of penguins and other birds.
- The unique legal status of Antarctica makes enforcement of any code of behaviour difficult.

Key question

How are the key threats to Antarctica likely to change over time?

7.7 The protection of Antarctica

Key terms

Adaptation – Any alteration or adjustment in the structure or function of an organism or system which enables it to survive better in changing environmental conditions.

Mitigation – Includes any actions, strategies, measures or projects undertaken (by mankind) to offset the known detrimental impacts of a process.

Resilience (ecological) – The amount of disturbance that an ecosystem can withstand without changing existing structures and processes.

The concepts of **resilience**, **mitigation** and **adaptation** all relate to the survival of species and maintenance of biodiversity. All three are carefully researched and monitored by scientific researchers from bodies such as the **Scientific Committee on Antarctic Research (SCAR)**.

Ecosystem resilience is the ability of an ecosystem to recover from a severe disturbance by resisting damage and returning to an equilibrium position of functionality. It includes the ability of species to recover from 'shock' events, disturbances or ongoing change, whether they be natural or induced by human activity. The ability to 'bounce back' from such events may also be linked to adaptation to changing conditions and this may apply to ecosystem change as a whole.

Natural disturbances and shock events that test the level of resilience of a system happen in Antarctica just as they do elsewhere in the world. Endemic species are generally capable of surviving shock events as they evolve strategies to allow their populations to rebuild following mass mortalities. Longevity among seabirds and the ability of plant seeds to survive for long periods are examples of this. Intense storms which affect large areas or more localised events such as 'scouring' of the **benthic** environment (ecological region at the bottom of a body of water) by icebergs will both impact on species. Providing such events are rare, it seems that species can recover, but increases in the frequency and magnitude of these events will present challenges to the resilience and adaptability of marine communities.

Human threats such as sealing, whaling and hunting for penguins have pushed species to the brink of extinction in the past. Once these activities ceased, a number of species recovered and in some cases, such as the king penguin, thrived. However, the unsustainable fishing of krill causes a dramatic change to an underpinning element of the ecosystem that may prove impossible to recover from.

The other major threat that exposes Antarctica to rapid environmental change is that of climate change. The following components are changing too quickly for many species to adapt:

- increasing sea temperatures
- ocean acidification
- expanding sea ice cover in some areas
- loss of sea ice and land ice cover in other areas
- higher intensities of ultraviolet radiation.

Some organisms may benefit, at least in the short term, while many others will be vulnerable as their ability to adapt is at a slower rate than the changes.

Mitigation is intervention by humans to eliminate or reduce the risk and hazards presented by a natural or human-induced phenomenon. It is often used in the context of discussing responses to climate change. The IPCC recognise that both mitigation and adaptation are terms which are fundamental to a response to the many challenges brought about by climate change. They are different (though not mutually exclusive) strategies to deal with environmental change. Mitigation measures for global climate change include energy conservation, reducing carbon emissions, etc. On a regional or local level in Antarctica, mitigation can only be undertaken by protecting the existing environment as much as possible and by monitoring any change.

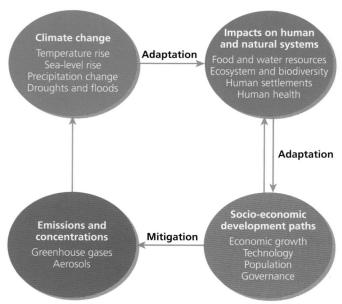

Figure 7.46 Linking adaptation and mitigation in an integrated framework of climate change

Source: Adapted from IPCC report

Antarctic Treaty System (ATS)

The Antarctic Treaty System (ATS) is the main remedy for the international governance of Antarctica. Improved scientific knowledge and more advanced equipment have allowed greater access to Antarctica. As with the high seas, this poses the threat of unchecked resource exploitation in Antarctica.

By the 1950s, permanent stations were established to commence the first substantial multi-nation research program during the International Geophysical Year (IGY) of 1957–58. Territorial positions were also asserted, though not agreed, creating a tension that threatened future scientific co-operation and potential conflict. Because of disputes over ownership, the **Antarctic Treaty** agreement was signed by the nations that had been active on the continent during the IGY in order to:

- avoid disagreements and conflict
- resolve disputes over ownership and mining rights
- establish guidelines to limit development on the continent.

The Antarctic Treaty 1959

The issue of sovereignty was resolved in December 1959, when 12 nations (Argentina, Australia, Belgium, Chile, France, Japan, New Zealand, Norway, South Africa, United Kingdom, United States and Soviet Union) signed up to the Antarctic Treaty. This formalised and guaranteed free access and research rights so that all countries could work together for the common causes of scientific research and exchange of ideas.

The Treaty, which applies to the area south of 60° south latitude, is surprisingly short, but remarkably effective. In its fourteen articles it:

- stipulates that Antarctica should be used exclusively for peaceful purposes: military activities are specifically prohibited
- guarantees continued freedom to conduct scientific research
- promotes international scientific co-operation including the exchange of research plans and personnel; it also requires that results of research be made freely available
- sets aside the potential for sovereignty disputes between Treaty parties, stating that no activities will enhance or diminish previously asserted positions with respect to territorial claims; no new or enlarged claims can be made
- prohibits nuclear explosions and the disposal of radioactive waste
- provides for inspection by observers (designated by any party) of ships, stations and equipment in Antarctica to ensure compliance with the Treaty
- requires parties to give advance notice of their expeditions; provides for the parties to meet periodically to discuss measures to further the objectives of the Treaty
- puts in place a dispute settlement procedure and a mechanism by which the Treaty can be modified.

Since coming into force on 23 June 1961, the Treaty has been recognised as one of the most successful international agreements. Problematic differences over territorial claims have been effectively set aside and as a disarmament regime it has been outstandingly successful.

The wider ATS system

The process of consensus has allowed the Antarctic Treaty to evolve into a system with a number of components that meet the needs of managing activities in the Antarctic, while still protecting national interests. This regime is now known by the broader title of the Antarctic Treaty System, which is governed by the annual **Antarctic Treaty Consultative Meetings (ATCM).**

The **Antarctic Treaty System** comprises a complexity of arrangements to regulate relations between nations

and to control their activities on Antarctica. The system includes the Antarctic Treaty at its core together with a number of related agreements:

- the **Protocol on Environmental Protection to the Antarctic Treaty** (Madrid, 1991) known as the **Madrid Protocol**
- two separate conventions dealing with the **Conservation of Antarctic Seals**
- the **Conservation of Antarctic Marine Living Resources.**

The system now includes a number of other organisations, including NGOs and scientific institutions that contribute to the decision making regarding activities taking place on the continent. These supplementary agreements have added to the scope of protection which now includes:

- protection of the Antarctic environment
- conservation of plants and animals
- preservation of historic sites
- designation and management of protected areas
- management of tourism
- collection of meteorological data and hydrographic charting
- logistic co-operation.

The Treaty also provides that any member of the UN can accede to it, thus membership continues to grow. The Treaty now has 46 signatories, 28 of whom are 'Consultative Parties', either as original signatories or as countries conducting substantial research there.

The Protocol on Environmental Protection to the Antarctic Treaty 1991 (the Madrid Protocol)

The Protocol (otherwise known as the Madrid Protocol) was negotiated by the UN and treaty members at an Antarctic treaty conference in in 1991. Its purpose is to give extra protection to the environment of Antarctica, especially against mineral exploration. The negotiation of the Protocol followed many years of international talks on controlling mineral resource activities in Antarctica. The Antarctic Minerals Convention was proposed in 1988 as part of the ATS but was never adopted. The Convention proposed that it might be possible for mining to go ahead in Antarctica providing it was consistent with the protection of the environment. However, Australia and France refused to ratify the Minerals Convention and instead made a separate proposal that eventually led to the Environmental Protocol.

The Protocol:

- designates Antarctica as a 'natural reserve, devoted to peace and science'
- establishes environmental principles for the conduct of all activities
- prohibits mining or any mineral resource exploration including exploration of the continental shelf
- subjects all activities to being assessed for their environmental impacts
- establishes a Committee for Environmental Protection responsible for advice, inspection and reporting
- requires that operators in Antarctica develop contingency plans to respond to environmental emergencies
- elaborates on the rules relating to liability for any environmental damage
- requires that waste of all kinds be returned to the country of origin wherever possible.

The protocol was ratified by all members and came into force in 1998. The ban on mining is of indefinite duration and strict rules for modifying it are in place. The prohibition can only be modified if all parties agree. Otherwise, the next review conference on the mining prohibition will not take place for 50 years. Any modifications to the mining prohibition are unlikely until 2048, and only then if 75 per cent of the current consultative parties agree.

Systems for inspection

Inspection and observation of all operations on Antarctica is compulsory under the terms of the Treaty.

Antarctica has no permanent population and so no citizenship or government. All personnel present on Antarctica are citizens of a nation outside of Antarctica and under that nation's jurisdiction, so there is no single unifying 'legal system'. Prosecutions against Treaty rules are the responsibility of individual nation states though the ATCM.

The Madrid Protocol introduced a more rigorous regime. Observers are now designated by the ATCM. **Environmental audits** are now carried out around bases, on land and in the sea to assess the impact that bases and their activities are having on the surrounding area. The scope of inspections under the Protocol includes inspection of stations, ships, aircraft and loading areas,

unfortunately with a lesser focus on maritime areas, which are equally important. Any new activities by operators are subject to **environmental impact assessments.**

International Whaling Commission (IWC)

The IWC is the global body responsible for the conservation of whales and the management of whaling. It was established in 1946, its purpose being to provide for the proper conservation of whale stocks and the orderly development of the whaling industry.

Figure 7.47 Part of the Southern Ocean Whale Sanctuary

The main duty of the IWC is to keep under review the measures laid down by the Whaling Convention, which governs the conduct of whaling throughout the world. These measures provide for:

- complete protection of certain whale species
- designated specified areas as whale sanctuaries (for example, The Southern Ocean Whale Sanctuary, Figure 7.47)
- established limits on the numbers and size of whales which may be taken
- prescribed open and closed seasons and areas for whaling
- prohibition on the capture of suckling calves and female whales accompanied by calves
- compilation of catch reports and other statistical and biological records.

In addition, the Commission encourages, co-ordinates and funds whale research, publishes the results of scientific research and promotes studies into related matters such as the humaneness of killing operations.

The International Whaling Moratorium

In 1982 the IWC decided that there should be a pause in commercial whaling on all stocks from the 1985/1986 season onwards. This pause is often referred to as the commercial whaling moratorium, and it is still in place today. This does not affect 'aboriginal subsistence whaling' carried out by indigenous cultures in Alaska, Greenland and parts of Canada, which is not viewed as commercial whaling. Japan continues to evade the moratorium by 'special permit' whaling allowing them to take whales.

Only Norway and Iceland take whales commercially at present, either under objection to the moratorium decision or under reservation to it. Both nations only take North Atlantic common minke whales within their own Exclusive Economic Zones and neither take whales from the Southern Ocean. These countries establish their own catch limits but have to provide information on those catches and associated scientific data to the Commission. The moratorium is binding on all other members of the IWC except for Russia who registered an objection to it, though it does not exercise this objection.

Non-governmental organisations in the Antarctic

Antarctic and Southern Ocean Coalition (ASOC)

In order to carry more influence with governments on the world stage, a group of over 30 different NGOs from around the world joined and formed a coalition under one umbrella group – the Antarctic and Southern Ocean Coalition. The coalition is comprised of a number of well-known environmental campaign groups including Friends of the Earth, Greenpeace (Figure 7.48) and the Worldwide Fund for Nature (WWF). It was formed in response to the concern that members of the Atlantic Treaty were meeting to negotiate a framework for mineral and gas exploitation

Figure 7.48 Greenpeace logo

in Antarctica. The group's initial objectives were to convince governments to:

- conclude negotiations of the world's first 'ecosystem as a whole' treaty on fishing
- prevent oil, gas and minerals exploitation by blocking ratification of the proposed Minerals Convention
- to open up the ATS to include participation by NGOs and specialist international bodies such as the International Council for Science (ICSU).

A major victory for ASOC campaigns came when a precautionary ecosystem approach was embedded into the Antarctic Treaty. ASOC was also successful in blocking the Minerals Convention and was partially instrumental in the development of the 1991 Madrid Protocol.

In 1991, ASOC was finally granted observer status in the ATS and is now able to attend annual meetings. It has expanded its scope of activities and currently focuses campaigns on the following:

- negotiating a legally binding Polar Code covering all vessels operating in the Southern Ocean
- establishing a network of marine reserves, including Marine Protected Area status for the Ross Sea
- managing Southern Ocean fisheries, including krill, sustainably
- regulating Antarctic tourism and biological prospecting
- strengthening the Southern Ocean Whale Sanctuary
- mitigating the impacts of climate change
- monitoring implementation of the Madrid Protocol.

ASOC is the only NGO working full time to preserve Antarctica and the surrounding Southern Ocean and continues to be the strong and powerful NGO lobby for environmental protection.

For more information on ASOC visit the website at www.asoc.org.

Scientific Committee on Antarctic Research (SCAR)

SCAR is an inter-disciplinary committee of the ICSU. Its role is to initiate, develop and co-ordinate the scientific research efforts taking place on Antarctica. It has a more holistic approach to co-ordinating research, that considers the role of the Antarctic region as part of the Earth system.

In addition to carrying out its primary scientific role, SCAR also operates in an advisory capacity to the ATCM and other organisations such as the **United Nations Framework Convention on Climate Change** (UNFCCC) and the IPCC. It advises on the science and conservation affecting the governance of Antarctica and the Southern Ocean and makes recommendations on a wide range of related issues. Some have been adopted by the international agreements of the ATS to provide protection for the ecology and environment of the Antarctic.

For more information on SCAR, visit the website at www.scar.org.

Fieldwork opportunities

Assess the level of globalisation affecting your local area:

1 Visit a local car park when full and take a systematic sample of 50 cars, making a note of the manufacturer.
   a) Identify the country of origin of the named manufacturer.
   b) Identify the place of final assembly (you will have to check this online).
   c) Plot the data graphically.
2 Visit a local supermarket. Pre-select 20 grocery items to include at least five items of fresh fruit or vegetables.
   a) Identify the countries of origin of each item.
   b) Identify the percentage of items certified as either i) fair trade or ii) organic.
   c) Plot the data as a pie chart.

Review questions

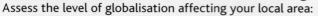

1 What is meant by globalisation?
2 Outline the main factors that have contributed to the process of globalisation.
3 Assess the extent to which international trade and variable access to markets have impacted upon people's lives across the globe.
4 Discuss the following statement using evidence from this chapter.
   'The benefits of globalisation (economic growth, development, integration and stability) outweigh the costs of globalisation. (inequality, injustice, conflict and environmental impact.)'

5 What is meant by global governance?
6 To what extent do the various international organisations and global summits contribute to a more sustainable future?
7 How successful has the Antarctic Treaty System been in protecting Antarctica from its many threats?

Further reading

General:

www.wto.org – World Trade Organisation

www.un.org – United Nations

www.unctad.info – UN Conference on Trade and Development

www.economist.com – The Economist

Globalisation:

McKinsey Global Institute – *Global flows in a digital age*

OECD report – *A global or semi-global village? (1990s to today)*

www.itseducation.asia/globalisation.htm – *Globalisation, wealth, poverty and sustainability*

www.polity.co.uk – *Global transformations*

Trade and TNCs:

www.ey.com – *Changes in geography, supply, sectors*

www.prople.hofstra.edu – *Transportation, globalisation and trade*

www.globalissues.org – *The Banana Trade War*

Global systems and governance (including Sustainable Development Goals):

Department for International Development (DFID) – *Sovereignty and Global Governance*

http://wealthofthecommons.org/essay/common-heritage-mankind-bold-doctrine-kept-within-strict-boundaries – *The wealth of the Commons*

UN System Task Team on the post-2015 Agenda – *Global governance and governance of the global commons*

UNEP report – *Environmental governance*

Antarctica:

www.discoveringantarctica.org.uk – educational website for The British Antarctic Survey

www.asoc.org – Antarctic and Southern Oean Alliance

Question practice

A-level questions

1. Explain the role of **one** major global institution and its contribution to global systems.   (4 marks)

3. Using Figure 7.25 (page 307) and Figure 7.50 assess the extent to which the data shown reflects the international way in which large transnational corporations operate.   (6 marks)

Table 7.8 Tourist visitors to Antarctica 2002–2014

Year	02–03	03–04	04–05	05–06	06–07	07–08	08–09	09-10	10-11	11-12	12-13	13-14
Landed	13,571	19,771	22,926	25,191	29,576	33,054	27,206	21,622	19,445	22,122	25,284	27,735
All	17,543	27,537	27,950	29,823	37,552	46,069	37,858	36,875	33,824	26,509	34,354	37,405

Note: The Antarctic tourist season is in the austral (southern hemisphere) summer, from November to March, and spans part of two calendar years, so seasons are referred to as 2012–2013.

Source: coolantarctica.com

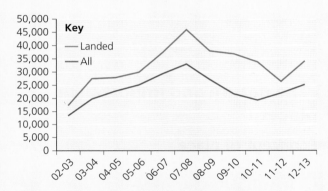

Figure 7.49 Tourist visitors to Antarctica 2002-2014

Source: coolantarctica.com

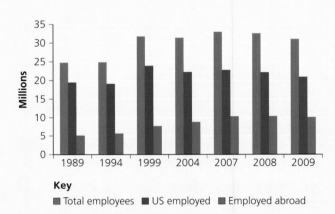

Key
■ Total employees   ■ US employed   ■ Employed abroad

Figure 7.50 US TNCs employment

2. Using Table 7.8 and Figure 7.49, analyse the trends shown for tourist visitors to Antarctica over the 12 seasons and outline the threats posed by these visitors to the Antarctic ecosystem.   (6 marks)

4. 'The process of globalisation brings more benefits than drawbacks for social and economic development in developing countries.'

How far do you agree with this view?   (20 marks)

Changing places

Place is an everyday word and a key term within the study of Geography. Place can be seen as a definite location on a map or it may refer to the description of the human and physical characteristics of a particular location. Place differs to the abstract notion of space because places have meaning to people. Space becomes place as we get to know it better. For many, the most familiar example of place is their home as it is where they feel most attached and can be themselves. Place is more than just location, however, and academic geographers increasingly discuss the different aspects of place – the so-called multidimensionality of place. This may include subjective aspects such as emotional responses to place, or film, photography, art and media representations of place. The Nobel peace prize winner Malala Yousafzai conveys an emotional response to place when she compares life in Birmingham to that of her homeland of the Swat Valley in Pakistan:

> 'When I stand in front of my window and look out, I see tall buildings, long roads full of vehicles moving in orderly lines, neat green hedges and lawns and tidy pavements to walk on. I close my eyes and for a moment I am back in my valley – the high snow-topped mountains, green waving fields and fresh blue rivers – and my heart smiles when it looks at the people of Swat.'
> Prologue, *I am Malala*, 2013

Malala refers to two contrasting places above. As part of this topic you will be expected to investigate two places, one local and one distant.

As geographers, we traditionally focus on the description of place and its locality using maps and statistics but a more accurate reading and writing of a place must also include experience of place, whether through direct interaction or indirect representation.

In this chapter you will study:
● the different meanings and representations of place
● how humans perceive, engage with and form attachments to places
● the character of place and how this can change over time
● how external agencies seek to improve perceptions of place.

8.1 The nature and importance of places

The concept of place

Key questions

How do we define place?

How do we distinguish place from location?

The first place you may think of is your home, where you live and where your physical possessions and memories are gathered together. The place where you live is a particular place. However, the meaning of the term place has been hotly debated and varies according to discipline. To those in planning, place may refer to the built environment; artists and writers attempt to evoke place in their work; and to a philosopher, place may be a way of being-in-the-world. Broadly speaking, geographers refer to three aspects of place: **location**, **locale** and **sense of place** (Figure 8.1).

Location is clearly the starting point here but places take on a significance far greater than simply points on a map and this is where the other two dimensions of place begin to form the wider picture. We can recognise these three aspects in the example of the town of Glastonbury.

Location: Glastonbury is in the county of Somerset (Figure 8.2). It is located 23 miles south of Bristol.

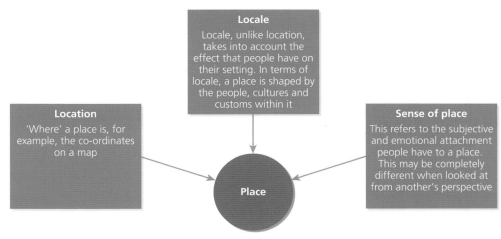

Figure 8.1 The different aspects of place

Latitude: 51.1456N and Longitude: 2.7144W. It is situated at a dry point on the low-lying Somerset Levels.

Locale: Glastonbury has its own unique character. It is home to a number of visitor attractions including Glastonbury Abbey and Glastonbury Tor (Figure 8.3). The *Visit Britain* website describes it as:

> 'alive with a history that sits comfortably alongside colourful myth and legend. The Abbey was an early centre of pilgrimage and many believe King Arthur is buried amid the ruins. It is said that a young Christ also visited Glastonbury and that the Holy Grail is hidden nearby. Quirky independent shops abound selling everything from healing crystals to cakes.'

The National Trust describes Glastonbury Tor as being 'one of the most spiritual sites in the country. Its pagan beliefs are still very much celebrated. It's a beautiful place to walk, unwind and relax.'

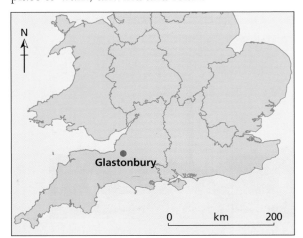

Figure 8.2 Location of Glastonbury

Sense of place: Glastonbury is a place of great spiritual importance for people interested in paganism, religious

connections or the King Arthur affiliation. For many others, Glastonbury evokes emotions about the internationally famous music festival which takes place most years at Worthy Farm in Pilton on the edge of Glastonbury (Figure 8.4).

Figure 8.3 Glastonbury Tor is considered a spiritual site. Excavations at the top of the Tor have revealed the plans of two superimposed churches of St Michael, of which only a fifteenth-century tower remains

Figure 8.4 The first Glastonbury Festival took place in June 1970. It was attended by 1,500 people and cost £1 entry with free milk from the farm

Key terms

Locale – This is the place where something happens or is set, or that has particular events associated with it.

Location – 'Where' a place is, for example the co-ordinates on a map.

Perception of place – This is the way in which place is viewed or regarded by people. This can be influenced by media representation or personal experience.

Place – Defined as a location with meaning. Places can be meaningful to individuals in ways that are personal or subjective. Places can also be meaningful at a social or cultural level and these meanings may be shared by different groups of people.

Placemaking – The deliberate shaping of an environment to facilitate social interaction and improve a community's quality of life.

Sense of place – This refers to the subjective and emotional attachment people have to a place. People develop a 'sense of place' through experience and knowledge of a particular area.

Theoretical approaches to place

The nature and meaning of place has long been debated in geography but there are three main approaches into which the study of place has been divided.

1. **A descriptive approach** is the idea that the world is a set of places and each place can be studied and is distinct.
2. **A social constructionist approach** sees place as a product of a particular set of social processes occurring at a particular time. For example, Trafalgar Square was built to commemorate a British naval victory in the 1800s and, using a social constructionist approach, could be understood as a place of empire and colonialism.
3. **A phenomenological approach** is not interested in the unique characteristics of a place or why it was constructed. Instead it is interested in how an individual person experiences place, recognising a highly personal relationship between place and person. The work of geographers Yi-Fu Tuan and Edward Relph has been particularly important here. Tuan developed the term 'topophilia' to describe the affective bond between people and place and argued that it is through human perception and experience that we get to know places. Relph argues that the degree of attachment, involvement and concern that a person or group has for a particular place is critical in our understanding of place.

There is clearly some overlap between these three approaches and it would be inaccurate to say that any one approach is more important than another. For A-level geography it is important that you consider all three approaches. Places are not simply geographical locations but should be understood as dynamic areas where people, ideas and information come together. Furthermore, places hold meaning: different groups of people may see places differently and have different scope to change these places.

Trafalgar Square, London

The cultural geographer Jon Anderson argues that places can be defined and given meaning by the traces that exist within them. Material traces are physical additions to the environment and include things such as buildings, signs and statues; non-material traces include events, performances or emotions which occur in that place. In Trafalgar Square the immediate traces are the statue of Admiral Lord Nelson, the column on which this stands, a large public square and two fountains. However, you can look behind these traces to try and understand the meanings behind them and the aims of the 'trace-makers' who constructed them.

The statue of Admiral Lord Nelson is at the top of a column that bears his name in a square commemorating

Figure 8.5 Trafalgar Square is a popular London tourist attraction

Nelson's naval victory over combined French and Spanish fleets in 1805. It has been suggested that the city planners of the 1800s wanted to commemorate British leadership and victory, and this place can be understood as a space of empire. Furthermore, it sought to inspire pride and patriotism in the country. Trafalgar Square is still used to celebrate victory and is often the focus for national celebrations such as sporting victories and London's successful Olympics bid.

Traces change over time and this is evidenced by The Fourth Plinth Programme in Trafalgar Square. It was initially intended to hold an equestrian statue of William IV, but had remained vacant due to lack of funding. Since 1999 however, it has become a centrepiece for some of the world's most provocative contemporary public art in a bid to showcase London as the cultural and artistic capital of the world. Commissions have included a marble torso of a pregnant artist, a large blue cockerel and the controversial Antony Gormley plinth entitled 'One and other' which involved 2,400 selected members of the public spending one hour each on the plinth doing whatever they wanted. Art installations like this reflect the notion that although London is a place of tradition, it is also not afraid to take risks and embrace innovation and diversity.

Trafalgar Square has sometimes been used for protest, when people have rallied against the British state rather than supporting and celebrating it. Anti-Iraq war demonstrations took place in 2004 and the traces left by the protestors made Trafalgar Square into a very different place at that time.

The importance of place in human life and experience

Key questions

How is place important to you?

How is your identity defined by place?

It is not hard to find examples of the ways in which place is important. People define themselves through a sense of place and by living in places and carrying out a range of everyday practices there. A person–place relationship is developed. The promotion of place is crucial in the marketing of holiday destinations. Food items are increasingly marketed in terms of the place from which they came and the popularity of particular events may be linked to the reputation of the place at which they happen. The Glastonbury Music Festival would be an example here. People may 'buy into' or 'consume' place. For example, those who like the countryside tend to holiday in rural locations, enjoy books and television programmes about these areas, spend money on walking gear and maps and even furnish their houses in a rustic country style. Numerous products are marketed so that people can buy into the notion of the rural idyll.

The importance of place can be explored by looking at its impact on three aspects: identity, belonging and well-being. The **placemaking** movement, which has expanded rapidly in recent years, places great emphasis on all three aspects.

Identity

Place can be critical to the construction of identity. Probably first and foremost is the sense of place developed in relation to our own home and local geographical area. Reading local newspapers, playing sport for a local team or attending a local fayre or event all foster a sense of local place.

Identity can be evident at a number of scales:
- Localism: An affection for or emotional ownership of a particular place. Localism rarely manifests itself in a political sense but can be demonstrated in 'nimbyism' (not in my backyard) which occurs when people are reluctant to have their local area affected by development.
- Regionalism: Consciousness of, and loyalty to, a distinct region with a population that shares similarities.
- Nationalism: Loyalty and devotion to a nation, which creates a sense of national consciousness. Patriotism could be considered as an example of a sense of place.

Historically, people have identified more with their local place or community because they have greater knowledge of this area and people. In some areas of the UK, this has led to calls for more regional government. In Cornwall, for example, the Mebyon Kernow party has been leading the campaign for the creation of a National Assembly for Cornwall. The party believes that the county of Cornwall, with its own distinct identity, language and heritage, has the same right to self-rule as other parts of the UK, such as Scotland

and Wales which have already achieved a degree of devolution.

Many people identify with place at a national level and this is usually strengthened by a common language, national anthem, flag and through cultural and sporting events. A resurgence in the Welsh language and culture has highlighted a stronger national identity among the Welsh in recent years.

Religion, too, can be used to foster a sense of identity in place. At a local level, churches, mosques and synagogues are places where people from the same religious identity come together to worship. There may also be larger sacred places such as Bethlehem or Mecca where people go on pilgrimages.

The power of place in political protest has arisen recently in reactions to unpopular political regimes and problems associated with capitalism. In 2011, Tahrir Square in Cairo was the focal point of the Egyptian Revolution against former president Hosni Mubarak and became a symbol for the ongoing Egyptian democracy demonstrations (Figure 8.6).

Figure 8.6 Tahrir Square also known as 'Martyr Square', is a major public town square in downtown Cairo, Egypt. The square has been the location and focus for political demonstrations in Cairo that saw the resignation of President Mubarak in 2011, and the ousting of President Morsi in 2013

In London, the Occupy movement, campaigning against social and economic inequality around the world, camped outside St Paul's Cathedral in the financial heart of the city. Similarly recognisable sites were chosen in other parts of the world as the Occupy movement relies on the power of place to attract attention and lodge itself in people's memories.

A global sense of place

The economic and social geographer Doreen Massey wrote about a global sense of place, in which she questioned the idea that places are static. She argued instead that places are dynamic, they have multiple identities and they do not have to have boundaries. She used her own local area to illustrate that place is influenced by constantly changing elements of a wider world.

> Take, for instance, a walk down Kilburn High Road, my local shopping centre. It is a pretty ordinary place, north-west of the centre of London ... Thread your way through the ... traffic diagonally across the road from the newsstand and there's a shop which as long as I can remember has displayed saris in the window. Four life-sized models of Indian women and realms of cloth. On the door a notice announces a forthcoming concert at Wembley Arena: Anand Miland presents Rekha, live, wih Aamir Khan, Jahi Chawla and Ravenna Tandon. On another ad, for the end of the month, is written, 'All Hindus are cordially invited.' In another newsagents I chat with the man who keeps it, a Muslim unutterably depressed by events in the Gulf.
>
> *Massey*, 1994

Massey argued that the character of a place can only be seen and understood by linking that place to places beyond. She concluded, 'What we need, it seems to me, is a global sense of the local, a global sense of place.'

Key questions

To what extent is your local place influenced by external factors?

How has globalisation affected your local place?

Globalisation of place

Some argue that globalisation has made place less important as the forces of global capitalism have eroded local cultures and produced identical or **homogenised** places (Figure 8.7). This can be seen through the increased presence of global chains such as Starbucks in high streets all over the world. The American novelist

James Kunstler has talked of a geography of nowhere – where processes such as urban sprawl have led to community-less cities covering huge areas of countryside with identical shopping malls, car parks and roads. He argues that 'every place is like no place in particular'. In the UK, the term **'clone town'** has been used to describe settlements where the high street is dominated by chain stores. The term **placelessness** has also been used to describe such places.

Figure 8.7 Airports, hotels and high street shops can look the same wherever they are and can lead to the loss of place identity. This Hilton hotel is in Quito, Ecuador but how different does it look to the ones in New York, Beijing and London?

Some local places and cultures are resisting the power of globalisation, as shown by the anti-Costa campaign in Totnes, Devon in 2012 (see text box). Multinational companies are also increasingly having to adapt to the local marketplace. This is known as **glocalisation.** One example of this is the McDonald's franchise. There are currently more than 36,000 McDonald's restaurants in over 100 countries around the world, but the aim to increase profits has led to the company adapting its brand and product to the local market place. In Hindu countries for example, beef has been removed from the menu while in Muslim countries, pork has been removed. The number of McCafes has increased in countries with a coffee culture.

'The independent coffee republic of Totnes' (so called in a BBC article)

Figure 8.8 Clonestopping poster displayed in Totnes

In 2012, the coffee chain Costa set about trying to open an outlet in the South Devon town of Totnes. Within weeks of the proposal, three-quarters of the town's population had signed a petition saying that they support the independent high street and would boycott any coffee shop chain that came to Totnes. The opposition was not anti-capitalist, rather a community fiercely proud of its independently-owned outlets and one eager to prevent Totnes becoming a 'clone town'. After an eight-month battle, Costa dropped their plans, announcing: 'Costa has recognised the strength of feeling in Totnes against national brands and taken into account the specific circumstances of Totnes.' This decision said something about a company actually coming to visit a place and understanding that place, rather than simply sticking a pin in a map, but also about the strength of local feeling about maintaining place identity.

Localisation of place

Place has become a political symbol for people fighting against global capitalism. One particular response has been a greater focus on 'local' place and the promotion of local goods and services. Some places, such as Totnes in South Devon, have introduced a local currency with the aim of encouraging people to shop locally and keep money in the local economy (Figure 8.9, page 342). The idea is that, because people are being encouraged to spend locally, less money will leak out of the local area and get lost in global financial systems. The introduction of the Bristol Pound in 2012 proved so successful in the city that since April 2015 residents have been able to use the local currency to make council tax payments.

Figure 8.9 The Totnes Pound was first launched in 2007 as a project of Transition Town Totnes. The aim is to foster stronger community connections, helping to bring together local consumers, businesses and suppliers who share a common interest: putting people and place over profits

Belonging

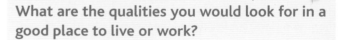

Key questions

What does it mean to belong to a place?

What influences the feeling of belonging, or not belonging to a particular place?

In the context of place, to belong means to be part of the community. Belonging is increasingly seen as one of the key factors that makes a place sustainable and successful. Regeneration schemes now often focus as much on the social environment as on the built environment and the work of different agencies and individuals can have a positive impact on how people feel about the place where they live. The Transition Town movement detailed below is an initiative that places great emphasis on community involvement and has been successful in developing a clearer sense of belonging in places all over the world.

The extent to which one might feel a sense of belonging to a place can be influenced by a number of factors, such as age, gender, sexuality, socio-economic status, religion and level of education. Race and ethnicity can also be linked to the feeling of belonging. Race is based on biological and physical characteristics such as skin colour. Ethnicity is a broader concept which can be defined as belonging to a social group that has a common national or cultural tradition such as language and ancestry. As globalisation and migration have increased, many places, particularly cities, have become much more ethnically and culturally diverse. London is often referred to as one of the world's most multicultural cities and hosts an enormous number and variety of festivals which demonstrate this. It has the largest Chinese New Year festival outside China and other celebrations include Vaisakhi (Sikh New Year), Diwali (the Hindu Festival of Lights) and the Shubbak Festival of Arab culture. In spite of its multicultural status, there are still ethnic clusters in parts of London, including China Town in Soho and Banglatown in and around Brick Lane. These have tended to develop, with dedicated shops and services, for reasons of mutual support and cultural preservation.

Well-being

Key questions

What are the qualities you would look for in a good place to live or work?

In 2014, the *Sunday Times* judges deemed Skipton in North Yorkshire as the best place to live in Britain due to its proximity to the Yorkshire Dales, its great schooling and independent shops along a 'buzzing' high street. Newnham came a close second for its offering of 'country living in the heart of Cambridge' – retaining a 'genuine village feel' despite its location near the centre of the university town. Earning a bronze medal, Monmouth, near the border between Wales and England, is said to have a charming feel and a great selection of independent schools, while the coastal town of Falmouth in Cornwall completed the top four for its combination of wonderful sea views, beaches

The Transition Town movement

Transition Network was founded in 2007, as a response to the twin threats of climate change and peak oil (the point in time when the maximum rate of extraction of oil is reached, after which the rate of production is expected to enter terminal decline). Since then, it can be said to have gone on to tackle many other issues including some associated with globalisation, such as dilution of place identity and loss of community and economy. One of the founders, Rob Hopkins, says:

'It's about what you can create with the help of the people who live in your street, reimagining and rebuilding your neighbourhood, your town. If enough people do it, it can lead to real impact, to real jobs and real transformation of the places we live, and beyond.'

There are now over 1,200 Transition initiatives worldwide and increasing numbers of places are embracing a life where communities come together to share skills, grow food, provide care for dependents and fight inequality.

Learn more about the Transition Network at www.transitionnetwork.org/

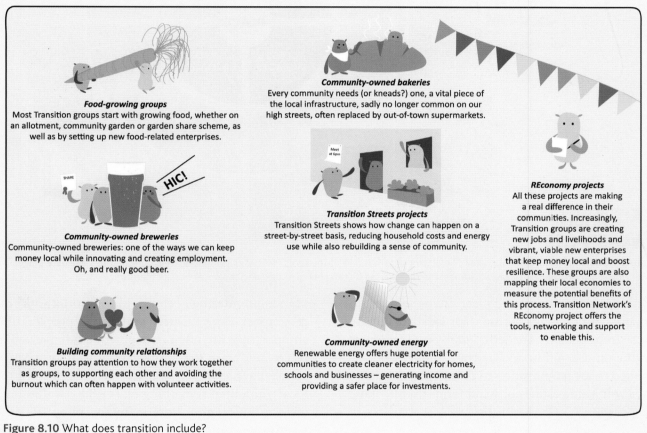

Figure 8.10 What does transition include?

and countryside and its newly earned reputation as a cultural hub.

Individuals may have different views about what makes a place great. Is it likely that all the residents of Skipton, Newnham, Monmouth and Falmouth agree with the *Sunday Times* survey? There are certain features which are generally accepted to be more important in promoting happiness and **well-being** within a place, as shown in Figure 8.11 (page 344), but different factors will be important to different groups of people. As with sense of belonging, age, gender, socio-economic status, religion, level of education and sexuality will also influence people's feelings towards and perceptions about different places.

Insider and outsider perspectives on place

People have a stronger relationship with the places they are familiar with. 'To be inside a place is to belong to it and identify with it, and the more profoundly inside you are the stronger is the identity with the place.' (Relph, 1976) This is the main reason

WHAT MAKES A GREAT PLACE?

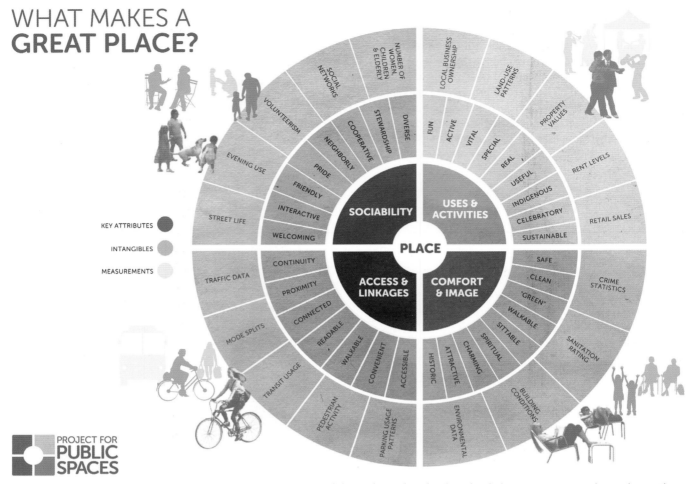

KEY ATTRIBUTES

INTANGIBLES

MEASUREMENTS

PROJECT FOR PUBLIC SPACES

Figure 8.11 What makes a great place? The Place diagram is one of the tools PPS has developed to help communities evaluate places. The inner ring represents a place's key attributes, the middle ring its tangible qualities, and the outer ring its measurable data

why people living within a place are more likely to oppose developments within their local area than those from outside. The term Nimby is an acronym for the phrase 'not in my backyard' and can be applied to local opposition to developments such as new housing estates or wind farms and fracking proposals.

There is also the notion here of people, activities and events being seen as 'in place' or 'out of place'. Why for example may a businessman feel 'in place' in Canary Wharf in London, while a homeless person may not, and what does this mean for how the place is defined? If certain people feel 'out of place', why is this, and who has created this feeling? As the politics of the UK and other countries becomes more focused on immigration, there is a greater need to understand the notion of place in the production of insiders and outsiders.

'A place for everything and everything in its place' is a well-known phrase that suggests there is a

particular ordering of things in the world. Cultural geographer Tim Cresswell argues that people, things and practices are often strongly linked to particular places and when these links are broken, or when people have acted out of place, they are deemed to have committed something of a crime. Graffiti on historic buildings or litter in an area of outstanding beauty are examples of this but increasingly we are seeing groups of people being treated as outsiders. In the past, this has included travellers, protestors and lesbian, gay and bisexual people. Gender has also been important. 'A woman's place is in the home' was the stereotypical societal view held by many in the twentieth century and this affected the type of places in which women felt comfortable.

Different people will perceive places in very different ways. **Positionality** refers to factors such as gender, race, ethnicity, age, religion, politics and socio-economic status, which influence how we perceive different places. The significance that an individual or group attaches to a particular place may be influenced by feelings of belonging or alienation, a sense of being an insider or an outsider. Some places and regions hold great significance for some groups of people but not for others. For example, the city of Mecca in Saudi Arabia is the most holy of religious places for Muslims, but for non-Muslims it is viewed more as an historical and cultural place. Place attachment also develops through positive experiences associated with a place, but not everyone will have the same experiences. Experiences of place change as we get older. Would you feel the same excitement going to the local playground now as you did when you were younger?

One group of people who are often referred to as 'out of place' are migrants. Deaths of migrants in the Mediterranean have highlighted the plight of hundreds of thousands of people trying to escape conflict and reach a 'better life' in Europe, but as people-without-place, such refugees present a crisis to state power. In the United Kingdom, **media** reports and anti-migration groups have used metaphors associated with water, blood and disease to describe the influx of refugees. Such negative terminology has invariably meant that the presence of such migrants has met with great resistance and calls from some to protect 'our place' and 'our culture' against people who do not 'belong' here.

Categories of place

Near places and far places

The terms 'near' and 'far' have several potential meanings when it comes to place. They could refer to the geographical distance between places. Equally, they could describe the emotional connection with a particular place and how comfortable a person feels within that place. Some places feel more familiar than others partly due to personal experience, but also because of frequent representational exposure. The key point is that geographically 'near' places do not automatically foster identities of familiarity and belonging and that in these days of globalised culture, travel and media, far-off places are not automatically strange, uncomfortable and different.

Experienced place and media place

Experienced places are those places that a person has spent time in, whereas **media places** are those that the person has only read about or seen on film. The 'reality' of a place can be far different to that put across by the media and this is most clearly seen through the portrayal of rural places. For a long time, the countryside in the UK has been 'idyll-ized' and countryside living has been stereotyped as involving a happy, healthy and close-knit community experiencing few of the problems of urban life. Geographer Paul Cloke has looked extensively at rurality and argues that magazines such as *Country Life,* television series such as *Doc Martin* and children's books such as *Postman Pat* seek to reinforce these images by focusing on more nostalgic images of the countryside. Stereotypical images of rural living permeate but the idyllic image of the countryside put forward by the media and advertising companies hides a host of problems. Unemployment and underemployment, the scarce availability of affordable housing and the reduction in public transport services have all sought to disadvantage low-income households in rural areas. Rural homelessness has also been hidden from the media glare.

In contrast, cities are often stereotyped in a negative way. Economic and social deprivation, homelessness, crime, vandalism and pollution are just some of the images routinely ascribed to British cities by the media. It is true that some of these problems are worse in urban areas, but it is wrong to assume that all cities are the same. Successful regeneration of urban areas has made city-living far more attractive in recent decades.

Factors contributing to the character of places

Key questions

What are the characteristics of your local place and how have these changed over time?

How can different organisations bring about change in places?

The **character** of a place refers to the physical and human features that help to distinguish it from another place. This character may be linked strongly to the natural environment but it is more often a combination of natural and cultural features in the landscape, and generally includes the people who occupy the place. These characteristics are known as **endogenous factors**.

Physical geography such as relief, altitude, aspect, drainage, soil and rock type.

Socio-economic factors such as employment opportunities, amenities, educational attainment and opportunities, inocme, health, crime rates, local clubs and societies.

Demographic factors: population size and structure (age and gender), ethnicity.

Cultural factors such as heritage, religion, language.

Factors affecting the character of place

Mobility of the population for work and leisure pursuits

Location: urban or rural, proximity to other settlements, main roads and physical features such as rivers, the coast etc

Political factors such as the role and strength of local councils and/or resident groups

The built environment: land use, age and type of housing, building density, building materials

Figure 8.12 Factors affecting the character of places

Endogenous factors are those that originate internally and may include aspects of the site or land on which the place is built, such as height, relief, drainage, soil type, geology and the availability of resources. They also include the demographic and economic characteristics of the area as well as aspects of the built environment and **infrastructure**.

Key terms

Agents of change – These are the people who impact on a place whether through living, working or trying to improve that place. Examples would include residents, community groups, corporate entities, central and local government and the media.

Endogenous factors – In the context of place, this refers to the characteristics of the place itself or factors which have originated internally. This would include aspects such as location, physical geography, land use and social and economic characteristics such as population size and employment rates.

Exogenous factors – This refers to the relationship of one place with other places and the external factors which affect this. The demographic, socio-economic and cultural characteristics of a place are shaped by shifting flows of people, resources, money and investment.

Infrastructure – Infrastructure relates to the services considered essential to enable or enhance living conditions. These primarily consist of transport communications (roads, railways, canals, and/or airports), communications infrastructure (broadband and phone networks) and services such as water supply, sewers and electrical grids. They may also include infrastructure such as the local education system, healthcare provision, local government and law enforcement, as well as emergency services.

As Figure 8.12 shows, there are clearly a wide range of factors which affect the character of places.

Some places may have an industrial story; others may have developed as agricultural places or tourist resorts. When places first started to grow, most had only one distinct **function**. This might be a good defensive position, a bridging point, availability of natural resources, natural routeways or trading centres. As the place developed, exogenous, external factors became more important and the importance of the initial functions diminished as technological advances occurred. **Exogenous factors** are those that have an external cause or origin. This might include links to or influences from other places, distances from and routeways to other places and the accessibility of the place.

De-industrialisation has brought about wholescale change in the economic structure of places and this has led to unemployment and urban decline in cities with a traditional manufacturing base. Likewise, mining, steel and shipbuilding towns have had to adapt to the challenges posed by globalisation. The increasing mobility of people has also affected place and this can sometimes lead to conflict. On a local scale, the construction of a new housing estate may be seen to affect the character of a place as land use changes and 'newcomers' move into an area. The purchase of second homes in seaside resorts and gentrification in cities are also thought by many to bring about change in the character and community of places. Finally, we might consider the impact of international migration as people from all over the world have settled in places around the UK, sometimes forming **diasporas** and definitely creating a more multicultural society.

8.2 Changing places: relationships and connections

Places change over time. They are dynamic and should be understood as a coming together of people, ideas, wealth and information rather than simple locations. Consider, for example, the economic rise and fall of British industrial cities in the nineteenth and twentieth century and the impact this had on their population and environment. British seaside resorts have also undergone significant change in the last fifty years as they have had to adapt to increasing numbers of British people travelling abroad for their holidays. Economic factors tend to have the biggest impact on the character of place but changes may also result from migration and conflict (Damascus, Syria), terrorism (New York City, 9/11), industrial accidents (Chernobyl), natural disasters (New Orleans) and climate change (The Maldives). Factors like these affect the way places are perceived and go on to affect the lives of the people living and working there. External forces can also bring about change, as shown in Table 8.1. This can be both positive and negative.

Table 8.1 Some impacts of external forces on place

Agent of change	Example	Impact
Government policies	Regeneration schemes and financial incentives for industries such as subsidies, tax breaks and enterprise zones	These can attract businesses to places and stimulate a positive multiplier effect
The decisions of multinational corporations	In 2010, Mondelēz International closed the Cadbury factory near Bristol and moved production to Poland	Job losses for employees Factory converted into housing
	In 2016, Tata Steel announced UK job cuts in response to difficult global market conditions	Major job losses at Port Talbot, Hartlepool and Corby – all highly dependent on the steel industry
The impacts of international or global institutions (for example, IMF, World Bank, UN, WHO)	In 2015, the World Bank was running 15 development projects in Haiti	Post-earthquake reconstruction of both homes and communities
	Millennium Development Goals	Varied level of success around the world

Conflict may arise when people resist changes forced upon their place. The redevelopment of areas of East London for the 2012 Olympic Games was not welcomed by everybody and this can be seen with other redevelopment projects currently planned or taking place in London. Proposed housing estates, landfill sites, wind farms and bypasses can all create tensions between different stakeholders. Think about how your local place has been affected by such developments? Why do people often resist change to their place and how can such conflicts be resolved?

Examples of continuity and change at a variety of scales

Places can be investigated at a range of scales. In 2012, the BBC in collaboration with the Open University researched the past and present conditions of six London streets (*The Secret History of our Streets*). They used the poverty maps of Charles Booth (c.1886) as a base and investigated the changing social and economic conditions of the residents who had lived there. Portland Road, Notting Hill, considered in 2012, 'the most gentrified street in the UK' and home to some of London's wealthier residents, was in 1899 the worse slum in London. In contrast, Deptford High Street has gone from being the 'Oxford Street of South London' to 'one of the poorest shopping streets in London, marooned amid 1970s housing blocks.'

The following three examples look at changing communities on a progressively larger scale. They illustrate how the characteristics of places can be shaped by a very different range of factors, including people, resources, money and investment. They also serve to show how past and present connections within and beyond localities can help shape places and the lives of the people who live there.

Bournville village, Birmingham

Bournville is an example of a place shaped almost entirely by the beliefs and ideals of one industrial family, the Cadburys. It was built as a garden village in the late nineteenth century after the family moved to the rural Bournbrook Estate, on the outskirts of Birmingham, in order to build new premises for their expanding cocoa and chocolate business. The area provided ample space for the construction of tree-lined roads and housing with front and back gardens, to ensure spacious and sanitary living conditions for their employees (Figure 8.13). No public houses were constructed because the Cadbury family were temperance Quakers but the Bournville Village Trust set up schools, hospitals, museums, public baths and reading rooms for the benefit of the Cadbury workers. They also laid out a tenth of the estate as parks, recreation grounds and open space and this proved to be a key driver in the later development of the 'garden city' movement.

Figure 8.13 Artist's impression of Bournville

As a company, Cadburys has achieved significant global success and, in 2010, it became part of the multinational operation known as Mondelēz International. In contrast, the 'place' of Bournville has changed very little. It is no longer exclusively inhabited by employees of the Cadbury factory but the physical appearance and community feel of the place have remained largely the same.

In 2003 the Joseph Rowntree Foundation officially found Bournville the 'nicest place in Britain'. *The Guardian* said, 'It's easy to see why. People do old-school community things like talk to each other and grow veg; flowers blossom in the municipal borders; houses are, in the main, hobbity to look at, and trees and bluebells are a few footsteps away.'

The city of Birmingham has grown well beyond the boundary of Bournville but the 'garden suburb' of Bournville still encapsulates the positive ethos and values of the Cadbury family. Large areas of green space remain, planning and building is tightly controlled, the area has little crime and there is a long waiting list for its affordable homes.

Devonport, Plymouth

In contrast to Bournville, Devonport in Plymouth has changed considerably in the last fifty years, largely as a result of external factors. It was originally established as a naval dockyard town due to its location on the sheltered, natural deep-water harbour known as the Plymouth Sound. By the eighteenth century, it had become one of the fastest growing towns in the country and enjoyed great prosperity.

However, Devonport's fortunes have fluctuated with those of the dockyard, particularly since the Second World War. In 1952, the Navy requisitioned Devonport Town Centre as a storage enclave and enclosed it with a three metre high wall. The community was effectively split in two and the displaced residents were rehoused in flats and apartments ill-suited to family living. Naval jobs continued to decline as a result of military cutbacks and this led to significant social and economic problems within Devonport in the 1980s and 1990s.

From 2001–11, Devonport benefited from the New Deal for Communities initiative which provided ten years of funding to improve some of the most deprived areas of the UK. The scheme specified place-related outcomes such as addressing crime, community and housing, and people-related outcomes such as education, health and employment. Community groups such as the Pembroke Street Estate Management Board have been heavily involved. The physical environment of Devonport has changed considerably. The dividing naval wall was removed, inter-war housing and flats demolished and historical landmarks incorporated into the redevelopment of the area. The regeneration has also tried to attract a wider range of people (Figure 8.14) with options for private or shared ownership and the availability of social housing.

Figure 8.14 The construction of Georgian-inspired homes in the 'Village by the sea' development has attracted people from higher socio-economic groups to Devonport

There are still pockets of deprivation within Devonport, but both quantitative and **qualitative data** show that the New Deal programme and other redevelopment projects have significantly changed and improved the area, with benefits to local residents.

Medellin, Colombia

The city of Medellin in Colombia, affectionately nicknamed 'The City of Eternal Spring', and the second largest city in Colombia, was for a long period associated with drugs and violence and dubbed 'the most dangerous city in the world.' The notorious drug Lord Pablo Escobar wielded enormous power in the city until his death in 1993. Unemployment, crime and poverty were widespread and this in turn created social inequality.

Today, Medellin, with a population of 2.2 million, has become a model for urban regeneration and sustainable city planning through long-term investments in infrastructure and education. As part of a plan to re-brand Medellin, city planners recognised the need to make the city equally accessible to all its citizens, rich and poor. Through this planning philosophy, the city's long-divided social classes are now more able to integrate in everyday economic and educational activities. The city's poorest, many of whom reside in shanty houses in the Aburra Valley, can now access the city's booming economic centre courtesy of a series of outdoor escalators and a gondola system that carries people up and down the valley.

Additional innovations include a bus rapid transit system named Metroplus, with dedicated bus lanes; an extensive above-ground tram system and a city-wide ride-sharing program. Emission-free transport has been promoted, and this has been helped by the EnCicla initiative, a free bike-sharing programme that offers an integrated alternative to the city's public and mass transportation systems.

Education, social programmes and the public arts and culture budgets have all been increased to transform the lives of the most underprivileged residents in this city. There are still problems in Medellin. Poverty rates have fallen but inequality between rich and poor has increased and cultural and geographic barriers continue to limit social integration. Crime rates and gang violence remain high. However, change takes time and there is a great feeling of optimism within the city.

8.3 Changing places: meaning and representation

The notion that places change because of the constant movement of people, ideas, wealth and information within them has already been discussed. A further dimension relates to the meaning and representation of place. **Meaning** relates to individual or collective perceptions of place. **Representation** is how a place is portrayed or 'seen' in society. Both of these may change over time. They also vary between people and communities as shown by the example of Belfast (see text box below) and the place study of Brick Lane and Spitalfields (page 363). In the latter, it is clear that different groups of people attach different values and

Belfast: one place, two representations?

Belfast is the capital city of Northern Ireland and home to 286,000 inhabitants. It is the region's economic powerhouse, an industrial city where the Titanic was built and which was once the largest producer of linen in the world. The city suffered greatly during the 30-year period of conflict called 'the Troubles', but has undergone a sustained period of calm and substantial economic and commercial growth since the Good Friday Peace Agreement was signed in 1998.

The city centre has seen large-scale redevelopment and different parts of the city have been rebranded as 'quarters', emphasising their unique history and culture. The Titanic Quarter is one of the largest brownfield redevelopment sites in Europe and includes the Titanic Studios and more than 100 companies, including Citi, Microsoft and IBM. This is the creative and cultural Belfast promoted by the City Council and Tourist Board, looking outwards to build a cosmopolitan city which is open and welcoming.

The other representation links to the complex political history of Belfast, the high degree of religious segregation and the image of 'bombs, bullets and balaclavas' which dominated newspaper front pages in the late twentieth century. Peace walls and political parades are an important part of the history of Belfast but some argue that they reflect a city struggling to move beyond the disputes and arguments which have shaped its past.

Figure 8.16 The new: The Titanic Quarter is home to the Titanic Belfast Centre, visited by more than two million people from 145 countries, and the set of the world's most successful television series, Game of Thrones, filmed in a former shipyard paint hall

Tourism has increased significantly in Belfast in the last decade, attracting more than 6.5 million visitors and contributing £450 million annually to the local economy. The evidence suggests that Belfast has successfully managed to change its international image and visitors from all over the world are attracted by the city's culture and vibrancy rather than being put off by events that have gone before.

To find out more about Belfast and the changes which are happening there, go to: www.belfastcity.gov.uk or www.belfasttelegraph.co.uk/ There are useful pages detailing ethnic diversity and segregation in Belfast at: www.geographyinaction.co.uk/

Figure 8.15 The old: Peace lines or walls were constructed in Belfast to keep neighbouring Protestants and Catholics apart. Many of these walls feature murals, which tourists come to see. In some cases, tourists write their own messages of peace and hope. It is hoped that the walls will be taken down as community relations strengthen

meanings to the area and hold different views on its present and future development.

This part of the chapter will explore how humans perceive, engage with and form attachments to places and how they present and represent the world to others.

Meanings of place

Sense of place is the meaning attributed to a place as influenced by our interactions with it. We experience place, for example, by living and working there or by visiting places. From these 'real-life' experiences and memories, we develop a sense of place. Think of the places that were important to you as a child. Was there a certain place you loved exploring? A favourite park, for example? What memories are attached to these places? What significance do they hold for you? Places can hold historical, spiritual or cultural significance, but for children it is primarily the emotional attachment associated with places that give them meaning and this is why 'home' is often the most important place.

Developing a sense of place is important. Research suggests that connecting to one's surrounding environment establishes knowledge of and appreciation for its resources; a sense of place supports the development of personal identity; having a strong sense of place can inspire stewardship and understanding; sense of place is said to nurture empathy. Could we argue therefore that our own experience and enjoyment of place is critical to our understanding of place? Making a link to geography fieldwork is relevant here. Fieldwork – that is, learning directly in the untidy real world outside the classroom – has long been an essential component of geography education and studying Geography without fieldwork has been likened to studying science without experiments. Practical skills are clearly important in the fieldwork experience but it is our engagement and interaction with place beyond the classroom which ultimately affects our understanding of that place.

Places can create memories. **Place-memory** refers to the ability of place to make the past come to life in the present. This can occur through material artefacts such as old photographs or place souvenirs. The preservation of buildings, monuments, museums and plaques are all examples of the 'placing' of memory, which can then be used to create a public memory. Attractions such as the Beamish Living Museum of the

North or the Ironbridge Gorge Museum enable visitors to experience a sense of place by standing inside and walking around recreated historical places.

Sense of place can be contrasted to **perception of place**, which is developed through what people have heard, seen or read about a place. Dartmoor National Park in Devon is closely linked with ideas of nature and wilderness in different artistic and literary works, including Arthur Conan Doyle's *The Hound of the Baskervilles* (1902). Many popular contemporary images of Dartmoor continue to associate the moor with wilderness, but these hide important human activities and conflicts that include mining, farming, quarrying and military training.

When it comes to researching place, daily life is the fieldwork. Cresswell argues that, 'the world itself is the best kind of resource for thinking about place and a lot can be learned from reflecting on everyday experience.' Walking down the local high street, reading the local newspaper or watching the local news are different ways in which you can experience your place. Traditionally, geography fieldwork has focused on explaining and analysing. Cultural geography encourages a more emotional, poetic and spiritual approach to fieldwork – one in which we get more of a 'place experience' and one which enhances our awareness of what it means to 'be in place'.

Management and manipulation of the perception of place

Key questions

How do your perceptions of different places compare to those of your friends and family?

How and why do organisations manipulate perceptions of place?

Perceptions of **international** places tend to be influenced more by the media than by personal and direct experience. What are your current perceptions of places in countries such as Afghanistan and Syria, for example, and how have these developed? Other influences, such as historical and political relationships or trading links, may also influence how a country is perceived. Governments are keen to attract trade and investment into their countries and a positive place perception is therefore important at an international level. Organisations

Figure 8.17 The components of urban rebranding

such as the British Council help to promote the UK through educational and cultural links. The government and monarchy are also seen to play a pivotal role in promoting international relations for the UK.

Perceptions of place are also important at the local scale and increasingly organisations are being employed to promote place, build up a place brand or improve perceptions of place. Investment in place is crucial for its survival and people are more likely to want to live or work in a place with a good reputation and positive image.

Different **agents of change** will aim to manage the perception of place. These may include national and local government, corporate bodies, tourist organisations and community groups.

Government

At both national and local level, strategies have been adopted to manage and manipulate perception of place in order to attract people and investment to a place. These include: **place marketing**, **rebranding** and **reimaging**.

Place marketing

Marketing or public relations (PR) companies may be employed by national and local government to improve or create positive perceptions of place. Using the example of Weston-Super-Mare in Somerset, strategies have included:

- advertising campaigns, including social media marketing through Facebook
- an official Weston-Super-Mare website and newsletter
- a Weston-Super-Mare logo
- creation and promotion of the first ever Love Weston Winter Wonderland, a festive attraction incorporating the annual Christmas lights switch-on. The aim was to increase perceptions of Weston-Super-Mare as a destination point for Christmas shopping.

Think about how your local place is marketed? Who is responsible for this? What are the key objectives of the marketing?

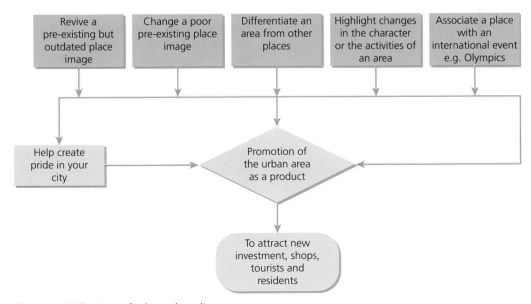

Figure 8.18 The aims of urban rebranding

Rebranding

Rebranding is used to discard negative perceptions of a place (Figure 8.18, page 351). Its main aim is to try to make a location a desirable place in which to live (as well as to invest and develop industrial activity) and one that people will want to visit for social and recreational purposes and to shop and spend money. In 2013, the slogan 'People Make Glasgow' was introduced as the new brand name for Scotland's largest city. It was chosen following a **crowdsourcing** social media campaign involving more than 1,500 people from 42 countries. The campaign emphasises that people are the heart and soul of this particular place – 'People Make Glasgow.'

Many argue that place (re)branding must start from the inside and involve local residents with 'insider' experiences. Geographers Edward Relph and Yi-Fu Tuan have emphasised the importance of being inside a place to truly understand it and likewise, it has been argued that without a thorough understanding of place, one would find it difficult to regenerate and rebrand a place.

Rebranding is not without its problems. Different stakeholders may include pre-existing residents, local businesses, potential investors, local government and potential home-owners or visitors and the challenge is to satisfy as many of these groups as possible. Pre-existing residents often want to protect and project their local distinctiveness while development agencies seek to establish place brands based on government incentives, available technology and an area's international links. Some city regeneration schemes have actually driven out the locals they originally intended to help, as rising property prices and rents have favoured more affluent people.

Re-imaging

Linked to rebranding, re-imaging seeks to discard negative perceptions of a place and generate a new, positive set of ideas, feelings and attitudes of people to that place. This may include the revival of a pre-existing but outdated place image. More commonly, it seeks to change a poor pre-existing image of a place. This has been well documented through the example of Liverpool in the 1980s and 1990s. Deindustrialisation had caused an economic downturn in the city and riots in 1981 had dominated newspaper headlines. Large-scale regeneration began and the Tate Liverpool art gallery was one of a number of projects aimed at re-imaging the city's industrial heritage through culture. The Merseyside Development Corporation used the term 'There's life in the old docks yet.' The exterior of the Grade I-listed warehouses of the Albert Docks remained untouched but the derelict interiors were transformed into modern art galleries. The Tate's presence in the city was seen as a key factor in Liverpool winning the title of European Capital of Culture in 2008, a far cry from the negative imagery of the city in the early 1980s.

Rebranding Amsterdam

Figure 8.19 The motto 'I Amsterdam' originated from a photography exhibition in 2004 promoting the city of Amsterdam. It was subsequently adopted for the rebranding and promotion of the city

In the late twentieth century, Amsterdam's reputation as a major international cultural centre had been threatened by a number of factors, including:

- greater competition from other cities both within and outside of the Netherlands
- social and economic decline in some areas
- the city's reputation for being liberal towards soft drugs and prostitution, which was seen as inappropriate for attracting new investors and enterprises
- a failed bid to host the Olympic Games.

A number of rebranding strategies were adopted but the most successful was the 'I Amsterdam' slogan, seen to be clear, short, powerful and memorable. The large three-dimensional 'I amsterdam' letters were positioned in front of the city's famous Rijksmuseum in 2005 and the sculpture is now the city's most photographed item, being photographed over 8,000 times a day on sunny days. The use of smartphones and social media has seen the image spread all over the world and Amsterdam has become one of the most successful destination brands on social media. Given the last ten years to imprint the new identity, the city of Amsterdam has experienced increased tourism and is one of the top five European cities based on its brand strength and cultural assets.

Other European cities such as Barcelona have undergone successful rebranding programmes in recent decades. Investigate the different strategies used to 'rebrand' these places and evaluate how successful they have been.

Corporate bodies

A **corporate body** is an organisation or group of persons that is identified by a particular name. Examples include institutions, businesses, non-profit enterprises and government agencies. Many corporate bodies will have an interest in place but some will want to manipulate perceptions of place. For example, tourist agencies aim to 'sell' place to potential visitors and marketing positive perceptions of place makes this easier. In the UK, tourist organisations range from Visit Britain, the non-departmental public body, funded by the Department for Culture, Media and Sport, to the individuals responsible for promoting a specific tourist attraction. The strategies are similar, to make a place look as good as it can and attract as many visitors as possible. Promotional materials such as brochures, videos, websites, magazine advertisements, slogans and logos are used and places may adopt a unique selling point. In 2012, the Pembrokeshire Coast National Park Authority put vintage-inspired designs featuring nostalgic images of the Pembrokeshire coast on show at Cardiff Airport, UK railways stations and across the London Underground in order to increase tourism to the area (Figure 8.20). The posters won numerous awards but they were also successful in attracting more people to the area.

Airlines and train companies also seek to manage perceptions of place but they do so in order to get people to use their travel services to visit these places. Figure 8.21 is one of several railway posters marketing a positive image of Torquay in Devon. Artworks like this emerged at the beginning of the twentieth century as railway companies commissioned posters, sometimes by famous artists, to sell the delights of the British coast and countryside and therefore boost the number of train passengers wanting to get there.

Figure 8.20 One in a series of retro-style railway posters used to promote the Pembrokeshire Coast National Park in 2012

The role of community and local groups

Community or local groups may take an active role in managing and improving the perception of their place to attract investment and improve opportunities and services within the area. Regeneration and rebranding strategies have increasingly involved local people, since they have the 'insider' experience of place and will be the people most affected by any changes. Residents associations and heritage associations play an important role and social media is increasingly being employed to engage and involve local people in planning and place-making schemes.

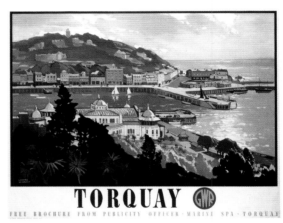

Figure 8.21 Great Western Railway poster advertising Torquay c.1947

8.4 Representations of place and the use of quantitative and qualitative sources

Places can be represented in a variety of different forms or media that may give contrasting images. It is therefore

Key terms

Meaning – Meaning relates to individual or collective perceptions of place.

Media – Means of communication including television, film, photography, art, newspapers, books, songs, etc. These reach or influence people widely.

Objective – Not influenced by personal feelings or opinions in considering and representing facts.

Qualitative data – Information that is non-numerical and used in a relatively unstructured and open-ended way. It is descriptive information, which often comes from interviews, focus groups or artistic depictions such as photographs. Some types of qualitative data, such as interviews, can be coded and may then be subject to quantitative analysis.

Quantitative data – Data that can be quantified and verified, and is amenable to statistical manipulation.

Representation – Representation is how a place is portrayed or 'seen' in society.

Subjective – Based on or influenced by personal feelings, tastes or opinions.

important to investigate and present your two place studies using a variety of quantitative and qualitative sources while at the same time acknowledging their limitations. This part of the chapter combines two parts of the specification. Firstly, it looks at how places can be represented using formal and informal sources and, secondly, it introduces the different types of quantitative and qualitative skills you can use to build up your own place studies.

Statistics

The UK census has been used to detail the social and economic characteristics of the population of the country for over 200 years. Smaller geographical units can reveal basic information about places such as age and gender structure, ethnicity and levels of economic deprivation. The census provides large-scale, **quantitative data**, which has been used by national agencies to better understand and plan for population growth and other demographic changes. The Office for National Statistics (ONS) is the UK's largest independent producer of official statistics and is responsible for collecting and publishing statistics related to the economy, population and society at national, regional and local levels. It also conducts the census in England and Wales every ten years.

Skills focus

The ONS website is a useful starting point for finding out about your local place. You can download data on characteristics such as population, health and crime and use these to produce maps and graphs. Go to www.ons.gov.uk/

The use of quantifiable data such as statistics is not as **objective** as it may first appear. This is because people selectively choose the data they wish to use for their particular purpose. Their use therefore becomes **subjective**. Another criticism of using statistics when studying place is that they tell us very little about the human experience of a place and what it is like to live there.

Maps

Maps have long been used to locate places but they can also influence how we think or feel about a place and as such play a very important role in both our sense of place and perception of place. It is important to cast a critical eye over the reliability and accuracy of maps as throughout history they have distorted

reality. For example, early world maps such as the Mappamundi (c.1300) depicted the world as a flat disk with the Holy Land and Jerusalem at the centre. During the period of colonial expansion, maps exaggerated area size and resources (and thus strategic importance). Early cartographers (map-makers) very much 'promoted the cause of empire' (Figure 8.22).

Figure 8.22 This early map of Brazil shows an interior landscape rich in resources but supposedly inhabited by 'less civilised' people. A boat of slaves is shown arriving from Africa

Maps can include hidden bias and influence. Google Maps is a search engine which allows people to search for and find out about places. While undoubtedly an invaluable tool in researching place, it is worth noting that it is not done in a strictly objective way. In fact, Google Maps, like any search engine filters place – directing people towards businesses that have engineered their appearance on the first page of a Google search.

Counter-mapping describes a bottom-up process by which people produce their own maps, informed by their own local knowledge and understanding of places. You could try this for your own local place. Figure 8.47 (page 371) is an example of this: the artist and cartographer Adam Dant illustrated this map of Spitalfields in East London with fifty portraits of the people who make the area distinctive. On the reverse of the map, stories of all those portrayed on the front are provided, plus a guide to the essential Spitalfields landmarks and destinations. The result is that the map not only provides factual information but it also conveys a sense of place.

Analysing the Mercator projection

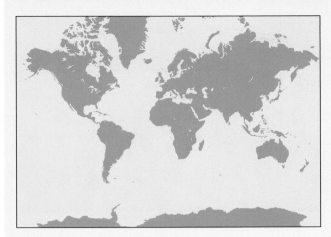

Figure 8.23 The Mercator projection

The Mercator map projection was first used in the sixteenth century as a navigational map for sailors. The northern landmasses look much bigger on the map than they are in reality.

You will commonly see the Mercator map projection in geography classrooms and unthinkingly accept it as an accurate representation of the world. However, educationalists have been denouncing its use in classrooms for a long time because it does not accurately depict the size of the Earth's northern and southern latitudes.

1 It distorts the relative size of land masses. For example, Greenland looks the same size as Africa but in reality Africa is 14 times larger.

2 It is based on the convention that the northern hemisphere is at the top when in truth the world does not come with a label saying 'this way up'.

3 The map is Eurocentric. Not only does it exaggerate the size of Europe but it puts it in the middle of the map.

Skills focus

The expansion of Geographical Information Systems (GIS) means that an increasing range of geographical data is available for research and analytical purposes. The Consumer Data Research Centre was established by the UK Economic and Social Research Council and provides a comprehensive set of geographical data which has been mapped and is easy to access. Go to http://maps.cdrc.ac.uk/

Biomapping

One method of trying to measure sense of place is **biomapping** or emotional cartography. This is the mapping of emotions shown by people to certain places through use of a device which records the wearer's Galvanic Skin Response (GSR). This is a simple indicator of emotional response in conjunction with a geographical location. A map can be created which visualises points of high and low feelings.

In a 2015 episode of the series *Coast*, a series of 'stress tests' were carried out on presenter Nicholas Crane by geographer Jon Anderson to measure the electro dermal activity (or nervous energy) of the skin while walking in London, cycling a Boris-bike around Trafalgar Square, and negotiating the London Underground system. The results showed very little variation when he was walking around London but there were notable spikes in nervous energy when he caught the underground train or crossed a busy road. The peak in stress level was recorded as he cycled around Trafalgar Square and had to contend with heavy traffic. Conducting the same experiment at the coast produced contrasting results. When the presenter took an early evening stroll along the beach at Southend, his stress levels were at their lowest. This was put down to the physical environment of the coast – a combination of natural elements such as the hypnotic and soothing sound of the waves, the feel of the sand under his feet and the sea breeze. Quantifiable data was produced in this experiment to back up the widely-held cultural notion that going to the seaside is good for you.

Figure 8.24 TV presenter Nicholas Crane preparing to cycle around Trafalgar Square as part of an emotional stress test

Figure 8.25 The Toshogu shrine in the UNESCO World Heritage site of Nikko in Japan

Figure 8.26 The Toshogu shrine in the middle of the day when visitors numbers are at their highest

Interviews

Interviews can generate detailed insights about a person's sense of place or perception of place. They are first-hand or direct reports of experiences, opinions and feelings. Interviews can be structured or unstructured, the latter being more like an everyday conversation rather than having pre-set questions. They tend to be more informal, open ended, flexible and free flowing. Disadvantages of using interviews include interviewer bias, in which the interviewer may affect the responses of the interviewee by using leading questions. The fact that people like to present themselves in a favourable light and therefore may not be honest should also be taken into consideration. Going beyond the spoken word, **ethnography** is a research method that explores what people do as well as say. For example, to understand different places as they are experienced and understood from the 'inside', the ethnographer would participate in the daily life of a person or group of people, watching what happens, listening to what is said and generally becoming part of that world.

Media representation of place

Knowledge and understanding of place can be influenced very strongly by a diverse **media** including television, film, photography, art, books, newspapers and the internet. These are increasingly reaching a larger, global audience and are therefore very important in shaping wider perceptions of place. In addition, this has meant that geographical distance has become less of an obstacle for learning about places when people can gain instant knowledge about them from media sources. The notion of near and far places is becoming increasingly blurred as people develop attachments to places they have never visited, through the media.

Photographs

At a time when photo-editing of people is commonplace, so places can easily be 'photoshopped' to make them appear different and, in the case of tourist areas, much more attractive. The difference between a photograph you may have seen of a particular place and the reality is often down to differences in weather, the time or the season, but with advanced technology, people are increasingly editing photographs to improve the image or perception of that place. Similarly, photographs can be selective in what they show. Marketing images tend to focus on the natural beauty or landscape of places without disturbance from humans; the reality for many tourist places may be very different, as shown in Figures 8.25 and 8.26.

Figure 8.25 is a tourist image of the Toshogu shrine in the UNESCO World Heritage site of Nikko in Japan. The shrine is set in a beautiful forest and is famed for its ornate architecture including woodcarvings and gold-leaf decoration. Figure 8.26 is taken at the shrine in the middle of the day when visitor numbers are at their highest. The 'reality' of Nikko here contrasts with the representation of the shrine in the tourist image.

Skills focus

Flickr is a photo-sharing platform which can be used to find visual images for your local and distant place. Google Earth is also a source of geotagged photographs. When analysing the value of a photograph, it is important to consider why it was taken and who it was taken for. A photograph can be selective in what it shows. What has been missed out? Why has this not been photographed? The conditions in which it was taken are also important. The website www.mesogo.com has some interesting comparisons of photographs of places before and after being photoshopped.

Textual sources

Novels may evoke a sense of place – a feeling that the reader knows what it is like to 'be there'. Some places have become so strongly associated with particular authors and stories that they are now promoted or advertised as such. Think for example of Thomas Hardy's Dorset or the area of West Yorkshire and East Lancashire known as Brontë country. While such associations can play a positive role in representation of place, the reverse can also be true and the role of the popular media (newspapers, television, film and the internet) can be instrumental in creating 'place prejudice'. An example of this would be the city of Liverpool, one of a number of (mainly northern) British cities that has historically suffered from a negative portrayal in the British media. Unemployment, economic deprivation, gangs, rioting and drugs problems were all characteristic of 'Scouse' depictions in the press in the late twentieth century, and the city authorities adopted an aggressive re-branding of the city to distance it from this poor media image. In 2008 the city was awarded the status of European Capital of Culture, reflecting Liverpool's more positive architectural, musical and sporting heritage.

Poetry

Poetry has long been used to describe and evoke a sense of place. There are many famous poets associated with particular places. William Wordsworth is linked to the Lake District, Seamus Heaney has written extensively about his Irish roots and William Blake famously described the poverty and despair of industrial London. Poets may refer to specific places in personal and responsive ways but they also enable the reader to sense and imagine what it is like to be in that place.

Daljit Nagra is a contemporary British poet who was born in England to parents who are traditional Sikh Punjabis. He has written extensively about the Britain where Indians came and settled. His poetry enables the reader to be immersed in the Sikh community and experience Sikh Britain from the 'inside'. Nagra has sought to explore the thoughts, feelings and cultural attitudes of first, second and third generation descendants towards their own community, other ethnic communities and the indigenous white population. His poem, *Our Town with the Whole of India!* describes the district of Southall in London which migrants had turned into a 'Little India' in the 1970s.

> Our town in England with the whole of India
> sundering
> out of its temples, mandirs and mosques for the
> customised
> streets. Our parade, clad in cloak-orange with
> banners
> and tridents, chanting from station to station for
> Vasaikhi
> over Easter. Our full-moon madness for Eidh
> with free
> pavement tandooris and legless dancing to be
> boostered
> cars. Our Guy Fawkes' Diwali – a kingdom of
> rockets
> for the Odysseus-trials of Rama who arrowed
> the jungle
> foe to re-palace the Penelope-faith of his Sita.
> Our Sunrise Radio with its lip sync of Bollywood
> lovers
> pumping through the rows of emporium cubby
> holes
> whilst bhangra beats slam where the hagglers roar
> at the pulled-up back-of-the-lorry cut price stalls.
> Sitar shimmerings drip down the furbishly columned
> gold store. Askance is the peaceful Pizza Hut ...
> A Somali cab joint, been there for ever, with smiley
> guitar licks where reggae played before Caribbeans
> disappeared, where years before Teddy Boys jived.

Daljit Nagra, 2014, Extract from *Our Town with the Whole of India!*

The poem describes the changes which have occurred in the community over the years, celebrating the multicultural nature of the area while at the same time highlighting potential threats accompanying it.

Skills focus

To illustrate how the same place can be represented in different ways, compare Blake's poem below with the more romantic, idealised images of London described by Wordsworth. Consider the imagery, the choice of vocabulary and the overall mood of each poem.

London

I wander thro' each charter'd street
Near where the charter'd Thames does flow,
And mark in every face I meet
Marks of weakness, marks of woe.
In every cry of every Man,
In every Infant's cry of fear,
In every voice, in every ban,
The mind-forg'd manacles I hear:
How the Chimney-sweeper's cry
Every black'ning Church appals,
And the hapless Soldier's sigh
Runs in blood down Palace walls;
But most thro' midnights streets I hear
How the youthful Harlot's curse
Blast the new-born Infant's tear,
And blights with plagues the Marriage hearse.

William Blake, 1794

Composed upon Westminster Bridge, September 3, 1802

Earth has not anything to show more fair:
Dull would be of soul who could pass by
A sight so touching in its majesty;
This City now doth, like a garment, wear
The beauty of the morning; silent, bare,
Ships, towers, domes, theatres, and temples lie
Open unto the fields, and to the sky;
All bright and glittering in the smokeless air.
Never did sun more beautifully steep
In his first splendour, valley, rock, or hill;
Ne'er saw I, never felt, a clam so deep!
The river glideth at his own sweet will;
Dear God! The very houses seem asleep;
And all that mighty heart is lying still!

William Wordsworth, 1802

Music

Music can help to evoke a sense of place. Different types of music may be associated with geographical areas, such as reggae with Jamaica. Song lyrics may also help to portray particular places. In 2010, Newport rapper Alex Warren and singer Terema Wainwright became an internet sensation after producing a parody of the Jay-Z and Alicia Keys song about New York. The spoof version of **Empire State of Mind** received almost 200,000 hits within just two days of being placed on YouTube. Like the music video for the original track, the video is shot in black and white with the singers referring to famous landmarks and cultural clichés. However, rather than the Manhattan skyline as a backdrop, the local duo celebrate the sights of the Welsh city. Rather than a grand piano in Times Square, Wainwright plays a battered synthesizer on the riverfront and sings in a thick Welsh accent: 'In Newport / concrete jumble nothing in order / not far from the border / When you're in Newport / Chips, cheese, curry makes you feel brand new / washed down with a Special Brew / Repeat the word Newport, Newport, Newport'.

Similar parodies have been added to YouTube for a number of other UK places.

Television and film

Places are a vital backdrop to most dramas. We associate certain places with different types of stories, such as derelict houses in horror films or space ships in science fiction. However, TV shows and films also play a significant role in representing place and this can be both positive and negative. The 2015 television dramatisation of the Poldark novels had a positive impact on the Cornish tourist trade as viewers, inspired by the shots of the dramatic coastline and beautiful beaches, flocked to the county to soak up the atmosphere. Hits on Visit Cornwall's website soared by 65 per cent after the first episode alone and the property website Rightmove reported that house-hunting enquiries more than doubled. It is anticipated that the 'Poldark effect' will be felt well into the 2020s.

On a global scale, *The Lord of the Rings* film trilogy has become synonymous with New Zealand and tourist numbers are 50 per cent higher than they were prior to the release of the first film in 2001. In 2012, *Tourism New Zealand* reported that *Lord of the Rings* contributed NZD$33 million (currently £15 million) a year to the New Zealand tourist economy. The government of New Zealand recognises the importance of the films and have capitalised on the positive representation of their country. The New Zealand postal service issued stamps with the faces of characters from the movie and Air New Zealand, the airline of Middle Earth, has two 777 airplanes decorated in a Tolkien-esque theme.

These two examples have shown how TV and film can represent place in a positive way. The reverse can also be true. Many crime dramas are located in urban areas, but not all crime occurs in cities. Equally, the sites chosen for filming can portray the same place in very different lights. Compare, for example, the representation of London in *EastEnders* (BBC, 1985–present) with the more glamourous skylines and buildings of *The Apprentice* (BBC, 2005–present) or the settings for crime dramas such as *Sherlock* (BBC, 2011–2014) or *Luther* (BBC, 2010–2013).

When researching media representations of your chosen place studies, you could have a look at the British Pathé website and YouTube channel. British Pathé is a multimedia resource with film, documentary and newsreel archive material. Over the course of the twentieth century, British Pathé reported on everything 'from armed conflicts and political crises to the hobbies and eccentric lives of ordinary British people.' The resulting archive is a useful audio-visual source which can be used to document changes in place and population over a period of time. The archive is available to view online for free.

Art

Art has long been used to represent place and this is most famously seen in the landscape paintings of the eighteenth and nineteenth century. Painters became synonymous with geographical places. Gainsborough and Constable (Figure 8.27) are known for their landscapes of Suffolk, now sometimes referred to as 'Constable Country' while French artist Paul Cezanne produced numerous paintings of Provence in southern France. The common criticism of such paintings was that they were pastoral fantasies giving the impression of a rural idyll, which did not exist for the majority of people living in the countryside at that time. These paintings reflected a romantic vision which still shapes many people's mental images of the countryside and is perpetuated through tourist brochures, chocolate boxes and jigsaw puzzles. Such constructs of rural places are powerful because they shape views on what the countryside is actually like and what it should be like.

Figure 8.27 Constable's painting entitled *The Hay Wain* is based on a site in Suffolk, near Flatford on the River Stour. The hay wain, a type of horse-drawn cart, stands in the water in the foreground. The location of the painting at Flatford Mill is now a popular tourist location run by the National Trust

Paintings may be considered less reliable than photographs because there is more scope for individual interpretation and selection. However, they can also show a deeper understanding of place because they allow the painter to show more of the character of what is there.

An artistic representation of an urban place can be seen in the work of L.S. Lowry, famous for his matchstick human figures and depictions of life in a northern mill town.

Figure 8.28 The iconic Lowry Centre at Salford Quays opened in 2000 on the site of the old docks. It now sits at the heart of Media City UK and has had a major influence on the cultural landscape of the northwest

Lowry became interested in the industrial landscape after he moved to Salford with his family. Although he made many drawings there, his mill scenes (a combination of the real and the imagined) have remained his most famous. Lowry's legacy and association with Salford have continued into the twenty-first century and been used for place regeneration. The Lowry building has been a key part of the redevelopment of the Salford Quays in Greater Manchester (Figure 8.28).

Skills focus

Where art has been used to depict place, it is important to ask a number of questions and view it as a cultural product:

- What and who are shown in the picture, and why?
- What and who are not shown, and why?
- Why did the artist paint this scene and who commissioned it?

We have seen then that art has traditionally represented place but it can also make a place distinctive and be used in the regeneration of an area. A second example is the Kelpies sculpture near Falkirk in Scotland (Figure 8.29). The Kelpies are the landmark feature of The Helix Environmental Regeneration Scheme on the Forth and Clyde Canal. They are modelled on heavy horses and represent the industrial history of Scotland and the Falkirk/Grangemouth area. Heavy horses would once have been the powerhouse of the area, working in the foundries, the fields, farms and the canal itself. The Kelpies sculpture has been used to regenerate place here but their national and international prominence has also served to develop a sense of pride and ownership in the local place. Indeed, they are the front cover image of a 2014 book entitled *Scotland: The Best 100 Places*.

Figure 8.29 The Kelpies sculpture

Graffiti

Graffiti is writing or drawing that has been put illicitly on a wall or other surface, often in a public place. It has traditionally been associated with youth cultures claiming ownership of a place but the famous UK graffiti artist Banksy argues that the importance of graffiti is also to give a voice to people who aren't normally heard in the mainstream (Figure 8.30). Many consider graffiti as a type of vandalism and authorities are keen to remove it from public areas. Increasingly however, graffiti is being accepted into mainstream culture and

art galleries may now stock graffiti images. In these places, graffiti is seen not as being 'out of place' but as something which can be bought and sold. It is also being used as a type of street art in the regeneration of places.

Figure 8.30 The streets of London have been a canvas for guerrilla artist Banksy's work for more than a decade. Some of the art makes a hard-hitting political point, while other works are intended to be funny

Dismaland, Weston-Super-Mare, Somerset

An unusual type of art installation brought thousands of visitors and greater place recognition to the seaside resort of Weston-Super-Mare in the summer of 2015. Based in a derelict seafront lido and billed as a 'bemusement park', Banksy's Dismaland featured artworks on themes including the apocalypse, anti-consumerism and celebrity culture. Unlike other British seaside resorts like Margate in Kent, which has built up a more positive representation following the opening of the Turner art gallery, Weston-Super-Mare relies on its beaches and traditional seaside attractions for visitors. Banksy's work brought significant benefits to a resort which has experienced significant decline in the last few decades. It is estimated that an additional 150,000 people visited the town as a result of the attraction and they added £20 million to the local economy. More importantly, local tourist chiefs were delighted that it helped to put the place of Weston-Super-Mare more firmly on the map.

Place and architecture

As well as designing buildings, architects are responsible for planning places and this planning is seen as increasingly important in terms of creating sustainable and healthy places. Architecture can also play a pivotal role in the redevelopment and rebranding of a place. Some redevelopment schemes utilise the existing style or heritage of buildings in an area; others involve more radical change as part of a rebranding process. Compare, for example, the redevelopment of the SouthGate shopping centre in Bath to the redevelopment of the Bull Ring in central Birmingham. The former mimics the city's original and popular Georgian architecture with a Bath-stone façade, while the Bull Ring was a much more radical redevelopment incorporating the construction of the iconic Selfridges department store with a façade of 15,000 aluminium discs mounted onto a blue background. The latter has been seen as one of several architectural landmarks, including the city library, which have made a major contribution to the regeneration of central Birmingham.

Redevelopment of place comes at a cost to some. The process of gentrification has been strongly promoted by local government in many British cities as a method of economic regeneration. Gentrification is the improvement of housing in an area that was formerly poor and run-down. It is mainly carried out by middle-class residents or newcomers who purchase the properties at cheaper prices and make repairs and improvements which increase the housing value. The nature of the whole area may improve and this leads to greater wealth in this particular place. Unfortunately, not everyone benefits and the original poorer residents may find themselves 'displaced' as they can no longer afford to live in this area. A similar process has occurred in some British coastal resorts where second-home owners have effectively priced less affluent locals out of the housing market.

Digital or 'augmented' place

The rise in the use of digital technology such as smart phones has led to much discussion about the notion of digital place. Parallel interactive places such as Second Life and Runescape have existed for sometime but these are seen as virtual places rather than real places. The development of GPS-capable mobile devices has meant that real places have become 'layered' and 'augmented' by technological advances and applications and this can have a huge impact on our sense of place. Smartphones know where we are and are linked to data that knows where other people or things are too. They can provide information and reviews about place in seconds.

GIS and software developers have been engaging in place for some time. Supermarkets employ GIS to map shopping habits and where customers live. Police forces want to know about the links between crime and place and politicians want to know about place so that they can focus their time and money at voters. GIS sources are increasing in number and becoming more readily available as free online resources. They could be used in place studies to illustrate changing demographic and cultural characteristics and economic change and social inequalities.

8.5 Place studies

> The single story creates stereotypes and the problem with stereotypes is not that they are untrue but that they are incomplete, they make one story become the only story ... It is impossible to engage properly with a place or person without engaging with all of the stories of that place or person.
>
> *Chimamanda Adichie*, TED Talk *The Danger of a Single Story*, 2008

The final section of the Changing Places topic asks you to explore the developing character of a place local to your home or study area and of a contrasting and distant place. A place for study should be about the size that you could comfortably walk around in a few hours. It may be that your local place is the place where you live or go to school – a village, small town, community or area of a city. It can be either urban or rural. The contrasting and distant place may be in the same country or a different country but it must show significant contrast in terms of economic development and/or population density and/or cultural background and/or systems of political and economic organisation. In some cases, the geographical distance between your local and distant place may be relatively small. It is also important to remember that your place studies must incorporate the knowledge acquired from the rest of the specification and thereby further enhance your understanding of the way your life and that of others are affected by continuity and change in these places.

Before choosing your place studies, it is worth looking at the availability of data and the different sources of information before committing to one distant place. For places abroad, charities and non-governmental organisations are useful sources of information, as are government websites and international or global institutions such as the World Bank.

Context for my place studies

Using the examples of central Torquay in South Devon and the area of Brick Lane and Spitalfields in East London, this section will suggest resources and activities which will enable you to build up your own two place studies. I have chosen to present Torquay as my local place study because it is the town in which I work and a place which I have got to know well through fieldwork and recreational activities. I have less 'insider' knowledge about my chosen distant place of Brick Lane and Spitalfields in East London, but it is a place I have visited and one which I know contrasts significantly with Torquay in terms of both economic development and cultural background.

Both Torquay and Brick Lane have undergone considerable change in the last few decades. The seaside resort of Torquay has experienced a decline in tourist numbers since the 1970s resulting in subsequent social and economic problems and a less favourable place identity. The Brick Lane and Spitalfields area has undergone social and economic changes of a different nature. Already contrasting with Torquay in terms of its cultural background (Brick Lane and Spitalfields is home to a large Bangladeshi diaspora), this area of East London is undergoing social and economic change as a result of gentrification and redevelopment projects.

1. Investigating location and locale

Maps are an important tool for any investigation of place. They can show the location of a place in relation to others and, depending on the type of map chosen, can also display physical and human features of the local area. For a rural place, you could start with an Ordnance Survey map, 1:25,000 scale. Look at the physical geography of the area: relief, height of the land, aspect and drainage. How have humans impacted the area? What are the main land uses? You could see how the area has changed by looking at older OS maps. Look for changes in the size of the place, types of land use and infrastructure.

Street maps are more suited to urban areas. Goad maps are detailed street maps which show individual buildings and their uses. These can be updated or annotated as part of your fieldwork. Google Maps also zoom in to street level. Along with Google Earth, these provide geographical coverage of places all over the world. They also provide aerial and satellite images. A useful exercise is to compare the image of your local place from the OS map to that from Google Earth. What characteristics are shown? Which are not? How do these images compare with the place as you know it? For your distant place, Google Street View is a useful tool for seeing what the place is like. The horizontal and vertical street-level images enable you to see land use and building type, which you might not otherwise get to see.

Maps can be used to highlight economic and demographic changes in an area as well as changes in lived experience. In London, for example, Charles Booth's c.1890 Poverty Map is viewed as an important step in the development of geodemographics (characterising people based on where they live). Over a period of several years, Booth and his researchers knocked on doors all around London, interviewing and characterising the people they met into seven categories, such as 'Well-to-do' or 'Vicious, semi-criminal'. The houses were then coloured on the map according to their category. The resulting map shows the characteristics of large parts of London at the turn of the twentieth century (Figure 8.31).

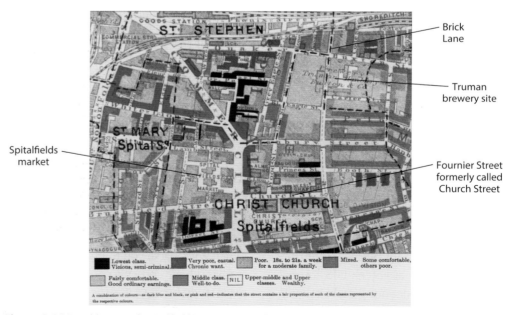

Figure 8.31 Booth's map of Spitalfields area c.1890. The original notebooks from Charles Booth's survey of life and labour in London (1886–1903) contain interviews with Londoners from all walks of life, eyewitness descriptions of the city, street by street, and the raw data that was later used to compile statistical reports and the famous maps of London poverty

Charles Booth's poverty maps for Brick Lane indicate that the area had mixed economic fortunes at that time. Some parts were classed as poor or very poor but equally there were areas of greater affluence too. This contrasted with large areas of nearby Whitechapel and the East End where poverty was much more widespread.

Literary sources such as books, atlases and newspapers can be used alongside maps to provide historical information about places. Your local library will be a useful source of information and there may be a Local History society in your area. Investigate how your chosen place has changed over time and consider the role of past connections on current characteristics.

The **internet** provides a wealth of information about different places. Official government websites provide geographical information such as the census data or local health statistics. An internet search on your chosen places will also result in other statistics, facts, opinions, stories, etc. Remember that this information is created for many purposes (to inform, to persuade, to sell, to present a viewpoint or to create or change an attitude or belief) and unlike more traditional media such as books and magazines, no one has to approve the content before it is made public.

Local place study: Torquay

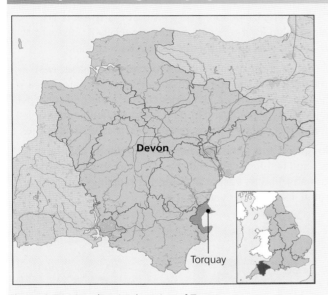

Figure 8.32 Map showing location of Torquay

Torquay is a seaside town in Devon, part of the unitary authority area of Torbay. It lies 18 miles south of Exeter and 28 miles east-north-east of Plymouth. It is situated on the north of Tor Bay, adjoining the neighbouring town of Paignton on the west of the bay and across from the fishing port of Brixham. Some of the material presented in this place study applies to the area of Torbay, but the main focus is the central tourist area of Torquay.

Constrasting place study: Brick Lane

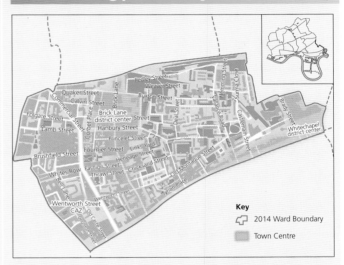

Figure 8.33 Map showing the location of Spitalfields and Banglatown within the borough of Tower Hamlets, London

Brick Lane is a street in the London Borough of Tower Hamlets. It runs from Swanfield Street in the northern part of Bethnal Green, passes through Spitalfields and is linked to Whitechapel High Street to the south by the short stretch of Osborn Street. Today, it is the heart of the city's Bangladeshi-Sylheti community and is known to some as Banglatown. This place study will focus on the southern half of Brick Lane and its immediate surroundings. Some of the data applies to the wider area known as Spitalfields and Banglatown.

Torquay

During the nineteenth century, Torquay grew from being a small fishing village of 800 people into a fashionable seaside resort. John Bartholomew's *Gazetteer of the British Isles,* 1887, described Torquay as '... among the first of English watering places. Its mild and equal climate, its excellent water supply and drainage system, and its bright and pleasant surroundings make it a great resort of invalids and convalescents in winter and spring. It has a museum, public gardens, baths and all the attractions of a great watering place.'

The *Western Morning News* described it in 1864 as 'the most opulent, the handsomest and the most fashionable watering place in the British Isles.' A slight local bias there perhaps but the resort attracted wealthy families, famous writers of the time and royalty.

The arrival of the railway in 1848 made travel more affordable and widespread and Torquay became more popular as a result. The term 'English Riviera' was adopted in the Victorian period as visitors likened the climate and topography of Torquay to the French Riviera resorts. This term has been an important strap line in the marketing of the area ever since. Torquay continued to expand as a tourist resort throughout the nineteenth and twentieth centuries, but like so many British seaside resorts, the 1970s saw the beginning of a decline in tourist numbers.

Figure 8.34

Brick Lane

Brick Lane was originally called Whitechapel Lane but it is thought that it was renamed because local earth was used by brick and tile manufacturers in the fifteenth century. By the seventeenth century, the street had also become a popular location for breweries.

In the seventeenth century, the area saw an influx of French Huguenots who had been driven out of France. The street and the surrounding area became well known for its weaving and tailoring. Like much of the East End, this area was also a haven for immigrants moving into London to escape persecution abroad or looking for a better life. During the nineteenth and twentieth centuries, it was best known for its Irish and Jewish population.

More recently, the area has become known as Banglatown as this has become a popular place for immigrants from Bangladesh, particularly Bengalis from the Sylheti region. The weekly Bangladeshi paper, *The Sylheter Dak* has an office there. Guidebooks suggest that the street is THE place to go for a curry in London, especially if you want to try traditional and authentic Bangladeshi cooking.

Evidence of the people and communities that have given the Brick Lane and Spitalfields area its unique character can be seen all around – a Huguenot church, a Methodist chapel, a Jewish synagogue and Muslim mosque stand among traditional and new shops, restaurants, markets and homes.

In recent times, the area has also developed a vibrant art and fashion scene and the Old Truman Brewery has transformed its vacant and derelict buildings into office, retail, leisure and event spaces.

Figure 8.35

2. Demographic characteristics

A key source of geographic data comes from the census, which is conducted every ten years in the UK. Other socio-economic data on factors such as employment, crime and health is generated on a more regular basis. The Office for National Statistics is the UK's largest independent producer of official statistics and its website is a useful starting block for finding out about the social and economic characteristics of your local and distant place. Go to www.neighbourhood.statistics.gov.uk. The Local Government Association also provides up-to-date published data about geographic areas and LG Inform Plus can be used to generate reports on places at a census ward level. Go to http://about.esd.org.uk/. At an even smaller scale, www.checkmyarea.com/ enables you to check how residents of a postcode have been profiled, and what their behaviours might be like.

For geospatial data from the 2011 census, go to http://maps.cdrc.ac.uk/. Type in the postcode for the area you want to investigate in more detail.

Torquay

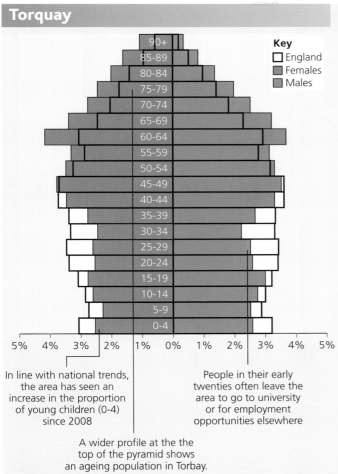

In line with national trends, the area has seen an increase in the proportion of young children (0-4) since 2008

People in their early twenties often leave the area to go to university or for employment opportunities elsewhere

A wider profile at the the top of the pyramid shows an ageing population in Torbay.

Figure 8.36 Population pyramid for Torbay, 2011. The solid bars represent Torbay's resident population and the hollow bars represent the population structure for England.

According to the 2011 census, Torquay is home to approximately 65,000 people. A high proportion of these, 21 per cent, are over the age of 65. This is significantly higher than the national figure of 16.3 per cent. Torbay is a popular retirement location, which may explain this trend.

There is very little ethnic diversity in Torquay. The 2011 census revealed that 97 per cent of the population of Torquay considered themselves white British.

Historically, the tourist-based economy attracted few migrants to Torbay. There has been a greater in-migration of Eastern Europeans in recent years but the area is still characterised by limited ethnic variation.

Brick Lane

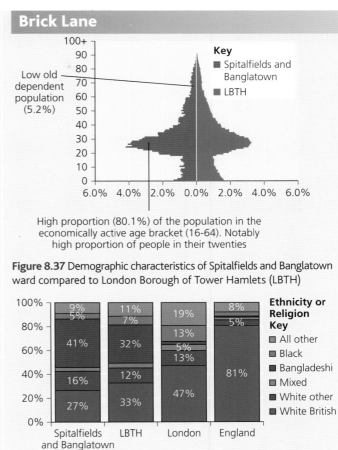

Low old dependent population (5.2%)

High proportion (80.1%) of the population in the economically active age bracket (16-64). Notably high proportion of people in their twenties

Figure 8.37 Demographic characteristics of Spitalfields and Banglatown ward compared to London Borough of Tower Hamlets (LBTH)

Figure 8.38 Compound bar graphs showing ethnicity and religion in Spitalfields and Banglatown from 2011 census.

Like many areas of inner London, Spitalfields and Banglatown ward has a notably high proportion of its population in the 20–40-year-old bracket. Access to a range of jobs in London attracts economically active people including immigrants. The area is also perceived as a trendy youthful hub.

Residents of Bangladeshi origin accounted for 41 per cent of the population of Spitalfields and Banglatown, which is higher than the borough average. There is a lower proportion of residents who are white British than the borough average. In terms of religion, Muslims make up the highest proportion of the population at 41.5 per cent. This area has a rich ethnic and religious mix as a result of a series of migratory moves by different groups since the nineteenth century.

3. Economic characteristics

To build up an economic profile of your two places, you could research the following:

- levels of employment and unemployment
- the balance between different economic sectors: primary, secondary, tertiary
- Gross Disposable Household Income (GDHI) estimates
- house prices (property websites and the Land Registry)
- access to services for different economic groups (Go to www.phoutcomes.info/).

The Index of Multiple Deprivation (IMD)

The Index of Multiple Deprivation is a UK-government qualitative study measuring deprivation at small-area level across England. It can be used to show economic inequality between and within different places. The English Indices of Deprivation 2015 are based on 37 separate indicators organised across seven distinct domains of deprivation, which are combined using appropriate weights. The seven different dimensions of deprivation are:

- Income
- Employment
- Health Deprivation and Disability
- Education Skills and Training
- Crime
- Barriers to Housing and Services
- Living Environment.

Note that these statistics are a measure of relative deprivation, not affluence. Not every person in a highly deprived area will themselves be deprived. Likewise, there will be some deprived people living in the least deprived areas. The IMD data for 2015 has been mapped at http://maps.cdrc.ac.uk/. Select the map showing the Index of Multiple Deprivation data for 2015.

Torquay

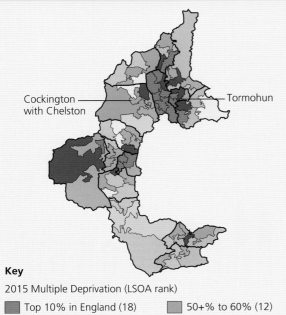

Cockington with Chelston

Tormohun

Key

2015 Multiple Deprivation (LSOA rank)

Top 10% in England (18)	50+% to 60% (12)
10+% to 20% (15)	60+% to 70% (6)
20+% to 30% (15)	70+% to 80% (6)
30+% to 40% (10)	80+% to 90% (3)
40+% to 50% (10)	90+% to 100% (0)
— LSOA boundary	— Ward boundary

Figure 8.39 Map showing the Index of Multiple Deprivation data for Torbay, 2015

In 2015, Torbay was within the top 20 per cent most deprived local authority areas in England for the rank of average score and the rank of local concentration. A number of areas within Torquay fell into the top 10 per cent most deprived areas, as shown by Figure 8.39.

Although the IMD data highlights significant areas of deprivation in Torquay, it is clear that, like most urban areas, it also has areas of great affluence.

The data in Table 8.2 (page 368) has been collected from a number of sources and highlights the significant income inequality found in Torquay. In doing so, it also emphasises the need to look closely for variation *within* areas as well as between areas.

Strategies such as the redevelopment of Torquay harbour and the designation of Torbay as an urban GeoPark have had some success in reviving the town's fortunes in recent years. Future research could focus on the social and economic changes brought about by strategies like these, as well as looking at the impact of the new South Devon link road which has brought greater accessibility to Torbay.

Brick Lane

Key
Index of Multiple Deprivaton 2015
The indices of deprivation for England in 2015 - showing the overall rank, split into deciles.

■ Most deprived decile	■ 2nd ■ 3rd ■ 4th □ 5th □ 6th □ 7th ■ 8th
■ 9th ■ Least deprived decile	□ Data missing □ Data not available
AREA INFORMATION	Most deprived decile

Figure 8.40 Map showing the Index Of Multiple Deprivation data for Tower Hamlets 2015.

Deprivation is widespread in Tower Hamlets and the borough is one of the most deprived areas in the country. However, the 2015 Index of Multiple Deprivation found fewer areas of Tower Hamlets to be among the most deprived 10 per cent in England than in 2010. Gentrification accounted for some of this trend as middle-class families moved into traditionally poorer parts of London while the poor moved to boroughs in the outskirts where housing costs and rents are lower. Spitalfields and Banglatown was ranked as the most deprived Lower Super Output Area (LSOA) in all of London in 2010 but there has been relative improvement since then. In 2015, it was ranked as being in the 20 per cent most deprived neighbourhoods in the country. Figure 8.40 shows the considerable variation even within this area.

Table 8.2 Statistical comparison of least and most deprived areas in Torquay

Measure	Least deprived small area (Cockington with Chelston)	Most deprived small area (Tormohun)
Life expectancy at birth (2011–2013)	88.7 years	74.6 years
Index of multiple deprivation score (2015)	5.9	67.4
Percentage income deprived (2015)	4.6%	38.8%
Percentage employment deprived (2015)	6.6%	35.0%
Jobseekers Allowance Claimant Rate (2011–2013)	1.1%	11.6%

Source: 2015 Indices of Multiple Deprivation, ONS, PCMD

Social characteristics and inequalities

The social aspects of the Index of Multiple Deprivation can be looked at in greater detail. Local educational data such as school performance can be found at www.education.gov.uk/schools/performance/ and crime data is available at www.police.uk. For health-related data, go to www.phoutcomes.info/ (see Figure 8.42 and Figure 8.43). This excellent website details the wider determinants of health including social and economic indicators. The data is available at a range of geographical scales and the website enables you to map the data and compare different areas.

Health profiles are also useful. These give a snapshot of health for each local authority in England and are published annually. Go to: www.apho.org.uk/.

Within Torquay, life expectancy at birth is not evenly distributed across the population. As Figure 8.41 shows, the gap between the most deprived and least deprived communities can be as much as 14 years, although this averages out to between 8 and 10 years when comparing the same gender group.

Figure 8.41 Torbay life expectancy at birth by gender and deprivation quintile 2009/11

The 2015 health summary for Torbay shows that there are a number of health indicators in which Torbay performs worse than the England average. Malignant melanoma can be linked to the higher number of sunny days per year and the nature of Torquay as a tourist resort, but many of the other indicators are linked to economic deprivation.

Compared with benchmark [Better] [Similar] [Worse] [Not compared]

Indicator	Period	England	Southwest region	Bath and northeast somerset	Bournemouth	Bristol	Cornwall	Devon	Dorset	Gloucestershire	Isles of Scilly	North somerset	Plymouth	Poole	Somerset	South Gloucestershire	Swindon	Torbay	Wiltshire
1.01i - Children in poverty (all dependent children under 20)	2012	18.6	14.4	11.4	17.9	23.1	16.3	12.2	11.7	13.0	0.0	13.4	20.2	14.8	13.3	10.8	15.1	21.2	10.6
1.01ii - Children in poverty (under 16s)	2012	19.2	15.1	12.0	18.4	23.6	16.9	12.7	12.3	13.8	0.0	14.1	20.9	15.3	14.1	11.4	15.9	22.1	11.2
1.02i - School readiness: The percentage of children achieving a good level of development at the end of reception	2013/14	60.4	62.4	62.5	67.4	58.4	59.0	67.8	67.5	57.1	*	69.9	58.3	60.4	61.4	72.3	60.6	61.3	58.7
1.02i - School readiness: The percentage of children with free school meal status achieving a good level of development at the end of reception	2013/14	44.8	43.5	33.0	50.7	43.8	40.8	50.0	44.9	35.4	*	46.2	45.6	46.7	40.8	52.1	44.2	48.1	35.7
1.02ii - School readiness: The percentage of year 1 pupils achieving the expected level in the phonics screening check	2013/14	74.2	74.4	74.0	73.5	73.7	72.0	78.6	75.7	74.9	*	81.1	74.1	71.3	73.2	74.3	72.7	75.2	70.9
1.02ii - School readiness: The percentage of year 1 pupils with free school meal status achieving the expected level in the phonics screening check	2013/14	61.3	58.5	54.7	60.4	60.6	57.6	63.9	55.1	56.0	*	62.1	65.2	56.5	54.6	61.0	54.5	63.6	48.0
1.03 - Pupil absence	2013/14	4.51	4.57	4.50	4.47	5.05	4.67	4.27	4.63	4.57	4.77	4.59	4.45	4.65	4.58	4.52	4.25	5.12	4.56
1.07 - First-time entrants to the youth justice system	2014	409	4.28	463	432	809	362	332	251	362	*	478	525	408	450	517	581	585	322
1.05 - 16–18-year-olds not in education, employment or training	2014	4.7	4.5	3.5*	5.9	6.3	4.2	4.2	4.1	4.5	0.0	3.1	6.2	6.1	4.4	3.0*	5.6	4.1	3.8*

Figure 8.42 Snapshot of public health outcome data for a range of places in Southwest England. (Data for Torbay has been highlighted)

Indicator	Period	England	London region	Hammersmith	Haringey	Harrow	Harvering	Hillingdon	Hounslow	Islington	Kensington and	Kingston upon Thames	Lambeth	Lewisham	Merton	Newham	Redbridge	Richmond upon Thames	Southwark	Sutton	Tower Hamlets	Waltham Forest	Wandsworth	Westminster
1.01i - Children in poverty (all dependent children under 20)	2012	18.6	23.5	26.0	26.9	17.0	18.5	19.6	21.2	34.5	21.7	12.1	29.0	27.3	15.8	27.8	19.3	8.8	28.4	14.7	39.0	24.9	19.5	31.3
1.01ii - Children in poverty (under 16s)	2012	19.2	23.7	25.6	26.8	17.0	19.6	20.1	21.5	34.4	20.9	12.3	29.0	27.6	15.9	27.2	19.3	8.8	28.6	15.3	37.9	25.0	19.2	30.7
1.02i - School readiness: The percentage of children achieving a good level of development at the end of reception	2013/14	60.4	62.2	60.8	61.3	61.3	65.5	52.5	58.3	57.8	56.7	64.9	55.9	75.3	59.9	65.1	63.8	64.2	65.6	59.6	55.0	63.0	63.7	57.9
1.02i - School readiness: The percentage of children with free school meal status achieving a good level of development at the end or reception	2013/14	44.8	52.3	50.7	52.2	47.9	49.0	39.4	49.1	48.7	43.7	43.9	46.7	68.1	44.4	59.9	50.1	36.1	55.0	40.4	50.7	57.6	52.5	52.3
1.02ii - School readiness: The percentage of year 1 pupils achieving the expected level in the phonics screening check	2013/14	74.2	77.4	81.0	73.8	81.5	76.5	76.6	79.9	73.7	79.9	78.9	78.8	79.8	75.8	80.4	73.4	82.5	77.0	78.6	75.5	73.1	80.3	79.5
1.02ii - School readiness: The percentage of year 1 pupils with free school meal status achieving the expected level in the phonics screening check	2013/14	61.3	68.6	72.9	66.5	71.5	62.1	67.0	71.2	65.7	75.1	66.2	72.4	70.4	62.7	76.4	61.2	67.0	70.4	63.2	74.3	65.6	70.0	76.8
1.03 - Pupil absence	2013/14	4.51	4.33	4.53	4.53	4.19	4.76	4.51	4.32	4.46	4.47	4.03	4.20	4.24	4.25	3.99	4.52	3.97	3.98	4.26	4.37	4.47	4.35	4.37
1.07 - First-time entrants to the youth justice system	2014	409	426	529	450	346	235	423	397	642	325	180	707	603	309	527	290	239	648	291	458	399	351	420
1.05 - 16–18-year-olds not in education, employment or training	2014	4.7	3.4	2.5	3.5*	1.5	4.0	2.4	3.2	5.2*	3.6*	3.9	2.2*	3.5*	4.3	4.3	3.3	4.3	2.0*	3.2	3.4	3.0*	2.9*	2.2*

Figure 8.43 Snapshot of public health outcome data for the borough of Tower Hamlets compared to other parts of London

The chart below shows how the health of people in this area compares with the rest of England. This area's result for each indicator is shown as a circle. The average rate for England is shown by the black line, which is always at the centre of the chart. The range of results for all local areas in England is shown as a grey bar. A red circle means that this area is significantly worse than England for that indicator; however, a green circle may still indicate an important public health problem.

- ● Significantly worse than England average
- ○ Not significantly different from England average
- ○ Significantly better than England average

Domain	Indicator	Local No per year	Local value	Eng value	Eng worst	England Range	Eng best
Our communities	1 Deprivation	24,740	18.7	20.4	83.8		0.0
	2 Children in poverty (under 16s)	4,945	22.1	19.2	37.9		5.8
	3 Statutory homelessness	56	0.9	2.3	12.5		0.0
	4 GCSE achieved (5A*-C inc. Eng & Maths)	840	56.6	56.8	35.4		79.9
	5 Violent crime (violence offences)	2,526	19.2	11.1	27.8		2.8
	6 Long-term unemployment	622	8.1	7.1	23.5		0.9
Children's and young people's health	7 Smoking status at time of delivery	219	16.8	12.0	27.5		1.9
	8 Breastfeeding initiation	n/a	-	73.9			
	9 Obese children (Year 6)	211	18.2	19.1	27.1		9.4
	10 Alcohol-specific hospital stays (under 18)	20.0	79.1	40.1	105.8		11.2
	11 Under 18 conceptions	65	29.5	24.3	44.0		7.6
Adults' health and lifestyle	12 Smoking prevalence	n/a	17.5	18.4	30.0		9.0
	13 Percentage of physically active adults	242	52.8	56.0	43.5		69.7
	14 Obese adults	n/a	24.0	23.0	35.2		11.2
	15 Excess weight in adults	227	66.8	63.8	75.9		45.9
Disease and poor health	16 Incidence of malignant melanoma	42.7	34.7	18.4	38.0		4.8
	17 Hospital stays for self-harm	316	259.2	203.2	682.7		60.9
	18 Hospital stays for alcohol related harm	1,184	858	645	1231		366
	19 Prevalence of opiate and/or crack use	814	10.2	8.4	25.0		1.4
	20 Recorded diabetes	7,831	6.5	6.2	9.0		3.4
	21 Incidence of TB	8.7	6.6	14.8	113.7		0.0
	22 New STI (exc Chlamydia aged under 25)	654	828	832	3269		172
	23 Hip fractures in people aged 65 and over	197	535	580	838		354
Life expectancy and causes of death	24 Excess winter deaths (three year)	107.8	20.2	17.4	34.3		3.9
	25 Life expectancy at birth (Male)	n/a	79.1	79.4	74.3		83.0
	26 Life expectancy at birth (Female)	n/a	82.8	83.1	80.0		86.4
	27 Infant mortality	6	3.8	4.0	7.6		1.1
	28 Smoking-related deaths	280	280.8	288.7	471.6		167.4
	29 Suicide rate	15	11.7	8.8			
	30 Under 75 mortality rate: cardiovascular	111	83.6	78.2	137.0		37.1
	31 Under 75 mortality rate: cancer	185	137.8	144.4	202.9		104.0
	32 Killed and seriously injured on roads	41	30.9	39.7	119.6		7.8

Figure 8.44 Health summary for Torbay, 2015

Representations of place

Particular weight must be given in your place study to the use of qualitative sources. These will vary significantly depending on the place you are investigating but may include novels, poetry, nature writing and travel writing. Use the representations of Torquay and Brick Lane as ideas for possible sources you could investigate.

Artistic representations

Early artistic images of Torquay focused on the attractive coastal scenery of the area and this has been used to great effect in marketing tourism within the area. Railway posters from the 1930s used idyllic images of sunny seaside holidays to inspire the public to travel to these destinations by train. More recently, the resorts of Torquay, Paignton and Brixham have been actively marketed as the English Riviera, evoking a sense of the Mediterranean, and the brand has used a strong graphic image of a palm tree The original English Riviera palm tree logo was created by award-winning designer John Gorham in 1982 and quickly became the identifying icon for the English Riviera. The icon has been used in a more recent representation by award-winning Devon artist Becky Bettesworth (Figure 8.45).

The area of Brick Lane and Spitalfields has developed a reputation for fashion and art, particularly street art

Figure 8.45 This recent representation of Torquay by award-winning Devon artist Becky Bettesworth features the icon of the Torquay palm

Source: www.beckybettesworth.co.uk

Figure 8.46 This doorway off Brick Lane shows the work of street artist Stik, depicting a burka-clad Stik figure holding hands with a non burka-wearing Stik figure. It is one of the few pieces of street art highlighting the cultural diversity of the area.

(Figure 8.46). The *Ravish London* website suggests that this area constitutes 'the Mecca of London street art; a de facto open-air gallery, a square mile of street art to compare with the financial square mile.' The nearby area of Shoreditch saw an influx of artists such as Tracy Emin and Damien Hirst in the 1990s and their success seemed to attract more artists to the area. Brick Lane and Spitalfields have also witnessed a huge increase in the number of galleries opening, helped financially by the gentrification of the area and increasing tourist numbers.

In 2011, local artist and cartographer Adam Dant unveiled his Map of Spitalfields Life (Figure 8.47). This map details key locations within the Spitalfields area today, but also the historic background of a variety of places. It includes characters who have made this area special, ranging from artist Tracey Emin to Fred, the chestnut seller. Dant's map provides a very strong sense of place.

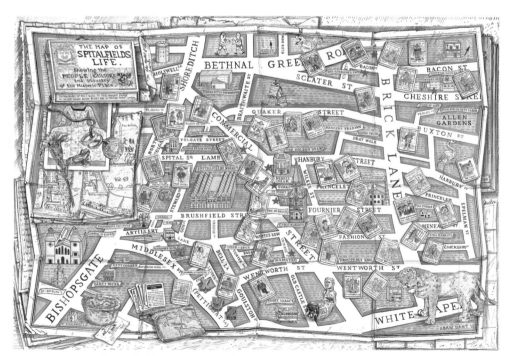

Figure 8.47 Map of Spitalfields Life produced by cartographer Adam Dant

Media representations

Media representations of Torquay have been mixed. For many people, Torquay will always be associated with the 1970s comedy series *Fawlty Towers* (BBC, 1975-1979) loosely based on an experience the actor John Cleese had while filming in Torquay and staying at the Gleneagles Hotel. The twelve episodes of *Fawlty Towers* were not actually filmed in Torquay. A more positive representation of Torquay is that of the crime writer Agatha Christie, who was born in Torquay. The connection with Agatha Christie has been promoted to bring a 'wider sense of glamour and style' to the Riviera brand and, according to the English Riviera website, the International Agatha Christie Festival held in Torquay, is 'fast becoming one of the UK's leading literary festivals.'

Public perceptions of Torquay are mixed. A survey carried out for the *Towards 2015* tourism report concluded that Torquay was perceived as having a 'classy past but an unclear present' while a range of positive and negative comments can be found on online discussion forums. Visitors to Torquay tend to be more complimentary. Typical posts from the *TripAdvisor* website refer favourably to the beautiful coastline, clean beaches and wide range of tourist attractions.

Media representations of Brick Lane are more numerous and more varied. The geographical area of Brick Lane and Spitalfields has numerous websites devoted entirely to it, detailing everything from the best curry houses to walking tours of the street art. The area has also become a popular tourist destination for visitors to London and features prominently in most guidebooks. The *Lonely Planet* writes: 'Full of noise, colour and life, Brick Lane is a vibrant mix of history and modernity, and a palimpsest of cultures. Today it is the centrepiece of a thriving Bengali community in an area nicknamed Banglatown. The southern part of the lane is one long procession of curry and balti houses intermingled with fabric shops and Indian supermarkets.' (Figure 8.48 and Figure 8.49)

In contrast, the website of the Old Truman Brewery located in the northern half of Brick Lane describes itself as being 'home to a hive of creative businesses as well as exclusively independent shops, galleries, markets, bars and restaurants.'

Figure 8.48 Brick Lane road sign written in both English and Bengali

Figure 8.49 View up Brick Lane

Other representations of Brick Lane come from those actually living or working there. Local poet Sally Flood has written extensively about Brick Lane. She was an embroidery machinist for most of her working life and much of her writing was done at the factory, from her seat with its window facing Brick Lane. Sally's memories of life in and around Brick Lane are documented on a number of websites and she has also published a collection of poems relating to this area.

Brick Lane is a mixture

of aromatic spices

Curries, onions and bad drains,

Pakistani restaurants

Jewish trimming shops

and betting shops,

Down-at-heel workers

and hopeful prostitutes,

cars and vans add to the pollution

with exhaust fumes.

Pavements and gutters

are littered with overspill

from dustbins and workshops.

This is where the immigrant

looks for fulfillment!

This is the breeding ground

for discontent,

Where the Meths drinker mixes

with the down-and-out,

Where Workers are exploited

and small-time drug peddlers

sell their dreams!

This is where the thug

dons the crown of King

and bullies thrive,

Where do-gooders

salve their consciences,

This is Brick Lane.

Sally Flood 'The Brick Lane I see', *Window on Brick Lane,* 1980

One of the more famous texts of recent times has been Monica Ali's novel *Brick Lane,* published in 2003 and subsequently made into a film (2007). The story follows the life of Nazneen, a Bangladeshi woman who moves to London at the age of 18 to marry. The fictional couple live in Tower Hamlets and the novel explores her new life in London and adaptations she makes as a Bangladeshi. The following extract gives an impression of the main features of Brick Lane and the changing prosperity of the area as it becomes a more fashionable neighbourhood.

> 'The bright green and red pendants that fluttered from the lamp posts advertised the Bangla colours and basmati rice. In the restaurant windows were clippings from newspapers and magazines with the name of the restaurant highlighted in yellow or pink. There were smart places with starched white tablecloths and multitudes of shining silver cutlery … 'You see,' said Chanu. 'All this money,

money everywhere. Ten years ago there was no money here.'

Monica Ali, Brick Lane, Page 256

Sources detailing the lived experience of place

Lived experience of place can come from a range of sources. In 2008, for example, the Department for Communities and Local Government conducted a place survey to gauge people's perceptions of their local area and the local services they received. In Torbay, the survey found that 82 per cent of Torbay residents were very or fairly satisfied with their local area. This compared favourably with the national average of 80 per cent at this time and was linked to the high quality of life afforded by the natural local environment and geography. Since 2011, the Personal Well-being Survey has monitored personal well-being and life satisfaction. Geographical patterns have emerged from this survey but the questions do not relate specifically to place.

You might want to produce your own questionnaires or carry out your own interviews with local people about their lived experience or perceptions of place. Ask your family and friends to take part or you could post a questionnaire online. Social media sources such as Twitter, Facebook, Instagram and Pinterest can provide lived experience of place as people increasingly communicate online about places they have visited. Make sure you acknowledge the strengths and weaknesses of each source.

Finally, blogs can be an excellent source of information about an area. In East London, for example, *The Gentle Author* writes a daily blog about the culture of the Spitalfields area and documents the lives of local people and places. 'How can I ever describe the exuberant richness and multiplicity of culture in this place to you?' @The Gentle Author, (www.Spitalfieldslife.com). The *Spitalfields Life* blog was born in 2009 with the words, 'In the midst of life I woke to find myself living in an old house beside Brick Lane in the East End of London.' Today, thousands of people from around the world read each daily posting.

Economic and social change

Statistics, maps, newspapers, poems and paintings are just some of the sources which can be used to show change in a place over time. Film or television footage may also be available for your local or distant place.

Figures 8.50a and 8.50b shows how photographs can be used to document change over time. The photographs from the official Spitalfields Market website show the early market as a collection of sheds and stalls. The second photograph shows the exterior of the market after a major regeneration programme. The website also documents the changes in text, noting the market's evolution from selling fruit and vegetables in the seventeenth century to being a major tourist attraction today, where visitors 'will find designers/makers and artists selling fashions, homewares and accessories or a treasure trove of vintage and antique clothing, furniture and other wondrous oddments.'

Figure 8.50a The early Spitalfields market

Figure 8.50b Spitalfields market, 2015

Photographs have also been used to document gentrification in East London. Figure 8.51 is a contemporary photograph showing improvements being made to housing on Fournier Street. Older photographs are available from the Tower Hamlets Local History Library and Archive and these have been used to create a series of before and after photographs for the same street. Search for these at http://spitalfieldslife.com/ which is regularly updated and provides fascinating information about the changes in this area.

Figure 8.51 Ongoing gentrification on Fournier Street. The design of the houses on the street demonstrates the importance of silk weaving to the lives of the original French Huguenot craftsmen in the 1700s. The second floor of the houses were known as weaver's garrets: they had huge attic windows with tiny attached frames designed to let in maximum light for the weaver to work by

Gentrification can bring great wealth to an area but can also lead to the displacement of original residents as house prices rise and shops and services change to suit the tastes of the evolving socio-economic composition of the area. National newspapers reported on a series of anti-gentrification protests in and around Spitalfields in September 2015 and a series of letters written to *The Guardian* newspaper highlights major concerns about the impacts of gentrification in East London at this time:

I was born in Whitechapel in 1974, to a family based in Shoreditch since the 1700s, and the stories and people, the spirit of that 1/4 square mile. Where historically class lines merged, cultures crossed, the rich brushed shoulders with Jack London. The hipsters have my approval for saving the architecture of the area, although I lay that more in the hands of the late 80s British art wave. But it's not the East End: inclusive, classless, tough, oozing character and fear in equal measure, pleasure and peril greeted every turn, never a yard from poverty or riches, sickness or health, life or death.

The current occupants have no verve or spirit, the survival with a smile has gone. It has always been driven by its association with the docks, the clay pits, the silk weavers; a hot bed of immigration and integration. But it no longer has a fraction of the character ... only the scenery remains for us Londoners to easier imagine the feeling of being on the edge of something dangerous or amazing at every turn. Takes a bit more than a beard for that.

Another,

I'm getting the impression that the 'them and us' mentality is coming from both sides. I live in Hackney with flatmates. None of us are born-and-bred Londoners, but have been here for some 3–8 years. They are British. I'm just a foreigner here for a while. None of us is a 'hipster' either ... I live in East London because I like the diversity. Not sure where else I'd live, because whenever I go west, it seems so white and it's all coffee chains.

The Guardian website, 24 March 2015

Careful evaluation of sources like this is required but these comments do highlight some of the key issues concerning in-migration and gentrification in this area at this time. In addition, they serve as another record of (some) people's lived experience of place.

Finally, we return to the poetry of long-time East London resident Sally Flood. We have already experienced a sense of place in her depiction of Brick Lane. She provides a similar emotional response to place in her writing about the changes occurring in her local landscape as a result of large-scale redevelopment. Many residents welcome such redevelopment schemes as they can improve the physical and economic landscape of a place. Others lament the loss of community and argue that many developments fail to address the capital's housing shortage or help local businesses.

This small street is fast receding

Growing smaller, day by day

Over years I watch it crumble

Watch the houses swept away,

Streets that share my childhood moments

Saw the war years, heard the bombs

Proudly held itself together

Now prepares itself to die.

An extract from 'Mount Terrace', *Tales by Eastenders,* 2015

The sources outlined here are by no means comprehensive but do enable an holistic account of these two places. Take care to evaluate all the sources you use and go through the checklist below to ensure you have used as many different sources of information as possible. Finally, do not forget your own lived experience of place. To be inside a place is to belong to it and identify with it more strongly. Use this to your advantage.

Table 8.3 Data sources

Data source	Examples
Statistics	Census data, employment data
Maps	Ordnance Survey, Google Maps
Geo-located data	Interactive transport app
Geospatial data	Index of Multiple Deprivation, Public Health Determinants of Health
Photographs	Photographs showing change over time
Text, from varied media	Poetry, novels, social media sources including blogs
Audio-visual media	Film archives
Artistic representations	Paintings, sculpture
Oral sources	Interviews, reminiscences, songs

Further reading

The first three sources are a particularly good starting place for the study of place:

Anderson, J. (2015) *Understanding Cultural Geography: Places and Traces* (Routledge)

Cresswell, T. (2014) *Place: An Introduction* (Wiky-Blackwell)

Phillips, R. (2016) 'Changing Place; Changing Places' A-level subject content overview from http://www.rgs.org/

Other useful texts include:

Cloke, P. Crang, P. and Goodwin, M. (2014) *Introducing Human Geographies*, Third edition (Routledge)

Horton, J. and Kraftl, P. (2014) *Cultural Geographies: An Introduction* (Routledge)

Jackson, P. (1989) *Maps of Meaning* (Routledge)

Kunstler, J. (1995) *The Geography of Nowhere* (Simon r Schuster)

Massey, D. (1994) *Space, Place and Gender* (Polity)

Rawling, E. (2011) 'Reading and writing place: a role for geographical education in the twenty-first century' in Butt (Ed.) *Geography, Education and the Future* (Bloomsbury)

Relph E. (2008) *Place and Placelessness*, SAGE

Tuan, Yi-Fu (2001) *Space and Place* (University of Minnesota Press)

Fieldwork opportunities

- Investigate how different places are experienced by different people by carrying out a walking tour of a particular place with fellow students. Record your individual feelings and observations as you walk through different areas. Back in class, discuss the reasons why individual students experience places in different ways.

- Geographical areas which have experienced significant immigration or in-migration can be investigated in terms of their changing demographic and cultural characteristics. Follow the example of Doreen Massey and walk down your local high street observing and recording the different groups of people, the shops and buildings within your locality.

- Investigate whether your local place has become a 'clone town'? How has your local place evolved over time and is there evidence of globalisation?

- The ways in which different groups of people experience and perceive places differently can be investigated through the use of surveys and interviews. Examples include the ways in which people with disabilities experience places, but there are other social groups who will have particular perspectives on place.

- Explore the extent to which the views of local people match those of the local government and corporate bodies involved in marketing and regenerating places.

- Compare the ways in which different places are marketed and represented. Investigate the different strategies used to improve perceptions of place.

- Investigate the impacts of industrial decline on different places in the UK. How does the character of these places change? How does it affect peoples' lived experience of that place?

- There is a wealth of information available for different parts of London and the impacts of the 2012 Olympics could be a focus of fieldwork here. 'The East End of London is a world in itself,' wrote Charles Dickens in the nineteenth century. Little has changed today and the economic and cultural diversity of London lends itself well to further research.

Review questions

1 What is place and how do humans form attachments to place?

2 How and why do different groups of people experience and perceive places differently?

3 How is place represented and how does this affect our perception of place?

4 What are the advantages and disadvantages of quantitative and qualitative sources for investigating place?

5 What are the key characteristics of your local and distant place and how have these changed over time?

Question practice

A-Level questions

1. 'Experienced place' and 'media place' are two categories of place. Explain what is meant by the terms 'experienced place' and 'media place' and illustrate how they may be different. (4 marks)

Figure 8.52 This photograph of the market stalls of A. Morris and C. & A. Ivory Ltd in Old Spitalfields Market was taken on 23 November 1928 when Queen Mary visited the East End of London to open the new extension at Spitalfields Market

2. Figure 8.52 is a photograph taken of Old Spitalfields Market, London in 1928 and the extract below is taken from a website advertising Old Spitalfields Market in London in 2015. Using these two sources of evidence, evaluate the reliability of the two sources of evidence in showing the nature and extent of change at Old Spitalfields Market between 1928 and 2015. (6 marks)

Old Spitalfields Market in London

Located just five minutes' walk from Liverpool Street Station, Old Spitalfields Market is the perfect shopping destination. Open seven days per week, the impressive array of shops and stalls draws in shoppers from all over the Southeast.

Our main market days on Thursday, Friday, Saturday and Sunday offer shoppers something different each day.

The Thursday antiques market offers a breathtaking array of collectable vintage and antique gems. Friday, is the destination for clothes shoppers and art lovers alike.

On the first and third Friday of the month, a record fair joins the market, with an eclectic mix of different musical styles on offer, including rare and collectable vinyl.

Saturdays offer a themed market each week from affordable vintage, the finest of Old Spitalfields traders, designer-makers and many more.

Our Sunday stallholders sell a little bit of everything, offering a market that has something for all the family.

www.oldspitalfieldsmarket.com

3. Name a quantitative source that you have used in one of your place studies (your source could be census data, maps or any geospatial data). Assess the usefulness of that source in helping you understand the development of the human geography of that place. (6 marks)

4. With reference to a place you have studied, assess the extent to which the socio-economic and/or demographic and cultural character of that place has been shaped by shifting flows of people, resources, money and investment. (20 marks)

AS level questions

1. The following is a list of methods from which one can acquire a sense of a place. Which combination of methods consists totally of representations by others?

   1 Reading a newspaper article about a place
   2 Following a 'town trail'
   3 Researching census data of that place
   4 Studying a map of the place

   a) 1, 2 and 3
   b) 2, 3 and 4
   c) 1, 3 and 4
   d) 1, 2 and 4 (1 mark)

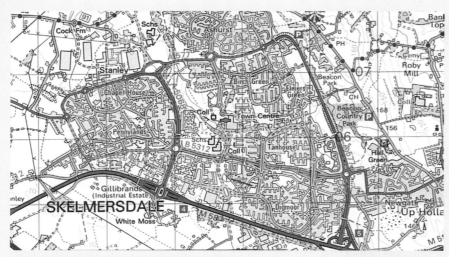

Figure 8.53a Current OS map of Skelmersdale in West Lancashire

Figure 8.53b OS map of Skelmersdale in West Lancashire in 1845

2. Which of the following statements is correct?

   a) Maps are a reliable source showing change in place over time because they are not subject to manipulation.

   b) If geographers want to really understand the character of a place they should study a range of sources of information, including census data.

   c) Qualitative data, such as interviews with people who have lived in a place for 50 or more years, is more reliable than using census data, which only tells you about that place at one moment in time.

   d) The census provides geographers with quantitative data, which is the only reliable and objective way for geographers to study the character of places.   (1 mark)

3. Name one place you have studied and outline how local or national government has tried to manage and manipulate perception of that place. (3 marks)

4. Figure 8.53a is from the current Ordnance survey 1:50,000 series and shows the town of Skelmersdale in West Lancashire. Figure 8.53b is taken from an Ordnance Survey map of the same area in 1845. The scales are slightly different.

   Using evidence from Figure 8.53a and Figure 8.53b analyse the main changes to the human and physical geography of the area that have occurred in the period shown.   (6 marks)

5. Describe how you used qualitative data to study change in your two contrasting places. To what extent was that data useful?   (9 marks)

6. In your study of two contrasting places, assess the extent to which **either** there has been a change in the demographic and cultural characteristics of those places **or** there have been economic changes over time that have affected those places.   (20 marks)

Chapter

9

Contemporary urban environments

Population levels are rising and nowhere is this more evident than in urban areas. In 2014, 54 per cent of the total global population were living in urban areas, up from 34 per cent in 1960. By 2050 it is expected that 66 per cent of the global population will be urban with close to 90 per cent of this increase concentrated in Asia and Africa. Cities can be efficient. It is easier to provide basic services such as water and sanitation to people living closer together. Likewise, access to health, education and other social and cultural services is much more readily available. However, as towns and cities expand, the cost of meeting basic needs increases and a greater strain is put on the environment and natural resources.

Urban areas are also experiencing the pressures of migration. Globalisation and the impacts of climate change have transformed agricultural practices and in low and middle-income countries increasing numbers of people are leaving rural areas to find employment in cities. Unemployment rates are high and the proportion of people living in slum areas has rocketed. Issues of social cohesion are evident as variations in wealth and ethnicity can sometimes lead to hostility.

The future survival of cities depends on sustainable growth and their ability to tackle the major issues such as provision of affordable housing, unemployment, pollution and waste disposal, transport and social inequality.

In this chapter you will study:
- global patterns of urbanisation and urban growth
- social, economic and environmental issues associated with urban growth
- the impact of the urban environment on local climate, weather and drainage
- features of sustainable urban growth.

9.1 Patterns of urbanisation since 1945

Urbanisation is defined as the process by which an increasing proportion of a country's population lives in towns and cities.

The urban population has grown rapidly from 746 million in 1950 to 3.9 billion in 2014. The most urbanised regions include Northern America (82 per cent living in urban areas in 2014), Latin America and the Caribbean (80 per cent) and Europe (73 per cent). In contrast, Africa and Asia remain mostly rural, with 40 and 48 per cent of their respective populations living in urban areas. This is likely to change over the next 50 years as the fastest growing urban areas are currently found in Asia and Africa. Just three countries – India, China and Nigeria – together are expected to account for 37 per cent of the projected growth of the world's urban population between 2014 and 2050. India is projected to add 404 million urban dwellers, China 292 million and Nigeria 212 million. The total world urban population is expected to surpass six billion by 2045 and much of the expected growth will occur in low income countries. The fastest growing urban areas are medium-sized cities and cities with less than one million inhabitants but some reports also highlight the fact that many areas projected to be urban in 2040 have not yet actually been built. In India alone, it is predicted that 70 per cent of cities have yet to be built.

Some cities have experienced population decline. Economic contraction led to population losses in the American cities of Buffalo and Detroit between 2000 and 2014, while New Orleans experienced population decline in the wake of the 2005 Hurricane Katrina natural disaster.

One striking feature of the last 30 years has been the rapid development of megacities – urban areas with a population of more than 10 million people. In 1990 there were ten megacities; in 2014, there were 28; and by 2025 the UN predicts there will be 37 megacities – housing

Table 9.1 Table showing past and predicted urban population growth

Country	Urban population (%) 1990	Urban population (%) 2014	Urban population (%) 2050	Average annual rate of change (%) 2010–2015
Australia	85	89	93	0.2
Brazil	74	85	91	0.3
China	26	54	76	2.4
Germany	73	75	83	0.3
India	26	32	50	1.1
Japan	77	93	98	0.6
Kenya	17	25	44	1.7
Nigeria	30	47	67	1.9
Russian Federation	73	74	81	0.1
United Arab Emirates	79	85	91	0.4
United Kingdom	78	82	89	0.3
United States of America	75	81	87	0.2

over 13 per cent of the global population. As Figures 9.1–9.3 show, the development of megacities is largely concentrated in Asia. In 2015, Tokyo was the world's largest city with over 38 million inhabitants but it was closely followed by the likes of Delhi, Mumbai and Shanghai, all with populations in excess of 20 million and rising rapidly. Such settlements can be defined as **metacities**, the term given to a conurbation (continuous built-up area) of more than 20 million people. The Chinese government has plans to merge nine cities in the Pearl River Delta creating an urban area 26 times larger than Greater London.

Key terms

Megacity – A city or urban agglomeration (urban area incorporating several large towns or cities) with a population of more than 10 million people. According to the UN, London achieved megacity status in 2013. This classification included residents in the Greater London area.

Metacity – A conurbation with more than 20 million people.

Urban growth – An increase in the number of urban dwellers. Classifications of urban dwellers depend on the census definitions of urban areas, which vary from country to country. They usually include one or more of the following criteria: population size, population density, average distance between buildings within a settlement and legal and/or administrative boundaries.

Urbanisation – An increase in the proportion of a country's population that lives in towns and cities. The two main causes of urbanisation are natural population growth and migration into urban areas from rural areas.

Urban sprawl – The spread of an urban area into the surrounding countryside.

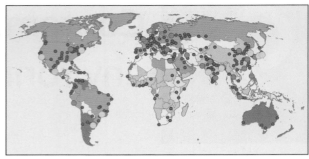

Key

Percentage urban

☐ 0–20%   ■ 20–40%   ☐ 40–60%   ■ 60–80%
■ 80–100%

City population

● 1–5 million   ● 5–10 million   ○ 10 million or more

Figure 9.1 Percentage of urban population and city size, 1990

Source: UN Population Division

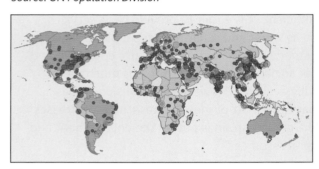

Key

Percentage urban

☐ 0–20%   ■ 20–40%   ☐ 40–60%   ■ 60–80%
■ 80–100%

City population

● 1–5 million   ● 5–10 million   ○ 10 million or more

Figure 9.2 Percentage of urban population and city size, 2014

Source: UN Population Division

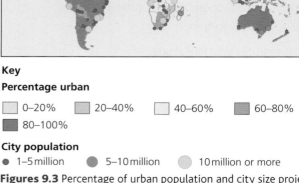

Key

Percentage urban

☐ 0–20%   ■ 20–40%   ☐ 40–60%   ■ 60–80%
■ 80–100%

City population

● 1–5 million   ● 5–10 million   ○ 10 million or more

Figures 9.3 Percentage of urban population and city size projected to 2030

Source: UN Population Division

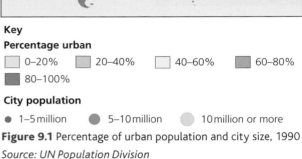

The causes of urban growth

The process of urbanisation plays an important role in human affairs. It has historically been linked to other important economic and social transformations which have brought about greater geographic mobility, lower fertility and longer life expectancy. Cities also play an important role in reducing poverty. They hold much of the national economic activity, government institutions, business and transportation, and have higher levels of education, better health, easier access to social services, and greater opportunities for cultural and political participation. In 2015, São Paulo, Brazil's economic and financial capital, accounted for 10 per cent of the population but 25 per cent of national Gross Domestic Product (GDP). Likewise, in Kenya, Nairobi, with 8.4 per cent of the country's population, accounted for almost 20 per cent of the country's GDP.

Natural population growth

Urban areas tend to have relatively young age profiles. Across the world, it has traditionally been young adults (15–40 years) migrating, lured by the prospect of higher paid jobs, better educational opportunities and greater social and cultural diversity. Between 2001 and 2011 the population of large city centres in England and Wales more than doubled, with the number of residents aged 22–29 nearly tripling to make up almost half of their total population. These migrants are in their fertile years – the years during which people have children – and so the rates of natural increase are higher in cities than in the surrounding rural areas. In London, an area stretching from Clapham, south of the River Thames, westwards to Fulham, north of the river has been termed 'Nappy Valley' due to the high proportion of young families living here. In the past, professional couples with young children would have moved out to the suburbs when they could afford it but the rising costs and time involved in commuting has encouraged more young families to remain in the city.

Rural – urban migration

The reasons for rural–urban migration are often divided into 'push' and 'pull' factors. **Push factors** cause people to move away from rural areas, whereas **pull factors** attract them to urban areas. In low-income countries, push factors tend to be more important than pull factors.

Push factors are largely due to poverty caused by:

- population growth, which means the same area of land has to support increasing numbers of people, causing over-farming, soil erosion and low yields
- agricultural problems, including desertification because of low rainfall, systems of inheritance that cause land to be subdivided into small plots, systems of tenure and debt on loans taken out to support agricultural change
- high levels of local diseases and inadequate medical provision
- agriculture is increasingly being organised globally. Land previously used to grow food for local people is now used to produce cash crops for sale to higher income countries. Many traditional rural communities have been driven off their land and into cities
- natural disasters such as floods, tropical storms and earthquakes – people flee rural areas and do not return
- wars and civil strife cause people to flee their land.

Pull factors include the prospect of:

- employment in factories and service industries (for example, hotels) which is better paid than work in rural areas. There is an increasingly high demand for unskilled labour in cities
- earning money from the informal sector, for example, selling goods on the street, providing transport (taxi/rickshaw driver) or prostitution
- better quality social provisions, from basic needs such as education and healthcare to entertainment and tourism
- a perceived better quality of life in the city, fed in part by images in the media.

Consequences of urbanisation and urban growth

Key questions

What are the key issues caused by urbanisation in the twenty-first century?

How do these issues vary between low and high income countries?

Problems with housing, traffic, waste disposal, crime and pollution can be found in cities all over the world irrespective of their economic status. Instead, these issues tend to be linked to the unique geographical circumstances of the city, such as topography, climate and function. The following section details the key consequences of urbanisation and characteristics of cities in the twenty-first century.

A. Urban sprawl

Urban sprawl is defined as the spread of an urban area into the surrounding countryside. It has been linked to the processes of urbanisation and **suburbanisation** and has traditionally occurred in an uncontrolled and unplanned fashion.

"We're waiting for the city to come to us..."

Figure 9.4 Urban sprawl

Urban sprawl has many negative impacts:

- Urban sprawl requires more roads and infrastructure such as pipes, cables and wires. It is less economically efficient to service low-density rural areas compared to compact urban developments with the same number of households.
- The reach of urban sprawl into rural areas ranks as one of the main causes of wildlife habitat loss.
- Urban sprawl causes more commuting from the suburbs to the city and thus more fuel consumption and traffic congestion.
- Urban sprawl can increase air pollution since a more car-dependent lifestyle leads to increases in fossil fuel consumption and emissions of greenhouse gases. The areas may also experience higher temperatures in line with the urban heat island effect.
- Sprawl has contributed to the loss of farmland and open spaces, which in turn has led to the loss of fresh local food sources with greater food miles as a result.

- Urban sprawl can have a serious impact on water quality and quantity. Covering the countryside with impermeable surfaces means that rainwater is unable to soak into the ground and replenish the groundwater aquifers. In addition, it can lead to greater water run-off and increased flood risk.
- In addition to the movement of people to the suburbs, one other important component has been the accompanying movement of industry and businesses, including retail companies. This is referred to as **decentralisation** and this outward movement has been blamed for the decline of retail in some city centres and an increasing **homogenisation** of the landscape, where cities become indistinct from one another. American cities in particular have witnessed the huge growth of large edge-of-city complexes including shopping malls and leisure areas. In some cases, new self-contained settlements have developed beyond the original city boundary. These are known as **edge cities**.

B. Shortage of housing in lower-income countries

Population density tends to be high in urban areas and one of the consequences of this is a shortage of accommodation, leading to the presence of large areas of informal and often inadequate housing. These normally develop on the edge of the city or in areas of low land value prone to environmental hazards such as flooding or landslides. They may also be found adjacent to transport networks or in areas suffering high levels of air, noise or water pollution. These settlements tend to have limited access to basic infrastructure such as water, electricity and waste disposal and a lack of services such as health centres and schools.

In 2013, UN Habitat reported that the number of people living in slum conditions was estimated at 863 million, up from 760 million in 2000 and 650 million in 1990. However, slums do not tend to have detailed enumeration (population counts) and this is why many feel that the actual number of slum dwellers is more likely to be in excess of one billion. Mumbai alone had more than nine million slum dwellers in 2015, up from six million in 2005.

Informal settlements take on different names and different forms depending on their location. The settlements lining the hills of Rio de Janeiro in Brazil are called favelas; in parts of India they are known as

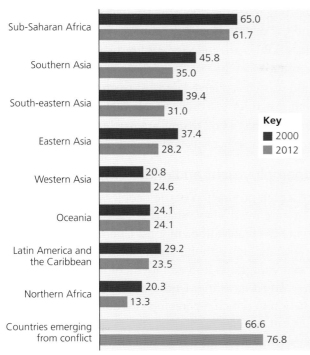

Figure 9.5 Proportion of people living in slums, 2000–2012
Source: UN Habitat

bustees and in West Africa, bidonvilles. The use of the term 'slum housing' has been criticised by those who see it as a political label. In Mumbai for example, settlements such as Shivaji Nagar and Dharavi are referred to by some as 'homegrown neighbourhoods'. The argument is that they were developed gradually, by the people who live there, with the help of local artisans of construction, and usually with little or no support from the authorities.

Figure 9.5 shows that the proportion of people living in slums fell in many areas between 2000 and 2012. This should be viewed in a positive light alongside the relative success of Millennium Development Goal 7 (Ensure environmental sustainability) to improve the lives of slum dwellers. Between 2000 and 2014, for example, the UN reports that more than 320 million people living in slums gained access to improved water sources, improved sanitation facilities or durable or less crowded housing. However, the total number of people living in slums has continued to increase and one of the targets in the post-2015 sustainable development goals is to 'ensure access for all to adequate, safe and affordable housing and basic services, and upgrade slums by 2030.'

In the past, the most extreme strategy adopted by authorities was to eradicate slums. However, this simply moved the problem elsewhere and a more common approach became to acknowledge the presence of slums

and provide help in the form of materials or services. These self-help or 'site and service' schemes have proved remarkably successful in some cities but the quantity and/or quality of housing remains inadequate in most urban areas around the world. More recent initiatives have been **slum upgrading programmes.** These seek to improve the slums in partnership with local NGOs and development organisations. They focus on securing rights for dwellers, formalising land tenure rights and providing basic amenities such as electricity, water and waste disposal. **Shack/Slum Dwellers International (SDI)** is an organisation which gives a voice to those living in informal settlements and links up poor communities across Africa, Asia and Latin America. The idea is for slum dwellers to share their knowledge and expertise so that they are not excluded from the economic and political processes happening in their cities.

The Tower of David, Caracas, Venezuela: the tallest slum in the world

Almost 70 per cent of the population of Caracas lives in informal housing; however, the residents of the 52-storey Tower of David building live in a much more unique environment than most. The building was constructed in downtown Caracas in 1990 to serve as the headquarters for a bank but the developer went bankrupt and it was never completed. It remained an empty concrete shell until 2007, when squatters moved in and built their own homes on each floor. A thriving community developed, including shops, hairdressers, an internet cafe and a gym. The squatters also installed electricity, completed the walls with bricks or zinc sheets, and got running water. Some commentators have praised the ingenuity of the squatters for constructing a self-contained community out of nothing, but in 2014 the process of relocating the 1,200 families began.

Figure 9.6 The bottom 28 storeys of the building are occupied by the squatters; the rest of the building remains empty.

C. Shortage of affordable housing in higher-income cities

Shortage of affordable housing is a key problem in many wealthy cities. In the UK, the rising demand for accommodation in cities has led to a dramatic increase in both house prices and rental costs. In some parts of London, average house prices rose by 50 per cent between 2010 and 2015 fuelled partly by in-migration, **gentrification** and by the purchasing of properties by wealthy foreign investors. The city is a major global hub and overseas investors buy properties in London to diversify their international portfolio.

D. Lack of urban services and waste disposal

Financial restraints in lower-income countries can result in a lack of basic services such as water and electricity. Maintenance of infrastructure such as roads, sewers and drainage is also limited and this can result in traffic congestion, polluted watercourses, flooding and the rapid spread of disease. In India, old pipeline infrastructure has not kept pace with urbanisation, resulting in large urban areas devoid of planned water supply and sewage treatment. In 2015, only five per cent of piped water reached the slum areas in 42 Indian cities and towns, including New Delhi.

Waste disposal poses a further problem. The amount of waste increases year-on-year, but there are economic, physical and environmental restraints on its disposal. In Addis Ababa, Ethiopia, the city authority is only able to deal with two-thirds of the rubbish created by the rapidly growing urban population. The rest is left to private contractors to collect or is simply dumped on streets and in rivers creating a huge health hazard. At Koshe, the huge landfill site on the edge of the city, families live in makeshift housing and search for anything they can use for themselves, or resell (Figure 9.7).

In other cities too, refuse is seen as an opportunity to make money. In Nairobi, recycling is an important part of the everyday economy:

- old car tyres are cut up and used to make cheap sandals

- washing machine doors are used as kitchen bowls, and the drums as storage units
- glass bottles are collected and returned to stores for refilling
- food waste is collected and fed to animals or composted for use on vegetable plots
- tin cans and old oil drums are used to make charcoal stoves, lamps, buckets and metal tips for ploughs.

E. Unemployment and under employment

Since a high proportion of the people who move to cities are relatively young, there is considerable pressure to create sufficient jobs. Unemployment rates are typically high although official data is hard to find and many migrants find employment in informal work such as street hawking. **Under-employment** refers to a situation in which a person is not doing work that makes full use of their skills and abilities. This may occur when a migrant moves to a new city.

F. Transport issues

The processes of urbanisation and suburbanisation have led to increased traffic in cities across the world (Figure 9.8). This has created more congestion and pollution, damaging human health and wasting billions of pounds in lost productivity. The spread of houses into the suburbs and beyond has created surges of morning and evening commuters. Traffic flows for shopping, entertainment and other commercial services add to the problem.

Figure 9.7 Metals, wood, plastic and clothing all have a value at the Koshe dump. Others come to find discarded food which they can eat

Figure 9.8 Lagos, Nigeria, suffers regular gridlock as seven million passengers move around in private cars, transit buses and motorcycle taxis known as 'okada'. It is hoped that the seven-line urban rail project under construction will ease traffic

During the car boom of the 1960s, city planners built more and wider roads as a solution. This didn't work. The more roads created, the more cars they attracted. A 1997 study in California found that new, additional traffic will fill up to 90 per cent of any increase in road capacity within just five years. No matter how much money is spent on traffic infrastructure, congestion and parking problems seem to get worse.

Contemporary urban processes

Key questions

How have urban processes changed over time and how do these vary across cities?

Why are many cities in high-income countries now seeing a resurgence in their urban areas?

Key terms

Counter-urbanisation – The movement of people from large urban areas into smaller urban areas or into rural areas, thereby leapfrogging the rural–urban fringe. It can mean daily commuting, but can also require lifestyle changes and the increased use of ICT.

Decentralisation – The movement of population and industry from the urban centre to outlying areas. The term may encompass the processes of both suburbanisation and counter-urbanisation.

Deindustrialisation – This refers to the loss of jobs in the manufacturing sector, which occurred in the UK in the second half of the twentieth century.

Gentrification – Gentrification is the buying and renovating of properties, often in more run-down areas, by wealthier individuals.

Suburbanisation – The movement of people from living in the inner parts of a city to living on the outer edges. It has been facilitated by the development of transport networks and the increase in ownership of private cars. These have allowed people to commute to work.

Urban resurgence – Urban resurgence refers to the regeneration, both economic and structural, of an urban area which has suffered a period of decline. This is often initiated by redevelopment schemes but is also due to wider social, economic and demographic processes.

Higher-income countries have seen a much slower rate of urbanisation in the last few decades and some cities have even witnessed a decline in numbers. Rather than moving into urban areas as seen in the industrial periods, a more significant trend has been the move outwards.

Suburbanisation: characteristics, causes and effects

Suburbanisation has resulted in the outward growth of urban development that has engulfed surrounding villages and rural areas. During the mid to late twentieth century, this was facilitated by the growth of public transport systems and the increased use of the private car. The presence of railway lines and arterial roads also enabled wealthier commuters to live some distance away from their places of work.

The towns and cities of the UK demonstrate the effects of past suburbanisation. In the 1930s there were few planning controls and urban growth took place alongside main roads – this was known as ribbon development. By the 1940s this growth, and growth between the 'ribbons', became a cause for concern. This led to the creation of **green belts** – areas of open space and low-density land use around towns where further development was strictly controlled.

Since 1950, suburban expansion has increased and has been better planned. During the 1950s and 1960s large-scale construction of council housing took place on the only land available, which was the suburban fringe. In the 1970s, there was a move towards home ownership, which led to private housing estates being built, also on the urban fringe. Building in these areas allowed people to have more land for gardens and more public open space.

As car ownership grew, the edge of town, where there is more land available for car parking and expansion, became the favoured location for new offices, factories and shopping outlets. In a number of cases, the 'strict control' of the green belt was ignored.

In recent years new housing estates have been built in suburban areas, along with local shopping centres and schools. People continue to move to the suburbs because of their desire for a quieter, less congested and less polluted environment. Suburbs are perceived as relatively crime-free environments and they also demonstrate other key benefits of the rural–urban fringe, such as woodlands and parks, golf courses and playing fields. Many are now well-established housing areas, highly sought after in the property market.

The negative effects of suburbanisation relate to urban sprawl and the environmental impacts of this process, discussed earlier. However, suburbanisation can also lead to:

- increasing social segregation within cities as the wealthy move out to the suburbs and the poor remain in the inner city. This issue is particularly acute in American cities, where segregation has occurred as a result of both wealth and ethnicity
- diversion of funding away from inner city areas to the suburbs to pay for new infrastructure and services.

Counter-urbanisation: characteristics, causes and effects

Counter-urbanisation is the migration of people from major urban areas to smaller urban settlements and rural areas. Counter-urbanisation does not lead to suburban growth, but to growth in rural areas beyond the main city. The difference between rural and urban areas is reduced as a consequence of this movement.

A number of factors have caused the growth of counter-urbanisation. One is that people want to escape from the air pollution, dirt and crime of the urban environment. They aspire to the 'rural idyll' – what they see as the pleasant, quiet and clean environment of the countryside, where land and house prices are cheaper. Car ownership and greater affluence allow people to commute to work from such areas. Indeed, many employers have also moved out of cities. Improvements in technology have allowed more freedom of location. The spread of broadband and high speed internet access means that someone working from a home computer can now access the same global system as a person in an office block in the centre of a city.

At the same time there has been a rising demand for second homes and earlier retirement. The former is a direct consequence of rising levels of affluence. Alongside this is the need for rural areas to attract income. Agriculture has faced economic difficulties and one straightforward way for farmers to raise money is to sell unwanted land and buildings.

Counter-urbanisation affects the layout of rural settlements. Modern housing estates are built on the edges of small settlements, and small industrial units on the main roads leading into the settlement. Former open areas are built on, old properties and some agricultural buildings are converted and modernised.

As with gentrified areas in inner cities, there may be tension between the newcomers and the locals. One of the main areas of conflict is that, despite the influx of new people, local services often close down. Bus services to many rural communities have disappeared, schools and post offices have closed, and churches have closed as parishes are amalgamated into larger units. The main reason for these changes is that the newcomers have the wealth and the mobility to continue to use the urban services some distance away.

The evidence for counter-urbanisation in an area includes:

- an increase in the use of a commuter railway station in the area, including car parking for commuters
- increased value of houses in the area
- the construction of more executive housing in the area, often on newly designated building land, following the demolition of old properties
- conversions of former farm buildings to exclusive residences.

Counter-urbanisation is one of a number of processes contributing to social and demographic change in rural settlements, sometimes referred to as the rural turnaround. This may include:

- the out-migration of young village-born adults seeking education and employment opportunities elsewhere
- the decline of the elderly village-born population, through deaths
- the in-migration of young to middle-aged married couples or families with young children
- the in-migration of younger, more affluent people, which results in increased house prices.

These changes do not take place uniformly within all rural settlements. There are considerable variations between and within parishes. The ones with the most change are key settlements that have a range of basic services and good access to commuter routes. Such settlements are called **suburbanised villages**.

Urban resurgence: characteristics, causes and effects

Urban resurgence refers to the regeneration, both economic and structural, of an urban area which has suffered a period of decline. An urban resurgence has been seen in many cities in recent years as redevelopment schemes have made city living more attractive. This is particularly the case for the former industrial cities of the UK which suffered from the manufacturing decline in the 1970s and 1980s but have reinvented themselves as cities of culture and commerce. London, Birmingham, Manchester and Leeds have all bounced back after severe **de-industrialisation** in the second half of the twentieth century. The cities have revived their fortunes by developing strong financial, business and consumer service industries and have also attracted more university students, young professionals and immigrant workers.

Urban resurgence is evident in the changing landscape of a city. Areas may still contain the industrial architecture of the past, including factories and warehouses, but increasingly these have been converted for housing or commercial use and modern infrastructure and services added. Many urban redevelopment schemes have successfully transformed run-down areas, rebranding them as fashionable districts or 'quarters' which then attract more newcomers, often young professionals, with a high disposable income.

The Jewellery quarter in Birmingham is one example of an area which has experienced bust and boom. In the early 1900s, it employed over 20,000 people in jewellery making, metalworking and hallmarking but a combination of foreign competition, reduced demand and the bombing of the area during the Blitz led to a sharp decline in the area's fortunes. Decline continued throughout the twentieth century and in spite of a small number of regeneration attempts, it was not until the early 2000s that large-scale improvements began to be seen. Former warehouses and factories have been converted into loft-style apartments and townhouses, and more than 30 restaurants, bars and cafés have helped create a vibrant hub for young businesses and professionals (Figure 9.9).

Figure 9.9 The Birmingham Mint redevelopment is found on the edge of the Jewellery quarter in Birmingham. The Grade II listed former minting factory was converted into flats with many of its traditional features still intact.

Further evidence of urban resurgence in Birmingham can be found in the Gas Street Basin canal area. Close to the city centre, this area was once the hub of a thriving canal transport network moving heavy goods such as coal and glass. Today, the canals have been cleaned up and are part of an attractive area where bars and restaurants line the waterways and traditional narrowboats navigate past large arts and entertainment complexes.

Urban resurgence is often driven by government-led regeneration schemes but there are wider economic, social and demographic processes which are also important. Redevelopment by private companies has led to the wholesale transformation of parts of UK cities in recent years and this has served to attract further investment. City living has also become more attractive as urban areas are improved and people choose to live closer to work, entertainment and leisure facilities, rather than face long and costly commutes. Globalisation and technological change have facilitated the resurgence of some urban areas. Parts of inner East London have experienced a huge in-migration of people, attracted by its reputation for creative and digital start-up businesses. East London Tech City is home to a cluster of independent start-up companies as well as global organisations such as Facebook, Google, Amazon and EE. This area has also become fashionable for its independent shops, galleries, markets, bars and restaurants.

Major sporting events can act as a catalyst to changing the fortunes of an area. The London Olympics brought much needed investment to former industrial parts of East London while the 2014 Commonwealth Games encouraged urban regeneration and business investment in parts of Glasgow depressed by the decline of shipbuilding, steelmaking and heavy engineering. Resurgence has a positive multiplier effect, initiating further improvements and attracting greater investment into an area. However, as more people are attracted back into a city, greater pressure is put on the urban infrastructure and some people may find themselves displaced as house prices rise in line with demand. There are also concerns that not everyone benefits from resurgence and this has led to increasing inequality between rich and poor.

Many American cities including New York, Boston and Los Angeles have experienced an urban resurgence in recent decades. The second half of the twentieth century had seen a huge population decline in urban areas as families increasingly moved out to the suburbs. This was accompanied by the loss of manufacturing and retail businesses and soon the traditional 'downtown' areas – the major retail centre of the city, were in decline along with neighbouring residential areas. The term '**dead heart syndrome**' was used to describe this process.

The 1990s saw the beginning of a resurgence. A sustained period of national economic growth, successful regeneration schemes and more attractive urban design has helped revive the fortunes of many American cities. In addition, young people are remaining in cities even when they start a family while many older people whose children have left home are moving back to the city to be closer to urban services. This population revival increases demand for services and has fuelled a prosperous urban economy in many cities.

Urban change

De-industrialisation, decentralisation and the rise of the service economy

De-industrialisation refers to the loss of jobs in the manufacturing sector, which occurred in the UK in the second half of the twentieth century. Prior to this, the Industrial Revolution and the rise of manufacturing industry had been a key development in the growth of many urban areas. Cities became synonymous with particular types of industry, such as textiles (Manchester), iron and steel (Sheffield) and shipbuilding (Glasgow). Thousands of jobs were created and people migrated into urban areas. By the 1980s, many of the older industrial cities were experiencing severe economic problems associated with the decline of manufacturing. This has been attributed to three main factors:

- mechanisation: most firms can produce their goods more cheaply by using machines rather than people
- competition from abroad, particularly the rapidly industrialising countries of the time such as Taiwan, South Korea, India and China
- reduced demand for traditional products as new materials and technologies have been developed.

Table 9.2 shows an overall downward trend in manufacturing employment in the UK. Urban areas bore the brunt of these job losses but unemployment figures varied significantly between cities and depended upon the size of the city, the composition of the urban economy and the actions of local government. Cities in the manufacturing heartlands such as Manchester, Liverpool

The New York City High Line

Urban resurgence is evidenced in New York with the changing nature of the New York City High Line – a 1.5 mile-long section of elevated rail track built originally to carry goods to and from Manhattan's largest industrial district. It was abandoned in the 1980s as the Lower West Side underwent a period of manufacturing decline but has been successfully redeveloped in the 2000s as an elevated park and walkway lined with trees, grasses and shrubs.

The walkway has become a site for artistic commissions and cultural events and the five million annual visitors have increased spending in local shops and cafes as well as encouraging real estate development in the neighbourhoods that line its route.

The High Line has given new life to a piece of industrial infrastructure as a public green space and functions

essentially like a green roof. Porous pathways contain open joints so water can drain between planks, cutting down on the amount of storm-water that runs off the site into the sewer system. The planting design is inspired by the self-seeded landscape that grew on the elevated rail tracks during the 25 years in which they were derelict.

Figure 9.10 New York City's High Line

Table 9.2 Trends in manufacturing employment in the UK, 1978–2015

Year	Number of employees (thousands)
1978	6,711
1980	6,403
1985	5,071
1990	4,908
1995	4,201
2000	3,991
2005	3,113
2010	2,563
2015	2,658

Source: ONS

and Sheffield suffered more extensive job losses than cities with more diverse economies.

There was also significant variation *within* cities. Inner city areas contained many of the old types of workplace most likely to be closed – old plants with the oldest production techniques, lowest productivity and most unionised workforces. Inner city areas also lacked suitable land for the expansion of existing manufacturing and, as a result, new investment tended to focus on the edge of urban areas or more rural locations. This movement of industry away from the inner city was part of a wider process known as decentralisation, which also affected residential and retail land use in the late twentieth century.

The decline in manufacturing employment in the late twentieth century was accompanied by the rise of the service economy in urban areas. The **service economy** covers a wide range of activities including:

- **tertiary activities** such as financial services (for example, banking, accountancy and insurance), retailing, leisure, transport, education and health
- **quaternary activities** where knowledge or ideas are the main output, such as advertising, computer programming and software design.

Population growth fuels the service sector but it has also grown because:

- financial services are needed to support manufacturing industries, which are still important in many cities today
- as societies become more technologically sophisticated, they need a larger range of specialised services to keep them running
- as societies become wealthier, they demand more leisure and retail services.

The expansion of the service sector has been evident in cities all over the world and for many urban areas there has been a dramatic shift in their economic core from manufacturing to service-based activities. The major financial centres are all located in world cities while corporate headquarters cluster in urban areas where they can access national and international markets, a highly skilled labour force and specialist support services.

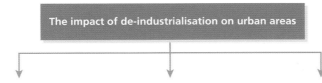

The impact of de-industrialisation on urban areas

Economic impacts
- Loss of jobs and personal disposable incomes
- Closure of other businesses which support closing industry
- Loss of tax income to the local authority and potential decline in services
- Increase in demand for state benefits
- Loss of income in the service sector as a result of falling spending power of the local population
- Decline in property prices as out-migration occurs
- De-industrialisation led to the **de-multiplier effect** in the urban areas affected

Social impacts
- Increase in unemployment
- Higher levels of deprivation
- Out-migration of population, usually those who are better qualified and more prosperous
- Higher levels of crime, family breakdown, alcohol and drug abuse and other social problems
- Loss of confidence and morale in local population

Environmental impacts
- Derelict land and buildings
- Long-term pollution of land from 'dirty' industries such as dye works and iron foundries remains a problem because there is no money for land remediation
- Deteriorating infrastructure
- Reduced maintenance of local housing caused by lower personal and local authority incomes
- Positive environmental impacts have been a reduction in noise, land and water pollution and reduced traffic congestion

Figure 9.11 The impacts of de-industrialisation on urban areas

Adapted from Drake and Lee (2000) The Urban Challenge, (Hodder Education)

Table 9.3 A summary of urban policy in the UK since 1979

Urban policy	Details	Examples
1979-1991 Emphasis given to property-led initiatives and the creation of an entrepreneurial culture	Greater emphasis placed on the role of the private sector to regenerate inner city areas. Coalition boards were set up with people from the local business community and they were encouraged to spend money on buying land, building infrastructure and marketing to attract private investment	Enterprise Zones Urban Development Corporations Urban Land Grants Derelict Land Grants City Action Teams
1991-1997 Partnership schemes and competition-led policy	A greater focus on local leadership and partnerships between the private sector, local communities, voluntary sector and the local authority. Strategies focused on tackling social, economic and environmental problems in run-down parts of the city, which now included peripheral estates	City Challenge City Pride Single Regeneration Budget
1997-2000s Area-based Initiatives	Many strategies in the 2000s focused upon narrowing the gap in key social and economic indicators between the most deprived neighbourhoods and the rest of the country. Local authorities were set targets to improve levels of health, education and employment opportunities and funding was allocated to assist them in delivering government objectives	Regional Development Agencies (RDAs) New Deal for Communities National Neighbourhood Renewal Strategy The Housing Market Renewal Programme
The Future?	There have been calls for greater devolution of powers (devolution deals) to English cities, of the type granted to Greater Manchester in 2014. Some feel this will lead to more effective place-based urban policies	Go to https://www.gov.uk/government/collections/future-of-cities to look at current research in Future of Cities

Table 9.4 An evaluation of three regeneration policies which have taken place in the UK, 1980-2010

Regeneration policy	Details	Successes	Failures	Case studies
Urban Development Corporations (1980s)	Urban Development Corporations (UDCs) were set up in the 1980s primarily to regenerate inner-city areas. The boards of UDCs were mostly made up of people from the local business community and they were encouraged to spend money on buying land, building infrastructure and marketing to attract private investment. Funding came direct from central government	Effective in attracting new businesses to run-down areas and improving the environment of UDC areas. By the mid-1990s, they had attracted over £12 billion in private-sector investment and created 190,000 jobs nationally	The property-led approach did little to tackle social problems Local people complained they had little involvement in the process and, in the London Docklands, locals did not tend to benefit from the new housing and jobs created	London Docklands Development Corporation (LDDC) Read: 'London Docklands: Post LDDC Developments' in *GeoActive*, 18:373 (OUP)
City Challenge (1990s)	This was a scheme where cities had to compete with each other for government regeneration grants. The cities with the 'best' schemes were awarded the grants. This was a local authority led scheme which formed partnerships between the private sector, local communities and the local authority. Strategies focused on tackling social, economic and environmental problems in run-down parts of the city	The fact that local authorities had to bid for funding was judged to have resulted in more successful regeneration schemes City Challenge gave equal importance to buildings, people and values 1997 data revealed that City Challenge had improved over 40,000 houses, created 53,000 jobs and reclaimed 2,000 ha of derelict land	Resources were thinly spread over large areas Areas which had previously received government funding based on need no longer received funding because their bid was unsuccessful Money was lost preparing bids by local authorities who did not win funding	Hulme City Challenge partnership, Manchester Read: Barker, Redfern and Skinner, *A2 Geography*, (Hodder) p. 157
New Deal for Communities (2000s)	NDC Partnerships were established to carry out 10-year strategic programmes designed to transform the 39 most deprived neighbourhoods and improve the lives of those living within them Local partnerships of residents, businesses, community organisations and local authorities were established but the focus was very much on communities being 'at the heart of the regeneration'	Between 2002 and 2008 NDC areas saw an improvement in 32 of 36 core indicators spanning crime, education, health, worklessness, community, housing and the physical environment Evidence found that gaps with both national and local authority levels had generally narrowed	The NDC strategy delivered greater positive change for place-, rather than people-, related outcomes Relatively little net-change was achieved for education and worklessness	Devonport Regeneration Company, Plymouth (see Chapter 8) You can find the final evaluation of the NDC strategy at http://extra.shu.ac.uk/ndc/ndc_reports.htm

City bars, restaurants and clubs provide important environments for social networking, which is heavily drawn upon in business.

While the growth in the service sector has gone some way to reducing the unemployment caused by de-industrialisation in urban areas, a number of problems still exist.

- Many of the men who lost jobs through de-industrialisation have continued to suffer from long-term unemployment.
- Many of the service jobs created are part-time or temporary.
- The number of service jobs created has not always made up for the loss of manufacturing jobs.
- Inner city locations have been avoided by both service industries and newer manufacturing companies leading to continued inner city decline.

Urban policy and regeneration in Britain since 1979

Urban policy relates to the strategies chosen by local or central government to manage the development of urban areas and reduce urban problems. Regeneration has been a key element of urban policy in the UK since the 1980s and while early strategies focused on 'top-down' economic regeneration, subsequent policies have recognised the need to adopt a more holistic approach, tackling economic, social and environmental problems from the 'bottom-up'.

Table 9.3 provides a summary of 'explicit' urban policies adopted in the UK since 1979. The use of the term 'explicit' is important because other more general strategies concerned with education, health and employment have helped in the regeneration of urban areas. Table 9.4 looks at three of the main regeneration strategies in more detail.

9.2 Urban forms

Urban form refers to the physical characteristics that make up built-up areas, including the shape, size, density and organisation of settlements. It can be considered at different scales: from regional to urban, neighbourhood and street and is evolving continually in response to social, economic, environmental, political and technological developments. A government report on 'Urban form and infrastructure in the UK', (2014) reported that the UK's urban form is characterised by 64 'primary urban areas', including one built-up megacity region (London, and the Greater South East), six large metropolitan areas (Birmingham, Leeds, Liverpool, Manchester, Newcastle-Upon-Tyne and Sheffield), and 56 towns and cities with more than 125,000 people.

This section of the chapter will focus first on the largest urban forms: those of megacities and world cities found in different parts of the world. It will then go on to explore patterns of land use within cities and finally it will consider how urban landscapes are changing in the twenty-first century.

The characteristics of megacities and world cities

Globalisation and economic competition between countries and cities has led to the rapid rise of the **megacity**. Faced with poorer economic prospects in rural areas and the perception of a better life in the city, mass migration has fuelled large-scale population growth. Added to this, government policies, such as the establishment of Special Enterprise Zones in Chinese cities, can encourage greater financial investment. Problems occur because this growth in population is not matched by a growth in resources and infrastructure. Megacities in lower-income countries have tended to sprawl in a haphazard fashion and this has inevitably led to the challenge of providing employment, housing and basic services. In addition, there are concerns about how city authorities can effectively govern such large cities.

Megacities can be a force for good. On average they produce two to three times more GDP than other cities and a United Nations report on urbanisation in 2014 highlighted a number of other characteristics and benefits:

1. They offer opportunities to expand access to services, such as healthcare and education, for large numbers of people in an economically efficient manner.
2. It is less environmentally damaging to provide public transport, housing, electricity, water and sanitation for a densely settled urban population than a dispersed rural population.
3. Urban dwellers have access to larger and more diversified employment markets.
4. Better levels of education and healthcare can improve the lives of the poor and empower women in countries where they do not have equal status.
5. Megacities are centres of innovation where many solutions to global problems are being trialled.

The growth of large urban areas can also fuel political pressure for change. Political protests are more common in urban areas where large numbers of younger people are brought together.

Key terms

Edge city – A self-contained settlement which has emerged beyond the original city boundary and developed as a city in its own right.

Fortress landscapes – This term refers to landscapes designed around security, protection, surveillance and exclusion.

World city – These are cities which have great influence on a global scale, because of their financial status and worldwide commercial power. Three cities which have traditionally sat at the top of the global hierarchy – New York, London and Tokyo – are now being joined by the likes of Beijing, Shanghai and Mumbai. These cities house the headquarters of many transnational corporations (TNCs), are centres of world finance and provide international consumer services.

Historically, the greatest global cities were generally the largest. However, today size is not so important. Of the world's most populous cities, only Tokyo, New York and Beijing are in the top rankings of the world's most important cities. It has been argued that what matters today is influence and the term **world city** (or **global city**) has been given to the cities which have greatest influence on a global scale. This can be measured in a number of ways and it is worth casting a critical eye over the different methods adopted.

One of the more commonly cited rankings is carried out by the Globalization and World Rankings Research Network (GaWC) based at Loughborough University. The Alpha, Beta and Gamma rankings, shown in Table 9.5, were based initially on the connectivity of cities through four 'advanced producer services': accountancy, advertising, banking/finance and law. New indicators were added in 2004 but economic factors are still deemed more important than political and cultural indicators.

Below these rankings are cities with 'sufficiency of services'. These are not world cities but have sufficient services so as not to be dependent on world cities. They tend to include smaller capital cities and traditional centres of manufacturing regions.

Figure 9.12 The characteristics of world cities

Table 9.5 GaWC city rankings, 2012

City ranking	Description	Examples from 2012 ranking
Alpha ++ cities	More integrated than all other cities and constitute their own high level of integration	London and New York
Alpha + cities	Other highly integrated cities that complement London and New York, largely filling in advanced service needs for the Pacific Asia	Tokyo, Hong Kong, Paris, Shanghai, Singapore, Beijing, Sydney, Dubai
Alpha and alpha – cities	Very important world cities that link major economic regions and states into the world economy	Chicago, Milan, Mumbai, Moscow, Sao Paulo, Frankfurt, Toronto, LA, Madrid and more
Beta level cities	These are important world cities that are instrumental in linking their region or state into the world economy	Bangalore, Lisbon, Copenhagen, Santiago, Guangzhou, Rome, Cairo and more
Gamma level cities	These can be world cities linking smaller regions or states into the world economy, or important world cities whose major global capacity is not in advanced producer services	Zagreb, Lahore, St Petersburg, Durban, Islamabad, Bristol and more

Spatial patterns of land use in urban areas

Key questions

How has urban land use changed over time?

What are the main factors influencing land use in different cities around the world?

The term **urban morphology** refers to the spatial structure and organisation of an urban area. Traditionally this may have been affected by physical factors such as relief and drainage. Early industrial areas developed close to rivers where they could harness the power of water for energy and transportation. Flat land was also important for transportation of goods via road or rail. Relief still plays an important role today because flat land is

easier to build on and may attract a higher land value. Conversely, flat land close to rivers may pose a flood risk. In poorer cities, informal settlements are often found on undeveloped steep land. Brazil's largest shanty town, Rocinha, is built on a steep and rugged hillside overlooking Rio de Janeiro. The poorer parts of the shanty town are found higher on the hilltop, with many houses only accessible on foot.

Skills focus

To what extent do physical factors still affect the pattern of land use in towns and cities today? Investigate different cities around the world to consider the impacts of relief, drainage and other factors such as soil and rock type and natural hazards on their land use.

As humans have been able to overcome the limitations imposed by physical factors, urban form today is more strongly influenced by human factors.

The main factor affecting land use in high-income countries is **land value** and this is traditionally higher in the centre of a city where accessibility is greatest. The **Peak Land Value Intersection (PLVI)** is the point with the highest land value and from here, land prices decline in line with the theory of **distance-decay**.

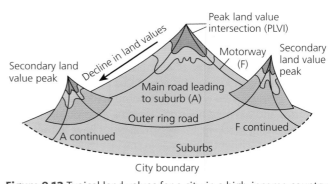

Figure 9.13 Typical land values for a city in a high-income country

It is usually only very profitable businesses such as large retailers that can afford the high prices of the PLVI. In UK cities, the site has often been occupied by the likes of Marks and Spencer.

Other retailers and commercial enterprises tend to occupy most of the **central business district (CBD)** but cannot all afford to pay the high prices required for the most accessible locations. Consequently, smaller retailers and businesses are more likely to be found towards the edge of the CBD.

Taking a transect from the CBD to the suburbs, land values fall significantly as the different land users are less reliant on accessibility and unable to pay the higher prices associated with this. This is known as the **bid-rent theory**. As Figure 9.14 shows, traditionally there has been a move from retailing to industrial and commercial and then to residential areas.

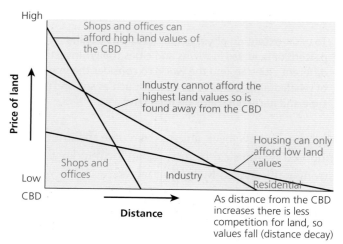

Figure 9.14 The bid-rent theory

The 1980s trend towards out-of-town shopping centres affected land values in some cities and this can be seen by the presence of secondary land-value peaks shown in Figure 9.13 (page 393). An increase in traffic meant that the city centre was no longer always the most accessible part of the city and a lack of space for expansion encouraged some of the large retailers to look elsewhere for potential new sites. For the consumers, the growth of edge-of-town developments provided free parking and other land uses such as cinemas and restaurants, which proved to be highly attractive. Land values subsequently rose in such areas.

New urban landscapes

Moving from the centre outwards, there have been a number of significant changes to the urban landscape in recent years.

Changes in the city centre

Some city centres in the UK have experienced decline in the last 30 years. In the 1980s and 1990s this was largely due to the development of out-of-town retail parks and the decentralisation of business and residential areas,

Main land-use zones in a city

Central Business District (CBD): This central area contains the major shops, offices and entertainment facilities.

Inner city: This is an area of old housing and light manufacturing industry. This area dates back to the Industrial Revolution when it was comprised mainly of factories and terraced housing providing accommodation for the factory workers. Many British cities have witnessed regeneration within these areas in the last three decades.

Residential: These areas consist of housing from a range of periods, which has traditionally increased in both size and price as one moves towards the outskirts. However, urban regeneration schemes and gentrification has meant that some of the most expensive property can now be found in traditional 'low class residential' areas while council estates on the edge of cities are now among some of the most depressed parts of British cities.

Green areas: Such as parks tend to be dotted throughout an urban area. They range from large botanical gardens of national importance down to playgrounds within a housing estate.

Out-of-town retail developments: Originally developed by large supermarkets, these spaces soon expanded to include non-food retail units and entertainment

complexes. They have had a negative economic impact on some town and city centres. In 1994 the UK government started to actively discourage their development.

Business or science parks: These tend to be found on the edge of urban areas where there is good access to major roads. Some science parks are also located near to universities.

Industry: Manufacturing industries often require large areas of land and tend to locate towards the edge of cities where cheaper land is available. De-industrialisation brought about the large-scale decline of manufacturing industry in British cities and former factory sites have either been demolished or converted into other land uses. The latter may still occupy their historical location in the inner city.

Informal settlements: Also known as slums or shanty towns, these are a feature of cities in low-income countries. They have traditionally developed on the edge of cities although they are also found adjacent to transport routes or in areas of the city unpopular with residents such as rubbish dumps. Physical factors such as steep slopes, unstable land and areas prone to natural hazards may also encourage their development.

which served to pull people away from the CBD. High parking costs, congestion and perceptions of the city centre as dirty and unsafe were further disincentives for shoppers. More recent competition has come from the phenomenal growth of internet shopping.

A number of strategies have been devised to help reverse the decline of city centres including the provision of a more attractive shopping environment, the construction of all-weather shopping malls, improvements in public transport links and the establishment of business and marketing teams to co-ordinate management of the CBD and run special events. In addition to this, the 2000s have seen two other notable strategies.

Town centre mixed development

Many cities are encouraging the development of functions other than retailing to increase the attractions of the city centre. These include:

- a wider range of leisure facilities including cinemas, theatres, cafés, wine bars, restaurants and other cultural and meeting places. Where these offer services of different kinds and at varying prices and degrees of quality, a greater range of people will be attracted
- the availability of spaces, including gardens and squares or plazas, to enable people-watching and other activities
- the promotion of street entertainment such as at Covent Garden in London
- developing nightlife, such as clubbing. (There are negative issues associated with this, including the high level of policing that is necessary)
- developing flagship attractions such as the At-Bristol Science Centre and M Shed museum and gallery in Bristol
- constructing new offices, apartments, hotels and conference centres to raise the status of the CBD for business and encourage tourists to remain near the city centre
- encouraging residential areas to return to city centres by providing flats, redeveloping old buildings (a form of gentrification) or building new upmarket apartments.

The combination of strategies above and the stricter planning controls placed on out-of-town developments has meant that many large cities in the UK have successfully attracted shoppers and visitors back to the city centre. However, decision makers are still worried about the decline of the CBD in smaller cities and urban areas.

In 2007, the city of Exeter in Devon saw the opening of a mixed-use city centre redevelopment scheme which replaced a post-war development that had become dated and unattractive for modern retailers. The Princesshay redevelopment contains more than 60 retail units, 122 flats, a unique visitor attraction (the city's mediaeval underground passages) a tourist information centre and 10 cafés/restaurants. It also includes new public art pieces and hosts events and festivals.

The developers were keen to instil a greater sense of security and vibrancy to the area and a key element of this was the promotion of city centre living and a night-time economy.

Figure 9.15 shows part of the City Centre Quarter of Exeter. The Roman walk signposted recognises the heritage of the city and runs along the original Roman city wall. In this photo, you can see shops, a restaurant and two stories of flats with cedar-clad walkways. These residential units help to create a sense of place, community and safety in the city centre, as well as meeting a local housing need.

The development of cultural and heritage quarters

Many cities across the UK have initiated the planning and development of cultural or heritage quarters as a deliberate model for urban regeneration of declining inner urban areas. Culturally-led urban development first began to appear in the 1980s and early UK examples are the Sheffield Cultural Industries Quarter and the Manchester Northern Quarter. A prerequisite for a cultural quarter is the presence of cultural production (making objects, goods, products) or consumption (people going to shows, visiting venues and galleries). Heritage quarters focus more on the history of the area based around small-scale industries. The most successful quarters tend to be those actually making something or those associated with a product, such as the Birmingham Jewellery quarter. Areas like these have built up a regional, and in some cases national, reputation which attracts visitors and tourists from further afield and brings financial benefits to the wider area.

Some critics have argued that not all towns and cities need cultural quarters and that in some areas they have simply created higher property values. Experiences of different 'quarters' has shown that some are more successful than others. However, as a tool for regeneration, improving perceptions of place and preserving history and culture, 'quarters' tend to be viewed in a positive light.

Gentrified areas

Gentrification is defined as the buying and renovating of properties often in more run-down areas by wealthier individuals. It is an important process of housing improvement supported by groups such as estate agents and local authorities and it has helped to regenerate large parts of British inner cities in the last few decades. Unlike the regeneration schemes described in Table 9.4 (page 390), gentrification involves the rehabilitation of old houses and streets on a piecemeal basis and is carried out by individuals or groups of individuals rather than large organisations. Gentrification can happen for a number of reasons:

- The Rent Gap: This refers to the situation when the price of property has fallen below its real value, usually due to lack of maintenance or investment, and there is a 'gap' between actual and potential price. Such properties are attractive to builders, property developers or individuals who can afford to renovate the properties and then sell them on to make a profit.
- Commuting costs: Commuting can be time-consuming, expensive and stressful. Moving closer to the city centre can eliminate the need to commute.
- The 'pioneer' image: This refers to the trend of creative individuals such as artists and designers moving into more 'edgy' neighbourhoods. These groups are not interested in the conformity of suburban living but are drawn to the diverse cultural opportunities of the urban centre. The gentrification of areas such as Hoxton and Shoreditch in London and SoHo in New York City has been linked to their notoriety as the location of vibrant arts scenes.
- The support of government and local decision makers: Both groups are keen to improve the economy and environment of inner city areas and gentrification is seen as an important part of this.
- Changing composition of households: Many cities have seen the growth of single or two-person households without children. These households are more likely to see the benefits of inner-city living.

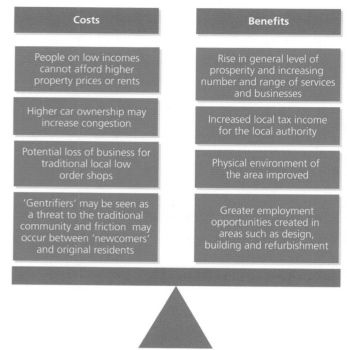

Figure 9.16 The costs and benefits of gentrification

Gentrification has both costs and benefits, as shown in Figure 9.16. Some commentators have emphasised the importance of gentrification in inner-city regeneration; others have raised concerns about the displacement of low-income families and small businesses. In many parts of London, gentrification has contributed significantly to the lack of affordable housing and prices have been pushed up even beyond the level of well-paid professional workers. Anti-gentrification protests are becoming more common.

Homes sold for a pound in Liverpool

One innovative way of regenerating inner city estates without displacing low-income families has been a scheme adopted in a number of British cities where run-down housing is sold off to individuals or families at very low prices. In 2013, Liverpool city council sold 20 derelict homes in Kensington, Granby and Picton for £1 each. In order to prevent people abusing this scheme, there were a number of basic rules. First, buyers had to show they were capable of doing the house up to a reasonable standard and secondly, they had to sign an agreement to live in the property for five years and not sublet it. There were about 1000 applicants for each house. The scheme has now been extended to other houses and shops in the area with the overall aim to improve the built environment and make it a thriving community again.

Fortress landscapes

The term **fortress landscapes** refers to landscapes designed around security, protection, surveillance and exclusion. In the UK, a number of relatively simple strategies have been adopted to reduce crime in urban hotspots such as the city centre and inner city estates. These include:

- greater use of closed-circuit television (CCTV)
- railings and fencing around private spaces
- 'mosquito' alarms which emit a high-pitched sound heard only by young people, to discourage loitering around certain buildings
- effective use of street lighting
- speed bumps to prevent joyriding.

More recent strategies have focused on the concept of 'designing out crime' through better urban architecture. In Manchester for example, the redevelopment of housing in parts of the Greenheys and Wythenshawe estates has included more windows to provide better natural surveillance, provision of front gardens with fences or hedges to mark a clear boundary between private and public space and bins in gated compounds rather than open alleyways. Features which have been avoided are recessed doorways for people to hide in, projecting window sills or exposed rainwater downpipes, which would make it easier for anyone to climb on to the roof and dark alleys and dead ends associated with muggings and drugs deals.

There is also evidence in some UK cities of the exclusionary tactics adopted by North American cities to segregate people from 'others' perceived as threatening or undesirable. The use of anti-homeless spikes fitted into the ground in shop doorways or outside upmarket apartments has been heavily criticised but the high pitched 'mosquito' alarm, sloped bus shelter seats and special benches which deter skateboarders raise few eyebrows. Gated communities are also becoming a feature of some urban landscapes. These are not as common in British cities as they are in places like the USA and South Africa but the use of guards or the electronic control of access into housing complexes is increasing. The notion of 'insiders' and 'outsiders' discussed in Chapter 8 is relevant here.

Fortress LA

Surveillance and exclusion measures are seen to be at their most extreme in American cities. Los Angeles has developed a reputation as a city built on fear. The author Mike Davis has detailed how paranoia and fear of gangs, minorities and the homeless has led to a refashioning of the urban landscape which includes:

- gated communities
- armed-response security units in residential areas
- shopping malls surrounded by staked metal fencing and an LAPD surveillance tower.

Edge cities

Edge cities are self-contained settlements which have emerged beyond the original city boundary and developed as cities in their own right. They are associated with the urban landscape of North America and have been viewed by some as a feature of postmodern urban living. Edge cities are largely the result of urban sprawl. This has happened on a much larger scale in the USA as a result of higher car ownership, greater willingness to travel long distances for work, shopping and entertainment and limited planning restrictions in the suburbs. Los Angeles is the classic example of a sprawling urban settlement. In 2015, the core city of Los Angeles which is about 30 kilometres wide, had a population of just under 4 million. However, this is surrounded by a metropolitan area of nearly 18 million people which is more than 100 kilometres at its widest point. Within this area there are more than 20 edge cities.

Edge cities develop close to major roads or airports and tend to be found in close proximity to shops, offices and other businesses which decentralised from the original city. Edge cities may lack a clear structure but they do have a wide range of amenities including schools, shopping malls and entertainment facilities. Residents may rarely go back to the original core city.

Edge cities have been linked to extreme **social segregation**, where the wealthy have moved to the new suburban settlements leaving only the poor and disadvantaged sections of society in the original city boundary.

The concept of the postmodern western city

The term **postmodernism** is used to describe the changes that took place in Western society and culture in the late twentieth century. It mainly concerned art and architecture and marked a departure from the conformity and uniformity of modernism. Postmodernism is characterised by the mixing of different artistic styles and architecture.

Both fortress landscapes and edge cities have been viewed as features of the **postmodern western city** but other key features include:

- a more fragmented urban form comprising independent settlements (such as edge cities), economies, societies and cultures
- a greater emphasis on producer services and knowledge-based industries rather than industrial mass production
- eclectic and varied architecture as seen in the London cityscape
- spectacular flagship developments such as the Guggenheim museum in Bilbao, Spain
- greater ethnic diversity but heightened economic, social and cultural inequalities and polarisation.

While elements of postmodernism can be seen in cities all over the world, it is important to acknowledge that the notion of the postmodern western city has been based on the experiences of a small number of (mainly American) cities. Los Angeles is often referred to as the archetypal postmodern city but the experiences of cities like Los Angeles are clearly not representative of cities elsewhere in the world.

9.3 Social and economic issues associated with urbanisation

Key questions

How do social and economic issues vary across and within cities?

How are different cities tackling issues such as inequality and segregation?

Economic inequality

Some argue that a key feature of cities in Asia, Africa and South America is the increasingly large wealth gap between rich and poor residents, known as **economic inequality**. In Mumbai, this stark contrast can be seen where the world's most expensive home towers over one of the largest slum areas of the city. Antilla is the 27-storey home of Mukesh Ambani, chairman of a

global energy conglomerate and estimated to be worth about $21 billion. Within a short distance of the house is Byculla, an area inhabited by some of Mumbai's nine million slum dwellers. Most people here survive on less than $2 a day.

Key terms

Cultural diversity – The existence of a variety of cultural or ethnic groups within a society.

Diaspora – A group of people with a similar heritage or homeland who have settled elsewhere in the world.

Economic inequality – The difference between levels of living standards, income, etc., across the whole economic distribution.

Social segregation – When groups of people live apart from the larger population due to factors such as wealth, ethnicity, religion or age.

Urban social exclusion – Economic and social problems faced by residents in areas of multiple deprivation.

Figure 9.17 Antilla, the world's most expensive home, is made of steel and glass, has spectacular ocean views, swimming and gym facilities and three helipads. It is seen here towering above a more modest part of Mumbai

Inequality exists in all urban areas. Enormous contrasts in wealth can be found over relatively small distances as seen in Figure 9.17. When you do fieldwork in an area, you can identify which areas are more affluent and you can sense if a neighbourhood is improving or deteriorating. The wealthy and the poor seem to concentrate spatially – a form of **social segregation**. There are a number of reasons for this:

- **Housing**: Developers, builders and planners tend to build housing on blocks of land with a particular market in mind. The requirement to include a proportion of 'affordable housing' may affect housing value in some areas but wealthier groups can choose where they live, paying premium prices for houses well away from poor areas, with pleasing environments and services such as quality schools and parks. Poorer groups typically have less choice.

- **Changing environments**: Housing is only a partial explanation for inequality since housing neighbourhoods change over time. Houses that were built for large families in Georgian and Victorian times are now too big for the average UK family. Many have been converted into multi-let apartments for private rental to people on low incomes. Conversely, former poor areas are being gentrified. The 'right to buy' legislation of the 1980s transformed many council estates, as houses were bought by their occupants and improved.

- **The ethnic dimension**: Ethnic groups originally come to the country as new immigrants. When they first arrive they may suffer discrimination in the job market and may be either unemployed or employed in low-paid jobs. They are only able to afford to buy cheap housing or they have to rent privately. Therefore, newly arrived migrants concentrate in poor areas in the city, often clustered into multicultural areas. Such ethnic groupings tend to persist into later generations.

Measuring poverty and inequality

The Index of Multiple Deprivation is a UK government qualitative study measuring deprivation at small-area level across England. The English Indices of Deprivation 2015 are based on 37 separate indicators, organised across seven distinct domains of deprivation which are combined, using appropriate weights, to calculate the Index of Multiple Deprivation 2015 (IMD 2015). This is an overall measure of multiple deprivation experienced by people living in an area and is calculated for every Lower layer Super Output Area (LSOA), or neighbourhood, in England. Every such neighbourhood in England is ranked according to its level of deprivation relative to that of other areas. The seven different dimensions of deprivation are:

1. Income
2. Employment
3. Health deprivation and Disability
4. Education, Skills and Training
5. Crime
6. Barriers to Housing and Services
7. Living Environment.

Each of these domains has its own scores and ranks, allowing users to focus on specific aspects of deprivation. The 2010 IMD data found that 98 per cent of the most deprived areas in England were in cities. Likewise, in 2015, the concentrations of deprivation were mainly found in large urban conurbations, areas that have historically had large heavy industry, manufacturing and/or mining sectors, coastal towns, and large parts of East London. The 20 most deprived local authorities in 2015 were largely the same as found for the 2010 Index, but the London Boroughs of Hackney, Tower Hamlets, Newham and Haringey have become relatively less deprived and no longer feature in this list. It has been suggested that gentrification is largely responsible for this change.

It is important to note that these statistics are a **measure of relative deprivation**, not affluence, and to recognise that not every person in a highly deprived area will themselves be deprived. Likewise, there will be some deprived people living in the least deprived areas.

GIS focus

Go to http://maps.cdrc.ac.uk/ and select the map showing the Index of Multiple Deprivation data for 2015. Type in the postcode for the area you want to investigate in more detail. The website also maps geodemographic data from the 2011 census and classifications for different types of urban dweller.

Urban social exclusion refers to the problems faced by residents in areas of multiple deprivation. These people are excluded from full participation in society by their social and physical circumstances. They cannot access a decent job because of poor education or obtain decent housing because of poverty. They often suffer from poor health and from high levels of crime in an unattractive physical environment. In a city, inequality can cause lack of social cohesion and in extreme cases lead to civil unrest.

Inner city areas have traditionally been the most deprived urban neighbourhoods. The characteristics of this decline, shown in Figure 9.19 (page 401), were caused by de-industrialisation in the second half of the twentieth century when unemployment became a major problem. Population loss followed and the movement of younger,

Skills focus

It is possible to measure the quality of life in an area using primary data, such as the quality, density and condition of housing and the nature of the physical and social environment (Figure 9.18). It is also possible to use secondary data from a census to assess deprivation levels. This may include poverty in terms of low income, or shown by poor health or the lack of possessions, such as cars. It is common for the poorest parts of an urban area to suffer from **multiple deprivation** (a combination of social, environmental and economic deprivation).

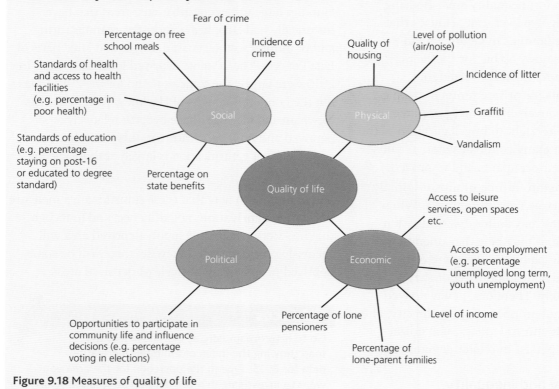

Figure 9.18 Measures of quality of life

more affluent and more skilled residents left behind a population who were older, less skilled and poor.

There is no clear geographical pattern to urban poverty today. It clearly exists but the image of a troubled inner core surrounded by affluent suburbs is out-dated. Some of the highest levels of urban poverty are now found in peripheral estates while many inner city areas have been transformed by regeneration schemes. Rich and poor areas today are found across city and suburb alike.

Inequality remains a major challenge for cities in the twenty-first century. Indeed, the gap between rich and poor can be far greater *within* a city than across a whole country. The inequalities tend to exist in terms of access to job opportunities, education, housing and basic public services such as water and sanitation. In many poorer cities for example, adequate water and sanitation services are primarily channelled to upper- and middle-class

neighbourhoods, while low-income neighbourhoods often depend on distant and unsafe water wells and lack any form of waste disposal. The poor also tend to live in overcrowded informal accommodation lacking basic infrastructure and services. The knock-on impacts of this are poorer health, higher unemployment and a lack of social mobility. The poor get stuck in a cycle of poverty from which it is hard to escape.

It is important to distinguish between poverty and inequality. Poverty is an absolute term, referring to a level of deprivation that does not change over time. Inequality is a relative term referring to the differences between people, usually economic, over a geographic distribution. In practice, poverty and inequality often rise and fall together but this need not necessarily be the case. Inequality can be high in a society without high levels of poverty due to a large difference between the top and the middle of the income spectrum.

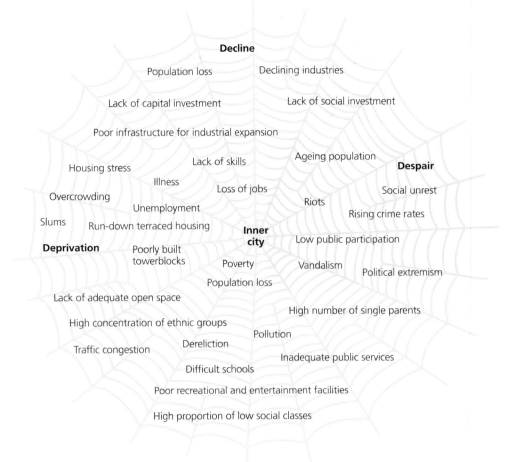

Figure 9.19 The web of inner city decline, despair and deprivation

Tackling poverty and inequality in cities

The main driver of urban inequality is wealth. Richer cities like London tend to have much higher rates of inequality while the more equal cities are those that are smaller, have lower average wages and are coming out of the end of years of industrial decline. On the national stage, a redistributive tax system can help to reduce inequality, but at the local level the effects of this are less clear. General strategies to reduce poverty and inequality are described below.

- Enforcing a living wage or providing an urban subsidy: In China, a government subsidy paid to urban dwellers to bring incomes up to a minimum level of 4,476 yuan ($700 or £446) has had extraordinary success in reducing urban poverty. The London Living Wage has also achieved notable success with over 10,000 families lifted out of working poverty since 2005.
- Provision of schools: Education may be taken for granted in many high-income countries but schooling beyond primary school level helps lift millions of people out of poverty in lower-income countries.
- Supporting low-skilled workers who want to develop their abilities: Cities need to ensure labour markets are inclusive, giving all residents the skills and opportunities needed to enter the workforce.
- Access to affordable housing: The presence of slum housing in poorer cities should not hide the fact that access to affordable housing is a major problem in wealthy cities too. In London, the lack of affordable housing has been blamed for breaking up families, reducing employment prospects and mobility and slowing the economy.
- Greater provision of public transport: Affordable public transport schemes improve mobility for the city poor, enabling them access to employment, education and services that could improve their lives. The Rio de Janeiro sky-high transport system, which connects six hilltops and covers 3.5 km, was

installed in 2011 to service 13 favelas and provide access to the main part of the city (Figure 9.20). The scheme was designed to give mobility to a once-stranded population and 12,000 people ride it daily

Figure 9.20 Cable cars overlook the cluttered rooftops of Rio's Complexo do Alemão favela

- Enforce minimum environmental standards: Poor health is strongly linked to poor environmental conditions. This can be improved through effective legislation.
- A number of British cities, including Liverpool and Sheffield, have established 'Fairness Commissions' looking at how local areas can address inequality.

Cultural diversity

Cultural diversity refers to the existence of a variety of cultural or ethnic groups within a society. Culture can relate to nationality, race, age and traditions. Immigration is a key influence. Urban areas, especially large cities, are places where cultural diversity tends to flourish. Cities like London, New York and Amsterdam have received migrants from all over the world and this has led to the creation of multicultural urban societies. London is considered to be one of the most diverse cities in the world. More than 300 languages are spoken by the people of London, and the city has at least 50 non-indigenous communities with populations of 10,000 or more.

Globalisation has increased movement around the world and it is now common to find people from different parts of the world living in major urban areas. The term **diaspora** is commonly used to describe a large group of people with a similar heritage or homeland who have settled elsewhere in the world. In some global cities, these diasporas make up a larger proportion of the

population than the indigenous residents. The United Arab Emirate state of Dubai is one such example. The majority of the city's population is made up of expatriates. Two-thirds of the population is of Asian descent, with people coming from India, Pakistan, Bangladesh and Sri Lanka. Arabic is recognised as the official language, but many languages are readily spoken, including English, Urdu, Punjabi, Tagalog, Bengali, Hindi, Persian and Chinese. In Los Angeles, 57 per cent of the population is multilingual. Cultural enclaves such as Chinatown, Koreatown and Thai Town are just some of the areas reflecting its cultural diversity.

Why is there greater cultural diversity in cities?

Cultural diversity results largely from immigration and immigrants are more likely to choose to live in urban areas. There are a number of reasons for this:

- cities tend to offer a greater range of employment opportunities
- cities are the first point of entry into the country for many immigrants
- cities tend to house earlier immigrant groups with the same ethnicity
- established cultural diversity in cities means there are specialist ethnic shops and religious centres located there
- urban populations tend to be more tolerant of immigrants.

In the UK, there have been a number of significant migrations which have shaped the cultural diversity of major cities (Figure 9.21). The port city of Liverpool attracted many Irish migrants in the nineteenth and twentieth centuries and there are large concentrations of the Indian ethnic minority in cities such as Leicester and Greater Manchester, where labour-intensive industries such as clothing were traditionally located. More recently, the influx of Eastern European migrants has led to changes in the demographic make-up of many British cities. In Southampton for example, more than 10 per cent of the population are now Polish and specialist Polish supermarkets and restaurants have opened to cater for them.

Cultural diversity brings many benefits to an urban society. The most commonly cited advantages relate to the greater exposure people get to different foods, music, language and religion. Events such as the Notting Hill

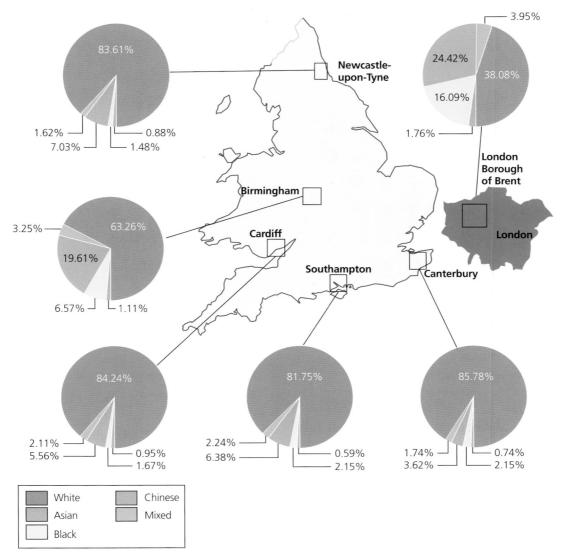

Figure 9.21 The ethnic breakdown of population for a selection of British cities, 2009

Carnival in London and Mela in Newcastle have become part of the British calendar and serve to illustrate an acceptance of and interest in different cultures.

Issues associated with cultural diversity

Cultural diversity can put extra pressure on already stretched urban services. Where language differences exist, local authorities may need to provide English lessons or bilingual literature. Hospitals may need to cater for specific illnesses and schools may alter their curricula and holiday patterns to cater for different ethnic groups. Variations in educational attainment have also been noted and it is the responsibility of local authorities to ensure that all children have the same opportunities.

GIS focus

You can use the 2011 census data to investigate cultural differences in your local urban area. Go to www.datashine.org.uk and select the Origins & Beliefs data. Compare the patterns shown with some of the other statistics available, such as employment rates, educational attainment, housing and health indicators.

Many countries have adopted a multicultural policy to protect and celebrate cultural diversity but some argue that at an urban level, this can encourage culturally and spatially distinct communities leading 'parallel lives'. This is known as segregation and there is evidence of this in cities around the world. In Dubai for example, British migrant workers tend to live in expat (sometimes

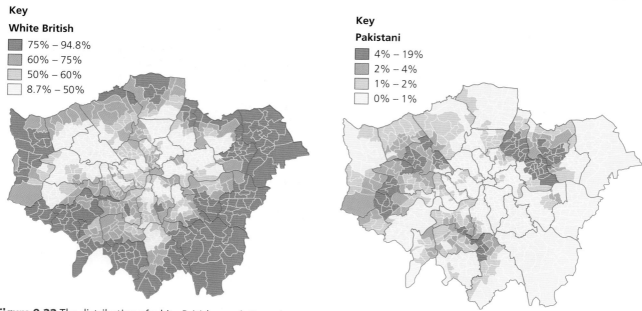

Figure 9.22 The distribution of white British populations shows a concentric pattern around the edge of London while those of Pakistani descent are concentrated in three main areas of the city (2011 census)

gated) communities and integrate little with local society. London has also witnessed an influx of very wealthy immigrants in recent years and these have tended to cluster together. The more affluent boroughs of Mayfair and Knightsbridge are home to a multitude of millionaires from the Middle East and the former Soviet Union and the price of housing in these areas now far exceeds the reach of even wealthy British people.

Social segregation based on ethnicity

In some European and North American cities, different ethnic communities have become isolated from wider society as they have maintained their own language and beliefs and limited their interaction with others.

Local schools can sometimes become dominated by a particular group and this can lead to suspicion and hostility as younger people do not get to know each other.

In American cities, the term **ghetto** has been used to describe an area of a city where the population is almost exclusively made up of an ethnic or cultural minority. As in many European cities, these are often located in the poorer parts of the city, which wealthier residents have left and where unemployment rates are high.

The place study on Brick Lane and Spitalfields in Chapter 8 highlights the large concentration of people of Bangladeshi descent in East London. There are

Figure 9.23 Reasons for ethnic segregation in urban areas

geographical patterns for other ethnic groups in the capital too (Figure 9.22).

Such geographical patterns tend to be the result of self-segregation, but there are also external factors which have encouraged ethnic minorities to live in particular areas. These are shown in Figure 9.23.

In spite of rising tensions between ethnic groups in some cities, analysis of the 2011 census data revealed that many large cities in the UK, including Leicester, Birmingham, Manchester and Bradford, recorded a decrease in segregation for most ethnic groups between 2001 and 2011. Figure 9.24 shows the changes for Indian and Chinese groups across England and Wales. In London too, even the most diverse wards of Brent and Newham experienced a further decrease in segregation.

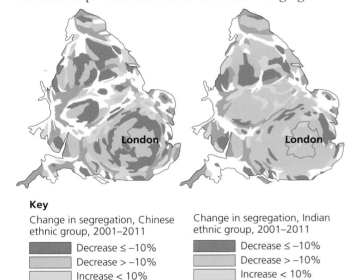

Key

Change in segregation, Chinese ethnic group, 2001–2011
- Decrease ≤ –10%
- Decrease > –10%
- Increase < 10%
- Increase ≥ 10%
- < 200 residents

Change in segregation, Indian ethnic group, 2001–2011
- Decrease ≤ –10%
- Decrease > –10%
- Increase < 10%
- Increase ≥ 10%
- < 200 residents

Figure 9.24 Changing rates of segregation for British Asian populations 2001–2011

To tackle the negative issues associated with ethnic segregation, a policy of **interculturalism** has been promoted by some. This emphasises interaction and the exchange of ideas between different cultural groups. Areas of mutual interest are found and community engagement is conducted in 'intercultural spaces' such as libraries, schools, sports clubs and community centres.

In the UK, there is legislation on anti-racism, employment rights and opportunities to combat discrimination, prejudice and racism. Encouraging greater political involvement of different cultural groups can also encourage greater integration and provide a voice for those who may feel under-represented.

9.4 Urban climate

How do climatic conditions vary across a city and what are the key factors influencing this?

Urban areas create their own climate and weather or **microclimate**. This is sometimes referred to as a 'climatic dome' within which the weather is different from that of surrounding rural areas in terms of temperature, relative humidity, precipitation, visibility, air quality and wind speed (Table 9.6, page 406). For a large city, the dome may extend upwards to 250–300 m and its influence may continue for tens of kilometres downwind.

Key terms

Albedo – The reflectivity of a surface. It is the ratio between the amount of incoming insolation and the amount of energy reflected back into the atmosphere. Light surfaces reflect more than dark surfaces and so have a greater albedo.

Microclimate – The small-scale variations in temperature, precipitation, humidity, wind speed and evaporation that occur in a particular environment such as an urban area.

Particulate air pollution – A form of air pollution caused by the release of particles and noxious gases into the atmosphere. Emissions of particles can occur naturally but they are largely caused by the combustion of fossil fuels.

Photochemical pollution – A form of air pollution that occurs mainly in cities and can be dangerous to health. Exhaust fumes become trapped by temperature inversions and, in the presence of sunlight, low-level ozone forms. It is associated with high-pressure weather systems.

Temperature inversion – An atmospheric condition in which temperature, unusually, increases with height. As inversions are extremely stable conditions and do not allow convection, they trap pollution in the lower layer of the atmosphere.

Urban heat island – The zone around and above an urban area, which has higher temperatures than the surrounding rural areas.

There are two levels within the urban dome. Below roof level there is an urban canopy where processes act in the space between buildings (sometimes referred to as 'canyons'). Above this is the urban boundary layer. The dome extends downwind and at height as a plume into the surrounding rural areas.

Table 9.6 The effects of urban areas on local climate

Element of climate	Effect of urban area (compared to nearby rural areas)
Temperature	
Annual mean	0.5–0.8°C increase
Winter minimum	1.0–1.5°C increase
Precipitation	
Quantity	5–10% increase
Days with less than 5 mm	10% increase
Relative humidity	
Annual mean	6% decrease
Winter	2% decrease
Summer	8% decrease
Visibility	
Fog in winter	100% increase
Fog in summer	30% increase
Wind speed	
Annual mean	20–30% decrease
Calms	5–20% increase
Extreme gusts	10–20% decrease
Radiation	
Ultraviolet in winter	30% lower
Ultraviolet in summer	5% lower
Total on horizontal surface	15–20% lower
Pollution	
Dust particles	1,000% increase

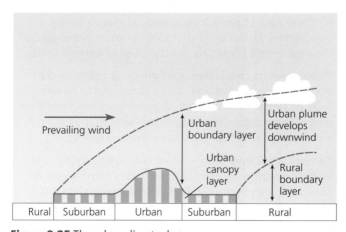

Figure 9.25 The urban climate dome

Temperatures: the urban heat island effect

Urban and suburban areas experience higher temperatures than the surrounding rural areas and this difference is what constitutes an **urban heat island (UHI)** (Figure 9.25). The annual mean air temperature of a city with one million or more people can be 1 to 3°C warmer than its surroundings and on a clear, calm night, this temperature difference can be as much as 12°C. Temperatures will fluctuate in urban areas dependent on season, weather conditions, sun intensity and ground cover. Surface urban heat islands are typically largest in the summer. Smaller urban areas will produce heat islands, but the effect tends to decrease as city size decreases.

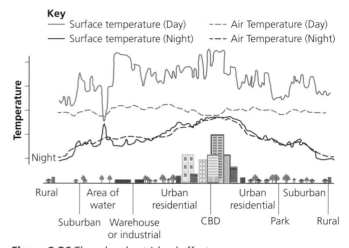

Figure 9.26 The urban heat island effect

Source: Adapted from US Environmental Protection Agency

Figure 9.26 shows a transect from the rural urban fringe through to the high-rise characteristics of the central business district and back to the rural-urban-fringe. It can be seen that typically the urban temperatures are at their highest in the mid-afternoon over the CBD. Secondary peaks of high temperature appear over other built-up areas such as the suburban residential areas. The temperature range from rural to city centre is often greatest at night due to the high heat storage capacity of building materials compared to vegetation. There is very little variation in surface temperatures over areas of water. This is because water maintains a fairly constant temperature over a 24-hour period due to its high heat capacity.

Figure 9.27a Landsat satellite image of Atlanta, Georgia, 2000 – aerial photo

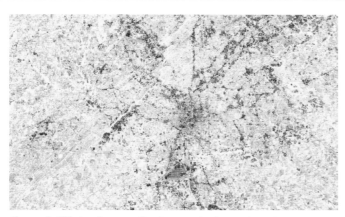

Figure 9.27b Landsat satellite image of Atlanta, Georgia, 2000 – land surface temperature map

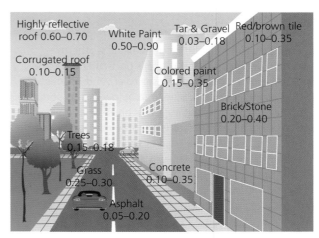

Figure 9.28 Urban environment albedos tend to be much lower than those of rural areas. Rural surface albedos may include deciduous forest (0.17), coniferous forest (0.14) and sand (0.37)
Source: Adapted from US Environmental Protection Agency

The Landsat satellite images shown in Figures 9.27a and 9.27b provide two views of urban Atlanta, Georgia, USA on 28 September 2000. The top image is an aerial photo-like view of the area, where trees and other vegetation are green, roads and development appear grey, and bare ground appears tan or brown. The bottom image is a land surface temperature map, in which cooler temperatures are yellow and hotter temperatures are red. Because vegetation cools the surface through evaporation of water, the most densely vegetated areas (darkest green in top image) are the coolest areas (palest yellow in bottom image). Where development is densest, the land surface temperature is near 30° C.

Cities tend to be warmer than rural areas for the following reasons:

- Surfaces in the city tend to be much less reflective than those in rural areas. Building materials such as concrete, bricks and tarmac have a much lower **albedo** and act like bare rock surfaces, absorbing large quantities of heat during the day (Figure 9.28). Much of this heat is stored and slowly released at night. Some urban surfaces, particularly buildings with large windows, have a high reflective capacity, and multi-storey buildings tend to concentrate the heating effect in the surrounding streets by reflecting energy downwards. In winter, rural areas keep snow for a much longer period and therefore have a greater albedo ranging from 0.86–0.95.

- Air pollution from industries and vehicles increases cloud cover and creates a 'pollution dome', which allows in short-wave radiation but absorbs a large amount of the outgoing radiation as well as reflecting it back to the surface.

- In urban areas, water falling on to the surface is disposed of as quickly as possible. This changes the urban moisture and heat budget – reduced

evapotranspiration means that more energy is available to heat the atmosphere.

- Heat comes from industries, buildings and vehicles, which all burn fuel (Figure 9.29). Although they regulate the temperature indoors, air conditioning units release hot air into the atmosphere. Even people generate heat and cities contain large populations in a small space.

Figure 9.29 This infrared thermal image shows that buildings and roads emit much more heat than lighter surfaces and vegetation.

Why is the urban heat island a matter of concern?

Considerable work has been done on the UHI in London in recent years and a number of concerns have been highlighted:

- As temperatures rise in the summer months, conditions can become uncomfortable in buildings and on city transport systems. During extreme heat island events, the cases of heat stroke, asthma, organ damage and even death increase. Vulnerable groups such as babies and the elderly are most likely to be affected.

Fieldwork opportunities

Urban temperatures can be mapped using isotherm maps as shown in Figure 9.30. In this particular case, the temperature in central London rises to around 3°C higher, while Richmond Park (dark blue area below left) is about 1°C cooler than its surroundings. It is possible to investigate the presence of an urban heat island in your local urban area by taking temperature readings for a transect from the rural-urban fringe through the urban centre and out to the other side of the settlement. Some UHI studies have been done by individuals on a bicycle but it is better to have a larger group of students who can record the temperature

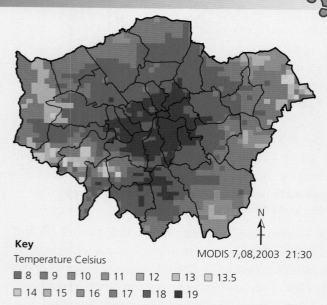

Key
Temperature Celsius   MODIS 7,08,2003 21:30
■ 8 ■ 9 ■ 10 ■ 11 ■ 12 ■ 13 □ 13.5
□ 14 ■ 15 ■ 16 ■ 17 ■ 18 ■ 19

Figure 9.31 Surface temperatures across London recorded by the MODIS satellite on 7 August 2003 at 21:30 hrs during the August 2003 heat wave. During this heat wave, the UHI intensity reached 9°C on occasions

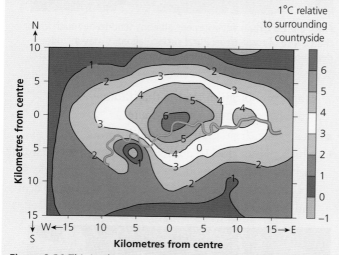

Figure 9.30 This isotherm map shows temperature differences between a rural reference climate station and a number of urban climate stations located across London, under calm and dry conditions at 02.00–03.00 hrs for six urban heat island events during the summer of 2000 (1 July to 30 September).

at allotted points at exactly the same time of day or night.

An alternative view of London's UHI can be gained from the analysis of surface temperatures as measured by infrared cameras located on satellites. Figure 9.31 shows there is a positive correlation between high surface temperatures and high density urban areas. The relatively cool areas to the southwest coincide with the large open and green spaces of Richmond Park.

- The hot and still, anticyclonic weather conditions responsible for intense urban heat island events also produce higher air pollution levels. This is because the chemical reactions that produce ozone and smog are accelerated by high temperatures while the lower wind speeds keep the heat and pollution trapped in the city.
- Excessive heat puts an increased strain on the supply of energy for cooling and air conditioning.
- In warmer periods, the added heat from the urban heat island can lead to increased water consumption by residents and businesses. This places extra strain on the water supply infrastructure and can lead to water-use restrictions. Evapotranspiration rates will also be higher with the result that plants and trees will potentially extract water from the soil at greater rates than normal.

- The earlier flowering times of plants and trees in cities and a prolonged growing season may cause discomfort for city residents who face a longer allergy season.
- The prolonged survival and higher reproduction rates of some animal and insect pests can be problematic and there is a greater potential for algal blooms in water courses as a result of rising temperatures.
- There is an increased risk of the deterioration of historical monuments and buildings through increased rates of temperature-related chemical weathering.
- Climate change is expected to increase the intensity of the urban heat island effect in most urban areas.

Strategies for managing the urban heat island

In light of the largely negative consequences of extreme urban heat island events, urban planning and design has increasingly focused on strategies to reduce the urban heat island effect:

- **Cool surfaces**: Cool roofs built from materials with high solar reflectance or albedo absorb and store less solar energy during the day and thus are not major emitters of heat into the urban atmosphere at night. Cool pavements are also being trialled around the world.

- **Green roofs**: Like cool roofs, green roofs, which consist of a growing medium planted over a waterproof membrane, can reduce rooftop temperatures by 20–40°C on a sunny day (Figure 9.32). In addition, green roofs reduce rainwater run-off, act as insulators and increase urban biodiversity by providing habitat space for birds and small animals.

- **Urban greening**: Planting trees and vegetation provides shade (surface peak temperature reductions of 5–20°C may be possible) and can have a natural cooling effect as seen by lower temperatures in urban parks around the world. Urban trees also act as carbon stores, can reduce urban flooding by intercepting rainfall and filter pollutants from the air.

Figure 9.32 The curving green roof structure of the School of Art, Design and Media at Nanyang Technological University in Singapore

- **Sky view factor**: Sky view describes the relative openness between buildings in an urban area. A restricted sky view, as found for narrow streets and tall buildings, will reduce the escape of heat from street and building surfaces. This can contribute to the accumulation of heat within 'street canyons' and lead to the increase of air temperatures. In addition, if streets are angled perpendicular to the prevailing wind, during intense urban heat island events this will reduce the chances of ventilation and removal of heat and pollutants that accumulate between the buildings.

- **Cool cars**: A lighter coloured car shell reflects more sunlight than a traditional dark car shell. This cools

the inside of the car and reduces the need for air conditioning. A recent study found that after parking in the sun for an hour, a silver Honda Civic (shell SR 0.57) had a cabin air temperature about 5–6°C lower than an otherwise identical black car (shell SR 0.05). Cars in cities contribute significantly to the higher temperatures and pollution levels experienced there and so the use of cool cars would benefit cities (and car drivers) significantly.

London skyscraper responsible for melting car parts

Figure 9.33 The 'walkie-talkie' building, 20 Fenchurch Street, London. The brise soleil is evident at the top of the building.

Individual buildings can sometimes have unexpected impacts on their immediate surroundings. In September 2013, the 525 ft tall building at 20 Fenchurch Street in London, nicknamed the 'walkie talkie' was blamed for melting car parts. The concave design and mirrored glass caused the sun to shine intense rays of light on to the pavement below and this led to damaged wing mirrors and plastic panelling on parked cars. A special architectural sunshade known as a *brise soleil* was subsequently fitted to prevent the problem.

Precipitation

Rainfall can be higher over urban areas than rural areas. This is partly because higher urban temperatures encourage the development of lower pressure over cities in relation to the surrounding area. Convection rainfall tends to be heavier and more frequent, as does the incidence of thunder and lightning. There are several possible reasons for this:

- The urban heat island generates convection. As ground surfaces are heated, rapid evapotranspiration takes place and can result in cumulus cloud and convectional weather patterns.
- The presence of high-rise buildings and a mixture of building heights induces air turbulence and promotes increased vertical motion.
- Due to the low pressure caused by rising air, surface winds are drawn in from the surrounding rural area. This air then converges as it is forced to rise over the higher urban canopy. A similar process occurs as the prevailing wind moves over the city. Friction from the urban boundary creates an orographic process similar to a mountain barrier but the moving air may split apart due to the barriers created by high-rise buildings. As the air comes back together downwind of the high buildings, they are thought to converge and rise upwards forming rain clouds.
- City pollution can increase cloud formation and rainfall. Pollutants act as hygroscopic (water attracting) nuclei and assist in raindrop formation. There is also some suggestion that city pollution enhances the chance of lightning as the cloud droplets take on different electrical charges.
- Cities may also produce large amounts of water vapour from industrial sources and power stations.

Studies have shown that rainfall downwind of major urban areas can be as much as 20 per cent greater than it is in upwind areas. The heating of the surface and the overlying air creates instability in the atmosphere that encourages air to rise. As it rises, it cools, and water vapour condenses into rain that falls downwind of the city.

Fog

In cities, the occurrence of fog increased along with industrialisation. Records of London weather show that in the early 1700s there would have been about 20 days of fog every year but by the end of the 1800s this had risen to over 50 days. It was discovered in the 1950s that the average number of particles in city air in the more developed world was much greater than in rural areas. The particles acted as condensation nuclei and encouraged fog formation at night, usually under high-pressure weather conditions.

In the UK the Clean Air Acts of the 1950s resulted in a dramatic reduction in smoke production and particulate emissions, and a decrease in the number of foggy days. In contrast, cities undergoing more recent industrialisation events are experiencing more fog.

Figure 9.34 People wearing face masks in Beijing. In December 2015, the first ever air pollution red alert was declared in the city as heavy winter fog combined with polluted air.

Cities such as New Delhi and Beijing (Figure 9.34) suffer regular winter fogs and the term 'airpocalypse' has been used to describe the high death toll which occurs when they trap pollutants to create a toxic smog.

Thunderstorms

Thunderstorms develop in hot humid air and are characterised by violent and heavy precipitation associated with thunder and lightning. In urban areas the chance of thunderstorms is increased, particularly during the late afternoon and early evening in the summer months.

Thunderstorms are produced by convectional uplift under conditions of extreme instability. Cumulonimbus clouds may develop up to the height of the tropopause, where the inversion produces stability. The updraught of air through the central area of the towering cloud causes rapid cooling and condensation. This leads to the formation of water droplets, hail, ice and super-cooled water, which coalesce during collisions in the air. During condensation, latent heat is released that further fuels the convectional uplift. As raindrops are split in the updraught, positive electrical charge builds up in the cloud. When the charge is high enough to overcome resistance in the cloud, or in the atmosphere, a discharge occurs to areas of negative charge in the cloud or to Earth. This produces lightning. The extreme temperatures generated cause a rapid expansion of the air which develops a shock wave. This is heard as thunder.

Wind

Urban structures and layout have an effect on wind speed, direction and frequency. Buildings can exert a powerful frictional drag on the air moving around them and this can cause changes in wind speed and direction. There are three main types of effects:

- The surface area of cities is uneven because of the varying height of the buildings. Buildings exert a powerful frictional drag on air moving over and around them. This creates turbulence, giving rapid and abrupt changes in both wind direction and speed. Average wind speeds are lower in cities than in the surrounding areas and they are also lower in city centres than in suburbs.
- High-rise buildings may slow air movement but they also channel air into the *canyons* between them. Winds in such places can be so powerful that they make buildings sway and knock pedestrians off their feet.
- On calm and clear nights when the urban heat island effect is at its greatest, convectional processes can draw in strong localised winds from cooler surrounding areas.

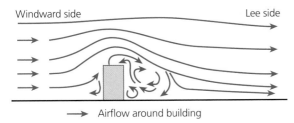

Figure 9.35 The effects of large urban buildings on winds

Winds are therefore affected by the size and shape of buildings. Figure 9.35 shows how a single building can modify an airflow passing over it. Air is displaced upwards and around the sides of the building and is also pushed downwards in the lee of the structure.

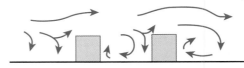
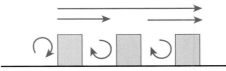

Figure 9.36 Airflow modified by a single building

Figure 9.36 shows the windward side of a building in more detail. The air pushes against the wall here with relatively high pressures. As the air flows around the sides of the building it becomes separated from the walls and roof and sets up suction in these areas. On the windward side the overpressure, which increases with height, causes a descending flow. This forms a vortex when it reaches the ground and sweeps around the windward corners. This vortex is considerably increased if there is a small building to windward.

In the lee of the building there is a zone of lower pressure, causing vortices behind it. If two separate buildings allow airflow between them, then the movement may be subject to the **Venturi effect** in which the pressure within the gap causes the wind to pick

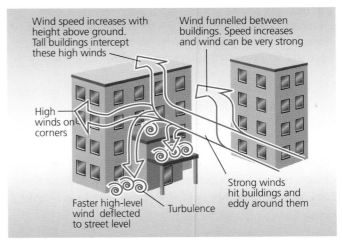

Figure 9.37 Airflow in urban areas modified by more than one building

Coping with the wind: the Burj Khalifa, Dubai

Figure 9.38 The Burj Khalifa in the United Arab Emirate state of Dubai is the world's tallest building

At over 828 m (2,716.5 feet) high, the Burj Khalifa is the tallest building in the world. At such heights, a major concern for any structure is wind stress and, in particular, wind vortexes pulling the building from side to side. The Burj Khalifa has to withstand gusts of over 240 kilometres per hour and a series of aerodynamics improvements had to be made to ensure the building could withstand these. The softened edges of the building deflect the wind around the structure and prevent it from forming whirlpools or vortices, and the entire tower was orientated relative to the prevailing wind direction. The 206-storey Burj Khalifa may still sway slowly back and forth by about 2 metres at the very top but it does not suffer any of the problems experienced by skyscrapers in the past. In the 1970s, for example, people working in the 240-metre tall John Hancock building in Boston, USA suffered motion sickness even in moderate winds and over 5,000 glass panels blew out, shattering on to the pavements below.

up speed and reach high velocities. Some buildings have gaps in them, or are built on stilts or podiums, to avoid this problem. However, a reasonable flow of air at street level is essential to remove pollution. Building design can also reduce ground level wind nuisance. For example, porches over doorways avoid pedestrians being down-blasted by wind.

Usually buildings are part of a group and the disturbance to the airflow depends upon the height of the buildings and the spacing between them. If they are widely spaced, each building will act as an isolated block, but if they are closer, the wake of each building interferes with the airflow around the next structure and this produces a complex pattern of airflow (Figure 9.37, page 411).

When buildings are designed it is important that pollution emitters (chimneys) are high enough to ensure that pollutants are released into the undisturbed flow above the building and not into the lee eddy or the downward-flowing air near the walls.

Utilising the wind: Masdar City, Abu Dhabi

The reliance on oil in the Middle East is not sustainable and the Masdar Initiative in Abu Dhabi therefore focuses on renewable energy in an attempt to diversify the future energy mix. It is hoped that the zero-carbon, zero-waste Masdar City will be home to 40,000 residents when it is completed in the early 2020s, providing an ultra-modern example of sustainable urban living.

The power of the sun has been harnessed to provide plentiful solar energy in Masdar City but the local winds have also been used to save energy. In the central atrium of the Masdar Institute, a barajeel, or wind tower provides cool air and a fine water mist reducing the need for air conditioning. Meanwhile, the Siemens headquarters is positioned to face the direction of the prevailing wind, making use of a Venturi tunnel underneath the building to maintain a cooling airflow through specially designed wind channels. Thanks to the Venturi effect, a breeze flows up to the roof of the building through atria in the buildings structure, cooling public spaces without any energy costs.

Figure 9.39 The orientation of the buildings at the Masdar Institute was designed to optimise street shading. The city grid was also angled to help capture and funnel prevailing winds to aid cooling

Table 9.7 Types of urban air pollution

Pollutant	Cause	Impacts
Carbon monoxide (CO) is a colourless, tasteless, odourless, poisonous gas produced by incomplete combustion of fuel	It is estimated that road transport is responsible for almost 90 per cent of all carbon monoxide emissions in the UK. Concentrations tend to be highest close to busy roads	Carbon monoxide affects the transport of oxygen around the body by the blood. Breathing in low levels can result in headaches, nausea and fatigue
Nitrogen dioxide reacts with hydrocarbons in the presence of sunlight to create ozone, and contributes to the formation of particles	Road transport is estimated to be responsible for about 50 per cent of total emissions of nitrogen oxides	Nitrogen dioxide can inflame the lining of the lung and impacts are more pronounced in people with asthma Oxides of nitrogen can also cause accelerated weathering of buildings and acid rain
Particles or particulate matter (PM) are tiny bits of solids or liquids suspended in the air	Particles originate mainly from power stations and vehicle exhausts. Other particulate matter includes small bits of metal and rubber from engine wear, dust, ash, sea salt, pollens and soil particles	Particles smaller than about 10 micrometers are referred to as PM10 and can settle in the airway and deep in the lungs, causing health problems
Sulphur dioxide (SO2) is a colourless gas with a strong odour produced when a material or fuel containing sulphur is burned	In the UK the major contributors are coal and oil burning by industry such as power stations and refineries	Short-term exposure may cause coughing, tightening of the chest and narrowing of the airways. Sulphur dioxide can also produce haze, acid rain, damage to lichens and plants and corrosion of buildings

Air quality

Air quality in urban areas is often poorer than in rural areas (Table 9.7). **Particulate air pollution** is caused by the release of particles and noxious gases into the atmosphere. Emissions of particles can occur naturally but they are largely caused by the combustion of fossil fuels. A combination of dust, soot and gases are produced from vehicles and industrial processes and this has had a negative impact on human health in many cities around the world.

Air pollution varies with the time of year and with air pressure. Concentration of pollutants may increase five- or six-fold in winter because **temperature inversions** trap them over the city (Figure 9.40).

The mixture of fog and smoke particulates produces **smog**. This was common in European cities through

the nineteenth and first half of the twentieth centuries because of the high incidence of coal burning. Britain suffered particularly badly, many of the smogs being so thick they were known as 'pea-soupers'. In the month of December 1952, smog in London was responsible for more than 4,000 deaths.

Smogs remain a feature of many cities around the world today and headlines regularly focus on the negative impact they have on human health. A particular concern has been the increase in **photochemical smog**. Photochemical oxidants (ozone and peroxyacetyl nitrate (PAN)) are associated with damage to plants and a range of discomforts to people including headaches, eye irritation, coughs and chest pains. The action of sunlight on nitrogen oxides (NO_x) and hydrocarbons in vehicle exhaust gases causes a chemical reaction which results in the production of ozone. (Do not confuse this low-level ozone with the high-level ozone in the atmosphere, which protects the Earth from damaging ultraviolet radiation.) Los Angeles has had a serious problem with photochemical smog for decades because of its high density of vehicles, frequent sunshine and basin topography that traps photo-oxidant gases at low levels (Figure 9.41, page 414). The chemical PAN, linked to vehicle emissions, has been deemed particularly hazardous and linked to the famous Los Angeles 'eye-sting'. Numerous attempts have been made to improve public transport in the city but Los Angeles is a large,

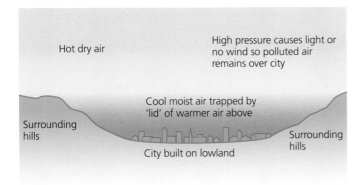

Figure 9.40 An urban temperature inversion

Figure 9.41 The skyline of Los Angeles, USA, with a layer of brown haze or smog hanging over the city

sprawling city and it is much easier to get around with a car. Consequently more recent strategies have included stricter city vehicle emission standards and the Clean Air Action Plan (CAAP) for the port of Los Angeles.

Photochemical smog is a particular hazard during anticyclonic conditions because once the air has descended it is relatively static owing to the absence of wind. Such weather systems tend to be stable and can persist for weeks during the summer months.

Pollution reduction policies

Reducing air pollution in cities has become increasingly important for national and local government and a number of different strategies have been adopted.

Clean Air Acts

After the catastrophic London smog of 1952, the British government decided legislation was needed to prevent so much smoke entering the atmosphere. The Act of 1956 introduced smoke-free zones into urban areas and this policy slowly began to clean up the air. The 1956 Act was reinforced by later legislation. In the 1990s, for example, tough regulations were imposed on levels of airborne pollution, particularly on PM10s. Local councils in the UK are now required to monitor pollution in their areas and to establish Air Quality Management Areas where levels are likely to be exceeded.

In London, air quality standards have improved in recent years but in 2015 NOx emissions were still higher than UK and European law recommended. Measures to clean up construction sites have been introduced, since these are responsible for around 12 per cent of London's NOx emissions. The use of dust suppressants at industrial sites has also been increased.

Vehicle control and public transport

Figure 9.42 shows the type of strategies adopted by different cities to reduce the number of vehicles in urban areas. Greater provision of public transport and general restrictions on polluting vehicles can be very effective.

In London, an Ultra-Low Emission Zone (ULEZ) is being introduced from September 2020. This is where exhaust emissions standards are set and a daily non-compliance charge introduced to encourage cleaner vehicles to drive in central London. It is hoped that almost all the vehicles running in central London during working hours could be zero or low emission.

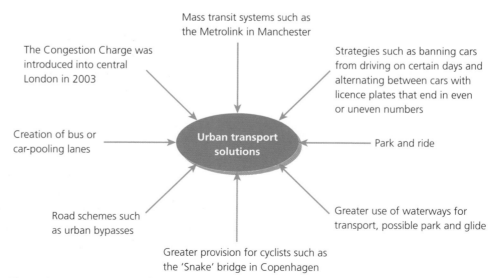

Figure 9.42 Strategies to reduce the number of cars in urban areas

Transport for London's bus fleet will also be upgraded so that all double decker buses operating in central London will be hybrid and all single deck buses will be zero emission.

Zoning of industry

Industry has been located downwind in cities if at all possible and planning legislation has forced companies to build higher factory chimneys to emit pollutants above the inversion layer.

9.5 Urban drainage

Built-up areas need to be drained to remove surface water run-off. Traditionally this has been achieved by using underground pipe systems to convey the water away as quickly as possible. They were not designed to take into account the amenity aspects of drainage systems, such as water resources management, community facilities, landscaping potential and provision of varied wildlife habitats. Water quality issues have become increasingly important; pollutants from urban areas are being washed into rivers or into the ground and once polluted, groundwater is extremely difficult to clean up.

Modern approaches such as **sustainable urban drainage systems (SUDS)** deliver a more holistic approach to managing surface water and wherever possible mimic natural drainage.

Urban hydrology

As already stated above, precipitation falls in greater amounts and with greater intensity in towns and cities than in the surrounding rural area. Natural landscapes like forests, wetlands and grasslands trap precipitation and then allow it to infiltrate slowly into the ground. In contrast, impermeable urban surfaces like roads, car parks and rooftops prevent precipitation from infiltrating; most of it remains above the surface where it runs off rapidly in unnaturally large amounts.

Urban areas are also designed to shed water quickly. Sloping roofs, smooth rounded guttering and cambered roads all contribute to the rapid movement of water away from the surface. Water then is gathered in smooth storm sewer systems which act like a high

density drainage system. It gathers speed and erosional power as it travels underground. When the water leaves the storm drains and empties into streams, they fill rapidly. Figure 9.43 shows how the water cycle is affected when a rural landscape is altered to an urban one.

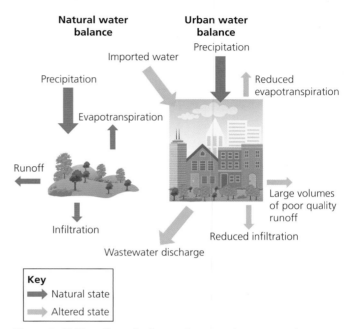

Figure 9.43 The effect of urban surfaces on the water cycle

Because much of the precipitation is unable to infiltrate the impermeable urban surfaces, groundwater and soil water levels are reduced. Since this is the water that feeds streams during dry periods, base level flows are reduced. The resultant storm hydrograph for an urban river (Figure 9.44, page 416) shows a river with a **flashy** discharge but low base flow.

The result of this is that urban areas are more likely to have flooded rivers after heavy rainfall. The combination of population and urban growth along with the predicted increase in the occurrence of severe weather as a result of climate change has meant that many more people are now at risk from flooding.

According to the Asian Development Bank (ADB), the Asian population vulnerable to inland flooding is expected to reach 350 million by 2025 while the Red Cross has stated that almost half of natural disasters they dealt with in 2014 were caused by floods.

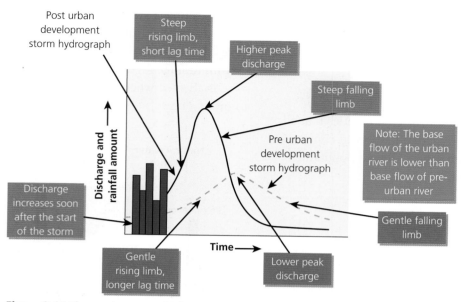

Figure 9.44 Changes in a storm hydrograph for a stream following urban development of its catchment

There are an increasing number of homes in the UK with paved-over front gardens. What are the reasons for this and what might the consequences be on both the urban hydrology and urban climate?

Catchment management in urban areas

While flooding is important when managing catchment in urban areas, it is not the only thing to be taken into consideration by planners.

Figure 9.45 summarises these issues. Pollutants can harm fish and wildlife populations, kill native vegetation, foul drinking water supplies, and make recreational areas unsafe and unpleasant. Sediment

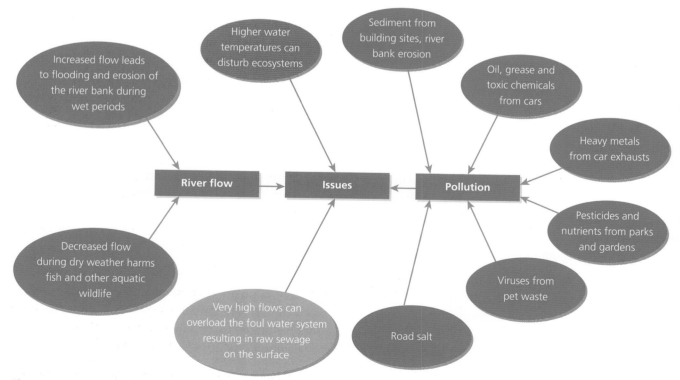

Figure 9.45 Issues associated with urban river catchment management

from erosion can fill spaces between rocks on the stream bottom, thus reducing living space or habitat for the biological communities.

One way to deal with the flood problem is shown by the way the city of Los Angeles has managed its river. The Los Angeles River channel (Figure 9.46) is designed to be **fail-safe**. Devastating flash flooding of the LA River led the city authorities in the 1930s to convert it from natural and meandering to cement and controlled.

While directing the river through this built channel has helped control flooding, it has removed the **ecosystem services** that a river typically provides. In addition, the entire system could be paralysed if one part of the structure sustains significant damage, such as from an earthquake.

Figure 9.46 The channelised Los Angeles River

Key question

What are the advantages and disadvantages of having a drainage system such as the Los Angeles River? Group these into: environmental, social and economic.

An alternative to this hard engineering is a series of measures that are described as **safe-to-fail.** They fall under a general title of (SUDS):

Sustainable urban drainage systems (SUDS)

SUDS is a relatively new approach to managing rainfall by using natural processes in the landscape to reduce flooding, control flooding and provide amenity for the community (Figure 9.47).

Roofwater is collected in water butts for use in gardens or flows to grass channels called swales. It then travels on to grass basins where it is stored before release into local ditches.

Figure 9.47 This bioswale rain garden is a sloped retention area designed to capture and convey water while allowing it to infiltrate the ground slowly over a 24- to 48-hour period. Some of the water is soaked up by the vegetation, thereby reducing flooding by natural means

Rain falling on roads or paths soaks through a permeable block paving where it is filtered and stored in the stone below, or it flows into grass channels, which have a stone filter drain underneath, before it joins the rest of the SUDS system.

Usually there will only be a little water in the detention ponds and the swales when the rainfall is light. If it rains heavily the swales and basins fill for a short period protecting those areas downstream. Water is collected, cleaned and stored in the local landscape, providing an attractive place for play and wildlife.

One example of SUDS being put into place is at the Moor Park Centre in Bispham, North Blackpool.

A case study of this can be found at www.susdrain.org/case-studies/case_studies/moor_park_blackpool.html. Susdrain, the community for sustainable drainage, have many other, mainly English, case studies. A lot of work has also been carried out in the USA. The link below contains more information on the use of such systems in Phoenix, Arizona, which has a desert climate with typical infrequent but heavy rainstorms. www.ecology.com/2015/07/28/changing-way-think-urban-infrastructure/

Key terms

Sustainable urban drainage systems (SUDS) – This is a relatively new approach to managing rainfall by using natural processes in the landscape to reduce flooding, control flooding and provide amenity for the community.

The Cheonggyecheon River project, Seoul, South Korea: an example of river restoration and conservation in a damaged urban catchment

Seoul is one of the world's largest cities and has a population of over 10 million people – 24.5 million in the larger metropolitan area. It is also one of the most densely populated, with over 17,000 people per square kilometre. One central area in the city, Cheonggyecheon, was built over the course of a small seasonal stream and had become overcrowded and blighted by traffic.

The Japanese colonial administration had first begun dredging the Cheonggyecheon stream in 1918, aiming to cover it over – it was seen as a sanitation and flood risk. The Second World War and the Korean War halted all conversion and maintenance work. Refugees and returning populations then established squatter communities along the stream's banks. The stream was fully covered between 1958 and 1961, and a four-lane overpass was built in 1971. This continued until 2003 when an ambitious two-year, $281-million scheme was launched to restore the river. This was transformed into a 5.8 km ecologically sensitive green pedestrian corridor.

Reasons for and aims of the project

- The City of Seoul is in the process of changing from a development-oriented urban landscape dependent on the car to one that values the quality of life of its people and the importance of functioning ecosystems. This mirrors changes in South Korea's planning priorities. In the second half of the twentieth century the country strove for accelerated industrialisation and modernisation. Now, however, there is a different emphasis from both city authorities and residents on health, sustainability and social responsibility.
- The ageing elevated freeway and concrete deck covering the Cheonggyecheon stream posed safety risks and needed to be repaired or removed.
- The government wanted to improve connectivity between the city's north and south sides, which the freeway divided.
- The areas next to the overpass housed a dense warren of over 100,000 small businesses. The overpass had a serious impact on their competitiveness. By the end of the twentiety century, buildings were disintegrating, the area was polluted and there was little green space.
- The project aimed to create both ecological and recreational opportunities along a 5.8 km corridor in the centre of Seoul.

Description of work carried out

- The elevated freeway and concrete deck covering the stream were dismantled.

- Twenty-two bridges – 12 pedestrian bridges and 10 for cars and pedestrians were built to improve movement north-south across the area.
- Car use was discouraged in the area, rapid bus lanes were added, and improved loading and unloading systems were implemented.
- To address the variable flow of the stream, water from the nearby Hanang River is pumped to the area to create a consistent flow with an average depth of 40 cm. This water is treated to make sure that it is not polluted.
- The corridor runs from Seoul to an ecological conservation area outside the city, and is split into three zones which mark the transition from an urban landscape to a natural environment:
  - In the central historic zone, underground waterways were redirected to create a new stream bed with landscaped banks, and the foundations and stones of earlier bridges were included as decorative elements. There is seating throughout to encourage the public to use the space.
  - The middle zone has recreation areas, waterfront decks and stepping stones that bridge the two banks. Its design focused on using environmentally friendly materials. Other features include fountains and waterfalls.
  - The stream widens as it enters the final zone, which is designed to look overgrown and untamed. Sections of the pier and overpass were left as industrial mementoes. Where the stream flows into the Hanang river is a wetlands conservation area.

Figure 9.48 Restoration of the Cheonggyecheon area

Figure 9.49 The Cheonggyecheon area following restoration

The attitudes and contributions of parties involved

- South Korean president Lee Myung Bak included the restoration of the stream in his successful bid to become the Mayor of Seoul in 2001, and it became a priority for his administration.
- Transportation experts were concerned that removing the elevated highway would increase traffic congestion and chaos in the northern end of the city since it carried 169,000 vehicles per day.
- The Cheonggyecheon Restoration Citizens' Committee of professional and citizen groups took responsibility for gauging public opinion. It communicated the project's goals and organised public information sessions.
- Local businesses saw the project as either an interruption or threat to their livelihoods. The Citizens' Committee surveyed the markets in detail, carried out interviews with those likely to be affected, and held regular consultation meetings. To minimise inconvenience and stimulate business activity, the Seoul Metropolitan Authority:
  - provided extra parking
  - reduced parking fees
  - improved the loading and unloading systems
  - promoted Cheonggyecheon businesses
  - provided support, subsidies and grants for business restructuring
  - struck special arrangements with those businesses that had to relocate.

Evaluation of the project

Since the restoration Cheonggyecheon has become popular with residents and visitors alike for rest and relaxation. The stream has become a tourist attraction, drawing an estimated 18.1 million visitors by the end of 2008.

Public facilities, such as the Dongdaemun plaza and various seating schemes, have encouraged a diversity of cultural programming. Parts of the project have become well known as lively and accessible venues.

The Cheonggyecheon Museum, situated on the embankment, has permanent and temporary exhibitions on the area's history and reconstruction.

Economic sustainability

Despite its central location the area had fallen behind the rest of Seoul in terms of economic viability. Since the end of the project, development capital has been invested in residential construction and property prices have risen at double the rates elsewhere in the city. The number of businesses in the areas closest to the restoration work has risen, compared with decreasing trends in most other city districts.

Traffic

Speeds in the central business district slowed by only 12.3 per cent. Bus and subway train usage has increased. The stream-side walk is an attractive alternative to street-level journeys, and there is more pedestrian activity in the area.

The removal of large numbers of cars from this artery has created a 2.5°C reduction in average temperatures. Air flows freely along the path of the stream, creating a cooling wind corridor. A 2005 public survey showed respondents overwhelmingly noticed improvements in air and water quality, noise and smells.

Environment

The stream has re-established lost habitats, plant and animal species have reappeared, and it has become an urban wildlife haven. Schoolchildren have access to a valuable educational resource through the ecology embedded seamlessly within their urban environment.

Inclusive design

The design was non-inclusive. There was limited consideration of certain user groups, for example older people, people with visual impairments and people with mobility problems. In September 2005, a group led a protest march demanding the right to access the new pathways alongside the stream.

In response, lifts were provided at seven locations, together with free wheelchairs for users with mobility problems. Despite this, irregular surfaces are uncomfortable for people who use a wheelchair and poorly-lit, congested tunnels are difficult for people with visual impairments.

Key question

What problems might planners face in the UK if they wished to carry out a scheme such as Cheonggyecheon in a major British city?

9.6 Urban waste and its disposal

Key questions

What are the key issues surrounding urban waste in the twenty-first century and how do these vary in cities around the world?

What are the best strategies for tackling urban waste disposal?

Waste creates a number of big issues for urban authorities and planners. Inadequate waste disposal can be linked to air and water pollution, both of which have a negative impact on human health. It is also becoming increasingly expensive to deal with waste as space for landfill is running out and treatment of waste through incineration is costly. Environmentally, waste is estimated to account for almost 5 per cent of total global greenhouse gas emissions, while methane from landfills represents 12 per cent of total global methane emissions. Recycling rates are increasing but it is clear that much more needs to be done to tackle the growing waste crisis in cities all over the world. Some argue that the first step in waste management is to *stop* calling it 'waste'.

Waste generation

Globally, waste increases by about 7 per cent, year on year. Population growth accounts for much of this increase but economic development also plays a role since greater personal wealth increases consumption of goods and services and this leads to more waste. In urban areas, where there are large concentrations of people, the amount of municipal solid waste (MSW) is particularly high and is set to increase significantly over the next decade as a result of urbanisation and rising

living standards (see Figure 9.50). In 2002, there were 2.9 billion urban residents who generated about 0.64 kg of MSW per person per day. In 2012, this had increased to about 3 billion residents generating 1.2 kg /capita/ day. By 2025 this is predicted to increase to 4.3 billion urban residents generating about 1.42 kg/capita/day of municipal solid waste. Solid waste is very much seen as an 'urban issue' because urban residents produce about twice as much waste as their rural counterparts. Globally, rural dwellers tend to be poorer, purchase fewer store-bought items (which results in less packaging), and have higher levels of reuse and recycling.

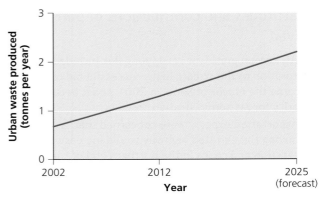

Figure 9.50 Urban waste generation 2002–2025

Source: 'What a Waste: A Global Review of Solid Waste Management', (2012), World Bank

Waste generation varies significantly between cities. As Figure 9.51 shows, 2010 rates of waste production were much higher in cities in high-income countries. This is because waste generation tends to be greater where disposable incomes and living standards are higher. However, it is the cities in low and middle-income countries which are set to see the biggest increase in waste generation over the next decade as a result of rapid urbanisation and continued industrialisation. The amount of municipal solid waste is growing fastest in

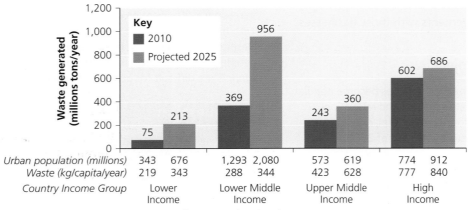

	Lower Income		Lower Middle Income		Upper Middle Income		High Income	
Urban population (millions)	343	676	1,293	2,080	573	619	774	912
Waste (kg/capita/year)	219	343	288	344	423	628	777	840

Figure 9.51 Urban waste generation by income level and year

China. It overtook the US as the world's largest waste generator in 2004.

A tipping point?

One interesting finding to come out of studies on urban waste management is that both richer and poorer cities in a region often outperform middle-income ones. The average resident of Kuala Lumpur, for example, uses 497 litres of water per day and produces 816 kilograms of waste per year. This contrasts starkly to richer Singapore – 309 litres of water and 307 kilograms of waste, but also poorer Delhi – 209 litres of water and 147 kilograms of waste. Somewhere in the midst of economic growth, consumption appears to overtake sustainability and it is not controlled until a city becomes comparatively wealthy. The 'tipping point' in Asia appears to be a per capita GDP of around US $20,000.

Sources of waste in urban areas

Urban waste is made up of millions of separate waste items, the sources of which are detailed in Table 9.8. Some of the larger waste items such as organics (food and horticultural waste) and paper are easier to manage, but wastes such as multi-laminates, hazardous (for example, syringes) and e-waste, pose disproportionately large problems for disposal. This is a particular concern

for low-income countries where they may not have the facilities to properly dispose of them safely. Annually, 50,000 tonnes of hazardous waste is deposited in Cairo.

The composition of waste varies according to a number of factors such as level of economic development, cultural norms, geographical location, energy sources and climate. As a country urbanises and populations become wealthier, consumption of inorganic materials (such as plastics, paper and aluminium) increases, while the relative organic proportion decreases. Generally, low- and middle-income countries have a high percentage of organic matter in the urban waste stream, ranging from 40 to 85 per cent of the total.

Impacts of increasing waste generation

- The costs of collecting and treating waste are high. In lower-income countries, solid waste management is usually a city's single largest budgetary item and it is common for urban authorities to spend 20–50 per cent of their budget on solid waste management.
- Environmentally, waste is a large source of methane, a powerful greenhouse gas. Waste also contributes to water, ground and air pollution.

Table 9.8 Generators and types of solid waste

Source	Waste generators	Types of solid waste
Residential	Households	Food wastes, paper, cardboard, plastics, textiles, leather, yard wastes, wood, glass, metals, ashes, special wastes (e.g. bulky items, consumer electronics, white goods, batteries, oil, tyres) and household hazardous wastes (e.g. paints, aerosols, gas tanks, waste containing mercury, motor oil, cleaning agents) and e-wastes (e.g. computers, phones, TVs)
Industrial	Light and heavy manufacturing, fabrication, construction sites, power and chemical plants	Housekeeping wastes, packaging, food wastes, construction and demolition materials, hazardous wastes, ashes, special wastes
Commercial	Stores, hotels, restaurants, markets, office buildings	Paper, cardboard, plastics, wood, food wastes, glass, metals, special wastes, hazardous wastes, e-wastes
Institutional	Schools, hospitals (non-medical waste), prisons, government buildings, airports	Same as commercial
Construction and demolition In some cities this can represent as much as 40 per cent of the total waste stream	New construction sites, road repair, renovation sites, demolition of buildings	Wood, steel, concrete, dirt, bricks, tiles
Urban services	Street cleaning, landscaping, parks, beaches, other recreational areas, water and wastewater treatment plants	Street sweepings; landscape and tree trimmings; general wastes from parks, beaches, and other recreational areas, sludge

Source: Adapted from 'What a Waste', (1999), World Bank

- Untreated or uncollected waste can lead to health problems such as respiratory ailments, diarrhoea, cholera and dengue fever.
- Many city authorities are struggling to collect increasing quantities of urban waste. The 2012 World Bank report on waste found that 30–60 per cent of urban solid waste in lower-income countries is uncollected. In Cairo, only 40 per cent of daily waste is collected or disposed of in an appropriate way. Much is simply dumped in the desert.
- Cities are running out of landfill space. In 2015, Beirut was plunged into a political crisis after the overflowing landfill site of Naameh, southeast of the city, was closed. The closure led rubbish collectors to pile mountains of untreated waste underneath bridges, by rivers and on the side of roads, leading to civil protests.

Approaches to waste management

Waste management generally follows an accepted hierarchy as shown in Figure 9.52. The key target is to reduce the amount of waste produced in the first place and this can be done through a combination of waste-related legislation, education and financial incentives. At the other end of the spectrum, waste disposal occurs through incineration or landfill.

Methods of waste management and disposal

Landfilling and thermal treatment of waste (incineration) are the most common methods of waste disposal in high-income countries. Most low- and lower middle-income countries dispose of their waste in open dumps. Some of this disposal may be **unregulated.** This means it is not controlled or supervised by regulation of law. Solid waste that is not properly collected and disposed of can be a breeding ground for insects, vermin and scavenging animals, and can thus pass on air- and water-borne diseases. A survey conducted by UN-Habitat in 2009 found that in areas where waste is not collected frequently, the incidence of diarrhoea is twice as high and acute respiratory infections six times higher than in areas where collection is frequent. Environmental threats include contamination of groundwater and surface water by leachate, as well as air pollution from burning of waste that is not properly collected and disposed of.

Recycling and recovery

Resource recovery is the selective extraction of disposed materials for a specific next use, such as recycling, composting or energy generation. **Recycling** is carried out when materials from which the items are made can be reprocessed into new

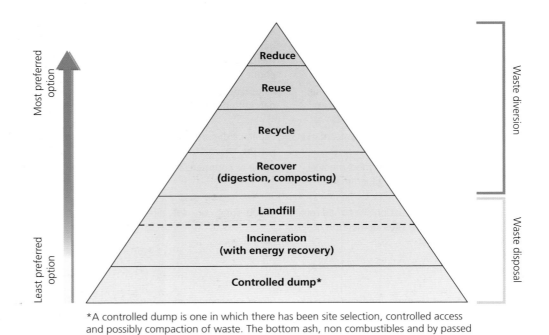

*A controlled dump is one in which there has been site selection, controlled access and possibly compaction of waste. The bottom ash, non combustibles and by passed waste from incineration all go to landfill.

Figure 9.52 The waste hierarchy

products. In recent years, the global market for recyclables has increased significantly. The world market for post-consumer scrap metal is estimated at 400 million tonnes annually and around 175 million tonnes annually for paper and cardboard (UN-Habitat 2009). This represents a global value of at least $30 billion per year. Recycling, particularly in low- and middle-income countries, occurs through an active, although usually informal, sector. Estimates suggest that about 1 per cent of the urban population – at least 15 million people – survive by salvaging recyclables from waste.

Manufacturing new products using recycled materials can save significant energy. For example, producing aluminium from recycled aluminium requires 95 per cent less energy than producing it from virgin materials. **Urban mining** is the name given to the process of recovering compounds and elements from products, buildings and waste which would otherwise be left to decompose in landfills. By collecting and salvaging valuable components to be reused and recycled, there is a greater chance of reducing landfill waste.

The key advantages of recycling and recovery are reduced quantities of disposed waste and the return of materials to the economy. However, there are some negative environmental issues. Notably, energy may be required for the operation of material recovery from waste and this leads to greenhouse gas emissions. Informal recycling by waste pickers will have little greenhouse gas emissions, except for processing the materials for sale or reuse, which can be relatively high if improperly burned (for example, metal recovery from e-waste).

Trade

Waste may be moved between and within countries. The **global waste trade** is the international trade of waste between countries for further treatment, disposal, or recycling. Toxic or hazardous wastes are often exported from high to low-income countries as seen in the example of e-waste in the textbox. However, these countries often do not have safe recycling processes or facilities and hazardous wastes are not properly disposed of or treated, leading to contamination of the surrounding environment. International laws such as the Basel Convention have been introduced to prevent transboundary movement of hazardous waste but evidence suggests it still happens.

Guiyu: The town of electronic waste

Electronic waste, also known as e-waste, refers to discarded electrical or electronic devices. A rapidly growing surplus of electronic waste around the world has resulted from quickly evolving technological advances and an estimated 50 million tons of e-waste are produced each year. According to the UN Step report (2013), e-waste is now the world's fastest growing waste stream. Much of this e-waste comes from the USA and Europe but it is shipped to poorer countries in Asia and Africa to be processed and recycled, often informally. Electronic goods often contain toxic substances such as lead, mercury, cadmium, arsenic and flame retardants. Once in landfill, these toxic materials seep out into the environment, contaminating land, water and the air. Those who work at these sites suffer frequent bouts of poor health.

The town of Guiyu in Guangdong province in China has long been known as one of the world's largest electronic waste dump sites. At its peak, some 5,000 workshops recycle 15,000 tonnes of waste daily including mobile phones, computer screens and computers shipped in from around the world. Large quantities of pollutants, heavy metals and chemicals are released into the air and local water supplies are heavily contaminated by toxic metal particles. Locals suffer respiratory and stomach problems and children living in the area have abnormally high levels of lead in their blood.

Reduction (incineration)

Incineration of waste (with energy recovery) can reduce the volume of disposed waste by up to 90 per cent. General waste can be safely burned at high temperatures and under carefully controlled conditions to produce electricity and heat. This is referred to as **energy from waste** and an increasing number of cities are processing their waste in this way. Incineration without energy recovery is still common but it is not a preferred option due to costs and pollution. The open-burning of waste, which is common in poorer countries, is particularly discouraged due to severe air pollution associated with low temperature combustion.

Burial (landfill)

Burial is the placement of waste in man-made or natural excavations, such as pits or landfills. Landfill sites are a common final disposal site for waste from urban areas. In lower-income countries there may simply be a hole in the ground where open dumping occurs. In higher-income countries, there are much stricter regulations and the types of material that can be sent to landfill are often defined by law. In the UK, most landfill sites now control and collect the gas that is released by the decomposing waste, often using it to generate electricity through turbines.

The environmental problems caused by landfills can be numerous. The greenhouse gas methane is produced by rotting organic matter and other chemicals like bleach and ammonia can produce toxic gases that negatively impact the quality of air in the vicinity. Dust and other forms of non-chemical contaminants can also make their way into the atmosphere. Landfills can also affect groundwater and river quality because toxic chemicals can leach out and contaminate the water.

Since 1999, Singapore has been incinerating much of its waste and sending the incineration ash and non-incinerable waste to an offshore landfill called Semakau Landfill. The Semakau site has been formed by the construction of a dyke in the shallow sea. It is lined with an impermeable membrane to stop leakage into the sea below. Each cell is covered with a layer of earth once it has been filled to ground level. Subsequently, grass and trees take root to form a green landscape. The actual **submergence** of waste in oceans is banned by international convention but, according to the United Nations, some companies have been dumping radioactive waste and other hazardous materials into the coastal waters off Somalia, taking advantage of the fact that the country lacks strong governance.

Table 9.9 A comparison of incineration and landfill

Type of disposal	Advantages	Disadvantages
Landfill	Facilities are properly sited with necessary controls Different types of waste accepted and ordered	Unsightly Often opposed by neighbouring residents Potential leaching of chemicals threatens groundwater supply Decaying matter produces methane, a strong greenhouse gas which is also explosive Landfill takes up a lot of space High transportation costs
Incineration	Can reduce volume of waste needing disposal by 90 per cent Can inactivate disease agents Can reduce toxicity of waste Can be used to produce energy Incinerator bottom ash can be recycled as a secondary aggregate	Expensive Not all waste is combustible Poses challenges of air pollution and incinerator bottom ash disposal Capacity limitations Unpopular with local residents

Landfill vs incineration in Amsterdam

Amsterdam is the cultural capital of The Netherlands – a densely populated low-lying country which has a growing population and a diminishing amount of spare land. In the late twentieth century, rising prosperity in the country had led to the greater production and consumption of goods and a consequent increase in waste generation. Lack of space and growing environmental awareness forced the Dutch government to take measures to reduce their reliance on landfill.

The Dutch approach is simple: avoid creating waste in the first place, recover the valuable raw materials from it, generate energy by incinerating residual waste, and only then dump what is left, but do so in an environmentally acceptable way. This approach – known as 'Lansink's Ladder' – was incorporated into Dutch legislation in 1994.

Landfill

- The increasing level of material consumption and the significant lack of physical space, together with environmental deterioration of the land, forced the Dutch government to take measures early on to reduce the landfilling of waste.

- There were increasing numbers of objections to waste disposal sites from the public due to smell, soil pollution and groundwater contamination.

- In 1995, the government introduced a landfill tax on every tonne of material landfilled. This gave waste processing companies the financial incentive to look for other methods such as recycling and incineration. Landfill tax was increased year on year until 2012 when it was repealed because the low level of landfilling had rendered the tax unnecessary.

- A landfill ban covering 35 waste categories was introduced in 1995.
- The amount of waste sent to landfill decreased significantly in the late 1990s and early 2000s. By 2006, the country had already reached the targets of the Landfill Directive set for 2016.

Waste to energy incineration in Amsterdam: the Afval Energie Bedrijf (AEB) incineration plant

- In a strategy referred to as waste-to-energy (W2E), Amsterdam has created the Afval Energie Bedrijf (AEB) incineration plant capable of producing 1 million MWh of electricity annually. Beyond the energy factor, the plant is also being used to create heating for several communities around Amsterdam, and produces 300,000 gigajoules of heat annually.
- Annually, 1.4 million tons of waste is brought to the W2E plant. This amounts to 600 trucks and 1 freight train per day of refuse from the Amsterdam metropolitan area.
- Sixty-four per cent of the waste that ends up at the plant is recycled.
- Aware of the environmental effects of the gasses from this process, the AEB plant have installed a complex process of scrubbing the flue gases. Attempts are being made to close the loop for other by-products so that the material can be used in other industrial processes—from trace elements for manufacturing, to fly ash for construction. Whatever material is left becomes landfill.
- Next door to the W2E plant is the Waternet water treatment plant. The two plants work together: the incineration plant supplies energy and heat for water treatment processes; the water treatment plant injects its sludge and biogas into the incineration plant as an additional fuel source.

- As the plant performs several functions (elimination of waste, generation of electricity and heat) simultaneously, it compares positively to other disposal methods, actually avoiding 438 kilotons of CO_2 per year.

The environmental comparison

If the equivalent amount of waste was put into landfill, the CO_2 emissions per year would be 1036 kilotons. This would mainly be as a result of methane gases developing at the landfill. If these gases were either captured or burned off, it would reduce the total emissions to 404 kilotons—still higher than the 438 kilotons of CO_2 'saved' by the AEB incineration plant.

Conclusion

The waste to energy strategy in Amsterdam is considered both economically and environmentally better than landfill. However, incineration is still a controversial waste disposal strategy and in 2014 the Dutch government set the target to reduce waste-to-incineration by 50 per cent and focus on improving rates of reduction and recycling. Like other twenty-first century cities aiming for greater sustainability, the movement away from waste disposal to overall waste reduction is the ultimate goal.

> ### Further research
>
> Singapore is another example of a city where it is possible to investigate landfill and incineration approaches to waste disposal. The National Environment Agency of Singapore is a good starting point for research here. Go to http://www.nea.gov.sg/energy-waste/waste-management

9.7 Other contemporary urban environmental issues

Environmental problems tend to be worse in poorer cities experiencing rapid growth and impact most severely on the more vulnerable groups within that urban society. The main problems concern pollution of the air and water and waste disposal but urbanisation is also leading to a loss of land and therefore habitats. By 2030, it is estimated that the urban land area will have expanded by as much as 3.3 million square kilometres. The Cities and Biodiversity Outlook (CBO)

project says that this will result in 'a considerable loss of habitat in key biodiversity hotspots' in cities such as floodplains, estuaries and coastlines. Noise pollution can also be an issue. In Shenzen, China, three-quarters of the complaints about pollution concern noise pollution rather than air or water pollution. Noise pollution is linked mainly to traffic and industry.

This section of the chapter focuses on three of the biggest environmental threats facing cities in the twenty-first century: atmospheric pollution, water pollution and dereliction.

Atmospheric pollution

Atmospheric pollution is caused by the release of particles and noxious gases into the atmosphere and this can have a negative effect on human health. In 2014, the WHO found that urban air pollution was 2.5 times higher than the recommended levels in about half of the urban populations being monitored. This puts urban dwellers at risk of serious, long-term health problems including heart disease, stroke, respiratory illnesses and cancers. An estimated 9,400 Londoners died prematurely from air pollutants in 2010, principally from exposure to nitrogen dioxide and fine PM2.5 particles.

Human activity produces the pollutants but it is primarily the weather that determines what will happen once they are released into the air. During wet or windy conditions pollution concentrations remain low, either blown or washed away. During periods of still, hot weather, pollution is able to build up to harmful amounts, leading to what are known as **pollution episodes**.

Dhaka, the capital of Bangladesh, is the world's fastest growing megacity with as many as 400,000 new migrants arriving each year. It stands on the bank of the Buriganga River and developed as an important trading centre. It has a diverse economy, a high literacy rate and rapidly developing infrastructure, but the city faces major urban challenges such as poverty, crime, congestion and political violence. However, one of the biggest problems the city faces is air pollution caused mainly by vehicle emissions and industrial practices. The two-stroke engines used in auto rickshaws (baby-taxies), tempos, mini-trucks and motorcycles have been identified as highly polluting and there are estimates that about 90 per cent of the vehicles that ply Dhaka's streets daily are faulty and emit smoke far exceeding the prescribed limit. Industrial pollution comes from factories in and around the periphery of Dhaka manufacturing products including textiles, fertilisers, pesticides and pharmaceuticals. However, according to a 2011 World Bank report, it is the brick-making kilns which account for about 40 per cent of Dhaka's fine-particle air pollution (Figure 9.53). The kilns cause 750 premature deaths a year from cancer and cardiopulmonary disease. In 2010, the national government ordered a shutdown of fixed-chimney kilns by July 2013. Facing opposition from kiln owners, the government has extended that deadline several times.

Figure 9.53 Air pollution from the brick kilns of Dhaka. The biggest culprit is the 'fixed-chimney' kilns, which use outdated technology to burn coal imported from India and firewood chopped from nearby forests.

Managing air pollution

A number of strategies used to improve air quality are described on page 414. Further measures include ensuring that houses are energy efficient, that urban development is well served by public transport routes, that street design is appealing and safe for pedestrians and cyclists, and that waste is well managed. More park and woodland areas also help. The environmental company AECOM has calculated that London's 8.3 million trees provide £95 million worth of air filtration every years in terms of health costs avoided. Together, these strategies serve as a catalyst for local economic development and the promotion of healthy urban lifestyles.

Figure 9.54 Air pollution monitoring station in London

GIS focus

City pollution levels are monitored continuously (Figure 9.54) and you can use this data to examine geographical patterns. Go to www.londonair.org. uk/london/asp/nowcast.asp for a 'nowcast', which is a service to show current pollution levels in detail across London in comparison with the government's Air Pollution Index. The Daily Air Quality Index (DAQI) details levels of air pollution in the UK and provides recommended actions and health advice. For European cities, the Urban Air Quality Index is updated every hour and allows comparison of air quality over a 24-hour period. Other forms of social media are increasingly used to convey information to the general public. For example, UK-AIR provides automated tweets about current and forecast air quality including episodes of poor air quality.

There are greater concerns about air pollution in low- and middle-income countries which lack the finances and legislation to adequately tackle the problem. Over 90 per cent of air pollution in these countries is attributed to vehicle emissions due to the high number of older vehicles, poor vehicle maintenance and low fuel quality. More recently, schemes such as the UN Partnership for Clean Fuels and Vehicles aim to improve air quality; there has been greater investment in improving road quality; and commitments have been made to promote non-motorised journeys.

Water pollution

Water pollution is the contamination of water sources including rivers, lakes, oceans, aquifers and groundwater. It occurs when pollutants are directly or indirectly discharged into water without adequate treatment to remove harmful compounds. Indirect sources of water pollution include contaminants that enter the water from soils or groundwater and from the atmosphere via rain. The high concentration of impermeable surfaces in urban areas increases run-off from roads and can carry numerous pollutants such as oils, heavy metals, rubber and other vehicle pollution into waterways and streams. The reduction in water percolation into the ground can also affect the quantity and quality of groundwater. Stormwater run-off in urban areas can overwhelm combined stormwater and wastewater treatment systems when high volume flows exceed treatment capacities. Reports suggest that urban stormwater can be as polluted as untreated domestic wastewater.

Causes of water pollution in urban areas

- Surface run-off from streets carrying oil, heavy metals and other contaminants from motor vehicles.
- Industrial waste.
- Untreated or poorly treated sewage which is low in dissolved oxygen and high in pollutants such as nitrates, phosphorus and bacteria. Treated sewage can still be high in nitrates.
- Rubbish dumps, toxic waste, chemical and fuel storage, which can all leak pollutants.
- Intentional dumping of hazardous substances.
- Air pollution can lead to acid rain, nitrate deposition and ammonium deposition, which can alter the water chemistry of an area.

Consequences of water pollution

Over 1.2 billion people lack access to clean water and waterborne infections account for 80 per cent of all infectious diseases. Increased water pollution creates breeding grounds for malaria-carrying mosquitoes as well as damaging ecosystems, leading to species extinction. For humans, different forms of water pollutants affect health in different ways:

- Heavy metals from industrial processes can slow development, result in birth defects and may be carcinogenic.
- Industrial waste often contains toxic compounds that damage the health of aquatic animals and those who eat them. They can cause immune suppression, reproductive failure or poisoning.
- Microbial pollutants from sewage often result in infectious diseases that infect aquatic life and terrestrial life through drinking water. Microbial water pollution causes diseases such as cholera and typhoid fever, which are a major cause of infant mortality in low-income countries.
- Organic matter and nutrients can cause an increase in aerobic algae and deplete oxygen from the water. This can lead to the suffocation of fish and other aquatic organisms.
- Suspended particles in freshwater reduces the quality of drinking water for humans and the aquatic environment for marine life. They can also reduce the amount of sunlight penetrating the water, disrupting the growth of photosynthetic plants and micro-organisms.

Water problems in India

India suffers chronic water problems. In 2015, a report by the Central Pollution Control Board (CPCB) found that over half of India's rivers were polluted. A primary cause of this was the quantity of domestic sewage generated by rapidly expanding towns and cities. A survey of 27 cities found that untreated sewage flowing in open drains was causing serious deterioration of groundwater quality with knock-on impacts to human health. Vector-borne diseases such as cholera, dysentery, jaundice and diarrhoea are widespread and water pollution was also found to be a major cause of poor nutrition and under development in children. According to the World Health Organization, more than 87 per cent of people in India's cities (compared with 33 per cent in rural areas) now have access to a toilet, but it is the leaking and incomplete sewage systems that are leading to the contamination of rivers and lakes.

Steps are being taken to tackle India's water problems. The Prime Minister Narendra Modi has made cleaning the Ganges, the river that is holy to Hindus, a key policy goal, while there are also plans for infrastructural improvements and wastewater recycling in cities across the country. In the city of Ahmedabad, the state council introduced a scheme in which children are being paid

to use public toilets. Open defecation is a problem where people do not have access to proper sanitation. The number of public toilets has also been increased to prevent human waste running into wells and streams, contaminating water that may then be used for drinking or bathing.

Figure 9.55 The Dharavi slum area in Mumbai – severely polluted with solid and liquid wastes generated by Mumbai

Managing water pollution

Improving water quality requires strategies to prevent, treat and remediate water pollution. Ideally pollutants are prevented from entering watercourses in the first place but the reality is that potential pollutants are treated before they are discharged. The final and often most expensive strategy is when polluted watercourses are restored through remediation.

In most high-income countries, water quality improvement focuses on two approaches:

1. The construction of water-treatment facilities and wastewater plants
2. Regulations aimed at 'point source' polluters such as industries which discharge water pollution into receiving waters or sewer systems that flow into treatment plants

The most difficult water-quality challenge is dealing with 'non-point source' pollution which is the result of precipitation run-off from a wide range of sources including fertilizers and pesticides from agriculture, and chemicals and toxins from urban settlements. These pollutants are difficult to regulate.

Lack of money and inadequate technology in low-income countries has resulted in much lower water quality standards. Effective legislation is often absent and enforcement of pollution controls limited.

Key water pollution strategies

- **Low-impact development (LID)** is a stormwater management approach that can help to reduce stormwater run-off. This is done primarily through the use of vegetation and permeable surfaces to allow infiltration of water into the ground. Permeable streets and pavements, 'green' roofs, rain gardens and more urban parks allow water to infiltrate into soils rather than flow directly into sewers. Filtering stormwater through vegetation and soil has been shown to reduce organic pollutants, oils and heavy metals by more than 90 per cent
- **Legislation, regulation and enforcement:** There are many different anti-pollution laws and agreements in operation worldwide. However, these laws need to be enforced. Some cities have adopted incentive-based approaches charging polluters per unit. Charges start low but are increased if pollution continues, creating an incentive to reduce discharges and purchase

wastewater treatment technologies. Regulation works in a similar way. Factories are allowed to discharge only limited amounts of carefully controlled pollutants. By slowly reducing the levels of permitted discharges, year by year, pollution levels are reduced

- **Education and awareness:** The more people know about the causes and effects of pollution, the more likely they will be to avoid adding to the problem. In 2014, Wessex Water used mobile billboards in hotspot areas urging its customers to get behind a campaign encouraging people to bin wet wipes rather than flush them down the toilet. Wet wipes are a common cause of sewer blockages since they do not decompose like toilet paper.
- **Improvements in sewage and wastewater processing**.
- **Appropriate technology:** One example of this is the Janicki omniprocessor – a small-scale innovation aimed at providing clean water in low-income countries. The omniprocessor first boils raw 'sewer sludge', during which the water vapour is separated from the solids. Those solids are then put into a fire, producing steam that drives an engine producing electricity for the system's processor and for the local community. Finally, the water is put through a cleaning system to produce drinking water.

The Thames Tideway Tunnel

Much of London is served by a combined sewerage system, collecting sewage from homes as well as rainwater run-off from roads, roofs and pavements. The interceptor sewers, constructed by Sir Joseph Bazalgette following the 'Great Stink' of 1858, are still the backbone of London's sewer network today, but they are struggling to cope with the expanding population and the demands of modern-day living. During periods of heavy rain, the Victorian drainage system discharges raw sewage directly into the River Thames, killing large quantities of fish and other aquatic life and threatening public health. In 2013 alone, 55 million tonnes of raw sewage was washed into the river. The Thames Tideway Tunnel is under construction to upgrade London's sewerage system. A specialised tunnel boring machine (TBM) has been used to excavate a circular tunnel, seven metres in diameter, using a rotating cutterhead, while simultaneously creating a tunnel wall using concrete segments. The completed tunnel will be 25 kilometres long and up to 65 metres deep. For more information, go to www.thamestidewaytunnel.co.uk/.

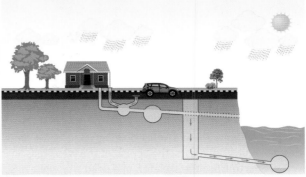

Now:
The low level interceptor sewers fill up and overflow into the River Thames.

After:
The overflow will be diverted into the tunnel instead of going into the river.

Figure 9.56 By intercepting the sewage before it enters the river, the Thames Tideway will help prevent the tidal River Thames from being polluted with untreated sewage, which can stay in the river for up to three months before the tide finally takes it out to sea

Dereliction

Dereliction refers to the state of having been abandoned and become dilapidated. In urban areas, derelict buildings are often associated with former industrial sites or run-down housing estates. In the UK, de-industrialisation led to many people leaving the inner city and industrial buildings being abandoned. Alongside this, services such as public houses and shops may also have become vacant as areas became subject to urban decline.

Dereliction has a negative impact on the surrounding area. Crime and vandalism rates tend to be higher, house prices fall and out-migration of residents takes place. The high costs involved in urban renewal often discourage authorities and individuals from renovation or rebuilding. In addition, the presence of listed buildings, which are subject to considerable planning regulation, can significantly increase the investment needed.

Derelict land can pose a risk to human health. Contamination from industrial processes lives on in an environment long after the industry that produced or used them is gone. In Glasgow, research has linked proximity to contaminated derelict sites with health problems such as low birth weight infants, heart disease, cancer and respiratory disease. Many of the industries prevalent in Glasgow during the industrial period were 'dirty' ones, with high usage of toxic and dangerous chemicals and environmental degradation of the surrounding areas. These industries included shipbuilding, dye works, tanneries, distilleries, cast iron foundries and chemical manufacturing.

Derelict London

For some, derelict buildings provide a fascinating look at our architectural past. The *Derelict London* website includes over 3,000 photographs of London sights not normally seen by tourists. The author Paul Talling clearly states that not all of the photographs are of derelict buildings but, as the *New Statesman* review suggests, the appeal of the website is 'in how it seems to trace the skeleton of a dead city while it is still in apparently rude health.'

The website is a useful resource for looking at how buildings have changed over time. Some properties have fallen into even greater disrepair but there are also photographs taken before and after regeneration has taken place.

Figure 9.57 Lambeth Hospital buildings, Kennington 2008 (left) and 2013 (right)

Lambeth Hospital shut in 1976. Much of the original workhouse was demolished, although the water tower survived but lay derelict. Some of the more modern hospital buildings remained in a derelict condition until being cleared for a new housing development. The former workhouse administrative block is now a cinema museum.

Source: www.derelictlondon.com

Key terms

Brownfield site – A term used in urban planning to describe land previously used for industrial purposes or some commercial uses.

Dereliction – The state of buildings having been abandoned and become dilapidated.

Greenfield site – An area of undeveloped land.

Land remediation – The removal of pollution or contaminants from the ground, which enables areas of derelict former industrial land to be brought back into commercial use.

Strategies to deal with derelict urban sites

One of the most common strategies for tackling urban dereliction is through **regeneration schemes**. Government-led schemes such as Urban Development Corporations, City Challenge and New Deal for Communities have been well documented and have had varying levels of success in tackling derelict areas. The government focus on using **brownfield sites** for new building developments rather than **greenfield sites** has also been successful. Between 1997 and 2009, the proportion of dwellings (including conversions) built on previously-developed land (brownfield sites) increased from 56 to 80 per cent, while the proportion of previously-developed land changing to residential use increased from 47 to 69 per cent. London had the greatest proportion of dwellings (including conversions) built on previously-developed land (98 per cent). The advantages of using brownfield sites in urban areas are numerous: it improves the physical environment; it revives older urban communities; existing infrastructure can reduce costs and encourage faster occupancy; it preserves historical landmarks and heritage architecture; it reduces urban sprawl and it preserves greenfield land. The disadvantages tend to focus on the greater costs of clearing contaminated land and the fact that most brownfield sites are in inner city areas, which have higher levels of traffic congestion and noise.

Land remediation

Land remediation is the term given to the removal of pollution or contaminants from the ground. This enables large areas of derelict former industrial land to be brought back into commercial use. In preparation for the 2012 London Olympics games, a 350-hectare area of East London was 'cleaned up' and has now become the largest new urban park in Europe, with 100 hectares of open land and 45 hectares of new habitat. Over 2.2 million square metres of soil was excavated, of which nearly half was treated by soil washing, chemical stabilisation, bioremediation or sorting. Eighty per cent of the excavated material was re-used on site. A total of 235,000 m³ of contaminated groundwater was successfully treated.

Community action

Community action has been at the heart of Detroit's recent strategy to tackle its derelict land areas. Once the centre of global car production, Detroit's population has fallen from nearly two million in the 1950s to less than one million today. The city now has 40,000 vacant and derelict lots, comprising around one third of its area. Between 1970 and 2000, over 150,000 buildings were demolished and large commercial developments built in order to try and revitalise the area, but the decline continued. Detroit community groups are now using areas of the city for a range of small, community-based activities, including urban farming. By 2010, they had converted over seven hectares of unused land into more than 40 community gardens and microfarms yielding over six tonnes of produce a year, including hay, alfalfa, honey, eggs, milk, beef, flowers, vegetables and herbs. Much of it is produced by volunteers and students, and sold on to other community organisations such as soup kitchens. Derelict buildings have been converted into community centres, cafés and greenhouses.

9.8 Sustainable urban development

Cities pose a threat to both the local and global environment. Cities consume three quarters of the world's resources and generate a majority of the world's waste and pollution. We have already seen how the consequences of waste disposal, pollution and dereliction are felt most severely in the cities in which they are created but, increasingly, the effects of urban growth are also being felt globally. Cities now rely on energy and resources from all over the world while the pollution and waste they generate is also dispersed globally.

The environmental impact of cities can be measured using the **ecological footprint** calculation. This is defined as the total area of productive land and water required to produce the resources a population consumes and absorb the waste produced. In 2007, the average person's footprint globally was 2.7 global ha. However, the variation both between and within countries is huge. Cities tend to have a higher ecological footprint than rural areas and wealthier cities have a higher footprint than poorer ones. The

City Limits survey in London in 2003 found that London's ecological footprint covered an area twice the size of the UK, and that if the entire population of the world made such demands, we would need at least three planets to sustain this level of activity. More recent research by the Global Footwork Network on other cities found that San Francisco's footprint was about 6 percent higher than the average American's (2011) while the average footprint of residents of the 'green' city of Curitiba was more than 40 per cent higher than the Brazilian average (2010). In all these cases, the higher footprint is attributed to the greater affluence of city residents correlating with increased consumption and waste production. The concept of the ecological footprint illustrates the disproportionate impact cities have on the environment and it is clear that if greater global sustainability is to be achieved, reducing the ecological footprint of cities is a priority.

Dimensions of sustainability

One of the post-2015 Sustainable Development Goals is to make cities inclusive, safe, resilient and sustainable and the city as an entity is viewed as a key factor in building a more sustainable world. The idea of sustainability was first brought to wider public awareness in the 1987 Brundtland Report entitled 'Our Common Future' by the UN World Commission on Environment and Development (WCED). **Sustainable development**, it stated, was 'meeting the needs of the present without compromising the ability of future generations to meet their own needs.' The term incorporates social, economic and environmental concerns and a **sustainable city** is therefore one which provides employment, a high standard of living, a clean, healthy environment and fair governance for all its residents, as shown in Figure 9.59 (page 433). People-centred planning has increasingly been incorporated into sustainable urban design and the notion of **liveability** is important here. Liveability may mean different things to different people. For some it is tied to natural amenities such as parks and green space; for others to cultural offerings, career opportunities, economic and political stability or some degree of safety within which to raise a family. In the context of the global liveability rankings, it relates to which cities provide the best or worst living conditions for their residents.

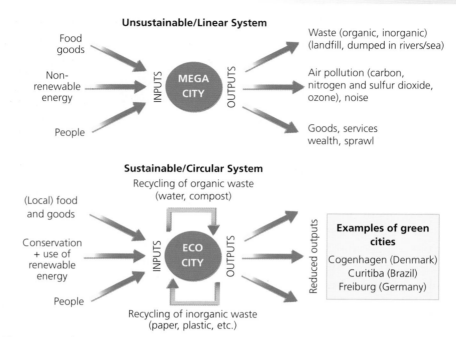

Figure 9.58 The city as a system

Features of sustainable cities

Thinking of the city as a system enables a comparison between the characteristics of a typical megacity and that of a green city. The linear system of the megacity is unsustainable. Uncontrolled use of inputs and outputs leads to resources becoming exhausted and extremely high pollution and waste levels. The circular nature of the sustainable city system shows how some of the outputs are recycled, which reduces both the demand for new input resources and pollution and waste levels (Figure 9.58).

Key terms

Ecological footprint – The total area of productive land and water required to produce the resources a population consumes and absorb the waste produced.

Liveability – The characteristics of a city which improve the quality of life for the people living there.

Sustainable city – A city which provides employment, a high standard of living, a clean, healthy environment and fair governance for all its residents.

Sustainable development – Meeting the needs of the present without compromising the ability of future generations to meet their own needs.

Urban resilience – The capacity of individuals, communities, institutions, businesses and systems within a city to survive, adapt and grow, no matter what kinds of chronic stresses and acute shocks they experience.

Sustainability can be measured in a number of ways and different organisations will use different criteria to rank cities in terms of their sustainability. Terms such as 'eco-city' and 'green city' are often used to describe cities with a good environmental record, but the range of indicators used for measuring urban sustainability needs to be far greater and incorporate the social, economic and political elements highlighted in Figure 9.59 and Figure 9.60.

Another twenty-first century buzzword which can be linked to the idea of sustainability is that of resilience. **Urban resilience** is defined as the capacity of individuals, communities, institutions, businesses and systems within a city to survive, adapt and grow, no matter what kinds of chronic stresses and acute shocks they experience. **Chronic stresses** include day-to-day challenges such as high unemployment, inefficient public transportation systems, endemic violence and chronic food and water shortages. **Acute shocks** are sudden events that may threaten a city, including earthquakes, floods, disease outbreaks and terrorist attacks. A sustainable city is one which can cope with the physical, social, and economic challenges that are a growing part of the twenty-first century.

Go to www.100resilientcities.org to look at some of the proposed ideas for developing greater urban resilience.

Social development	Economic development	Environmental management	Urban governance
• Adequate provision of schools and health services • Availability of food supplies • Green housing and buildings • Clean water and sanitation • Green public transport • Green energy access • Recreational areas and community support	• Decent employment opportunities • Production and distribution of renewable energy • Investment in green technology and innovation	• Waste and recycling management • Energy efficiency • Water management • Air quality conservation • Adaption to and mitigation of climate change • Forest and soil management	• Adoption of green urban planning and design strategies • Strategies to reduce inequalities • Strengthening of civil and political rights • Support of local, national, regional and global links

Figure 9.59 Pillars for achieving urban sustainability

Contemporary opportunities and challenges in developing more sustainable cities

The main obstacle to cities becoming more sustainable tends to be financial. In the context of rapidly growing urban populations and limited budgets, city authorities often choose short-term 'cheaper' solutions over long-term planning. Likewise, in cities where large numbers of people still lack access to water and electricity, providing basic infrastructure tends to take priority over investment in environmental projects and sustainable development. Further challenges to developing sustainable cities include poor infrastructure, weak institutions and lack of enforcement of planning regulations.

Table 9.10 on page 435 highlights some of the main challenges facing cities in low- and high-income countries. Drawn from the United Nations World Economic and Social Survey 2013, the table also suggests possible opportunities for overcoming these problems. Well-documented examples such as Curitiba in Brazil, Singapore, Freiburg in Germany and Shanghai in China could be researched further to illustrate how many of the challenges have been overcome.

Figure 9.60 Key features of sustainable cities

Copenhagen: Europe's most sustainable city?

One European capital city which regularly tops the list of the world's most sustainable cities and was awarded the title of European Green Capital in 2014 is Copenhagen in Denmark.

Social development

- Only two per cent of employees in Copenhagen work more than 40 hours a week. This allows more time for family, friends, hobbies and recreational activities, all of which are associated with lower stress levels.

- The city has 249 miles of cycle lanes, which makes cycling an easy and safe option. More than half of commuters travel to work or school by bike each day. The government actively promotes cycling through projects such as the elevated *Cykelslangen* cycle bridge. It is estimated that the residents of Copenhagen have one million fewer sick days simply due to the fact that they cycle so much.

Figure 9.61 The elevated *Cykelslangen,* or Cycle Snake opened in Copenhagen in 2015. It is a two-way 220-metre bridge. You can get across the harbour in less than a minute creating a less stressful atmosphere for cyclists, pedestrians and vehicle users alike

Economic development

- Copenhagen is the economic and financial centre of Denmark. It is home to a number of international companies and has successful business clusters in IT, biotechnology and pharmaceuticals.

- Copenhagen has some of the highest wages in the world and unemployment rates are low.

- Danes have an entrepreneurial spirit. In Copenhagen they have held on to their independent shops and cafes. Small boutiques and local coffee shops are the norm rather than global chains.

- Reasonable rents in the city allow young start-up companies to flourish.

Figure 9.62 Added to its green credentials, Copenhagen regularly tops the lists of the most desirable cities in which to live

Environmental management

- The Climate Plan 2011 set a target to be carbon neutral by 2025, which will involve halving its emissions and offsetting its remaining carbon use by producing more renewable energy than it consumes.

- The harbour that bisects the city has been transformed from an industrial zone into a cultural and residential hub. Where once 100 overflow channels carried wastewater into Copenhagen harbour after heavy rains, investment in reservoirs and conduits has meant that the harbour water is now so clean that residents can go swimming in it.

- Copenhagen is served by an integrated public transportation network: a driverless Metro, regional trains and buses.

- In 2001, a large offshore wind farm was built just off the coast of Copenhagen at Middelgrunden. It produces about 4 per cent of the city's energy.

- Copenhagen's Finger Plan 2007 includes protection of its green belt and limits urban sprawl through better use of existing city land.

Urban governance

- As a country, Denmark has one of the highest taxation levels in the world but this provides very generous state welfare provision which reduces inequality.

- The Citizen's Dialogue Project is a scheme financed through Copenhagen's annual budget, which involves direct public participation in planning legislation for the city.

- Denmark boasts very high income equality and gender equality and is perceived to be a very fair and egalitarian society.

Table 9.10 Challenges to and opportunities for building sustainable cities

Main urban trends	Developing countries		Developed countries	
	Challenges	Opportunities	Challenges	Opportunities
Social				
By 2025, urban population will live mainly in small (42 per cent) and medium-sized (24 per cent) cities	Improve access to housing, water and sanitation; improve public infrastructure; foster institutional capacity	Investment in public infrastructure (including transportation); construction of compact buildings in middle-income cities; strengthen links between cities and rural areas	Social cohesion	Investment in compact urban development and decentralisation
Number of urban people living in slums continues to grow	Reduce number of urban poor and disease risk; improve social cohesion; reduce youth unemployment	Investment in universal access to affordable water and sanitation; public transportation, and creation of jobs to reduce growth of slums; employment of the 'youth' dividend in low-income countries.	Reduce urban unemployment due to economic crises (of youth in particular); provide adequate housing in poor neighbourhoods	Strengthening and widening of social safety nets; upgrading investments in social protection for an effective response to crises and their aftermath
Inefficient use of public services (water, electricity)	Improve waste and recycling management; support consumption of local produce; change overconsumption patterns of high-income households	Subsidies to households and small firms to reduce non-saving water systems and waste; incentives to local communities to improve recycling systems	Change overproduction and overconsumption styles; improve waste and recycling management	Investment in retrofitting in buildings of water- and energy-saving devices; upgrading of public infrastructure
Ageing	Create productive employment for older persons	Investment in universal pensions; extension of working age; support for family networks	Fiscal pressure to reduce health costs; improve productivity	Investment in retraining older persons, and extending the working age
Economic				
Inequality and financial fragility	Create policy space for inclusive development; reduce underemployment; promote economic diversification	Investment in green industry, adaptation to climate change, structural economic change (industrial and service 'leapfrogging' for least developed countries); strengthening regional cooperation	Reduce unemployment; boost economic growth; strengthen international cooperation	Investment in green infrastructure; policy coherence and coordination
Food insecurity	Improve access to food; increase productivity	Investment in urban agriculture, local crops, storage facilities; R&D	Reduce food waste	Investment in storage infrastructure; reducing food subsidies; policy coordination
Environmental				
Energy access	Provide access to clean energy and reduce use of 'dirty' energy in poor households (e.g. least developed countries); discourage high energy consumption in high-income households)	Investment in capacity development, energy-saving devices, production and use of renewable sources of energy; subsidies and incentives for efficient energy use and water use for middle- and high-income households	Reduce overproduction and overconsumption to sustainable levels	Investment and incentives to produce and use renewable energy sources; decentralization of energy production
Climate change	Reduce impact on livelihoods; reduce carbon emissions; generate financial resources for adaptation	Investment in health and education infrastructures and facilities; adaptation and mitigation technology and early warning systems; green public transportation; strengthen regional cooperation for green technology transfer	Upgrade disaster risk prevention systems; reduce carbon emissions to sustainable levels	Investment in mitigation, industrial green transformation; retrofitting of buildings; policy coordination

Source: UN/DESA, Development Policy and Analysis Division

Strategies for developing more sustainable cities

There is no 'one scheme fits all' approach towards achieving sustainability because the challenges faced by individual cities are diverse and depend on their population size, economic status, technological capacities and development priorities. However, some of the key strategies are summarised below:

- **Investment in infrastructure such as roads, water, sewers and electricity and services such as schools and healthcare**: Curitiba's credentials as a sustainable city are well documented. Its integrated bus system is just one part of the Curitiba Master Plan introduced in the 1970s. An extensive network of dedicated bus lanes provides a service comparable to underground or subway systems but at a cost estimated to be about 200 times less. The associated reduction in traffic has led to a significant reduction in carbon emissions.

- **Green investment in low-income countries can help poorer cities 'leapfrog' from high-carbon energy use to a zero-carbon development path**: This could provide employment for the 'youth bulge' within these cities.

- **Investment in the production and use of renewable energy sources as well as the renovation of infrastructure, retrofitting of buildings and improved electricity and water efficiency**: In the city of Freiburg in Germany, solar investment subsidies are given to residents installing solar panels and 'plus energy' homes have been constructed which create more energy than they consume. Low-energy construction standards have also been introduced in a bid to reduce CO_2 emissions by 40 per cent by 2030.

- **Investment in the reduction of waste production and improvement of waste collection and recycling**: The 'Garbage that is not garbage' scheme in Curitiba promotes recycling through the separate collection of different waste components, while the 'Garbage purchase' programme encourages residents in the favelas to sell their rubbish back to the city in exchange for food, bus tokens and football match tickets. The scheme has helped clean up densely populated areas that the rubbish vans cannot reach.

- **Provision of more 'green' areas**: The British environmental charity Groundwork published a report in 2012 entitled 'Grey places need green spaces' in which they outlined the benefits of green spaces in cities. These include greater public health, better personal well-being and economic prospects and reduced violence and aggression.

- **Investment in more sustainable and affordable housing**: Low carbon housing developments include the experimental BEDZED development and Greenwich Millennium Village in London. Environmentally sustainable, recycled and local materials were used in their construction and the accommodation comprises a mix of social housing and private units.

- **Adoption of a local currency**: Local currencies such as the Bristol Pound serve the needs of local people because they keep money within the local economy. Research by the New Economics Foundation has found that for every local currency pound spent in a local business, £1.73 is generated through the multiplier effect. In contrast, for every pound spent in a chain store, only 35 pence is re-spent in the local economy. Local currency can also encourage a sense of community and can include a mechanism to generate donations for local schools and social services.

- **Active participation of different city stakeholders including government, residents and local businesses in urban planning.**

- **Disaster risk reduction**: Schemes such as tidal barrages and early warning systems can help mitigate the impacts of floods, storm surges and other hazards to which some cities are vulnerable.

Finally, it has been argued that greater investment in rural areas is important to reduce the rural – urban migration that has put increasing pressure on cities in the last few decades.

Case study of an urban area: London

Figure 9.63 London's skyline

London is the capital and most populous city of the UK. In 2015, the Greater London Authority reported that the population had topped 8.6 million. This figure is expected to reach 11 million by 2050. London has a history stretching back to Roman times when it was named Londinium. From then it developed as a port around the navigable River Thames and eventually became the seat of political power and government. It has been suggested that London currently holds the role as the most important world city. This is due to a number of factors:

- Economically, London is a global financial centre with a growing reputation as a technological hub and top rankings for software and multimedia development. Its reputation as an economic powerhouse means that it is particularly successful in attracting direct foreign investment.
- London performs very well globally in terms of the number of people in higher education, the quality of universities and access to libraries.
- More than 300 languages are spoken by the people of London, and the city has at least 50 non-indigenous communities with populations of 10,000 or more. Virtually every race, nation, culture and religion in the world can claim at least a handful of Londoners.
- According to the World Cities Culture Forum, London is seen as one of the most cosmopolitan and tolerant cities in the world, attracting a large diversity of people – from activists to business leaders, intellectuals to fashionistas. The city is also a major centre for art forms including music and dance, while its leading museums and galleries are among the most visited in the world.
- London is one of the most visited cities in the world due to a combination of history, heritage, art and culture.

- The hosting of the 2012 Olympic Games further raised the profile of London as a world city and led to the huge redevelopment of formerly run-down areas.

Cultural diversity in London

London is often referred to as 'the world under one roof' because of its multicultural population. Its function as a port has resulted in a long history of immigration and this has led to a rich ethnic and cultural diversity. Many see the arrival of hundreds of West Indian men aboard the *Empire Windrush* in 1948 as the start of mass immigration into the UK. Many of these men were answering adverts offering employment and intended only to stay a few years. In fact, the people of the *Windrush* and subsequent immigrant groups have played a vital role in creating a new concept of what it means to be a Londoner.

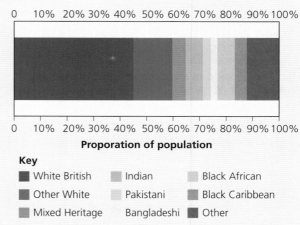

Key

- White British
- Other White
- Mixed Heritage
- Indian
- Pakistani
- Bangladeshi
- Black African
- Black Caribbean
- Other

Figure 9.64 Ethnicity in London, 2011

In 2015, just over 3.8 million of London's residents (44 per cent) were of a black and minority ethnicity origin. This is expected to increase to 50 per cent by 2038. In spite of some instances of hostility towards immigrant groups, London is regarded as a welcoming city and residents generally recognise the positive social and economic contributions immigrants make to life in the capital.

Economic and social well-being

In spite of its status as an international financial centre, London has huge areas of poverty and the gap between rich and poor continues to widen.

In 2015, the London Fairness Commission reported that for every £1 of wealth owned by the bottom 10 per cent of London households, the top 10 per cent own £172.

Inability to get on to housing ladder (ever) due to low income and inflated London house prices

Long commutes on buses because of inability to afford the tube fare

Jobs, sometimes paying below the minimum wage

Multimillion pound houses

Fine dining and theatre experiences

Holidays abroad

Twice as likely to die from chronic lower respiratory illness

Lack of leisure time due to holding down more than one job to make ends meet

Overcrowded high-rise flats

Well-paid jobs with six-figure bonuses for some

Use of private schools and hospitals

Fear of crime and gang violence

Large proportion of monthly income spent on renting low-cost accommodation a long distance from work

Figure 9.65 A city of two halves

Key points from London's Poverty Profile, 2013

- Incomes in London are more unequal than in any other region, with 16 per cent of the population in the poorest tenth nationally and 17 per cent in the richest tenth.

- The richest 10 per cent by financial asset wealth have 60 per cent of all assets. The richest 10 per cent of households by property wealth have 45 per cent of that wealth.

- The top tenth of employees in London earn around four-and-a-half times as much as the bottom tenth. This ratio is an increase over the last decade and higher than any other English region.

- Among London's boroughs, Kensington and Chelsea has the greatest imbalance between high and low earners. The top quarter earn at least £41 per hour, three-and-a-half times the level of the lowest quarter at £12 per hour or less, which is in turn higher than the lowest quarter for England as a whole.

Skills focus

To see how these figures have changed over time and how different boroughs compare, go to www.londonspovertyprofile.org.uk/key-facts/ It is also worth looking at the 2015 Index of Multiple Deprivation data for London which measures relative deprivation. This can be viewed at http://maps.cdrc.ac.uk/.

Inequality impacts on everyday life. Compare the map showing average incomes in Figure 9.66 with life expectancy in Figure 9.67. The figure varies by about 5 years between the extremes. The difference between Hackney and the West End is the same as the difference between England and Guatemala.

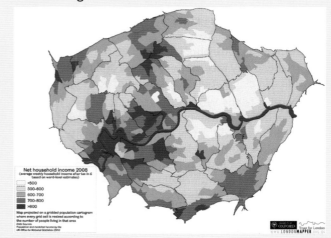

Figure 9.66 Average household income after tax (before housing costs) based on ward-level estimates for 2008 projected on a gridded population cartogram (Data source: Office for National Statistics, 2011)

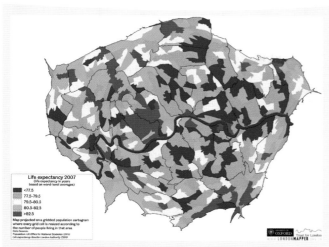

Figure 9.67 Average life expectancy based on ward-level estimates for 2007 projected on a gridded population cartogram (Data source: Greater London Authority, 2011)

The nature and impact of physical environmental conditions

In the first half of the twentieth century, 'smog' events became so strongly linked to the city of London that the term 'London particular' was and still is used in reference to smogs. Since then, air quality has improved significantly. The introduction of the Clean Air Act in 1956 and more recent legislation, along with the introduction of greener buses, taxi age limits and the Low Emission Zone (LEZ), has reduced CO_2 emissions in particular. However, nitrogen dioxide levels still breach EU legal limits and Figure 9.68 shows that London still suffers higher levels of air pollution than surrounding rural areas. The 2014 Public Health England report on air pollution says that 5.3 per cent of all deaths in people age 25 and over are now linked to air pollution. These figures vary geographically but the highest percentage of deaths linked to air pollution are in London. More specifically, the boroughs of Kensington & Chelsea and Westminster have the highest rates, where 8.3 per cent of deaths can be attributed to air pollution.

Linked to air quality is the impact of London's urban heat island, shown in Figure 9.30 (page 408). With the centre of London already up to 10°C warmer than the surrounding rural areas and average summer temperatures predicted to rise further, summer heat waves pose a threat to homes, workplaces and public transport. They have a negative effect on health, particularly that of vulnerable people, and lead to greater consumption of water and energy.

The London Climate Change partnership has suggested that extremely high demands on London's power supply network may lead to 'brownouts', due to the high cooling demand, and increases in electricity demand for cooling could negatively affect London's sustainability.

One further concern linked to London's physical environment is the threat of flooding. This comes from five sources – tidal, fluvial (from rivers and tributaries), surface (from rainfall), sewer and groundwater flooding. Climate change will bring wetter winters and more frequent heavy downpours, as well as rising sea levels and higher tidal surges, all of which pose a major threat to London. Fifteen per cent of London is on the flood plain, protected by flood defences. Residential areas are located within this area but it also includes much of the infrastructure Londoners rely on daily: 49 railway stations, 75 underground stations and 10 hospitals. The Thames Estuary 2100 (TE2100) project is one response to the risk of flooding but other strategies are being adopted. Go to http://climatelondon.org.uk/ for more information on these.

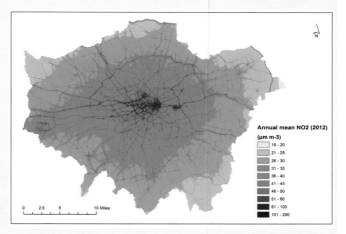

Figure 9.68 Concentrations of nitrogen dioxide are higher along the roads in London and at key transport hubs such as Heathrow Airport and City Airport. Over the course of the day, NO_2 levels are particularly high during rush hour as they predominantly come from vehicle emissions

Moving towards greater urban sustainability

The London Sustainable Development Commission was established in 2002 to advise the Mayor of London on making London a 'sustainable world city'. Various strategies have since been implemented to tackle key issues such as inequality, air pollution and urban deprivation. A snapshot of strategies is shown below. There are many more to investigate and evaluate.

As the urban area of London continues to expand, greater pressures will be placed on the services and infrastructure. Strategies encouraging greater urban sustainability will be key to the city's future viability. Proposals to make London into the world's first **National Park City** are being discussed.

Further research

There are numerous areas of London's geography to research. The Olympic legacy, inequality and the success of strategies to tackle environmental problems are just three areas which merit more attention. Look also at the place study of East London in Chapter 8 for information on the Bangladeshi diaspora, and the impacts of gentrification and urban resurgence. Useful websites include www.london.gov.uk and www.tfl.gov.uk. The London mapper website visualises data in a variety of different ways: www.londonmapper.org.uk/ Likewise, the 2014 book, *London: The Information Capital* includes 100 maps and graphics on all aspects of London living. Both sources are excellent examples of GIS. Finally, a useful resource for looking at the extent and impact of inequality in London is geographer Danny Dorling's *The 32 Stops: the Central Line*.

Table 9.11 Urban sustainability strategies adopted in London

Social developments	Economic developments
Urban renewal has brought economic investment and growth to parts of London previously suffering urban decline. For example, Olympic regeneration in Newham has provided additional social housing, leisure facilities and greater investment in education. The Queen Elizabeth Olympic Park provides parklands, waterways and leisure activities. From 2016, the Olympic Stadium will be home to West Ham United FC	The London Living Wage is a calculated hourly rate of pay which gives the wage rate needed for a worker in London to provide their family with the essentials of life, including a cushion against unforeseen events. It is not compulsory but a voluntary commitment made by employers. It has been estimated that over 10,000 London families have been lifted out of working poverty as a direct result of the Living Wage
A focus on improving education in the capital through the London Challenge initiative has continued to improve outcomes for pupils in London's primary and secondary schools at a faster rate than nationally. Pupils in London do better in school than the rest of the country	When Crossrail opens in 2018, it will increase London's rail-based transport network capacity by 10 per cent. It is hoped this will cut journey times across the city, ease congestion and encourage regeneration
In Newham, the Every Child programme offers children the chance to learn a musical instrument or take part in cultural events. Children in this area also benefit from universal free school meals, which have been linked to greater attainment and better health. Similar schemes operate in other boroughs	At a borough level, schemes such as Workplace in Newham have been introduced to help people find jobs. Workplace provides residents with free advice about employment options and training. Since 2007, Workplace has supported over 20,000 Newham residents into employment, and helped more than 900 businesses fill their vacancies
Environmental developments	**Urban governance**
The Congestion Charge was introduced in 2003 to discourage driving in central London to reduce congestion and pollution	The role of the Mayor of London was created under the Greater London Authority Act 1999 as part of the government's commitment to restore a city-wide government for London
More recent 'green' transport policies have included Routemaster buses with green electric hybrid engines, a fleet of eight hydrogen fuel buses, more than 1,400 charge points to support the use of electric vehicles, zero emission taxis from 2018 and a central London Ultra Low Emission Zone (ULEZ) from 2020 with the objective of reducing air pollutant and CO2 emissions from road transport	Mayors are important to ensure that a city has a strong voice and can attract investment from home and abroad. Devolved powers include planning, transport, employment, economic development, health and policing
Cycle superhighways and the cycle hire scheme more commonly known as 'Boris Bikes' have helped to increase cycling in London by 173 per cent since 2001. The aim is for a 400 per cent increase by 2025, although there are continued concerns about cycling safety	
In response to the UHI effect in London, targets have been set to increase green cover in central London by 5 percent by 2030	

Case study of a contrasting urban area: Mumbai

Although not its capital, Mumbai is India's largest city with a population in excess of 20 million. It is the financial and commercial centre of India as well as home to the popular Bollywood movie industry.

Figure 9.69 The location of Mumbai

Mumbai is located on the west coast of India and is the capital of the state of Maharashtra. It developed as a trading centre selling local goods such as gold, jewellery and textiles. The arrival of the East India Company in the early seventeenth century led to the establishment of the British Raj (Empire) and goods such as raw cotton were regularly shipped to England for manufacturing. Originally, Mumbai was a series of seven islands separated by swamps, but by 1845 these had been filled in and Mumbai occupied one large island. Mumbai has a natural deep-water harbour and has been the main port in the Arabian Sea since the opening of the Suez Canal in 1869.

Figure 9.70 Mumbai's skyline resembles that of a wealthy city. The Bandra-Worli Sea Link shown in the photo is a cable-stayed bridge that connects central Mumbai with its western suburbs.

Following India's independence from the British in 1947, Mumbai developed rapidly. High-rise, modern architecture, the Bombay Stock Exchange, tarred roads and a boom in manufacturing and services have changed the city's status and brought it on to the world stage. In 2015 it accounted

for 33 per cent of India's income tax, 6.16 per cent of GDP (the largest single contributor in India), 25 per cent of industrial output and 40 per cent of foreign trade. Mumbai is now a megacity and while its population growth may be slowing, by 2020 it will have an estimated 24 million people with the highest population density of any city in the world. Such numbers place an inevitable strain on the urban infrastructure.

Economic and social well-being

Mumbai's population has nearly doubled since 1991 and this is largely due to the influx of migrants from other parts of India seeking better employment opportunities. The resulting population is very diverse and 16 major languages of India are spoken here. Poverty and inequality are two of the big issues, however. It is estimated that around 60 per cent of Mumbaikars live in 'slums' and the average Indian would need to work for three centuries to pay for a luxury home in Mumbai.

Dharavi slum

There are many informal slum areas in Mumbai but Dharavi in central Mumbai is perhaps the most famous, brought to the attention of many by the Oscar-winning film *Slumdog Millionaire* in 2008. Until the late nineteenth century, this area was mangrove swamp inhabited by Koli fishermen. When the swamp filled in (with coconut leaves, rotten fish and human waste) the Koli people lost their fishing ground, but there was more land area for others. The Kumbhars came from Gujarat to establish a potters' colony, Tamils opened tanneries and thousands more travelled from areas such as Uttar Pradesh to work in the rapidly expanding textiles industry. The result was a very diverse neighbourhood in a very diverse city.

Cottage industries have thrived in Dharavi. It is home to thousands of micro-industries, including garment-makers, tanners, welders and potters, which produce over $650m annually. However, the living and working conditions remain very poor. Years of government neglect have resulted in inadequate hygiene standards and it is said to have the highest population density in the world at over 300,000 people per square kilometre. Housing quality is poor and the slum lacks basic infrastructure. Each toilet is shared by over 1,000 residents and services such as water and electricity are not always available.

Due to the northward expansion of the city of Mumbai, Dharavi found itself occupying an area of prime land in the new business district of India's richest city. This made it a key target for developers eager to make money from the construction of luxury apartments, while the government has been keen to improve the appearance and reputation of the area. The government-led Dharavi Redevelopment Project will see all residents who can prove residency since

2000 provided with a new, 300-square-foot house for free. The scheme has not been without controversy. The main concern has been the potential loss of the community networks and businesses which have built up there.

The nature and impact of physical environmental conditions

Mumbai has a tropical climate. The south-west monsoon brings heavy rainfall to the city between June and September and although Mumbaikers adapt to the wet conditions, the rains can be devastating for the city, which is mainly built on low-lying land. On 26 July 2005, Mumbai received 944 mm of rainfall – the average amount for the entire season and a 100-year high (Figure 9.71). This, combined with high tides, caused a devastating flood. Electricity, water supply, communication networks and public transportation were totally shut down; more than 400 people died; and over 10,000 homes were destroyed. The city suffered losses amounting to £1.2 billion. Urban growth was of course partly to blame. There was nowhere for the rainwater to go as rapid and often uncontrolled development had replaced most of the public parks, private gardens, beaches, mangrove swamps and wetlands with a built environment.

Figure 9.71 Mumbai commuters walk through floodwaters in 2005 after torrential rains paralysed the city

The Greater Mumbai Disaster Management Action Plan was created in response to the 2005 flooding. It identified the risks and vulnerabilities the city could face in the future, including earthquakes and cyclones; it created the Disaster Management Cell to co-ordinate relief and rescue efforts; and it widened and deepened the Mithi River which drains out into the Arabian Sea. The plan is clearly a positive strategy for Mumbai and shows the government is keen to address the major flood issue, but environmentalists are concerned that there are still too many factors which make Mumbai vulnerable to flooding again: continued construction on the floodplain, removal of mangrove forests and the clogging of storm drains and waterways with plastic rubbish. In addition to this, the Intergovernmental Panel on Climate Change (IPCC) has predicted that the increase in rainfall, heat, humidity and sea-level rise associated with climate change will make Mumbai the second most at-risk city in the world.

The future?

Economic growth has clearly brought wealth to many in Mumbai and new housing projects are trying to address the shortage of housing. However, this megacity is struggling to cope with its rapid growth. Services are stretched or non-existent, air and water supplies are polluted, inequality is growing and over half the population still lives in slums. The Indian government has pledged to make cities like Mumbai 'smarter' in terms of the economy and environment. In addition, non-government organisations have been working towards improving the lives of slum dwellers in Mumbai. However, as Figure 9.72 shows, Mumbai faces considerable environmental pressures and much greater investment and long-term planning are essential if the city is to become truly sustainable in the twenty-first century.

> ### Further research
>
> Mumbai has been well-documented in geographical magazines. Further useful information can be gathered from the World Bank (www.worldbank.org/) which funds a number of projects in Mumbai. In March 2015, *The Guardian* newspaper's Guardian Cities Team did a five-day feature on Mumbai.

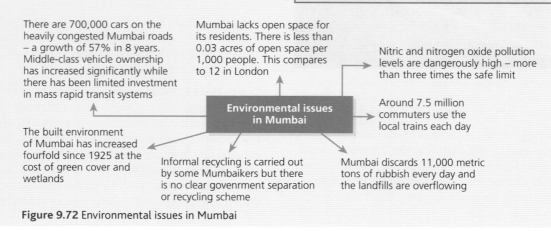

There are 700,000 cars on the heavily congested Mumbai roads – a growth of 57% in 8 years. Middle-class vehicle ownership has increased significantly while there has been limited investment in mass rapid transit systems

Mumbai lacks open space for its residents. There is less than 0.03 acres of open space per 1,000 people. This compares to 12 in London

Nitric and nitrogen oxide pollution levels are dangerously high – more than three times the safe limit

Environmental issues in Mumbai

Around 7.5 million commuters use the local trains each day

The built environment of Mumbai has increased fourfold since 1925 at the cost of green cover and wetlands

Informal recycling is carried out by some Mumbaikers but there is no clear govenrment separation or recycling scheme

Mumbai discards 11,000 metric tons of rubbish every day and the landfills are overflowing

Figure 9.72 Environmental issues in Mumbai

Review questions

1 What are the key problems facing urban areas in the twenty-first century and in what order should they be tackled?

2 Which are the most effective strategies for tackling inequality in urban areas?

3 What are the main influences on climate and drainage in a city? How can humans limit their impact on both of these?

4 Which aspects of urban sustainability do you think are most important and why?

5 How realistic is it to expect all cities, rich and poor, to adopt sustainable practices?

Fieldwork opportunities

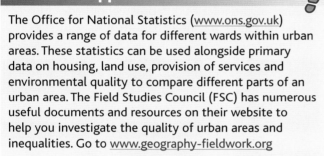

The Office for National Statistics (www.ons.gov.uk) provides a range of data for different wards within urban areas. These statistics can be used alongside primary data on housing, land use, provision of services and environmental quality to compare different parts of an urban area. The Field Studies Council (FSC) has numerous useful documents and resources on their website to help you investigate the quality of urban areas and inequalities. Go to www.geography-fieldwork.org

It is also worth looking at the 2015 Index of Multiple Deprivation data which measures relative deprivation. This can be viewed at http://maps.cdrc.ac.uk/ or http://dclgapps.communities.gov.uk/imd/idmap.html.

The redevelopment of King's Cross Station in London provides a new fieldwork venue in the UK capital. Disused railway land on the 27-hectare site is being transformed into a mixed-use site complete with homes, shops, offices, galleries and restaurants. The website below provides both historical and current information about the site as well as brochures and videos. Visit www.kingscross.co.uk/discover-kings-cross

Changes in temperature and/or pollution levels can be measured and mapped across an urban area. Traffic counts at strategic points may show a correlation with such data.

Infiltration rates can be measured across a variety of urban surfaces. This could then be linked to a Sustainable Urban Drainage System.

Further reading

Two relatively recent urban geography books:

Hall, T. and Barrett, H. (2011) *Urban Geography* (Routledge)

Pacione M, (2009) *Urban geography: a Global Perspective of* (Routledge)

See also Drake, G. and Lee, C. (2000) *The Urban Challenge* (Hodder)

The United Nations publishes a number of useful reports on cities. In 2014, a report entitled *World Urbanisation Prospects* was published outlining the key issues resulting from urban growth.

UN–Habitat – an organisation concerned with global urban issues has also published a number of reports entitled *The State of the World's Cities* which focus on sustainability issues. A new *World Cities Report* is due to be published by the UN in 2015/2016 (www.unhabitat.org/)

The Urban Geography Research Group (UGRG) of The Royal Geographical Society/Institute of British Geographers has a website with useful links and resources. (www.urban-geography.org.uk/)

Hans Rosling's brilliant Gapminder website contains a wealth of data on socio-economic trends and variations around the world (www.gapminder.org/)

For current urban trends and processes it is worth reading newspaper websites such as *The Guardian* and *The Independent,* which often have up-to-date features on cities. *The Guardian* has a twitter feed @guardiancities. It also has a resilient cities page supported by the Rockefeller Foundation.

Magazines such as *Geographical* magazine and *Geography Review* regularly include articles on specific cities and urban issues.

Question practice

A-level Questions

1. Which of the following would not be considered a desired goal of sustainable cities?
   a) A reduction in household waste with a corresponding increase in recycling and re-use
   b) Greater use of open space and protection of wetlands, woodlands, stream valleys, habitat, etc.
   c) Increased movement on foot, bicycle and public transport
   d) Low-density spread-out residential developments   (1 mark)

2. Which statement most accurately describes photochemical pollution?
   a) Run-off water from buildings, streets and pavements carries many pollutants, including sediment, nutrients, bacteria, oil, metals, chemicals, road salt, pet droppings and litter, into rivers.
   b) The accumulation of particulates (for example, smoke from car exhausts) in the atmosphere increasing the likelihood of fog and the reduction of visibility.
   c) The poisoning of urban rivers with industrial effluent so that light cannot penetrate to the bottom and many species are unable to survive.
   d) The type of pollution that leads to the formation of ozone (O_3) and other oxidising compounds (from primary pollutants, for example, nitrogen oxides (NO_x) in the lower layer of the atmosphere.   (1 mark)

3. Which of the following statements describes a situation that could lead to a decrease in cultural diversity in a city?
   a) As immigrants become established, with increasing wealth, they often move to the suburbs.
   b) Cities are the first point of entry into the country for many immigrants.
   c) Cities tend to offer a greater range of employment opportunities.
   d) Urban populations tend to be more tolerant of immigrants.   (1 mark)

4. What is an edge city?
   a) An area of housing on the edge of a city from which residents must commute for work and services
   b) An area of shopping and entertainment found on the edge of a city, such as Cribbs Causeway in Bristol or the Trafford Centre in Manchester
   c) A city designed around security, protection, surveillance and exclusion
   d) A self-contained settlement which has developed beyond the original city boundary close to major roads or airports   (1 mark)

5. Analyse the roles of differing technologies in the handling of solid urban waste as shown in Figure 9.73.   (6 marks)

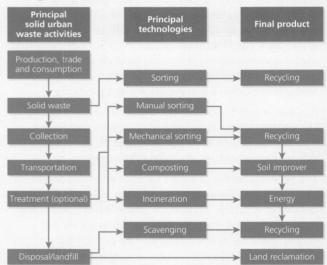

Figure 9.73 Solid waste handling and treatment system components

6. Assess the impacts of social segregation and cultural diversity on contrasting urban areas you have studied.   (9 marks)

7. To what extent has the emergence of megacities and world cities affected global and regional economies?   (9 marks)

8. 'Environmental problems vary from city to city.' Using examples, discuss the extent to which you agree with this statement.   (20 marks)

AS Questions

1. What is gentrification?

   a) The buying and renovating of properties, often in more run-down areas, by wealthier individuals

   b) A government-led scheme to attract private investment into inner city areas

   c) The development of cultural and heritage quarters

   d) A publicly-funded regeneration scheme aimed at tackling social, economic and environmental problems in run-down parts of the city   (1 mark)

2. Which of the following is an argument in favour of the incineration of urban waste?

   a) Incineration is unable to remove or reduce greenhouse gases, especially as waste is something that generates its own methane supply.

   b) Flue gas and residual ash can be highly toxic and difficult to dispose of. Ultra-fine particles have also been known to get past many flue gas filtration systems.

   c) Incinerators generate more CO_2 per unit of energy then other more conventional methods.

   d) Incinerators produce electricity; for example, incineration in Denmark contributes to 5 per cent of the country's electricity generation. (1 mark)

3. Outline reasons for the rise of the service economy.   (3 marks)

4. Figure 9.74 shows the location of different-sized cities around the world. Assess the extent to which there is a relationship between city size and location.   (6 marks)

5. Assess the success of two or more strategies for developing more sustainable cities.   (9 marks)

6. Using examples, to what extent do differing physical and environmental conditions have differing impacts on contrasting urban areas?   (20 marks)

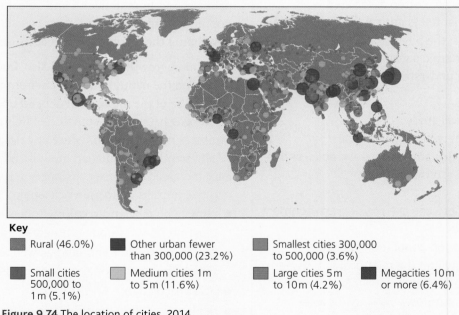

Key

■ Rural (46.0%)   ■ Other urban fewer than 300,000 (23.2%)   ■ Smallest cities 300,000 to 500,000 (3.6%)

■ Small cities 500,000 to 1m (5.1%)   ■ Medium cities 1m to 5m (11.6%)   ■ Large cities 5m to 10m (4.2%)   ■ Megacities 10m or more (6.4%)

Figure 9.74 The location of cities, 2014

Chapter 10

Population and the environment

Human population growth is the driver of many environmental, socio-economic and political challenges that we face in the world today. This chapter will examine the relationship between some key aspects of the physical environment and population numbers as well as people's health and well-being. It will in turn consider the impacts of population change on the environment.

The ability of a place to support a population depends upon whether it can provide an adequate food supply. This can be achieved by local food production and/or trade in food with other areas that have a surplus. This in turn depends upon the climate, soils and other aspects of the environment. Population change is also often associated with economic development. It is critical to consider the relationship between the physical environment, population and economic development at all scales of study, in order to understand how the future of our planet will look.

In the chapter you will study:
- the relationship between the physical environment, particularly climate and soils, and food production systems
- food security
- the relationship between the physical environment and (human) health
- natural and migration population change
- population ecology and the relationship between population and resources
- global population futures – varying possible scenarios of future population growth.

10.1 Introduction – the relationship between population and the environment

Appreciating the many facets of the relationship between population and the physical environment is critical to understanding the geographical characteristics of **population distribution** and change, as well as the impact of people on the environment.

Features of the physical environment such as the climate, geology, topography and ecosystems will determine the nature of soils, drainage, water supply and potential environmental hazards. These will in turn affect food production, energy supplies, settlement patterns and human health and so contribute to population size, distribution and its success in developing the resources it has at its disposal.

Population growth and size will also, however, have an impact on the natural physical environment. The degree and nature of the impact will depend upon the level of resource consumption, standard of living and mitigation measures put in place to reduce any negative impacts. Recent trends in population growth and development have increased resource consumption leading to many negative impacts on the physical environment, including:
- climate change and resulting increases in extremes of temperature, floods, droughts and sea level rises
- pollution of water and land as well as atmospheric pollution on a regional scale, such as acid rain
- ozone depletion
- depletion of finite natural resources
- damage to wildlife and their habitats leading to increased extinction rates and consequent threats to species interdependence

Figure 10.1 outlines the different facets of this reciprocal relationship between population and resources. This is further examined in the population

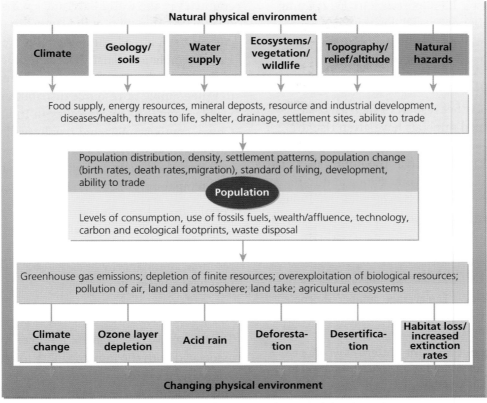

Figure 10.1 The inter-relationship between human population and the physical environment

ecology section of this chapter, particularly by the Population, Resources and Pollution Model.

Elements in the physical environment

Climate

Of all the elements in the physical environment, this is arguably the most important. Rainfall, temperature, wind velocity and levels of solar insolation all determine food productivity, the type of farming system adopted and which species are selected for cultivation. For example, many crops need temperatures of at least 5 °C to grow and pasture-fed livestock need to be supplemented with fodder crops in temperatures below 5 °C. High concentrations of population are determined by adequate rainfall and temperatures that are suitable for growth of crops and rearing of livestock, with sufficient sunlight for photosynthesis. Characteristics of climate can also drive the level and nature of diseases. For example, tropical diseases such as malaria, yellow fever and Ebola will directly affect death rates and life expectancy in the human population. High infant and child mortality rates in tropical and subtropical areas can also influence fertility rates as families seek to

compensate for their loss. Climate change arising from global warming is the aspect of the environment giving greatest concern to human population at the start of the twenty-first century.

Soils

The most important feature of soil is its fertility. This depends upon soil structure, texture, acidity (pH), organic matter and nutrients. These will determine agricultural output and the type of farming system employed. Fertility can be maintained with artificial chemical fertilisers in areas of high population density but this may have unsustainable outcomes, such as contributing to water pollution, eutrophication and increased greenhouse gas emissions. Areas with fertile soils are often associated with densely populated areas. Some, however, such as rich volcanic or alluvial soils, are also prone to hazards (such as volcanic eruptions or floods), which may negatively affect the population.

Water supply

People need access to water to survive. It is important for human hydration but its key use is for irrigation to maintain food production. It has other key uses including hygiene and sanitation as well as in many industrial processes. Its importance in determining population distribution is well illustrated in Egypt, where 95 per cent of its population of 80 million people live within 12 miles of the River Nile.

Geology and other resource distributions

Concentrations of other resources such as fossil fuels or other valued minerals have given rise to industrialisation and consequently densely populated conurbations in Europe (such as the Ruhr Valley), parts of the USA and increasingly now China and India. Even when these resources become depleted, industrial inertia leaves a legacy of large, dense populations with new tertiary industries emerging to serve them.

Key terms

Population density – The average number of people living in a specified area, usually expressed as the number of people per km².

Population distribution – The pattern of where people live. This can be considered at all scales from local to global, in an area or country.

Population parameters

Elements of the physical environment clearly determine distribution patterns and **population density**. For example, on a national scale, Egypt's population is very unevenly distributed, determined primarily by water supply and specifically by access to the River Nile. Thus the Nile Valley has areas with a population density of 1,500+ people/km² while the average for Egypt as a whole is 80 people/km² because the desert areas have fewer than 10 people/km². Most

countries in the world have uneven distributions due to a combination of the factors in the physical environment already discussed. Figure 10.2 shows global population density estimates for 2015. Even at this scale, the densely populated areas relating to water supply or resource distribution are clearly visible along the Ganges Valley in India, in eastern China, southeast Asia and northwest Europe. Areas of population sparsity can be correlated to deserts, continental interiors and where there are extremes of cold temperatures in Siberia and the Canadian Shield.

Population change considers population from a temporal as well as a spatial perspective. Over time people have migrated to areas of fertile soils and resource abundance, where they have thrived and numbers have grown.

Role of development processes

The process of development has been a narrative associated with the human ability to acquire and make use of the resources made available by the physical environment. Development surges have resulted from some trigger to progress; the most notable of these have been the Neolithic Agrarian Revolution in Mesopotamia, around 12,000 years ago and the Industrial Revolution in Europe in the eighteenth and nineteenth centuries. These have sparked technological developments which have enabled specific areas (and the Earth as a whole) to support a larger human population. Also important has been the increasing ability of the human race to control many of the infectious diseases which have threatened survival and kept death rates high. This has been one of the main

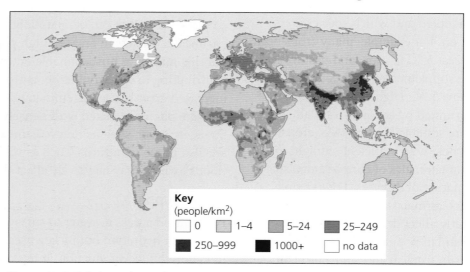

Figure 10.2 Global population density estimates, 2015

reasons for the rapid rise in world population from around 1 billion in 1800 to 7 billion by 2011.

Rapid population growth and the process of development have had a largely negative impact on the environment itself. The impact of individuals will depend on where they live on the planet, their individual lifestyles and the resources and services available to them. These themes are further examined in the section on population ecology in this chapter, which considers ecological footprints and the global challenges resulting from continued population growth.

10.2 Environment and population

The amount and type of food produced varies from place to place. That variability depends upon both environmental and human variables. The two key environmental variables are **climate** (temperature, growing season, water availability) and **soils**. The variation in production can occur at all scales from local to global.

Key terms

Agricultural productivity – The ratio of agricultural outputs to agricultural inputs.

Agriculture – The science or practice of farming, including cultivation of the soil for the growing of crops and the rearing of animals to provide food, wool and other products.

Climate – A region's long-term weather patterns. This is measured in terms of average precipitation (that is, the amount of annual rainfall, snow, etc.), maximum and minimum temperatures throughout the seasons, sunshine hours, humidity, the frequency of extreme weather and so on.

Food security – Food security exists when all people at all times have access to sufficient, safe, nutritious food to maintain a healthy and active life.

Salinisation – The build-up of salts in soil, eventually to toxic levels for plants.

Soil – The upper layer of earth in which plants grow, a black or dark brown material typically consisting of a mixture of organic remains, clay and rock particles.

Topography – The relief and drainage of an area.

Zonal soil – A soil which has experienced the maximum effect of climate and natural vegetation upon the parent rock, assuming there are no extremes of weathering, relief or drainage.

Global and regional patterns of food production and consumption

Food production

In the early 1960s, global food supplies for humans stood at only 2,300 calories per person per day; this was very unevenly distributed. In high-income countries the average was 3,030 calories per day whereas in low-income countries of the developing world it was below 2,000. Probably more than half the people in these low-income countries suffered from chronic under-nutrition.

By 2010, despite the fact that global population had increased by almost two and half times, the world could produce enough food to provide every person with more than 2,800 calories per day. This should have been sufficient to ensure that all had enough to eat but unfortunately, food production and availability was uneven, meaning that 800 million people still suffered from under-nutrition. This, however, represents a drop in overall percentage terms from 50 to 20 per cent.

These gains resulted from a number of factors:
- The package of technologies referred to as the **green revolution**, including increased use of new, high-yielding crop varieties and technologies. Irrigation, fertiliser and pesticides increased yields for millions of farmers, although with some negative social and environmental impacts.
- An increased reliance on global trade. During the 1970s alone, net imports of cereals by low-income countries more than tripled – from 20 million to 67 million tonnes.

Figure 10.3 (page 450) shows how food supply has increased in all regions of the world between 2000 and 2009. The rates of increase vary.

Skills focus

Describe the changes in food supply for the six regions of the world shown in Figure 10.3 (page 450). Suggest reasons why these changes vary from region to region.

In low-income countries, production and consumption of the main agricultural products have been growing at much higher (and increasing) rates than in the higher-income countries. This is a result of higher

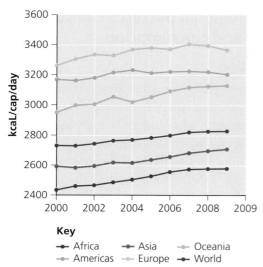

Figure 10.3 Regional variations in food supply over time

Source: FAO

population growth rates, increasing wealth and a greater responsiveness of demand to income growth in these poorer countries. In contrast, a slower growth of demand has occurred in the richer world because high per capita consumption and slow growth of population have had a dampening effect on the growth in demand for many commodities. Countries such as China and Brazil have high growth rates, while North America, Europe and Russia have lower growth rates and, in some cases, production has declined.

Figure 10.3 also shows that the global food production increased from 2,740 kcals per capita per day to 2,830 kcals per capita per day between 2000 and 2009. The resulting 2009 global pattern is shown in Figure 10.4.

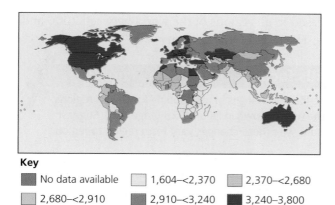

Key

■ No data available	▢ 1,604–<2,370	▨ 2,370–<2,680
▢ 2,680–<2,910	▨ 2,910–<3,240	■ 3,240–3,800

Figure 10.4 Global food supply, 2009

Source: FAO

This increase is made up of all farm produce. Two examples of the increase in production are shown in Figures 10.5 and 10.6, which show changes in all crop production and all meat production.

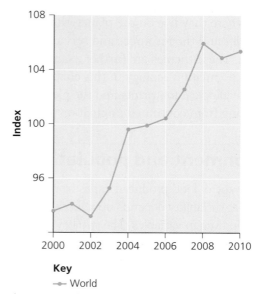

Key
— World

Figure 10.5 The change in the amount of crops produced globally, 2000–2010

Source: FAO

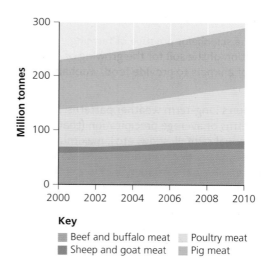

Key

▨ Beef and buffalo meat	▢ Poultry meat
▨ Sheep and goat meat	▨ Pig meat

Figure 10.6 Changes in global meat production 2000–2010

Source: FAO

Skills focus

Figure 10.6 is a compound line graph. Describe the changes in the production of the different sources of meat between 2000 and 2010.

Food consumption

Food consumption expressed in kilocalories (kcal) per capita per day is a key variable used for measuring the changes in the global and regional food situation. Analysis of **United Nations Food and Agriculture Organisation (FAO)** data shows that dietary energy has been steadily increasing on a worldwide basis; availability of calories per capita from the mid-1960s to the late 1990s increased globally by approximately 450 kcal per capita per day and by over 600 kcal per capita per day in developing countries (see Table 10.1). The per capita supply of calories has remained almost stagnant in sub-Saharan Africa and has recently fallen in the countries in economic transition.

Table 10.1 Global and regional per capita food consumption (kcal per capita per day)

Region	1964–1966	1974–1976	1984–1986	1997–1999	2015	2030 (est)
World	2,358	2,435	2,655	2,803	2,940	3,050
Developing countries	2,054	2,152	2,450	2,681	2,850	2,980
Near East and North Africa	2,290	2,591	2,953	3,006	3,090	3,170
Sub-Saharan Africa	2,058	2,079	2,057	2,195	2,360	2,540
Latin America and the Caribbean	2,393	2,546	2,689	2,824	2,980	3,140
East Asia	1,957	2,105	2,559	2,921	3,060	3,190
South Asia	2,017	1,986	2,205	2,403	2,700	2,900
Industrialised countries	2,947	3,065	3,206	3,380	3,440	3,500
Transition countries	3,222	3,385	3,379	2,906	3,060	3,180

Source: WHO

According to the FAO, undernourishment, or hunger, is dietary intake below the minimum daily energy requirement. Dietary energy requirements differ by gender and age, and for different levels of physical activity. Accordingly, minimum dietary energy requirements, the amount of energy needed for light activity and minimum acceptable weight for attained height, vary by country, and from year to year depending on the gender and age structure of the population. It ranges between 1,700 and 2,000 kcals per day.

The amount of undernourishment is calculated using the average amount of food available for consumption, the size of the population, the relative disparities in access to the food, and the minimum calories required for each individual. According to the FAO, 868 million people (12 per cent of the global population) were undernourished in 2012. This has decreased across the world since 1990, in all regions except for Africa, where undernourishment has steadily increased.

The decrease in undernourishment has not quite been able to reach the **Millennium Development Goal** of halving hunger between 1990 and 2015. The global financial, economic and food-price crisis in 2008 drove many people to hunger, especially women and children. The sudden increase in food prices prevented many people from escaping poverty, because the poor spend a larger proportion of their income on food and subsistence farmers are net consumers of food.

Agricultural systems

Agriculture can be studied in a systematic way and farms can be considered as open systems (see Figure 10.7, page 452). The inputs include physical, cultural, economic and behavioural. Generally, as an area develops the physical factors become less important as the human inputs increase.

Figure 10.8 (page 452) summarises how farming can be carried out. Within that general pattern there are the following types:

- **Commercial farming** is carried out so that the majority of the produce is sold and the income generated can provide a livelihood for the farm workers as well as be invested back into the farm.
- **Subsistence farming** means that the majority of the produce is consumed by the landowner and the farm workers, though a little surplus may be sold to buy other living requirements and/or be invested in the farm.
- **Intensive farming** is usually relatively small scale and can either be:
  - **capital intensive**: money is invested in soil improvement, machinery, buildings, pest control, high-quality seeds/animals. There are few people employed and so output is high per hectare and per worker, for example, tomato production in the Netherlands.
  - **labour intensive**: the number of farm workers is high and so there is a high output per hectare but a low output per worker, for example, rice cultivation in the Ganges Valley.
- **Extensive** farming is carried out on a large scale over a large area. This varies greatly. There are areas where, although the labour force is low, there

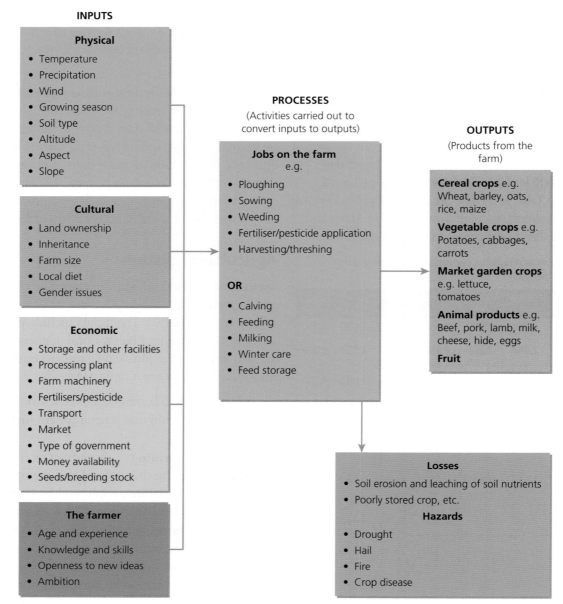

Figure 10.7 Aspects of the farming system

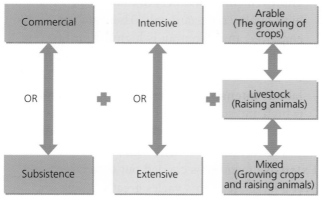

Figure 10.8 Farming types

is a high capital input, for example, on quality seeds/animals or pesticides and insecticides, such as for wheat farming in the Canadian Prairies. Other areas still have a low labour force but rely on the sheer amount of land they are farming to provide sufficient output for their needs, for example, sheep ranching in Australia.

Agricultural productivity

Productivity is a key measure of the economic performance of agriculture and an important driver of farm incomes. It represents how efficiently the agricultural industry uses the resources that are available to turn inputs into outputs. It is a relative measure which enables economists to see if improvements are made by comparing one year to another. Understanding and assessing the potential for greater productivity is a critical first step on the pathway to producing more with less.

Those looking at the results need to consider the longer-term trends as changes from year to year are often shaped by factors outside the farmer's control. Factors such as weather, animal disease, policy interventions and general economic conditions can have short-term effects on productivity.

Agricultural productivity is typically measured in terms of yield: how many kilograms of grain per hectare, kilograms of meat per animal or litres of milk per cow, etc. The most commonly used measurement of productivity is **total factor productivity (TFP)**.

TFP is the ratio of agricultural outputs (gross crop and livestock output) to inputs (land, labour, fertilizer, machinery and livestock). As producers use inputs more effectively and precisely, or adopt improved cultivation and livestock raising practices, their TFP grows while using a fixed or even reduced amount of inputs.

For crops, TFP improves with:
- higher yielding, disease resistant and drought or flood tolerant crop varieties
- more efficient and timely cultivation and harvesting practices

- using technologies that indicate precisely when and how much water and fertiliser to apply.

For raising livestock:
- breeding animals for favourable genetic qualities and behaviour
- using better animal care and disease management practices
- adoption of high quality feeds contribute to greater productivity.

While Figure 10.9a indicates a promising global trend in producing more output with fewer resources, Figures 10.9b and 10.9c show there is considerable variation across countries, particularly when considering per capita income and development level.

Low-income countries have boosted their agricultural output dramatically since the mid-1980s, and a growing share of their agricultural output is now attributable to TFP, the efficiency of production. Raising productivity in low-income countries requires massive investments in agricultural research and development, rural infrastructure and support for the special needs of smallholder farmers, women producers and co-operative producer associations.

In high-income countries, decades of public and private investment have had a powerful impact on agricultural research and development, rural infrastructure and extension of productive technologies and innovations to the farm level.

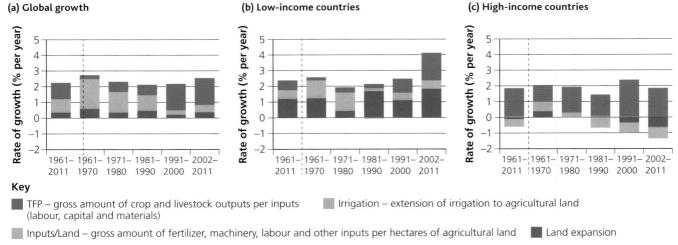

(a) Global growth   **(b) Low-income countries**   **(c) High-income countries**

Key

■ TFP – gross amount of crop and livestock outputs per inputs (labour, capital and materials)   ▨ Irrigation – extension of irrigation to agricultural land

▨ Inputs/Land – gross amount of fertilizer, machinery, labour and other inputs per hectares of agricultural land   ■ Land expansion

Figure 10.9 Growth in global agricultural output, 1961–2011
Source: US Department of Agriculture

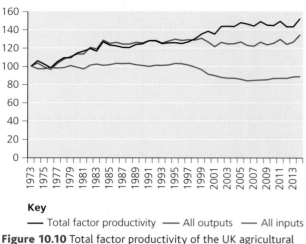

Key

—— Total factor productivity —— All outputs —— All inputs

Figure 10.10 Total factor productivity of the UK agricultural industry (1973 = 100)

Key question

Figure 10.10 shows how UK agriculture has become more efficient. How is it possible that the productivity has increased by 53 per cent whereas all outputs have only increased by 35 per cent?

Key physical environmental variables

As we have seen from Figure 10.7 (page 452) farmers make decisions on how to use their land based on:

● their own strengths, knowledge and skills
● cultural factors
● economic variables
● physical factors.

Two of the most important physical factors are the climate and the soils; they are closely related because soils are often determined by the climate and the natural vegetation of an area.

Climate

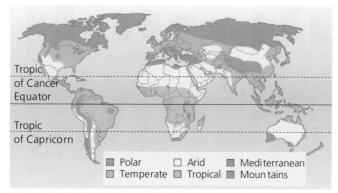

Figure 10.11 Climate zones of the world

By studying Figure 10.11 and Figure 10.2 (page 448), a correlation can clearly be seen between the patterns of world population density and types of climate. Two examples of how climate has an effect on human activities and numbers are given below.

Skills focus

You must be able to describe patterns on a map. Simply naming places/regions is not a pattern. To what extent do the patterns on Figure 10.2 (page 448) and Figure 10.11 match?

Climate 1: Polar climates

The polar climate is one of long and intensely cold winters, with very low temperatures (falling below −40°C) accompanied by snow and strong winds. The cold is such that:

● there is a permanent covering of the land by glacial ice with a surface layer of snow, or
● it freezes the land from the surface downwards to produce permanently frozen ground (permafrost). In summer, the top layer thaws for a short time but the ground underneath remains frozen. In winter there is a covering of snow. This area is known as the tundra.

According to the Arctic Human Development Report, the polar areas are occupied by approximately 13.1 million people spread between eight countries with a population density of less than four people per square kilometre. Most of these people live in the tundra climatic area of continental North America and Eurasia. Glacial areas are mostly devoid of permanent human settlement. They are visited by nomadic hunters and oil and mineral prospectors.

In the second half of the twentieth century the number of Arctic people started to grow rapidly because of improved healthcare for indigenous populations and the discovery of vast natural resources, which led to a large influx of immigrants. Recently population growth has slowed down in general and in some cases (for example, the Russian North) the total population has been declining. It is estimated that two-thirds of the total Arctic population lives in relatively large settlements, however the indigenous peoples tend to be found in small, widely scattered communities.

The only form of arable farming in such a place is one where the farmer creates an artificial environment. For example Tim Meyers farms seven hectares of land in the Yukon-Kuskokwim Delta of Alaska (Figure 10.12). The nutrient-rich soil is frozen and so, to get at it, Meyers first had to develop a method for thawing the ground. It was a two-year process of clearing vegetation and then spreading manure, composted tundra and a 'slurry' of salmon. Added to this was some lake water and a small amount of dry molasses to 'increase biological life'. Raised beds and high polytunnels help mitigate the low temperatures and short growing season, and a vast underground root cellar stores and extends the lifespan of the harvest. He grows and raises root vegetables, brassicas, gourds, strawberries, poultry and eggs.

Figure 10.12 Tim Meyer's farm in the Alaskan tundra

In North America, some Indian and Inuit groups hunt caribou (the American reindeer) but in Eurasia, many Arctic tribes herd them. The reindeer are not totally domesticated, as they generally roam free on the Arctic pastures. A single owner can possess hundreds or even thousands of animals. Traditionally, the herders followed the animals northwards to their breeding grounds. They return south in autumn to escape the harsh polar winter. Reindeer have never been bred in captivity, and individuals are just tamed for milking and as draught animals. The reindeer is the source of everything from milk, meat and skin (clothing and tents) to antlers and the fatty marrow of the bones.

Further details about the climate and examples of human activities can be found in Chapter 4 (page 137).

Climate 2: Tropical monsoon climate

The monsoon climate occurs in much of India and Bangladesh. It involves the seasonal reversal of winds.

In the winter, they are from the north and northeast, blowing outwards from central Asia. These winds are very dry and there is a winter drought. In summer the winds reverse direction and blow from the southwest, bringing with them hot and wet air that originated over the equatorial area of the Indian Ocean. This causes high and intense amounts of rain that is increased by the uplift of the air over the Western Ghats of southwest India and the foothills of the Himalayas as well as by intense convection caused by the hot land surface. An example is Mumbai on the west coast of India. It has an annual mean rainfall of 1,811 mm with all but 120 mm falling in just four months (June–September). Temperatures range from 30°C in summer to 19°C in winter.

Rice seedlings are grown in nurseries until the monsoon rains begin and then they are transplanted into flooded fields (10–12 cm deep water). The fields are level and have low mud walls to retain water. Although rice is best suited to lowland flood plains with deep fertile alluvial soils, it can be grown in terraces cut into the slopes of hilly areas. Rice growing in India is mostly labour intensive where a large cheap labour force using simple tools and oxen does all the work (Figure 10.13).

The Indian economy gains due to good monsoon rains in the country. It can support large numbers of people in both rural areas and the rapidly growing cities of the Indo-Gangetic Plain. On the other hand, weak monsoon rains result in crop failure, which affects the economy in a negative manner due to lower production. Later on, this translates into rising prices, low industrial output, and other issues. There has been a drive by the Indian government to increase the amount of land that is irrigated but it still stands at less than half the total land under cultivation.

Figure 10.13 Labour intensive rice farming in the monsoon area of India

Key question

Some farmers in India describe the monsoon as a 'rain god'. Why is it of such importance?

Climate change and agriculture

According to the FAO, climate change is not hypothetical and the effects of global warming are already becoming obvious. The world is beginning to witness floods, droughts, shifting of monsoon patterns and more frequent and intense weather events. These will have wide-ranging effects on the environment, especially on agriculture and **food security**.

More than any other region, the people in the Asia-Pacific region are likely to be hardest hit as a result of climate change. Agriculture is likely to be the most vulnerable economic sector because of its dependence on climate and weather. People in the Asia-Pacific region are mainly agrarian, with almost 60 per cent of the population living in rural areas. Almost a billion people will face the direct impacts of climate change, with consequences to their livelihoods and ways of life. In addition, the region's population is expected to increase by another 850 million people by 2050; this will severely test its ability to produce enough food for everybody.

The predicted impacts of climate change in the Asia-Pacific region are varied. Overall, the region is expected to become warmer. In the Pacific region, rising sea levels may not only affect the growing conditions but also the ability to live on some of the low-lying islands. Coastal areas of south and southeast Asia are likely to face the combined threats of changing rainfall, temperature and sea levels. The cooler northern regions of Asia are likely to become warmer. Changes in rainfall patterns may result in severe water shortages or floods. Rising temperatures can cause changes to crop growing seasons or even reduce their yields.

In 2012, the FAO held a Global Conference on Agriculture, Food Security and Climate Change. At this conference, the concept of **Climate-Smart Agriculture (CSA)** was introduced (Figure 10.14).

CSA is an integrative approach to address the interlinked challenges related to food security and climate change through:

- **economic** means by sustainably increasing agricultural productivity, to support equitable increases in farm incomes, food security and development

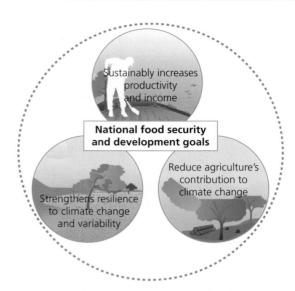

Figure 10.14 Climate Smart Agriculture (CSA)

- **social** change by adapting and building the resilience of agricultural and food security systems to climate change at multiple levels
- **environmental** means by reducing and/or removing greenhouse gas (GHG) emissions from agriculture (including crops, livestock and fisheries).

Although the ideas behind CSA have been welcomed by global organisations such as the FAO and the World Bank, it does not seem to have taken roots in the Asia-Pacific region at all. So far, only Japan, the Philippines and Vietnam are members of the Global Alliance for Climate-Smart Agriculture (GACSA).

Example of adaptation to climate change

Changing climatic conditions and extreme weather events are particularly treacherous for small, resource-poor farmers who do not have any savings, alternative income or insurance. The **Indian Agricultural Research Institute (IARI)** carried out **Climate Change Adaptation Projects** to test technologies and strategies for sustainable livelihood security in rural communities that are vulnerable to climate risks. One such community is the Mewat district of the Indian state of Haryana, about 100 km south of Delhi.

Mewat is a low rainfall, drought-prone area. Analysis of past weather data showed that the mean minimum temperature increased at the rate of 0.18°C every 10 years during the monsoon (Kharif) season, and by 0.47°C every 10 years during the dry (Rabi) season. Between 2020 and 2050, temperature change is expected to accelerate, with minimum temperatures increasing by 1.87°C during the monsoon season and by 2.73°C during the dry season. Based on these findings IARI developed a customised set of interventions to conserve resources, adapt to climate changes and improve livelihoods:

- Superior seed varieties were tested and the successful ones were made available through village seed banks.
- Heat-stress tolerant varieties of the major crop, wheat, were introduced, increasing yields by 12 to 18 per cent.
- Conservation farming practices were also introduced, as well as integrated pest management and soil nutrient management.
- Farmland was levelled using lasers and this improved water-use efficiency by about 15 to 20 per cent.
- Underground pipelines were laid in farmers' fields and used for delivering drip irrigation; this resulted in an additional 40 per cent saving. Overall, the irrigated area increased by 45 per cent while the labour hours required for irrigating crops were reduced by 28 per cent.
- Greater crop diversification was introduced with high yielding varieties of chilli, aubergine, tomato and onion, accompanied by improved production technologies (for example, raised-bed planting, starting seedlings in a nursery, and micro and sprinkler irrigation) which increased household profits by £320 per hectare. Households that planted diversified crops had incomes that were 44 to 86 per cent higher than households that maintained conventional cropping of just pearl millet, wheat and mustard.
- Farmers were also given access to an information and communications technology platform, **mKRISHI®**, which is operated by **Tata Consulting Services**. Using their mobile phones, farmers and farmers' groups were connected to weather forecasting and agricultural production advisory services.

Soils

According to Moujahed Achouri, Director of the United Nations Food and Agriculture Organisation's (FAO) Land and Water Division in his address to the Global Soils week in April 2015, healthy soils are the foundation of global food production and ought to become a key agenda item in public policy. He went on to say that *'soils are essential for achieving food security and nutrition and have the potential to help mitigate the negative impacts of climate change.'*

In addition to sustaining 95 per cent of food production, soils host more than a quarter of the planet's biodiversity, are a major source of pharmaceuticals and play a critical role in the carbon cycle. Unfortunately, the level of soil degradation – estimated at 33 per cent globally – is alarming and has the potential to threaten food security and send many people into poverty.

Sustainable soil management can contribute to the production of more and healthier food. FAO wants the international soil community and policy makers to work together to reduce soil degradation and restore already degraded land.

Zonal soils

Soil is made up of a mixture of solids (minerals and organic matter), liquids and gases that occur on the land surface and is characterised by one or both of the following:

- **horizons**, or layers, that are distinguishable from the initial material as a result of additions, losses, transfers and transformations of energy and matter
- the ability to support rooted plants in a natural environment.

The upper limit of soil is the boundary between soil and air, shallow water, live plants or plant materials that have not begun to decompose. The lower boundary is less well defined and separates soil from the weathered rock layer underneath. Soil consists of horizons near the Earth's surface that, in contrast to the underlying parent material, have been altered by the interactions of climate, relief and living organisms over time.

Soil is an ecosystem that can be managed to provide nutrients for plant growth. They absorb and hold rainwater for use during dryer periods, filter and buffer potential pollutants from leaving fields, serve as a firm foundation for agricultural activities and provide habitat for soil microbes to flourish and diversify to keep the ecosystem running smoothly.

Soil performs five essential functions:

- **Cycling nutrients**: Carbon, nitrogen, phosphorus and many other nutrients are stored, transformed and cycled in the soil.
- **Regulating water**: Soil helps control where rain, snowmelt and irrigation water goes. Water and dissolved solutes flow over the land or into and through the soil.
- **Sustaining plant and animal life**: The diversity and productivity of living things depends on soil.

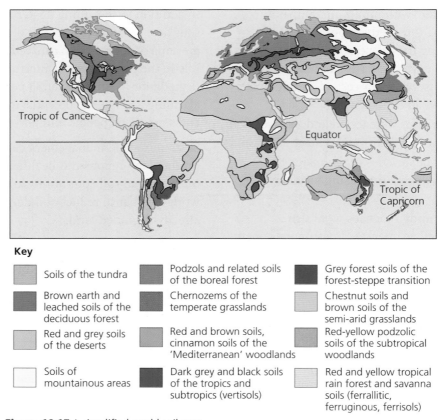

Key

Soils of the tundra	Podzols and related soils of the boreal forest	Grey forest soils of the forest-steppe transition
Brown earth and leached soils of the deciduous forest	Chernozems of the temperate grasslands	Chestnut soils and brown soils of the semi-arid grasslands
Red and grey soils of the deserts	Red and brown soils, cinnamon soils of the 'Mediterranean' woodlands	Red-yellow podzolic soils of the subtropical woodlands
Soils of mountainous areas	Dark grey and black soils of the tropics and subtropics (vertisols)	Red and yellow tropical rain forest and savanna soils (ferrallitic, ferruginous, ferrisols)

Figure 10.15 A simplified world soil map

- **Filtering and buffering potential pollutants**: The minerals and microbes in soil are responsible for filtering, buffering, degrading, immobilising and detoxifying organic and inorganic materials, including industrial and municipal by-products and atmospheric deposits.
- **Physical stability and support**: Soil structure provides a medium for plant roots.

A **zonal soil** is a major soil group often classified as covering a wide geographic region or zone and embracing soils that are well developed and mature, having taken a long time to develop from the parent material. They are in a state of dynamic equilibrium with the climate, vegetation and parent matter. Within the area of a zonal soil there are variations caused by local factors.

Figure 10.15 shows the distribution of the major global soil types.

Zonal soil 1: Chernozem

Chernozems are deep black soils rich in organic matter, their name deriving from the Russian for black earth (Figure 10.16). Their mineral content mainly derives from wind-blown (loess) sediments. Chernozem has a black colour and contains a high percentage of organic matter

as well as high natural percentages of nutrients such as phosphorus and ammonia. It is over one metre deep and has a clay-like structure which is good for retaining water.

These soils are found in regions with a continental climate (cold winters and hot summers) and flat or gently undulating plains with tall-grass natural vegetation. They cover an estimated 230 million hectares worldwide, mainly in the middle latitude steppes of Eurasia and North America.

Figure 10.16 A chernozem soil agricultural landscape in late autumn in the Ukrainian steppes

In earlier times, the chernozems were covered by tall grass natural vegetation. The people who inhabited the region (for example, Native Americans on the Prairies of North America and Eurasian Nomads on the Steppes) were nomadic herdsmen or hunter-gatherers. The high natural fertility of chernozems and their favourable **topography** attract modern agriculture because they permit a wide range of agricultural uses, including arable cropping (with supplemental irrigation in dry summers) and cattle ranching.

The black earth region of the Steppes of Russia has a climate that allows planting in both winter and spring. The single harvest is carried out between July and October. This time range allows planting and field work periods to be spread out, as well as helping with crop rotation.

Crops include cereals (wheat, barley and maize), oilseed plants (sunflower, oilseed rape and soybean) as well as potatoes.

Zonal soil 2: The red/yellow latosols of the tropical rainforest.

Figure 6.45 (page 264) describes latosols. They are the product of a hot, wet climate and thick forest cover. Although they appear to be very fertile, most of the organic nutrients are stored in the vegetation and not the soil.

One successful way of farming in this area is by shifting cultivation. Indigenous peoples (such as the Quicha and the Kayapo of the Amazon Basin) clear small areas of vegetation. They then burn it because the ash provides nutrients for the infertile soil. The land is then farmed for two to three years before they move on to another area of the rainforest. This allows both the forest and the soil to recover. This only supports very low numbers of people and the rainforests have typically been sparsely populated.

In more recent times both poor settlers and ranchers have moved into the rainforest areas and vast tracts of forest have been cleared. The causes and impacts of this clearance on both the environment and the population are covered in Chapter 6 (pages 265–69).

Soil problems and their management

Soil erosion

Soil erosion is defined as the wearing away of the top layer of soil (**topsoil**). It is the most fertile layer because it contains the most organic, nutrient-rich materials. It is this layer that farmers want to protect for growing their crops and for their animals to graze on. Soil is eroded by both water and wind (Figure 10.18 and Figure 10.19, page 460). Water erosion is judged to be the more serious of the two types. It has been estimated that up to 75 billion tonnes of topsoil are lost every year, which equates to approximately 9 million hectares of productive land lost.

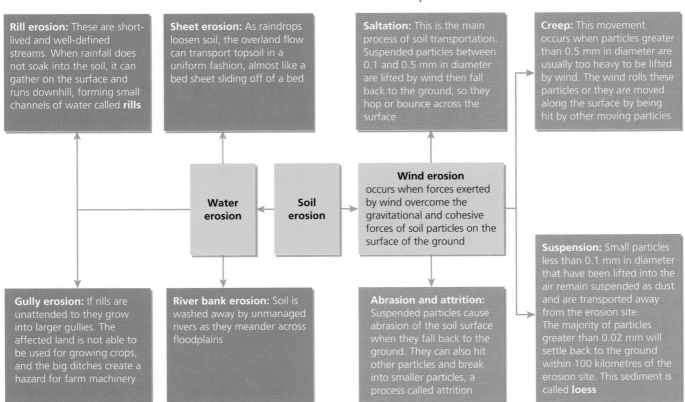

Rill erosion: These are short-lived and well-defined streams. When rainfall does not soak into the soil, it can gather on the surface and runs downhill, forming small channels of water called **rills**.

Sheet erosion: As raindrops loosen soil, the overland flow can transport topsoil in a uniform fashion, almost like a bed sheet sliding off of a bed

Saltation: This is the main process of soil transportation. Suspended particles between 0.1 and 0.5 mm in diameter are lifted by wind then fall back to the ground, so they hop or bounce across the surface

Creep: This movement occurs when particles greater than 0.5 mm in diameter are usually too heavy to be lifted by wind. The wind rolls these particles or they are moved along the surface by being hit by other moving particles

Water erosion

Soil erosion

Wind erosion occurs when forces exerted by wind overcome the gravitational and cohesive forces of soil particles on the surface of the ground

Gully erosion: If rills are unattended to they grow into larger gullies. The affected land is not able to be used for growing crops, and the big ditches create a hazard for farm machinery

River bank erosion: Soil is washed away by unmanaged rivers as they meander across floodplains

Abrasion and attrition: Suspended particles cause abrasion of the soil surface when they fall back to the ground. They can also hit other particles and break into smaller particles, a process called attrition

Suspension: Small particles less than 0.1 mm in diameter that have been lifted into the air remain suspended as dust and are transported away from the erosion site. The majority of particles greater than 0.02 mm will settle back to the ground within 100 kilometres of the erosion site. This sediment is called **loess**

Figure 10.17 Characteristics of soil erosion

Figure 10.17 (page 459) summarises the ways in which soil is eroded and transported away from fields.

Repeated erosion reduces the fertility of the soil by:

- removal of topsoil that is rich in crop nutrients and organic matter
- reduction of the depth of soil available for rooting and water storage for crop growth
- reducing infiltration of water into soil, thereby increasing run-off and erosion.

This in turn can lead to:

- loss of seeds, seedlings, fertiliser (nitrates) and pesticides
- young plants being 'sandblasted' (wind erosion)
- increased difficulty (therefore fuel consumption and man-hours) of field operations.

Damage to the environment can include:

- deposition of sediment onto roads, neighbouring land and into drains
- damage to the quality of water courses, lakes and rivers through excess inputs and increased chemical loading
- increased run-off and sedimentation causing a greater flood-hazard downstream
- sediment in rivers damaging the spawning grounds of fish.

Water erosion control

Figure 10.18 Sheet erosion on red sandstone soils in south Devon

The key to reducing soil erosion by rainwater is to reduce the amount of surface flow of water. This is done by:

- installing and maintaining field drains (see Figure 1.32, page 20) and ditches. Sediment should be removed from ditches and replaced in the fields where it came from
- reducing the amount of water running off roads and farm track onto fields

- the judicious use of farmyard manure to stabilise the topsoil
- protecting soil in winter by early sowing or the use of cover crops
- work across slopes whenever possible. Contour ploughing reduces overland flow and the formation of rills and gullies.

Wind erosion control

Figure 10.19 Wind erosion of sandy soil in East Anglia

Wind erosion control is carried out by:

- increasing soil cohesion by applying organic matter (for example, farmyard manure) to the soil; this improves its structure
- increasing the roughness of the soil surface or by leaving crop residues or stubble in fields and not ploughing them into the soil. In Burkina Faso, for example, when millet and sorghum stubble is cut at a height of one metre and left vertical to the soil surface, it traps a large amount of dust, together with leaves that tornado winds have blown off the trees
- increasing plant cover so that surface wind speed can be cut

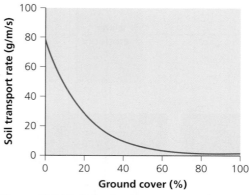

Figure 10.20 Relationship between ground cover and wind erosion

- increasing plant cover to about 50 per cent gives fields adequate protection from wind erosion (Figure 10.20)
- planting lines of trees or hedgerows cuts wind speed which reduces both evaporation (by up to 20 per cent) and wind erosion. A tree line has a wind speed reduction effect for up to 12 times the height of that tree line, both before and after the barrier. This means that the cropped area between windbreaks can be as wide as 100 m if the trees are over 5 m high.

Waterlogging of soil

Soil is considered to be waterlogged when enough of the pore spaces in the soil are occupied by water rather than air for there to be insufficient oxygen (**anaerobic conditions**) for plants to respire. Plants differ in their demand for oxygen and so there is no universal level of soil oxygen that can identify waterlogged conditions for all plants. In addition, a plant's demand for oxygen in its root zone will vary with its stage of growth.

Key questions

Rice and wheat require different amounts of water at different times in their growth cycle. How might this impact on where they are grown?

Lack of oxygen in the root zone of plants causes their root tissues to decompose. The consequence is that a plant's growth and development is stalled. If the anaerobic circumstances continue for a considerable time the plant eventually dies. Fortunately, waterlogged conditions often do not last long enough for the plant to die. Once a waterlogging event has passed, plants recommence respiring. As long as soil conditions are moist, the older roots close to the surface allow the plant to survive, though it may be stunted with low yield.

Waterlogging occurs under two conditions:
- Surface-fed waterlogging happens when precipitation, irrigation water or river floodwater exceeds the combination of evapotranspiration and percolation so that the water stays in and on the surface of the soil.
- Groundwater-fed waterlogging is caused when the rate of rising groundwater is not matched by the rate of evapotranspiration. This may be a natural rise in groundwater or be caused by see page from irrigation canals.

The methods and impacts of soil drainage are discussed in Chapter 1 (see Figure 1.33, page 21).

Salinisation

Salinisation refers to a build-up of salts in soil, eventually to toxic levels for plants (3,000–6,000 ppm of salt results in trouble for most cultivated plants). Salt in soils can be directly toxic to plants but its biggest effect is that it decreases the osmotic potential of the soil (the soil has greater concentrations of solute than does the root) so plants cannot get water from soil.

Problems with salinisation are most commonly associated with excessive water application, rather than with too little. All irrigation water contains dissolved salts derived as it passed over and through the land. This is particularly true of water that has come from groundwater sources. Rainwater also contains some salts. Evaporation of this water from the dry surface of the soil leaves the salts behind. It is also likely to become a problem on soils where the groundwater is within three metres or less of the surface. In such cases, water rises to the surface by capillary action rather than percolating down through the entire soil profile. It then evaporates from the soil surface.

Figure 10.21 Salinisation of soil in the Indus Valley, Pakistan

Salinisation is a worldwide problem and estimates of how much farmland is affected vary from 10 per cent to 20 per cent globally. It is particularly acute in hot semi-arid areas which are poorly drained and use a lot of irrigation water; there is never sufficient rainwater to flush the salt out of the soil. These conditions are found in China's Northern Plain, Central Asia, the San Joaquin Valley of California and the Indus Plain of Pakistan (Figure 10.21, page 461). Salt concentrations are high enough in much of this irrigated land to decrease yields significantly. In extreme cases, land is abandoned because it is too salty to farm profitably.

The 'treatment' for salinisation is to flush the soil with lots of water. However, this results in the salinisation of rivers and groundwater where the flush water goes. For example, following the flushing of soil in the Lower Colorado River valley, the river became too salty downstream for the Mexican farmers to use for irrigation. The Mexican government forced the USA to construct a desalinisation plant near the Mexican border so the water would be useable.

In extreme cases, when the salt crust is too thick, it cannot be flushed, as water just runs off the salty surface.

Structural deterioration of soil

Soil structure is the arrangement of soil particles into groupings. These groupings are called **peds** or aggregates, which often form distinctive shapes typically found within certain soil horizons. For example, granular soil particles are characteristic of the topsoil. Soil aggregation is an important indicator of the workability of the soil. Soils that are well aggregated are said to have 'good soil tilth.'

Soil structures can be classified as follows:

- **Granular and crumb structures** are individual particles of sand, silt and clay grouped together in small, nearly spherical grains. Water circulates very easily through such soils. They are commonly found in the top, A-horizon of the soil profile.
- **Blocky structures** are soil particles that cling together in nearly square or angular blocks having more or less sharp edges. Relatively large blocks indicate that the soil resists penetration and movement of water. They are commonly found in the lower (B-horizon) layers of the soil where clay has accumulated.
- **Prismatic and columnar structures** are soil particles which have formed into vertical columns or pillars separated by miniature, but definite, vertical cracks. Water circulates with greater difficulty and drainage is poor. They are also found in the lower, more clay-rich layer of soil.

- **Platy structures** are made up of soil particles aggregated in thin plates or sheets piled horizontally on one another. Plates often overlap, greatly impairing water circulation. It is commonly found in forest and high latitude soils.

Under natural conditions, the breakdown of soil structure is rare but where agriculture is practised, changes in soil structure are common. They can be divided into two main categories:

- Those associated with a net reduction in soil organic matter in the topsoil. There are two main reasons for this. Firstly, cultivation of the soil causes physical fracturing and mixing of soil and increased aeration, and therefore stimulates breakdown of soil organic matter. Secondly, less organic material is returned to the soil in the form of decaying plant material because much of the plant matter is removed with arable crops through harvest of grain and burning of stubble.

When soil organic matter content declines there is also a breakdown in soil structure involving a splitting-off of individual particles of soil from the aggregates because the bonds holding aggregates together become so weak (because of the low organic matter content) that they can no longer withstand disruptive forces (such as wetting and drying or raindrop impact). When exposed to heavy rain these soils may form a surface crust that impedes movement of water into the soil and/or emergence of seedlings through it.

- **Compaction** of soil is created by a combination of pressure and sliding forces as they are applied to the soil from a wide range of sources. These forces include driving and trailer wheels, plough soles, disc edges, rotary blades and grazing livestock. The way a soil reacts to application of pressure depends on the texture of the soil, how wet it is, how hard it is, its depth below the surface and also on the shape of the contact area.

Farm vehicles are often very heavy and repeatedly travel over the same ground. The concentrated weight compacts the wheel ruts which then become impermeable to water. Rainwater gathers in the ruts and travels downslope causing, in extreme circumstances, gully erosion. The bottom (sole) of ploughs can smear the soil immediately below the depth of a plough furrow creating an impermeable layer called a **plough pan**.

Strategies to ensure food security

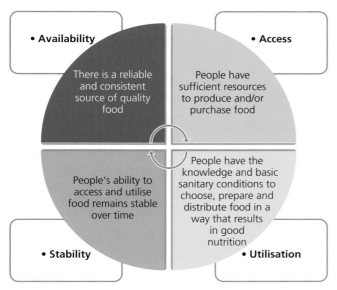

Figure 10.22 The four main components of food security

Food security depends upon (Figure 10.22):

- **Food availability:** Sufficient quantities of food available on a consistent basis.
- **Food access:** Having sufficient resources to obtain appropriate foods for a nutritious diet.
- **Food use:** Appropriate use based on knowledge of basic nutrition and care, as well as adequate water and sanitation.
- **Food stability:** That the availability, access and utilization components remain sustainable for future generations.

Opinions about food security vary. Discussion points include:

- There is enough food in the world to feed everyone adequately, but it is not necessarily in the places where it is needed.
- Future food needs can – or cannot – be met by current levels of production.
- National food security is very important – or no longer necessary – because of global trade.
- Globalisation may – or may not – lead to the persistence of food insecurity and poverty in rural communities.
- Households get enough food, but distribution within the household and whether that food fulfils the nutrition needs of all members of that household is unclear.

Some of the differences in opinion impact on the approaches taken to ensure food security.

The **Global Agriculture and Food Security Program** (GAFSP) in partnership with the **World Bank**, believes that the world will need to produce at least 50 per cent more food by 2050 to feed everyone. GAFSP was established in 2010 to finance medium- and long-term efforts to improve agricultural productivity, increase incomes and ensure food and nutrition security in low-income countries.

At the start they wanted a program that could raise both public and private money to generate a large and sustainable impact on agriculture in the world's poorest countries and improve the lives of smallholders. By the end of 2014 their efforts have helped improve agricultural productivity, increased incomes and improved food security in 31 of the poorest or most malnourished countries across the globe.

On the other hand, according to the **Food and Agriculture Organisation** of the United Nations (**FAO**) the rate of global food production has increased faster than the greater demand for food from population increase. They say that the world produces more than one and a half times enough food for everyone on the planet, enough to feed the predicted 2050 population.

They go on to say that a combination of storage losses after harvest, overconsumption and waste mean that some 800 million people in developing countries are malnourished.

In 2011, the FAO reported annual food losses in sub-Saharan Africa exceeding 30 per cent of the total crop production; this represents more than US$4 billion in value every year. These annual food losses far exceed the total amount of international food aid provided each year to sub-Saharan African countries.

Strategy 1: Increasing food production

One example is in Rwanda where GAFSP funding is contributing to the Rwandan government's **Land Husbandry, Water Harvesting, and Hillside Irrigation Project (LWH)**. This works to increase the productivity and commercialisation of small farmers. After just 30 months the project had already reached more than 92,000 people and yields of maize, beans and potatoes in treated areas were 30 per cent, 167 per cent and 219 per cent above national averages, respectively. The project also looked at nutrition and gender by:

- improving access to nutrition-rich vegetables like iron-fortified beans
- training communities in the construction and management of kitchen gardens

● partnering with local institutions to create innovative financial products for smallholder farmers, particularly women.

Within Rwanda there are several LWH sites, one of which is called **Gatsibo-8.** The total area of the site to be treated is 985 hectares with 45 hectares of that being irrigated. The project aims to help 8,661 people where 4,513 are male (52 per cent) and 4,148 female (48 per cent).

The farmers have been organised into 87 self-help groups in order to move from individual small holdings to a more co-operative system. These self-help groups were formed based on location. There was a year of basic training, for example, compost making, pest management and liming, followed by further education on co-operative management, book-keeping, marketing and entrepreneurship.

There has also been investment in crop drying and storage facilities to reduce losses after the harvest and to increase the quality of the produce. Eighty per cent of produce is now collected in these stores and then sold collectively by the co-operative.

Measuring the success of the scheme:

● By the middle of 2015 maize and beans yield rose from 0.8 tons per hectare to 3.6 tons per hectare and 0.6 tons per hectare to 2.5 tons per hectare respectively.
● Seventy per cent of the land treated was marginal and has now been converted into productive land.
● Seventy-five farmers have been able to build new houses.
● Farmers have been able to buy cows and goats.
● Farmers have been able to send their children to school.
● The introduction of kitchen gardens and vegetables in the site has improved the balance of people's diet and overall nutrition.
● The job opportunities created by the scheme have attracted people from other parts of Rwanda who have now settled in Gatsibo.

Strategy 2: Improving post-harvest practices

In 2013/14 the World Food Programme (WFP) carried out research trials in Uganda and Burkino Faso which aimed to:

● reduce the post-harvest food losses of grains, pulses and legumes of participating farmers by over 70 per cent, leading to increased household food security, nutrition and income
● increase the ability of participating low-income farmers to decide on the percentage of their harvest to retain and the timing of the sale of surplus product
● increase the ability of smallholder farmers and small- to medium-scale traders to link to markets looking for high-quality produce, thereby increasing the overall marketable grain quantities, individual financial returns and improving the food security of participating communities.

This was done in three stages: training, equipping and field support. It involved 400 low-income farmers and their families.

The challenges

Losses of harvested crop

From the time that crops ripen in the field a variety of functions are performed, including harvesting, assembling, drying, threshing, storage, transportation and marketing. Inefficient management practices which allow crops to be unnecessarily exposed to contamination by micro-organisms, chemicals, excessive moisture, fluctuating temperature extremes, mechanical damage and ineffective storage practices contribute greatly to food losses.

Food safety

Adding to the losses caused by biological deterioration are the serious health risks which arise when damage caused to the external pods of legumes or husks/kernels of grains during pre- and post-harvest stages, contribute to contamination and mould growth.

Improving post-harvest management has the potential to impact on the health and well-being of all people living in the region. The most serious of food-related health risks is the constant threat of food poisoning caused by aflatoxin contamination. Aflatoxins are naturally occurring, highly carcinogenic poisons produced by a fungus that is particularly prominent in maize. It can develop when produce comes into contact with soil during harvesting, threshing and drying. Contamination of crops can also occur after grain has been placed into storage, due to pest infestation and poor storage conditions that lead to accelerated growth rates of fungi. The problem has become so widespread throughout

Africa, particularly in the East African region, that aflatoxin poisoning has become an epidemic.

The solutions

- **Pre-harvest** instructions on land preparation and the correct timing of planting and harvesting to reduce a plant's susceptibility to aflatoxins, as well as guidance on controlling moisture content and avoiding direct crop contact with exposed soil was provided.
- Farmers learned the importance of properly **drying crops** to reduce the chance of fungal growth, and ways of creating **low humidity storage** conditions.
- The traditional practice of stockpiling dried crops either directly on the floor, in baskets or in polypropylene sacks on the floor of their houses (due to fear of theft) was strongly advised against, regardless of the length of storage.
- Training in **harvesting** so that the harvest was carried out at the optimum time:
  - to avoid losses (too early and crops are moist and grains unfilled/too late and attacks by insects, birds and rodents begin)
  - to understand the impact of weather conditions at the time of harvest, where rain causes the dampening of crops leading to mould growth and the risk of aflatoxin contamination.
- Training in **drying**:
  - minimising damage by reducing the moisture content below the level required for mould to grow during storage
  - never allowing crops to have direct contact with the soil during drying
  - limiting aflatoxin contamination (using tarpaulins to reduce the risk of contamination and to provide cover when exposed during damp weather).
- Training in **threshing**:
  - precautions to avoid damage to grain during threshing/shelling.
- **Solarisation:**
  - an additional step to kill all life stages of insect pests prior to storage by placing the grain into a solar oven (dark plastic base with clear plastic cover) for 1–5 hours.
- **Improving** on farm storage by introducing hermetic storage units or new storage technologies to protect crops from insects, rodents, birds, rain and temperature fluctuations. A variety of containers were trialled but they all were capable of being sealed in

such a way that a build-up of carbon dioxide in the container eventually reaches a level of toxicity where it is impossible for insects and moulds to survive.

- **On-farm support** took place where support workers were present to make sure that storage instructions were carried out correctly.

Figure 10.23 shows how the metallic silo storage reduced crop losses the most, but also that all the hermetic types of crop storage greatly reduced crop loss. Farmers were able to choose when they went to market with their produce rather than getting rid of it before losses proved the harvest worthless.

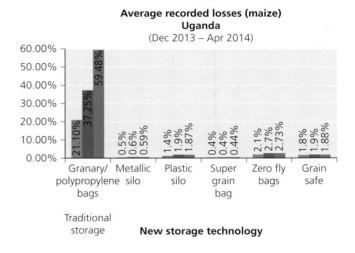

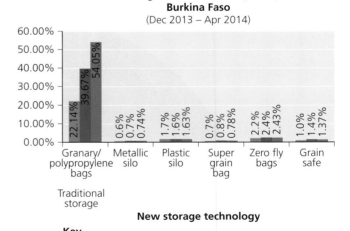

Figure 10.23 The losses of stored maize before and after the introduction of new storage technology

Figure 10.24 (page 466) shows the difference in prices at different times of the year in Uganda. Farmers received more money if they sold the same crop just three months later.

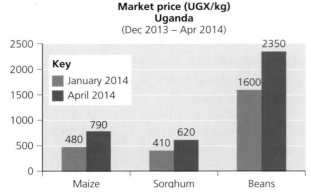

Figure 10.24 The difference in market price between the time immediately after harvest and 90 days later

10.3 Environment, health and well-being

Few things matter more to people than their **health**, **mortality** and **well-being** and these are directly affected by the environment in which they live. As Figure 10.25 shows, this can be the physical environment where factors such as air and water quality are important, the economic environment which affects the type of housing they live in and the jobs that they do or the social environment determining where an individual was educated or lives, and the people and institutions with whom they interact.

Key terms

Disability-adjusted life years (DALYs) – A measure of morbidity within a society. They measure the number of years of healthy life lost by being in poor health or a state of disability.

Epidemiological transition – This describes changing patterns of population age distribution, mortality, fertility, life expectancy and causes of death. It assumes that infectious diseases are replaced by chronic diseases over time due to expanded public health and sanitation.

Health – Defined by the World Health Organization (WHO) as a state of complete physical, mental and social well-being and not merely the absence of disease or infirmity.

Morbidity – Morbidity relates to illness and disease. It can also be used to describe the incidence of a disease within society. Some diseases are so infectious that by law they must be reported. These include malaria, rubella and tuberculosis.

Mortality – Mortality relates to death. It can be measured by death rate, infant mortality, case mortality and attack rate. Figure 10.27 shows the causes of death in England and Wales between 2000 and 2012. Non-communicable diseases such as heart disease and cancer have claimed the most lives.

Non-communicable disease – A medical condition or disease that is by definition non-infectious and non-transmissible among people.

Well-being – The state of being comfortable, healthy or happy.

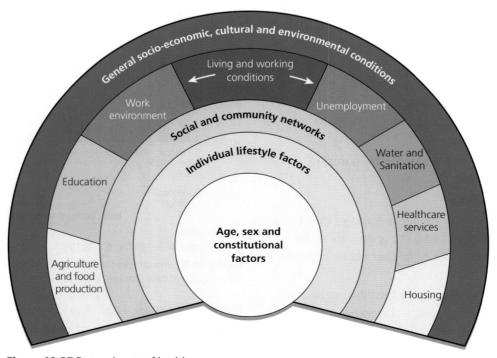

Figure 10.25 Determinants of health

Source: Dalgren G and Whitehead M (1991) Policies and strategies to promote social equity in health, Institute for Future Studies, *Stockholm*

Global patterns of health, mortality and morbidity

What are the geographical patterns of health?

Health is unevenly distributed. If all the people who died in 2012 were imagined as a group of 1,000 people, Figure 10.26 shows how **mortality** rates are considerably higher in low- and middle-income countries. More than half (514) of these 1,000 deaths would have been caused by just 10 conditions, shown in Figure 10.27.

Non-communicable diseases (NCDs) were responsible for 68 per cent of all deaths globally in 2012, up from 60 per cent in 2000. The four main NCDs are cardiovascular diseases, cancers, diabetes and chronic lung diseases. In comparison, deaths from infectious diseases are declining. Historically, NCDs have been associated with more deaths in high-income countries, but in 2012, 28 million (almost three-quarters) of the 38 million of global NCD deaths occurred in low- and middle-income countries.

Figure 10.28 (page 468) shows that the number of people dying globally from NCDs has increased between 2000 and 2012, while deaths from communicable diseases have fallen. This is the result of improvements in sanitation, diet and healthcare.

Economic and social development

As societies and nations develop, the economic and social conditions in which people live should improve. As part of this development, improvements in health and healthcare provision will also be witnessed. For example, increases in food productivity and supply and in transport infrastructure mean a population will be fed more reliably and is less prone to famine or diseases related to malnutrition. Similarly, developments in sanitation and public hygiene reduce the chances of water-borne infections such as typhoid and cholera. With advances in medical technology and vaccination programmes, the risk of infectious diseases has been reduced in many parts of the world.

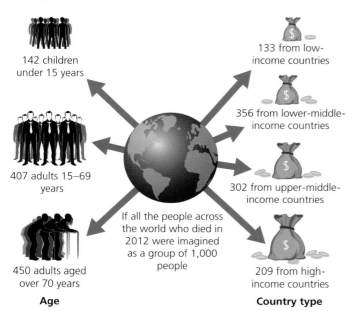

142 children under 15 years

407 adults 15–69 years

450 adults aged over 70 years

Age

133 from low-income countries

356 from lower-middle-income countries

302 from upper-middle-income countries

If all the people across the world who died in 2012 were imagined as a group of 1,000 people

209 from high-income countries

Country type

Figure 10.26 Distribution of mortality across age and country type in 2012

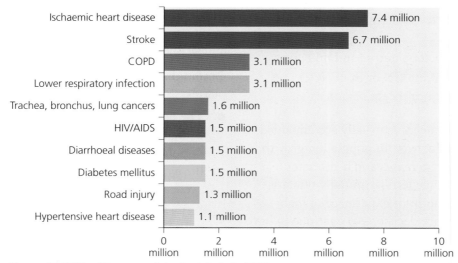

Figure 10.27 Top 10 causes of death globally in 2012

Source: WHO

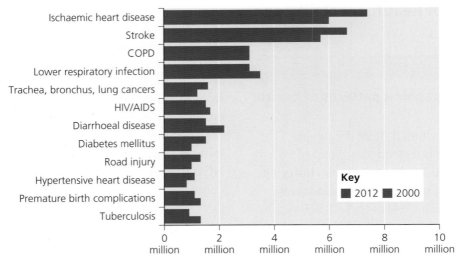

Figure 10.28 Comparison of leading causes of death between 2000 and 2012

Source: WHO

Economic developments linked with improved health include:

● technology to improve food productivity and supply (for example, the Green Revolution)
● improved transport infrastructure to distribute food and medical supplies
● investment in drainage and sewage systems especially in urban areas
● trading of resources or manufactured goods in exchange for a wider variety of foods and medicines.

Social developments improving health include:

● improved sanitation
● better education about sanitation and disease transmission
● advances in medical technology such as antibiotics and vaccines
● better training for doctors, nurses and midwives
● aid programmes from the UN or from NGO provision to improve healthcare resources.

Socio-economic development is usually accompanied by a sudden and stark increase in population growth rates as life expectancy increases and death rates decrease. At the same time fertility initially remains high as social norms and prior experience of high infant mortality influence decisions to continue to have large families. Healthcare developments eventually reduce infant mortality rates and with greater survival, parents start to have fewer children, particularly in

areas where development is accompanied by improved access to contraception. This leads to subsequent declines in fertility rates.

Epidemiological transition (see also the Demographic Transition Model)

The idea of **epidemiological transition** was posited by Abdel Omran (1971) who suggested that the process of socio-economic development accounts for a transition, over time, from infectious diseases to chronic and degenerative diseases as the most important causes of death (Figure 10.29).

The transition occurs as a country undergoes the process of modernisation from developing to developed nation status. It is the element of demographic transition (examined later in this chapter) explained by advances in medical technology, disease prevention and sickness therapy and treatment.

Omran divided this transition into three distinct phases:

● **The age of pestilence and famine**: Mortality is high and fluctuates; this prevents sustained population growth. There is low and variable life expectancy of between 20 and 40 years. The cyclical low growth patterns are associated with wars, famine and epidemic outbreaks, interspersed with periods of relative 'prosperity'.
● **The age of receding pandemics**: The rate of mortality declines as disease epidemics occur

Table 10.2 The 4 stages of epidemiological transition

Phase	Life expectancy	Change in socio-economic conditions	Causes of morbidity and mortality
Age of infection and famine	20–40	Poor sanitation and hygiene; unreliable food supply	Infections; nutritional deficiencies
Age of reducing pandemics	30–50	Improved sanitation; better diet	Reduced number of infections; increases in occurrence of strokes and heart disease
Age of degenerative and man-made diseases	50–60	Increased ageing; lifestyles associated with poor diet, less activity and addictions	High blood pressure, obesity, diabetes, smoking-related cancers, strokes, heart disease and pulmonary vascular disease
Age of delayed degenerative diseases	c. 70+	Reduced risk behaviours in the population; health promotion and new treatments	Heart disease, strokes and cancers are main causes of mortality but treatment extends life. Dementia and ageing diseases start to appear more

less frequently. Average life expectancy increases and population growth is sustained and begins to rise exponentially. This phase is associated with advances in medicine and the development of healthcare systems. For example, half of the deaths prevented during the nineteenth century may be due to clean water provided by public utilities. One breakthrough of note was the discovery of penicillin in 1928, which led to widespread and dramatic declines in death rates from previously serious diseases.

● **The age of degenerative and man-made diseases**: Mortality continues to decline and eventually approaches stability at a relatively low level. In this phase, infectious disease pandemics are replaced as major causes of death by non-communicable degenerative diseases. Infectious agents as the major contributor to morbidity and mortality are overtaken by anthropogenic causes. The average life expectancy increases to more than 50 years. Fertility becomes a more important contributory factor to population growth.

The theory was revisited and modified in the 1980s by other public health researchers. Olshansky and Ault added a fourth stage to the transition:

● **The age of delayed degenerative diseases:** Declining death rates are concentrated at advanced ages. Life expectancy increases to 70-80. The causes of mortality remain the same as in the third phase but the distribution is delayed until older ages because of new treatments, prevention and health promotion.

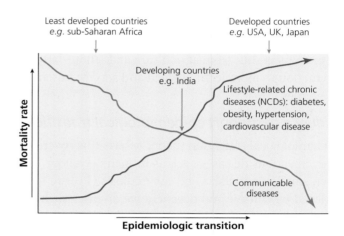

Figure 10.29 Line graph outlining epidemiological transition

Omran discovered variations in the pattern, pace and the determinants of population change within the transition and so identified three types of model:

● **Classical/western model:** (for example, Western Europe) – a slow decline in death rates followed by lower fertility.

● **Accelerated model:** more rapid transition; falls in mortality take place over a shorter period of time (for example, many parts of Asia; some parts of Latin America).

● **Contemporary/delayed model:** recent decreases in mortality are not accompanied by a decline in fertility because infant and maternal mortality remain relatively high (for example, sub-Saharan Africa).

In Omran's third phase fertility rates decline from highly positive replacement numbers to stable or even negative replacement rates. This represents the effect of individual choices on family size and the ability to implement those choices through the use of contraception. He suggested three sets of factors encourage reduced fertility rates:

- Bio-physiologic factors: reduced infant mortality and the expectation of longer life in parents
- Socio-economic factors: childhood survival and the economic perceptions of large family size
- Psychological or emotional factors: society changes its rationale and opinion on family size; parental energies are redirected to qualitative aspects of child-rearing.

This aspect of the transition could also be linked to the adaptations associated with migration to urban areas; for example a shift from agriculture and labour-based production to more technological and service-based economies.

Evidence to support epidemiological transition

Epidemiological transition theory asserts that, over time, infectious diseases such as influenza, yellow fever, smallpox, malaria and tuberculosis (TB) are replaced by chronic and degenerative diseases such as heart disease, cancer and dementia as the more important causes of death.

Evidence to support the idea of transition can be seen in Figure 10.27 on page 467 which shows that NCDs now account for seven of the ten most common causes of death globally. Figure 10.28 on page 468 shows that even in the period from 2000–2012 numbers dying from infectious diseases are in decline while those dying from chronic disease are increasing. Figure 10.30 shows a clear shift in the major causes of death in the USA between 1900 and 2010.

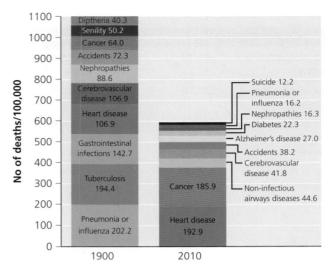

Figure 10.30 Epidemiological transition in the USA (deaths per 100,000 of population)

This shift in demographic and disease profiles is caused by factors such as the use of antibiotics and increased public sanitation. The transition is evident in most developing nations but some critics question whether the transition has actually taken place during the twentieth century. They argue that the increase in chronic disease may be an illusion resulting from more advanced medical technologies being able to diagnose diseases that previously remained undiagnosed, which gives the appearance of an emergence of new chronic illnesses.

Table 10.3 shows differences in causes of death by rank between low-income countries and high-income countries and provides further evidence to support the concept of epidemiological transition.

According to WHO, in the next two decades there will be dramatic changes in the world's health needs as a result of epidemiological transition. At present, lifestyle and behaviour are linked to 20-25 per cent of the

Table 10.3 Differences in the top six causes of death between low-income and high-income countries (deaths in rank order, per 100,000 population)

Low-income countries		High-income countries	
Cause	Deaths/100,000	Cause	Deaths/100,000
Lower respiratory infections (e.g. pneumonia)	91	Ischaemic heart disease	158
HIV/AIDs	65	Stroke	95
Diarrhoeal diseases	53	Cancer (trachea/lung)	49
Strokes	52	Alzheimer's disease	42
Ischaemic heart disease	39	Chronic obstructive pulmonary disease (COPD)	31
Malaria	35	Lower respiratory infections	31

Source: WHO

global burden of disease. This proportion is increasing rapidly in poorer countries. By 2020, chronic diseases such as depression and heart disease, as well as road traffic accidents, will replace the traditional enemies such as infectious diseases and malnutrition as the leading causes of disability and premature death. NCDs are expected to account for seven out of every ten deaths in the developing regions. Epidemiological transition is well advanced, suggesting that public health policy in poor countries, with its traditional emphasis on infectious disease, will need to adapt.

Environmental variables and their links to the incidence of disease

Climate

Links between the natural environment and disease tend to focus on climatic conditions. Drought leads to crop failure, reduction in food consumption and the potential for famine. At the other extreme, flooding caused by heavy rains or tropical storms can lead to water-borne disease and respiratory infections.

Seasonal affective disorder (SAD) is a type of depression that has a seasonal pattern. The episodes of depression tend to occur at the same time each year, usually during the winter, and have been linked to reduced exposure to sunlight during the shorter days of the year (Figure 10.31).

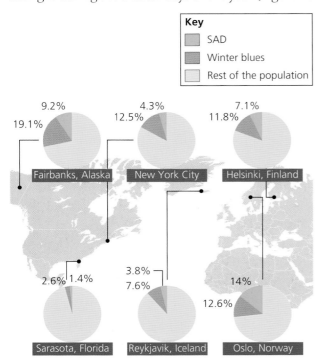

Figure 10.31 The changing incidence of seasonal affective disorder with latitude

In the UK, common everyday ailments linked to the natural environment include allergies such as **hay fever** and **asthma,** where the body reacts to pollen, dust and other substances in the air during the spring and summer.

Large-scale seasonality of mortality has declined in the UK possibly due to the increased use of central heating. The number of people affected by respiratory problems in cities has declined due to better air quality.

Other small-scale aspects of the environment can also have an impact on morbidity. A relationship has been made between soft water and high levels of heart disease. High concentrations of aluminium in water have been suggested as an explanation for the geographic distribution of Alzheimer's disease. Excesses of nickel, cadmium, mercury and lead are known to be hazardous to health while in the southwest of England, high amounts of radon and radioactivity from granite rocks has been linked to some cancers.

Topography, drainage and health

In many low-income countries of Asia, high densities of people occupy the flood plains of major rivers (for example, The Ganges, the Yangtze and the Mekong). They are attracted by the flat land and the seasonal flooding that enables them to grow rice. Flooding, for them, is a good thing; that is, until they have exceptionally wet years. Unplanned flooding accounts for 40 per cent of all natural disasters worldwide and causes about half of all deaths from natural disasters. Most floods occur in low-income countries and tropical regions where the impact on health is significant, the number of people displaced is often large, and the death toll is high. Increased rates of diarrhoea (including cholera and dysentery), respiratory infections, hepatitis A and E, typhoid fever and leptospirosis have been described as occurring after floods in developing areas. In addition, floods contaminate freshwater supplies, increase the risk of water-borne diseases and create breeding grounds for disease-carrying insects such as mosquitoes. They also lead to drownings and physical injuries, damage homes and disrupt medical and health services.

The landscape of Nepal is dominated by high mountains and steep slopes. The 2015 earthquake caused avalanches on Mount Everest where at least 19 died and at least 120 others were injured or missing. In the Langtang Valley 329 people were killed by a landslide of weathered rock and soil. After the earthquake, one of the first concerns among health officials was that hospitals in four of the worst-affected districts were completely destroyed or too badly damaged to function. Landslides had damaged roads and tracks and although trained medical personnel were able to treat patients on the ground, emergency aircraft and aid vehicles were required to transport people to health centres and hospitals further away.

Other natural hazards can have an indirect link to disease too, as shown by the Haitian earthquake of 2010. Even before it struck, the country lacked a public sewage system, and less than half the population had access to drinking-water services. Malnutrition was rampant, nearly 200,000 people lived with HIV/AIDS, and only half of the childhood population was vaccinated against basic diseases like diphtheria. In the cramped and unhygienic conditions of the refugee camps infectious diseases spread very quickly, especially cholera. Between 2010 and 2014, UNICEF reported that there had been nearly 700,000 cases and over 8,500 deaths from cholera in Haiti.

Air quality and health

Ambient (outdoor) air pollution is a major environmental health problem with the potential to affect everyone in both developed and developing countries. According to the WHO, air pollution is now the world's largest single environmental health risk. Terms such as 'airpocalypse' have been used to describe the increasing number of deaths and poor health associated with polluted air.

Polluted air affects the cardiovascular and respiratory health of the population and was linked to 3.7 million premature deaths worldwide in 2012. The WHO estimates that around 80 per cent of air pollution-related premature deaths were due to heart disease and strokes, while 14 per cent of deaths were due to chronic obstructive pulmonary disease (COPD) or acute lower respiratory infections and 6 per cent of

deaths were due to lung cancer. People living in low and middle-income countries are disproportionately affected by air pollution, with 88 per cent of premature deaths occurring in developing countries; the greatest burden being in the Western Pacific and Southeast Asia regions.

A 2014 study by the World Health Organization found that Delhi had the dirtiest atmosphere of 1,600 cities around the world for **particulate matter (PM)** – solid particles or liquid droplets in the air (Figure 10.32). These are created by things such as vehicle emissions, forest fires and industrial processes. They are able to penetrate deep into the lungs.

Figure 10.32 Children protecting their faces from the Delhi smog in 2014

Most sources of outdoor air pollution are well beyond the control of individuals and demand action from authorities and policymakers (from city to international level). Policies and investments which support cleaner transport, power generation and industry, energy-efficient housing and better municipal waste management would reduce key sources of urban outdoor air pollution. Rural air pollution in developing regions can be tackled by reducing emissions from biomass energy systems, agricultural waste incineration, forest fires and certain agro-forestry activities (for example, charcoal production). By reducing air pollution levels, countries would reduce the burden of disease from stroke, heart disease, lung cancer and both chronic and acute respiratory diseases, including asthma.

In addition to outdoor air pollution, the use of fuelwood for indoor cooking and heating produces harmful pollutants which negatively affect the health of millions of people around the world. Exposure to smoke increases the risk of respiratory infections, lung cancer, cardiovascular disease and cataracts. In low-income countries, 2.5 billion rural households are primarily based on traditional sources for cooking purposes and use solid fuels, with a high preference for fuelwood. The WHO reports that in 2012, 4.3 million deaths globally were attributable to household air pollution. This amounts to 7.7 per cent of global mortality and almost all of it is in low- and middle-income countries. Women and children tend to be most at risk because they spend more time in fuelwood-burning environments.

Water quality and health

Water pollution and poor water quality have important impacts on health and can lead to many diseases. The WHO claims two million deaths annually are attributable to unsafe water and poor sanitation and hygiene.

Water-related diseases and morbidity include the following.

- Diarrhoeal diseases such as cholera are caused by bacteria and chemicals in water that people drink. Inadequate drinking water, sanitation and hygiene are estimated to cause 842,000 diarrhoeal disease deaths per year (WHO 2014). Diarrhoea is a leading cause of malnutrition and the second leading cause of death in children under five years old. More than 50 countries still report outbreaks of cholera to the WHO.
- *Schistosomiasis* is an acute and chronic disease caused by parasitic worms which have part of their lifecycle in water; people are infected during exposure to infested water. An estimated 260 million people are sufferers.
- Diseases like malaria have water-related vectors.
- *Legionellosis* (or Legionnaire's disease) is a respiratory disease caused by bacteria which infect the lungs and cause pneumonia. The most common form of transmission is inhalation of contaminated aerosols produced in conjunction with water sprays, jets or mists. Infection can also occur by aspiration of contaminated water, particularly in susceptible hospital patients.
- Cancer and tooth/skeletal damage: millions are exposed to unsafe levels of naturally-occurring arsenic and fluoride.

Human sewage is one of the main pollutants of water. Toilets may be a simple hole in the ground (latrine) or a drop over a water source. The same water may then be used for drinking, washing and watering animals. In 2015, WaterAid estimated that nearly 800 million people live without safe water and over 500,000 children die every year from diarrhoea caused by unsafe water and poor sanitation.

Much of the burden of water-related diseases (water-related vector-borne diseases in particular) is caused by methods of water resource development and management. In parts of the world, attempts to manage water supplies through dam construction, irrigation development and flood control have led to adverse health impacts from water pollution causing significant preventable disease. For example, stagnant waters created by man-made reservoirs in sub-Saharan Africa create the perfect conditions for malaria-spreading *Anopheles* mosquitoes to breed and multiply. A recent study found that cases of malaria in high risk areas across the region are associated with people living in the vicinity of a dam. Even attempts at more sustainable strategies such as increasing the use of wastewater in agriculture, which is important for livelihood opportunities, are also associated with serious public health risks.

Most of the diseases and health problems related to water quality are preventable and relatively affordable and give plentiful opportunities to improve health and reduce disease, even in the world's poorest countries. The WHO claims that 4 per cent of the global disease burden could be prevented by improving water supply, sanitation and hygiene. Much progress has been made over the past decade because of the focus provided by the Millennium Development Goals (MDGs). The number of children dying from diarrhoeal diseases has fallen steadily since 1990 and 2.3 billion people gained access to improved drinking water over the same period. Practical measures have included:

- better tools and procedures to improve and protect drinking water quality at the community and urban level, for example through Water Safety Plans (including better education and awareness to reduce faecal contamination and disinfecting supplies with chlorine)
- availability of simple and inexpensive approaches to treat and safely store water at the household-level (for example, boiling and covered storage).

As the world focuses its attention on the post-2015 Sustainable Development Goals (SDGs), controlling human exposure to waterborne pathogens associated with faecal waste remains a priority in attaining the goal of safe drinking water and sanitation for all (SDG 6).

There are many other contaminants which are detrimental to human health. Oil spills contaminate land and water, leading to digestive problems and diarrhoea. In the Niger Delta (see the case study in Chapter 11, page 608), oil spills, waste dumping and gas flaring are endemic and this type of pollution has damaged the soil, water and air quality for decades. Hundreds of thousands of people have been affected, particularly the poorest and those who rely on traditional livelihoods such as fishing and agriculture.

The global prevalence and distribution of malaria

Malaria is a tropical **vector-borne disease**, biologically transmitted by insects, which has a devastating impact on people's health and livelihoods around the world. According to the latest available data from the WHO, about 3.4 billion people live in areas at risk of malaria transmission, in 106 countries and territories. In 2013, an estimated 198 million cases occurred, killing about 584,000 people. As Figure 10.33 shows, sub-Saharan Africa has the highest concentration and approximately 82 per cent of malaria cases and 90 per cent of malaria deaths in 2013 occurred in Africa, with children under the age of five and pregnant women most severely affected. Malaria is the sixth biggest killer in low-income countries and it was deemed significant enough to be part of Millennium Development Goal 6 in the year 2000.

Malaria is caused by parasites of the *Plasmodium* family and is transmitted by the female Anopheles mosquito. If the mosquito bites a person, the parasite may be injected into the person's blood stream. The parasites multiply in the liver, and then infect red blood cells. Symptoms of malaria include fever, headache and vomiting, and usually appear between 10 and 15 days after the mosquito bite. If not treated, malaria can quickly become life threatening by

disrupting the blood supply to vital organs. It can either kill a person directly or weaken the immune system so that other diseases affect the sufferer. The disease also contributes greatly to anaemia among women and children – a major cause of poor growth and development, and can lead to low birth weight among newborn infants.

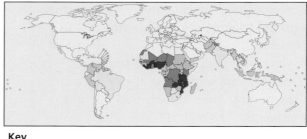

Key
Confirmed malaria cases per 1000 population

■ >100	■ 10–50	□ 0.1–1	□ No ongoing malaria transmission
■ 50–100	■ 1–10	□ 0–0.1	■ Not applicable

Figure 10.33 Countries with ongoing transmission of malaria, 2013. Higher latitudes are unaffected by malaria as are areas of higher altitude, even within tropical areas.
Source: WHO

Links to physical environment

Malaria is strongly linked to the natural environment. Mosquitoes breed in stagnant water and transmission is greatest in areas during and just after the rainy season. Figure 10.34 shows how, in Africa, the length of the malaria transmission season is directly related to the length of the rainy season. At altitudes above 1,500 m and where rainfall is below 1,000 mm, malaria transmissions fall. A high temperature is also important. The parasites require temperatures of between 16°C and 32°C to develop inside the mosquito and this is one reason why the disease is largely concentrated in the tropics and subtropics.

On a smaller scale, coastal areas have much less seasonal variation in temperature, are lower in altitude and generally have higher relative humidity. Providing the temperatures lie within the transmission vector's range of tolerance, coastal areas will often show a higher prevalence of the disease.

Land use is another important factor which can determine malaria risk. Studies in India have shown that those living in proximity to forested areas (again with higher humidity) are more susceptible to infection.

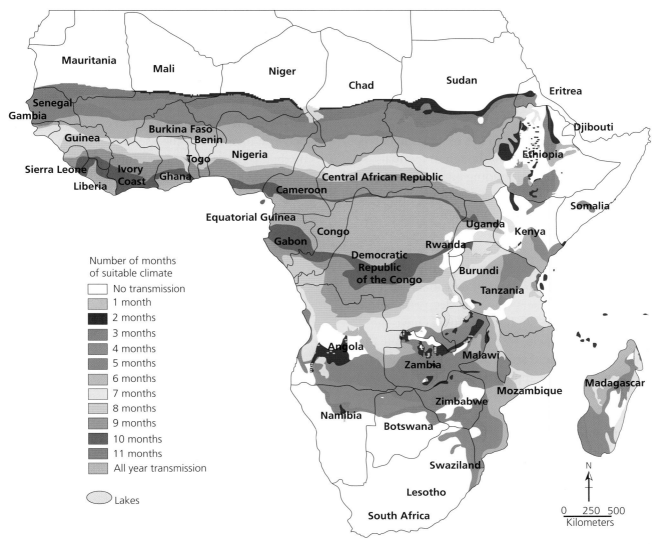

Figure 10.34 The duration of the malaria transmission season

Source: MARA/ARMA (Mapping Malaria Risk in Africa/Atlas du Risque de la Malaria en Afrique)

Links to socio-economic environment

While environmental variables are the overriding determinant of malarial mosquito presence, socio-economic factors will also influence the level of risk of contracting the disease. Malaria is frequently referred to as a disease of poverty and there is sufficient evidence to affirm that the burden of malaria is greatest among the world's poorest countries. When the burden is measured as **disability-adjusted life years (DALYs)**, 58 per cent of the total global burden due to malaria is concentrated among the poorest 20 per cent of the global population, while only 0.2 per cent of total global DALYs are lost by the richest global 20 per cent.

Malaria is a common cause of morbidity in many countries but is concentrated in the world's poorest continents because of the lack of investment in projects for developing new drugs and vaccines and the limited number of existing programmes to control the disease. Studies undertaken in many African nations, as well as those in Asia and South America, have attempted to identify key socio-economic, demographic and geographic factors that affect potential vulnerability to malaria. A number of common socio-economic variables prevail in many of these studies that increase the incidence of malarial infection:

● **Housing quality and occupancy:** Homes are those with earth/sand floors and using materials such as mud, bamboo cane or wooden trunks for walls and grass or palm leaves to provide a roof; those with poorly-fitted windows or doors, or windows without glass, screens, curtains or shutters. Densely clustered built-up areas and overcrowded rooms (for sleeping) is another important factor in increasing risk. Areas with high malaria incidence are associated with high infant mortality and subsequent high fertility rates. A consequence of this is that children (and sometimes parents) often share a room with at least five occupants.

- **Unsanitary conditions:** Studies of malaria 'hotspots' in Chennai identified that although individual houses were clean, surrounding areas were dirty and polluted by rubbish and waste outflows. Spitting and open defecation are also commonly associated with attracting mosquitoes.

- **Occupation:** Agricultural workers are more exposed to the mosquito vector. This particularly applies to those living near to stores of irrigation water and to migrant workers who may sleep outdoors at night. For example, studies in Ethiopia have shown that highland farmers, who seasonally migrate to find work in lowland areas, are prone to malarial infection. The same applies to miners in some areas, for example, gem miners in Sri Lanka.

- **Rural vs urban environments:** Results from studies are more mixed and inconsistent about the nature of the environment. Generally, those in rural areas are seen to be more at risk, however contamination rates in urban slums and squatter settlements are high due to density of households and unsanitary conditions.

- **Age and gender:** These are not significantly associated with the risk of infection, though children aged under 5 are more likely to suffer the extreme consequences of the disease. However, studies in both The Gambia and Tanzania indicate a shift in malaria burden from the under-fives to older children aged 5 to 14 years. This is thought to be as a result of the focus on prevention in the younger age groups through better education, awareness and provision of nets.

- **Ethnicity:** This only influences risk because of different attitudes towards health and education. For example, in Thailand non-Thais searching for work and housing have greater exposure to malaria threats.

It is important to distinguish between the overall risk of infection, determined by the mix of physical and socio-economic environmental factors listed above, and the utilisation of prevention and treatment methods. These have a much clearer association with poverty and socio-economic status (SES) such as the level of income and educational attainment as well as accessibility to intervention methods. There is more consistent evidence that vulnerability to the consequences of malaria is higher among lower socio-economic groups.

- **Income:** There is a strong positive correlation between income and the use of prevention methods. Those with higher incomes spend more on repellents, insecticide-treated nets (ITNs) and mosquito coils, to reduce their risk of infection. Higher income is associated with better nourishment and studies show that cerebral malaria is less common in well-nourished children. Again, this disproportionately affects agricultural workers, many of whom receive seasonal incomes at harvest time and cannot afford nets or treatment during the main malarial season.

- **Education:** Those with a clearer understanding of the link between malaria and surrounding environmental conditions, including hygiene and sanitation, are more likely to use prevention strategies. For example, studies in Malawi found that net ownership was largely absent in homes where the head of the household had not completed primary education.

- **Distance and accessibility:** Greater distance to the nearest clinics or hospitals is associated with fewer seeking treatment for symptoms and less expenditure on prevention methods. For example, repellents and coils are less available and so used less in rural areas; instead there is a reliance on burning medicinal plants as a repellent.

Most of the variables listed are proxies for poverty and evidence suggests that malaria infection, and particularly death resulting from malaria, is strongly associated with low socio-economic status. It is children from the poorest income groups living in poor housing who are most at risk.

The impacts of malaria

Well-being can be defined as 'the state of being comfortable, healthy or happy'. Clearly if an individual is ill then they cannot be any of those things. If groups of people are ill then it has an effect on the economic well-being of a community.

Impacts on health and well-being

Malaria kills a child somewhere in the world every minute. Ninety per cent of malaria deaths occur in Africa, where malaria accounts for about one in six of all childhood deaths. The disease also contributes greatly to anaemia among children – a major cause of poor growth and development.

Malaria causes the patient to develop a high fever, which comes and goes. The pattern of fevers (if there is one) may vary according to the species of malaria. Initially, it feels like the flu with high fever, fatigue, and body aches, with hot and cold stages. Signs and symptoms in children may be nonspecific, leading to delays in diagnosis. People also may have headache, nausea, shaking chills (rigors), sweating and weakness. Anaemia is common in patients with malaria, in part due to the effects of the *Plasmodium* parasite on the red cells.

Plasmodium falciparum causes a particularly severe form of malaria. In addition to fever, patients may experience complications such as destruction of red blood cells, yellow skin discoloration, kidney failure, fluid in the lungs, cerebral malaria, convulsions, coma or death.

People who have lived for years in areas with malaria may develop a partial immunity to new infections.

Malaria infection during pregnancy is associated with severe anaemia and other illness in the mother and contributes to low birth weight among newborn infants – one of the leading risk factors for infant mortality and sub-optimal growth and development.

As well as the burden on local healthcare systems, malaria illness and death cost Africa approximately US$12 billion each year in lost productivity. The effects permeate almost every sector. Malaria:
- increases school and work absenteeism
- decreases tourism
- inhibits foreign investment
- affects crop production.

Impacts on economic well-being

Malaria has serious and far-reaching impacts, slowing economic growth and development and prolonging the vicious cycle of poverty. While it is true that poverty and lack of development may be a key cause of the presence of malaria, there is a strong argument that malaria causes underdevelopment.

- Costs to individuals and their families include the purchase of drugs for treating malaria; expenses for travel to, and treatment at, clinics; lost days of work; reduction in crop production; absence from school; expenses for preventive measures; expenses for burial in cases of deaths.
- Costs to governments include maintenance, supply and staffing of health facilities; purchase of drugs and supplies; public health interventions against malaria, such as insecticide spraying or distribution of insecticide-treated bed nets; lost days of work with resulting loss of income; and lost opportunities for joint economic ventures and tourism.
- It has been estimated that in some areas, malaria accounts for up to 40 per cent of public health expenditures, 30–50 per cent of inpatient hospital admissions and 60 per cent of outpatient health clinic visits.
- Direct costs (for example, illness, treatment and premature death) have been estimated to be at least US$12 billion per year. The cost in lost economic growth is many times more than that.

Management and mitigation strategies

How can malaria be prevented?

The WHO suggests that the key interventions to control malaria are:

1. prompt and effective treatment with artemisinin-based combination therapies (ACT)
2. use of insecticidal nets by people at risk, and
3. indoor residual spraying with insecticide to control the vector mosquitoes.

In many parts of the world, the parasites have developed resistance to malaria treatments so this holistic approach offers better protection. Sleeping under insecticide-treated bed nets (ITNs) is seen as a particularly effective low-cost strategy (Figure 10.35, page 478). These nets work by creating a protective barrier against deadly malaria-carrying mosquitoes that typically bite at night. The insecticide woven into each net kills the mosquitoes so that they cannot go on to bite others who may not be protected. Bed nets can reduce malaria transmissions by as much as 90 per cent in areas with high coverage rates.

Figure 10.35 Insecticide-treated bed nets are seen as a cost-effective strategy in the fight against malaria

Burning mosquito coils is favoured in some parts of the world, particularly in urban areas in Asia and South America where they are readily available and relatively affordable. The burning coils generate smoke that can control mosquitoes effectively (Figure 10.36). However, the practice has been discouraged by some health authorities as in some coils the smoke may contain pollutants which are of health concern.

Figure 10.36 Mosquito coils when burnt emit a repellent smoke

Despite being preventable and treatable, malaria continues to have a devastating impact on people's health and livelihoods around the world. In most countries where malaria is endemic, the disease disproportionately affects poor and disadvantaged people, who have limited access to health facilities and cannot afford the recommended treatment.

Management

Great progress has been made in tackling malaria in the last fifteen years. Adoption of the Millennium Development Goals in the Year 2000 and the Roll Back Malaria partnerships have helped to reduce cases of malaria significantly. Between 2000 and 2012, the substantial expansion of malaria interventions led to a 42 per cent decline in malaria mortality rates globally and in the decade from the year 2000, 3.3 million deaths from malaria were averted, and the lives of three million young children were saved. The WHO has developed a new global malaria strategy for the 2016–2030 period. It sets the target of reducing the disease burden by 40 per cent by 2020, and by at least 90 per cent by 2030. It also aims to eliminate the disease in at least 35 new countries by 2030. The strategy provides a comprehensive framework for countries to develop tailored programmes that will sustain and accelerate progress towards malaria elimination. Alongside Goal 3 of the post-2015 Sustainable Development Goals: 'to ensure healthy lives and promote well-being for all at all ages', there is certainly hope for the future.

The global prevalence and distribution of coronary heart disease (CHD)

Coronary heart disease (also known as ischaemic heart disease) is the leading cause of death worldwide, accounting for around 7.5 million deaths annually. CHD rates vary considerably among populations and the risk of heart attacks within populations from across the world can be ascribed to the varying levels of a number of risk factors among individuals. A heart attack occurs when the blood vessels supplying the heart muscle become blocked, starving it of oxygen and leading to heart muscle failure or death. A wide range of risk factors can be responsible for a heart attack, often acting in combination. Many of these factors are significant in all populations but their incidence varies, as does the occurrence of the disease globally.

Links to physical environment

Air quality: In a historical context, the changing physical environment has been a contributory factor to increases in heart disease. Industrialisation has brought a number of negative effects. Studies in the US have shown that increased exposure to airborne pollutants, especially particulate matter (PM) emanating from industry and transport systems, increases the risks of CHD. This suggests that rural environments, relatively clean of air pollution, would indicate a lower risk, though there is no clear evidence that this is the case in the developed world.

Climate: Studies in the UK have shown that air quality was not significantly associated with CHD mortality rates, but temperature and sunshine were both significantly associated. CHD mortality rates were higher in areas with lower average temperature and hours of sunshine (this was after adjustment for deprivation and the unhealthy lifestyle of populations). The results suggest that climate has a small but significant association with CHD mortality rates. Climate is important in that cold and damp winters in temperate climates have a negative impact on the cardiorespiratory system, which increases the risk of heart attack.

Relief and topography: There are no clear links between landscape and CHD risk other than the notion that more challenging relief requires more physical effort when walking, which can be both advantageous in increasing exercise and activity (which reduces risk) but can be a threat for individuals with other underlying risks. The attraction of landscape, whatever the topography, may encourage more physical activity but the same might be said of gym provision in built-up areas.

Links between CHD and the physical environment, with the possible exception of climate, are limited and minimal. Lifestyle choices within a variety of physical environments are more important risk factors. For example, it could be argued that the Japanese diet of fish (which reduces CHD risk) is influenced by Japan's physical environment as much as it is by culture. However, studies have shown that societal factors can be more influential. For example, Japanese migrants to the USA have, over generations, adopted American values, diets, lifestyle habits and ultimately CHD rates.

Links to socio-economic environment

In developed countries, socio-economic factors and lifestyle choices are much more important determinants of CHD prevalence. For example, poor diet and other negative lifestyle factors are estimated to account for one third of all deaths associated with CHD in England.

There is an increasing risk of CHD with age but gender is a less important determinant, despite it traditionally being thought of as a disease of men. CHD is multifactorial and it is the interaction between some of the genetic, lifestyle and other social factors which increases risk:

- **Social deprivation:** There is a positive correlation between deaths from circulatory diseases and deprivation (for example, within London, those living in Tower Hamlets have a three times increased risk of dying prematurely from CHD than those living in Kensington and Chelsea).
- **Tobacco use:** Mortality from CHD is 60 per cent higher in smokers.
- **Alcohol use:** World Health Report estimates that 2 per cent of CHD in men in developed countries is due to excessive alcohol consumption.
- **High blood pressure:** 22 per cent of heart attacks in Western Europe were due to hypertension, which doubles the risk of heart attack.
- **High cholesterol:** 45 per cent of heart attacks in Western Europe are due to abnormal blood lipids.
- **Poor nutrition:** A WHO report stated that a diet high in saturated fat, sodium and sugar and low in complex carbohydrates, fruit and vegetables increases CHD risk.
- **Overweight and obesity:** Obesity is an independent risk factor for CHD but is also associated with other risk factors such as high blood pressure, high cholesterol and diabetes.
- **Diabetes:** Men with Type 2 diabetes have two to four times greater risk of CHD; women have three to five times greater risk.
- **Infrequent exercise:** Physical activity reduces risk; a WHO report estimates that over 20 per cent of CHD in developed nations is caused by physical inactivity.
- **Ethnicity:** South Asian people moving to more developed regions have a higher premature death rate from CHD (though this may be associated with deprivation); the premature death rate from CHD in people from West Africa and from the Caribbean is much lower.
- **Family history:** First degree relatives of patients with premature heart attacks have double the risk themselves.

In developing countries with low mortality, such as China, the same risk factors apply, with the additional risks of under-nutrition and communicable diseases. In developing countries with high mortality, such as those of sub-Saharan Africa, low vegetable and fruit intake are also important factors. The relative impact of these risk factors is shown in Table 10.4. Some major risks are modifiable in that they can be prevented, treated and controlled. There are considerable health benefits at all ages, for both men and women, in stopping smoking, reducing cholesterol levels and blood pressure, eating a healthy diet and increasing physical activity.

Table 10.4 Relative importance of risk factors

Risk factor	Developed countries (%)	Low mortality developing countries (%)	High mortality developing countries (%)
Tobacco use	12.2	4.0	2.0
High blood pressure	10.9	5.0	2.5
Alcohol use	9.2	6.2	< 1.0
High cholesterol	7.6	2.1	1.9
Obesity	7.4	2.7	< 1.0
Under-nutrition	No importance	3.1	14.9
Unsafe sex	0.8	1.0	10.2

Source: WHO

Urbanisation and cardiovascular diseases

According to the World Heart Federation, one of the major factors increasing the risk of CHD in developing societies is the rapid rate of urbanisation taking place (Figure 10.37). They cite the following negative impacts of urbanisation, which increase risk, particularly for children growing up in fast-developing cities:

- areas with insubstantial housing conditions and poor access to healthcare services, healthy foods and safe, green places that are free of environmental toxins and pollutants for outdoor activity
- crowded living environments can spread diseases such as rheumatic fever, which if left untreated can cause rheumatic heart disease
- city dwellers are more likely to be exposed to marketing and advertising for unhealthy foods, tobacco and alcohol
- higher levels of particulate matter air pollution
- urban environments may discourage physical activity and encourage sedentary habits
- individuals turn to prepared and heavily processed convenience foods that are often high in sugar, salt and saturated fats
- city dwellers are more frequently tobacco users than rural dwellers
- children in cities may be particularly susceptible to second-hand smoke given the number of smokers in urban areas, along with crowded living conditions.

One World, One Home, One Heart.

Urbanization and cardiovascular disease
Raising heart-healthy children in today's cities

WORLD HEART FEDERATION®

Figure 10.37 World Heart Federation poster highlighting the link between urbanisation and CHD in cities in the developing world

Impacts of CHD

Impacts on health and well-being

The most common symptom of CHD is angina; a low level but fairly constant chest pain which can spread to other parts of the upper body. The pain can increase when the heart is put under more stress, for example from physical activity. These symptoms can be relieved with the use of nitrate tablets or sprays.

As well as angina, the other main symptoms of CHD are heart attacks and heart failure. Heart attacks can permanently damage the heart muscle and, if not treated straight away, can be fatal. Heart failure can also occur when the heart becomes too weak to pump blood around the body. This can cause fluid to build up in the lungs, making it increasingly difficult to breathe.

It is possible to recover from heart attacks and lead a normal life. Those who are at high risk or have suffered a heart attack can undergo heart bypass surgery to reduce the risk, but this in itself can be stressful and risky.

Rehabilitation programmes for those recovering from heart attacks or surgery focus on exercise; education about lifestyle choices; and relaxation and emotional support to build confidence. Those suffering from heart attacks can lose confidence in undertaking physical activity. CHD sufferers at all stages of the disease may be required to take continued medication for the rest of their lives. This may include warfarin (an anticoagulant which 'thins' the blood) which may have subsequent side effects. In some cases, electronic devices are fitted to patients to regulate heart beat and blood flow.

Impacts on economic well-being

The impacts of heart disease include the cost to the individual and to the family of healthcare and time off work, the cost to the government of healthcare and the cost to the country of lost productivity. All of these are difficult to quantify but the following are some estimates about costs.

- In 2009, CHD is estimated to have cost the UK healthcare system around £8.7 billion and the cost to the UK economy was estimated at £19 billion (NHS).
- 'Healthcare costs associated with smoking-related illnesses result in a global net loss of US$200 billion per year, with one-third of those losses occurring in developing countries' (WHO).
- 'The cumulative Medicare costs of treatment of heart diseases in people aged 65 and over in the USA amounted to US$76 million in 2000' (WHO).
- 'The number of people who die or are disabled by coronary heart disease could be halved with wider use of a combination of drugs that costs just US$14 a year per patient' (WHO).

Management and mitigation strategies

Prevention

Significant health gains in the treatment of heart disease can be made within a short period of time through public health and treatment interventions (Table 10.5). Governments are stewards of health resources and have a fundamental responsibility to protect the health of citizens. They can do this by educating the public, making treatments affordable and available and advising patients on healthy-living practices. Some examples of prevention strategies are:

- In the UK, dieticians promote the benefits for heart health of eating oily fish, more fruit and vegetables, and less saturated fat.
- In Finland, community-based interventions, including health education and nutrition labelling, have led to population-wide reductions in cholesterol levels closely followed by a sharp decline in heart disease.
- In Japan, government-led health education campaigns and increased treatment of high blood pressure have reduced blood-pressure levels in the population.
- In New Zealand, the introduction of recognisable logos for healthy foods has led many companies to reformulate their products. The benefits include greatly reduced salt content in processed foods.
- In Mauritius, a change from palm oil to soya oil for cooking has brought down cholesterol levels, but obesity has been unaffected.

Health education

The above strategies are not effective without public understanding, support and demand. Health education is essential to promote healthy choices. Schools are an ideal venue for health education as they can provide a healthy diet, prohibit smoking and allow opportunities for exercise. The WHO has initiated a number of activities to assist schools around the world, and since 2000 has co-ordinated World Heart Day events and activities on 29 September each year (Figure 10.38, page 482) including:

- medical activities such as blood-pressure testing
- activities to engage the public in physical activity

Table 10.5 Approaches to coronary heart disease prevention and treatment

Region	Availability of equipment for treatment of high blood pressure (%)	Availability of suitable drugs (%)	Number of medical professionals working in disease control per 100,000 people	Countries with tobacco legislation (%)
Africa	81	70	111	22
Americas	96	88	310	50
Eastern Mediterranean	93	92	259	75
Europe	97	100	772	80
Southeast Asia	88	71	151	70
Western Asia	96	96	516	69

Source: WHO

● scientific conferences
● activities to promote a heart-healthy diet.

The number of countries taking part in World Heart Day increased from 63 in 2000 to over 120 in 2015.

Figure 10.38 World Heart Day in Kolkata: students hold posters to raise awareness about healthy heart issues

Policies and legislation

Only governments can legislate for the prevention and/or control of disease. The most common legislation involves reducing tobacco smoking, which has clear links to reducing heart disease. Legislation can include advertising bans, smoke-free areas, health warnings on packets, taxation and outright bans in public places. A smoking ban was first introduced in Singapore in 1970 and 37 years later the idea was implemented in the UK.

Role of international agencies and NGOs in promoting health and combating disease
International agencies

Leading the effort within the UN system to promote and protect good health worldwide is the World Health Organisation (WHO), which came into force in 1948.

World Health Organisation

The WHO constitution states that, 'Health is a state of complete physical, mental and social well-being and not merely the absence of disease or infirmity.' Its primary role is to direct and co-ordinate international health within the United Nations' system. At the outset, the WHO's top priorities were to address malaria, women's and children's health, tuberculosis,

venereal disease, nutrition and environmental sanitation. Many of these remain on the WHO's agenda today, in addition to such relatively new diseases as HIV/AIDS and Ebola.

There are six regional WHO offices globally and staff work in 147 member countries. Key facts about the WHO and their main areas of work:

● responsibility for the International Classification of Diseases, which has become the worldwide standard for clinical and epidemiological purposes
● advising national ministries of health on technical issues and providing assistance on health systems and care services
● advising on the prevention and treatment of both communicable and non-communicable diseases
● working with other UN agencies, NGOs and other partners on international health issues and crises.

The WHO has been criticised for being over-bureaucratic and lacking the practical 'front-line' application to health issues. However the organisation's ability to focus, promote and co-ordinate efforts to tackle health problems on an international scale has achieved undeniable progress and success. Most notably:

● In the 1970s, WHO efforts led to the eradication of smallpox – the only major infectious disease that has been completely eradicated (probably WHO's greatest achievement).
● In 1988, the WHO launched its Global Polio Eradication Initiative and by 2006 the number of cases was reduced by more than 99 per cent.
● It has given special attention to adapting global HIV/AIDS policies to fit the specific needs of different regions. For example using a different approach in sub-Saharan Africa, where the epidemic is largely spread through heterosexual sex, compared with Eastern Europe, where injecting drug use is the primary mode of transmission.
● It has worked in partnership with other agencies to achieve the health-related Millennium Development Goals (MDGs).
● In 2010 the WHO launched a global effort to mobilise resources for a collective Strategy for Women's and Children's Health to save the lives of more than 16 million women and children.

WHO interventions cover all areas of the global healthcare spectrum, including crisis intervention; response to humanitarian emergencies; establishing international health regulations, which countries must follow to identify disease outbreaks and stop them from spreading; and the prevention of chronic diseases.

Other international organisations

It would be misleading to suggest that the entire work of supporting global health rests with the WHO. Other UN bodies are also engaged in this critical task, for example, the Joint United Nations Programme on HIV/AIDS (UNAIDS); the United Nations Population Fund (UNFPA) in support of reproductive, adolescent and maternal health; and the health-related activities of the United Nations Children's Fund (UNICEF).

- **UNAIDs**: an innovative partnership that leads and inspires the world in achieving universal access to HIV prevention, treatment, care and support.
- **UNICEF**: works with various partners towards achieving the MDGs and have mobilised both human and capital resources in their focus to reverse the spread of HIV/AIDS and to reverse the incidence of malaria and other diseases affecting children.
- **World Bank:** A vital source of financial and technical assistance to developing countries around the world to fight poverty, including making investments to improve health.

Health-related decades are declared by the UN General Assembly, including the decade to 'Roll Back Malaria' in developing countries, particularly in Africa (2001-2010) and the 'Water for Life' international decade for action (from 2005).

Non-governmental organisations (NGOs)

A non-governmental organisation (NGO) is any non-profit association that operates independently of both government and of profitable businesses. They are organised on a local, national or international level and are primarily funded by public donations. NGOs have a long history of involvement in the promotion of human well-being. They have increasingly been promoted as alternative healthcare providers to the state, especially in developing countries; governments cannot always cope with the enormity of some health issues such as HIV/AIDs, malaria or viral outbreaks without support. NGOs are increasingly instrumental in the implementation of international health programs, as they possess attributes enabling them to function as effective agents creating links between the community, private sector and government. The attributes of NGOs that increase their potential effectiveness include:

- their ability to reach areas of severe need
- their promotion of local involvement
- their relatively low cost of operations
- their adaptiveness and innovation, independence, and sustainability.

They have more flexibility and freedom to respond in creative ways to a range of situations and to promote the same health goals, but are less hampered by bureaucracy and resource constraints.

The role of NGOs in health promotion and provision can be divided into four main areas:

- **Service providers:** front line providers of clinical healthcare: treating illness in the absence of, or complementary tó, any existing government or private provision.
- **Social welfare activities:** providing or ensuring that the health infrastructure is in place, for example, food supply, clean water provision, public hygiene, sanitation and shelter for the unwell.
- **Support activities:** working with and training locals as health workers.
- **Research and advocacy:** involvement in research to improve efficacy of prevention campaigns for both communicable and non-communicable diseases.

NGOs, despite their good work, need to be careful that their efforts are not seen as undermining public sector provision and health systems, by not diverting health workers and managers into parallel operations. The NGO Code of Conduct for Health Systems offers guidance on how international NGOs can work in host countries in a way that respects, and supports, the primacy of the government's responsibility for organising health system delivery.

Médecins Sans Frontières (MSF)

Médecins Sans Frontières (MSF) (also known as Doctors Without Borders) is an NGO that exists to save lives by providing medical aid where it is needed most. Since 1971, MSF has cared for millions of people caught up in crises such as natural disasters, conflicts, famines, refugee exoduses or people excluded from healthcare. They have projects in over 60 countries (mainly in Africa, Latin America and South Asia) requiring both rapid response and longer-term specialist medical help to tackle health crises and diseases including cholera, Ebola, malaria, meningitis and HIV/AIDs. Eighty-nine per cent of MSF's funding comes from 6 million individual donors around the world, which helps to ensure operational independence and flexibility. The remaining funds come from governments and international organisations.

Figure 10.39 Distribution of aid by Médecins Sans Frontières in Port au Prince, Haiti following the earthquake of Jan 2010

MSF also treats malnutrition with ready-to-use therapeutic food (RUTF) which includes all the nutrients needed by children and can be stored for long periods and used in a variety of settings. MSF healthcare teams work with locally trained midwives, and traditional birth attendants, to set up programmes so that complicated births can be identified quickly to help prevent maternal deaths.

As well as providing vital medical assistance, MSF undertakes invaluable field research to generate evidence that improves the effectiveness and quality of the clinical care they provide. Research topics include treatment of multidrug-resistant TB, HIV/AIDS, neglected tropical diseases (such as Chagas and kala-azar) and mental health.

MSF acts as a watchdog to protect public health needs against corporate interests. The MSF 'Access Campaign' lobbies both governments and pharmaceutical companies, challenging the high cost of existing medicines and the absence of treatment for many of the diseases affecting patients in developing countries. The campaign pushes for price cuts to medicines, vaccines and diagnostic tests by encouraging the production of more affordable generic products. It also aims to steer the direction of research towards new drugs and vaccines.

10.4 Population change

There are **two** main components of population change:

- **Natural change:** The change in population brought about by the difference between **crude birth rate** and **crude death rate**. If the birth rate exceeds the death rate, then the population size will grow – this is called a **natural increase**. If the number of deaths exceeds the number of births there is a **natural decrease**.

- **Migration change:** Migration of people into or out of an area or country will also cause the population size to change. Those people moving into an area are called **immigrants** and those moving out are **emigrants**. The difference between the number of immigrants entering and emigrants leaving an area or country is known as the **net migration change.**

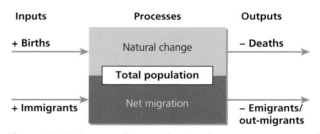

Figure 10.40 Systems diagram showing the two components of population change

Source: After Waugh D., Geography: An Integrated Approach

Figure 10.40 shows that these two components are different and distinctive but not necessarily mutually exclusive and will affect each other. For example, many international migrants are of **reproductive age**, consequently an increase of immigrants into a population will increase the child-bearing potential of that population, leading to an increase in birth rate. There will also now be a smaller proportion of elderly in the population so the movement of younger immigrants into a country will also reduce the death rate. If birth rates or death rates (or both) are changed by a migration event, then migration will have had an impact on natural change.

Natural population change – key vital rates

Natural population change will give either a natural increase or a natural decrease. Table 10.6 (page 486) shows birth rates and death rates for a number of countries. Two other **demographic** indicators (**total fertility rate** and **infant mortality rate**) are also included in the table.

Table 10.6 also shows there is a wide variation in birth rates and death rates in countries across the world and that both birth rates and death rates are able to rise or fall over time. The majority of countries in the world have a natural increase but some have a natural decrease (Figure 10.41, page 486).

Niger has the highest birth rate on the table, although it fell slightly to 49.9 per 1,000 in 2011. It also has a relatively low death rate for a low-income country. This might be explained by its youthful population structure, and the fact that healthcare has improved, largely because of the many **non-governmental organisations (NGOs)** that provide medical and other forms of aid. The difference between the birth rate and death rate figures in 2011 was 38.4 people per thousand, which is also the highest natural increase on the table. This means that for every 1,000 people in Niger there will be an extra 38.4 people per year. As Niger had a population of 17 million people in 2011, there will be an increase of about 650,000 (and climbing) each year. This makes it one of the highest natural population growth rates in the world with a population expected to double in the next 20 years.

Japan and **Russia** are the only nations in the table with a natural decrease. Japan did have a small natural increase of 1.6 per 1,000 in 2001, but as a result of falling birth rates and rising death rates (both associated with having an ageing population

structure) the situation has reversed. They now have a natural decrease of –1.6 people per 1,000. Russia had a larger natural decrease of –6.5 per 1,000 in

Key terms

Crude birth rate – The total number of live births per 1,000 of a population per year – also known as 'birth rate'.

Crude death rate – The average number of deaths per 1,000 of a population per year – also known as 'death rate'.

Demography (demographic) – Demography is the study of human population. A demographer is someone who studies the statistics and characteristics of populations.

Emigrant – A person leaving their native area or country in order to settle elsewhere.

Immigrant – A person moving into an area or country to which they are not native in order to settle there.

Infant mortality rate – The number of children who die before their first birthday per 1,000 live births per year.

Life expectancy (at birth) – The average number of years a person born in a particular year in a location is expected to live.

Natural change – The difference between birth rates and death rates. If birth rates are higher there is a natural increase in population; if death rates are higher, there is a natural decrease.

Net migration change – The difference between the total number or average rate of immigrants and emigrants in an area or country over a given period of time. More immigrants than emigrants will give a positive net migration and more emigrants than immigrants gives negative net migration.

(Net) replacement rate – The number of children each woman needs to have to maintain current population levels or give zero population growth by generation. It is a measured fertility rate. In richer developed countries the replacement rate is about 2.1 but ranges from 2.5 to 3.3 in less developed countries because of higher mortality. If fertility is above the replacement rate, population will grow and if it is below, population will decline.

Reproductive age – The age at which women can give birth. In official demographic data it is usually considered to be between 15 and 44.

Total fertility rate – The average number of children born per woman in an area or country if all women lived to the end of their childbearing years. It is considered to be a more direct and accurate measure of fertility than birth rate as it refers to the births per woman.

Table 10.6 Differences between vital rates of natural change in ten different countries

A	B		C		D	E
	2001		**2011**		**Total fertility rate 2011 (average births per woman)**	**Infant mortality rate 2011 per 1,000 < 1 year old**
Country	Birth rate – per 1,000 population	Death rate – per 1,000 population	Birth rate - per 1,000 population	Death rate – per 1,000 population		
Brazil	20.5	6.4	15.3	6.4	1.8	14
Canada	10.6	7.1	11	7	1.6	5
The Gambia	45.1	11.9	43.2	10	5.9	51
India	25.2	8.8	21	7.9	2.5	45
Japan	9.3	7.7	8.3	9.9	1.4	2
Mexico	23.6	4.6	19.2	4.5	2.2	14
Niger	52.7	16.2	49.9	11.5	7.6	64
Russia	9.1	15.6	12.6	13.5	1.6	10
Sri Lanka	18.5	6.7	18.3	7	2.3	9
United Kingdom	11.3	10.2	12.8	8.7	1.9	4

Source: World Bank – World Development Indicators

2001, but partly as a result of government **pro-natal policies** (incentives specifically designed to increase the birth rate), it now has a much smaller decrease of 0.9 per 1,000.

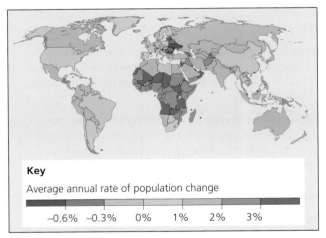

Key

Average annual rate of population change

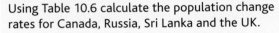

−0.6%   −0.3%   0%   1%   2%   3%

Figure 10.41 World map showing variations in natural population change in 2013

Key question

Is there a pattern of natural population change that occurs globally?

Growth rates and net replacement rates

From the natural change, demographers can calculate the growth rate of a population by multiplying by 100 to get a percentage. Niger has a natural increase of 38.5 per 1,000 of its population. This can be converted to a growth rate figure of 3.85 per cent.

Skills focus

Using Table 10.6 calculate the population change rates for Canada, Russia, Sri Lanka and the UK.

There is an apparent **correlation** between birth rates and fertility rates as they are both measures of the reproductive potential within a population. However they are different forms of measurement.

● **Birth rate** is a measure of the total number of births as a proportion of the total population (averaged down to the number per 1,000 of the population). Hence it is called 'crude' birth rate as it is an average measure based on the whole population.

● **Total fertility rate** is a more accurate measure of future population change as it is the average number of children that each woman of reproductive childbearing age will give birth to.

The fertility rate can be used as an indicator of future population growth. A fertility rate of 2 would mean that every woman gives birth to (on average) one girl. This means the current generation of women would replace itself with a younger generation of girls. This is known as the **net replacement rate**. In reality, the replacement rate is 2.1 because not all girls reach childbearing age (plus there are slightly more male births recorded globally). The replacement rate for populations with a high mortality is considerably higher than 2. Replacement rates for different countries range from below 2.1 in richer countries to almost 3.5 in Sierra Leone.

In Table 10.6, the sub-Saharan countries of Niger and The Gambia have the highest fertility rates. If they maintain these high fertility rates their populations will continue to grow rapidly as each new generation will be larger than the previous one. Many of the other countries in the table have lower than replacement fertility, which is an indication that their population will shrink in the long term. It is possible to have below net replacement rate but still have a natural increase because the populations consist of more than just the youngest two generations and they have more people living longer, for example, in the UK.

Infant mortality rates

The infant mortality rate is also shown in Table 10.6. The number of children dying before their first birthday forms part of the death rate statistics but is considered to be a particularly important measure of mortality because:

- it is **age specific** as it relates to one particular group. This group is vulnerable but arguably is the group that society should be taking most care of
- it gives an **indication of the level of healthcare** available in the population, particularly maternity and post-natal care, as well as the prevalence of, and ability to, combat diseases
- it gives **an indication of the wealth** of the country. Higher income countries will be able to afford the healthcare, medicines, healthy diet and clean water that will keep infants alive
- it will have an **impact on fertility rate**. Families in areas with high infant and child mortality will have little choice but to have more children to ensure that some survive into adulthood.

Key question

Suggest possible reasons for the change in the UK's natural population increase between 2001 and 2011.

Skills focus

Using Table 10.6, plot a scatter graph showing the relationship between infant mortality rates and birth rates for the 10 countries listed. Assume that infant mortality is the independent variable here. What conclusions can you draw from the graph? What other methods could you employ to see whether there is correlation between these two variables?

Cultural controls

Birth rates and fertility rates vary considerably on a global scale and are influenced by a number of cultural controls as well as other socio-economic and political factors, as shown in Figure 10.42.

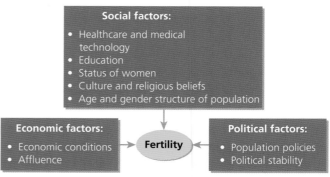

Figure 10.42 Factors affecting fertility

Cultural controls are particularly important in determining fertility rates and thus population growth.

Decisions about family size are influenced by socio-economic and sometimes by political factors, which in turn will have long-term effects on culture. For example, larger families were once the norm in the UK but as society has become more affluent parents tend to have fewer children, partly to give special attention to those children but also because having more children can be regarded as an expensive option when family planning methods and contraception are so readily available. Similarly, China's infamous one child policy has arguably dictated that Chinese culture accepts one child as being the social 'norm'. As China has relaxed this policy a counter-culture of wealthy Chinese choosing to have more than one child, because they can afford to, has developed. (However, after abandoning the one child policy in October 2015, a survey of Chinese suggested that around 40 per cent of parents would still only have one child, though this was mainly for economic rather than cultural reasons.) A number of cultural controls will set the tone of future population growth.

Religion

The doctrine of certain faiths such as Roman Catholicism and Islam proscribe the use of artificial contraception and abortion as methods of population control. With the exception of the Philippines, most countries where Catholicism is the main faith demonstrate low fertility and only very slow population growth (and in some cases decline). It may be that other 'Western' cultural values such as equal opportunities for women or materialism override religious doctrine. On the other hand, countries where the majority of the population are Muslim show some of the highest fertility rates and fastest growth, even though certain types of artificial contraception are permissible. Again, it may be other cultural or even environmental factors are a stronger determinant of fertility rates.

Gender

- **Gender preference:** For cultural and socio-economic reasons, some societies have a distinct preference for having male children. This is particularly true in rural areas of developing countries where the male will be expected to farm the family land and look after his parents in their old age. This is the case in parts of Pakistan, India and in sub-Saharan African countries such as Niger and Mali. An obvious consequence of this will be higher fertility rates and larger families as parents will continue to have children until they have one or two boys.

- **Status of women:** Women are still discriminated against in many cultures, often not receiving education and when married are expected to leave their own family and look after in-laws. The degree of equal opportunities afforded to females in a society in terms of education and careers will inevitably influence the number of children each mother will have in her lifetime.

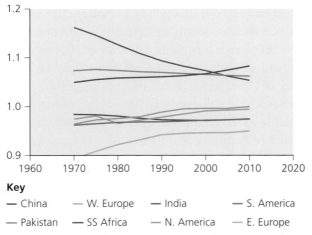

Key

— China   — W. Europe   — India   — S. America
— Pakistan   — SS Africa   — N. America   — E. Europe

Figure 10.43 Ratio of males to females resulting from preferences for male children

Marriage traditions

- **Child marriage:** According to UNICEF more than 60 per cent of women in five countries of sub-Saharan Africa and Bangladesh were married before they reached 18. Most African countries have legal minimum marriageable ages of between 16 and 18 but in certain countries (for example, Niger and Chad) the minimum age is 15. On a local scale, courts can give permission for girls to marry at 12. This will clearly have an impact on the number of children that a mother will have in these cultures.

- **Polygamy:** Polygamy (mainly polygyny, where a man takes more than one wife) is relatively common in some cultures. It is especially common across much of Africa and several Middle Eastern countries. Although often associated with Islamic faith it seems to have more correlation with tradition and culture than religion. Even in societies where polygamy is common, it occurs unevenly. In polygynous societies, having more than one wife is seen as a status symbol of power and wealth.

Population policies

Even in countries with less coercive population policies than China, government-led media campaigns and other strategies advocating population control will influence culture. For example. India's 'one family, two children' slogan has seen success in many Indian states where fertility rates have fallen significantly. Fertility rates in many Indian states are now even lower than in Kerala, which started the trend of smaller families with the introduction of compulsory female education in the 1980s.

Models of natural population change – the demographic transition model

A 'model' provides a structure which is representative of the characteristics, processes and relationships that occur in the real world.

The **demographic transition model** (DTM) was developed by American demographer Warren Thompson in 1929 and was based on recorded changes in birth rates and death rates in industrialised countries since the start of the industrial revolution. It traces a change from high

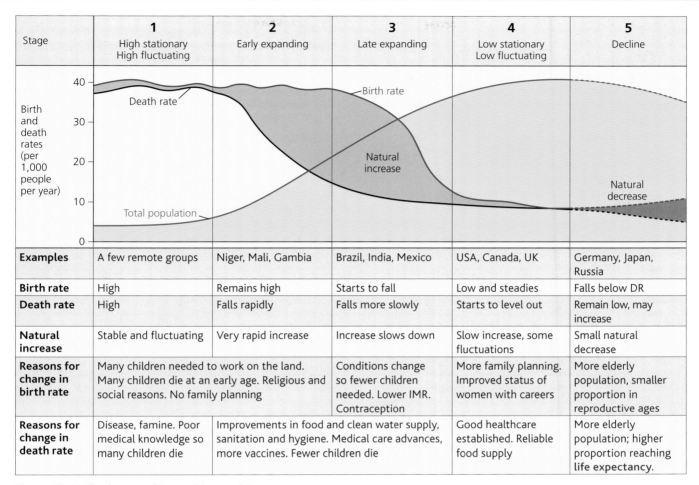

Stage	1 High stationary High fluctuating	2 Early expanding	3 Late expanding	4 Low stationary Low fluctuating	5 Decline
Examples	A few remote groups	Niger, Mali, Gambia	Brazil, India, Mexico	USA, Canada, UK	Germany, Japan, Russia
Birth rate	High	Remains high	Starts to fall	Low and steadies	Falls below DR
Death rate	High	Falls rapidly	Falls more slowly	Starts to level out	Remain low, may increase
Natural increase	Stable and fluctuating	Very rapid increase	Increase slows down	Slow increase, some fluctuations	Small natural decrease
Reasons for change in birth rate	Many children needed to work on the land. Many children die at an early age. Religious and social reasons. No family planning		Conditions change so fewer children needed. Lower IMR. Contraception	More family planning. Improved status of women with careers	More elderly population, smaller proportion in reproductive ages
Reasons for change in death rate	Disease, famine. Poor medical knowledge so many children die	Improvements in food and clean water supply, sanitation and hygiene. Medical care advances, more vaccines. Fewer children die		Good healthcare established. Reliable food supply	More elderly population; higher proportion reaching life expectancy.

Figure 10.44 The demographic transition model

birth rates and high death rates to low birth rates and low death rates as a country develops from a pre-industrial economy to an industrialised economic system. The rate of change will vary from country to country and countries will, at any one time, be located at different stages along the transition.

Figure 10.44 shows an updated version of Thompson's model. It has five stages, each identified by what is happening to birth and death rates or total population.

The DTM records the rate of total population growth, which is at its most rapid when natural increase is greatest during stage two and the early part of stage

three. It then starts to slow down during the latter part of stage three and stage four, when there is less difference between birth and death rates and so a smaller natural increase. In recent years a fifth stage has been added to the original model to account for those countries with declining population numbers.

Britain's demographic transition

The experience of demographic change in the UK, as one of the pioneering industrial nations, fits the model extremely well. Figure 10.45 (page 490) shows where there are distinct continuous falls in death rate followed, at a later stage, by a significant and continuous fall in birth rate.

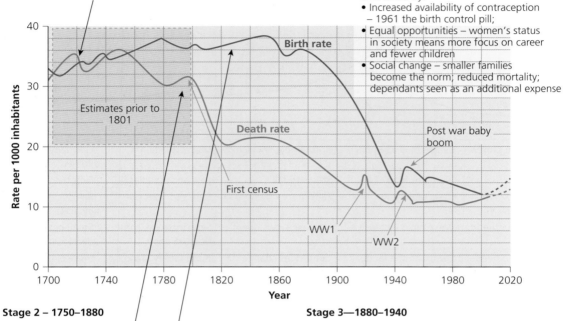

Stage 1 – Prior to the agricultural and industrial revolutions in Britain, birth rates were high and families large. These high birth rates were cancelled out by high fluctuating death rates resulting from periodic famines and disease epidemics. Infant mortality was particularly high, so population size remained stable

Stage 4 – 1940-present
Death rates start to level out at a low level thanks to:
• National Health Service (1946) – keeps death rates low, reduces infant and child mortality
Birth rates continue to fall to a similar low level aided by:
• Increased availability of contraception – 1961 the birth control pill;
• Equal opportunities – women's status in society means more focus on career and fewer children
• Social change – smaller families become the norm; reduced mortality; dependants seen as an additional expense

Stage 2 – 1750–1880

Fall in death rates caused by:
• Improved food supply – from new agricultural methods plus supplies from the New World
• Improved sanitation and hygiene – developing underground sewers and clean water supply
• Impact of disease reduced – partly because of improved sanitation and also medical advances

Birth rates remained high because:
• Lack of contraceptive methods
• Infant and child mortality persisted, especially in the mid-19th century due to poor housing
• Children still seen as an asset as workers and to care for parents in old age

Figure 10.45 Britain's demographic transition since 1700

Stage 3—1880–1940

Continued fall in death rates as a result of:
• Medical advances – 1907 training for midwives; 1929 – discovery of penicillin
• Rise in living standards – better nutrition, industrialisation and urbanisation (brought more shelter with warmth) and better wages

Fall in birth rates due to:
• Employment and education legislation – children no longer allowed to work; had to attend school
• Economic and social changes – more women in the workforce, higher incomes, rising middle class
• Growth of new ideas – challenges to accepted beliefs e.g. Darwinism

The model applied to countries with contrasting physical and human settings

Table 10.7 Countries at various levels of economic development and application to DTM

Country	Birth rate 2011	Death rate 2011	GNI per capita 2011 (US$)	Life expectancy
Niger	49.9	11.5	370	57.5
Sri Lanka	18.3	7	2,580	73.9
Botswana	24.1	17.3	6,930	46.7
Brazil	15.3	6.4	10,700	73.3
Japan	8.3	9.9	45,190	82.6
Canada	11	7	46,860	81.1

Source: World Bank – World Development Indicators

The countries in Table 10.7 have been ranked from lowest to highest according to their level of economic development, measured by **gross national income per capita**. The data gives an overview of whether there is a relationship between the level of economic development and progress in terms of demographic transition. Table 10.8 (page 491) outlines contrasts between Niger and Canada and considers whether physical conditions and other human factors play a part in the demographic transition experienced by these two countries.

Table 10.8 Contrasting physical and human settings and demographic transition in Niger and Canada

Niger	**Physical setting**	One of the Sahelian nations with a very arid, subtropical desert climate. Much of the north and east of the country is Sahara desert. The terrain is predominantly desert plains and sand dunes. The extreme south has a tropical climate near the edges of the Niger River basin. Non-desert areas in the south and west are threatened by periodic drought and desertification (Figure 10.46). Recurring drought events are a hazard and have occurred more frequently in recent years as a result of climate change
	Human setting	It is mostly sparsely populated, especially in the north. There is a higher concentration of population in a band of towns and cities in the south along the Niger basin. The economy is very reliant on its primary sector, especially agriculture and mining. Most people live in rural areas and are subsistence farmers or nomadic pastoral herders. Niger is rich in uranium deposits and also started producing oil in 2010. Over 90 per cent of the population are Muslim, coming from a number of different tribal groups. Droughts, failed crops, insect plagues and internal conflicts have led to food shortages, high food prices and hunger for many
	Application to Demographic Transition Model	High birth rates (largely as a result of religious and cultural beliefs) and relatively low and falling death rates put Niger in Stage 2 of the model but death rates are much lower than would be expected for this stage. This is partly due to their young population structure and also because the government has made great strides to reduce child mortality by reducing hunger and malnutrition and improving healthcare. They have been supported by NGOs such as Save the Children and the Eden Foundation
Canada	**Physical environment**	Comprises a wide range of climatic types including arctic, temperate continental and temperate maritime. The west of Canada is mountainous and the southern central area features rolling fertile plains (Figure 10.47). The north is a rugged and mountainous wilderness of taiga and tundra. Canada is extremely rich in mineral resources
	Human setting	It is a very large country with a relatively small population of around 35 million people, so on a national scale it is sparsely populated but highly urbanised with a concentration of population in the large cities in the southeast of the country, bordering the USA. Smaller concentrations appear in the southern parts of central provinces and on the west coast. The economy is based very much on its tertiary sector, particularly financial services and manufacturing, though it also has thriving mining and oil industries. It is a multicultural society, mostly welcoming to immigrants and tolerant of different cultures, languages and traditions
	Application to Demographic Transition Model	Canada has a low birth rate and death rate with a low natural increase. This places its demographics fairly clearly in Stage 4 of the model. Despite being the wealthiest of the countries considered, Canada has not progressed into Stage 5. Indeed, the natural increase is getting slightly larger. The main reason is that Canada is a nation which encourages some controlled immigration. Having plenty of space and being rich in resources, it is a country that can cope with increased population and wants to keep a balanced population structure that avoids becoming an ageing population, as in other equally rich countries

Figure 10.46 Environmental conditions make reliable food production in Niger difficult

Figure 10.47 A variety of landscapes in Canada provide various economic opportunities for its population

The model is not perfect but still provides a useful framework against which to compare real data. Its limitations can be used to explain some of the anomalies that occur between a nation's transition and what the model predicts. Countries do not all progress smoothly through the model; sometimes the transition falters or even reverses. The model might have more validity if social, cultural and political factors were given as much importance as economic wealth and development, however these are less predictable and may render the model unnecessarily complex.

Table 10.9 Summary evaluation of the Demographic Transition Model

Strengths	Limitations
The model can be: ● easily and universally applied ● used to compare stages of demographic development ● used by demographers to make predictions about future change	The model does not account for: ● migration and its impact on birth rates ● government policy which may influence birth rates ● wars and other conflicts which will impact on death rates ● the impact of major disease pandemics on death rates, for example, HIV/AIDS ● environmental limitations placed on future economic development, for example, meagre resources, harsh climatic conditions, natural disasters

Key questions

How do contrasting physical environments affect natural population change?

Why are lower than expected death rates seen at earlier stages in demographic transition than the model suggests?

Population structure – age and sex composition

Population structure is a term used to describe the age distribution and sex composition of a population. It is usually examined at a 'national' scale, by country, but it can be used at many different scales.

Population structure is usually depicted by the construction of a **population pyramid** as shown in an annotated version in Figure 10.48.

A population pyramid is a snapshot at any given time, but population structure is constantly changing as each age group moves up the pyramid over time. Pyramids provide insights into past trends in population, such as changes in fertility, mortality and international migration, as well as what is currently happening to population.

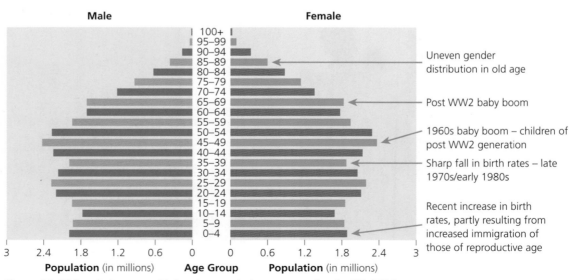

Figure 10.48 Population pyramid showing population structure for the UK, 2014

Interpreting population pyramids

Population pyramids can:

- show past changes in population
- be used to predict short-term and long-term future changes in population
- show the effects of wars/diseases and famine
- indicate life expectancy
- give an idea of the stage of demographic transition
- show effects of migration on a population structure (including the age and gender of migrants).

Dependency and its implications

For analytical purposes, population structures and their representative pyramids are divided broadly into three categories according to their economic productivity:

- **Ages 0–14** – known as the '**young dependent**' population.
- **Ages 15–64** – a group known as the '**economically active**' in a population.
- **Ages 65 plus** – usually known as the '**elderly dependent**' population.

The economically active tend to be the **working population** who earn an income and pay taxes. The two dependent groups, the young and elderly, are known as the non-economically active population. They are dependent on the economically active to make sure they have a means of living. Young dependants rely on their parents to support them while the elderly rely on pensions and/or family support depending on the development of social services where they live.

The **dependency ratio** is a measure of the level of dependency and is expressed as:

$$\frac{\text{Young dependants } (0-14) + \text{elderly dependants } (65 \text{ and over})}{\text{Economically active (those aged } 15-64)} \times 100$$

In 2014 the UK's population structure comprised of:
0–14 = 11.041 million (17.3 per cent of population)
15–65 = 37.904 million (65.2 per cent of population)
65+ = 11.172 million (17.5 per cent of population)

$$\frac{11.041 + 11.172}{37.904} \times 100 = 58.62$$

This means that for every 100 people of working age there are around 59 people dependent on their earnings. (The dependency ratio does not take into account those who are unemployed.) The dependency ratio in the UK is currently falling; in 2007 it was 60.68. However, the ratio of young to elderly has recently changed and there are now more elderly dependents than young dependents.

Impact of natural change on population structure

The shape of the population pyramid evolves based on changes in fertility and mortality. The demographic transition model gives a predicted sequence of change in birth rates and death rates over time. The

Table 10.10 Implications of different population structures for resources and development

Problems	Potential benefits
Youthful populations	
Increasing demands for improving maternal and child healthcare and education	Young people of today are the country's human resources of tomorrow
Providing food, water, energy and shelter for growing population	Can foster growth and development if there are favourable political and economic conditions
Governments need to keep up with demand for schooling	
Lack of attendance in schools (especially in rural areas) leads to low levels of literacy and poorly educated workforce	Can provide a cheap workforce and a growing market for foreign investors
Ageing populations	

Welfare and healthcare costs escalating as the elderly are more likely to need support	Some pensioners are healthy and affluent:
	  – growth in leisure/tourism industry
	  – growth in private health and residential care businesses
	  – companies able to target growing markets

Pensions will cost increasingly more in the future (the '**pensions time bomb**'):
  – costs have to be borne by a smaller economically active proportion
  – workers may have to pay higher taxes
	Fewer people of working age so there should be less unemployment in the population.

Smaller proportion of the population are 'economically active' – this may affect economic growth and the overall standard of living; may be skills shortages	Some elderly still 'work':
	  – looking after grandchildren, enabling parents to work
	  – volunteering in their local community

two concepts can be linked and a 'definitive' shape of population pyramid outlined for each stage of the model (Figure 10.49).

The model determines what stage of demographic development a country is at so we can suggest an appropriate population structure for countries at each stage of development.

The nature of these different population structures has different implications for the use of resources and services and for the economic potential of countries at different stages of development.

Implications of population structure for the balance between population and resources

The dependency ratio for most developed countries is between 50 and 70, whereas for low-income countries it is often more than 100. However, the composition of dependency will be quite different. In richer countries a greater proportion of the dependency comprises the elderly population whereas in less developed countries the higher overall dependency is largely of a youthful population. Both types of structure have problems resulting from dependency but they also have potential benefits.

Not only will population structure have an impact on resources, but the growth rate and actual size of population also has implications for the balance between population and resources. A larger population will usually make more demands on the resource base of an area and put at risk development or increases in standards of living for people. However, much will depend on other factors, such as the consumption patterns of that population or the levels of technology available. These arguments are extended in more detail in Section 10.5 Principles of population ecology applied to human populations (page 502).

The concept of the 'demographic dividend'

The **demographic dividend** refers to a period when the population structure of a country means there is low dependency.

Key term

Demographic dividend – The benefit a country gets when its working population outgrows its dependents, such as children and the elderly. A boost in economic productivity results from growing numbers in the workforce relative to the number of dependents.

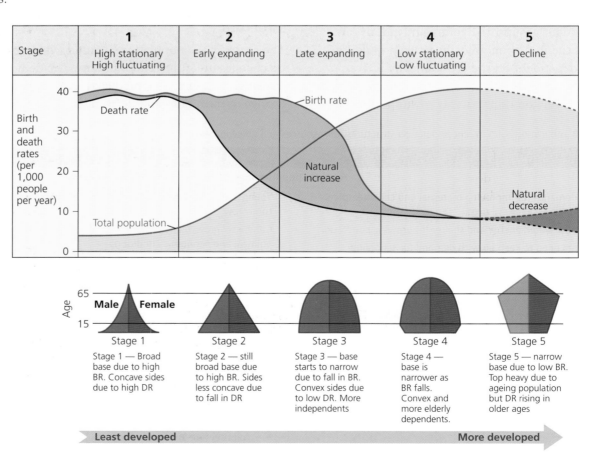

Figure 10.49 Population pyramid outlines for each stage of the demographic transition model

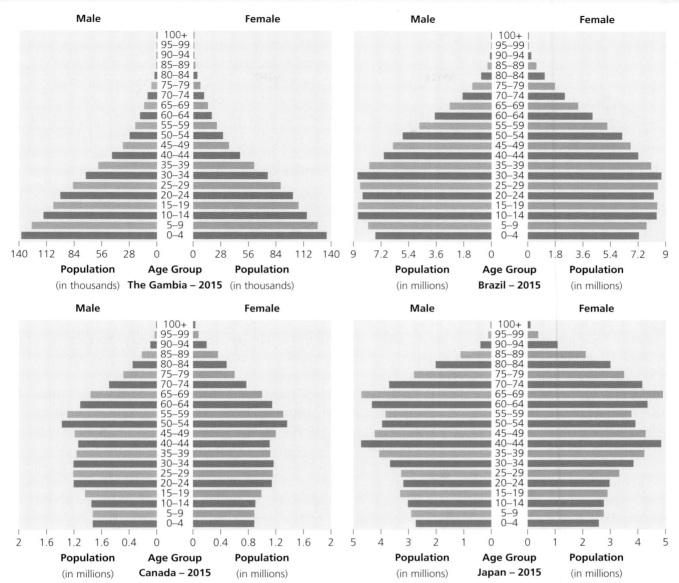

Figure 10.50 Population pyramids for countries at different stages of development

Source: US Census Bureau, International database

The lag between falling death rate and subsequent reduced birth rate may last for one or two generations before parents adjust to falling infant mortality by having fewer children. This lag period is one of rapid population growth and produces a 'generational bulge' that rises up the population pyramid, eventually into the economically active sector (as seen in the population pyramid for Brazil in Figure 10.50). With few elderly dependents because of previously low life expectancy, and fewer young dependents because of declining fertility, the dependency ratio will fall. For countries progressing through demographic transition relatively quickly, just as China, Brazil and India have done in the past 50 years, this 'bulge' will appear in Stage 3, just as fertility rates are starting to fall.

This large group of young and aspirational people will give rise to a period of higher productivity and a boom in economic growth and social development because:

- a large, young, educated **workforce attracts investment** from footloose global companies (TNCs)
- workers with fewer children **invest more of their income**, leading to financial stability and growth
- fewer children also means **more women join the workforce**, promoting more gender equality
- salaried workers provide a **growing market** for consumption of goods and services.

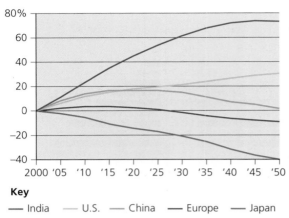

Key

— India — U.S. — China — Europe — Japan

Figure 10.51 Changes in working-age populations (actual and projected)

Source: UN Population Division

Countries can only take advantage of their demographic dividend providing there is investment in education and employment and that, politically, there is the scope for change. The Asian Tiger economies are good examples of where this has occurred. Countries may further maximise the dividend if there is a rich natural resource base to support growing demand, such as in Brazil and China. However, the dividend may be a wasted opportunity if there is poor educational attainment or a lack of transparency and democracy in government. Political instability will deter investment and lack of employment opportunities leads to the emigration of young people, thus jeopardising the dividend itself. Myanmar, for example, has not benefited fully from its dividend for these very reasons.

The boost in productivity will create extra demand in the economy and this may cause environmental degradation and put additional pressure on the use of natural resources, unless they are managed sustainably. The demographic dividend itself can only be sustained by managing fertility rates at replacement level. Even then, improved life expectancy will mean, in the longer term, the bulge will move up the pyramid to form an ageing population structure. This in turn will put pressure on specific services such as healthcare. Thus it could be argued that taking advantage of the demographic dividend is a 'one-time opportunity' for most countries.

Key question

Which type of population has the more challenging dependency issues, a youthful population or an ageing population?

International migration

The other major component of population change is migration. The geographical definition of migration is 'the movement of people across a specified boundary in order to establish a new permanent or semi-permanent residence.' Migration can take place at various scales and is the result of a number of factors. Whatever the scale and causes, it is likely to have an impact on population change at both the **origin** and the **destination** of the migrants.

International migration is an important force in development and a high-priority issue for both developing and developed countries.

Types of migrant – refugees, asylum seekers and economic migrants

In 2015, 244 million people, or 3.3 per cent of the world's population, lived outside their country of origin. The majority of migrants cross borders in search of better economic and social opportunities. Others are forced to flee crises such as wars, conflict, persecution or natural disasters. This suggests that the majority of migration is one directional from less developed to high-income countries. Migration to high-income countries has been rising for the past 50 years and in the first decade of the twenty-first century the number of immigrants living in OECD countries rose from around 75 million to more than 100 million.

However, aside from this general trend there is considerable diversity, both in the direction of flows to individual countries and between different immigration categories, such as **labour migration**, **family migration** and **humanitarian migration**. About one third of all international migration is from one developing country to another and a slightly lower proportion is between richer developed nations. Around 60 per cent of international migrants reside in countries designated as 'high income', but these include many developing countries such as Bahrain, Brunei, Qatar, Singapore, South Korea and the United Arab Emirates. Figure 7.7 (page 285) in Chapter 7 of this book shows the main flows of international migration.

During large-scale international migration events, such as the European migrant crisis of 2015, there can be confusion in the media and in public debate about the differences between **asylum seekers**, **refugees** and **economic migrants** and the complex legal distinctions of each status can be lost. The 1951 UN Convention on the Status of Refugees (and subsequent 1967 Protocol) is the key legal document in defining who is a refugee, their rights and the legal obligations of states. There are 147 signatories and the Convention is overseen by the UNHCR (the UN's Refugee Agency).

Key terms

Asylum seeker – A person who has fled their country of origin and applies for asylum under the 1951 Convention on the grounds that they cannot return to their country of origin because of a well-founded fear of death or persecution. While they wait for a decision on their application to be concluded, they are known as an asylum-seeker.

Economic migrant – A person who has voluntarily left their country of origin to seek, by lawful or unlawful means, employment in another country.

Refugee – In its broader context it means a person fleeing, for example, civil war or natural disaster but not necessarily fearing persecution as defined by the 1951 Refugee Convention. Legally however, a refugee is an asylum seeker whose application claim for asylum has been successful.

There is some overlap between the different categories. For example 'asylum seekers' and 'refugees' are frequently conflated, which gives rise to confusion and suggests that asylum seekers are automatically granted refugee status. The position of asylum seekers is similar to those who enter on short-term visas. They are granted temporary admission while their applications are pending, but if their claim is unsuccessful they have no lawful right to remain in their chosen country of destination.

Some migrants claiming to be asylum seekers are in fact economic migrants who hope to secure entry into a country by claiming asylum. This is relatively easy for migrants coming from areas of ongoing conflict or civil disorder.

Causes of migration – push and pull factors

Global migration is running at record levels and is predicted to increase further, though as a proportion of total population, only 0.60 per cent of world population switched country between 2005 and 2010, a smaller proportion than some of the great migrations of the late nineteenth and early twentieth centuries.

People migrate for a number of reasons. These can be categorised as **push factors** or **pull factors.**

- 'Push' factors are reasons which are based at the origin of the migration and which initiate the migrant's desire to move.
- 'Pull' factors are based at the intended destination of the migrant and attract people.

Figure 10.52 lists examples of push and pull factors. Observers often distinguish between '**forced**' migration, when the individual or family have little choice but to move, and '**voluntary**' migration. Forced migration is always caused by extreme push factors at the origin, such as natural disasters or wars. A voluntary decision to migrate may be based on factors present at both the origin and the destination as indicated by Lee's Push-Pull Model of migration shown in Figure 10.53 (page 498).

ORIGIN	DESTINATION
'Forcing' factors	**Associated with voluntary migration**
• war, conflict, political instability,	• better quality of life, standard of living
• ethnic and religious persecution	• varied employment opportunities, higher wages
• natural and man-made disasters such as earthquakes, tsunamis, drought, famines	• better healthcare and access to education services
	• political stability, more freedom
	• better life prospects
Socio-economic conditions	**For retirees:**
• unemployment, low wages or poor working conditions	• specific type of environment with a range of services to cater for their needs
• shortage of food	

Figure 10.52 Push and pull factors

A combination of war, religious and ethnic persecution and poverty is causing the movement of refugees and economic migrants on a global scale. UN figures claim that 1 in every 122 people globally is either a refugee, a displaced person or is seeking asylum. In 2014, UN figures suggested that wars and conflicts across the globe had forced 19.5 million migrants to flee their homes and that number was expected to increase in 2015.

Processes and impacts of migration

Lee's push-pull model of migration suggests that the process of migration from one area or country to another is not usually a straightforward one. The decision to migrate is made by individuals or families based on existing conditions at the origin. For example, during Europe's 'migrant crisis' (see text box) in 2015, the majority of the migrants were fleeing civil war in Syria. Innocent civilians were caught up in fighting between government forces and various rebel groups including the militia of the so-called Islamic State (of Iraq and Syria) or ISIS. Migrants genuinely feared for their lives and so the negatives at the origin far outweighed any positives for them to stay in Syria. Equally the prospect of higher living standards and the promise of better opportunities in the EU (Germany offered to take 800,000 refugees) made the positives at the destination outweigh any negatives. For many the decision to migrate was an obvious choice. However, both refugees and economic migrants still have to overcome the intervening obstacles, which for many involved negotiations with people smugglers in either Libya or Turkey and a hazardous sea voyage, often on unstable craft. As the year progressed an increasing number of refugees also faced long relentless walks through Balkan countries, where they were met with physical barriers at borders. People had to find new routes to reach their intended destination in Austria or Germany.

The intervening obstacles make the migration a 'step by step process'. Initial moves by migrants may be to seek refuge in 'safer' towns, countries or refugee camps before deciding on a longer and more permanent move.

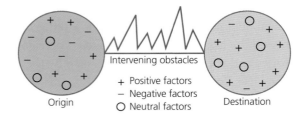

- For people to move, they need to be 'pushed' from their country of origin and 'pulled' to another country.
- The majority of migrants are voluntary movers, doing so for largely economic reasons.
- The negative factors at the origin are the 'push' factors; the positive at the destination are 'pull' factors.
- Migrant has to evaluate these factors and obstacles before they move.
- Intervening obstacles might include: travel costs, family pressures, language barriers, misinformation, immigration controls, bureaucracy, border controls.

Figure 10.53 Lee's push-pull model of migration

Migration as another feature of globalisation

Table 10.11 Benefits and costs of international migration

Benefits	Costs
+ Migrants can access new opportunities + Destination receives additional skills and labour + Origin country receives remittances, which many relatives increasingly rely on	− Richer countries gain at expense of poorer countries who lose their most talented people − Migrants to richer countries increase their consumption levels to match the unsustainable levels of their adopted country, which puts more pressure on global resources

International migration has become another feature of globalisation (see Chapter 7). Ever larger numbers of unemployed in poor countries are seeking better opportunities abroad. In the past, the movement of

European migrant crisis 2015

Syrian War – Over three million refugees have fled from civil war in Syria since 2011. More than one million went to neighbouring Lebanon, a small and relatively poor country; one in four of its current population are now refugees, many located in camps established by the UN. Local services and infrastructure are at breaking point. Over one million Syrian refugees have also crossed the border into neighbouring Turkey. In both cases these destinations may be temporary transit points as migrants seek a further move into European countries.

Europe – During the spring and summer of 2015, the migrant crisis and its resultant effects appeared across

TV news screens nearly every day. By the end of the summer it was estimated that a total of over 600,000 migrants had entered Europe from North Africa and the Middle East, their main access points being Italy (via Libya) and, as numbers from Syria and Iraq increased later in the summer, this switched to Lesbos, a Greek island off the coast of Turkey. Holding centres for migrants in Sicily (Italy) and in Lesbos were already full and the continued flow of migrants meant that 'squatter settlements' were springing up around both entry points and other border locations across Central and Western Europe.

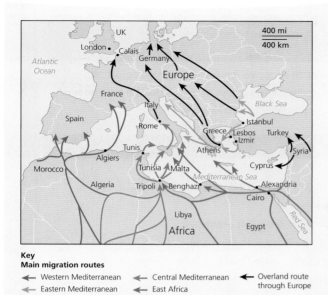

Key
Main migration routes

← Western Mediterranean   ← Central Mediterranean   ← Overland route through Europe
← Eastern Mediterranean   ← East Africa

Figure 10.54 Map showing main migration routes through North Africa and Middle East into Europe

Apart from Syrian refugees, migrants were travelling from Iraq, Afghanistan and from northern and sub-Saharan African countries to escape conflict, persecution or poverty or simply to seek a better life in a richer country. European governments blamed a rise in human smuggling and trafficking of migrants and sought to address this.

The situation in the Mediterranean was particularly acute; news reports showed migrants crossing the sea on overcrowded, makeshift boats supplied by smugglers. Many refugees were known to have drowned or died in the cramped conditions on board. EU Governments were torn between their humanitarian impulse to save lives and offer sanctuary and the struggle to provide aid with dwindling resources and pressure from electorates wary of admitting more migrants at a time of economic austerity.

UK – The UK attracts many migrants because their chances of gaining employment in the UK may be improved as many have learnt English in their home country. This has caused turmoil around the port of Calais, where at least 3,000 migrants are living in temporary camps. On a daily basis, hundreds try to illegally enter trucks and containers crossing to the UK. This has caused disruption to ferry crossings and Eurotunnel services, and increased political tensions between the UK and France.

The EU Commission has suggested a quota system so that each member state takes a certain proportion of the migrants.

Figure 10.55 Overcrowded migrant boat in the Mediterranean

Figure 10.56 Migrants at a refugee squatter settlement in Calais

poor people seeking better lives usually took the form of rural to urban migration in developing countries, with squatter settlements appearing in and around major cities. Alternatively refugees avoiding conflict would cross borders and settle in established refugee camps. While these migrations still persist, more recently there has been a 'scaling up' of migration taking place over longer distances. This is largely achieved by organised traffickers but also migrants are better informed and seem willing to take greater risks.

The implications of migration

Any notable migration event will have implications for the origin and for the destination area or country. In the first instance the population **size** will change as people move from the origin to the destination. The age and gender characteristics of the migrants will also have an impact on the **structure** of the population.

Table 10.12 Implications of migration for the origin and for the destination

Implications at origin (Home country)	Implications at destination (Host country)
Demographic implications	
Lower birth rates, people of childbearing age leave Population structure – ageing population remain; population unbalanced Loss of male population of working age	Balances population structure, if previously ageing population Migrants in reproductive age groups means increase in birth rates Increase in male population of working age
Social implications	
Advantages Reduced pressure on healthcare (see health) Reduced pressure on education **Disadvantages** Loss of traditional culture Break-up of family units Break-up of communities May lose qualified workers such as doctors, nurses and teachers	**Advantages** Cultural advantages of new foods, music, fashion, etc. **Disadvantages** Pressure on maternal and infant healthcare Pressure on schools (particularly primary) Young male migrants create social problems Can give rise to ethnic and racial tensions Increase in crime Segregation of migrants into certain areas
Economic implications	
Advantages Reduced pressure on food, energy, water, etc. Less unemployment Remittances sent back home by migrants Migrants develop new skills which they can bring back home **Disadvantages** Lose better educated/most skilled from workforce Creates dependency on remittances Less agricultural and industrial production Decline in services – not enough people to support them	**Advantages** Overcomes any labour/specific skill shortages May provide cheap labour who work longer Working migrants spend money/pay taxes Increases size of workforce – can provide economic boom and multiplier effect Reduced dependency – 'demographic dividend' **Disadvantages** Pressure on jobs/unemployment Resentment towards migrants in time of recession
Political implications	
Pressure to re-develop areas in decline May introduce pro-natal policies	Pressures to control immigration Rise of anti-immigration political parties Growth of right-wing racist organisations
Environmental implications	
Farmland, buildings and sometimes whole villages may be abandoned Less environmental management	Pressure on land for development – roads, housing, other infrastructure Increased demand for energy, water and food puts pressure on natural resources
Health implications	
Migrants leave areas where infectious diseases are endemic or sometimes epidemic Less pressure on limited health services but... ...demographics of migration mean that the most vulnerable (children, elderly and poor) remain at risk	Increase in infectious diseases transmitted by/to migrants from areas of different disease prevalence Increased pressure on health services because of rise in infectious diseases Increased pressure on health services to treat non-communicable/chronic diseases – the notion of 'health tourism'

Impact of migration change on population structure

Voluntary migration is age and gender specific, so inevitably the movement of a large number of people from one population to another will have an impact on the population structure of both the origin and the destination. This can happen at all scales from rural to urban migration occurring in developing countries up to an international scale. Migration causes a fairly **immediate impact** on the size and structure of the population, unlike natural change which tends to have a more gradual, long-term effect.

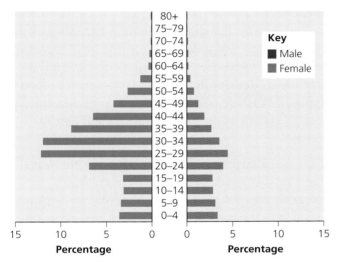

Figure 10.57 Population pyramid for United Arab Emirates. Some populations have experienced levels of immigration which have severely distorted their population pyramids

Source: UN Population Division

Specific impacts on structure:

- To bolster the economically active proportion of the population and therefore reduce dependency. Spending on accommodation, goods and services together with paying taxes will help support the economy. It will have the reverse effect at the origin where dependency increases and services close due to lack of support. Remittances sent back to relatives may offset these negative effects.
- Those of economically active age are also of reproductive age and there is a consequential effect where the birth rate will increase in the destination area or country. This has happened in the UK and has increased the numbers in the younger age groups.
- Migration events can be gender imbalanced. For example, the population pyramid for the United Arab Emirates (Figure 10.57) shows a large

protrusion of males of working age groups from 20 to 49. This will be shown as indentations in the pyramids of countries of origin for workers from, for example, Pakistan, India and the Philippines.

Skills focus

Using the population pyramid for the United Arab Emirates in Figure 10.57, estimate the percentage of young dependents (0–14) and elderly dependents (65+). From these figures, calculate the dependency ratio for the UAE.

Health implications of migration

Disease transmission

In much of the developed world, immunisation, antibiotics, improved healthcare and public hygiene have decreased the incidence of infections that were, historically, significant causes of morbidity and mortality. By the end of the twentieth century, domestic transmission of serious infectious diseases such as measles and polio had been eliminated in some developed countries. This is not the case in the developing world where many diseases are associated with poverty and inequality.

Such disparities in disease prevalence and healthcare often exist between a migrant population's place of origin and its destination. These can be among the driving push and/or pull factors initially triggering migration. There are a number of implications for increased migration across disease disparate regions, including the transmission of certain diseases globally and particularly in countries receiving migrants. For example, by the 1980s, tuberculosis (TB) was thought to have been virtually eradicated in the UK. However, the incidence of TB in the UK has gradually increased over the two decades since the mid-1980s, especially in London where the number of cases rose by 50 per cent between 1999 and 2009. This rise has largely been attributed to the increase in travel and migration. Three-quarters of the increase has been in non-UK born groups including black African (28 per cent) and Indian (27 per cent) people. However, 85 per cent were not recent immigrants and had lived in the UK for at least two years, which suggests that transmission may have occurred after they had arrived in the UK. Reports suggest that the poorer socio-economic status and living conditions experienced by certain population groups have led to a gradual re-emergence of TB as a public health problem.

Impact on health services

In 2011, the National Institute of Economic and Social Research (NIESR) reached the following conclusions regarding the impact of migration on UK health services:

- Economic migrants pose a disproportionately small burden on health services: They are likely to be relatively light to moderate users of health services as they are relatively young, healthy, employed and, disproportionately, in professional roles (many in healthcare).
- Some migrants do place greater demands on parts of the health service: this is associated with social deprivation (for example, a higher incidence of TB); poor English language skills and lack of knowledge of the health system.
- There is mixed evidence of health behaviours and lifestyles of migrants (for example, drinking, smoking and diet) and the impact of these on demand for services. The health of migrants deteriorates with age and length of stay but this may not affect healthcare use.

Health impacts during migration

Another feature worth consideration is the implications for health of the migration journey itself. This is particularly relevant for refugees and asylum seekers, who are likely to display a greater prevalence of illness resulting from trauma, abuse, injury, deprivation and exposure, especially if they have been victims of trafficking or smuggling.

Key questions

How and why are increasing numbers of international migrants overcoming barriers to migration?

What are the advantages and disadvantages for a developed country, such as the UK, in accepting large numbers of refugees from less developed countries?

10.5 Principles of population ecology applied to human populations

Population ecology and growth dynamics

In ecology, population growth is determined by birth rates and death rates. The number of births is controlled by the natural reproductive potential of the species. This is also known as the **biotic potential**, which has evolved over time for individual species and is related to the survival rates of the young of that species. Thus fish species such as herring produce many young because their chances of survival are very low. In contrast, most mammals (including humans) have a small number of offspring as their rate of 'survivorship' is high, though they can be vulnerable to increases in death rate. The number of deaths is controlled by environmental limiting factors that prevent survival. This group of factors is collectively known as **environmental resistance.** Some environmental limiting factors are '**density independent**' and are not influenced by the population size and density. For example, natural hazards such as drought, floods, volcanic eruptions, etc., will increase the death rate whatever the population size. On the other hand, '**density dependent**' factors such as food supply and disease will become more prevalent in limiting growth as the population size and density increases.

When the biotic potential is greater than environmental resistance, the population will grow. From these ecological concepts, model 'growth curves' can be plotted to show how the population of a species grows over time (Figure 10.58).

Key terms

Biotic potential – In population ecology, the natural reproductive potential of the species.

Environmental resistance – A term used in population ecology to explain mortality rates controlled by environmental factors that prevent survival, for example, disease or shortage of food. Sometimes known as 'limiting factors'.

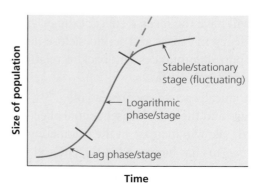

When **biotic potential > environmental resistance** = Rapid population growth during **log phase**

Figure 10.58 Model growth curve showing how populations grow over time

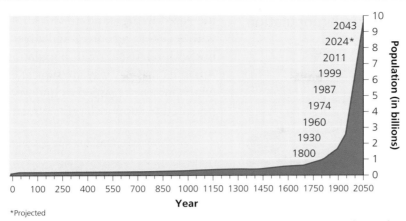

*Projected

Figure 10.59 J curve of human population growth showing exponential growth since 1800

Population ecology applied to the human population

For the greater part of human history, human population size remained relatively stable. Although birth rates were high and family sizes were larger in the past, most parents were only survived into adulthood by two of their children, who effectively replaced them in the population. The reason for the stabilisation of population was therefore the high death rates and low life expectancy brought about by the many diseases and lack of food security that threatened human lives.

Developments in technology have improved food supplies in many parts of the world. This, together with improvements in medical science, healthcare and sanitation, have been the main reasons behind the exponential growth in population over the past two hundred years. Medicines and vaccines have been developed that combat many of the infectious diseases, such as smallpox, measles and malaria, which were killing people well into the twentieth century. Recognising that improved sanitation would also reduce the spread of water-borne diseases such as typhoid and cholera, also helped to control the death rate and increase life expectancy. In 1800, the Earth's population was estimated to be around one billion, whereas in 2015, it is estimated to be 7.2 billion, largely due to the control of death rates. In effect, human population has overcome much of the 'environmental resistance' that was limiting growth (Figure 10.59). For the last 200 years, human population has been through a 'log phase' of exponential growth. World population is still growing rapidly though the rate of growth is starting to slow down.

Key question

According to demographer Nicholas Eberstadt, 'we didn't start breeding like rabbits, we stopped dying like flies.' Why does this statement summarise the population growth of the past 200 years?

Implications of population size for the balance between population and resources

Continued global population growth will put pressure on the Earth's resources and its environment. This situation will affect us all in some way. It raises the questions, 'at what point will breaking-point be reached?' and 'how many people can live on the planet sustainably?' (Figure 10.60)

Figure 10.60 A crowded Oshodi market in Lagos, Nigeria – the country's population is expected to soar from 200 million today to 900 million by 2100. How will this impact upon food resources, health and well-being?

Concepts of overpopulation, underpopulation and optimum population

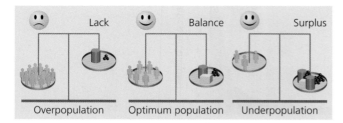

Figure 10.61 The balance between population size and available resources

The relationship between human population and the resources available to support that population is a complex one, which can be addressed at all scales. An ideal balance between population and resources is known as **optimum population** (Figure 10.61 and 10.62). This refers to a population size in an area which, working with all resources, will give the highest standard of living for the people of that area. It maximises the income per capita and is a balance which most nations will try to achieve through direct or indirect population or resource management. It is constantly changing, for example, the development of new technology may increase the resources available.

Overpopulation exists when there are too many people in the area relative to the available resources, putting pressure on those resources. It means that a continued increase in population will reduce the average standard of living for all. Conversely it implies that with no advances in technology and no increase in resources, only a fall in population would increase living standards. Some commentators mistakenly relate the idea to population density, stating that areas or countries which are densely populated are overpopulated. This is often incorrect because the measure applies only to the relationship between population size and available resources. Thus a densely populated country such as the Netherlands (448 people per km^2) can enjoy a relatively high average standard of living, while less densely populated Mali (12 people per km^2) struggles to provide the basic needs for its population because of a poor resource base.

Underpopulation occurs when there are too few people to use the resources efficiently for a given level of technology. It suggests that an increase in population would mean a more effective use of resources and increased living standards for all. Developed countries with access to resources and advanced technology but with relatively small populations such as Canada, Australia and Norway are typical examples of countries which may be regarded as underpopulated and with potential to support larger populations.

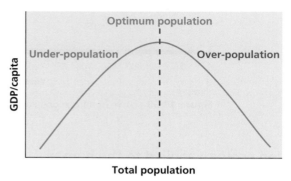

Figure 10.62 Relationship between population size and standard of living

Table 10.13 Characteristics of overpopulation and underpopulation compared

Overpopulation	Underpopulation
Low per capita income/poverty	High per capita incomes (not maximised)
Unemployment/underemployment	Low unemployment
Outward migration	Inward migration
Insufficient food, minerals and energy resources	Good living conditions
Prone to natural disasters	High levels of technology
War, conflicts, tribalism	

Carrying capacity and ecological footprint

Key terms

Biosphere – The biological component of Earth systems (the others being atmosphere, hydrosphere and lithosphere).

Carrying capacity – The maximum population size that an area or environment can sustain indefinitely.

Ecological footprint – A measure of the demand placed by humans on Earth's natural ecosystems.

Overshoot – An ecological term referring to a point when the population and its associated consumption of resources exceed the long-term carrying capacity of its environment.

Total productive bio-capacity – All the food, water and energy resources produced by the Earth's natural systems annually to sustain us.

Carrying capacity is an ecological concept usually applied to animal and plant populations. It is a calculation of how large a population any given environment can support. Human ecologist Professor William Rees has applied the concept to humans and suggests that the measure of carrying capacity can be flexible depending on the average lifestyle of the population concerned. His arguments about carrying capacity hinge on the varying levels of consumption of resources in different parts of the world. For example, the Earth may support a larger population of 10 billion at a modest level of consumption but it will have a much lower carrying capacity if we use and waste more resources in order to live the more comfortable lifestyles enjoyed in rich industrialised countries.

These ideas suggest that Earth has bio-productive limits; it can only accommodate so much consumption of its resources. The basis for Rees's calculation of carrying capacity is measuring the '**total productive bio-capacity** of the Earth' and dividing this by the total population. This would give us a measure of carrying capacity or how many people the Earth can support.

Calculating carrying capacity and ecological footprint

$$\frac{\text{Total productive biocapacity}}{\text{Total population}} = \text{Global hectares (gha) available per person}$$

Total productive bio-capacity = all the food, water and energy resources produced annually to sustain us.

One global hectare (gha) = a unit of measurement which represents the average productivity of all biologically productive areas (cropland, forests, fishing grounds etc.) on earth in a given year.

Global hectares per person = the amount of global hectares needed by each person to provide for their consumption of resources.

The calculation gives us the global hectares available for each person on the planet. This is the **ecological footprint** that we are each entitled to, with a population of 7 billion. The ecological footprint is a measure of the demand each of us places on the Earth's **biosphere**. One of the largest components is each individual's carbon footprint.

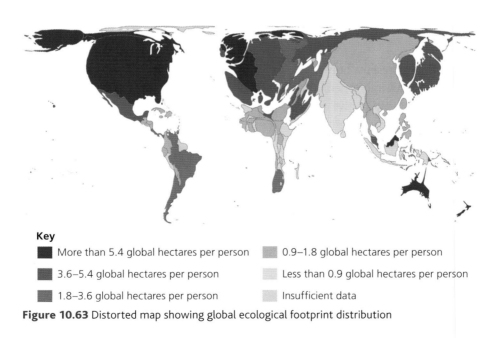

Key

- ■ More than 5.4 global hectares per person
- ■ 3.6–5.4 global hectares per person
- ■ 1.8–3.6 global hectares per person
- ■ 0.9–1.8 global hectares per person
- ■ Less than 0.9 global hectares per person
- ■ Insufficient data

Figure 10.63 Distorted map showing global ecological footprint distribution

In 2010, calculations show that if we share the Earth's productive bio-capacity **evenly**, each person has approximately **2 global hectares** (gha) available to them. In reality the share of ecological footprint is vastly uneven and although some parts of the world are at or below the 2 gha level, other areas are well above. Productive bio-capacity is also not evenly distributed, so some countries have a more generous share, suggesting that either they could accommodate a larger population or that their existing population can afford a larger ecological footprint.

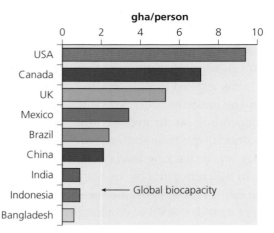

Figure 10.65 Bar graph showing per capita ecological footprints for selected countries

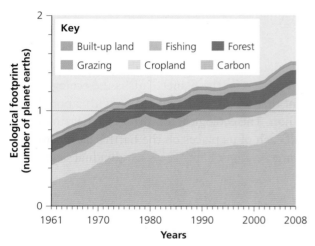

Figure 10.64 Compound graph showing components of ecological footprint and changes in their composition from 1961–2008
Source: WWF

Ecological footprint accounting represents human consumption of biological resources and generation of waste, compared to the biosphere's productive capacity in a year. Figure 10.63 (page 505) gives an impression by distortion of size of the total ecological footprint of countries globally. (The shading shows each country's average footprint.) Figure 10.65 also shows some calculations of ecological footprints for different parts of the world in 2010.

According to these calculations, if all humans had an ecological footprint similar to the average Indian or Indonesian, the Earth could support 15 billion people. However, if everyone lived as we do in the UK, then the Earth could only support 2.5 billion, and only 1.5 billion if we lived the lifestyle of the average US citizen.

These figures are based on rates of consumption that are already unsustainable and evidence suggests that we are already in a state of **overshoot**. Each year our levels of consumption are exceeding the productive bio-capacity of the planet, not only in terms of its ability to produce but also in its ability to assimilate our pollution and waste. Rees believes we are living beyond the means of the planet to sustain us and we will need 1.5 planets to support our current way of life (Figure 10.66).

An alternative way of looking at this states that there is an 'Earth Overshoot Day' each year, designated as the day it is thought we have used the productive bio-capacity of the planet for that year; after that day we go into 'ecological debt'. In 2000, it was 1 November but in 2014 it was 19 August.

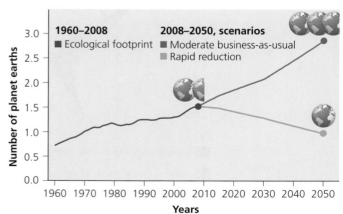

Figure 10.66 Number of planets needed to support world population now and in the future

Implications of carrying capacity and ecological footprint

The concept of carrying capacity introduces a 'limit' to population growth, so the exponential 'J curve' growth of human population shown in Figure 10.59 (page 503) needs to be considered in terms of what will happen to population if it exceeds this limit. Some observers would argue that the carrying capacity has already been exceeded in some parts of the world, such as in the Sahel where famines are regular occurrences. However, on a global scale, population growth is starting to slow down and an alternative view is that population growth is gradually forming a **'sigmoidal'** or 'S' curve as it adjusts to carrying capacity (Figure 10.67). This adjustment is known as the **logistic model** of population growth which suggests that as population size increases, the rate of increase declines as more environmental resistance is encountered (for example, dwindling food and water resources), leading eventually to an equilibrium population size fixed by the carrying capacity. It could also be the result of the greater awareness of an intelligent species of negative implications of continued rapid growth and the need for sustainable development.

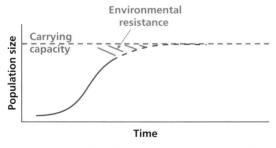

In population ecology this pattern is characteristic of long-lived species with relatively few predators, *e.g. primates*

Factors slowing down the growth rate are known as **limiting factors** – they create something known as **environmental resistance**

Figure 10.67 The 'sigmoidal' or 'S' curve of population growth

The negative environmental implications of growing ecological footprints include:

● climate change, exacerbation of global warming
● more land taken for settlement, industry and transport
● degradation of natural ecosystems (for example, 10 per cent of coral reefs are already degraded beyond recovery)
● increased threat of species' extinctions
● over-cultivation and overgrazing reducing land and soil quality
● depletion of fish stocks beyond recovery.

The population, resources and pollution model

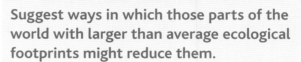

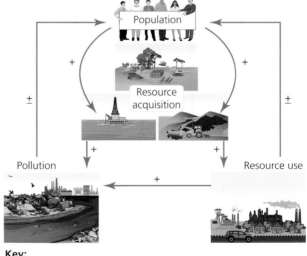

Key:

+ signs indicate a 'positive feedback' loop in which one activity increases another

− signs indicate a 'negative feedback' loop in which one activity reduces another

Figure 10.68 The population, resources and pollution model (after Dr D. D. Chiras). The model outlines how all organisms interact with their environment.

At the beginning of this chapter we considered the relationship between humans and their environment; how humans exploit and utilise resources provided which, in turn, has an impact on the environment. The population, resources and pollution (PRP) model (Figure 10.68) represents a fundamental ecological relationship true to all organisms including humans, and so the model:

- provides an insight into the human–environment interaction
- illustrates several important relationships between the two
- adopts a 'systems' approach (as seen in Chapter 1) – understanding that a consequence of changing one variable in the model is that others will be affected
- promotes 'systems thinking' which is vital to sustainable development
- provides an insight into sustainable solutions – for example, to control resource depletion population control measures might be adopted; encouraging less demand for resources will reduce levels of pollution, for example in energy generation
- uses the concepts of positive and negative feedback (also seen in Chapter 1).

The model shows that populations acquire resources from the environment. This acquisition of resources will alter the existing biotic and abiotic conditions of ecosystems. For example, open-cast coal mines damage and destroy wildlife habitat; they can also lead to deforestation and soil erosion that pollutes nearby streams. On the diagram this is represented by the arrow between resource acquisition and pollution.

Resources that are extracted from the environment are utilised; for example, coal may be used in power stations to generate electricity, or mineral deposits are crafted into finished consumer products such as cars. The conversion of raw materials into energy or into finished products results in pollution and this is indicated on the diagram by the arrow connecting resource use to pollution. Pollutants are classified by the medium they contaminate, usually land, water or atmosphere. Recent evidence suggests that many forms of pollution cross these boundaries – a phenomenon known as **cross-media contamination**. For example, acid rain is an atmospheric pollutant but when it falls from the sky, acid is deposited contaminating forests, rivers and lakes.

Positive and negative feedback mechanisms

Positive and negative feedback either enhance or counter changes that occur in a system (see Chapter 1).

- **Positive feedback** enhances or amplifies changes – moving a system away from its equilibrium state and making it more unstable.
- **Negative feedback** is an opposing force which counters any change, holding the system in a more stable equilibrium.

Negative feedback loops are the 'default' for controlling biological systems such as population growth and regulating homeostasis, as shown in Figure 10.69.

These ideas are applied in population ecology and can equally be applied to human population growth. Population regulation uses negative feedback to keep plant and animal populations within the carrying capacity of the environment.

In the PRP model, the negative feedback loops from resource use and pollution are of great concern. For

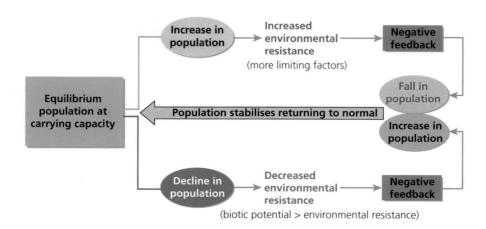

The mortality rate changes depending on whether the population is above or below the carrying capacity – to return the population to the carrying capacity

Figure 10.69 Homeostatic regulation of population growth

example, continued soil erosion from overgrazing, over-cultivation or pollution of land and water courses would lead to a decline in food production that could have devastating effects on human population. This is already seen in parts of sub-Saharan Africa. At the very least, the negative feedback from increased resource depletion means that people may have to expect a levelling out, and in many cases a decline, in their standard of living as fewer resources are available to them.

The sigmoidal curve and the PRP model both suggest that negative feedback will be the consequence of continued population growth. As population nears or exceeds the carrying capacity then negative feedback will mean that there are insufficient resources and death rates will overtake birth rates, causing a population decline back to a stable of equilibrium. This is also known as homeostatic regulation.

Positive feedback on the model is shown by resource acquisition enhancing survival and promoting human population growth. One way to view positive feedback

is to consider agriculture and population together. They can be viewed as being in a **positive feedback spiral**, in that one drives the other with increasing intensity. Population growth has led to an increase in agricultural yields and production through technology, which in turn means that more population can be supported and so population growth continues at an exponential rate.

Positive feedback mechanisms may cause serious problems, such as the devastating cycles of depletion and environmental destruction. Eventually, in agricultural production, despite technological advances, soils are depleted, the carbon cycle is disrupted and yields diminish leading to instability. Another problem is the use of fossil fuel energy, which has increased our capacity to produce and distribute food but is creating a rise in global temperature because of greenhouse gas emissions. The resulting shift in rainfall and drought patterns may devastate crop production in many areas.

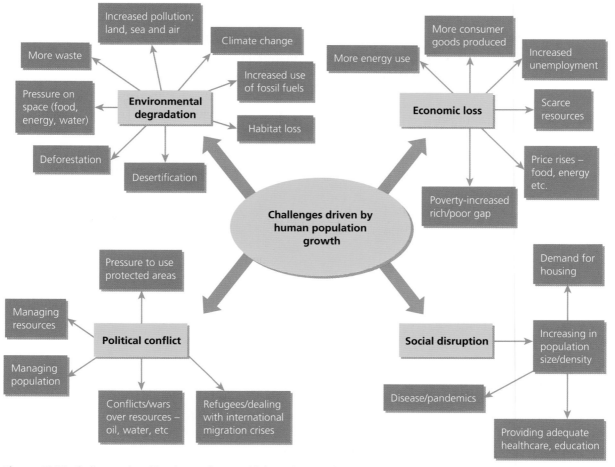

Figure 10.70 Challenges faced by the modern world driven by population growth

Contrasting perspectives on population growth

The negative implications of continued population growth are all too obvious. As countries develop economically and take advantage of their demographic dividend, then it is likely that the consumption of resources will continue to grow. An increasing population with an ever-decreasing supply of resources is an unsustainable position.

In addition to the environmental pressures, Figure 10.70 (page 509) identifies a range of social, economic and political implications. There are, however, some positive implications to consider:

● Increasing education and awareness can influence and change lifestyles, thereby reducing ecological footprints – though more needs to be done.

● Biotechnologies and appropriate technologies could increase the carrying capacity in specific areas and globally.

These contrasting views concerning the balance between population growth and resources are outlined by theorists such as Thomas Malthus and Ester Boserup and their respective followers.

Malthus and neo-Malthusians

Just as human population was starting its unprecedented exponential growth curve around the beginning of the nineteenth century, Thomas Malthus, an English clergyman, published *An Essay on the Principle of Population* in 1798. Malthus was one of the first writers to make an observation about the relationship between humans and resources and he used his observations to predict the future. He pointed out that 'the power of population is infinitely greater than the power in the earth to produce subsistence for man'. In other words, food production cannot increase as rapidly as human reproduction.

The principle behind Malthus's theory is that food production can, at best, only increase arithmetically (so from an index value of 10 to 20, 30, 40 and so on) so at any one time there is a fixed 'carrying capacity' that can only support a given population. However, population growth will occur geometrically (so from a given index value of 1 to 2, then 4, 8, 16, 32 and so on). Eventually the population value will exceed the carrying capacity value and there will be a population 'crash' caused by his predicted catastrophe (Figure 10.71).

Time periods (years)	25	50	75	100	125	150	175	200
Population	1	2	4	8	16	32	64	128
Food Supply	1(0)	2(0)	3(0)	4(0)	5(0)	6(0)	7(0)	8(0)

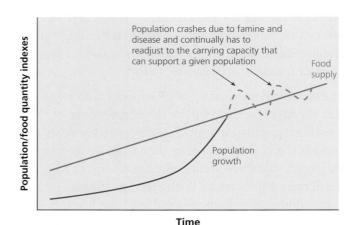

Figure 10.71 Malthusian diagram showing the relationship between population and resources

Malthus's views on the population/resource balance are **pessimistic**. He believed that there is an ultimate limit to how much human population can grow and this limit is determined by the environment and its ability to produce food. He also believed that the balance is maintained by various checks:

● **Positive checks:** Increased deaths through war, famine and disease; increased incidence of abortion, infanticide, etc.

● **Negative checks:** He advocated moral restraint, for example, celibacy and later marriages (his theory was written before the advent of modern contraception).

Malthus predicted that unless population growth was controlled life would end in misery, leading to what became known as a 'Malthusian catastrophe'.

Although his ideas originated 200 years ago, they received much support in the mid-twentieth century from demographers known as neo-Malthusians. There is some evidence to support neo-Malthusian ideas about population growth:

● the regular famines that occur in countries such as Sudan and Ethiopia

● the wars that are often fought over food, water and energy resources

● water scarcity – especially in the Middle East.

Malthusian predictions were also examined by a group known as The Club of Rome, a global think tank formed in 1968 which used a computerised model to investigate the state of a range of resources as well as population growth. Their 'Limits to Growth' report predicted that economic growth could not continue indefinitely because of the limited availability of natural resources.

and enable it to extend upwards in line with population growth (Figure 10.73).

Time periods (years)	25	50	75	100	125	150	175	200
Population	1	2	4	8	16	32	64	128
Food Supply	1(0)	2(0)	3(0)	4(0)	5(0)	6(0)	7(0)	21(0)

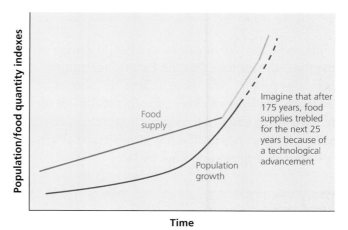

Figure 10.73 Boserup's view of the population–resource relationship

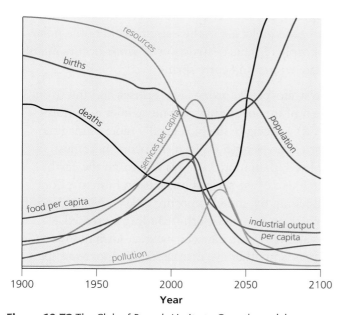

Figure 10.72 The Club of Rome's Limits to Growth model

The model (Figure 10.72) predicted:
- population would continue to rise until 2050 after which it starts to decline
- rapid resource depletion in the first half of the twenty-first century
- food per head starts to decline in the early twenty-first century
- industrial output per head starts to decline
- pollution increases before falling in 2030.

Alternative ideas of Boserup and Simon

There are alternative, more optimistic views of our future such as those expressed by Danish economist Ester Boserup who also recognised that environments have limits that restrict production. Unlike the 'determinism' (limits determined by the environment) claimed by Malthusians, however, Boserup suggested a 'possibilistic' philosophy, where human ingenuity could alter the carrying capacity

In 1965 Boserup suggested that population growth often stimulates innovation and limits can be increased by improvements in technology. These would increase resource availability so much that the carrying capacity could sustain a much larger population than Malthus had suggested. It seems that our history is such that when there is a crisis for survival in an area, we find a solution that eventually averts that crisis, thus proving that 'necessity is the mother of invention'. This is shown in agricultural innovations such as:
- intensification of farming in Europe in response to the EU Common Agricultural Policy (CAP)
- the 'Green Revolution' in Mexico, South and Southeast Asia
- recent developments in biotechnologies including genetic modification (GM).

Julian Simon, an American economist, argued that every important long-term measure of human material welfare shows improvement in all parts of the world, demonstrating that despite rapid population growth humans are measurably better off. He uses facts to support the idea that raw materials have become less scarce and that the relative costs in accessing them have also reduced. Using other facts to support his claims, Simon suggests:

- the air in rich countries is safer to breathe
- water cleanliness has improved
- the condition of cropland is improving rather than worsening
- historically, food production increases have always been at least as fast as population increases.

In his book *The Ultimate Resource* (1981) he argues that the only resource that the Earth was running short of is people, even though there are more of us. There are challenges to his arguments and more recent trends suggest flaws in his claims, but the underlying premise, that as population has grown resources have increased, is largely supported by fact.

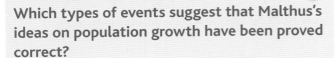

Key questions

Which types of events suggest that Malthus's ideas on population growth have been proved correct?

Malthus had forecast the end of civilisation and mankind by 1900 yet since then the world's population has increased sevenfold and, for many, their standard of living has improved not diminished. So why were the 'doom and gloom merchants' so wrong; or were they?

Consider reasons why Boserup's theory of population growth is more feasible.

10.6 Global population futures

There has been much speculation about what will happen to global population size during the twenty-first century following its rapid increase in the twentieth century. This is something that will affect us all, particularly if rates of consumption of global resources are maintained. With more people on the planet the demand for resources will undoubtedly increase, especially if more of those people become more affluent and desire a higher standard of living.

At the same time, the earth's physical environment is undergoing change, particularly with regard to its atmosphere and climate, which will impact on human health.

Health impacts of global environmental change

Ozone depletion

Depletion of ozone (otherwise known as triatomic oxygen or O_3) is a relatively recent phenomenon in human history and is known to be caused by halogenated chemicals such as chlorofluorocarbons (CFCs) used in refrigeration, insulation and spray-can propellants. These chemicals react with and destroy the ozone molecule in the extreme cold of the polar stratosphere. This destruction of ozone occurs mainly in late winter and early spring.

The stratospheric ozone layer utilises and filters out much of the incoming solar ultraviolet radiation (UVR), especially the biologically more damaging, shorter wavelength UVR. If the ozone becomes depleted, then more UV rays will reach the earth, which can cause serious impacts on humans, flora and fauna. Extended exposure to UV rays has harmful impacts on human health including increased prevalence of skin cancers, cataract formation, increases in other eye diseases and an increase in infectious diseases due to excessive UV radiation weakening the human immune system.

Skin cancer

Most skin cancers are caused by exposure to the sun. This may be long-term exposure, or short periods of intense sun exposure and burning. The ultraviolet light in sunlight damages the DNA in the skin cells. Malignant melanoma incidence rates have increased more than fivefold since the mid-1970s. Melanomas are more likely to affect older people and those from more affluent areas. According to the Office for National Statistics (ONS), rates of newly diagnosed malignant melanoma skin cancer are highest in the South West and the South East regions of England (Figure 10.74).

Australia has the highest rate of skin cancer in the world. The combination of the predominantly light-skinned population, tropical latitude with high levels of ultraviolet radiation and cultural emphasis on out-door activities has contributed to this problem.

- Two out of three Australians will be diagnosed with skin cancer by the age of 70.
- Around 2,000 Australians die from skin cancer each year.

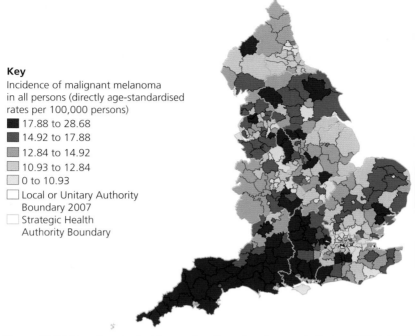

Key
Incidence of malignant melanoma
in all persons (directly age-standardised
rates per 100,000 persons)
- 17.88 to 28.68
- 14.92 to 17.88
- 12.84 to 14.92
- 10.93 to 12.84
- 0 to 10.93
- Local or Unitary Authority
  Boundary 2007
- Strategic Health
  Authority Boundary

Figure 10.74 Map showing incidence of malignant melanoma in England 2004-2006

- Skin cancers account for about 80 per cent of all new cancers diagnosed each year in Australia. Each year, Australians are four times more likely to develop a common skin cancer than any other form of cancer.
- It has been calculated that skin cancer costs the health system around AU $512.3 million in 2010 (diagnosis, treatment and pathology).

Figure 10.75 Skin cancer protection poster

Cataracts

Cataracts are a form of eye damage which causes a loss of transparency in the lens of the eye leading to cloudiness of vision and can eventually result in blindness. There are different types of cataracts and a number of factors can lead to their formation, including their being a natural part of the ageing process or as a result of diabetes. Smoke from burning fuelwood and cigarettes as well as poor nutrition (particularly a lack of vitamin A) can also cause cataracts.

Exposure to UV radiation can damage different parts of the eye, including the lens, cornea, retina and conjunctiva, and research shows it appears to be a major risk factor for cataract development. Cataracts are the leading cause of blindness in the world. Proteins in the eye's lens unravel, tangle and accumulate pigments that cloud the lens and eventually lead to blindness. Increased ozone depletion means anyone who spends a lot of their time outdoors is at risk of eye problems from UV radiation, but the extent of the risk depends on a number of factors including:

- **Geographic location:** UV levels are greater in tropical areas near the earth's equator.
- **Altitude:** UV levels are greater at higher altitudes.
- **Time of day:** UV levels are greater when the sun is high in the sky, typically from 10 a.m. to 2 p.m.
- **Setting:** UV levels are greater in wide open spaces, especially when highly reflective surfaces are present, like snow and sand. Exposure is less likely in urban settings, where tall buildings shade the streets.

UV levels are not significantly affected by cloud cover and risk of UV exposure can be quite high even on hazy or overcast days. This is because UV is invisible radiation, not visible light, and can penetrate clouds. Wearing good quality sunglasses blocks harmful UV radiation and close fitting 'wrap around' styles are better as they prevent sunlight entering the peripheries of the glasses.

The only effective treatment for cataracts is surgery to remove the cloudy lens and replace it with a clear plastic intraocular lens implant. Although curable, cataracts severely diminish the eyesight of millions of people in both the developing and developed world. A combination of variables means that those living in rural areas in developing countries are disproportionately affected by UV-induced cataracts and they are less able to afford prevention or treatment for the problem. A sustained 10 per cent depletion of the ozone layer is expected to result in nearly 2 million new cases of cataracts globally each year.

Climate change

A change in climatic conditions can have three kinds of health impacts:

- Relatively direct impacts, usually caused by increases in the frequency or severity of extreme weather events such as:
  - storms which increase the risk of dangerous flooding, high winds, and other threats

  - warmer average temperatures leading to hotter days and more frequent and longer heat waves.
- Consequences of environmental change and ecological disruption that occur:
  - increased concentrations of unhealthy air and water pollutants
  - changes in temperature and precipitation patterns could exacerbate the spread of some diseases.
- Consequences associated with demoralised populations in the wake of climate-induced economic dislocation, environmental decline, and conflict situations. For example:
  - greater frequency of infectious disease epidemics following floods and storms
  - substantial traumatic, nutritional or psychological health effects following population displacement from sea level rise or increased storm activity.

Although global warming may bring some localised benefits, such as fewer winter deaths in temperate climates and increased food production in certain areas, the WHO suggests that the overall health effects of a changing climate are likely to be overwhelmingly negative (Figure 10.76). Between 2030 and 2050, they estimate that there will be approximately 250,000 additional deaths per year from malnutrition, malaria, diarrhoea and heat stress.

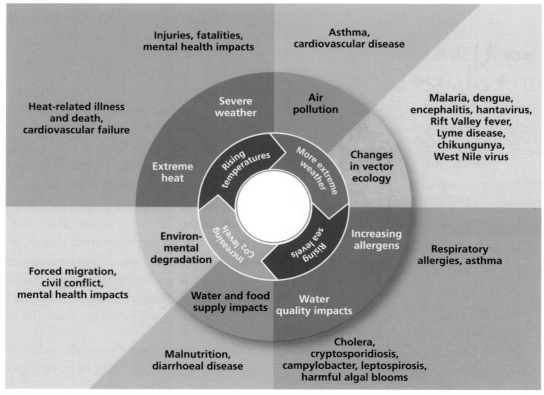

Figure 10.76 Impact of climate change on human health
Source: US Department of Health and Human Services

For each potential impact of climate change, certain groups will be particularly vulnerable. Vulnerability depends on factors such as population density, level of economic development, food availability, income level and distribution, local environmental conditions, pre-existing health status and the quality and availability of public healthcare. For instance, those most at risk of being harmed by thermal extremes include the elderly and the poor.

Thermal stress (heatwaves, cold spells)

One of the potential effects of human-induced climate change is a change in mortality related to thermal stress. Research by the Intergovernmental Panel on Climate Change (IPCC) concludes that climate change will cause increased heat-related mortality and morbidity but decreased cold-related mortality in temperate countries. A change in temperature at both extremes affects both cardiovascular and respiratory mortality.

Heatwaves

The main risks posed by a heatwave are dehydration, overheating, heat exhaustion and heatstroke. The very young, the elderly and the seriously ill are the groups who are particularly at risk of health problems when the weather is very hot. During the summer heatwave in northern France in August 2003, unprecedented high day- and night-time temperatures for a period of three weeks resulted in 15,000 excess deaths. The vast majority of these were among older people. The WHO reports that 70,000 excess deaths were recorded across the whole of Europe in the same period.

Heatwaves can lead directly or indirectly to other health risks. These include:

- smogs, which can lead to high concentrations of nitrogen dioxide and particulate matter in the atmosphere
- the growth of blue-green algae in water courses, which can cause problems for aquatic life, including fish, as well as toxic algal blooms, causing problems for public recreational water activities
- health and environmental problems including odour, dust and vermin infestation. Additional measures may be necessary to mitigate these problems, including more frequent waste collections and extra pollution control measures at landfills and other waste treatment facilities
- wildfires
- water shortages.

Heatwaves have a much bigger health impact in cities than in surrounding suburban and rural areas. The impact on mortality from heat stress may be more significant in developing-country cities (such as Mexico City or New Delhi) where populations are especially vulnerable as they lack the resources to deal with heatwaves.

Cold spells

In many temperate countries, there is clear seasonal variation in mortality. Death rates in winter can be as much as 25 per cent higher than in summer. Extremes of cold impact severely on those suffering with both cardiovascular and respiratory diseases and lead to an increase in mortality. However, annual outbreaks of winter diseases such as influenza, which have a large influence on winter mortality, are not strongly associated with colder temperatures.

Climate change is likely to bring milder winters in temperate regions. Results from research indicate that, for most of the cities studied, global climate change is likely to lead to a reduction in mortality rates due to decreasing winter mortality. This effect is most pronounced for cardiovascular mortality in elderly people in cities which experience temperate or cold climates at present. There is conflicting and limited evidence as to whether the decreases in winter mortality are greater or less than the increase in summer deaths due to heatwaves. The net impact on mortality is likely to vary between populations.

Emergence and changing distribution of vector-borne diseases

Important determinants of vector-borne disease transmission include:

- vector survival and reproduction
- the vector's biting rate
- the pathogen's incubation rate within the vector organism.

Vectors, pathogens and hosts each survive and reproduce within a range of optimal climatic conditions. Temperature and precipitation are the most important, while altitude, wind and daylight duration are also important. Any changes in temperature and rainfall regimes resulting from global climate change will alter the geographical distribution of optimal conditions for most vectors.

The once limited geographic ranges of many vector-borne diseases are expanding, spurred largely by anthropogenic factors. By 2100 it is estimated that average global temperatures will have risen by 1.0–3.5 °C, increasing the likelihood of many vector-borne diseases in new areas. The greatest effect on transmission is likely to be observed at the extremes of the range of temperatures at which transmission occurs. For many diseases these lie in the range 14–18 °C at the lower end and about 35–40 °C at the upper end.

Malaria and dengue

Malaria and dengue fever are the most important vector-borne diseases in the tropics and subtropics. Populations living at the present margins of malaria and dengue, without effective primary healthcare, will be the most susceptible if these diseases expand their geographic range in a warmer world. For example, the *Aedes* mosquito vector of dengue is highly sensitive to climate conditions and studies suggest that climate change could expose an additional 2 billion people to dengue transmission by the 2080s.

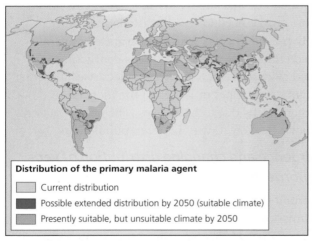

Distribution of the primary malaria agent

☐ Current distribution

■ Possible extended distribution by 2050 (suitable climate)

▨ Presently suitable, but unsuitable climate by 2050

Figure 10.77 Climate change and malaria

Figure 10.77 shows that malaria is likely to spread by 2050, based on the Hadley Centre climate change model's 'high' scenario. Areas shown in yellow indicate the current distribution of malaria. Areas shown in red indicate areas where climate will be conducive to malaria by 2050. Other areas may become free of malaria as climate changes. Climatic anomalies associated with the El Nino–Southern Oscillation phenomenon, resulting in drought and floods, are also expected to increase in frequency and intensity and have been linked to outbreaks of malaria in Africa, Asia and South America.

Lyme disease

Lyme disease (*Lyme borreliosis*) is the most common vector-borne disease in temperate climates of the northern hemisphere, including the USA and Europe. The *Borellia* bacteria is transmitted to humans by the bite of infected deer ticks of the *Ixodes* genus. It is an emerging vector-borne disease thought to be associated with warmer and more humid conditions.

West Nile Virus

West Nile Virus (WNV) is another emergent disease transmitted by mosquitoes. Human infections attributable to WNV have been reported in many countries in the world for over 50 years. Since 1997 it has spread widely and in 1999 the virus reached New York, resulting in a large and dramatic outbreak that spread throughout the USA in the following years. Since its introduction into the USA, the virus has spread and is now widely established from Canada to Venezuela. The WNV outbreak in the USA (1999–2010) highlighted the fact that establishment of vector-borne pathogens outside their current habitat represents a serious danger to the world.

Zika virus

Zika virus infection is caused by the bite of an infected Aedes mosquito, usually causing rash, mild fever, conjunctivitis, and muscle pain. Another emergent disease, the virus was isolated in 1947 in the Zika forest in Uganda. Since then it has remained mainly in Africa, with small and sporadic outbreaks in Asia. In 2014, Chile notified the WHO of the Zika virus on Easter Island. Outbreaks have been reported since in the Pacific region, and the virus has now spread to South and Central America and the Caribbean. In May 2015, the public health authorities of Brazil confirmed the transmission of Zika virus and other countries of the Americas have subsequently reported its presence. There is increasing evidence that pregnant women who contract the virus during pregnancy may have an increased risk of giving birth to a baby with microcephaly (an abnormally small head associated with abnormal brain development).

Agricultural productivity

The current impacts of climate change on agriculture (and adaptations to these impacts) are discussed earlier in this chapter, but it is important to examine how climate change is predicted to affect productivity into the future.

Direct impacts

Higher growing season temperatures will significantly affect agricultural yields, farm incomes and food security. There will be gains and losses depending on the location of the growing region. Some gains in productivity include:

- In mid and high latitudes, the crop yields are projected to increase and extend northwards, especially for cereals and cool season seed crops, such as oil seed rape.
- Crops grown in lower latitudes, such as maize, sunflower and soya beans, could become viable further north and at higher altitudes; yields could increase by as much as 30 per cent by the 2050s.
- There are potentially large gains in agricultural land for regions such as Russia, owing to longer planting seasons and more favourable growing conditions under warming, amounting to a 64 per cent increase over 245 million hectares by the 2080s.

However, without adaptations by farmers, climate change may not necessarily confer benefits to productivity. For example, an increase in the mean growing season temperature will bring forward the harvest time of current varieties and possibly reduce final yield without adaptation to a longer growing season. There are also likely to be losses in productivity:

- In areas where temperatures are already close to the physiological limits for crops, such as seasonally arid and tropical regions, higher temperatures will be detrimental, increasing the heat stress on crops and water loss by evaporation.
- Varying rainfall patterns will have a significant impact on productivity. The impact of climate change on regional precipitation is difficult to forecast but there is increasing confidence in projections of a general increase in high-latitude rainfall, especially in winter, and an overall decrease in many parts of the tropics and subtropics.

In conclusion, because different crops show different sensitivities to temperature change there are still uncertainties about yields for given levels of global warming. As an example, it is thought that a 2 °C warming in the mid-latitudes could increase wheat production there by 10 per cent, whereas at low latitudes the same amount of warming may decrease yields by around the same amount.

Indirect impacts

Climate change may also impact indirectly and adversely on crops through effects such as:

- **Pests and diseases:** Indications are that pests such as aphids and weevil larvae respond positively to higher levels of CO_2. Increased temperatures also reduce the overwintering mortality of many pests enabling earlier and more widespread dispersion.
- **Changes in water availability owing to distant climate changes:** Water for irrigation is often extracted from rivers which depend upon distant climatic conditions. For example, agriculture along the Nile in Egypt depends on rainfall in the upper reaches of the river such as in the Ethiopian Highlands.
- **Sea level rise:** Vulnerability of crop productivity is greatest where a large sea-level rise occurs in conjunction with low-lying coastal agriculture. Many major river deltas provide important agricultural land owing to the fertility of fluvial soils.

Nutritional standards

The impact of climate change on global food production will have uncertain and varying consequences for human health and nutrition. It can be viewed from different perspectives depending on levels of development.

Developed regions

Increasing food prices may lower the nutritional quality of dietary intakes, exacerbate obesity, and amplify health inequalities. Changes in agricultural production resulting from new crop and livestock species may lead to use of different pesticides and veterinary medicines, and in turn affect the transfer mechanisms through which contaminants move from the environment into food. This will have implications for food safety and the nutritional content of food.

The complementary promotion of healthier diets and climate change mitigation may increase consumption of foods whose production reduces greenhouse gas emissions. For example, reducing red meat consumption will have positive effects on saturated fat in the diet but negative impacts on zinc and iron intake.

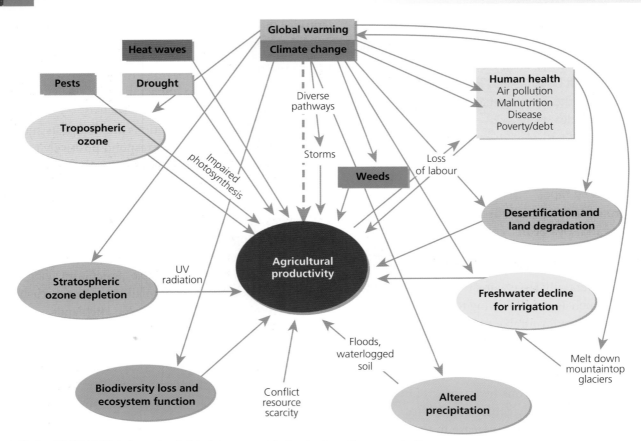

Figure 10.78 Multiple impacts of global warming and climate disruption on agriculture

Developed countries may be able to adapt to the food safety consequences of climate change, although the ability to respond to nutritional challenges is less certain.

Developing regions

Nations experiencing rapid economic growth often go through a 'nutrition transition' where increased affluence guides the population to a more 'westernised' diet. China, in particular, has seen a significant shift from primarily cereal consumption (mainly rice) to more meat consumption. This will have mixed nutritional results for populations and though it has not come as a result of climate change, it will exacerbate the effects through 'positive feedback'. Increased livestock production will increase methane emissions, require more land usage for fodder production and so more clearance of forests.

Least developed regions

These regions are the most vulnerable to loss of production due to climate change. A study by UNICEF found clear and alarming links between climate change and the nutritional status and migration patterns of populations. The predictability of rainfall is critical for rain-fed agriculture and climate change has given an unpredictable combination of extended periods of rainfall (leading to floods) and droughts, both of which have resulted in crop failure. When crops fail, farmers may sell livestock as an economic default, which can lead to human iron and zinc deficiencies. Another important aspect of climate-induced crop failure is an over-reliance on fewer crops which are less likely to fail. Exclusive diets of particular crops lead to specific forms of malnutrition.

Prospects for global population change

Over the course of the twenty-first century, world population is likely only to rise by 50 per cent, unlike the 400 per cent of the previous one hundred years. The reason for this is much slower growth rates. The peak of world population growth rate was reached in the 1960s and it has continued to fall from 2 per cent to approximately 1 per cent (Figure 10.79). In 2015, the UN's Department for Economic and Social Affairs (Population Division) produced a report, 'World Population Prospects: The 2015 Revision'. The key findings of this report are available at http://esa.un.org/unpd/wpp/publications/files/key_findings_wpp_2015.pdf.

Drivers of world population growth

The two drivers of world population growth are the world **fertility rate** and the world **life expectancy**.

Fertility rates

A country's total fertility is the key factor in the demographic transition that each country goes through as it develops. The global average fertility rate in the 1960s was five children per woman but this has now halved. The UN forecasts that this downward trend will continue during the twenty-first century from 2.5 to just over two children. The current UN projection for global fertility is that it will be less than two by 2100 (Figure 10.80).

As global fertility reduces to the replacement level, it ceases to be the main driver behind world population growth for the foreseeable future.

Life expectancy

As living standards rise and healthcare and medical science improve around the world, more and more children survive. The transfer of medical knowledge, often through NGOs, has improved the chances of

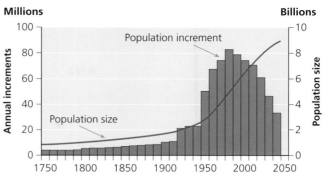

Figure 10.79 Long-term world population growth, 1750–2050
Source: UN Population Division

survival in even some of the poorest countries. This means that average life expectancy around the world is also increasing rapidly (Figure 10.81, page 520). This has become the main reason for population growth; more people are surviving into older age.

There have been a range of predictions offered. The most discussed forecasts of world population are those presented by the Population Division of the United Nations and they are constantly reviewing their predictions in the light of new information and new methods of forecasting.

In 2012 the UN made three possible forecasts for the 2050 and 2100 world population. The forecasts are made according to different modelled fertility rates (Figure 10.82, page 520).

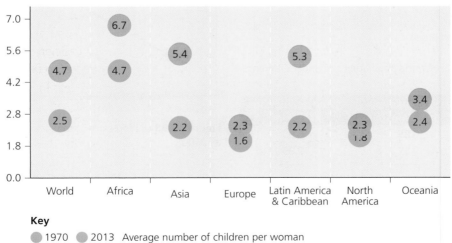

Key

⚫ 1970   ⚫ 2013   Average number of children per woman

Figure 10.80 Global decline in fertility rates from 1970 to 2013

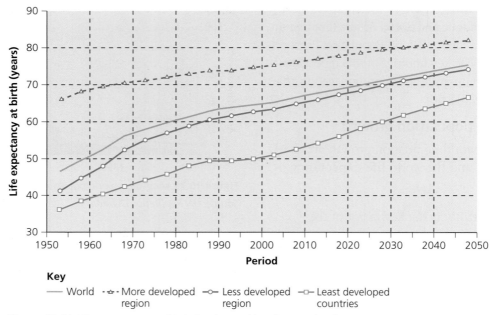

Key

— World ⋯▵⋯ More developed region —○— Less developed region —□— Least developed countries

Figure 10.81 Life expectancy at birth for the world and major development regions

Source: UN Population Division

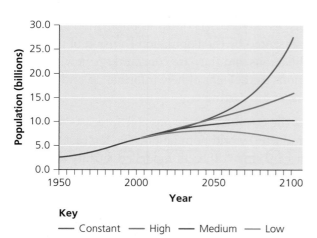

Key

— Constant — High — Medium — Low

Figure 10.82 UN estimates and forecasts of world population by different total fertility scenarios

Table 10.14 UN forecasts of world population size under different scenarios

Fertility scenario	2050 population in billions	2100 population in billions
High	10.85	16.65
Medium	9.55	10.85
Low	8.34	6.75*

Perhaps to underline the difficulty of making predictions, even for an organisation such as the UN, the 2050 forecasts display a considerable 2.51 billion range between the different modelled scenarios. A more dramatic divergence of 9.9 billion in forecasts appears for the end of the century. The extent of the spreads in both sets of forecasts

points to the opportunity for fertility reduction, but also to the threat of more unsustainable growth. *Note that the prediction for 2100 with continued reduced fertility could give a population lower than the Earth's current population. Figure 10.83 shows that the UN report claims there is an 80 per cent chance that the actual number of people in 2100 will be between 9.6 and 12.3 billion.

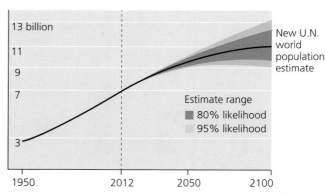

Figure 10.83 UN forecasted probabilities of future population size by 2100

Why has the UN revised estimates upwards?

Fertility in sub-Saharan Africa is falling much more slowly than it has done in other developing parts of the world. Between 2010 and 2012 the UN revised their estimates of fertility rates in Africa by one quarter of a child per women. Because of the **compound** nature of population growth, this has made an enormous difference to forecasts. One quarter child more per woman equated to 0.60 billion more people by 2100.

Nigeria is crucial to these calculations. With 175 million, it has Africa's largest population and this is expected to be five times greater by 2100, giving a population of 900 million. According to the UN, of all the people added to the planet in the twenty-first century, one in every five will be Nigerian!

More optimistic predictions

The International Institute for Applied Systems Analysis has used different methods to forecast. They argue that population will peak at 9.4 billion in 2075 and fall to below 9 billion by 2100. There are three main reasons why they suggest the world's population will stabilise before the end of the century.

- They disagree with the UN claim that China's fertility, which has already fallen to 1.6, will start to rise again.
- They project that Nigeria's population will only triple by 2100 (not quintuple) because of better access to education. In Nigeria, improvements in education are already happening.
- Their model incorporates a crucial variable not included in the UN projections – the level of education in a given population. Educating girls has been found to be one of the most successful methods of reducing fertility in the long term (Figure 10.84).

Figure 10.84 Girls in Kerala receive compulsory education, which has contributed to lower fertility rates there

Consensus views

Despite the variety of forecasts that have been made about the future of global population change this century, there are some common themes which can be regarded as certainties:

- World population growth rate will continue to slow down. The rate of deceleration is difficult to predict.
- The most rapid population increases will be in countries of sub-Saharan Africa.
- There will be continued population decline in many parts of Europe and Japan; China and other parts of Asia and the Americas are also likely to see population decline in the future.
- The global population will have an increased average age.
- Countries with declining populations will face new challenges with ageing populations.
- Life expectancy will continue to rise in the foreseeable future.
- Global total fertility rates will continue to fall towards or below replacement rate in most global regions.
- Just as China reaches its peak population size in around 2028, it will be overtaken by India as having the largest population size.

Key question

Why is there such an increasing divergence in forecasts of population size for the end of the century?

Projected distributions

Global population distribution will change significantly during the next century. According to the UN's 2015 revision of World Population Prospects, changes in the projected global population distribution among major regions are as shown in Table 10.15.

Table 10.15 Projected changes in the world's population distribution by region, 2015–2100

Region	Population (millions)			
	2015	2030	2050	2100
Africa	1,186	1,679	2,478	4,387
Asia	4,393	4,923	5,267	4,889
Europe	738	734	707	646
Latin America and the Caribbean	634	721	784	721
North America	358	396	433	500
Oceania	39	47	57	71
World	7,349	8,501	9,725	11,213

Source: UN Population Division (2015)

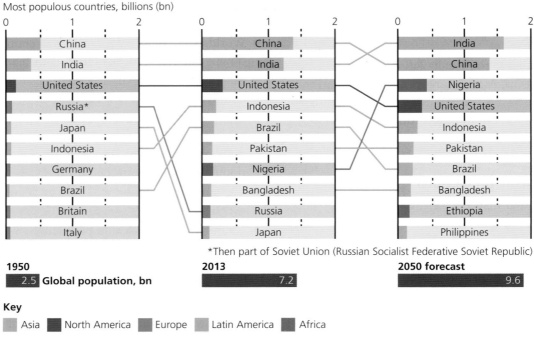

Most populous countries, billions (bn)

1950	2013	2050 forecast
China	China	India
India	India	China
United States	United States	Nigeria
Russia*	Indonesia	United States
Japan	Brazil	Indonesia
Indonesia	Pakistan	Pakistan
Germany	Nigeria	Brazil
Brazil	Bangladesh	Bangladesh
Britain	Russia	Ethiopia
Italy	Japan	Philippines

*Then part of Soviet Union (Russian Socialist Federative Soviet Republic)

1950   2.5 **Global population, bn**   **2013**   7.2   **2050 forecast**   9.6

Key
Asia   North America   Europe   Latin America   Africa

Figure 10.85 New world order of population sizes during twenty-first century

As with any type of projection, there is a degree of uncertainty and the results presented in the table are based on the medium projection variant. This assumes a decline in fertility for countries where large families are still prevalent, as well as a slight increase in fertility in countries with fertility rates of less than two. The key findings suggest the following:

● **Africa**: More than half of global population growth from 2015 to 2050 is expected to occur in Africa. Africa has the highest rate of population growth among major regions, growing at a pace of 2.5 per cent annually. Rapid population increase in Africa is anticipated even if there is a substantial reduction of fertility levels in the near future.

● **Least Developed Countries (LDCs)**: Population growth remains especially high in the group of 48 countries designated as the least developed countries (LDCs), of which 27 are in Africa.

● **Asia**: This is the second largest contributor to future population growth, but the size of increase is relatively small in comparison.

● **Europe**: This is projected to experience a shrinking population. Several countries, mostly in Eastern Europe, are expected to experience a population decline of more than 15 per cent by 2050.

● **Short list of countries**: Half of the world's population growth between 2015 and 2050 is expected to be concentrated in only nine countries: India, Nigeria, Pakistan, DR Congo, Ethiopia, Tanzania, USA, Indonesia and Uganda.

Figure 10.86 shows a distorted world map based on country population size, which demonstrates the change in global population distribution from 1800, when there were 1 billion occupants of the planet, to the projections for 2100. Note the rapid growth in size of Africa, some growth in the Americas and relative decline in both Europe and Asia. Similar interactive images tracing the projected changes in population distribution during the twenty-first century can be found at:

www.viewsoftheworld.net/wp-content/uploads/2011/10/WorldPopulationAnimation.gif, or http://i.imgur.com/zb25KZV.gif

Appraisal of the population-environment relationship

There is an ongoing debate as to whether '**population**' or '**consumption**' is a more significant threat to the environmental limits set by planet Earth. They are both significant threats and should not be set in opposition to each other. A growing and more affluent global population poses particular challenges for ecological footprints and carbon budgets. The main challenges are:

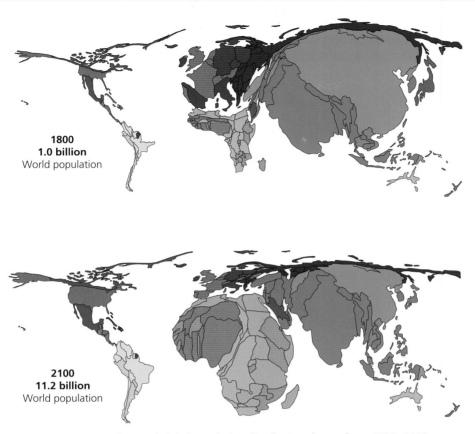

1800
1.0 billion
World population

2100
11.2 billion
World population

Figure 10.86 Distorted map of global population distribution change from 1800–2100

- to continue to supply the resources needed for survival: food, water and energy
- to do this using **sustainable** methods.

Water and energy supply

Clean drinking water is critical for survival. It provides sanitation and hygiene for millions of people. About 70 per cent of water globally is used in agriculture for irrigation and so it is intrinsically linked with the ability to produce food and other important crops such as cotton, rubber, palm oil, etc.

Energy demand is predicted to increase by 40 per cent by 2030. Population growth will accelerate the depletion of non-renewable energy supplies such as fossil fuels. Demand for energy will inevitably have to be met with more renewable energy and also with increases in nuclear energy. There will be increasing pressure on fuelwood supplies in developing countries.

The future of water and energy resources are discussed in more detail in Chapter 11 – Resource security.

Key question

What can governments do to manage population growth more sustainably?

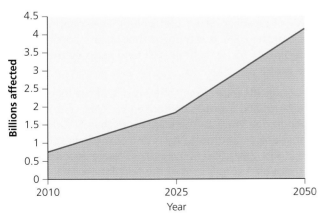

Figure 10.87 Increasing numbers will be affected by water scarcity

Case study of a country experiencing specific patterns of overall population change: Japan – decline and ageing

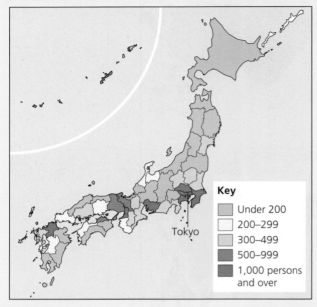

Tokyo

Key
- Under 200
- 200–299
- 300–499
- 500–999
- 1,000 persons and over

Figure 10.88 Population density map of Japan

Japan is one of the world's oldest populations with one of the lowest birth and fertility rates. The total population of Japan in 2014 was 127.9 million (tenth largest globally) and population density was 337 per km². This density is an average and doesn't give the full picture of how the population is distributed.

Physical environment and its influence on population size and distribution

Japan is a mountainous archipelago consisting of thousands of islands of which Honshu, Hokkaido, Kyushu and Shikoku are the largest. The islands contain fertile valleys and narrow coastal plains where much of the population is concentrated. As around 70 per cent of the land is mountainous, the people live in a relatively small proportion of the total land area, so the urban areas are very densely populated.

Due to the large north–south extension of the country, the climate varies strongly in different regions. It enjoys a mainly temperate climate but ranges from long cold winters in Hokkaido and the Sea of Japan coast in the north, to subtropical weather in the southernmost islands such as Okinawa, where the mean temperature in January is 17 ºC. The climate in most of the major cities, including Tokyo, is temperate to subtropic and consists of four seasons. Winters are mild and the summer is hot and humid. There is a rainy season in early summer, and typhoons hit parts of the country every year during late summer.

Located in an area where several continental plates meet, Japan experiences frequent hazards from earthquakes and volcanic eruptions. The volcanic activity in the islands has provided the population with fertile soils. These soils, together with the temperate and seasonal warmth, have fostered intensive agricultural production and so the population grew and thrived, despite the potential hazards of volcanic activity and typhoons. The population has partially overcome the limitations presented by the relief of the landscape by terracing to increase farmland availability.

The archipelagic nature of Japan caused its population to become a strong seafaring nation and so it still has a strong fishing industry which provides for its population.

Japan, however, has relatively small amounts of fossil fuel and mineral deposits. In the industrial age, it relied on expansion into neighbouring countries such as Korea, Manchuria in northeast China and Inner Mongolia to support its ever-growing population and its desire to become an industrial superpower. This, together with further expansion of 'Empire' into the Pacific, was ultimately one of the triggers for the Second World War.

Demography and culture

Japan has a distinctive culture which is very traditional and paternalistic – expecting loyalty from individuals who are extremely hard-working and intensely loyal to family, employer and nation. It is the third largest economy in the world and the gross national income per person is US$48,324 per annum. Traditionally males have been the main breadwinners in Japanese society but as the economy has moved more towards a post-industrial service economy, well-educated females have become an increasingly important part of the workforce. This may have had an impact on birth rates and fertility rates, which are 8.3/1,000 and 1.4 respectively. Death rates are 9.9/1,000. This means that Japan has a natural decrease and population has only stabilised because of migration (mainly of Japanese citizens returning). The country is in Stage 5 of the DTM. It is thought that Japan's population will decline by 30 per cent in 2050 and halve by the end of the century.

The average life expectancy of 83 is one of the highest in the world and over 65s comprised 23.9 per cent of the population in 2013. There are more than 50,000 centenarians. Infant mortality is one of the lowest in the world at just 2.2/1,000 of live births. Young dependents only make up 13.5 per cent of the population. The percentage of economically active people is therefore

62.6 per cent and the dependency ratio is 60. By 2025 it is estimated that 33 per cent of Japan's population will be over 65 and 38 per cent by 2050, so they face a 'demographic time bomb' where the population decline means that there are not enough young people to replace the older population who are dying out. The burden borne by the working and younger population in supporting the elderly may become unsustainable.

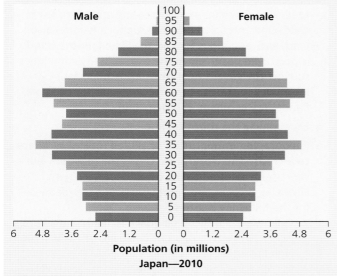

Figure 10.89 Japan's population structure

Figure 10.90 Japan – an ageing society. Japan is currently the only country with an older population of more than 30 per cent

Socio-economic implications of Japan's ageing population

The consequences arising from Japan's ageing population range from the serious – for example, the domestic economy is shrinking, to the more trivial – Japan's ageing problems means that more adult incontinence nappies are sold than baby nappies. All generations are affected by the population structure.

- 25 per cent of the government's budget is spent on pensions/social security for the elderly.
- 40 per cent of all government public spending is on the elderly.
- Since the 1960s retirees pay only a small part of their medical care through a universal system.
- The working population is shrinking fast, which will have a negative impact on domestic economic growth – affecting both production (fewer workers) and consumption (fewer people earning means less spending).
- A rise in home care by families, who may have to give up full-time jobs to become sole carers, takes more people out of the economic workforce.
- Pressure on care services for the elderly means more medical and care workers are needed.
- There is increasing pressure on NGOs (charitable organisations) to provide this care.

Reasons for the ageing population:

- Excellent **healthcare, sanitation and high living standards** mean high life expectancy.
- **Low fat diets** – unlike many western MEDCs, Japan has fewer obesity and heart disease problems.
- Reputedly the air is clean and it is true that Japan uses fewer fossil fuels to generate energy – relying more on nuclear and HEP. (They also have a tendency to use more public transport.)
- **Women** receive good education so as a result many are more **focused on careers**, so they have **fewer children**.
- **Contraception is accessible and widely used** – there are few religious or social taboos.
- **Price of schooling** in Japan has risen – many can only afford to send one child to school.

- It is estimated that 3.8 million elderly Japanese live alone with no family support.
- There are increasing numbers of elderly living on the streets, in poor quality housing or ending up in prison (especially males). Reoffending rates are high as prison offers better care for some.
- There is a rising number of cases of dementia.
- More elderly are having to return to work to make up their income – 3,500 over 65s work for McDonald's.
- Increasingly dysfunctional relationships between young couples and the growth of a new group of Japanese 'thirty-something' men – *otaku* (obsessive/nerds) – who indulge in fantasy relationships with 'animes' or virtual females on computer.

There are many positives as well:

- Due to globalisation, the Japanese manufacturing industry is still very successful and income from large TNCs such as Toyota, Sony, etc., brings in revenue to support the domestic economy.
- There is a growth in the number of private hospitals specialising in healthcare for the elderly and so creating jobs and income.
- There is an increasing demand for leisure activities and tourism among the affluent elderly – also generating income and creating jobs.
- One of the positive features specific to Japan is the embracing of new robot technology – to build 'partner robots' to assist the elderly including companion pets, bionic limbs that aid mobility, fetching, carrying and even robots that help when going to the toilet. This technology is being developed by some of the larger TNCs such as Honda. It is not without problems and is expensive so is only available to the more affluent, but already is worth US$1 billion to the economy.

Responses by the government:

The Japanese government has a duty to try to tackle some of the issues for a sustainable future but has to be cautious with its policies so that it does not alienate the elderly, an important lobby group (the 'grey vote'). Responses are as follows:

- **Raise taxes** – the government has already raised consumption tax by 5 per cent and intends to raise it by a further 5 per cent in 2015 – to maintain revenue to care for the elderly.
- **Raising retirement age to 70.**
- **Pro-natal population policies** – encouraging parents to have more children by increasing maternity leave, child allowances and day care support. Parents also enjoy tax reductions for each child and motherhood is portrayed positively in the media. Many companies support the government by offering financial incentives to staff to become pregnant (for example, Bandai Toy company and Daihatsu).
- Encouraging the **robot technology** revolution.
- **Encouraging immigration of skilled foreign workers to increase the workforce:** This is the most obvious short-term solution to the dependency problem. However, this has been problematic for the Japanese for the following reasons:
  - Japanese is a difficult language to learn and all immigrants are given a rigorous language test which has a high failure rate.
  - It is challenging for foreigners to integrate with Japanese culture and to cope with the hard work ethic that Japanese employers expect of their workforce. It is difficult to be accepted by local staff as a result of this.

Japan does have an immigrant population of around 2.2 million (about 2 per cent of total population), mainly from China and SE Asian countries who have an equally strong work ethic. For example, Japan accepts foreign nurses from Indonesia and the Philippines. Professor of Economics, Noriko Tsuya, has calculated that in order to replace the dwindling population and maintain it at current levels, by 2050 Japan will need a net migration of over 340,000 per year. Japanese society is unlikely to be ready for these numbers so the government has to be very careful with their immigration policy.

Case study of a specified local area: Place, health and well-being – Hart, Hampshire

There are a number of well-documented examples of places with a reputation for having poorer health but this case study will focus on good health and well-being using the Hart district of Hampshire. This area was voted the most desirable place to live in the UK from 2011–2014 (Halifax Quality Of Life Survey) and it consistently appears in the top ten UK districts with the highest life expectancy. This case study will focus specifically on the local area of Hook within the Hart district. It will look at the social, economic and environmental reasons why health and well-being is so much better here than in other parts of the UK.

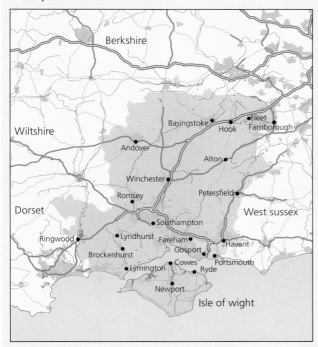

Figure 10.91 Map showing the location of Hook

Hook is a large village with a population of 8,334. It is situated in northeast Hampshire approximately 42 miles southwest of London. The village is bounded on the east side by the picturesque valley of the River Whitewater, and to the south by two areas of common land – Hook Common and Bartley Heath. It has a railway station on the South Western Main Line and is close to Junction 5 of the M3 motorway which links London and Southampton. This makes it a prime location for commuters. Hook is also home to a number of large IT companies such as Hewlett Packard and Lenovo. Virgin Media, BMW and Serco all have headquarters in the nearby Bartley Wood business park providing plenty of employment opportunities.

The health of people in the Hart district is very good. The key health statistics reveal much lower than average mortality rates, lower cases of key diseases and a higher life expectancy for both men and women. The only statistic reported to be worse than the national average in 2015 was the rate of people killed or seriously injured on roads.

The chart below shows how the health of people in this area compares with the rest of England. This area's result for each indicator is shown as a circle. The average rate for England is shown by the black line, which is always at the centre of the chart. The range of results for all local areas in England is shown as a grey bar. A red circle means that this area is significantly worse than England for that indicator; however, a green circle may still indicate an important public health problem.

● Significantly worse than England average

○ Not significantly different from England average

○ Significantly better than England average

Domain	Indicator	Local No Per Year	Local value	Eng value	Eng worst	England Range	England best
Our communities	1 Deprivation	0	0.0	20.4	83.8		0.0
	2 Children in poverty (under 16s)	1,035	5.8	19.2	37.9		5.8
	3 Statutory homelessness	22	0.6	2.3	12.5		0.0
	4 GCSE achieved (5A*-C inc. Eng & Maths)	710	67.2	56.8	35.4		79.9
	5 Violent crime (violence offences)	464	5.0	11.1	27.8		2.8
	6 Long-term unemployment	63	1.1	7.1	23.5		0.9
Children's and young people's health	7 Smoking status at time of delivery	85	9.2	12.0	27.5		1.9
	8 Breastfeeding initiation	850	86.6	73.9			
	9 Obese children (Year 6)	128	13.2	19.1	27.1		9.4
	10 Alcohol-specific hospital stays (under 18)	5.0	22.4	40.1	105.8		11.2
	11 Under 18 conceptions	20	11.9	24.3	44.0		7.6
Adults' health and lifestyle	12 Smoking prevalence	n/a	10.8	18.4	30.0		9.0
	13 Percentage of physically active adults	288	60.9	56.0	43.5		69.7
	14 Obese adults	n/a	16.7	23.0	35.2		11.2
	15 Excess weight in adults	142	61.0	63.8	75.9		45.9
Disease and poor health	16 Incidence of malignant melanoma	18.0	21.9	18.4	38.0		4.8
	17 Hospital stays for self-harm	143	158.8	203.2	682.7		60.9
	18 Hospital stays for alcohol related harm	393	443	645	1231		366
	19 Prevalence of opiate and/or crack use	132	2.2	8.4	25.0		1.4
	20 Recorded diabetes	3,757	4.8	6.2	9.0		3.4
	21 Incidence of TB	2.3	2.5	14.8	113.7		0.0
	22 New STI (exc Chlamydia aged under 25)	368	627	832	3269		172
	23 Hip fractures in people aged 65 and over	92	547	580	838		354
Life expectancy and causes of death	24 Excess winter deaths (three year)	10.8	5.4	17.4	34.3		3.9
	25 Life expectancy at birth (Male)	n/a	83.0	79.4	74.3		83.0
	26 Life expectancy at birth (Female)	n/a	85.9	83.1	80.0		86.4
	27 Infant mortality	1	1.3	4.0	7.6		1.1
	28 Smoking-related deaths	92	189.2	288.7	471.6		167.4
	29 Suicide rate	7	-	8.8			
	30 Under 75 mortality rate: cardiovascular	29	37.1	78.2	137.0		37.1
	31 Under 75 mortality rate: cancer	87	108.7	144.4	202.9		104.0
	32 Killed and seriously injured on roads	48	52.4	39.7	119.6		7.8

Figure 10.92 Health summary for Hart district, 2015

Reasons for good health and well-being in Hook and the surrounding area

- Hook and the surrounding area is affluent. Good accessibility and proximity to large local companies means that many local residents are employed in professional and managerial positions and as such enjoy high incomes. In 2014, the average weekly earnings for the area were £839 – more than a third higher than the UK average of £629.

- Deprivation levels are low. Figure 10.93 shows that the geographical area of Hook is among the 20 per cent least deprived areas of the country.

- Residents tend to be fit and well – over 97 per cent reporting good or fairly good health in the 2011 census.

- Inhabitants live in relative security with one of the lowest crime rates in the country.

- Residents enjoy a relatively good climate – less rainfall per year than the national average (736 mm against 879 mm) and more weekly sunshine hours (32.5 hours against the national average of 29.5 hours).

- Residents have easy access to areas of outstanding natural beauty such as the New Forest and the Test Valley. More locally, Bartley Heath and Hook Common contain miles of protected habitats open for the public to explore.

- Sporting and leisure opportunities in the immediate vicinity include football, squash, badminton and cricket. The village also has a sports centre and bowling club.

- Local websites show that Hook is a lively, friendly place which has a wide range of local community groups including a choral group, the Hook Players theatre group and a local history group. The Base youth centre runs activities for young people and charity groups such as the Lions and local Rotary group raise thousands of pounds each year for good causes. Such organisations and activities help to create a strong community bond which has been shown to make people feel safer and happier.

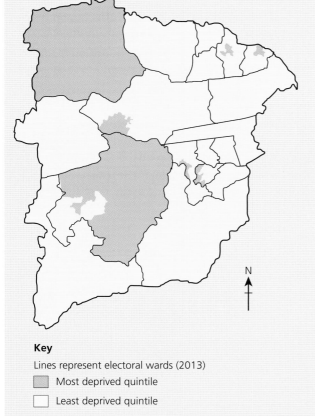

Key

Lines represent electoral wards (2013)

▢ Most deprived quintile

▢ Least deprived quintile

Figure 10.93 Index of multiple deprivation data for Hart district, 2010

- The employment rate for Hook is 77 per cent, significantly higher than the national average of 71 per cent. Only 2 per cent of the population is classified as unemployed.

Figure 10.94 The River Whitewater and other local beauty areas provide walking and other recreational opportunities for locals

Fieldwork opportunities

Census data from the Office for National Statistics (ONS) website www.ons.gov.uk can be used to support fieldwork investigations of population change at different scales throughout the UK. For example, in order to investigate the health of your local area, start by looking at the annual public health profile and then focus in on a smaller area utilising parish profiles and local government websites. The ONS provides analysis and summaries of census data and you could then conduct your own survey or questionnaire looking at different aspects of health. Remember that health is affected by a range of social, economic and environmental factors. Look for provision of green spaces and recreational opportunities, look at employment data and crime figures and then try and gauge the level of community engagement and opportunities, which can all lead to improvements in health and well-being.

Review questions

1 Summarise the changes in the global production and consumption of food during the late twentieth and early twenty-first centuries.

2 Using an example show how one food production system has developed as a result of environmental variables.

3 Summarise the effects that climate change may have on food production in the next 25 years.

4 Outline the ways in which soil problems can cause problems in agriculture.

5 Explain how economic and social development causes an epidemiological transition.

6 Outline the arguments in favour of the following statement:

   'Physical environment has as big an effect upon the incidence of disease as the human environment'.

7 How relevant is the Demographic Transition Model to the population dynamics of the twenty-first century?

8 What are the key causes of migration in the twenty-first century?

9 What are the major social, economic and environmental challenges presented by human population growth in the twenty-first century?

10 Outline technological developments that would make continued global population growth more sustainable.

Further reading

Environment and population:

'Climate-Resilient Agriculture: What small-scale producers need to adapt to climate change'

http://www.christianaid.org.uk/images/climate-resiliant-agriculture-briefing-july-2015.pdf
This offers one view of the direction that the world needs to take in order to ensure food security for all in the face of changing climate. It focuses on the needs of individual farmers rather than those of national or global institutions.

FAO Statistical Yearbook 2013: World food and agriculture
www.fao.org/docrep/018/i3107e/i3107e00.htm
This is an overview of global and regional changes in food production.

'Meat Atlas: Facts and figures about the animals we eat'
Jointly published by the Heinrich Böll Foundation and Friends of the Earth
www.foeeurope.org
Although this document has been funded by the European Commission, the publishers are both pressure groups that have their own agenda and will use the information to further their own points of view. Just remember, there are other ways of looking at this topic.

Environment, health and well-being:

www.who.int/ – the World Health Organization has numerous publications relevant to specific topics such as the Ebola virus, supported with a range of data and statistics.

'Geographies of Health (2nd Edition)' by AC Gatrell and SJ Elliot (2009)

Population change and migration:

Available online, all of the following general resources have data and other sources of information regarding population change:

www.ons.gov.uk – the Office for National Statistics

www.prb.org – Population Reference Bureau

www.unfpa.org/world-population-trends – United Nations Population Fund

www.data.worldbank.org – The World Bank – has a wide range of 'development and demographic indicators'

www.populationmatters.org – Population Matters

www.migrationobservatory.ox.ac.uk – The Migration Observatory at the University of Oxford

www.migrationpolicy.org – the Migration Policy Institute (US)

www.migrationwatchuk.org – MigrationWatchUK monitors migration flows in and out of the UK

Population structure:

As above, plus

www.ec.europa.eu/eurostat/statistics – EU Eurostat provides information and data on ageing population trends in particular

www.policyproject.com – Policy project report on 'Understanding the demographic dividend'

www.gatesinstitute.org – 'Creating and Capitalising on the Demographic Dividend for Africa' from the Bill and Melinda Gates Institute for Population and Reproductive Health at John Hopkins University

Population ecology:

www.footprintnetwork.org – The Global Footprint Network is an international think tank with a focus on reducing ecological footprints and increasing sustainability

'Living Planet Report' – published by the Worldwide Fund for Nature (WWF), Global Footprint Network, Water Footprint Network and London Zoo
www.footprint.wwf.org

'Environmental Science' by DD Chiras (2014) – for the Population, resources and pollution model

Global population futures:

As Population change and Migration, plus

'Global Population to 2050 and beyond: sources analysis and discussion' – a briefing from the Anthony Rae Foundation

'Overpopulation may lead to conflict' from the Population Institute at www.populationinstitute.org

'A World with 11 billion people? New population projections shatter earlier estimates' by Robert Kunzig, published in *National Geographic*, September 2014.

'Future World Population Growth' by Max Roser (2015), published by Our World in Data at www.ourworldindata.org/population-growth

'UN World Population Prospects, The 2015 Revision' available at http://esa.un.org/unpd/wpp/publications/files/key_findings_wpp_2015.pdf

Question practice

A-level questions

1. In agricultural systems, what is meant by capital-intensive farming?

   a) There are numerous inputs, such as machinery, fertilisers, pesticides and other equipment, into a farm relative to its size. As a result, the output yields from the system per unit of area and per worker are both high.

   b) The farming system is large scale and uses a lot of land so needs many workers to make sure output yields are high.

   c) There are few workers relative to the size of a farm, which is large scale and used mainly for livestock grazing. The output per worker is high but for each unit of area farmed is relatively low.

   d) There are a large number of workers relative to the size of an area being farmed, so the output per unit area farmed is high but the output per worker is relatively low.   (1 mark)

2. What is epidemiological transition?

   a) A fall in infant mortality rates is accompanied by a rise in birth rates.

   b) Infectious diseases are replaced by chronic diseases as the main cause of morbidity and mortality, due to improved public health and sanitation.

   c) An increase in infectious disease resulting from a viral pandemic causes increased mortality, measured by death rates.

   d) A fall in birth rates is accompanied by an increase in life expectancy and lower death rates.   (1 mark)

3. When does a country start to experience a demographic dividend?

   a) Dependency ratios are high because there is a very youthful population structure.

   b) Measures are taken by government to resolve issues resulting from an ageing population structure.

   c) There are high rates of net immigration leading to an increase in the birth rate.

   d) Dependency ratios are low as birth rates fall and large numbers of a previously youthful population enter the economically active sector.   (1 mark)

4. What is positive feedback in the population, resources and pollution model?

   a) An increase in human population leads to an increase in the ability to acquire resources. There is also an increase in agricultural production as having more people engenders and enables advances in technology.

   b) A decrease in population in a country, because death rates are higher than birth rates, means that there are more resources available for each person and because there are fewer people there is less pollution.

   c) An increase in population leads to an increase in over-grazing and over-cultivation in agriculture, which in turn leads to soil erosion and degradation of farmland so eventually less food is produced.

   d) A decrease in resource use means that there is less pollution, so the population enjoys better health and population starts to increase rapidly.   (1 mark)

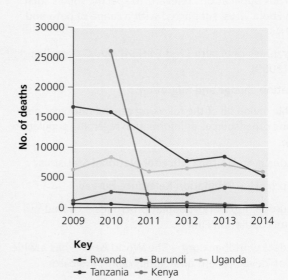

Figure 10.95 Reported deaths caused by malaria in five East African countries from 2009 to 2014.

5. Analyse the trends in deaths caused by malaria illustrated by the line graph shown in Figure 10.95.   (6 marks)

If you don't have migration, you won't be able to fill important jobs to keep the economy going.

Begin to work out programs for regular migration. Begin to work on better co-operation between law enforcement and social ministries on the welcome of migrants, the integration of migrants and enforcement of mechanisms against smugglers and traffickers.

Brunson Mckinsey, Former US diplomat and Director General of the International Organisation for Migration(IOM) from 1998 to 2008

6. Using the extract above and your own knowledge, assess the impacts of international migration as it affects both developed countries and developing countries.　**(9 marks)**

7. Table 10.16 shows changes in crude birth rates, crude death rates and life expectancy over a 33-year period from 1981 until 2014.

   Assess the importance of the population statistics shown in Table 10.16 in explaining the development of these countries and their position in terms of demographic transition.　**(9 marks)**

Table 10.16

Country	Crude birth rate/1000		Crude death rate/1000		Life expectancy at birth	
	1981	2014	1981	2014	1981	2014
Argentina	24	18	9	8	70	76
Bangladesh	43	20	14	5	54	72
Italy	11	8	10	10	74	83
Sudan	45	33	13	8	54	63
USA	16	13	9	8	74	79

Source: World Bank

8. 'The current strategies used to ensure food security for a growing global population are generally effective but will have to be adapted in the future as a result of environmental change.'

   How far do you agree with this statement?　**(20 marks)**

Chapter

11

Resource security

One of the major challenges resulting from continued population growth in the twenty-first century is the large-scale exploitation of natural resources. This chapter looks at the increasing demand for **energy**, **water** and **minerals** as they are all critical for human survival and development. The uneven distribution of resources is partly resolved by geographical transfers from areas of surplus to areas of shortage or demand. For water this is generally on a local or regional level but for energy and mineral resources it takes place on a global scale.

Understanding the nature of resource exploitation and demand requires an appreciation of the relationship between the human and physical environment, how the resources are used and why they are so important to people in different regions. The increasing scarcity of all three resources means that ensuring a secure and sustainable supply of each has become a major objective for most governments.

In this chapter you will examine the following ideas:
- concepts of resource development
- global patterns of availability, production and consumption of resources
- geopolitical issues concerning energy, water resources and mineral distribution and trade
- sources of energy, water and minerals and their relationship with the physical environment
- environmental impacts of resource development
- strategically managing the supply of and demand for resources
- sustainability issues associated with the exploitation, trade and consumption of energy, water and mineral resources.

11.1 Resource development

A **resource** is defined as any aspect of the natural environment that can be used to meet human needs.

As they are useful or necessary, natural resources such as fossil fuels or fresh water are also a material source of wealth. They have economic value and can be used to improve a country's wealth and economic development. Geographically, they are unevenly distributed around the Earth. Some nations have an abundance of certain resources but experience a scarcity of others that are equally important. For example, Saudi Arabia is an 'oil-rich' nation with around 25 per cent of the world's conventional oil **reserves**, but at the same time it has a **scarcity of water** supplies. It therefore has a greater problem ensuring water security than energy security.

Resource security is usually determined at a national level though it can be applied as a global concept. It refers to the ability of a country to safeguard a reliable and sustainable flow of resources to maintain the living standards of its population, while ensuring ongoing economic and social development. Resource security relies on a country being able to develop necessary resources from its own physical environment or otherwise being able to secure supplies from a trading partner. To provide their populations with all vital resources, most nations need to employ both strategies.

Key terms

Exploitation – The action of using natural resources to the fullest or for the most profitable use.

Exploration – The process of searching an area with the intention of finding and mapping natural resources.

Flow resources – Resources that are renewable and can be replaced, examples include fresh water and timber. They are commonly expressed in terms of the annual rates at which they are regenerated.

Reserves – That part of a resource that is available for use under existing economic and political conditions and with available technology.

Key terms

Resource frontier – A newly colonised region where resources have been discovered and are brought into production for the first time.

Resource peak – This marks the point in time when the maximum production rate of a resource occurs (at all scales from individual reserves to global) with production declining in subsequent years.

Stock resources – Non-renewable resources which can be permanently expended. Their quantity is expressed in absolute amounts rather than rates, for example, oil.

Classification of resources

Natural resources include a wide range of materials and organising them into groups helps us to understand the differences in how they are used.

Stock and flow resources

Stock resources are compound deposits of materials which have taken millions of years to form, usually on or in the Earth's crust. From an economic perspective stock resources have a 'fixed' and finite supply as once they are used they cannot be replenished. Their quantity can be measured in absolute amounts (for example, in tonnes).

Flow resources are those which can be naturally renewed within a sufficiently short time span to be relevant to decision makers. It is useful to make a further distinction between flows which are not dependent on human activity and those which are only renewable if human use remains at or below their capacity to reproduce or regenerate. The latter are known as **critical flow resources** and can be exhausted if not carefully managed.

This distinction between these different types of resources is best exemplified with energy resources:

Energy resources are classed as either renewable or non-renewable.

- **Non-renewable energy resources** (also known as finite or stock resources) are those that have been built up, or have evolved, over time. They cannot be used without depleting the stock because their rate of formation is so slow that it is meaningless in terms of human lifespan. Non-renewable energy resources are primarily fossil fuels (oil, natural gas and coal) but they also include uranium used in nuclear energy.

Figure 11.1 A wind farm near Wellingborough in Northamptonshire. Wind power is a flow resource and a renewable source of energy

- **Renewable energy resources** (also known as flow resources) yield a continuous flow that can be consumed in any given period of time without endangering future consumption, as long as current use does not exceed net renewal during the same period. Renewable energy resources include solar power, hydroelectric power, geothermal energy, wave and tidal power, wind power (Figure 11.1) and biomass sources. Renewable resources can be subdivided into:
  - **critical** – sustainable energy resources from forests, plants and other biomass; critical resources may be depleted by overuse, for example, if they are exploited at a faster rate than they are replaced, so they require prudent management
  - **non-critical** – 'everlasting' resources such as tides, waves, running water, wind and solar power.

Resource management involves controlling the exploitation and use of resources in relation to the associated economic and environmental costs. A key element of this is the concept of **sustainable development**, which requires a controlled system of resource management to ensure that the current level of exploitation does not compromise the ability of future generations to meet their needs.

Water is a renewable resource within the confines of the closed global hydrological system. However, water is effectively **finite** since only about one per cent of freshwater is easily available for human use, so it is also important to consider what limitless exploitation

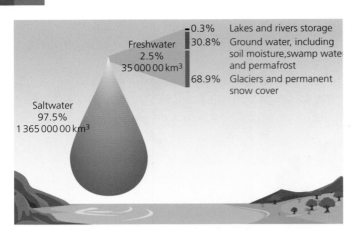

Figure 11.2 Distribution of Earth's water by store

of this resource might lead to. Water can only truly be considered renewable at any location when there is carefully controlled usage, treatment and release.

The majority of human uses require fresh water. Figure 11.2 indicates that of the 1.39 billion km³ of water on Earth, only about 2.5 per cent of this is fresh water (97.5 per cent is salt water) and 70 per cent of this is locked up in ice caps, glaciers and permanent snow. **Groundwater** represents about 90 per cent of the world's readily available freshwater resource so only a small fraction of the Earth's fresh water is available to us on the surface or in the atmosphere. Around 1.5 billion people depend upon groundwater for their drinking water but the supply is steadily decreasing. Depletion is most prominent in Asia and North America where rates of **abstraction** exceed the rates of **recharge**.

Key terms

Abstraction (water) – The removal of water available in the environment, either permanently or temporarily, from rivers, lakes, canals, reservoirs or from underground aquifers.

Groundwater – The water stored underground in the cracks and spaces in soil, sand and rock. It moves slowly through geologic formations called aquifers.

Recharge (groundwater) – A hydrological process involving the downward movement of water by infiltration and percolation causing the replenishment of natural groundwater. It occurs naturally but can be done artificially through human intervention and management.

Key question

Why are some renewable resources described as 'critical'?

Stock resource evaluation

Stock resources are mainly mineral deposits found in or on the Earth's crust. Mineral resources are a concentration of naturally occurring, solid, liquid or gaseous, inorganic or fossilised organic material, including precious metals and gemstones, mineral ores (such as iron and copper) and fossil fuels such as oil, gas and coal. Mineral deposits are potentially valuable, especially when discovered in sufficient quantity and of such grade or quality that reasonable prospects exist for their economic extraction.

There is a distinction between 'resources' and 'reserves'. The term 'resources' includes all deposits of mineral resources, whether undiscovered, discovered or discovered but unviable economically. Alternatively, 'reserves' are the parts of the resource base constrained by economic (and legal) viability and by being technically feasible to extract. The distinction between reserves and resources can easily become blurred because commodity prices are so volatile and technological advances are relatively rapid. Thus a combination of continued increases in price of a particular mineral and improved technology (which reduces costs) to access and exploit it, will convert more 'resources' into actual 'reserves'.

A McKelvey box or diagram (named after US geologist Vincent McKelvey) is a visual representation that helps to distinguish between resources and reserves and the differences that exist within these two broad categories. In Figure 11.3a the larger box represents the entire 'resource base' of a given mineral such as copper ore. The smaller cube within the resource base is the actual copper reserve which is determined by economic recoverability, based on access and cost against market price and the degree of geological assurance or certainty of its existence.

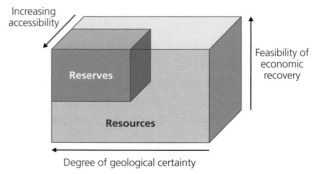

Figure 11.3a McKelvey box showing the distinction between resources and reserves based on geological assurance and economic feasibility

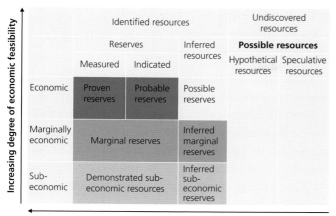

Figure 11.3b A McKelvey rectangular diagram is a useful visualisation for classifying resources by their degrees of geological assurance and economic recoverability

Figure 11.3b shows mineral resource classifications in relation to two distinct parameters. The first of these is the degree of **geological certainty** that a deposit actually exists, which is based on initial survey information about its characteristics including the quality of the deposit, the amount of material and its location and depth. The second parameter is the **profitability** of the deposit, which is based on estimates of the amount of money received for the material balanced against the costs of extraction. This can vary widely with upgrades in technology, shifts in world prices and levels of competition existing for that specific resource. Transport costs will also have a bearing on profitability.

Mineral resources and reserves are sub-divided, in order of increasing geological confidence, into **inferred**, **indicated** and **measured** categories. These categories are used as international standards to evaluate stock resources. Although there are slight variations in terminology between US, Canadian and Australian standards, these three large mining nations use the same categories based on the McKelvey concept. 'Inferred' mineral resources rate a lower level of confidence than that applied to 'indicated' mineral resources. An indicated mineral resource has a lower level of confidence than a 'measured' mineral resource.

Measured reserves

A 'measured reserve' is that part of a mineral resource for which quantity, grade or quality, densities and physical characteristics are so **well established** that they can be estimated with confidence. The quantity is computed from dimensions revealed in outcrops, workings and drill holes followed by detailed sampling to establish the grade or quality of the deposit. Inspection sites are so closely examined that the information gathered allows declaration of tonnages and grades with enough confidence to be called **'proven reserves'**.

A proven reserve is the economically mineable part of a measured mineral resource demonstrated by at least a preliminary feasibility study. They constitute mining assets with the greatest degree of confidence.

Indicated reserves

An 'indicated reserve' is that part of a mineral resource for which quantity, grade or quality, densities and physical characteristics, can be **estimated** with a level of confidence sufficient to allow further evaluation of the economic viability of the deposit. Estimates of quantity and grade are computed from similar information to that used for measured resources, but the inspection sites are spaced further apart, thus reducing the level of confidence that the reserve is continuously viable. The degree of assurance, although lower than that for measured reserves, is high enough to assume continuity between testing points and to establish a reliable tonnage grade and estimate.

An indicated reserve estimate is sufficient to support a preliminary feasibility study, which can serve as the basis for major development decisions and allows for conversion to **'probable reserve'**. Probable reserves constitute mining assets with a lesser degree of confidence than proven reserves. However, the viability is sufficiently supported by information on engineering operations and legal factors to justify further major expenditure on developing the resource.

Inferred resources

An 'inferred resource' is that part of a mineral resource for which quantity and grade or quality can only be estimated on the basis of limited geological sampling. Estimates are based on assumed continuity of the deposit but in areas beyond those which have been measured and tested, so are not verified by continuous samples or measurements. Due to the uncertainty attached to inferred resources, it cannot be assumed that they will be upgraded to an indicated or measured mineral resource as a result of continued exploration. There is insufficient information to declare tonnages or grades with any confidence and their viability for mining has not been validated and so they remain only as **'possible reserves'**. Possible reserves do not stand alone but are an extension of proven and probable reserves. They are, by definition,

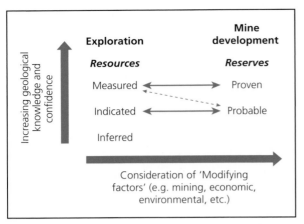

Figure 11.4 The relationship between exploration of resources and mine development of reserves

based to a large extent only on reasonable extrapolations, assumptions and interpretations of existing reserves.

Possible resources

The diagram in Figure 11.4 distinguishes between possible reserves and **possible resources.** Estimates of possible resources are based on broad geological knowledge of the existence of other, mostly undiscovered, deposits of the resource base and an expectation that these may become economically viable in the long term. There is less confidence of these resources becoming available and viable than for inferred resources and so the timelines for possible future production are much longer. They consist of:

- **Hypothetical resources**: undiscovered materials that are reasonably expected to exist in known mining regions under known geological conditions
- **Speculative resources**: undiscovered materials that may occur in known types of deposits in geological settings where no previous discoveries have yet been made.

Resource development over time

Security of supply of resources relates to a combination of **physical risks** and **geopolitical risks**. Physical risks include the accessibility of the resources available in an area or country. Geopolitical risks include the concentration of production in a relatively small number of countries, and the confidence any individual country has in trading with producers who will seek to exert their market power.

Physical risks depend upon:

- the quantity of the resource that has been found
- the quality of the resource in that reserve

- its physical location and accessibility
- the technology available to access the resource economically.

All of these factors need to be considered before deciding whether to exploit a resource in an area and only if it is deemed economically viable will production of the reserve go ahead. Advancing technology has a key role to play in the future availability of and access to more remote resources. Equally importantly, the price or value of the resource may increase, prompting the **exploration** for reserves in more remote locations, as they become more viable for production. Exploration for mineral deposits that have the potential to become an economic resource is time consuming and expensive, and unlocking the economic value of embedded mineral deposits is even more so. For example, the cost of building a modern ore mine starts at a minimum of $50 million and can easily reach $1 billion or more. Thus reserves can be classified as **recoverable** or **possible**.

- **Recoverable reserves** are the amounts of a resource likely to be extracted for commercial use within a certain time period and at a certain level of technology. (**Proven reserves** are those claimed to have at least a 90 per cent chance of being recoverable under existing conditions.)
- **Possible reserves** are deposits that are thought to exist because the geological terrain is similar to areas that have yielded comparable deposits but no exploration has yet taken place.

Cycle of natural resource development

The development of each resource will vary according to its type and location. Figure 11.5 gives the chronology in which resource development takes place. Organisations such as transnational corporations (TNCs) have to apply for licences from the appropriate authorities (for example in the UK, from the government's Department for Energy and Climate Change) to give them the right to explore for resources in any given area.

- Exploration of potential sites can take a number of years. For underground reserves, geologists are employed to use satellite imagery, conduct geophysical surveys and gather field data before they can assess the extent and potential quality of the reserve.
- Discoveries have to be fully **evaluated** to determine their viability for production given existing market conditions and costs of production. In developed

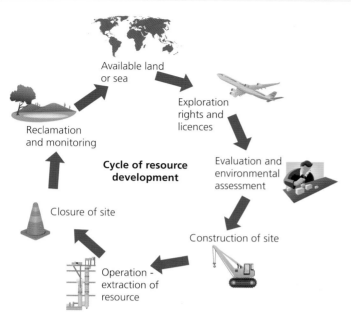

Figure 11.5 Cycle of resource development

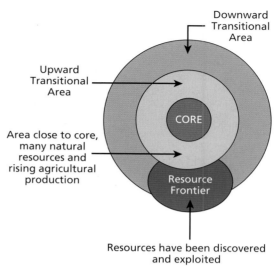

Figure 11.6 Friedman's core-periphery development model including the idea of a 'resource frontier'

countries a further assessment of the environmental impacts of the operations is usually required before extraction can take place.

- Once extraction licences have been awarded, the necessary infrastructure needs to be **constructed** to access the resource and start operations. This includes road building, laying pipelines (if transporting oil or gas is involved) and improving port facilities.

- Eventually production of the resource will commence and the reserve will continue to be **exploited** until all the viable reserves are recovered. The life span of the operation will vary depending on the type and quantity of the resource. (Exploration programmes will continue, to find further viable reserves in adjacent areas during operations; this will extend the life of the operation.)

- Closure of an operation does not necessarily mean that all the reserve has been exhausted. Any residual stock may be uneconomical to recover. In the case of valuable resources such as oil, secondary and tertiary recovery operations are put in place, for example using steam or detergent solutions under pressure to force oil to the surface.

- In developed countries, where environmental laws are more stringent, environmental management has become an essential part of resource extraction. Operating organisations are responsible for rectifying any environmental damage resulting from the resource development. This is not always the case for resource exploitation in developing countries.

Resource frontier

A **resource frontier** refers to an area where resources are brought into production for the first time.

In Friedman's core-periphery model (1963), he suggested a differential pattern of development based on a core region of a country, which attracts all the wealth and investment because of its advantageous location and existing resources. As development spreads from the core area, surrounding areas also start to benefit. In contrast, the periphery remains largely undeveloped, failing to attract investment and growth and lagging behind in standards of living and other development indicators. Within the periphery, however, is the possibility that resources such as minerals or fossil fuels will be discovered, prompting a rush of investment to the area, creating jobs directly or indirectly related to the development of the resource (Figure 11.6). Social conditions in resource frontiers may still be poorly developed but economically the region experiences rapid growth, with an influx of workers.

Resource frontiers can be identified at a variety of scales. For example, on a national level in the UK, Scotland could be identified as a 'peripheral' region far away from the core in southeast England. However, the development of North Sea oil reserves from the 1980s saw the northeast of Scotland become a resource frontier. The local economy boomed as it became a base for the oil industry and its associated technology and ancillary industries. Similarly, overcoming the transportation problems of crude oil in the Arctic region of North Alaska with the Trans-Alaskan oil pipeline caused Alaska to become a 'resource frontier' in the USA (Figure 11.7, page 540).

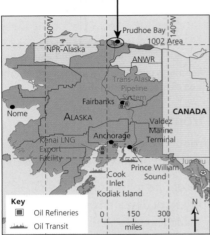

Figure 11.7 A resource frontier – the industrial petroleum extraction centre at Prudhoe Bay on north Alaska's Arctic Ocean. Its location is shown on the map

On a global scale, the depletion of resources in 'core' areas of high-income countries in North America and Europe prompted a search for mineral and fossil fuel reserves elsewhere. The increasing influence of energy and mining TNCs on the world stage, together with the process of globalisation, has led to the discovery of many new reserves of oil and gas in lower-income countries, particularly in Asia, Africa and Central America. Countries such as Azerbaijan, Chad and South Sudan (all with considerable oil reserves) have become 'resource frontiers' on a global scale.

Resource peak

Key terms

Conventional oil and gas – Refers to petroleum, or crude oil, and raw natural gas extracted from the ground by **conventional** means and methods.

Unconventional (oil and gas) reserves – Hydrocarbon reservoirs that have low permeability and porosity and so are difficult to produce. They require enhanced recovery techniques such as fracture stimulation. For example, shale deposits, tar sands and heavy oils.

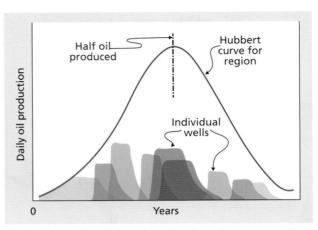

Figure 11.8 Hubbert's bell-shaped peak production curve for oil

The **resource peak** refers to the time of the maximum rate of production of a resource, either from a given reserve or for the resource as a whole. The idea developed from the concept of 'peak oil' originally suggested by Shell geophysicist M.K. Hubbert in 1956, and has since been applied to other finite energy resources and minerals. Hubbert accurately predicted that **conventional oil** production from the lower 48 mainland states (excluding Alaska) of the USA would peak between 1965 and 1970 and then enter a decline. His theory suggests that at any scale, from an individual oil-producing reserve to the planet as a whole, the **rate** of oil production tends to follow a bell-shaped curve (Figure 11.8). US mainland oil production followed Hubbert's predictions, actually peaking at 9.6 million barrels per day (mbpd) in 1971, and by the mid-2000s had fallen to 1940s levels. Despite advances in technology and the increasing value of oil, the US oil industry could not produce at its former rate because (with the exception of Alaska) the reserves were not available. More recently, the successful application of hydraulic fracturing ('**fracking**') technology to exploit **unconventional reserves** in shale rock has caused US oil production to rebound, producing 9.7 mbpd in April 2015.

Peak production curves for oil are also known as 'Hubbert curves' and can be plotted for all reserves as they are discovered. The concept of peak oil is sometimes confused with oil depletion; it is important to understand that peak oil is the point of maximum production, while depletion refers to a period of falling reserves. Viable production can usually continue for many years after peak production has been reached.

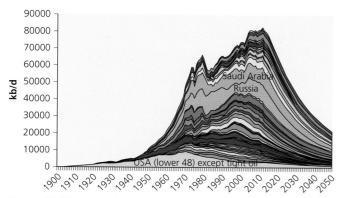

Figure 11.9 Peak production curves for different oil-producing regions

Reports on global fossil fuel peaks vary considerably because they are dependent on so many factors, especially those concerning levels of consumption. In 2009, the **International Energy Agency's (IEA)** World Economic Outlook report stated that oil and gas liquid outputs would continue to rise until peaking in 2030. However, until 2015, newly emerging economies such as China and India were pushing demand higher, which may shorten the IEA's predicted peak.

The **Association for the Study of Peak Oil and Gas (ASPO)** suggests that regular conventional oil reached a peak in 2005 (Figure 11.9). The **Organisation of Petroleum Exporting Countries' (OPEC)** production has been flat since then, although this could be because of voluntarily limiting output to keep prices high, rather than due to an enforced production decline. Other researchers also suggest that peak oil occurred globally between 2005 and 2011 and that oil has passed its maximum production rate and that both gas and coal will do so by 2030 to 2040. However, according to the US Energy Information Administration, unconventional supplies such as shale oil and gas and oil sands may extend the new peak oil to 2019.

Sustainable resource development

Resources are the keystone of every economy; when resources are used and processed they add to the wealth of present and future generations. However, the dimensions of our current resource use present major problems:

- the chance of future generations, particularly in low-income and developing countries, to have access to their fair share of scarce resources is endangered

- the consequences of our resource use in terms of environmental impacts will cause damage that goes beyond the carrying capacity of the planet
- these effects will be exacerbated once the developing world has attained levels of growth and resource use in line with the industrialised countries.

Sustainability can be achieved by different methods of managing resources, which can be broadly classified into two categories – supply side and demand side management. The two ideas are different approaches to the problems outlined above, but they are not in conflict. They are more likely to achieve success if they are used as complementary strategies.

Reducing depletion

Resource depletion is the consumption of a resource faster than it can be replenished and is mostly applied to our use of non-renewable fossil fuels such as coal, oil and gas and to critical renewable resources such as biomass. It equally applies to water consumption, especially in areas which rely on supplies from aquifers, when the rate of abstraction from these large underground stores of water exceeds the rate of recharge. The ability to reduce depletion depends on the nature of the resource but can generally be achieved by using demand management strategies as outlined in Table 11.1.

Table 11.1 Supply and demand management strategies for sustainable resource development

Supply side management	Demand side management

Involves seeking methods of increasing the supply of resources:
• increasing exploration efforts for existing non-renewable resources
• increasing research efforts to develop:
  – more sustainable alternative or substitute resources to replace unsustainable ones
  – new technologies that are more sustainable and cause less environmental impact
	Involves reducing consumption of resources, individually and at all other geographical scales:
	• changing individual behaviour and lifestyle to discourage wasteful and/or extravagant use of resources
	• developing technology to enable more efficient use of resources
	• recycling after use
	• reducing population growth with population control methods so there is less pressure on resources
	• regulatory controls and frameworks as part of global governance, for example:
	  – Agenda 21
	  – Kyoto Protocol

Minimising environmental impacts

Another key aspect of sustainable development is minimising the environmental impact of resource use. Technological advances have enabled a more sustainable approach to our polluting habits, such as catalytic converters on vehicles, reducing greenhouse gas emissions. Similarly, flue-gas desulphurisation (FGD) plant and carbon capture and storage (CCS) technology have reduced the emissions of sulphur and carbon into the atmosphere from power stations. There is still much to be done to minimise environmental harm but these and other technologies are lowering the risk of acid rain and reducing carbon emissions.

Seeking alternatives/substitutes

A whole array of renewable energy alternatives has evolved to take the place of non-renewable resources as they start to run out. However, commentators argue that the rate at which these are being introduced has not been sufficient to significantly reduce the use of fossil fuels. While non-renewable energy is still the main source of energy the negative impacts on the environment will continue.

Key question

What factors affect the rate of resource use?

Environmental Impact Assessment (EIA)

An EIA is a process of evaluating the likely **environmental consequences** (both beneficial and adverse) of a proposed project, such as resource development, prior to the decision to go ahead with the proposed action. The assessment also considers all interrelated socio-economic, cultural and human health impacts. In the UK, the EIA process derives from European law based on an EU Directive and it is now an integral feature of many large-scale planning applications that are likely to have environmental effects. EIAs have become a key aspect of development projects in many countries across the world and are implemented by planning or environmental authorities. The objective is to equip decision makers with an indication of the likely consequences of their decisions. In essence the EIA:

- assesses the impact of changing land use on the environment
- attempts to put environmental factors on an equal footing with economic factors
- quantifies potential environmental effects of land use change
- enables decisions to be taken with full knowledge of environmental consequences
- suggests modifications or alternatives to the proposal that would reduce its impact.

The main stages in the EIA process are:

1. Outline of the proposed development
2. Description of the existing environment including all biotic and abiotic features (humans, flora, fauna, air quality, soils, water, landscape and cultural heritage)
3. Assessment of the likely impacts of the development on the environment (often using a recording method called a **Leopold Matrix**)
4. Consideration of modifications to the development that would reduce the environmental impacts (remediation or mitigation)
5. Publication of an **environmental statement** which is communicated to the planning authority and in most cases is legally required to be made available to the public
6. A decision is made for or against the proposal – there is often the right to appeal against the decision by both those in favour of and those against the development.

EIA in relation to resource development projects

A typical application of EIA in relation to resource development projects would be for the open-pit (also known as open-cast) mining of coal for energy or for mineral ores such as copper. For example the assessment would consider a range of impacts caused by such a project including:

- **aesthetic problems:** degradation of landscape
- **air pollution:** dust and particulate matter (PM) from mining and vehicle movement
- **noise pollution:** from blasting and from heavy vehicles

- **water pollution and turbidity:** increased sediments in drainage water
- **dereliction:** abandonment of site when mine exhausted and closed.

The EIA would identify the magnitude and importance of each of these problems and suggest remedial action to mitigate against them so that the project could still go ahead. For example, landscaping embankments during the operations and restoring the site for appropriate agricultural, recreational or conservation use after final closure may be something the planning authority insists upon.

Lumwana Copper Project in northwest Zambia

An EIA was prepared for the Zambian government to assess the environmental impacts of the Lumwana Copper Project. In poor regions of less developed countries like Zambia, social and economic gains from investment and employment may have to be balanced against environmental damage, though remediation efforts to minimise environmental impacts may still be needed to tip the balance in favour of development in the **cost–benefit analysis.**

The developer was Equinox Copper Ventures (ECV), a Zambian-registered joint venture company and subsidiary of Equinox Minerals Ltd (a joint Canadian/Australian corporation). The method of extraction was to use conventional open-pit mining (Figure 11.10) in the Lumwana East River basin. The major social and environmental impacts were assessed and are shown in Table 11.2.

Figure 11.10 Long-haul mine truck in a dusty open-cast mine in Zambia

Table 11.2 Environmental impact assessment for Lumwana Copper Project, Zambia

Environmental impacts identified	Offsetting considerations or mitigation effects	Economic and social costs and benefits identified
Disturbance of nearly 9,000 ha of land	Despite large footprint, actual area where flora and fauna is impacted is relatively small	Beneficial multiplier effects in the local and regional economy
	Impacts reduced by progressive revegetation	Project will promote local suppliers and contractors providing services to the mine
Contamination of surface water from spillage and accidental release	Control of direct run-off from operational areas into local watercourses	
Seepage through the base of waste rock dumps and **tailings** could impact groundwater quality	Tailings and waste rock are non-acid forming so risk of acid rock drainage is low but will be monitored	
Radiation from plant effluent and metallurgical discharge	Radiation levels comply with Zambian standards and US Environmental Protection Agency limits	
	ECV committed to employing local people from Lumwana and Solwezi and will provide infrastructure including clinics, schools, power and water supply	Attracts workers from surrounding areas seeking employment putting the local population in competition with outsiders
Road traffic will increase. Heavy vehicles transporting concentrate will increase noise and dust pollution	Upgrading the highway and incorporating a pedestrian area adjacent to the road	
	Proactive approach through education and awareness raising will be used with the local population	Key health issues of the region are the prevalence of malaria and HIV/AIDs

Key terms

Cost-benefit analysis – A systematic analysis of the advantages and disadvantages likely to result from a development project, where an objective value is allocated to all economic, social and environmental aspects affected.

Tailings – Tailings consist of the slurry mix of crushed rock and effluents generated by the mechanical and chemical processes used to extract the desired mineral from the ore. (It produces a waste stream known as mine tailings.)

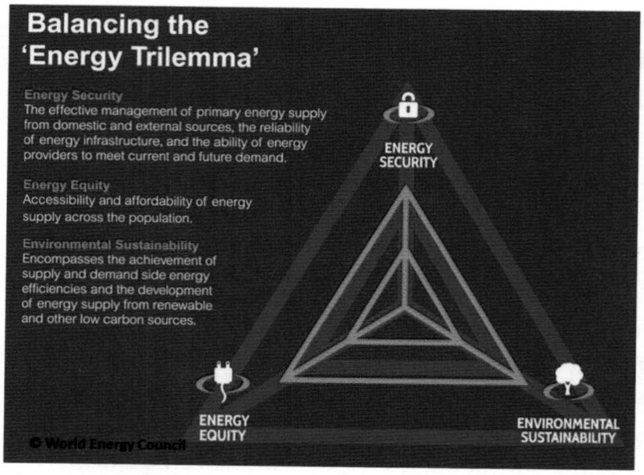

Balancing the 'Energy Trilemma'

Energy Security
The effective management of primary energy supply from domestic and external sources, the reliability of energy infrastructure, and the ability of energy providers to meet current and future demand.

Energy Equity
Accessibility and affordability of energy supply across the population.

Environmental Sustainability
Encompasses the achievement of supply and demand side energy efficiencies and the development of energy supply from renewable and other low carbon sources.

ENERGY SECURITY

ENERGY EQUITY

ENVIRONMENTAL SUSTAINABILITY

© World Energy Council

Figure 11.12 The World Energy Council's 'Energy Trilemma' outlines the key challenges facing world energy.

11.2 Natural resource issues – Energy

Energy is one of the main drivers of economic and social development. In the recent past, the main uncertainty concerning energy was the future availability and price of oil. With the development of unconventional oil reserves the oil endowment has quadrupled so that 'peak oil' is less of a concern, in the short term at least. The IEA predicts that global energy demand is set to grow by 37 per cent by 2040 (Figure 11.11). The majority of that demand will continue to be met by non-renewable energy sources, presenting an unsustainable picture of an energy system under pressure and in danger of falling short of the expectations placed upon it.

The global energy system faces the following strategic challenges in the twenty-first century:

● the continued depletion of non-renewable fossil fuels
● the growing risk of disruptions to energy supply
● the threat of environmental damage caused by energy use
● the persistence of energy poverty in less developed countries.

The **World Energy Council (WEC)** is a UN-accredited global energy body comprised of a network of energy leaders and practitioners from governments, private and public organisations, universities and NGOs. The WEC advocates that attaining sustainable energy for all can only be achieved by addressing three key issues. It calls this the **'Energy Trilemma'**: the triple challenge of providing **secure**, **affordable** and **environmentally sensitive** energy (Figure 11.12).

Energy resources – global patterns of production, consumption and trade

There is a marked **energy gap** between the rich and poor nations of the world.

● Around 2.5 billion of the world's population, living in low-income countries, have no electricity or other modern energy supplies and depend almost entirely on fuelwood or other biomass for their energy needs.
● In more developed countries oil provides the bedrock for modern life. Developed countries consume around 75 per cent of the total supply of the three major fossil fuels, though as China and India industrialise their consumption also continues to increase.

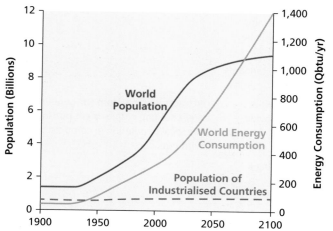

Figure 11.11 World population and energy demand projections

Source: US Department of Energy

The main energy resources produced, consumed and traded are fossil fuels, that is, coal, oil and gas.

- **Coal** is relatively abundant, though it is unlikely to grow in popularity as a future resource because it is so polluting. Carbon capture and storage and cleaner coal-burning technologies may reduce these concerns as the other fossil fuels run out.
- **Oil** industry experts predict that current reserves will only last for another 40 years or so and although **gas** supplies will last longer, they are both finite. Accessing unconventional 'tight reserves' is likely to extend the longevity of both gas and oil so there is now more uncertainty about predicting future supplies.

At current levels of global production and demand, there is enough oil in the world to last for at least a further 40 years, natural gas for 60 years, and coal for over 100 years. The geographical distribution of resources will also determine their future use as primary energy sources.

Patterns of energy production

Global energy supply is unevenly distributed. Some countries are '**energy rich**' because they have large reserves of energy resources such as coal, natural gas and oil. These three fuels constitute around 80 per cent of world energy production. Energy rich nations have the wealth and technology to exploit these reserves and many are also large consumers of their own energy resources, for example, China and the USA. As reserves of fossil fuels in richer countries go beyond peak production, two major trends have emerged that have influenced the pattern of production:

- more exploration and production of gas and oil reserves in less developed countries where governments have given exploration rights to TNCs, whose investment and expertise has helped resource development; for example in countries such as Indonesia, Chad, Angola and South Sudan
- increases in oil and gas prices, together with advances in technology, have prompted the development of **unconventional resources**, including tight reserves from shale rock in North America and Europe and oil sands in Canada and Venezuela.

Other countries produce little energy because they lack resources (even developed nations such as Japan and Spain have very few fossil fuel reserves) or are unable to secure the investment needed to develop the resources available. In some cases conflict or political instability prevents resource development and production.

Skills focus

Using Table 11.3 calculate proportions and draw three pie charts for coal, oil and gas to show the top ten producing countries of each.

Coal production patterns

- Coal production is scattered across the globe due to its relatively common availability (Figure 11.13, page 546).
- It is sometimes called the 'fuel of the past' because it is the most polluting of the fossil fuels. Despite this, production in 2013 was 2.5 times greater than in 1973.
- There has been a movement away from deep mining; the bulk of coal production is now from open cast mining.
- Geographically, global production of coal is dominated by China with 45.5 per cent. The USA is the second

Table 11.3 Top producing countries of the three main fossil fuels in 2012 *(measured in millions of tonnes of oil equivalent – Mtoes)*

Top coal producers 2012	Mtoes	Top oil producers 2012	Mtoes	Top natural gas producers 2012	Mtoes
China	1,890	Saudi Arabia	559	USA	559
USA	495	Russia	523	Russia	541
India	261	USA	407	Qatar	141
Indonesia	256	China	208	Iran	132
Australia	240	Canada	186	Canada	130
Russia	201	Iran	168	Norway	97
South Africa	146	Venezuela	165	China	90
Columbia	58	Kuwait	161	Algeria	73
Poland	58	Mexico	152	Indonesia	67
Kazakhstan	53	UAE	150	Saudi Arabia	66

Source: International Energy Agency

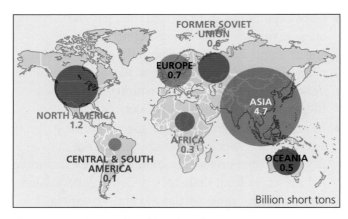

Figure 11.13 Pattern of world coal production, 2010
Source: US Energy Information Administration

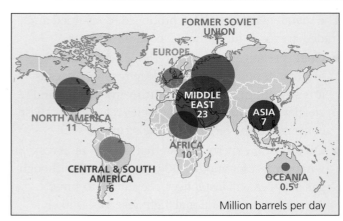

Figure 11.14a Global patterns of oil production, 2010
Source: US Energy Information Administration

largest producer (11.6 per cent) followed by India, Indonesia and Australia. These five countries together account for more than 75 per cent of world production.

- Coal extraction is less dependent on location than on a country's development and need for a coal-based energy strategy.
- The largest producers consume much of their own coal for power generation (China 57 per cent, USA 92 per cent, India 71 per cent and Australia 90 per cent).

Oil production patterns

- The global pattern of oil production is less dispersed as, geographically, oil reserves are more limited than coal (Figure 11.14a).
- Production is concentrated, with over 70 per cent of the world's oil production originating from the Middle East, USA and Russia
- Due to large reserves, Middle East countries contribute over 25 per cent of the world's production.
- Less conventional reserves such as oil sands and heavy oil in Canada and Venezuela have enabled these countries to become important producers.
- Geological factors influence the global pattern of oil supply. Extraction tends to be more viable in Middle Eastern countries which have large oilfields, such as Saudi Arabia, compared to relatively smaller reserves found in the North Sea.
- Other factors affect production, for example oil extraction in the physical environment of the Middle East is cheaper when compared to the harsh cold climates found in Alaska, where oil is deep below preserved permafrost layers. Conversely, the Middle East has a more politically volatile environment than other oil producing regions.

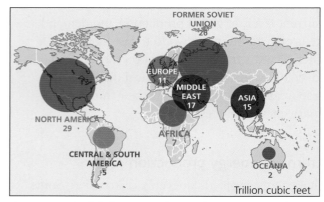

Figure 11.14b Global patterns of natural gas production, 2010
Source: US Energy Information Administration

Natural gas production patterns

- Natural gas has become an important energy resource within the last 50 years (Figure 11.14b). Previously, technology was insufficiently advanced to transport and store natural gas
- In 2013 world natural gas production reached a new record of almost 3,500 billion cubic metres
- The largest natural gas reserves are found in Russia.
- Production is dominated by a few countries, with the top five gas producers (see Table 11.3, page 545) representing over half of the global gas production.

Nuclear energy

The primary energy source for nuclear fission is the chemical element uranium. Mining uranium is dependent on the quality and quantity of discovered ores. Larger concentrations of higher grade ores tend to occur in continental shield areas of the Earth. The five largest producers of uranium in 2013 were Kazakhstan (38 per cent), Canada (16 per cent),

Australia (11 per cent), Niger (8 per cent) and Namibia (7 per cent).

Electricity is produced from uranium using nuclear fission. The highest producers are wealthy countries which lack fossil fuel reserves but have highly developed technology and are willing to invest in nuclear power for their domestic energy security. France and Japan are leading producers of nuclear power, together responsible for generating one quarter of the world's total production.

Renewable energy

The final category of production comes from renewable energy sources such as solar, wind, hydroelectric power (HEP), bio-fuels and geothermal energy, which together account for around 15 to 20 per cent of world energy production.

- Renewable energy sources can be established in many countries so the distribution of production is scattered globally.
- China is the largest producer of renewable energy (largely from HEP); India is the second largest using a mixture of wind and solar in urban areas, while rural communities still largely use biofuels. Brazil, using HEP and bioethanol to produce renewable energy, and Nigeria, using solar and biomass, are also in the top five producers.
- Around 65 per cent of the world's renewable energy is produced in low-income countries from solid biofuels (mainly fuelwood) and other forms of biomass.
- Renewable energy in general is found in countries which lack the natural resources, suitable conditions, technology and money needed to invest in finite energy production.

- Significant developments in solar and wind energy, due to supportive policies by national governments in developed countries, have raised the importance of renewable energy production, especially for electricity generation.

Patterns of energy consumption

Global energy use is increasing by approximately 2 per cent per year (although this rate of growth is starting to slow down) while population growth is about 1.1 per cent. The increase in energy consumption is therefore not due to population growth alone. The amount of energy used per person globally is increasing 'on average' by about 10 per cent per year.

Skills focus

Using Table 11.4 draw a comparative bar graph showing the changes in per capita energy consumption between 2000 and 2014 for each country.

Table 11.4 shows that emerging economies are increasing their consumption of energy more rapidly than developed nations, where the trend is towards declining energy usage in terms of both per capita and total consumption (even in countries with growing populations such as the UK and USA). Consumption in high-income economies has stabilised because of more energy conservation, energy efficient technologies and because there are fewer energy intensive industries. Energy producing countries such as Russia, Saudi Arabia and Iran are also large consumers of energy (Figure 11.15, page 548).

Table 11.4 Energy use by representative countries including growth

Country	Per capita energy use kg oil equivalent/capita (kgoe/capita)			Population (millions)			Total energy use (millions tonnes oil equivalent) (Mtoes)		
	2000	2014	Growth	2000	2014	Growth	2000	2014	Growth
Brazil	1,074	1,418	32%	175	202	15%	188	306	63%
China	920	2,143	132%	1,263	1,351	7%	1,161	3,034	161%
India	438	637	45%	1,042	1,237	19%	453	872	92%
Russia	4,224	5,283	25%	147	143	−3%	619	751	21%
Saudi Arabia	4,858	7,079	46%	20	28	40%	99	217	119%
Iran	1,866	2,873	54%	66	76	15%	123	238	93%
Nigeria	700	792	13%	123	169	37%	86	131	52%
Japan	4,092	3,546	−13%	127	127	0	519	437	−16%
UK	3,786	3,018	−20%	59	64	8%	223	178	−20%
USA	8,057	6,815	−15%	282	314	11%	2,269	2,224	−2%

Source: World Energy Council

Table 11.5 Reasons for changing patterns in energy consumption

High-income developed countries	Developing and emerging economies
High standards of living demanding more electrical and electronic appliances	**Rapid industrialisation** – as manufacturing moves to lower-cost economies – industrial development relies on energy supplies
High rates of **vehicle use**	**Urbanisation** – as more people live in urban areas, they have greater access to energy supplies
Growth has levelled because appliances and vehicles are becoming more **energy efficient**	**Rising incomes** – as populations become more affluent they aspire to the same living standards enjoyed in richer countries, so they use more appliances and vehicles
Heavy **energy-intensive industries are in decline** because of global shifts in manufacturing to developing countries	**Energy producing countries** are using their energy wealth as a trigger for development and as more reserves are exploited, some production is used by the home market
Population growth is more stable (this does not affect per capita figures)	
Increased **awareness** of environmental impacts of energy use (campaigns)	
Better technologies and incentives for **energy conservation**	

Consumption of fossil fuels

Coal: As well as being the leading producer, China dominates consumption of coal and together with the USA, India and Russia, these four countries account for 75 per cent of global coal consumption.

Oil: The USA is by far the largest oil consumer, followed by China, Japan, India and Russia. Oil's share of global primary energy consumption has declined as it has been displaced by natural gas and nuclear for electricity production. However, it still accounts for over 90 per cent of energy consumed in the transport sector.

Natural gas: The consumption of gas has tripled over the past 40 years and it now accounts for one-fifth of global energy consumption. This dramatic increase in gas consumption is a result of:

- large discoveries in many countries
- development of long gas pipeline connections
- development of liquefied natural gas (LNG) plants and carriers to transport it
- environmental issues (gas is less polluting; it emits only half of the carbon dioxide of coal)
- development of gas **combined cycle technology** for more efficient electricity generation.

Many countries have now developed a gas infrastructure but transport and storage difficulties mean that less gas is traded internationally than oil. The largest consumers in rank order are the USA, Russia, Iran, China, Japan and Canada, and together they account for half of world consumption.

Global trade in energy

Countries initially try to achieve energy security with their own sources, but many countries consume much more energy than they can produce so they have to import (for example, Japan and Spain). Conversely, some countries produce more energy than they need and for many low-income countries this provides a much needed source of foreign exchange.

By far the most traded energy resources are fossil fuels because they are needed by many countries for electricity generation and transport. Oil is the most traded because of the mismatch between major areas of production and areas of most consumption, so the fuel will travel vast distances to reach its consumers. These distances create many challenges:

- the environmental risks of long-distance pipelines
- oil-related problems such as oil spills at sea
- issues related to political instability in the Middle East.

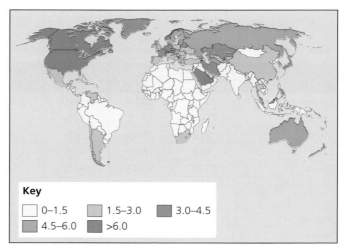

Figure 11.15 Global pattern of primary energy consumption per capita in 2014 (measured in tonnes of oil equivalent per person)

Source: BP

The largest net exporters of oil are Saudi Arabia, Russia, Kuwait, Venezuela and the United Arab Emirates. The recent lifting of trade sanctions by Western countries is likely to see the return of Iran as a major oil exporter. The largest net importers are the USA followed by China, Japan, India and South Korea. Even though they are major producers, the USA and China still rely on oil imports.

Gas exportation is hindered by the need to build pipelines, which requires huge investments and can also damage the environment. Location in proximity to large consuming economies means that gas pipelines are feasible to supply a demand. Russia is the largest exporter, able to export via pipeline to both Europe and to China, South Korea and Japan. Norway and

Algeria are also major exporters to the EU as is Canada to the USA. The main importers are Japan, Germany, Italy and South Korea.

As coal is a low value, bulky fuel, it is less economically viable for transportation so only a relatively small proportion of the coal produced enters the world market. Some European countries, such as the UK and Germany, import coal as they are still reliant on it for electricity generation.

Renewable energy resources are not usually traded as they are provided *in situ* within the country of production, though the electricity produced from renewable sources can be exported.

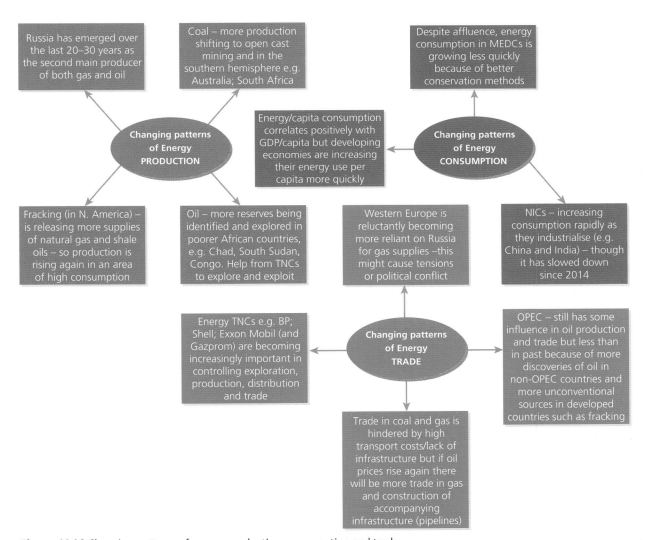

Figure 11.16 Changing patterns of energy production, consumption and trade

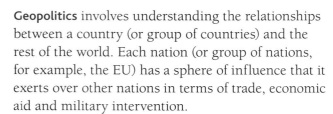

Key questions

How are the patterns of production, consumption and trade of energy related?

How and why are more unconventional reserves of oil and gas being developed?

Geopolitics involves understanding the relationships between a country (or group of countries) and the rest of the world. Each nation (or group of nations, for example, the EU) has a sphere of influence that it exerts over other nations in terms of trade, economic aid and military intervention.

From the 1990s the political alliances of countries that were based on the Cold War divisions started to dissolve, though some are still rooted in the old East/West divide. The geopolitics of energy created its own alliances, for example OPEC is an economic alliance of countries with a surplus of oil that they are able to export to developed countries which need it more. OPEC still has a major influence on the trade in oil as the organisation is responsible for about 50 per cent of crude oil exports.

Geopolitical issues – Energy

Key term

Geopolitics – The study of international relations, as influenced by geographical factors.

Organisation of Petroleum Exporting Countries (OPEC)

OPEC is a good example of a geopolitical relationship between countries. The organisation's 12 member countries (Algeria, Angola, Ecuador, Iran, Iraq, Kuwait, Libya, Nigeria, Qatar, Saudi Arabia, United Arab Emirates and Venezuela) are all developing countries located in the Middle East, Africa and South America.

OPEC operates as a 'cartel' by attempting to control supply to the world market in order to influence the price of oil. This requires co-operation between them as they jointly attempt to withhold supplies in order to keep oil prices high. Twice in the 1970s OPEC restricted the supply of oil on the world market to cause huge price hikes in oil and its derivatives. This strategy reduces the use and depletion of oil reserves and thus inadvertently increases sustainability.

The development of oil resources in many non-OPEC countries means that the cartel has less control of world markets than it used to but its influence is still strong (Figure 11.17). For example, the fall in oil prices in 2014–15 has been attributed to both the increase in production in North America because of fracking, and also a fall in demand as economic growth in China starts to slow down. However, another factor is that OPEC has released more oil supplies onto the world market in an attempt to undercut US unconventional oil production (which is more costly), therefore making it less viable to produce.

OPEC has not always been a united organisation and is influenced by external geopolitics, with members having different political relationships outside of the group. For example, Saudi Arabia and Kuwait have good trade relationships with the West (for example, USA and Western Europe) while Venezuela, Algeria and Iran look instead towards former communist states of the East for support and trade.

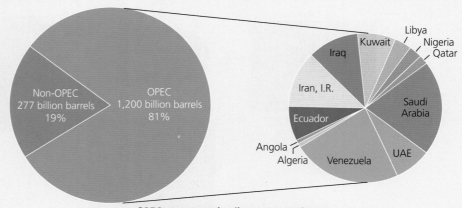

OPEC proven crude oil reserves, end 2012
(billion barrels, OPEC Share)

Venezuela	297.7	24.8%	Iraq	140.3	11.7%	Libya	48.5	4.0%	Algeria	12.2	1.0%
Saudi Arabia	265.9	22.1%	Kuwait	101.5	8.5%	Nigeria	37.1	3.1%	Angola	9.1	0.8%
Iran, I. R.	157.3	13.1%	United Arab Emirates	97.3	8.1%	Qatar	25.2	2.1%	Ecuador	8.2	0.7%

Figure 11.17 OPEC share of world proven crude oil reserves, 2012

Energy co-operation and conflict

Good political relationships concerning energy will lead to co-operation between countries in order to secure energy supplies for the future. There are also a number of potential threats and conflicts regarding the future security of energy supplies:

- Saudi Arabia at present has strong links with the USA and other Western countries, but this could change with leadership transition, economic reform and the threat of terrorism.

- The Arab Spring uprisings that have occurred in the Middle East and North Africa (MENA) region since 2010 have led to differing outcomes for individual oil producing states, such as Libya, which are likely to affect future supplies.

- In Iraq, the control of oil facilities by the extremist militant group, the self-proclaimed Islamic State (of Iraq and Syria) (ISIS) as well as internal economic problems, continues to disrupt supplies.

- Iran's relationship with the Western world, particularly with the USA, has been extremely difficult since the Islamic revolution in 1979. The UN Security Council, the EU and several other countries imposed trade and other sanctions on Iran in 2006 because of concerns that it was developing a military nuclear capability. These sanctions stopped oil trade between Iran and the West but an international agreement, signed in July 2015, promises to end these sanctions.

- Russia's President Putin has warned the West that its energy dependence on Russia could create obstacles if Moscow is not treated as an equal partner in future trade agreements. However, trade relations between Russia and Europe are seriously threatened because of Russia's support for separatists in Eastern Ukraine and its annexation of Crimea from Ukraine. A further escalation of the conflict in the Ukraine as well as Russia's involvement in Syria may lead to more sanctions and could put Europe's gas supply at risk. More European countries are considering fracking as an alternative to increase gas supplies so they are less reliant on imports from Russia.

Political developments in the Middle East and in the Ukraine are of great importance to the energy security of the rest of the world and the West has an obvious interest in supporting stability in both regions.

Key question

Who are the main 'players' in the geopolitics of energy?

11.3 Natural resource issues – Mineral ores

An ore is a type of rock containing minerals with a high concentration of a certain element, typically a metal, which can be extracted from the rock. For example cassiterite is an ore of tin and sphalerite is an ore of zinc. The ores are extracted from the earth by either underground or open-pit mining; they are then refined, often via smelting at high temperatures, to extract the valuable metal. Metals can be subdivided into a number of categories including:

- **Abundant metals:** such as iron, aluminium, magnesium and silicon
- **Scarce ferroalloy metals:** such as chromium, nickel and tungsten
- **Scarce base metals:** such as copper, lead, tin and zinc
- **Scarce precious metals:** such as gold, silver and platinum.

A number of key issues affect the geography and economics of mineral ores. These are interrelated and based on their value as stock and as finite resources. The key issues are:

- **Finite nature:** As we use more, so stocks will decrease.
- **Price volatility:** Scarcity influences price but so does fluctuating supply and demand for specific minerals; this can be harmful to development in emerging economies reliant on ore exports.
- **Geopolitics:** Ores can be so valuable that they can influence relationships between countries.
- **Globalisation:** Its influence is seen in the increased movement and trade of minerals and particularly the role of large transnational mining corporations.
- **Environmental impacts:** The exploration and exploitation of more mineral deposits will inevitably lead to greater impacts on the global environment.

Global patterns of production, consumption and trade

The global distribution of mineral deposits and the patterns of production and consumption of metal commodities are very uneven. For example, although the USA only contains around 5 per cent of the world's population, it is thought to consume around 20–30 per cent of the world's mineral resources. This demand is partly met by domestic production but most of their minerals are now imported.

Industrial societies are very dependent on mineral resources to produce finished goods such as appliances and vehicles, and to provide services such as gas and electricity supply. Some of these are used directly by individual consumers but some uses are more indirect.

Global patterns of mineral ore production

Production of many mineral ores has shifted significantly in the past 50 years as most of the developed nations such as the UK and USA have depleted their own reserves. This shift in production of ores has been to developing parts of the world, where it has enabled some nations to become important emerging economies and middle-income countries.

The gap between consumption and production of iron, steel and base metals (aluminium, copper, lead and zinc) has widened in the developed countries of North America, Europe, Japan and South Korea. More metals are consumed than extracted in these regions. Despite being a major producer of many metals, China has exhibited a similar trend during its industrialisation of the last 20 years, with rapidly growing consumption overtaking their extraction and production of metals.

Countries	Production (in million metric tonnes)	Countries	Production (in million metric tonnes)
China	900	Canada	35
Australia	420	Iran	33
Brazil	370	Sweden	25
India	260	Kazakhstan	22
Russia	100	Venezuela	16
Ukraine	72	Mexico	12
South Africa	55	Mauritania	10
United States	49	Other countries	50

Figure 11.18 World major producers of iron ore 2011

The largest reserves of many minerals are based in some of the larger countries and, of the developed countries, Australia and Canada continue to be large producers of metal ores (Figure 11.18). They have extended production overseas to developing countries, so many of the large mining corporations have Canadian and Australian bases. Much of the rest of the gap between consumption and production is being filled by emerging economies, particularly South American countries such as Brazil, Chile and Peru. India and Southeast Asian countries such as Indonesia and the Philippines are also major producers of some mineral ores. Finally, developing countries in central and southern Africa such as Guinea, Democratic Republic of Congo, South Africa and Zambia are also contributing to production to satisfy the growing demand. These developing regions produce more mineral resources than they consume. Much of this is achieved from foreign investment in their mineral resource base but this can lead to 'mineral booms' and the so-called 'resource curse' which do not always serve these economies well in the long term.

Table 11.6 List of top producing countries of metal mineral ores

Mineral ore	Uses	World measured reserves (thousand metric tonnes)	Top four producing nations
Bauxite	ore of aluminium	21,559,000	Australia, China, Brazil, Guinea
Chromium	alloys, stainless steel, electroplating	418,900	South Africa, India, Kazakhstan, Turkey
Copper	alloys, pipes, wire	321,000	Chile, China, Peru, USA
Iron ore	steel	64,648,000	China, Australia, Brazil, India
Lead	alloys, batteries	70,400	China, Australia, USA, Peru
Nickel	alloys, stainless steel	48,660	Canada, Philippines, Russia, Norway
Tin	alloys, containers	5,930	China, Indonesia, Peru, Bolivia
Zinc	steel alloys, galvanised metals	143,910	Australia, China, Peru, USA

Smelter and refinery production of metals remains located mainly in developed countries, although this balance has started to change with the growth of Chinese production of refined copper and aluminium.

Consideration should also be given to the contribution made by secondary resources, mainly recycled metals, to the overall production of new materials. This element of production is undertaken mainly in industrialised nations where discarded capital and consumer products are recycled to their metal base. This has increased in recent years due to better recycling methods and processes but will always lag behind the increasing demand for new resources.

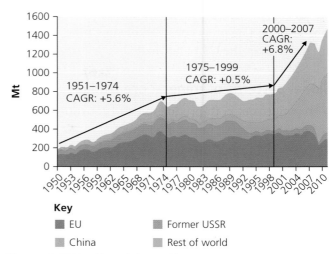

Figure 11.19 World steel demand by region 1950–2010

Global patterns of mineral ore consumption

A US Geological Survey (USGS) report studying the global consumption of metals since the 1970s revealed that aluminium demonstrated the most dramatic increase with a threefold expansion. Copper and zinc consumption doubled, while lead consumption increased the least by around 50 per cent. The consumption of ferrous metals used to produce steel increased slowly until the late 1990s from when it increased rapidly, largely due to the increased demand fuelled by China's industrialisation.

During the twentieth century, the US, Europe and Japan were the largest consumers of metals and minerals; the growth in consumption in these countries was relatively slow and intermittent, partly due to de-industrialisation in some regions. Around the turn of the twenty-first century, consumption patterns started to change with a marked shift towards Asia. By the early 2000s, China had overtaken the US as the major steel consumer and by the end of the decade was consuming 25 per cent of the world's steel. Growth in consumption was also high in South Korea and India, though the absolute quantities consumed were small compared to China (Figure 11.19).

The production, consumption and trade in mineral ores all played a crucial role in the rapid growth and industrialisation of the Chinese economy. More iron and steel was consumed as it undertook major construction programmes in urban areas and infrastructure projects such as the Three Gorges Dam. Industrialisation also demanded the development of reliable energy supplies and plans to have a national grid infrastructure for

electricity increased its consumption of copper from 1.5 million tonnes in 1999 to 6.5 million tonnes by 2010. By 2014, China's consumption of copper was 10 million tonnes, nearly half of the global demand (Figure 11.20). Aluminium consumption remains dominated by the US and the EU but, again, China's consumption has grown rapidly alongside its manufacturing industry. As China had become the main consumer of mineral ores globally, recent slowing of its economic growth has curbed demand for metals, steel in particular, which has caused repercussions in world markets.

The USGS report also examined consumption rates per capita, which provides a different picture than that shown by absolute consumption. Interestingly, on this measure, South Korea surpasses all other countries in the per capita consumption of both steel and copper

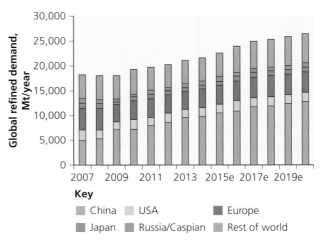

Figure 11.20 Copper demand by region 2007 onwards

and all except the US in aluminium. The per capita consumption rate grew considerably in China, but is still lower than the USA and the EU.

Global patterns of mineral ore trade

With patterns of both production and consumption shifting, the growth in the international mineral market between countries has been far from uniform over the past 40 years.

During the latter part of the twentieth century there was a general weakening in the international minerals market. Trade in the most widely used industrial minerals: iron and steel, aluminium, copper, nickel, tin, lead and zinc, saw a combination of supply surplus and a growth in demand that was less than the growth in total economic activity. This led to a marked downward trend in mineral prices. Two factors led to this decline in trade at the end of the twentieth century:

- reduction in military demand following the end of the Cold War
- recessions and decline in manufacturing in the USA, then later in Europe and Japan

Traditionally the industrialised economies of North America and Europe were significant mineral importers and provided the major market for consumption of metal ores. At the turn of the twenty-first century the geographical focus of trade shifted as markets moved away from the Atlantic towards Asia and the Pacific. China began to stand out as a leader in terms of the mineral ore trade, not only in the Asia–Pacific region, but on a global stage. The supply to demand balance for minerals in China is extremely vulnerable and the shortage experienced for some ores was met by importing large quantities from elsewhere in the world, for example iron ore from Australia, Brazil and India and other metals from parts of Africa. In 2009, against a backdrop of world recession, China increased imports of iron ore by 33 per cent, of copper by 14 per cent and aluminium by eight times in order to fill the gap in resource shortages.

The forecast trade flows shown in Figure 11.21 were made in 2012, at a time when Chinese demand for steel was expected to continue to grow. The forecast was accompanied by warnings that the Chinese economy might start to shift away from investment in

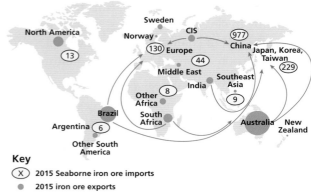

Figure 11.21 Forecast iron ore global trade flows for 2015
Source: BHP Billiton

infrastructure. Since 2014 the Chinese economy has started to slow down and in 2015 recorded a marked deceleration in growth at 6.9 per cent. This led to a rapid decline in the demand for minerals and some metals stockpiled in China. Although not shown on the forecast, China moved from being a net importer to a net exporter of iron and steel.

The combination of China's resource wealth, geographical position, large population and economic growth rates makes it the largest potential minerals market in the world. The direction of China's mineral trade is very significant for countries in the Asia–Pacific region and elsewhere in the world.

Trade in mineral ores is increasingly affected by a number of factors relating either to the demand side or supply side of the global market. These include:

- the inelasticity of demand for specific minerals needed for certain uses (determined largely by the availability of substitute materials)
- recession and decreasing growth rates which lead to falling demand and prices
- mining companies are reluctant to invest in new production with falling prices
- technological change has brought about cost reductions which have increased supply
- environmental concerns discourage new exploration and extraction but encourage recycling and alternative supplies.

Some of these factors are evident in the recent history of production, consumption and trade in rare earth elements (see textbox) which have been a rapidly growing part of the mineral ore market.

Rare earth elements (REEs)

Rare earth elements are a group of seventeen chemical elements that occur together in the periodic table. The group consists of yttrium, scandium and the 15 lanthanide elements. REEs are all metals and the group is often referred to as the 'rare earth metals'. They have many similar properties that often causes them to be found together in geological deposits. They are relatively abundant in the Earth's crust, but discovered viable reserves are less common than for most other ores.

REEs are used in many electronic devices such as TVs, computers, DVDs and mobile phones and also in catalytic converters, rare earth magnets and lighting. They are used in large quantities in batteries that power hybrid and electrical vehicles. Rare earth magnets are very strong and are used in alternative energy applications including wind turbines, making the turbines highly efficient. Consequently, there has been a rapid growth in demand for products that require rare earth metals.

Demand for REEs was ignited in the mid-1960s with the production of colour TVs. The US became the dominant producer over the next twenty years. In the late 1980s, China entered the market for REEs and began selling them at such low prices that US production was undermined and became inactive. By 2010, China controlled 95 per cent of production (Figure 11.22) but, in addition to being the world's largest producer of REEs, China is also the dominant consumer. In 2010 China started to significantly restrict exports of REEs to ensure a supply for domestic manufacturing. This move caused panic buying by manufacturers and some REE prices rose exponentially in just a few years. Japan and the United States are the second and third largest consumers of rare earth materials and together with the EU they complained to the World Trade Organization about China's restrictive trade policies.

China's restriction of exports shows the danger of the dominance of one producer in a market. It was an awakening for REE consumers and motivated mining companies in the US and other countries to re-evaluate old rare earth prospects and explore for new ones. Mines in Australia began producing rare earth oxides in 2011 and by 2013 they were supplying about 2–3 per cent of world production REE exploration and development is now being undertaken in Canada, South Africa, Thailand, Malawi and Sri Lanka; other reserves are found in Brazil, India, Vietnam and Malaysia.

High prices of REEs also prompted product manufacturers to do three things:

- seek ways to reduce the amount of REEs needed to produce each of their products
- seek alternative materials to use in place of REEs
- develop alternative products that do not require REEs.

In 2013, China still produced around 90 per cent of REEs but also started buying them from other countries, for example Australia.

The global demand for electric vehicles, consumer electronics and energy-efficient lighting is expected to rise rapidly over the next decade. REEs are also used in new developments in medical technology, such as surgical lasers and magnetic resonance imaging, so the demand for them should remain high.

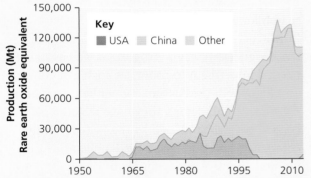

Figure 11.22 History of production of rare earth elements (REEs)

Geopolitics of mineral ores

The geopolitics of minerals arises from a number of issues including the following:

- the inter-dependency existing in the international trade of raw materials
- the potential for trade wars resulting from fluctuating prices of different minerals
- the potential future conflict concerning access to 'common' resources
- the issues arising from the environmental and socio-economic mine developments
- the role and dominance of large transnational mining corporations
- the role of China in the world's mining and metals industry (see trade in REEs in textbox).

International trade and development

Growth in exploration and mining in Africa, Latin America and parts of Asia has been spurred by:

- technological advances giving increasing viability to mining previously inaccessible deposits in remote, less developed regions

- the development of large ocean-going vessels able to carry large quantities of bulk mineral commodities such as iron ore and bauxite.

Trade in mineral ores is now largely one-directional, from developing to developed countries. It seems simplistic to assume that this will bring economic and then social benefits to the developing country as a result of foreign investment in the country's resources. This happens to an extent, but paradoxically, 'resource-rich' low-income countries are also subject to a phenomena known as the **resource curse** (sometimes referred to as the 'paradox of plenty'). The resource curse refers to when countries with an abundance of non-renewable resources, such as minerals, tend to have lower economic growth and development than resource-poor countries. This is thought to be due to a number of possible causes:

- the over-dependence on one resource which is subject to volatile global price fluctuations
- other sectors of the economy, particularly manufacturing, do not develop or prosper
- resource frontiers draw in skilled workers from other sectors so wages rise
- higher wages makes the national currency less competitive
- poor exchange rates make tradeable goods less competitive in world markets
- authoritarian rule and, additionally, government corruption results in unfair regulation of the industry and unfair distribution of the wealth gained from it
- control or partial control of the resources by large TNCs results in leakages.

It seems clear that many developing countries are not always maximising this newly-found wealth (this is true of many African nations). However, many experts believe that the resource curse is not universal or inevitable, so only affects certain types of countries under certain conditions.

Some developing countries have responded to this problem with specific measures. For example, in an attempt to increase the 'value' of their exports and encourage processing industries, Indonesia introduced new mining laws in 2009, prohibiting the export of unprocessed ores. This has resulted in falling supplies of both mined nickel and bauxite onto the world market since the ban. Production increases in other countries have been insufficient to compensate for Indonesia's reduced output, leading to a short-term price rise for these minerals. Similarly, it is possible that China's reluctance to sell rare earth elements (REEs) (see textbox) is a defence of their value-added manufacturing sector.

China's impact on mineral ores

The role of China in world trade and the geopolitics of the mineral ore trade cannot be ignored. This especially applies to the iron/steel and copper industries. China is well endowed with its own mineral resources including iron ore and copper, however rapid economic and industrial growth since the 1990s has meant that demand for metals has outstripped supply. Consequently, China has continued to increase domestic production as well as importing huge amounts from overseas. This has had three main geopolitical effects globally:

1. **Chinese investment in resource mining in developing countries**: The Chinese mining industry is state controlled and it has invested heavily in establishing mining operations in developing countries, especially in Africa. The search for and exploitation of resources has brought infrastructural benefits, such as restoration of the Benguela railway in Angola, as well as much needed investment to countries such as Botswana and Zambia in the African copper belt. In Zambia, with new technologies, they have restored old copper mines abandoned by the British mining companies and purchased the Baluba mine to maintain control of the supply of REEs.

2. **Trade with other developing countries**: Such is China's demand for steel that they have negotiated deals with other emerging economies, such as Brazil, to import iron ore. This has been worth millions of dollars to the Brazilian economy, though the recent slowdown in the Chinese economy has caused problems for Brazil because of an over-reliance on exports to China.

3. **Impact on developed countries**: The ability of China to manufacture large amounts of steel economically with cheap labour and state support has had major impacts on the global steel industry:
   - In the US steel belt of Ohio, American manufacturers of steel have moved their

investment to Chinese production resulting in the closure of many steelworks. These are the lifeblood of many communities, such as Youngstown, where economic decline has set in and rising unemployment levels are being recorded.

- The slowdown in the Chinese economy in the mid-2010s reduced demand for steel in China but they still had a large capacity to supply cheap steel to the world market. This undercut the ability of steel companies in developed countries to compete. For example in the UK, Tata Steel sold its UK operations and many large steelworks were closed or threatened with closure, causing increases in unemployment and devastation of communities.

Role of transnational corporations in mineral ore distribution

Mining became 'global' before most other branches of industry and the global distribution of mineral ores is still largely controlled by transnational mining corporations. They are competitors but will often work together to combine exploration efforts and develop mine operations. Most are multivariant in the materials they mine (in some cases this includes coal and oil). Table 11.7 shows a list of the major metal mining corporations and their operations.

These giants of the metals mining industry set criteria for the reserves that interest them, usually targeting projects that offer a lifespan of at least 20 years. They provide the expertise, technology and capital investment needed to explore and extract ores in developing countries and help them to develop their resources. They are usually responsible for:

- negotiating exploration rights and leasing land from national governments
- establishing subsidiaries or entering into joint ventures with mining companies from the host nation
- developing infrastructure such as access roads and nearby accommodation to support projects.

These corporations can bring economic benefits in terms of jobs and skills to local communities but they have also been criticised for negligence of the environmental and the social impacts of their operations. There is growing evidence that corporate social responsibility is now an important consideration for many of these companies, so mining operations are being managed more sensitively by these operators. However, there are still environmental impacts; in some instances, local disasters have devastated local communities as well as the environment.

Table 11.7 Top transnational and national mining corporations

Corporation	HQ/country of origin	Main mining operations
BHP Billiton	Australia	Operates some of the largest mines in the world, mainly South America and Australia; produces copper, iron ore, silver, aluminium, manganese, and nickel
Rio Tinto	UK/Australia	Mainly in Africa, South America and Australia; produces a wide range including iron ore, bauxite, copper, zinc (and coal)
Vale (CVRD)	Brazil	Primarily in Brazil; largest iron ore producer, second largest nickel producer; also copper, gold and platinum
Glencore Xtrata	Switzerland/UK	Operates globally producing copper, nickel, zinc/lead, aluminium and iron ore; dominates the mining industry but that includes revenues from coal and oil
Anglo-American	UK/USA	Mainly in Chile, Brazil and South Africa; produces nickel, platinum and iron ore, also coal and diamonds
Codelco	Chile	State-owned operators in Chile; largest copper producers with largest copper mines in the world
Freeport McMoRan	USA	Operate in North and South America, Africa and Indonesia; producers of copper, gold and cobalt
Grupo Mexico	Mexico	Operate mainly in Mexico; third largest copper producer in world (with the second largest mine in the world at Cananea)
Fortescue Metals Group	Australia	Iron ore in Pilbara region of Australia; exports large quantities to China
Barrick Gold	Canada	Operates in North and South America, Africa and Australia; largest gold producer; also produces copper

11.4 Natural resource issues – Water

In the global context, water is a renewable resource but, like energy, it will come under increasing pressure because of population growth. It is a vital resource for many aspects of life. In addition to providing drinking water, it is needed for irrigation to grow food, and in maintaining health and preventing disease through its use for hygiene and sanitation purposes. Water is also used in a variety of production processes, all of which contribute to its consumption (Figure 11.23).

A population increase of another two to three billion by 2050, combined with more industrialisation and urbanisation, will result in an increasing demand for water and will have serious consequences for the environment. Already there is more waste water generated and dispersed today than at any other time in the history of our planet.

The key issues concerning water security are:

- the variable access to clean, **safe**, **potable** or **improved drinking water**
- the need to conserve water supplies
- improving the supply and quality of drinking water
- reducing tensions and conflict caused by access to water supplies
- improving trans-boundary co-operation of shared water resources
- ensuring sufficient supply for food production
- reducing the environmental impacts of water use
- **water stress**.

Key terms

Improved drinking water – A source of water that by using intervening treatment is protected from outside contamination, particularly from faecal matter.

Potable (water) – Water that is suitable for drinking (it does not have to be completely pure but must not contain unacceptable levels of hazardous materials, nor look, taste or smell unpleasant).

Safe drinking water – Water that is safe for human consumption. The water must be free from harmful pollutants and bacteria that could make people ill.

Water stress – Water stress occurs when the demand for water exceeds the amount of water available during a certain period, or when poor quality of water restricts its availability for human use.

Water surplus (areas) – Areas where there is more than sufficient water available to meet human needs.

Key facts about water availability

- 770 million people around the world lack access to safe, clean water. That is approximately one in nine people: more than twice the population of the USA.
- 842,000 people die from a water-related disease each year – this can be caused by unsafe drinking water and/or inadequate sanitation and hand hygiene, all leading to severe diarrhoea.
- Water-borne diseases from faecal pollution are a major cause of illness in developing countries; every minute a child dies of a water-related disease.
- Polluted drinking water contributes to the death of 10 million children annually.
- 2.5 billion people lack access to basic sanitation (36 per cent of the world's population).
- Women and children around the world spend 140 million hours a day collecting water.

Source: Water.org and WHO

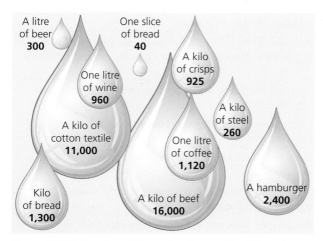

Figure 11.23 Number of litres of water used in different production processes

Even wealthier countries, which are more able to build reservoirs to store water, or afford borehole drills and pumps to tap into aquifers underground, will put more pressure on their sources of water as population increases. In the USA, the abstraction rates from the giant Ogallala aquifer, mainly used for irrigation in mid-west states, are already exceeding the recharge rates by 15 km^3 each year, which is not sustainable (Figure 11.24).

In April 2011, the Human Rights Council adopted a UN General Assembly resolution to recognise that having

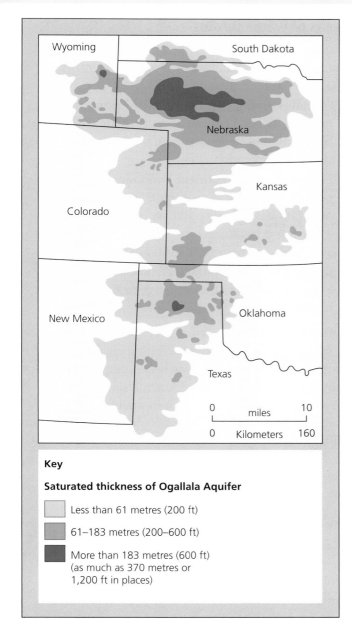

Figure 11.24 The Ogallala aquifer in the mid-west USA

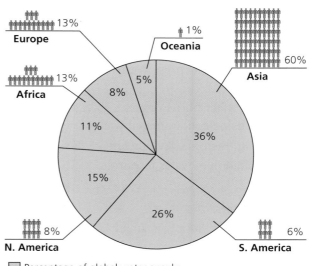

Percentage of global water supply
Percentage of global population

Note: Totals exceed 100 because figures have been rounded up.

Figure 11.25 Percentage of global water supply in comparison to population by continent

Table 11.8 Populations without access to safe water

Region	Population size without access to safe water
Developed countries in North America, Europe, Russia, Japan, Australia and New Zealand	9 million
South East and East Asia (including China) and Oceania	186 million
West, Central and South Asia (including India)	180 million
Africa	358 million
Latin America and the Caribbean	36 million
	Total = 769 million

Source: Water.org

Water resources – global patterns of availability and demand

Key question

Why does the supply of and demand for water vary on a global scale?

There is enough fresh water on the planet for seven billion people but it is distributed unevenly and too much of it is wasted, polluted or unsustainably managed.

The key pressures are again on poorer countries; a total of 450 million people in 29 of the poorest countries suffer from water shortages. The areas of the world most likely to suffer water shortages are those experiencing the most rapid population growth: the countries of sub-Saharan Africa. This growth will result in an increase in demand

access to sufficient water and sanitation is a human right. According to the World Health Organization (WHO), between **50 and 100 litres of water per person per day** are needed to ensure that most basic needs are met (including irrigation for food supply). Supplies must be safe, acceptable, affordable and physically accessible (the water source has to be within 1,000 m of the home and collection time should not exceed 30 minutes).

Most of the people categorised as lacking access to clean water use about five litres a day, one tenth of the average daily amount used in rich countries to flush toilets (Figure 11.25). Most people need at least two litres of safe water per day for drinking and for food preparation.

for water, for drinking for both humans and livestock, for irrigation to maintain food supplies and for hygiene to prevent the spread of disease. Shortages will result in:

- droughts which kill crops, livestock and eventually people
- increases in water-borne diseases such as cholera because of using contaminated water
- barriers to industrial development
- over-exploitation of water resources.

The availability of water is affected by both physical and human factors:

- **Physical factors:** Supply is largely determined by factors such as latitude, wind direction and proximity to coasts. These factors will influence rainfall and thus whether areas receive plentiful water supplies. Those regions receiving more than adequate rainfall to meet the needs of supporting the population are areas of **water surplus**.
- **Human factors:** Availability of water for the population is largely dependent on human intervention, such as treatment facilities to provide **improved drinking water** and the infrastructure to supply safe drinking water to a population through pipelines or wells (Figure 11.26).
  - Another key feature of water availability is the general discrepancy between urban and rural areas, especially in the poorest countries. 82 per cent of those who lack access to improved water live in rural areas, while just 18 per cent live in urban areas.
  - Availability may be adversely affected by human factors such as overuse, even in areas where nature provides a surplus.

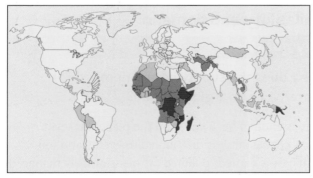

Key

Proportion of population (%)

■ <50   ■ 50–75   ■ 76–90   □ >90

▨ Not applicable   ▨ Data not available

Figure 11.26 Proportion of population using improved drinking water, 2011

Source: WHO

Areas of water surplus

Water surplus countries have certain geographical, water management and water usage characteristics that determine water availability:

- **Climates giving regular, plentiful rainfall** resulting in good run-off and stores such as aquifers, large bodies of freshwater sources (lakes, rivers, etc.) and low evaporation rates.
- **Effective water treatment and management** protects against water pollution and ensures premium water quality and supply. (Agreements for water resources shared with other countries are also an effective management strategy.)
- **Low water-usage characteristics,** for example, low population or efficient usage of available water supplies.

Countries with water surpluses are located in temperate or tropical areas, including much of South America, North America, Northern Europe (including Russia), South East Asia and Australasia, but there is not always a correlation with rainfall levels. For example, both Russia and the USA have areas with low rainfall but both have an overall water surplus. They have a surplus because of their ability to manage their water resources, but on a smaller scale both countries have regions with a water deficit.

Areas of water scarcity

Water scarcity is both a natural and a human-made phenomenon and already affects every continent. Around 1.2 billion people, or almost one-fifth of the world's population, live in areas of **physical water scarcity**, and 500 million more people are approaching this situation. Hydrologists assess scarcity by looking at the population–water equation. They offer a range of parameters for areas suffering from water deficit, as the degree of deficit varies.

Most countries with water scarcity are located in the Middle East and North Africa (MENA region), Central Asia and parts of the Indian sub-continent. Sub-Saharan Africa has the largest number of water-stressed countries of any region.

Arid regions are most associated with physical water scarcity and where this occurs the water resources cannot meet the demands of the population. There are, however, an increasing number of regions in the world where water scarcity is a situation brought about by human factors. The Colorado River basin in the USA

Table 11.9 Measures of water scarcity

Degree of scarcity	Characteristics	Example areas
Areas of water stress	Annual water supplies drop **below 1,700 m³/person**	India, parts of China,
	Demand for water exceeds supply over a period of time causing shortages	sub-Saharan African countries
Water scarcity	When annual water supplies drop below **1,000 m³/person**	South Africa, parts of Western Asia (Iraq)
'Absolute' water scarcity	When annual water supplies drop **below 500 m³/person**	North Africa–Algeria, Libya and Egypt Middle East–Saudi Arabia
Physical water scarcity	Arid areas receiving **less than 500 mm of rainfall annually**	MENA region and parts of North and Central America, parts of Australia, Central Asia and Northern China
	Also known as areas of water deficit	
	Where nature's provision of water is insufficient to meet population needs	
Economic water scarcity	Exists when a population does not have the monetary means to utilize an adequate supply of water	sub-Saharan Africa; parts of South America and South Asia
	Characterised by unequal distribution and poor infrastructure	

is an example of a seemingly abundant source of water being over abstracted and unsustainably managed, leading to serious physical water scarcity downstream.

There is an obvious correlation between areas of scarcity and rainfall levels but there are anomalies, with areas suffering physical water scarcity still being able to provide water to their population, for example, Australia. Conversely, some areas with an adequate natural supply of water are, for economic reasons, unable to fully utilise their sources because they cannot afford to abstract it, transport it or treat it if it is polluted. This is a situation known as **economic water scarcity** and it affects the poorest countries in the world, particularly much of sub-Saharan Africa, parts of South America and South Asia (Figure 11.27). Around 20 per cent of the world's population faces shortages because their countries lack the infrastructure to supply them.

Economic water scarcity is almost entirely due to a lack of investment and good governance. However, striving to meet the Millennium Development Goals (MDG) has meant a great deal has been achieved to alleviate the problem of water availability. The MDG 7 target, to halve the proportion of people without access to safe drinking water, was met five years ahead of schedule in 2010.

Pattern of demand for water

North America and Australia have the highest per capita water usage globally (Figure 11.28, page 562). They are both affluent regions of the world and therefore use water in a range of domestic appliances such as dishwashers, washing machines, etc. Personal hygiene levels are high, which also places a great demand on water withdrawals. Water is also abstracted for recreational purposes such as swimming pools, maintenance of golf courses, etc. Unlike much of Europe, the agricultural areas of the USA, Canada and Australia, are much drier and need more irrigation.

Sub-Saharan Africa has the lowest per capita usage, simply because rainfall is unreliable and they lack the infrastructure to supply water on tap. They have to be careful in using the resources they can access and this is used primarily for irrigation.

Geopolitics of water resource distributions

Key question

Why will water security be important in determining co-operation and conflict in the twenty-first century?

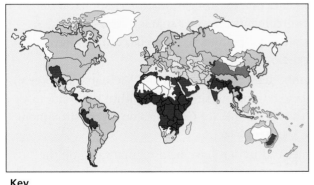

Key

- Little or no water scarcity
- Physical water scarcity
- Not estimated
- Economic water scarcity
- Approaching physical water scarcity

Figure 11.27 Areas of physical and economic water scarcity

Source: World Water Assessment Programme

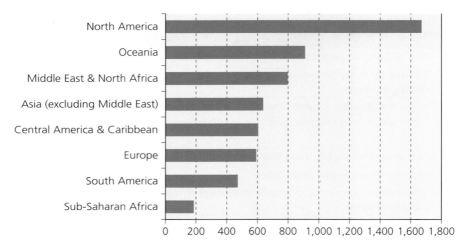

Figure 11.28 Per capita water consumption by region, 2000 (cubic metres per year)

Political negotiations that focus on conflicts over the shared use of water resources are sometimes called **'hydro-politics'**. As water resources become scarce, tensions among different users may intensify, both at the national and international level. Water is a less traded commodity than energy but the potential for conflict over access to water is in many ways more threatening and dangerous. This is largely because it is a resource which is shared by many countries in different regions and for which there is no alternative. For example:

- **276 river basins are transboundary** (shared by two or more countries). Without formal agreements, any changes made within a river basin can lead to transboundary tensions
- **200 transboundary aquifers** have also been identified and over-abstraction by one country can lead to tensions and conflict, particularly in areas prone to water scarcity
- **148 countries** have territory within at least one transboundary river basin (39 of these have more than 90 per cent of their territory within transboundary basins).

When water supplies are polluted, diverted, blocked or over-abstracted by one country, it can affect the life chances of those in neighbouring countries. If major projects such as dams or abstraction proceed without regional collaboration, they can heighten regional instability.

Potential for conflict

As the demand for water reaches the limits of supply in certain areas, conflicts may brew between nations that share freshwater reserves. The UN recognises the potential for up to 60 such conflict 'hotspots' globally;

more than 50 countries on 5 continents might be caught up in disputes unless agreements on how to share rivers, reservoirs and aquifers are made (Figure 11.29).

The main problem arises when one country unilaterally makes development plans without consulting those with whom it shares water resources. For example, such is China's demand for energy to fuel industrialisation that it has started building dams in the headwaters of some of Asia's major rivers, including the Indus, Ganges and Mekong, without agreements with downstream **riparians** (those with access rights to the river banks).

Improving transboundary cooperation

The Berlin Rules on Water Resources (2004) summarises international law customarily applied in modern times to freshwater resources. Instead of a drift towards war, water management can be viewed as an opportunity for regional economic development to be strengthened. There are around 450 transboundary agreements that exist regarding international waters. Legal agreements on water sharing have been negotiated and maintained even as conflicts have persisted over other issues.

- Cambodia, Laos, Thailand and Vietnam have been able to co-operate since 1957 within the framework of the Mekong River Commission, even throughout the Vietnam War.
- Israel and Jordan have held regular talks on the sharing of the Jordan River, even though they were until recently in a legal state of war.
- The Indus River Commission has survived two wars between India and Pakistan.
- In 1999, a framework for the Nile River Basin was agreed in order to fight poverty and spur economic

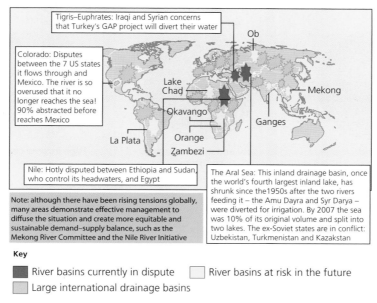

Figure 11.29 Present and potential water conflict hotspots

development in the region by promoting equitable use of common water resources. This was threatened by disagreement (now resolved) between Ethiopia and Egypt over the construction of the Great Ethiopian Renaissance Dam in the headwaters of the Blue Nile.

● The nine Niger River Basin countries have agreed on a framework for a similar partnership.

The cases demonstrate the need for institutions such as the UN to develop a process of engagement with countries to seek international water resources co-operation.

11.5 Water security

Water security is defined as the capacity of a population to safeguard sustainable access to adequate quantities of acceptable quality water for:

● sustaining livelihoods
● human well-being and socio-economic development
● ensuring protection against water-borne pollution and water-related disasters
● preserving ecosystems in a climate of peace and political stability.

Source: UN Water, 2013

Providing stable freshwater supplies is a priority for every country in the world. Yet stable supplies are increasingly hard to come by in many countries, as water-related risks increase. By 2025, 1.8 billion people will be living in countries or regions with absolute water scarcity, and two-thirds of the world's population could be living under water-stressed conditions due to overuse.

Sources of water

Key question

What physical conditions need to exist to supply reliable sources of water?

The water used to satisfy these demands can be obtained from the following sources:

● **Surface water supplies** – rivers and reservoirs
● **Underground stores** – groundwater aquifers
● **Seawater** (after desalination).

Components of water demand

The major uses and demand for water can be categorised into the following three main sectors:

● **Agricultural use:** Mainly irrigation of crops; also used in cleaning/washing and watering livestock.
● **Industrial and commercial use:** Used as a coolant in electricity generation; also used in heating, steam turbines, transport and various industrial processes especially textiles, paper and fibre manufacture, metal processing, food processing and the construction industry.
● **Domestic/household use:** Known as **public or municipal water** supply and used for drinking and food preparation, personal hygiene, sanitation and washing/cleaning.

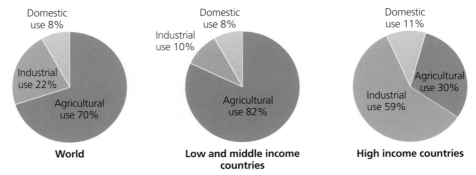

Figure 11.30 Variations in use of water in different parts of the world

Globally the world uses fresh water in the following proportions:

- **Agriculture (mainly irrigation) = 70 per cent**
- **Industry = 22 per cent**
- **Domestic = 8 per cent**

The composition of demand will vary in different parts of the world (Figure 11.30). For example, less developed countries in arid parts of the world use a higher proportion of their water for agriculture than western Europe, where plentiful rainfall means that less irrigation is needed. People here are more affluent and industry is more advanced, so a higher proportion is used in the domestic and industrial sectors.

The concept of water stress

Water stress results from an imbalance between water use and water resources available and leads to a situation where the lack of water is a major constraint on human activity. International organisations have assigned a measure to water stress which is, water availability of less than 1,700 m³ per person per year.

Another water stress indicator is the amount of water withdrawal as a proportion of the total water resource available. Figure 11.31 shows exposure to water stress in different parts of the world based on the ratio of withdrawals to supply. Areas with a higher percentage of withdrawal mean that water users are competing for limited resources.

Water stress causes a deterioration of freshwater supplies in terms of:

- **Quantity** – for example, over-abstraction from rivers, reservoirs or aquifers
- **Quality** – for example, organic pollution or eutrophication of surface water or saltwater incursion into aquifers.

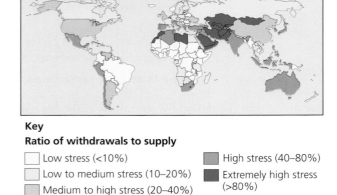

Key
Ratio of withdrawals to supply

☐ Low stress (<10%)
☐ Low to medium stress (10–20%)
☐ Medium to high stress (20–40%)
☐ High stress (40–80%)
☐ Extremely high stress (>80%)

Figure 11.31 Risks of water stress by country

Source: World Resources Institute

Relationship with aspects of physical geography

Physical factors are important in ensuring reliable sources of surface water and aquifers. The usual requirements include:

- reliable annual rainfall (this may be seasonal but storage can help with temporal discrepancies)
- large bodies of surface fresh water – lakes and rivers provide the most accessible source
- low evaporation rates to reduce surface losses
- suitable geological structures that provide for groundwater aquifers.

Rivers and lakes

In deciding to abstract river water for human use, the following factors about the river's regime need to be considered:

- river discharge – the annual flow must be adequate
- variations in flow over time should be included when determining abstraction rates

- water quality – the level of contamination and turbidity before treatment
- other uses made of the river (for example, uses such as transport or wildlife conservation).

Over-abstraction from rivers and lakes will have a number of negative impacts including:

- reduced volume of water following abstraction concentrates pollutants
- decreased velocity and discharge will increase sedimentation downstream
- reduced downstream flooding will adversely affect soil quality
- reduced flow downstream will impact on ecosystems, habitats and may kill aquatic life.

Reservoirs

Reservoirs are created by flooding valley floors; a dam is built across the river valley to allow water to be stored behind it in a reservoir (Figure 11.32). Creation of reservoirs for storing and abstraction of water is a controversial decision for two reasons:

- conflict with other uses in the valley – such as settlement (people are displaced), food production, etc.
- disruption to the hydrological cycle and impacts on the whole river system.

In addition, good sites for reservoirs are often those that have the highest value for scenery and wildlife. This makes decisions on where they should be located difficult.

Figure 11.32 Aerial view of the dam forming Lake Vyrnwy reservoir in mid-Wales and its location on the map. The reservoir was constructed in the late nineteenth century to supply water to the growing city of Liverpool, 70 miles away.

Table 11.10 Physical factors affecting the most suitable location of reservoirs

Physical characteristic	Preferred site and reason
Topography	A long narrow valley basin with steep sides that will provide a large volume and small surface area
Geology	Impermeable rock so that water is not lost to seepage. Geology must be stable as earthquakes or faulting may damage or destroy the dam. Avoiding areas where the rocks or minerals may have an economic value that exceeds the value of the reservoir
Catchment area	Large area to increase water volume for storage
Climate: Water supply	Adequate and sustainable supply of water from upstream requires regular and frequent precipitation. Greater volume gives more flexibility and reliability
Climate: Flow fluctuation	Rainfall needs to be reliable. Cooler and less windy to reduce losses due to evapotranspiration. Avoiding areas prone to freezing for long periods

Table 11.11 Evaluation of the potential benefits and environmental impacts of reservoirs

Potential benefits gained from reservoirs	Environmental effects of reservoirs
Flood control: Prevents flooding downstream	**Sedimentation/siltation:** Silt/sediment is deposited in the reservoir: reduces capacity and increases the erosive power of the river downstream
Stored water supply: For agricultural, industrial or public use	**Creates microclimate:** Smaller temperature fluctuations: higher wind speed, higher humidity downwind
Multiple purposes including: HEP generation Recreational use Fishing	**River regime downstream of dams:** Changes in flow, temperature, turbidity, erosion and fluvial landforms such as deltas
	Habitat change: Flooding removes habitat, creates a barrier to migration and dispersal of seeds

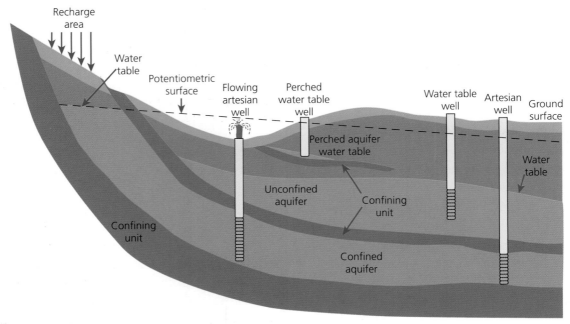

Figure 11.33 Geological structure of confined and unconfined aquifers

Aquifers

An aquifer is a layer of rock that holds groundwater. They occupy specific geological formations and need porous or permeable rocks such as chalk, limestone, sandstone or gravel to hold a 'reservoir' of underground water. Porosity is a measure of the proportion of rock which is 'space' and can hold fluids, while permeability is the rate at which fluids flow through a rock because of the connected pores or fissures. Permeable rocks enable aquifers to 'recharge' from natural sources.

Aquifers usually form in sedimentary basins and the different layers of permeable and impermeable rock will form confined and unconfined aquifers (Figure 11.33). The layer below the aquifer must be impermeable to prevent the escape of the water. Confined aquifers will have impermeable layers of rock both above and below them. They are called artesian aquifers and can be recharged where permeable layers meet the surface.

Water occurring below the ground is brought to the land surface by wells or springs. If the head of the well is below the natural water table level then it will flow naturally under atmospheric pressure, otherwise the water is pumped to the surface.

Groundwater is vulnerable to overuse and the improper disposal of chemicals at the land surface. The trend of falling groundwater levels may indicate climate change or prolonged over-abstraction.

Over-abstraction occurs when the rate of extraction of water from the aquifer is greater than the recharge rate. It can lead to a number of serious environmental impacts including:

- subsidence
- loss of wetlands/vegetation changes
- drying of soils leading to 'osmotic dehydration' of crop roots
- lower water table
- rivers drying up, leading to loss of habitat/loss of recreation
- saltwater incursion into groundwater.

Aquifers are naturally recharged by rainfall but it can take many years for water to percolate down and to top up aquifers, so they are also recharged artificially.

Strategies to increase water supply

Key question

What strategies need to be deployed to ensure that there is always sufficient water in the correct location?

Water supply and demand is prone to spatial and temporal mismatches. For example, in the UK the greatest demand for water is in south-east England, because of the higher population density and greater agricultural needs in arable farming. The south-east receives less rainfall and has less surface water than other areas, so is reliant on groundwater sources. It has the highest potential water stress of any part of the UK.

There is an increasing awareness that freshwater resources are limited in certain areas and at certain times. They need to be protected both in terms of quantity and quality. Measures can be taken to avoid a worsening crisis. The provision of adequate supplies involves:

- increasing the total amount of water available for use, for example, by increasing abstraction
- using water more efficiently, for example, by reducing waste use
- making better use of existing supplies by:
  - reducing pollution of existing sources
  - distributing it more effectively.

Catchment Abstraction Management Strategy (CAMS) is an integrated approach to manage the supply of water. It is used to assess the amount of water available for abstraction licensing, taking into account what the environment needs. Environmental flow indicators (EFIs) are used to assess whether river flows are sufficient to support a healthy ecology. Surface water management can best be achieved by storage to even out temporal differences in supply or by diversion and inter-basin transfers to even out spatial inequalities.

Storage by reservoir

New reservoirs may be built to trap and store surplus winter rainfall or regulate river flow, to maintain habitats and prevent flooding downstream.

Direct supply reservoirs, such as Lake Vyrnwy (Figure 11.32, page 565), store water for release via pipeline to the public supply. They may fill from rainfall, from natural river flow or via pumping from nearby rivers.

Reservoirs are an obvious way of increasing and managing water supply on a regional or national scale. Larger-scale multi-purpose reservoirs such as those created by the Three Gorges Dam (Yangtze River) and Aswan Dam (Lake Nasser on the River Nile) can also provide additional benefits such as hydro-electric power (HEP), bolstering the fishing and tourism industries, but they are not without their economic or environmental problems, as indicated by Table 11.11 (page 565). (See also the text box, Environmental impacts of a major water supply scheme – The Aswan High Dam and Lake Nasser, page 569)

Diversion and inter-basin transfer

River diversion is used to transfer water from one river catchment to another when there is a surplus of water in one area and too little in another. Water is drawn from rivers, reservoirs and aquifers, and pumped across the region enabling localised droughts to be tackled quickly. Transfer can take place by through aqueducts or by river diversion, canal or pipeline.

In the UK there is a huge variation in average rainfall between the different regions; Snowdonia receives over 400 cm per year, whereas parts of East Anglia receive only 55 cm per year. Demand will also vary due to population distribution. For example, heavy rainfall in sparsely populated mid-Wales is stored in reservoirs and then transported via rivers and pipelines to the West Midlands conurbation.

West to east transfers of water in the UK have been mooted. In particular, the suggestion of a Severn–Thames inter-basin transfer was proposed as long ago as 1973 but has never been implemented. It would necessitate some transfer by canal plus pumping of water over the Cotswolds. The notion of taking water from an area of water surplus in the west of England to an area of water stress in the south-east is attractive. However, the prohibitive construction costs, ecological concerns and public resistance to more reservoirs being built to capture and store the transferred water have meant that the project has never been given the go-ahead.

In water-stressed developing countries which are still heavily reliant on irrigation for agriculture, irrigation canals transfer water from major river basins to farming areas. Such is the case in the Sindh province of Pakistan, which has a network of irrigation canals abstracting water from the Indus basin. Unfortunately evaporation leads to large losses of water from these canals. Poor maintenance of linings and earth bunds

Figure 11.34 Water transfers in southwest USA to address shortages in southern California

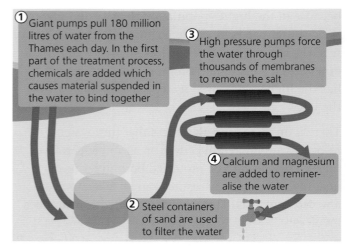

Figure 11.35 The desalination process at Thames Water's Beckton operation using reverse osmosis.

Table 11.12 Evaluation of inter-basin water transfer

Advantages of inter-basin transfer	Disadvantages of inter-basin transfer
Localised drought can be remedied quickly	Capital and running costs are very high – water is expensive to pump
Habitats can be protected	There are regional differences in water quality parameters (for example, chemical composition)
Addresses spatial mismatch	
	Requires development of storage reservoirs in receiving basin

leads to the dual problem of waterlogging and salinity, which destroys crops and renders the soil infertile.

On a larger scale, water is transferred between major river catchments in the south-western USA by means of large canals, pumping stations and tunnels. An enormous quantity of water is transferred 300 km from the Colorado River Basin to Southern California, where it is used for irrigation in the agricultural areas of the San Joaquin Valley and for public water supply in Los Angeles, San Diego and other cities (Figure 11.34).

Desalination

The removal of salt from seawater would provide a huge supply of water but desalination is an expensive process which is not considered sustainable because it is very energy intensive and a significant source of greenhouse gas emissions. There are two main methods used in desalination:

- **Reverse osmosis**: Involves filtration of seawater at high pressure (600–1,000 PSI) through a partially permeable membrane, using small polyamide tubes, to convert it to potable water. This process collects fresh water and rejects very saline water, which is pumped back into the sea.
- **Distillation:** Water is boiled by heating under reduced pressure to save energy (water boils at a lower temperature in a vacuum). The steam produced is condensed and collected separately and the salt is left behind in the boiler.

Desalination plants have generally been developed in wealthier countries with water stress or scarcity issues, especially those in the Middle East such as Saudi Arabia, Oman and the UAE, but also in the Mediterranean, the USA and Australia. In the UK, Thames Water's desalination plant at Beckton in East London provides water for 400,000 households by reverse osmosis membrane filtration (Figure 11.35).

Environmental impacts of a major water supply scheme – The Aswan High Dam and Lake Nasser

Background to the Aswan scheme

The Aswan Dam is situated across the Nile River in Aswan, southern Egypt. Since the 1950s, the name commonly refers to the High Dam, which is larger, newer and upstream of the Aswan Low Dam, which was first completed by the British in 1902. Following Egypt's independence from the UK, the High Dam was constructed between 1960 and 1970. It was partially funded from Egypt's takeover of the Suez Canal in 1956 but mainly with assistance from the Soviet Union (aiming to gain geopolitical support in the region). It is nearly 4 km wide across the valley. The reservoir of Nile waters stored behind the dam is called Lake Nasser and stretches back 550 km southwards to the Sudanese border (Figure 11.36). It is 35 km at its widest point and holds around 132 km³ of water. The scheme aimed to increase economic production by regulating the annual river flooding and providing storage of water for agriculture, and later, to generate hydroelectric power (HEP). The dam has had a significant impact on the economy and culture of Egypt.

Figure 11.36 Panoramic view looking south towards Lake Nasser from the Aswan High Dam

Before the dams were built, the Nile River flooded each year during late summer, as water flowed down the valley from its East African drainage basin. These floods brought high water and nutrients and minerals that enriched the fertile soil along the floodplain and delta; this made the Nile valley ideal for farming. However the floods vary: in high-water years the whole crop might be wiped out, while in low-water years widespread drought and famine occurred. As Egypt's population grew there was a desire to control the floods and thus both protect and support farmland and the economically important cotton crop. With the capability of reservoir storage provided by these dams, the floods could be reduced and the water could be stored for later release. When necessary, a maximum of 11,000 cumecs of water can pass through the dam. In emergencies, there are spillways for a further 5,000 cumecs to be released.

Socio–economic impacts

The dam had significant human socio-economic impacts in the immediate area. Lake Nasser flooded much of lower Nubia and around 100,000 Nubian people had to be resettled in purpose-built villages in Sudan and Egypt. However, the lake created a lot more land for people to farm on because the stored water was used to irrigate land around the lake. The lake also created a large fishing industry producing over 25,000 tonnes of fish per year.

Unfortunately, the Aswan Dam had a negative impact on the lives of farmers downstream. Previously, when the river flooded annually, it deposited fertile alluvial sediments on its downstream banks. Since the dam was built, the annual flood has been stopped and these sediments are held back in the reservoir. Farmers downstream now have to use fertilisers to grow their crops, which makes it more expensive and creates indirect environmental impacts.

The HEP capacity of the Aswan High Dam is 2.1 gigawatts (GW). By the 1980s this was providing half the country's electricity. This proportion has fallen dramatically in recent years as energy demands have increased. The dam now only provides around one tenth of the country's electricity.

Environmental impacts

Critics of the project in the 1970s pointed out potential failings of the dam (Figure 11.37).

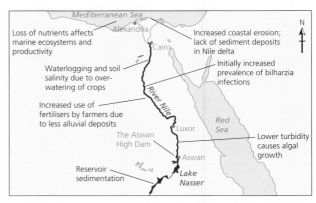

Figure 11.37 Outline map of the Lower Nile valley in Egypt labelled with some of the key environmental impacts

Coastal erosion

The erosion of coastlines has accelerated due to lack of sediment, once brought down to the Nile delta by the river. Coastal erosion is severe in some areas, especially at the Rosetta and Damietta promontories. The coastal erosion that occurred even before the construction of the High Dam is thought to be the result of the limited trapping of sediments behind the Low Dam. The High Dam accelerated erosion and made necessary the construction of expensive coastal protection works in the Nile Delta. These efforts to prevent overall coastline retreat have been largely unsuccessful.

Reservoir sedimentation

Sediment deposited in the reservoir is lowering the water storage capacity of Lake Nasser. The reservoir storage capacity is 162 km³, including 31 km³ dead storage at the bottom of the lake. The annual sediment load of the Nile is about 134 million tonnes. Sediment accumulates much faster in the upper reaches of the Lake where sedimentation has already affected the live storage zone. It would take about another 900 years until the live storage zone could be completely filled with sediment but operation of the dam would become impossible some time before that. Efforts to dredge the dam can go some way to reduce this problem but it is easier to dredge parts of the river before it enters the reservoir.

Marine ecosystems

Marine productivity in the Mediterranean has decreased because nutrients that used to flow down the Nile to the Mediterranean are trapped behind the dam. For example, the sardine catch off the Egyptian coast initially declined massively, though this now seems to have recovered to an extent but only to half the levels prior to the building of the dam.

The Aswan Dam also tends to indirectly increase the salinity of the Mediterranean Sea and this affects the Mediterranean's outflow current into the Atlantic Ocean. This current can be traced thousands of kilometres into the Atlantic.

Health and disease impacts

Schistosomiasis, also known as **bilharzia**, is an infection caused by a parasitic worm that lives in fresh water in subtropical and tropical regions. The standing water in irrigation canals is a breeding ground for snails carrying the bilharzia parasite. The incidence of bilharzia increased due to the dam inhibiting the natural fluctuations in water height. However, other important factors contributed to the prevalence of schistosomiasis including poor sanitation and limited awareness of how the disease was transmitted. These have largely been addressed and the provision of clean water, sanitation, health education and rural clinics has reduced the overall prevalence of the disease from more than 40 per cent during the pre-dam period to less than 2 per cent in 2010.

Waterlogging and increased soil salinity

Before the construction of the High Dam, groundwater levels in the Nile Valley fluctuated by as much as 8–9 m/year as a result of changing water levels. During summer when evaporation was highest, the groundwater was too deep to allow salts dissolved in the water to be pulled to the surface through capillary action. With the continuous irrigation enabled by the scheme, fields in the Nile Valley tend to get water logged as groundwater levels remain high, with little fluctuation. Soil salinity has also increased because the distance between the surface and the groundwater table is small enough (1–2 m depending on soil conditions and temperature) to allow water to be drawn up by evaporation so that relatively small concentrations of salt in the groundwater gradually accumulate on the soil surface over the years. Since the water is not flushed out annually by flooding, the soils are saltier and less fertile.

Most of the farmland did not have proper subsurface drainage to lower the groundwater table so salinisation gradually affected crop yields. It took twenty years after the completion of the Aswan scheme to address this problem by providing two million hectares of farmland with subsurface drainage (at a cost that exceeded the construction cost of the High Dam). Overall, agricultural production in Egypt has increased as a result of the dam. The more reliable and constant water supply allows multi-cropping during the year and Egyptian farmers have also adopted more successful seeds and harvesting methods.

Algal growth and water quality

The lower turbidity of the water downstream from the dam means that sunlight penetrates deeper in the Nile water. As a result of this and the increased presence of nutrients from fertilisers in the water, more algae now grow in the Nile. This has increased the costs of water treatment to supply potable water. The decrease in water quality was an unforeseen consequence of the High Dam scheme.

Strategies to manage water consumption

Where people are not facing scarcity, water is often taken for granted and wasted. Whatever the use of fresh water (agriculture, industry or domestic use), huge savings and improvement of water management are possible.

Domestic use

Probably the most effective way of avoiding wastage is to ensure households pay for what they use. The introduction of water meters in the UK ensures that households with a meter fitted pay for the amount of water used, rather than a standard rate. This gives householders an incentive not to waste water. Because the south of England has been declared an 'Area of Serious Water Stress', water companies such as Southern Water, Thames Water and South East Water are installing meters to encourage customers to use less water.

A number of simple methods are available to reduce water consumption:

- toilets – around 30 per cent of the water in homes is used to flush toilets. This can be reduced by:
  - installing a displacement device (for example, a 'hippo bag') in the cistern to reduce the volume of water used per flush
  - installing low flush or dual flush toilets
- personal hygiene – having showers instead of baths saves water use
- machinery – using newer technologies such as water-efficient washing machines and dishwashers
- filling kettles with just enough water for your needs; this will also reduce energy bills
- collecting rainwater to use on gardens.

Similar savings can be made in commercial and industrial settings. Newer buildings and complexes are designed to reduce water use by collecting rainwater, which is filtered and then sent into toilet and washbasin areas. This not only reduces abstraction from traditional sources but also saves the energy required in treating water from those sources.

Agricultural use

It is possible to reduce water consumption when irrigating by using water efficient methods. Micro-irrigation is the most efficient method. This employs low-flow technology to deliver water directly to plant

Figure 11.38 Micro-irrigation devices such as soaker hoses reduce water losses

roots at rates that prevent deep percolation, evaporation and run-off losses. Typical devices include:

- drip feed irrigation: small irrigation heads that deliver a drip or trickle directly at the plant base
- soaker hoses: hoses with small openings that allow water to seep out; they are placed at the plant base and are suitable for using with rows of crops (Figure 11.38)
- micro-sprayers: overhead sprinklers which deliver a fine mist that can be controlled.

Additional measures that growers can use to reduce water use include:

- using mulches wherever possible to retain water content
- collecting storm water and irrigation run-off in ditches or drains
- watering early in the morning to reduce evaporation losses
- minimal or no ploughing (leaving at least 30 per cent of the previous season's crop residue in the field)
- contour ploughing reduces run-off
- using organic fertilisers because organic matter retains soil moisture
- using cover crops in winter and fallow times to improve the soil and reduce water loss and soil erosion.

Sustainability issues

Sustainability issues associated with water management are concerned with achieving a greater long-term balance between supply and demand, particularly in areas of water shortages and stress. More sustainable management will help to maintain water supplies for future generations.

Virtual water trade

In recent years, the concept of 'virtual water trade' has gained weight. Professor John Allan introduced the idea, pointing out that countries with water scarcity, for example, those in the Middle East, can save water resources by relying more on the import of food. The water used in the production process of an agricultural or industrial product is called the 'virtual water' contained in the product. The idea is to estimate the unit requirement of water needed to produce a commodity, which gives the product a virtual water value. It is using comparative advantage in terms of water efficiency, as the water requirements will be less in the exporting country than in the importing countries. For example, water-scarce countries like Israel should discourage the export of oranges (relatively heavy water guzzlers) to prevent large quantities of virtual water being exported to different parts of the world.

Conservation measures
Land use, ecological management and afforestation

One of the strategies to manage water yield is to control land use in a river basin, as this will affect run-off, infiltration rates and groundwater storage. Impermeable surfaces in and around urban areas increase run-off, taking water to one of its shorter-term stores in the river channel. An increase in permeable land uses increases infiltration and percolation rates, taking water into longer-term groundwater storage, thus increasing supply for future use.

Vegetation cover such as forests and tree plantations also improves the water cycle in diminishing run-off and improving the replenishment of the water table. Providing that there is not excessive evapotranspiration from leaves, they will help to maintain groundwater stores at a more sustainable level. Afforestation can be a good tool for water conservation and forested catchments can supply a high proportion of water for domestic and agricultural use. On a micro-climatic to regional scale, tree planting has often been proposed as a way to increase rainfall. However, trees also consume water. The more the aerial system of trees is developed, the more water they transpire. The desirability of tree planting in arid lands is debated because trees may consume more water than they provide to the water cycle.

Recycling and grey water

Other conservation methods include:

- **Recycling of water:** This is done through extensive sewage treatment. In London 90 per cent of used water is recycled by being treated immediately after being returned to the drains
- **Greywater use:** Greywater is water that has been used for washing or cleaning but has not been in contact with faecal matter (black water). It can be re-used for purposes where non-potable water is adequate (for example, flushing toilets, gardening or irrigation). Greywater recycling is ideally suited to hotels, leisure centres and large office and residential blocks as it recycles water used in baths and showers to flush toilets
- **Leakage control:** Better maintenance of supply pipes, taps and appliances will prevent losses.

Groundwater management – aquifer recharge

Aquifers take a long time to recharge naturally but they can be recharged artificially by:

- artificial infiltration – pumping water underground
- diverting rivers to permeable surfaces
- diverting storm water into recharge wells. In Australia these are known as 'leaky' wells.

This is a good way of storing water that is surplus to requirements so that it is available when supplies are less plentiful. Maintaining aquifers may help maintain rivers and ensure surface stability.

Water conflicts

Key question

Why is there likely to be conflict at different scales over the access to and use of water?

Water conflict is a term describing a conflict between countries, states or groups with opposing interests over access to water resources.

Local-scale conflicts

Where demand for water exceeds supply and no effective management operates, conflicts between the various users of that water source may ensue. Conflict at a local level is likely to be caused by different users wanting a scarce resource; it can lead to sabotage, or escalate into violent protest.

For example:

- In 2007 in the north-eastern Indian state of Orissa, farmers clashed with police because the state government had allowed a large number of industries to source water from the Hirakud Dam, although the farmers depended on this water for irrigation.

- In northern Chile, the Atacama Desert is known as one of the driest places in the world, only receiving an average of 1 mm of rain a year. In such an arid environment there are competing demands for the only available water, provided by the Loa River and some aquifers. Agriculture has traditionally been one of the most important sectors of the economy but the discovery of vast copper reserves has made mining more important. The influence of the mining companies, who can afford to pay more, means water has been diverted away from vineyards (leaving them to die) and into the mines, where it is used to extract copper deposits. In some areas the water table has fallen to over 140 m below the surface because of over-abstraction by the mines.

National scale

Disputes at this scale are often the result of water scarcity and an uneven distribution of available resources. For example, although Israel and the Palestinian Authority are two separate states, they can be examined at a national scale for water management as much of the Palestinian territory is controlled by Israel. The climate of the region is very arid and with the growing populations of the countries that share the River Jordan there is increased tension concerning the availability and use of water.

Political and religious disputes in the region date back thousands of years, however many argue that access to water is at the root of the current disagreement, because of its importance for irrigation. Apart from the River Jordan, the mountain aquifer, mostly in Palestinian territory on the West Bank, is the main source of water. The West Bank is controlled by Israel, and they have built settlements on Palestinian territory there. The mountain aquifer in the occupied West Bank is largely used by Israel (80 per cent), only leaving 20 per cent for the Palestinians. Israel claims they have the right to use the aquifer, because some groundwater flows into Israeli territory (Figure 11.39).

International scale

There are 13 basins worldwide that are shared between 5 and 8 riparian nations. Five basins, the Congo, Niger, Nile, Rhine and Zambezi, are shared between 9 and 11 countries.

Examples of conflict regions:

- The Aral Sea (Kazakhstan and Uzbekistan) – due to the over-abstraction (for cotton irrigation) from

Figure 11.39 Pressure on limited groundwater resources in Israel and Palestine

the two rivers (Syr Darya and Amu Darya) that flow into the Aral Sea, half of its surface has disappeared, representing two thirds of its volume – 36,000 km² of land is now covered by salt.

- India continues to build new dams that are seen by its rival Pakistan as a threat to its 'water interests' and thus its national security.
- Turkey, from its dominant position upstream, has been diverting the Tigris and Euphrates rivers, thus increasing water stress in the already volatile states of Iraq and Syria.

These 'basins at risk' are just some examples of where tensions at an international scale could escalate into so called 'water wars'.

For more detailed information on water conflicts visit the following website: www.worldwater.org/water – conflict/.

11.6 Energy security

In 2014 Energy Ministers of the G7 group signed a joint statement in Rome on energy security. The fundamental principle they subscribed to was that energy security is a common responsibility.

Co-operative arrangements regarding energy security have been reached before but the fact that the issue was addressed again, illustrates the importance of geopolitics for energy security in all regions.

Key terms

Energy mix – The composition of different sources of energy used in a country or globally. Primary energy mix refers to those used to produce all secondary energy. Some energy mix data refers only to sources used to produce electricity.

Primary energy (resources) – Energy sources obtained in their raw material or natural form, such as oil, natural gas, wind or running water; they are not usable until they are converted into heat or mechanical action to produce secondary energy.

Secondary energy (resources) – Energy that is transformed or converted from primary energy sources into manufactured sources of power which are usable, such as heat, electricity or petroleum.

Primary and secondary sources of energy

Primary energy resources are potential sources of energy found in their natural form (Figure 11.40), such as wind or running water or raw materials such as coal, oil, wood and uranium. Primary energy sources need to be converted or processed into **secondary energy** sources such as petrol or diesel used to run vehicles, or into electricity which is then used to power domestic, industrial and commercial premises. In more developed countries energy sources such as oil, natural gas, coal, hydroelectric power and uranium provide the main sources of power. In contrast, many less developed countries rely on biomass sources such as fuelwood and animal waste.

Table 11.13 Main primary sources of energy and how they are converted into secondary energy

Primary energy resources			Converted by/energy systems	Secondary energy form
Non-renewable sources	**Fossil fuels**	**Crude oil**	Oil refinery	Petrol, diesel, fuel oils, etc.
			Thermal power station	Electricity
		Natural gas	Thermal power station (combined cycle)	Electricity
			Combustion after transfer by National Grid	Heat for cooking, central heating, etc.
		Coal	Thermal power station	Electricity
			Direct burning	Heat
	Mineral fuels	**Uranium**	Nuclear fission (nuclear thermal power station)	Electricity
Renewable sources		**Solar energy**	Photovoltaic cells (unit, array, farm)	Electricity
			Solar furnaces/tower and parabolic reflectors	Electricity
			Harnessing heat/light passive solar architecture	Heat
		Wind energy	Wind turbines (farm)	Electricity
		Biomass sources	Burning fuelwood directly	Heat (for example, for cooking)
			Fermentation/distillation and refining into liquid	Bioethanol or other alcohol-based fuels
			Anaerobic digestion to produce gas	Methane gas for burning – heat for cooking/heating or electricity generation
		Hydroelectric power (HEP)	Potential energy released and converted to kinetic energy to turn turbines	Electricity
		Tidal energy	Tidal turbines or tidal barrage (kinetic energy)	Electricity
		Wave energy	To drive hydraulic turbines or compressed air chambers	Electricity
		Geothermal energy	Thermal power stations	Electricity
			Directly harnessing underground heat	Heat

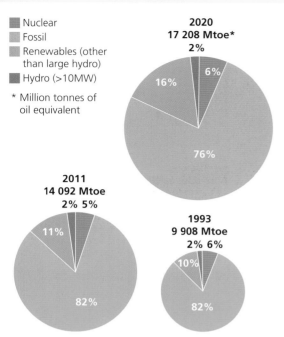

Nuclear
Fossil
Renewables (other than large hydro)
Hydro (>10MW)

* Million tonnes of oil equivalent

2020
17 208 Mtoe*
2%
6%
16%
76%

2011
14 092 Mtoe
2% 5%
11%
82%

1993
9 908 Mtoe
2% 6%
10%
82%

Figure 11.40 Global growth and changing composition of total primary energy supply (TPES)
Source: World Energy Council

Fossil fuels

Oil: Even though its share of total primary energy supply (TPES) decreased from 46 per cent in 1973 to 31 per cent in 2013, oil still accounts for the largest share, followed by coal and natural gas.

Natural gas: Accounts for 21 per cent of the energy budget and it is the fastest growing fossil fuel source because many countries now use it to reduce dependence on oil and reduce environmental problems.

Coal: Still accounts for around 29 per cent of the world's energy supply; solely in terms of electricity production it is even higher at 40 per cent. It is the most plentiful fossil fuel but concerns about greenhouse gas emissions and acid rain have reduced its use.

Nuclear energy

Nuclear power generates just over six per cent of the world's energy. Nuclear production requires advanced technology and huge investment, so is mostly found in developed, high-income countries (Figure 11.41). Developing countries, such as China and India, are starting to invest more of their wealth in nuclear power plants. Political factors have had negative impacts on nuclear energy development. Nuclear accidents at Chernobyl, Ukraine (1986), and more recently at Fukushima in Japan (2011) forced the evacuation of

Table 11.14 Advantages and disadvantages of each of the fossil fuels

	Advantages	Disadvantages
Oil	Indispensable in road transport and petrochemical industries Leading tradable commodity Flexible, easy to transport fuel	High price volatility Geopolitical tensions in areas with the largest reserves Market is dominated by leading oil-producing countries and large TNCs
Natural gas	Cleanest of fossil fuels Flexible and efficient fuel for power and heat generation Increasing proved reserves from unconventional sources (reassessments of speculative reserves and shale gas)	Reserves increasingly offshore or in more remote areas Large investment requirement for gas transport and distribution Increasingly long supply routes with high infrastructure costs
Coal	Wider geographical distribution and more plentiful reserves New technologies to improve environmental performance Stable prices	High emissions of carbon, particulates and other pollutants Environmental mitigation such as carbon capture (use) and storage (CCS/CCUS) have a negative impact on energy efficiency

Source: World Energy Council, World Energy Report 2013

the local populations. Large amounts of radioactive material escaped into the surrounding areas causing loss of lives and wide scale environmental damage. These incidents, together with the inherent problems of disposing of nuclear waste, have dampened the appetite of some advanced economies to seek security in nuclear energy and have resulted in a levelling off in new reactor construction.

Figure 11.41 EDF's Sizewell B nuclear power station in Suffolk

Advantages of nuclear energy:

- highly energy efficient fuel with newer reactors even more efficient
- relatively large reserves of uranium
- moderate and predictable cost of electricity over the life of a power station
- excellent replacement for fossil fuels in generating thermally sourced electricity
- atmospheric pollution is much less than with fossil fuels.

Disadvantages of nuclear energy:

- high investment and compliance costs in constructing nuclear power stations
- public concerns about operation and radioactive waste disposal
- potential dangers and impacts of nuclear accidents
- difficulty in finding suitable sites
- plutonium is a by-product which can be used in nuclear weapons – so can cause geo-political tensions or threats of terrorism.

Renewable energy sources

Over the last 40 years, the contribution of renewables to world TPES has been stable at around 12–15 per cent. Renewables have maintained their rank of third largest contributor to global **electricity production**. For some countries the proportion of electricity generation from renewable energy sources can be very high, for example 100 per cent of Iceland's electricity is generated from geothermal and HEP sources. In Paraguay and Norway, respectively 100 per cent and 98 per cent of their electricity is produced by HEP.

Solar energy

The sun contributes significantly to heating buildings through windows and walls. Features to maximise this energy are incorporated into new buildings using **'passive' solar architecture**, for example larger south-facing windows. The potential for solar energy in less developed tropical countries is excellent because of the reliability and intensity of the sunlight. It is already used in 'appropriate technologies' such as solar cookers, but more electricity could be generated from photovoltaic (PV) units/arrays. The principle of using **heliostats** (parabolic mirrors) to direct the sun's energy onto a focal point on solar furnaces and solar towers is utilised to produce extremely hot

Figure 11.42 The PS10 solar power tower, near Seville, Spain

temperatures for industrial uses or alternatively to generate steam for turbines to produce electricity (Figure 11.42).

Wind energy

The potential for wind energy varies enormously because of variations in wind speed and reliability. Turbines are used to generate electricity and are increasingly found in large numbers on wind farms connected to an electricity grid network. Conflicts with other land uses may restrict wind energy use in many parts of the world. Countries with frequent, prevailing winds and large coastlines are more likely to develop offshore wind farms.

Biomass sources

Biomass sources currently provide around 12 per cent of TPES. 'Bioenergy' is a broad category which includes the use of any plant or animal matter converted into solid, liquid or gaseous biofuels using one of a range of conversion technologies. The term 'traditional biomass' mainly refers to fuelwood, charcoal and agricultural residues which are used for household cooking, lighting and space-heating in developing countries. These are the main fuel sources for over a third of the world's population. Biomass can also be burnt in thermal power stations or used in the anaerobic digestion process producing a biogas that can be used for cooking or heat generation or burnt for thermal electricity generation.

Hydroelectric power (HEP)

Falling water generates around two per cent of total energy supply. It is thought to be under-exploited; however large dam projects are controversial and carry heavy environmental costs.

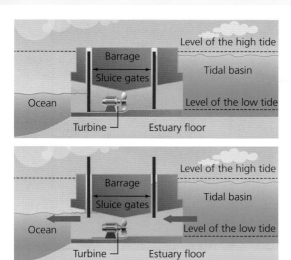

Figure 11.43 How tidal barrages work – this is a one-flow generation system

Tidal power

The energy potential of the oceans is massive but only a fraction is likely to be harnessed. Tidal power is harnessed either through tidal current water turbines (underwater versions of wind turbines) or by tidal barrages. Barrages operate in a similar way to HEP dams and turbines: the incoming tide is captured behind a barrage at high tide, giving a great deal of potential energy. At low tide sluice gates are opened and the captured water is released through turbines, thus maximising its kinetic energy (see Figure 11.43). Electricity can be generated through turbines, on both the incoming ebb tide and the outgoing flow tide. Viable sites for tidal barrages are limited as specific physical conditions are necessary, such as deep coastal bays, inlets or estuaries with a high tidal range. Tidal barrage construction is controversial because of the enormous cost and because of the environmental disruption they may cause by changing natural tidal flows.

Wave power

Various technologies are being developed to make wave power commercially viable, especially as energy is lost in transmitting the electricity to shore from offshore wave farms. The most common offshore technologies are based on harnessing the kinetic motion of surface waves to move floats up and down so that hydraulic rams drive an electrical generator (see Figure 11.44(a) – the Pelamis Wave Energy Converter). Onshore technologies are an alternative and the LIMPET device (on the Isle of Islay), for example, harnesses onshore wave action inside a

(a) Offshore

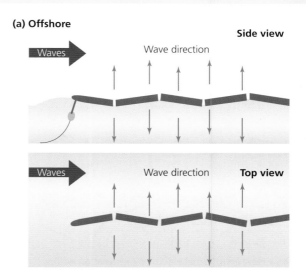

(b) Onshore

① The wave capture chamber is built into the rockface.
② Waves force water into the chamber.
③ This column of water causes the air to be compressed and decompressed.
④ Power is created by these rushes of air driving the turbine.

Figure 11.44 Offshore and onshore wave energy technologies

concrete chamber to compress air which is forced through a pair of turbines (Figure 44(b)).

Geothermal energy

Earth's temperature rises by approximately 1°C for every 30 m of depth within the crust, and this temperature gradient is even steeper in geologically active areas such as plate margins. Geothermal energy makes use of this heat, either directly by producing hot water or indirectly to produce electricity. The most common way of capturing geothermal energy is to tap into naturally occurring 'hydrothermal convection' systems where cooler water is pumped or naturally percolates into the Earth's crust, where it heats up and then rises to the surface. Once this heated water is forced to the surface, the steam is captured and used to drive electric generators (Figure 11.45a, page 578).

The main and overriding advantage of all renewables is that they are sustainable in the long term and reduce our dependency on fossil fuels. Some renewable energy forms have greater environmental impacts than others, either in their manufacture, construction or operation. With the exception of biomass sources, all renewables require certain physical factors to be present to make them viable, thus rendering them **geographically specific,** which could be a limitation for their further development.

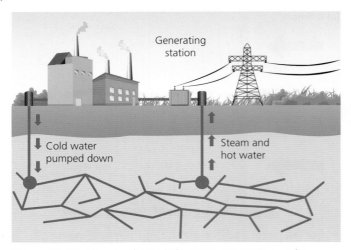

Figure 11.45a How geothermal electricity generation works

Figure 11.45b Iceland's Nesjavellir geothermal power station

Table 11.15 Advantages and disadvantages of renewable energy sources

	Advantages	Disadvantages
Solar	Pollution free (except for production and disposal of pv units) Plentiful silica resources (to make pv cells) High reliability, no moving parts Relatively cheap and efficient Good potential in less developed countries	Intermittency of solar intensity Ineffectiveness of storing the energy generated, so limited uses Use of some toxic materials in pv cell production Grid connection challenges
Wind	Quick installation and dismantling (onshore) Relatively simple technology No atmospheric pollution (except in production of turbines) Land use does not always conflict with other operations (for example, farming) Option for offshore combined with tidal energy	Some land use conflicts Conflicts such as visual intrusion, effects on wildlife (birds) Local objections to installation: visual, noise and radio interference Intermittent nature of wind thus relatively inefficient and unreliable Offshore technology and potential problems for sea navigation
Biomass	Proven simple combustion technology Alternative to oil for transport Energy crops are an efficient way of using 'marginal' land Mostly carbon neutral Relatively high energy density Flexibility of form – solid, liquid, gas	Growing energy crops conflicts with food production Some emissions (NOx and SOx) Fuelwood use can cause deforestation, soil erosion and desertification Carbon neutrality is questionable as CO_2 sink is lost
HEP	Low operating costs No waste or CO_2 emissions Proven technology Can top-up supply in periods of peak demand	Dams have high construction costs Requires large amounts of land so causes conflict, for example, relocation of populations Environmental impacts on drainage basin and micro-climate effects
Tidal	Generation periods are predictable Potential power output is large	Power generation is intermittent Ecological impacts (for example, on marine life)
Wave	Produces electricity regularly Less environmental impact (than tidal barrages, for example) Virtually pollution free Onshore sites relatively cheap	Difficulty in transmission from offshore Offshore 'farms' require large areas of sea Onshore technology is relatively inefficient Environmental impacts – reduces turbulence, increases sedimentation
Geothermal	Clean and sustainable Relatively cheap in suitable areas	Reliant on specific geology to be viable so restricted to certain locations

Energy mix and components of energy demand

Key question

Why do individual nations develop different compositions of resources to provide their energy?

The global TPES is in effect the global **energy mix** but this will vary considerably from country to country. The utilisation of this energy will also vary from a global average, especially between industrialised and non-industrialised countries.

Energy mix

Energy mix is the composition of different primary energy sources from which households and industries in a named area (usually at national level) get their energy. An important consideration when analysing energy mix is that some data refers only to the combination of energy sources used to generate electricity. This is not the primary energy mix of a country as it only includes one form of secondary energy and ignores transport or other uses such as gas or biofuels used in heating or cooking, etc.

Energy mixes in contrasting settings: Mali, Iceland and France
Mali

- **Energy mix:** With 78 per cent of a 3.500 mtoe annual primary energy supply, biomass, mainly in the form of wood and charcoal for domestic use, plays the dominant role in the Malian energy mix. Despite the substantial oil reserves in the north of the country, Mali depends on importing fossil fuels, mainly refined oil for transport (costing around 16 per cent of the national budget). The remaining 4 per cent of the primary energy supply is largely made up of renewably generated electricity, mainly HEP from the Félou project on the Senegal River. Mali is starting to develop its huge potential for solar energy, opening its first grid-connected solar power plant in 2011 (Figure 11.46).
- **Energy consumption:** Households consume 86 per cent of Mali's energy, (road) transport 10 per cent, industry (mainly mining) 3 per cent and agriculture 1 per cent.

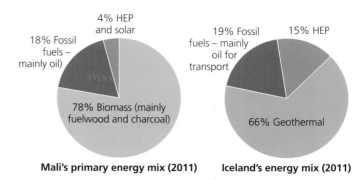

Figure 11.46 Pie diagrams showing contrasts between energy mix in Mali and Iceland, 2011

Mali's primary energy mix (2011): 18% Fossil fuels – mainly oil; 4% HEP and solar; 78% Biomass (mainly fuelwood and charcoal)

Iceland's energy mix (2011): 19% Fossil fuels – mainly oil for transport; 15% HEP; 66% Geothermal

- **Potential:** There is potential for harnessing more HEP from the Niger and Senegal rivers and for developing more solar energy. Undeveloped wind energy potential is also available, particularly in the Sahelian and Saharan zones, where annual average wind speed is estimated at 3 to 7 m/s.

Why this energy mix? There is little use for electricity except in urban areas. The main energy need is for cooking and heating, for which firewood is the cheapest and most readily available source. There are no coal reserves and oil reserves have not yet been developed. Mali is a very poor country and cannot afford to develop much of its energy potential without help from the World Bank.

Iceland

- Iceland has one of the highest energy consumption/capita of any country.
- It is one of the lowest carbon emitters as it only uses fossil fuels for transport.
- It straddles a plate margin which is geologically active so has abundant geothermal energy.
- It is a mountainous country with plentiful water supply and potential energy in falling water for HEP.
- As an advanced country with a relatively small population, Iceland can afford the technology to harness its geothermal potential and to construct dams to develop HEP for its electricity supply (Figure 11.46).

Table 11.16 Iceland's renewable energy, 2011

Source	Primary energy	Electricity only
HEP	15%	76%
Geothermal	66%	24%
Fossil fuels	19%	0

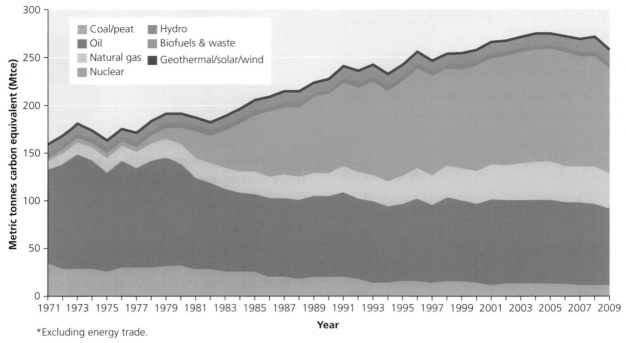

Figure 11.47 Compound line graph showing France's primary energy mix 1971–2009

France

- France gets 38 per cent of its primary energy from 59 nuclear power stations (Figure 11.47).
- Nearly 80 per cent of France's electricity is supplied by nuclear energy and it is able to export electricity.
- Government policy of generating energy from nuclear followed the OPEC oil crises of the 1970s.
- France has a shortage of fossil fuel reserves (especially oil/gas).
- Physical conditions were suitable for developing nuclear energy with many large rivers for cooling.
- Politically it has been a less controversial option for determining energy security.

Factors influencing a country's energy mix

- **Availability of energy sources within the country:** Nations seek to use available resources first.
- **Inertia:** Retaining a mix that already exists because of economic or technical difficulties in changing (even if the reasons are outdated). For example, the UK's continued reliance on coal.
- **Government energy policy:** Making decisions to strive for energy security (for example, France) or to abide by international treaties (for example, Kyoto).
- **Geopolitics:** Trading partners who are friendly or the country can co-operate with; some suppliers cannot be relied on or present potential conflict.

- **Level of development (economic/technological):** Has to be 'appropriate' for example, less developed countries cannot afford to develop nuclear energy or become over-reliant on fossil fuel imports.
- **Physical/locational conditions:** Suitable for certain types of renewable energy (for example, UK – wind; Sweden, Iceland and Norway – HEP; Spain – solar).
- **Diversity:** Governments may decide to diversify their energy mix so that countries are not too reliant on only one or two sources.

Components of energy demand

Energy that is generated by primary sources is used in a number of ways, but most of the data divides the components into four categories: industrial, commercial, residential (household use) and transport. Figure 11.48 shows the world energy consumption by

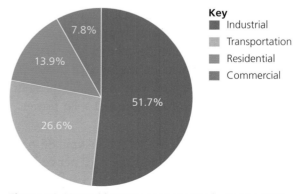

Figure 11.48 World energy consumption by sector, 2012

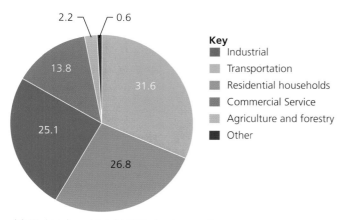

(*) Figures do not total 100% due to rounding.

Figure 11.49 EU energy consumption by sector, 2013

Table 11.17 China's energy consumption by sector

	Industrial use	Residential and commercial use	Transport
Electricity (end of use)	70%	20%	1%
Oil	38%	8%	34%
Natural gas	54%	32%	1%

these four sectors in 2012 and at this scale the majority of energy is used for industrial purposes.

Consumption by sector will vary from region to region and Figure 11.49 shows a slightly different composition for the EU in 2013. Here, in some of the most developed countries of the world, energy generated for transport, services (commercial) and household use (residential) is significantly higher and the proportion devoted to industrial use is lower.

Conversely, rapidly industrialising countries, such as China, show a higher than global average consumption by the industrial sector. For example, 70 per cent of China's end-of-use electricity is consumed by industry, while only 20 per cent is for residential and commercial utilisation (see Table 11.17).

Figure 11.50 shows differences in consumption by sector among selected countries and demonstrates

the importance of energy for industrial use in China. In India, residential, commercial and agricultural use is more dominant, contrasting with the USA where transportation is the major use of energy. The varying compositions of energy use generally reflect the importance of each sector to a country's overall economy.

Energy supply and physical geography

Key question

How do physical conditions influence the geographically specific nature of energy supply?

The availability of different energy resources is dependent on a number of physical limiting factors; for fossil fuels and geothermal energy, geological conditions are a major determining factor, while for many renewable sources of energy, such as wind and solar energy, climate is a more important constraint.

Geology

Coal

Coal is a sedimentary rock which started to form 360 million years ago in swamp conditions on the edge of sedimentary basins such as lagoons or lakes. Plant debris accumulated and was buried under layers of mud and sand where the decomposition process was slowed down due to the anaerobic conditions. The basin gradually sank under the weight of more sediment and the layers of dead vegetation were subjected to increasing amounts of heat and pressure in a process known as 'coalification' (Figure 11.51, page 582). Increased heat and pressure over time reduces the levels of moisture and volatile materials present in the sediment. At the same time the carbon content and quality of coal that is produced will increase.

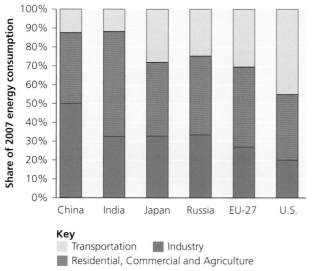

Key
Transportation   Industry
Residential, Commercial and Agriculture

Figure 11.50 Differences in energy consumption by sector in selected countries/regions, 2007

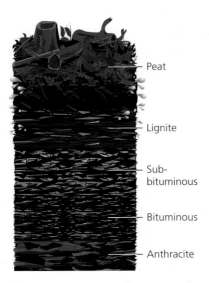

Figure 11.51 Stages in the process of coalification

The different types of coal formed are ranked according to their quality, which is dependent on carbon content:

- **anthracite:** (86 per cent to 98 per cent pure carbon) is an excellent fuel that is still used to heat homes
- **bituminous coal:** (containing 70 per cent to 86 per cent carbon) is used to make coke for furnaces
- **sub-bituminous coal:** (70 per cent to 76 per cent carbon) is burned in industrial boilers
- **lignite** (also known as '**brown coal**'): (65 per cent to 70 per cent carbon) is a low-grade fuel with a high moisture content that is also used in industrial boilers.

Thermal power stations generating electricity use a grade which is between anthracite and bituminous known as **steam coal**. The coal is first milled to a fine powder, which increases the surface area and allows it to burn more quickly.

Oil and natural gas

Oil and natural gas are hydrocarbons of organic origin (plants, algae and animals) which settled to the bottom of the sea as they died, were buried and fossilized in sedimentary 'source' rocks such as finely grained shale. The organic matter in the sediment changed into a substance called kerogen and when underground temperatures rose above 110°C the kerogen gradually changed into oil. Under hotter conditions it formed into natural gas.

After oil and gas have formed in the shale, pressure continues to rise, squeezing the oil and gas out, upwards or horizontally into more porous and permeable rocks (for example, sandstone) that have

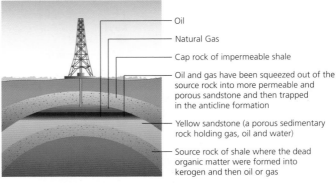

Figure 11.52 A conventional oil and natural gas reservoir: this illustrates a structural 'anticlinal trap' that contains oil and natural gas

more spaces within them for the oil and gas to flow into. Some eventually reaches the surface and escapes naturally, but most of it comes up against a layer of impermeable rock that it cannot pass through. This is known as a 'cap' rock because it forms a seal or trap, and slowly the oil and gas build up in the porous rock beneath it forming a reservoir.

Reservoirs are sedimentary rock formations that hold oil, natural gas or both within their pores, like a fossilised sponge. All pore spaces in the rocks are filled with water, gas or oil. Crude oil is a complex mixture of hydrogen, carbon and traces of other substances. Its texture varies, but it is generally liquid and in a geological trap it will accumulate above layers of trapped water. Natural gas is mainly made up of methane; as it is the lightest substance it moves to the top of the trap.

There are two types of traps – **structural traps** and **stratigraphic traps**. Structural traps such as the one shown in Figure 11.52 hold gas and oil because the Earth's crust has been deformed in some way, such as the dome-shaped anticline shown.

Unconventional availability of oil and gas

Hydraulic fracturing ('fracking'): Economic pressures on fuel prices have prompted technological advances enabling the **resource development** of other known reserves of oil and gas, such as those found in shale. The pore spaces in shale are so tiny that the oil and gas have difficulty moving through it into a conventional reservoir.

New technology has enabled engineers to penetrate down to the level of the shale and then drill horizontally through the shale rock (Figure 11.53). The permeability of the shale is increased by pumping

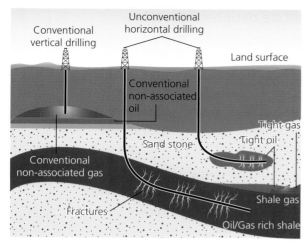

Figure 11.53 Unconventional fracking of 'tight' gas and oil deposits compared to conventional drilling

water down the well under sufficiently high pressure to fracture the rock. These fractures liberate some of the gas and oil from the pore spaces and allow them to flow to the conventional well or directly to the surface. This technology means that the previously '**possible resources**' of oil and gas are now accessible as recoverable '**measured reserves**'.

Oil and tar sands: These are bituminous sands and are a type of unconventional petroleum deposit. The largest discoveries and development of these reserves have been on the Athabasca River in Alberta, Canada and in the Orinoco basin in Venezuela (though technically the latter fall into the category of 'heavy oils'). These areas have thus become **resource frontiers**.

The bitumen is too thick to be pumped as a liquid from the ground so has to be mined by strip mining or open pit mining. The relatively low recovery rate of oil of around 10 per cent is compensated for by the massive scale of the reserves and the relatively low cost of extraction. A number of techniques have been developed to separate the bitumen from other constituents but it still takes around four tonnes of tar sands to produce one barrel of oil (see the text box on the Athabasca tar sand oil development in Alberta, Canada, page 585).

Climate
Solar energy

Solar energy can be harnessed anywhere that receives sunlight, so it is abundantly available but it is also variable and intermittent. Different locations and weather conditions will affect the amount of energy that can be harnessed for heating or electricity

production. Solar power cannot generate electricity at night without storage mechanisms, and is less effective in overcast or cloudy conditions.

Climatic considerations:
- Sunlight:
  - tropical and sub-tropical areas receive the most reliable and regular sunlight
  - the sun is higher in the sky so the days are longer
  - cloudy regions make solar systems harder to justify, though not necessarily uneconomical.
- Frequent fog, smog or air pollution: These will negatively affect the amount of sunlight received.
- Air density: Solar exposure is better in the mountains than near sea level because the air is thinner and scatters less sunlight.
- Snowfall: PV systems will not operate effectively if covered with heavy layers of snow.
- Rainfall: Electrical and metallic components are susceptible to corrosion in humid conditions.
- Wind: Frequent strong winds can damage solar installations and cool down surfaces on solar water heating panels quickly.

Wind energy

Wind energy can also be variable and intermittent. For economic reasons wind farms are most viably productive in areas where average wind speed is greater than 5.5 metres per second.

- **Wind speed**:
  - The minimum wind speed at which the wind turbine will generate usable power is between 7 and 10 mph for most turbines.
  - The minimum wind speed at which the wind turbine will generate its designated power (known as 'rated power') is between 25 and 35 mph for most turbines.
  - At very high wind speeds of between 50 and 80 mph, most turbines will shut down. This is the cut-out speed and is a safety feature which protects the turbine from damage.
  - Turbines have to be spaced well apart to prevent eddying or a reduction in wind speeds in the middle of wind farms, which means large areas of land are required.
- **Air density**: The greater the density of the air, the more energy is received by the turbine. Air is less

dense at higher elevations than at sea level, and warm air is less dense than cold air.

- **Prevailing winds**: Optimum conditions for wind energy generation exist when there is a prevailing wind direction.

The UK has the best wind energy potential in Europe because of its position in temperate latitudes on the northeast rim of the Atlantic. It has the long 'fetch' of uninterrupted air flow across the ocean from which prevailing south-westerlies blow.

Drainage systems

The size and shape of drainage basins, together with the volume of water in the channels, are important considerations in the development of large- or small-scale conventional HEP which involves dam construction. Two vital factors to consider are the **flow** and the **head** of the stream or river. The flow is the volume of water which can be captured and re-directed to turn the turbine generator, and the head is the height the water will fall (potential energy). The larger the flow and the higher the head, the greater will be the energy available for conversion to electricity.

The key equation to remember is:

$$\text{Power} = \text{Head} \times \text{Flow} \times \text{Gravity}$$

The topography of the drainage basin is another important consideration as dam construction is expensive. A long, steep-sided valley basin with a relatively narrow exit is ideal as this maximises the volume of the reservoir and makes dam construction more manageable and less costly. The geology of the rocks should be stable and impermeable, to secure the dam and to prevent water seepage.

Energy supply and globalisation

Key question

How and why has the energy industry become a feature of the globalisation process?

By 2025 energy consumption in less developed countries is set to surpass that of more developed countries. As countries such as China and India develop, so will their appetite for natural resources, leading to fierce demand for ever-diminishing supplies of non-renewable energy.

Competing national interests

The importance of oil for transport and the versatility of gas, combined with their uneven distribution and relatively short lifespan, make these the most important commodities. Oil is vulnerable to any event that impacts on its supply and demand. All countries that depend on oil imports are defenceless against external events affecting their supply.

As reserves of non-renewable fossil fuels continue to diminish, the following scenarios are likely:

- European dependence on Middle East oil will remain significant.
- Asian dependence on Gulf oil will increase.
- China will start to develop oil reserves in Africa (for example, Chad and Sudan).
- The USA will still rely on imports, despite developing more of their 'tight' reserves.
- Natural gas will continue to be the fastest-growing primary energy source.
- The growth of natural gas will require major infrastructure investments (pipelines).
- Russia will expand its exports of oil to China, Japan and South Korea through the East Siberian Pacific Ocean (ESPO) pipeline but will want to increase gas supplies to Europe.
- The use of sustainable renewable sources of energy will increase.

Except for the growth of trade between Russia and its eastern Asian neighbours, it is unlikely that the pattern of oil production and trade will change significantly in the near future.

Role of transnational corporations (TNCs)

Tremendously wealthy and powerful TNCs dominate the international oil trade. Oil companies have considerable power in today's globalised world; they help fund development projects and even support presidential campaigns in many countries. Because of their size, influence and wealth they can control and co-ordinate economic activities and influence governments, for example by being granted exploration rights in an area to initiate the cycle of natural resource development there. Energy TNCs are usually involved in all stages of exploration, production and distribution.

Three out of the top five largest TNCs in 2011 were oil companies — Royal Dutch Shell, BP and Exxon-Mobil.

Energy companies based in developing countries are also starting to exert influence on a global scale. They largely control production in their own country; some are totally or partially state-owned, such as Petrochina, Gazprom (51 per cent owned by the Russian state) and Petrobras (a semi-public Brazilian company).

Royal Dutch Shell

Shell is an Anglo-Dutch transnational oil and gas company with headquarters in The Hague in the Netherlands.

Shell is **vertically integrated** and is active in every area of the oil and gas industry, including:

- exploration and production – using geological expertise
- refining – it owns 47 refineries
- distribution and marketing – it has its own fleet of oil tankers and lorries
- petrochemicals
- power generation and trading
- some renewable energy activities in the form of biofuels.

Shell has in excess of 100,000 employees in over 100 countries, produces around 3.1 million barrels of oil equivalent per day and has 44,000 service stations worldwide. Its main production activities are in South East Asia and in the Niger Delta in Nigeria where it has been accused of damaging local communities and the environment by its activities. Shell does, however, provide jobs and income, as well as improving the skills, education and training of the local workforce. Like other energy TNCs, Shell will sometimes work in consortiums or joint ventures with other companies (for example, with Petronas, a Malaysian state company) to gain exploration rights, or with other TNCs to invest in mutually advantageous infrastructure such as oil pipelines. Like many other energy companies, Shell saw the need to invest in alternative, renewable sources of power. It withdrew its interest in wind and solar in the 2000s but has retained an interest in Brazilian biofuels and is also involved in research into large-scale hydrogen energy projects.

Figure 11.54 Shell tankers at the Stanlow oil refinery in Cheshire (owned by Shell until 2011)

Environmental impacts of the Athabasca tar sand development in Alberta, Canada

Background to tar sand development in Alberta

As conventional sources of crude oil are depleted, unconventional sources, such as the bitumen found in tar sands (also known as oil sands) play a role in offsetting declining conventional production. The Canadian tar sands rival the conventional oil in Saudi Arabia and unconventional heavy oils in Venezuela as the largest proven reserves of oil in the world.

The bulk of Canada's tar sands reserves are found in three major deposits in the province of Alberta, where they underlie 140,000 km² of relatively pristine boreal forest, an area that is approximately the size of the state of Florida (Figure 11.55). The largest and most viable of the deposits is the Athabasca River deposit. The sands are estimated to contain 175 billion barrels of recoverable reserves of crude bitumen. The portion of land which is currently mined is around 5,000 km². The land here is leased to companies for open-pit strip mining. For the remaining deposits, thermal 'in-situ'

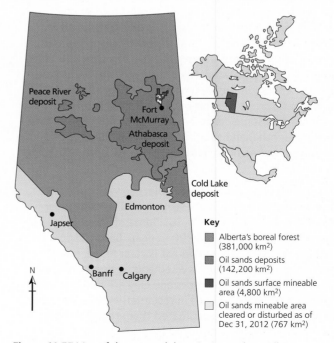

Figure 11.55 Map of the tar sand deposits in northern Alberta

methods have been developed to extract bitumen reserves which are too deep to be mined.

The oil sands have been developed commercially since the 1960s but advances in extraction technology combined with continued rising oil prices led to more extensive development from the start of the twenty-first century. Canada's so-called 'black gold' has come to be regarded as an abundant, secure and affordable source of crude oil. More than a million barrels of crude flow out of Alberta's oil sand plants every day and most of this is exported to the US where it provides around 7 per cent of their daily consumption. There are around 20 mine projects in operation, owned and operated by a number of different partner companies, including many large transnational oil corporations such as Chevron-Texaco, Shell and Total (and smaller specialist mining companies with various national affiliations); in addition to Canada there is investment from the US, UK, Japan, South Korea, France and Norway.

Environmental impacts

Development of this unconventional resource requires unconventional extraction and production processes with associated risks and consequences. To justify the investment, the scale of the project is enormous, arguably the largest industrial project in human history and possibly the most destructive.

Producing oil from the tar sands is complex. They consist of 85 per cent sand, clay and silt, 5 per cent water and 10 per cent crude bitumen (a tar-like substance which is converted to oil). Bitumen does not flow like oil and extracting it from the sands is an extremely dirty process. The technology is a hot-water-based separation process that requires large quantities of both water and energy. Once separated, the bitumen has to undergo further energy intensive processes to upgrade it into a synthetic crude oil. This occurs before it is sent to refineries. In some cases, however, raw bitumen is diluted (using naphtha) and sent by extensive pipeline networks to refineries (mainly in Midwest USA) for both upgrading and refining.

Most current production takes place in vast open-pit mines where the sands are strip-mined layer by layer from the surface (Figure 11.56). Some of these open-pit mines can be 150 km² and 90 m deep. Before strip-mining begins, the boreal forest is clear felled, rivers and streams have to be diverted and wetlands drained. The overburden of soil, rocks and clay are stripped away (and stockpiled to be used in reclamation) to access the bitumen deposits. It is estimated that around 4 tonnes of material has to be removed to produce each barrel

Figure 11.56 Open pit mining of Athabascan oil sands in northern Alberta, Canada

of bitumen. This leaves massive scars on the landscape as well as destroying habitats and natural ecosystems, leading to Canadian Geographic terming the mining projects as 'scar sands'.

Only a fraction of the bitumen is close enough to the surface to be strip mined and over 80 per cent of the established reserves are deeper and must be extracted **in situ** using steam. The bitumen is heated and softened by injecting high pressure steam into the ground enabling it to be pumped out, leaving most of the solids behind. In situ development could occur in an area 30 times greater than the current mining area. In-situ extraction methods have a much smaller footprint and require less land disturbance, but they impose another set of negative environmental impacts.

Table 11.18 Advantages and disadvantages of 'in-situ' extraction compared to open-pit mining

Advantages of in-situ bitumen extraction	Disadvantages of in-situ extraction
Smaller surface footprint – surface area for well heads is relatively small	**Linear disturbances** – from 3D seismic and core hole exploration, roads and pipelines; these negatively impact on wildlife
More efficient water usage – water used for steam production is mostly recovered and recycled	**More production well-pads** – smaller footprint but more of them and more scattered
No tailings pond – much of the sand is left in the ground	**Lower bitumen recovery rates** – only about 40–60 per cent; much lower than surface mining
Lower costs – with no mine and no tailings pond, projects require less capital	**Greater greenhouse gas emissions** – steam is produced by burning natural gas

Climate change

Tar sand development is extremely carbon intensive. The mining and upgrading procedure releases at least three times the CO_2 emissions of conventional oil production and is said to be the single largest industrial contributor in North America to climate change. The transition from conventional oil to tar sands actually worsens greenhouse gas emissions, as these processing emissions are added to the emissions from the end product.

The carbon emissions contribute to a rise in global temperatures, which have impacts on a global scale. Ironically, the effects of this can be seen visibly within the Athabasca basin with the rapid melting and retreat of the Athabasca Glacier upstream in Jasper National Park in the Canadian Rockies. The outlet stream from the glacial lake is the source of the Athabasca River, which eventually passes through the bitumen deposits of the oil sands on its way to the Arctic Ocean. Water is abstracted from the river, to be superheated for the in situ extraction of bitumen and also for separating the bitumen extracted by open-pit mining. The emissions from these processes, and from the vehicles used by tourist visitors to the glacier, contribute to the greenhouse gases that are melting the glacier. This is an example of the human-induced (Anthropocene) resource cycle with accelerating positive feedback effects.

Loss of forests and wildlife habitat

The tar sands are already considered to be the cause of the second fastest rate of deforestation on the planet after the Amazon Rainforest. Canada's boreal forest is globally significant as it is a complex ecosystem of trees, wetlands and lakes representing 25 per cent of the world's intact forests. It provides 1.3 billion acres of wild habitat for a vast array of species including wolves, grizzly bears, lynx, nesting migratory birds and thousands of plant varieties.

Planned tar sand development projects are expected to see at least 5,000 km² of forest cleared, drained and stripped for open-pit mining. The remaining 137,000 km² could well be fragmented into a web of seismic lines, roads, pipelines and well pads for in-situ drilling. These will negatively affect the populations of wildlife, such as the woodland caribou, which avoid linear features.

Water abstraction and pollution

Approved projects will see 3 million barrels of tar sands produced daily by 2018; for each barrel of oil produced as much as five barrels of water are used in extraction, separation and upgrading processes. The Athabasca River passing through the areas being cleared and strip-mined serves as the primary source of water used in these processes. Current operations are permitted to withdraw 350 million m³/year (equivalent to the amount used by a city with 2 million people). Initially, it was thought that the river had sufficient flows to meet the demands of tar sand operations but it is becoming clear that this might not be the case, particularly in winter when flows are naturally lower. Even in-situ operations, despite recycling, leave some water underground, so a continuous supply of new water is needed. As these operations are sometimes a distance from rivers, they rely on using groundwater aquifers, which lowers the water table in the region and threatens surface freshwater.

Water over-abstraction poses threats to the sustainability of fish populations in the Athabasca River and also to the Peace-Athabasca Delta, the largest boreal delta on Earth. It is a World Heritage Site and one of the most important waterfowl nesting and staging areas in North America.

Tar sand mining operations produce large volumes of waste in the form of **tailings** (around six barrels of tailings per barrel of bitumen extracted). Liquid tailings are a slurry mix of water, sand, clay and unrecovered bitumen but also contain naphthenic acids and trace metals making them toxic to aquatic organisms and mammals. Operators are required to store tailings waste in vast wastewater reservoirs on site in large containment dykes, because the water is too toxic to be returned to the river. These reservoirs are misleadingly referred to as tailings 'ponds', as they cover an area of around 100 km² and are thought to contain 720 billion litres of toxic tailings (Figure 11.57). There is growing evidence that toxic chemicals from tailings ponds are leaching into groundwater and into the nearby Athabasca River.

Air quality

Air pollution from tar sands operations is increasing at both local and regional scales. Criteria Air Contaminants (CACs) are the most common pollutants released by

Figure 11.57 Polluted tailings ponds near the Suncor tar sands mining operation in northern Alberta

heavy industry. They include lead, particulate matter (PM), carbon monoxide, nitrogen oxide and sulphur dioxide, all of which are emitted in large volumes by tar sand operations. These pollutants affect human health and contribute to acid rain.

Impacts on humans

Human health in many communities has taken a significant turn for the worse, allegedly caused by tar sands production. Production has led to many serious social issues throughout Alberta, from housing problems to the vast expansion of temporary foreign worker programmes that exploit so-called 'non-citizens'. Water abstraction and pollution of the Athabasca River also jeopardises subsistence and commercial fishing by local aborigines such as the Cree Indians.

Cumulative impacts and reclamation

In addition to the environmental devastation in northern Alberta, pipeline infrastructure to refineries and to supertanker ports crosses the continent to all three major oceans and the Gulf of Mexico.

Very little of the area directly affected by mining operations has been reclaimed and tailings ponds are expected to cover even greater areas over the next 20 years. UNEP has identified Alberta's tar sand mines as one of 100 global 'hotspots' of environmental degradation. According to Environment Canada, development of the tar sands presents 'staggering challenges for forest conservation and reclamation' and yet oil sands projects continue to be approved.

Management of energy security

In order to ensure security of energy supply, countries have to develop strategies which will increase the reliability of future supplies and simultaneously reduce consumption by maximising energy efficiency.

Supply management strategies

Decisions about energy security will often lead to countries diversifying their energy mix, so they are not too dependent on one source of energy, and thus vulnerable to economic change.

Oil and gas exploration

As the price of fossil fuels increases, extra exploration effort is invested in finding new reserves or reassessing speculative reserves to analyse whether they have become viable to develop. Countries are more willing to grant exploration rights to TNCs in these circumstances. For example, although North Sea oil and gas are past their peak, exploration continues and new discoveries are still being made. In August 2015, the biggest gas field discovery in a decade was made by Maersk, a Danish TNC. It has been approved for production by Britain's oil and gas authority and will trigger a £3 billion investment that could supply five per cent of the UK's gas consumption.

Oil and gas price rises prompt large TNCs to develop new technologies to access previously unviable reserves. Such has been the case with hydraulic fracturing in the USA and this seems likely to be emulated in the UK, by similar exploration for unconventional sources. In August 2015, government ministers announced that they would award fracking companies, such as Cuadrilla, licences to explore for oil and gas in areas spanning 1,000 square miles of Northern England and the East Midlands (Figure 11.58).

Nuclear power

Nuclear power provides a long-term, efficient and viable energy security alternative for many countries which have become over-reliant on fossil fuels. It currently provides around 20 per cent of the UK's electricity but all operating energy plants (with the

Figure 11.58 A Cuadrilla drilling rig in Lancashire

exception of Sizewell B in Suffolk) will be closed by 2023. No new reactors have been built since the 1980s. In the government's 2002–03 Energy Review nuclear expansion was put on hold for economic, political and environmental reasons.

An energy crisis in 2006 prompted a review of nuclear policy and an almost complete 'U' turn, with nuclear again being considered as a major component of energy security because of:

- increasing reliance on imports of gas
- insufficient energy being generated from renewables
- the need to meet CO_2 targets agreed at Kyoto.

The 2007 Energy Policy review announced that more nuclear power stations would be built and, together with renewables, nuclear would be an important part of the future energy mix. New nuclear reactors are planned, to be built on existing nuclear energy sites such as Sizewell, Hinkley Point (Somerset), Oldbury (Gloucestershire) and Wylfa (Anglesey). The government invited bids from companies to build and operate the new plants and there has been substantial international interest in the UK's nuclear programme. Most of the new installations are planned to be operational from the mid-2020s and it seems likely that they will be run by a combination of French, Chinese, Japanese and German-owned companies.

Development of renewable resources

UK energy policy since the 2008 Energy Act has been built around reducing CO_2 emissions rather than security of supply or cost, so there has been much support for the development of sustainable fuel sources. A variety of government subsidies have been offered to encourage renewable electricity generation by businesses, local authorities and individual householders. Financial support schemes include:

- **Renewables Obligations** (RO): Targeted to provide financial support for electricity suppliers with large-scale installations over 5 megawatts (MW).
- **Feed-in Tariff** (FiT) scheme: Aimed at small-scale installations up to 5 MW; generators are paid for every unit of electricity they produce.

These incentives have seen the rapid growth of renewable energy in the UK, particularly from wind,

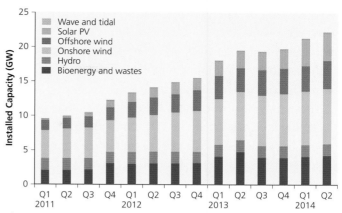

Figure 11.59 Increases and composition of renewable energy sources in the UK, 2011–2014

solar and biomass sources (Figure 11.59). The UK has been the world leader in offshore wind energy since 2008, with as much capacity already installed as the rest of the world combined.

Skills focus

Using the compound bar graph in Figure 11.59, estimate the percentages of each type of renewable energy resource used in i) the first quarter of 2011 and, ii) the second quarter of 2014.

Some notable case studies from the DECC report 'UK Renewable Energy Roadmap Update 2013' include:

- **solar – Blackfriars Bridge:** the world's largest solar bridge with 4,400 solar PV panels installed on the bridge spanning the Thames, the foundation for the newly built Blackfriars Station
- **wind – London Array:** the world's largest offshore wind farm opened in July 2013 off the Kent coast
- **biomass** – New Earth's **Avonmouth** Energy Recovery facility generates electricity from anaerobic gasification of recycled municipal and commercial waste from a neighbouring waste processing plant.

In 2015, the government decided to cut the subsidy schemes offered for renewable generation installations. The RO support for larger projects was cut in April 2015, followed by an announcement of cuts to the FiT scheme to take effect from 2016.

Demand management strategies – managing energy consumption

The tone is set to reduce energy consumption by international agreements such as Agenda 21 and the

Kyoto Protocol. Energy conservation efforts are mostly driven by:

- government incentives (for example, subsidising householders' loft and cavity wall insulation)
- media and other campaigns to raise awareness – these may be government led or run by energy-saving NGOs such as the Carbon Trust
- pricing strategies – the cost of fuel gives an incentive for energy saving but off-peak electricity pricing can also reduce pressure on generation.

Household energy saving

The EU Directive on the Energy Performance of Buildings is an important part of government strategy for tackling climate change. The directive requires an energy performance certificate (EPC) to be issued whenever a property is constructed, rented out or sold. The EPC shows the energy efficiency rating of a dwelling on an A–G rating scale similar to those used for refrigerators and other electrical appliances. To achieve good ratings homes will have to consider:

- improving thermal efficiency of walls, windows and roof
- draught-proofing of floors, doors and windows
- installing a high efficiency condensing boiler
- being part of a district heating system (for example, Southampton or Woking)
- using new building materials to reduce heat loss (thermal blocks)
- using low carbon technologies: solar panels, biomass boilers or wind turbines
- installing energy efficient appliances and lighting
- improved daylighting – fitting larger windows (passive solar heating)
- using environmentally friendly/recycled building materials.

Industrial and commercial energy saving

The Carbon Trust is a non-profit organisation whose primary aim is to help the business community move toward a low carbon economy. It helps large companies create climate change strategies and gives smaller businesses free energy audits and no-interest loans for energy efficient equipment. The energy savings strategies offered are similar to those in residential premises.

Many industrial processes generate heat, which is a source of energy that can be wasted unless it is captured. The Carbon Trust encourages industrial

energy consumers to install **heat recovery systems** which collect and re-use heat arising from any process. Heat recovery can help to reduce the overall energy consumption of the process itself, or provide useful heat for other purposes. This should not be confused with **combined heat and power systems**, which are also beneficial in reducing energy consumption in industrial, commercial and domestic settings.

Combined heat and power (CHP) systems

CHP generates electricity while also capturing usable heat that is produced in the process. Conventional electricity generation in coal- and gas-fired power stations generates heat but, because of their remote location, up to two-thirds of the overall energy is lost. Energy is also 'lost' in transmission costs from generation to consumption.

CHP enables consumers to generate their own power locally using renewable or non-renewable fuels. It captures the waste heat from generation so it can be used for space or water heating, thus reducing energy consumption and carbon emissions (Figure 11.60). It can be used in homes, businesses or public settings. For example, Center Parcs most recent site at Woburn Forest uses two CHP generators, which saves the company £200,000 and 900 tonnes of carbon emissions per year.

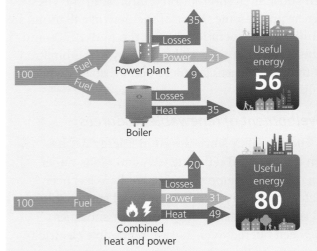

Figure 11.60 The principle of combined heat and power systems. Generating electricity locally can increase efficiency and reduce consumption

Sustainable transport strategies

Energy conservation can be achieved in transport use through technology, design and incentives to alter lifestyle and the type of transport used.

- **Technology:**
  - engine design: more efficient fuel usage (for example, fuel injection) with lower emissions

- ignition control: reduces wastage
- hybrid cars: using electric power – these will be more sustainable options as the proportion of electricity generated by renewable sources increases.
- **Methods and design:**
  - aerodynamic designs reduce fuel consumption
  - using more renewable fuels (for example, bioethanol)
  - using cleaner, more efficient fuels (for example, LPG, low sulphur alternatives).
- **Schemes and campaigns to discourage car travel or reduce congestion:**
  - government policy/legislation (for example, road tax based on emissions)
  - congestion charges (for example, London Congestion Charge Zone)
  - toll roads (for example, M6)
  - park and ride schemes in cities combined with high car park charges in city centres
  - campaigns and incentives to encourage car sharing
  - campaigns to encourage taking public transport/cycling/walking
  - planning homes, services and workplaces in close proximity to reduce travelling.

Energy sustainability issues

One of the main differences between renewable and non-renewable sources of energy is their impact on the environment. Although they have disadvantages, renewable sources of energy are 'cleaner' and less harmful to the atmosphere than non-renewables. Non-renewable sources of energy release harmful pollutants, such as carbon and sulphur compounds, into the atmosphere when they are burnt.

Environmental impacts

The environmental impacts of using energy resources take place during exploitation, transportation and when using the fuels to produce energy.

In developing countries, particularly in rural areas, people largely rely on biomass such as fuelwood to meet their energy needs for cooking. The use of biomass is not in itself a concern unless fuel is harvested unsustainably. The scarcity of fuel supplies has led to the widespread removal of woodland for firewood. This results in less interception of rainfall, reduced infiltration, faster run-off and greater soil erosion by water and wind often leading to desertification.

Table 11.19 Summary of environmental impacts caused by using energy resources

Stage of utilisation	Type of impacts	Examples
Exploitation	Fuelwood: deforestation leading to soil erosion and desertification	Sahel region of sub-Saharan Africa
	Oil/gas well leaks from drill points: damages ecosystems, kills wildlife, leaks methane	BP's Deepwater Horizon, Gulf of Mexico oil spill
	Open cast mining for coal and oil sands scars landscape and disrupts drainage	Athabasca oil sands, Alberta, Canada
	Hydraulic fracturing (fracking) may lead to subsidence and water contamination	
Transportation	Oil and gas leaks from ruptured pipelines	Niger Delta, Nigeria
	Oil spills from large oil tankers at sea	Exxon Valdez spill, Alaska
Utilisation	Combustion of fossil fuels leading to: Acid rain from carbon, nitrogen and sulphur compounds causing damage to wildlife, vegetation and to buildings	Scandinavia; Northeast USA; South-east China and Eastern Asia
	Enhanced greenhouse effect causing global climate change	Global effects

Acid rain

During the 1980s and 1990s a major environmental crisis confronting developed regions of Western Europe and Eastern North America was **acid rain**. The impacts were notable because the areas suffering mostly from the impacts were not the areas which were the source of the problem. Acid rain is a 'migrating' pollution. Scandinavia suffered the effects of acid rain from polluting sources in the UK and Germany. Acid rain is now prevalent in South East Asian countries such as Thailand and Laos with the pollution originating from industrialisation in India and China.

● **Causes:** The main sources of acid deposition are:
  – burning of fossil fuels in power stations
  – exhaust fumes from motor vehicles.

● **Formation:** Acid rain consists of the deposition of pollutants, such as sulphur dioxide and nitrogen oxide, released by burning fossil fuels. These chemicals react with precipitation, mist and clouds in the atmosphere to produce wet deposits of sulphuric acid (H_2SO_4), nitric acid (HNO_3) and compounds of ammonia when it rains (Figure 11.61).

● **Environmental impacts:** Acid rain:
  – damages and kills trees, particularly conifers (causing yellowing of needles)
  – leaches toxic metals (aluminium) from soils which then accumulate in rivers/lakes
  – kills fish stocks and damages ecosystems
  – damages buildings and monuments by accelerating weathering, particularly of limestone
  – has harmful effects on human health causing respiratory complaints (for example, bronchitis).

● **Reducing impacts:** Various ways of reducing acid deposition exist:
  – use of catalytic converters on cars reduces NOx emissions from exhausts
  – burning fossil fuels with a lower sulphur content
  – replacing coal-fired power stations with nuclear or other sources of electricity
  – use of 'scrubbers' and flue gas desulphurisation (FGD) plants that remove sulphur either before or after coal is burnt
  – energy conservation – reducing use of and therefore demand for electricity
  – 'liming' – adding lime to neutralise the acidity in acidic lakes – this is only a short-term measure and does not solve the problem (used in Norway and Sweden).

Acid deposition in South East Asia

Worsening air pollution, as a result of economic expansion, is leading to higher acidity in rainwater in cities across East Asia. In Thailand, monitoring showed that from 2001 to 2008 Bangkok was the city with the highest chemical deposition. It is caused by the increasing consumption of fuel by vehicles, combined with smoke from industrial plants and residue from other human activities, such as farmers burning their fields. Acid rain is polluting rivers and streams, threatening aquatic life and also poses threats to human health and to valuable heritage monuments through the corrosion of stone.

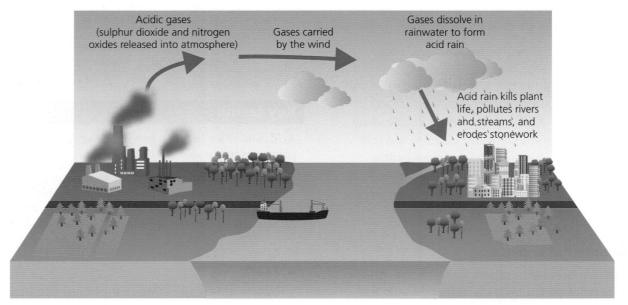

Figure 11.61 The process of acid rain formation

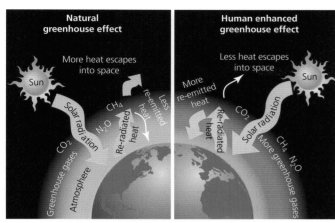

Figure 11.62 The natural and the enhanced greenhouse effect

Source: National Park Service, USA

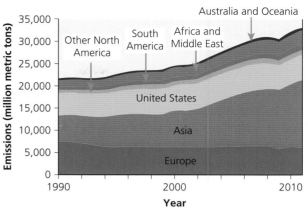

Figure 11.63 Rate of increase of CO_2 emissions by region in million metric tonnes

Source: US Environmental Protection Agency

Enhanced greenhouse effect

The Earth's natural greenhouse effect means that solar insolation from the sun is trapped in the lower atmosphere (troposphere) by greenhouse gases when it is re-radiated from the Earth's surface. This keeps temperatures at a level that makes the planet habitable. The rise in consumption of fossil fuels over the last 200 years has increased the quantity of greenhouse gases in the atmosphere, for example current CO_2 levels are 40 per cent higher than before the industrial revolution. As a result, the Earth's greenhouse effect is enhanced as more heat is trapped and global temperatures rise (Figure 11.62). Scientists think that global temperatures will rise on average by 1 to 2°C over the next century.

There is a growing consensus that the burning of fossil fuels has released CO_2, methane and nitrous oxides into the atmosphere and this is a major factor causing climate change. The main environmental impacts and concerns for the future are:

● melting of polar land ice
● rising sea levels
● changing ecosystems/biomes
● extreme weather events (droughts, floods, cyclones).

Most observers agree that the less developed world is more susceptible to the effects of climate change. Poorer countries do not always have the infrastructure and resources to counter the effects. It is not uncommon for single weather events, such as tropical cyclones and floods, to kill thousands of people in regions such as South Asia, Southern China and Central America. In Africa the UN

reports a 40–60 per cent decrease in total water in the large catchment basins of Niger, Lake Chad and Senegal. In a continent already struggling with poverty and famine, climate change is a matter of life and death.

Although emissions continue to rise, the rate of increase is at a noticeably slower rate (2 per cent) (Figure 11.63). In 2013, global CO_2 emissions from fossil fuel use and cement production reached a new all-time high. However, in a sign that efforts to tackle climate change may have been more effective than thought, the International Energy Agency found global emissions of CO_2 did not rise in 2014.

Nuclear waste management

There are a number of concerns regarding the development of nuclear energy:

● risks associated with nuclear accidents
● disposal of radioactive waste
● health and safety concerns (particularly leukaemia and cancer)
● potential contamination of water supplies
● concern over crops and grazing animals' safety.

Nuclear waste has a very long half-life (the measure of how long it takes to lose half its radioactivity). It remains highly radioactive for thousands of years, which must be considered when disposing of it safely. Nuclear power stations produce high-level radioactive waste in the form of spent fuel rods and fission products which have been removed from reactors. These are taken for nuclear re-processing where reusable uranium and plutonium are separated out to leave unusable radioactive waste. This is vitrified into solid blocks of glass and stored in steel-clad or lead-lined glass containers underground.

In the UK, the Nuclear Decommissioning Authority (NDA) has responsibility for disposing of radioactive waste. It has been unsuccessful in finding sites for the disposal of waste for the reasons shown below:

- **Physical conditions:** It needs to be buried 200–1,000 m deep in geologically stable rock types.
- **Economically:**
  - expensive to purchase land for this purpose
  - if located near to existing power stations the economy may become too dependent on the nuclear industry
  - negative impacts on tourism in the area.
- **Transport:** It requires safe transport links for the transport of waste from power stations.
- **Political and social barriers:**
  - strength of local pressure groups
  - concerns that the site may become the target of terrorism.

11.7 Mineral security

Metals and industrial minerals represent a major category of non-renewable resources that humans extract from, and return to, the natural ecosystem. Their use has increased dramatically as the population has grown and economies have developed. This rapid growth in metal and mineral use has serious implications for both the security of future resources and for the health of the environment.

Copper

Copper (Cu) is the third most-used metal in the world (after steel and aluminium). It is a scarce base metal with important properties that give it value for use in a range of industries.

Sources of copper

Copper ores are found primarily in **igneous rocks** as hydrothermal deposits but also in **sedimentary rocks**. Copper mines are only developed where there is more than 5 kg of copper per tonne of rock (0.5 per cent of mass – the **'cut-off ore grade'**) but ideally the content should be near to 2 per cent to establish a proven reserve.

Copper is found in many mineral ores which occur in deposits large enough to mine. The three main ores are bornite, **chalcopyrite** and malachite. Currently, the most common source of copper ore is the mineral chalcopyrite ($CuFeS_2$), which accounts for about 50 per cent of copper production.

The amount of copper believed to be accessible for mining on the Earth's land is 1.6 billion tonnes. In addition, it is estimated that 0.7 billion tonnes is available in deep-sea nodules. Mineral rich nodules of copper, magnesium and other metals are formed as a product of deep-sea volcanic activity. Retrieving these nodules from the sea floor is, as yet, too expensive to allow this to be a major source of copper.

Copper ores are formed when geothermal solutions (which are hot and under pressure) bring copper, dissolved from deep underground, to cool near surface environments. When the solution reaches favourable temperature and pressure conditions, the dissolved minerals are released in veins and **disseminations** (spread deposits) within the rock. This is a process known as **mineral precipitation**. Copper is usually deposited as copper sulphide or sometimes as native copper metal. (Native copper was the only source of copper until the turn of the twentieth century, when extraction methods were improved.) The most common copper minerals in the primary hydrothermal zone are bornite and chalcopyrite.

Copper can also be deposited in sedimentary rocks such as sandstones and shales. This is the result of copper-bearing fluids moving through permeable rock strata. It is thought that the metals precipitated when the fluids reached a 'chemical trap', where the rock changed in a way that made it impossible for the metals to remain in solution. Examples of copper-bearing sandstones can be found in Zambia and the Democratic Republic of Congo.

Distribution of copper reserves and resources

Copper deposits found disseminated as crystalline intrusions in igneous rock (**porphyry** deposits) account for about 60 per cent of the world's copper resources. **Strata bound** deposits, in which copper is concentrated in layers in sedimentary

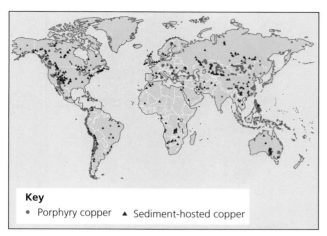

Figure 11.64 World distribution of known copper deposits (2008). The red indicates copper associated with igneous rocks (porphyry deposits) and the blue indicates copper contained in sedimentary rocks (sediment-hosted copper).

Source: USGS

rocks, account for about 20 per cent of the world's identified copper resources. These two types of deposit produce about 12 million tonnes of copper per year (Figure 11.64).

South America has the largest measured and undiscovered copper resources. The world's largest igneous deposits are mined in this region. Since the 1990s, the Andean region of South America has emerged as the world's most productive copper region, producing about 45 per cent of the world's copper by 2007. Chile is by far the largest producing country, although production has stabilised over the last few years, with Peru's production increasing rapidly. Eight of the top ten largest mine developments are in South America (six in Chile, including the world's largest mine, Escondida, and two in Peru).

Table 11.20 World's current top ten leading copper producers

Production (in million tonnes)		
Country	2006	2015
Chile	5.361	5.750
China	0.915	1.750
Peru	1.050	1.600
USA	1.220	1.250
DR Congo	0.131	0.990
Australia	0.875	0.960
Russia	0.675	0.740
Canada	0.607	0.695
Zambia	0.503	0.600
Mexico	0.129	0.550

North America has reserves of mineralized porphyry copper, including a massive deposit in northern Mexico where the world's second largest mine (the Cananea mine) is located at Sonora. Of the copper ore mined in the United States, the majority is produced in three western states: Arizona, Utah and New Mexico. There are also large deposits in western Canada. In Europe, Poland is the largest producer of copper and has the largest sedimentary deposit in the world. In 2006, it was the tenth largest producer.

China, whose production has increased rapidly, is the world's largest copper consumer and second largest producer. Increasing production costs mean Chinese mining operations overseas are more economically appealing, although some large porphyry deposits have been found in the Tibetan plateau.

Reserves in less developed regions are becoming increasingly important sources of copper ores. In particular:

- **Southeast Asia archipelagos** have significantly large reserves with huge ore deposits in Indonesia, Papua New Guinea and the Philippines.
- **Africa and the Middle East** have the world's largest accumulation of sediment-hosted copper reserves, with large deposits in the African 'copper belt' countries of the Democratic Republic of Congo and Zambia. Significant 'possible' copper resources remain to be discovered in Africa.

End uses of copper

Metals derive their usage from their physical properties and copper's most important properties make it popular for a range of end uses. These properties include that it is:

- ductile (capable of being drawn into wire)
- strong and malleable (capable of being hammered and moulded)
- a better conductor of heat and electricity than any other metal, except silver
- corrosion resistant
- easily joined by soldering or brazing
- can be combined with other metals to make alloys
- biostatic (does not sustain bacterial growth); can be used to keep drinking water safe
- decorative, with a sheen when polished.

It is thought that the average UK home has around 180 kgs of copper for electrical wiring, plumbing and electrical appliances. Most copper used in wiring and plumbing remains in use for at least 50 years.

- **Electricity**: Due to the reliability and efficiency of copper's conductivity, its main uses are in the electricity supply industry and electrical and electronic products. Copper is very ductile and can be drawn into very thin wires. It is used in power stations and substations and for electrical machinery such as electrical motors, electromagnets, generators and communication devices.

- **Water and other liquid vessels and pipework**: Copper is corrosion resistant and can be joined easily by soldering; it can also be shaped easily because of its ductility. It withstands extremes of heat with no long-term degradation. These qualities make it extremely useful for pipework, for making sealed copper vessels used in water supply and in the brewing and distilling industries.

- **Hygiene**: It is a naturally hygienic metal with antibacterial qualities that will slow down the growth of germs such as E. coli, MRSA and legionella. This is important for applications such as food preparation, coinage and in hospitals, as well as for plumbing systems.

- **Chemicals and agriculture**: Copper is also used in pigments, pesticides and fungicides, although it has of late been largely replaced by synthetic chemicals.

- **Alloys**: Copper can be combined with other metals to make alloys. The best known are brass (with zinc) and bronze (with tin), and it can be combined with nickel to form cupro-nickel. Alloys are harder, stronger and tougher than pure copper and can be used in industrial applications.

- **Ornamental use**: Copper and its alloys have been fashioned into ornamental objects and jewellery for centuries. They have an attractive golden colour which varies with the copper content. They have a good resistance to tarnishing, making them long-lasting.

- **Other applications**: As copper is non-magnetic and non-sparking, it is used for special tools and military applications.

Components of demand for copper

The equipment and construction industries use the largest percentage of copper (Figure 11.65).

- **Equipment**: This includes a range of capital equipment used in industry such as gears, bearings, tanks, pressure vessels and pipes. Its excellent heat

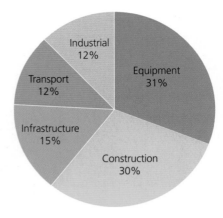

Figure 11.65 The major consumers of copper

transfer capabilities mean that it is widely used in producing heat exchange equipment. It is also used in power cables, transformers and other equipment for the electrical and renewable energy industries. This component includes its major use in motors to drive consumer electrical appliances.

- **Construction**: It is used in producing plumbing components such as taps, valves, pipes, fire sprinkler systems and other fittings in buildings. Brass is also used in these applications.

- **Infrastructure and communication**: Copper and copper alloys are a critical part of wires and cables used in the telecommunications industry. Fixed-line telephone cables and wires used for network connectivity contain copper although it is increasingly being substituted with fibre-optic cables for these uses. The $300 billion semi-conductor industry is also a major copper consumer.

- **Transport**: Copper contributes to several applications in the railway industry and in other forms of transportation. It is used for radiators, motors, brakes and bearings in the vehicle industry.

Role of copper in global commerce and industry

The world's production and consumption of copper have increased dramatically in the past 25 years. As large developing countries have entered the global market, demand for copper has increased. The risk of disruption to the global copper supply is considered to be low because copper production is relatively dispersed globally and not limited to a single country or region. However, because of its importance in construction and power transmission, the impact of any copper supply disruption would be high.

Codelco (Chile), Freeport-McMoRan, Rio Tinto and BHP Billiton are among the top copper producers. The Southern Copper Corporation and Newmont Mining Corporation Copper (both US TNCs operating in SW USA and Peru) are also major producers.

Copper is traded internationally in various forms from raw materials, copper concentrates, semi-finished products (copper semis) through to scrap. It is virtually 100 per cent recyclable, so copper is one of the most widely recycled of all metals; approximately 40 per cent of the world's demand is met by recycled copper. Recycled copper and its alloys can be used directly or further reprocessed to refined copper without losing any of the metal's chemical or physical properties. Copper is also traded as fabricated items and as part of end-use products. This is a form of indirect copper trade, for example, importing electronic equipment also means the import of copper used in its production.

With demand much less than its production, Chile is the biggest exporter across these product categories. China is the largest importer of copper ores and refined copper while Japan is the second largest importer of copper ore but the second biggest exporter of refined copper. Japan also indirectly exports copper through the sales of end products such as motor vehicles and electronic appliances.

Copper prices are the key driver for copper production; as they fall producers prefer to 'leave the copper in the ground' in the hope that supply shortages will raise prices. The diversified nature of copper's end usage in electronic products, construction and vehicles makes copper prices a good indicator of an economy's overall health. Consumption of the metal across such a wide range of markets has meant that it is also known as 'Dr. Copper' because analysts see copper prices as a reflection of the global economy's health. (Steel and aluminium have more concentrated end usage in the construction and vehicle industries; their prices are therefore regarded as less good indicators).

Figure 11.66 shows the movement in copper prices, which reached their peak in 2011. Since then copper prices have been falling mainly as a result of falling demand in the Chinese economy, which in 2015, grew at its slowest pace in 26 years. This has combined with a response lag by producers and recyclers, so supplies remained steady. Together this has sent copper prices on a downward trajectory since 2014.

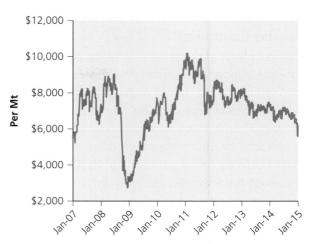

Figure 11.66 Fluctuating world copper prices reflecting the health of the global economy

Key aspects of physical geography associated with mineral ore occurrence

If all the mineral resources in the Earth's crust were evenly distributed then none of them would be sufficiently concentrated to become viable reserves for exploitation. Geological processes have provided localised concentrations of deposits to enable exploitation to occur.

Geological conditions

Based on the process of geological formation, mineral deposits may be classified into four broad categories:

1. **Magmatic deposits**: Named because they are linked with magma which is emplaced into the crust (either continental or oceanic) and are found within rock types derived from the crystallisation of such magmas. The process of crystallisation separates ore and non-ore minerals according to their crystallisation temperature (chromite and magnetite form in this way). These deposits also include ores containing nickel-copper deposits and platinum metals.

2. **Hydrothermal deposits**: These are formed by another igneous process associated with magma intrusions. The rocks around intrusions may be lifted and deformed so that cracks and fissures appear. These cracks allow hot solutions containing dissolved minerals to escape towards the surface and as these cool down, the minerals precipitate in a predictable order according to their solubility. They are found in concentrated deposits that are more viable for exploitation. Most metal ores,

including iron, copper, tin and lead are found as hydrothermal deposits.

3. **Metamorphogenic deposits**: Metamorphic processes change pre-existing mineral deposits and form new ones. The changes are brought about by intense heat and pressure over a long period of time. For example, low density minerals are replaced by minerals of high volumetric mass. These are found primarily in PreCambrian formations (for example, the iron ore deposits in Ukraine; manganese deposits in Brazil and India and the gold and uranium ores in South Africa).

4. **Sedimentary deposits** (including **placer deposits**): Not all metal deposits have igneous origins – sedimentary deposits can also be a valuable source of many metals, including copper. Sedimentary deposits are generally tabular and sheet-like, horizontal in form but frequently folded and faulted. Beds containing metallic ore are generally less than 20 ft thick. Beyond copper, other metals found in sedimentary deposits are lead and zinc. **Placer** deposits are alluvial; they are carried in flowing water and deposited when the water slows down. Dense minerals are the first to settle, including tin ore and gold.

Location and working

As mineral deposits can be found in all types of geological formations, traditionally the major physical geographical constraints on exploration and mine development were determined by remoteness, access to processing and markets and ability to exploit the resource. Thus areas with extremes of temperature or physical environments such as deserts, densely forested areas or high mountains were largely avoided and left unexplored.

Advances in mechanical and transport technology have released these constraints and geologists can feel free to explore suitable geological formations on any continent, with the exception of Antarctica. Trends in mine production have shifted to more remote areas such as the Atacama Desert in Chile (copper), under rainforests in Brazil (iron ore) and other locations in the continental interiors of Australia, Africa and Asia. The last remaining frontier for exploration and development is the hypothetical and speculative resources that lie in the deep seabed.

Physical conditions will, however, often determine the nature of the mine working itself. The shape, size, quantity and grade of the ore deposit found from exploration will define whether it is extracted through traditional deep underground mining or through open-pit (sometimes known as **open-cast**) mines.

Ore extraction: underground and open-pit mining

Mine working has undergone important changes during the twentieth century with a major shift from underground to open-pit mining (Figure 11.67), which has become more common as mining has evolved in emerging economies. Productivity increases in the past 50 years have been achieved through the ability to access lower grade ores by more efficient processing and using large-scale capital equipment. The main cost advantage of open-pit mining is that mechanisation has allowed the rapid removal of vast amounts of material using larger digging machines and trucks. Equipment is not restricted by the size of the opening it must work in, which allows faster production and the lower cost also permits lower grades of ore to be mined.

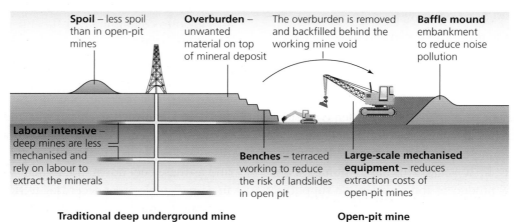

Spoil – less spoil than in open-pit mines

Overburden – unwanted material on top of mineral deposit

The overburden is removed and backfilled behind the working mine void

Baffle mound embankment to reduce noise pollution

Labour intensive – deep mines are less mechanised and rely on labour to extract the minerals

Benches – terraced working to reduce the risk of landslides in open pit

Large-scale mechanised equipment – reduces extraction costs of open-pit mines

Traditional deep underground mine   **Open-pit mine**

Figure 11.67 Features of traditional deep underground mine and open-pit mine

Extraction costs will depend on the size, shape and location of the reserve. These include:

- **Overburden**: Rock and soil lying above the deposit can be expensive to remove. It is blasted away in open-pit mines. An open-pit mine is the least expensive, the first choice when a deposit is close to the surface, big enough and has little overburden.
- **Water removal**: As depth increases, more water flows into the mine; this needs to be pumped out.
- **Depth**: Costs rise as depth increases; this applies in both types of mine. Underground mining includes a wide variety of added requirements such as ventilation systems, ground support and finding a safe and cost-effective way to get the mineral ore to the surface.
- **Form of the deposit**: Costs increase if the mineral is found dispersed irregularly or found in thin layers. This is less costly and safer to access in open-pit mines.

Costs have to be balanced against what can be extracted from the mine. This will depend on:

- **The 'stripping ratio'**: This is the amount of waste rock mined relative to the amount of ore mined. A stripping ratio of 3:1 means that there will be three times as much waste rock mined as ore. To be profitable, an open-pit mine must be designed so that the cost of mining the waste rock does not exceed the value of the ore.
- **The 'cut-off ore grade'**: Processing costs depend on the purity of the ore. The **cut off ore grade** is the lowest ore purity that can be exploited economically. This varies depending on the metal or mineral being processed.

Environmental impacts of the Carajás iron ore extraction scheme in Pará state, Brazil

The environmental impact of mining includes formation of sinkholes or surface subsidence (following the collapse of the surface layer), loss of biodiversity and contamination of soils, surface water and ground water by toxic chemicals or leachates resulting from the mining process. The latter can also affect the health of the local population. Some mining methods may have significant environmental and public health effects.

Background to the Carajás scheme

The Carajás complex (Figure 11.68) is the largest single iron ore mine in the world (rivalled in size only by the Hamersley Ranges iron ore project in the Pilbara region of Western Australia). It is located in the state of Pará in northern Brazil, in an area naturally covered by the Amazonian rainforest.

The mine complex boasts the richest reserves and concentrations of high grade iron ore anywhere in the world, with an iron content of around 65 per cent. It is now fully owned by Vale (CVRD) and holds 7.2 billion tonnes of iron ore in proven and probable reserves. Carajás is not only rich in iron ore but also ores for copper, tin, aluminium, manganese and gold. The rich ores are found within Archaean iron formations to a depth of between 100 m and 150 m. The upper 80 per cent comprises of a soft, friable enriched limonite passing down into hematite to a vertical depth of around 300 m.

The operation uses an open-pit mining system with four giant man-made chasms. In 2013 Vale's iron ore production at Carajás was 105 million tonnes. Around 3,000 workers are employed at the mine using a fleet of giant machines, such as diggers, crushers and loading trucks. In 2011, the mine was producing 300,000 tonnes (worth around $36 million) per day. Around 70 per cent of this was exported as ore to China using a dedicated 900 km railway line to the Ponta da Madeira deep sea terminal near São Luís in the north-eastern state of Maranhão. Carajás is also served with its own airport.

Figure 11.68 The Carajás open-cast iron ore mine complex

Environmental impacts

The biggest environmental impact of the mine development in the Carajás region has been the necessity to clear large tracts of pristine rainforest to make way for the massive open pit that forms the mine. As a 'resource frontier' further clearance of rainforest has been necessary for associated infrastructure:

- the railway to the seaport which is an integrated part of the scheme
- road access for large trucks and to convey other large machinery
- the airport
- mining settlements – accommodation and services for the miners
- temporary storage of overburden (debris and soil).

Deforestation has resulted in both direct and indirect impacts on landscape, wildlife habitats, the climate, soils, relief, vegetation and drainage. Furthermore, the access into the rainforest by roads has led to further forest clearance by non-related activities such as clearance for agriculture or illegal logging.

Landscape, wildlife and ecosystems

The mine is in an area of national forest but what was a landscape of dense vegetation is now a scene of bare cliffs and billowing dust. Noise pollution is another feature of the devastation with the sounds of birds and animals in the rainforest replaced by the constant roar of massive engines and the grinding of diggers and crushing machinery.

Habitat destruction has led to endangered species being threatened with extinction. It is estimated that around 25 per cent of wildlife in the region has been lost as a direct result of the project. The development of railways, road and to a lesser extent the airport has fragmented habitats, reducing migration of some fauna and creating smaller gene pools.

Climate, soils and vegetation

The loss of rainforest over such a large area has had an impact on microclimate. Trees are a regulator of both temperature and humidity and their loss means that there is less rainfall, less humidity and a higher diurnal temperature range with extremes of heat during the day baking the red ferralitic soils. In the tropics decomposition of organic matter is rapid and the nutrients released are taken up by the trees equally quickly. This cycle is broken, the soils lose any organic input or fertility and without the tree cover are prone to erosion. Consequently it will be difficult

for vegetation to regenerate naturally in such harsh conditions.

Relief and drainage

The open pit mining alters the relief and drainage of the area directly. In addition, large masses of soil can be eroded because of storm events which can cause landslides around the mine and the silt being washed into local rivers, which can kill some aquatic species and causes local flooding.

Global environmental impacts

Though not as energy intensive as the tar sands operation in Canada, the Carajás project contributes to global climate change by a net addition to greenhouse gases. In addition to burning fuel to operate the trucks and large machines on site, deforestation in the area removes a significant carbon sink.

The sprays and chemicals used on the site contribute to ozone depletion and the emissions from vehicles and from waste materials will also contribute to acid rain.

Human impacts

The area was once settled by native Amazon Indians from three different tribes. Some of these remain living in the rainforest but their traditional way of life has been disrupted and in many cases totally lost. Some native Indians work on the site and have been integrated into mainstream Brazilian culture.

Expansion and remediation efforts

There are plans to expand the Carajás mine into surrounding areas which will mean more of the same environmental degradation. However, Vale point out that the expansion will be good for the Brazilian economy and that in all their operations they attempt to adopt 'green' measures.

Like many transnational corporations, Vale is keen to point out that sustainability is embedded in its thinking and planning and refers to a series of measures designed to limit the mine's impact. It argues that the complex only covers around 3 per cent of the national forest. Before it starts any new digging it has to present a restoration plan to the state authorities and the national conservation agency ICMBio. These plans aim to restore the area to its original state by using spoil to fill in the mines and reshape the topography, then replant the area with the original native species, which the company have preserved and which are growing in a separate nursery.

Vale also supports an extensive monitoring operation in the forest run by ICMBio; they have paid for 80 additional forest rangers plus vehicles, boats and the use of a helicopter to guard against poaching of animals and illegal logging.

The company intends to implement a completely truckless, electrical, automated transport system to help move rock and ore around the mine site more efficiently and safely. Although one of the purposes is to remove fuel costs, it will also reduce emissions.

Vale are clearly making some effort to operate the mine more sustainably, however they are a profit-making corporation and their priority is iron ore extraction; the environment may only be a secondary concern.

Sustainability issues associated with ore extraction, trade and processing

The biggest issue for mineral ores is their finite availability, which raises their value and price as commodities. Advances in technology can improve the supply and extend the lifespan of many metals by:

- more cost-effective exploration for new deposits
- advances in machinery and extraction methods which can convert more resources into reserves
- improving resource efficiency through, for example, better recycling methods
- substitution with the use of alternative materials can extend the lifespan of many resources.

Despite this, there is still concern about the long-term sustainability of our use of many metals. How we manage their supply and demand in the future is of critical importance.

Ore extraction

In areas of wilderness, mining may cause destruction and disturbance of ecosystems; in areas of farming it may destroy productive arable or grazing land. In urban environments threats to sustainability come mainly from noise and visual pollution.

Unsustainable practices of mineral ore extraction lead to a number of negative impacts. These impacts are shown in Table 11.21 together with suggested measures that make operations more sustainable.

Despite mining companies being more aware of the potential for environmental impacts there have been a number of recent environmental disasters as a result of ore extraction:

- In August 2014, 40,000 m³ of copper sulphate acid spilt into the Sonora river and public waterways

Table 11.21 Impacts of unsustainable practices and measures to improve sustainability

Impacts	Measures to improve sustainability
Land take - land area needed is usually much larger than the mine itself whether underground or open-pit mining **Habitat loss** – results in loss of species	Restoration plans: using native species by capturing animals and releasing them or relocating them; habitat restoration using native plant species, (e.g. the Carajas project); creating new habitats
Pollution: **Noise pollution** – machinery disturbs wildlife **Dust pollution** – caused by blasting and vehicle movement	Baffle mounds around the periphery of open-pit mines help to absorb and deflect some noise Dust can be reduced with water sprays
Water turbidity – caused by particles and silt from mines; can block sunlight from aquatic plants and choke filter-feeding animals	Holding lagoons can help suspended solids to settle and clear water can then enter rivers
Toxic leachates – chemicals or toxic metals unearthed as part of mining can become soluble; drainage into nearby rivers kills aquatic life	Toxic metals are more soluble when acidic so passing mine drainage water through a filter of crushed limestone may immobilise some toxic metals
Spoil disposal – is often loosely compacted and unstable with potential for landslides	Drainage using pipes in the base of spoil heaps prevents waterlogging and reduces landslide risk

in Sonora state in northwest Mexico. This was the result of heavy rainfall causing the tailings pond to overflow at the large Buenavista copper mine operated by Grupo Mexico at Cananea. The massive spill poisoned the river and caused water shortages in an area already experiencing water stress. The contaminants are likely to persist in the river and could enter the food chain posing a risk to animals and people.

Figure 11.69 The Samarco mine disaster in 2015. The village of Bento Rodriguez, near the city of Mariana in south-east Brazil was flooded when an iron ore tailings dam failed, killing 17 people.

- In November 2015, an iron ore tailings dam burst at the Samarco mine (operated as a joint venture by BHP Billiton and Vale) in Minas Gerais state in southeast Brazil. The collapse unleashed a wave of waste water that flattened the village of Bento Rodriguez killing 17 people and polluting a 400-mile stretch of the Rio Doce (Figure 11.69).

Trade

The comparative advantages of different countries around the world can foster the development of a mutually beneficial and more sustainable mineral resources trade.

In 2009, the Word Economic Forum launched the **Mining and Metals Scenarios to 2030** project to examine the global economics, trade and geopolitical future of the minerals and metals sector. In their report they presented three scenarios for the future:

- **Green Trade Alliance (GTA)**: This proposes the most sustainable scenario, in which a GTA is formed to 'promote environmental sustainability without compromising competitiveness'. The alliance would include a mix of developed and developing countries and those which are resource-rich, wherever they are on the development continuum. The idea would be supported by a 'Sustainable Trade Organisation' overseeing GTA agreements. Environmental standards would be the basis for protectionist measures adopted by GTA members, resulting in limited cross-border flows between GTA and non-GTA countries. Initially there would be tension but there would also be scope for non-GTA members to be encouraged to join the alliance.

- **Rebased globalism**: This scenario suggests an evolution of the economic environment for mineral resources as it is in the 2010s. Economic power is held jointly between markets where there is strong demand and by those countries that control resources. Free market principles hold and there are free cross-border trade flows. This scenario may assist developing economies to hold more power and to overcome the 'resource curse' but much will depend on the role and mindset of large TNCs.

- **Resource security**: This is the least sustainable scenario and is based on the premise of national self-interest. The idea is that countries hoard domestic resources and enter trade alliances based on regional or ideological foundations. In this case, countries would adopt import substitution strategies and the trade environment would be filled with protectionist measures resulting in limited cross-border flows of resources.

Processing

Metal and mineral commodities are non-renewable and their total availability is finite. However, processes and mechanisms that enable recycling or reuse of these materials increase their long-term sustainability and contribute to the production of metals. Recycling sources are old scrap materials from discarded products and new scrap from industrial processes. These contribute to the 'end-of-life' (EOL) stocks of metals, most of which can be used in new materials. Recycling procedures have been established in many countries as a result of the Agenda 21 commitment. For example, the collection of lead, used predominantly in lead acid batteries, is particularly well-organised.

A United Nations Environmental Programme (UNEP) status report published in 2011 estimated the end-of-life recycling rates (EOL-RR) for the following metals:

- **Iron and steel**: 70–90 per cent is recycled, which contributes to around 35 per cent of new production.
- **Aluminium**: 40–60 per cent is recycled, contributing around 24 per cent towards new production.
- **Copper**: 40–50 per cent is recycled, contributing around 18 per cent of new production.
- **Lead**: Recycling estimates vary between 60–90 per cent but this contributes to around 55 per cent of new production.
- **Tin**: Around 90 per cent is recycled.

Table 11.22 Factors influencing recycling rates

Factors that make recycling more challenging	Factors that assist recycling
Mixed materials – for example, the separation of alloys	Media campaigns – to educate and raise awareness of the advantages of recycling
Dilution and dispersal – some uses are so dispersed that it is unviable to recover	Design for 'end of life' – easy dismantling to remove useful parts at point of discard
Transport – cost of transporting scrap material may make it unviable	Labelling when assembled – to state what goods are made of
Labour costs – recycling is much more labour intensive than raw material extraction	Reusing used parts – removing parts which are still of value
	Legislation – for example EU Directives on disposal of batteries and electronic equipment
	Financial incentives – for example, landfill tax

These figures show that although secondary resources can make a significant and potentially increased contribution to production, there will continually be the need to extract new raw materials.

Increasing the use of secondary resources as a source of production for new materials represents a considerable challenge because:

● A large proportion of metals end up in the built infrastructure where they will remain long term.

● Some metals end up in complex products combined with other materials that inhibit recycling.

11.8 Resource futures

The future of energy resources

Key question

What will the total primary energy supply of the planet be like by 2050?

There are a number of implications for global energy resulting from continued population growth:

● Global energy mix will change considerably, though it is uncertain exactly what the mix will be. By 2050, however, it is thought that non-carbon emitting sources will make up 40 per cent of the mix. Coal could remain important with increased use of clean carbon technologies.

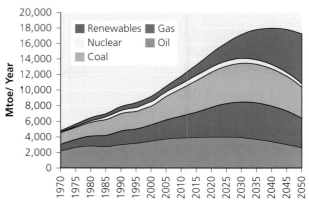

Figure 11.70 Forecast energy consumption by source for 2052

Oil and gas are likely to decline, simply because reserves become depleted (Figure 11.70).

● The average per capita energy use is set to decline as individuals become more attuned to energy efficiency.

● These changes will not be uniform and there is likely to be a growing energy gap between rich and poor countries.

Technology

With the foreseeable depletion of fossil fuels, new technologies are being developed to replace them:

● **Hydrogen:** Unlike many renewables, hydrogen is a high energy-density fuel and can be used to replace fossil fuels as the technology for conversion is available. The '**hydrogen economy**' is a system of delivering energy using hydrogen in different forms. The main uses being developed are:

  – to produce and store surplus energy in fuel cells by **electrolysis of water**: converting H_2O into H_2 and O releases energy

  – combustion – used as vehicle fuel, or for domestic/industrial heating

  – used as chemical energy – to make hydrocarbons – which can then be used as fuel in much the same way as oil and gas.

Hydrogen fuel cells could be the perfect alternative to fossil fuels. They generate electricity using only hydrogen and oxygen and are pollution free. The following issues hinder further breakthroughs in this important technology:

  – Electrolysis requires a lot of other energy use and if fossil fuels are used in the process, then the environmental benefit of using hydrogen is negated. Using renewable energy for the process would overcome this issue.

- Hydrogen cannot be compressed easily or safely, and requires large tanks for storage.

Developments of this fuel are ongoing and many large oil TNCs, such as Shell, and vehicle manufacturers (for example, Honda) are investing heavily in the technology as they envisage its future potential. Aspiring to become the first carbon-free nation, Iceland is attempting to develop a hydrogen economy, using geothermal energy to produce pure hydrogen for use in vehicles, so they no longer have to import oil.

- **Gasification of coal:** Coal is converted into gas below ground where the deposits are deep, thin or difficult to take out as solid fuel.
- **Secondary and tertiary oil extraction:** Using bacteria to digest heavy oil – producing lighter, easier flowing oil.
- **Nuclear:** The nuclear power industry has been working on safer technology solutions since the Fukushima nuclear disaster in 2011. Ideas include:
  - smaller reactors more evenly distributed to serve smaller populations, with safety mechanisms built in to eliminate the chance of radiation release
  - offshore nuclear reactors – like the floating platforms the oil and gas industry already uses; constructed to withstand hurricanes and with reactor cores submerged beneath the platform, so a fresh supply of cold seawater would always be available to cool the reactor
  - nuclear fusion – promises limitless carbon-free power without producing dangerous nuclear waste. Trying to replicate the same atomic reaction that occurs in the sun, in a controlled way, is still many years from maturity.
- **Offshore wind:** Promises to gain massive growth in 25 years as advances in turbine blade designs, developed from aeronautics technology, maximise energy from wind.
- **Geothermal energy:** Researchers in Iceland have spent several years drilling straight into volcanoes to access very hot water and magma deposits, with a view toward developing these high-temperature resources into geothermal power stations that can produce 10 times more energy.

- **Solar fuels:** Converting sunlight, water and carbon dioxide into usable chemical energy that can be stored like gasoline for extended periods of time.
- **Ocean thermal energy conversion (OTEC):** Aims to generate electricity by exploiting the temperature differences between surface water, heated by the sun, and water in the ocean's chilly depths.

Economic

- Energy demand growth will continue to move away from OECD countries.
- Energy trade flows towards Asian markets will continue to gather pace and despite a slowdown in the Chinese economy, India and Middle East countries will take up the reins. Issues concerning the trade in natural gas could be eased by the increased availability of LNG.
- By the mid-2020s, non-OPEC oil supply from Canada, Brazil and the USA may start to decline and the oil economy will rely on supplies from Venezuela and Middle East countries.
- The new geography of oil demand will present difficulties for the oil refinery sector, particularly in Europe which will have an excess of refining capacity.
- Natural gas production will increase in every region except in Europe. Unconventional gas will account for 60 per cent of the growth in production, though shale gas outputs are likely to decline by the 2030s.
- Despite a rapidly growing economy, sub-Saharan Africa will only account for 4 per cent of the growth in global energy demand.

Environmental

- **Carbon capture and sequestration:** Although not a power generation technology, carbon capture and sequestration enables the scrubbing of carbon from power plants for storage in underground reservoirs such as exhausted oil and gas traps. This means that carbon emissions from fossil fuels and biomass can be eliminated so they can be used with less environmental impact.
- Targets for carbon emissions will become more stringent as global climate change presents more problems around the world.
- The development of unconventional sources of fossil fuels will do little to reduce carbon emissions.

On a national scale, there is likely to be a move to more local energy generation from a combination of smaller power stations, combined heat and power plants and 'energy from waste' alternatives. There will be some energy efficiency gains and environmental benefits from this.

Political

Geopolitics will continue to be an important feature of the energy industry. A number of trends seem likely to emerge over the next thirty years until 2050.

- The influence of TNCs in resource development will continue to grow. They will have the wealth to control the technologies of the future.
- Developing countries will increasingly rely on TNCs to help develop their resources. The corporations will have a monopoly on the skills and capital needed to do this effectively.
- OPEC may re-emerge as a powerful cartel in influencing supply and pricing of both oil and gas.
- As traditional fossil fuel resources become depleted and remaining reserves more valuable, there will be a scramble to claim previously protected areas and exploit them. The Arctic is less well protected against development than Antarctica and there have been a number of territorial claims by the USA, Russia, Canada and Denmark in the race for fossil fuels in the Arctic Ocean. This may lead to conflict.
- Energy supply and security is even more likely to become a bargaining tool or threat in geopolitical relationships between countries.

The future of water resources

Information collated by organisations such as the FAO and the projections of the IPCC forecast grave situations for future water supply. By 2050, it is predicted that 15 per cent of the world's countries will be experiencing water deficiency, most in the MENA region. Water availability for the future rests upon good management of the water the world has now.

Technology

Unlike energy resources, water has no alternatives other than developing technologies that will give a greater supply from existing sources, or reducing demand. Much of the developing technology to increase the supply is focused on the largest source of water – the oceans, and the use of desalination techniques:

- **Osmotic distillation** and advanced membrane technology may, in future, require less energy intensity and thus become more sustainable.
- **Electrodialysis** is an alternative method to desalinate seawater which is being considered. This works in the opposite way to reverse osmosis by allowing sodium and chlorine ions to pass through a membrane in the presence of an electric field, leaving purified water on the other side.
- **Saltwater greenhouse technology** is well suited to arid parts of the world; it produces water for irrigation by pumping seawater into a greenhouse and piping it over honeycomb cardboard pads that provide a large area for evaporative cooling.
- Smaller-scale **appropriate technologies** such as plastic solar stills (marketed as 'watercones') are more appropriate for distilling salt water in smaller quantities in less developed countries.

Economic

There are many countries with an abundance of water and redistribution can occur in various ways, including water shipping, water management and virtual water trade. Trading water on a larger scale will take place as water becomes more valuable than oil in some regions.

Environmental

Increased water use by humans not only reduces the amount of water available for industrial and agricultural development but also has a profound effect on aquatic ecosystems and their dependent species. Any action to manage water resources more sustainably will need to be co-ordinated with efforts to mitigate the effects of climate change – they cannot be considered separately. In the future an integrated river basin management approach will be undertaken by planners, scientists and hydrological engineers.

- Impacts on water quality and drainage are key considerations when environmental impact assessments are conducted.
- Future-proofing river basin management strategies deals with these challenges.

In developed countries, the impacts of water-source development are already considered carefully before implementation goes ahead and it is thought that developing countries will follow this example.

Political

Water still presents the most likely cause of international conflict in the future. The potential hotspots of conflict will need to be monitored carefully and co-operation encouraged.

The future management of global water resources will feature strongly in the UN's new Sustainable Development Goals (SDGs), which will set targets to achieve by 2030.

- Proposed Goal 6 is to 'ensure availability and sustainable management of water and sanitation for all'.
- Sustainability of water resources is also relevant to:
  - Goal 3.3 relating to diseases and epidemics (including water-borne diseases) and Goal 3.9 which relates to reducing death from water contamination
  - Goal 12, which relates to the safe management of chemicals and wastes, and reduction of their release into water
  - Goal 15, which includes targets that focus on the role of water in wildlife conservation.

The future of mineral ores

Concern surrounding the future security of mineral supplies continues. New sources of minerals are needed to meet future demand. This is particularly so for the so-called 'critical' minerals, which are those elements that are essential for new and green technologies. In addition to rare earth elements (REEs), the EU has defined a list of 13 other elements as critical, including antimony, germanium and tungsten. To ensure that future shortages do not cause serious problems, new approaches will need to be developed.

Technological future

Exploration

Exploration holds the key to the future supply of mineral ores. A range of techniques continues to be developed to **explore** new mineral deposits more cost effectively:

- **Remote sensing**: This includes techniques that collect information about mineral deposits from a distance. The most commonly used are aerial

and satellite surveys, which enable large areas to be surveyed to identify topography and geological features. Infrared images may also indicate possible mineral content in surface rocks.

- **Magnetometry**: This can be used to identify magnetic rock content, such as iron ore.
- **Gravimetry**: This is used to detect the density of rocks and distinguish magmatic igneous deposits, which are usually denser than sedimentary and placer deposits.
- **Seismic surveys**: Echoes of surface vibrations are used to provide information on depth, angle and thickness of rock strata; increasingly used in underwater surveys.

Increasing deep sea exploration will be undertaken using a combination of these techniques.

Extraction and exploitation

- **Mechanisation**: Greater efficiency is found in many aspects of mining equipment. Larger excavators can extract material more rapidly and cost-effectively.
- **Electronic technologies**: The introduction of computer technology, remote-control interfaces, satellite communications and robotics has led to greater safety and productivity in mining, mineral processing, smelting and refining operations.
- **Exploiting low-grade ores**: Technological developments have made it possible to mine ores of declining grades and more complex mineralogy without increasing costs. Alternative methods have been developed to extract metals from low-grade ores:
  - Copper in leachate water from spoil heaps can be concentrated by evaporation and then separated by electrolysis.
  - Copper from low-grade copper sulphide spoil heaps can be extracted using bacteria. The sulphur is oxidised producing an acid solution that leaches out the copper, which can then be purified.

Economic future

Supply

Mineral resources are abundant but reserves of exploitable minerals are limited. High extraction costs or a lack of suitable technology may prevent 'possible' reserves being exploited. However, in spite of recent price volatility and weakness, especially for the staple ores of iron, copper and nickel, metal prices are likely to remain at a relatively high level compared to the last three decades of the twentieth century. This will encourage continued investment in technology and exploration, which will increase supply.

Demand

Demand could be more finely balanced. On the one hand, increased recycling efforts with regard to waste materials may reduce demand on reserves in the ground. The substitution of metals for more abundant materials will also dampen demand. On the other hand, this is likely to be more than countered by increasing demand from growing populations. An unprecedented demand growth will come from millions of aspiring individuals in emerging economies striving for a better material standard of living. Even with dramatic increases in recycling, newly-mined materials will be needed to meet this demand.

Environmental and political

These two aspects of mineral ores will be increasingly and intrinsically linked in the future. More exploration and development of new mines will continue to put pressure on the natural environment. Protection for the environment can only come from governmental controls on impacts, at all levels, from local to global. Insisting on EIAs before giving the go-ahead on development projects may offset some of the worst devastation brought about by mineral exploitation, but not all regions and countries will impose such restrictions on operators.

The trend of mine production moving to emerging economies will continue. Huge investments seen in Latin America, Africa and parts of Asia over recent years will escalate over the next 10 to 20 years, especially among the six resource-developing countries of Brazil, Chile, DR Congo, Peru, South Africa and Zambia. The perceived economic gains from foreign direct investment from TNCs and from other national governments, (including jobs and improvements in local infrastructure) may be seen as too good an opportunity for development to be missed. In many cases, this may obscure the concerns over environmental impacts.

On a global scale, some sites are likely to have large mineral deposits which remain unexplored. These are in areas that could be expensive to exploit, with transport and infrastructure difficulties or they are protected. Examples include Antarctica, the Arctic region (including Alaska, Greenland and Siberia) and the deep seas.

- Antarctica is protected from all forms of commercial mineral exploitation by the Montreal Protocol part of the Antarctic Treaty System.
- In Alaska, the Arctic National Wildlife Refuge is protected from mineral exploitation; though the Refuge comes under increasing political pressure from some quarters to permit drilling for oil.
- The right to exploit deep ocean floor deposits (such as manganese nodules) in international waters has only recently been established. Few companies have been willing to invest in expensive technologies to exploit them, but the first mining permits have recently been granted in Papua New Guinea for mining operations at a depth of 1,500 metres.

Case study of energy resource issues: Oil and natural gas production in the Niger Delta, Nigeria

Background to Nigerian oil production

Nigeria is the largest oil producer and exporter in Africa and has been a member of OPEC since 1971. It is the most populous member country with a rapidly growing population of over 180 million. Nigeria also holds the largest natural gas reserves in Africa (though Algeria is the largest producer), but has limited infrastructure in place to develop this sector. Natural gas is associated with oil production and is mostly 'flared' in order to access the oil. Development of gas pipelines and transport as liquefied natural gas (LNG) via tanker are expected to accelerate the growth of the gas industry both for export and for use in domestic electricity production.

As one of the four MINT economies showing rapid growth in the early twenty-first century, Nigeria is heavily dependent on its hydrocarbon industries, which are the mainstay of its economy. According to the IMF, the hydrocarbon sector accounted for more than 95 per cent of the country's export earnings in 2011.

Resource development and production

The oil industry is primarily located in the delta region of the River Niger (Figure 11.71). Oil was discovered there in 1956 by Shell-BP, who had the sole exploration rights at the time. Shell made further discoveries in shallow waters off the coast near Warri in the 1960s. Exploration rights in both onshore and offshore areas were given to other international companies such as Exxon in the 1960s. In 1977 the government established the Nigerian National Petroleum Company (NNPC), a state-owned company which is a major player in the industry. This gave Nigeria more control over its oil resources but it is still reliant on the expertise and investment of TNCs such as Shell, Chevron-Texaco and Exxon Mobil to explore and develop new reserves and to provide the infrastructure to distribute it for export or domestic use. These TNCs establish Nigerian-based subsidiaries (for example, the Shell Petroleum Development Company) which are 'joint ventures' with the NNPC.

The region produces between 2 and 2.5 million barrels per day (mbpd), though it is thought to have a production capacity of up to 3 mbpd. Production disruptions caused by local militant groups have compromised oil production in the region for many years. Peak production reached 2.5 mbpd in 2005 but fell back again because of militant action. A recent amnesty with militant groups saw a rise in production back to an average 2.15 mbpd in 2012. It is thought that light Nigerian crude oil is the best quality oil in the world and the least expensive to refine.

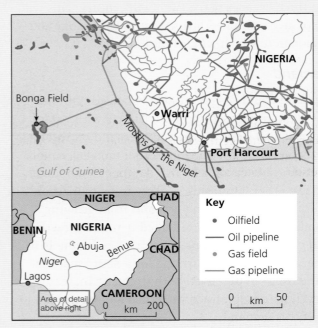

Figure 11.71 Map of Nigeria and the oilfields in the Niger Delta region

Trade

There are four oil refineries based in Nigeria: the two main ones are at Port Harcourt and Warri. Both only operate at 30 per cent capacity as most of the oil extracted is exported as crude by TNCs and refined overseas (Figure 11.72).

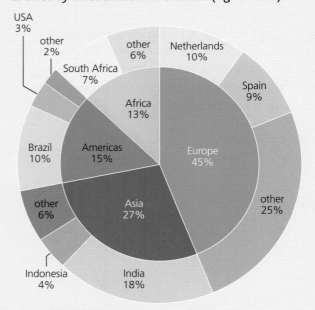

Note: Nigerian crude oil and condensate exports averaged 2.05 million barrels per day

Figure 11.72 Exports of Nigerian oil

Source: US Energy Information Administration

Nigeria exports its oil around the world. Until 2012, the US was the largest market for Nigerian oil exports but as America has moved towards domestic supply from fracking, it is now only tenth. Nigeria's largest regional markets are now in Asia and Europe, and India is currently the largest importer of Nigerian oil.

Nigeria's energy mix

Only a small proportion of the oil is supplied to the domestic market, although as the country develops consumption of petroleum is growing at a rate of around 12.5 per cent. Most of the demand is in urban areas; petrol is unavailable to most Nigerians because it is too costly. The majority of the population still live in rural areas and rely on fuelwood as their main energy supply for cooking and space heating (Figure 11.73).

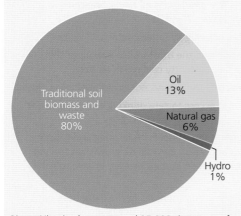

Note: Nigeria also consumed 35,000 short tons of coal in 2012.

Figure 11.73 Nigeria's energy mix, 2012

Source: US Energy Information Administration

Figure 11.74 Woman carrying fuelwood in southern Nigeria

Implications for resource security and human welfare in the Delta region

The region already has many ethnic and religious tensions as there are a number of local tribes living in different parts of the delta. Around 31 million people from up to 40 different ethnic groups live in the 'South-South' region though there are majority groups such as the Ijaw, Ogoni and Igbo tribes. Environmental devastation associated with the industry, combined with the uneven distribution of oil wealth, serve to escalate these tensions. This has given rise to numerous conflicting factions in the region, including recent guerrilla activity by the Movement for the Emancipation of the Niger Delta (MEND).

Local groups are largely excluded from the benefits of oil wealth and seek a share by attacking the oil infrastructure and staff. Oil theft (known as 'bunkering') and pipeline sabotage lead to damage, loss of production, pollution, forced shut down of production and even piracy in the Gulf of Guinea. The oil companies blame militant activity for the pollution but locals claim that companies such as Shell are involved in illegal activities such as 'flaring' (burning off gas), and that the pipeline infrastructure is not maintained and many leaks are due to corrosion (Figure 11.75).

Figure 11.75 Niger Delta residents pass a burning pipeline as they evacuate their homes by boat

The main implications for resource security and human welfare are summarised in Table 11.23 (page 610).

Attempts to manage the resource

Nigeria earns some $10 billion every year from oil but Niger Delta residents remain in poverty. Less than 20 per cent of the region is accessible by good roads, even in the dry season, and hospitals and schools are seriously under-funded. The government can use oil revenue to overcome some of these issues but wealth distribution in Nigeria is very inequitable. Economic growth means more people from rural areas are moving to the cities, where there is more demand for electricity for industrial and domestic use.

Table 11.23 Main implications for resource security and human welfare in the Niger Delta

	Positive implications	Negative implications
Resource security	Potentially high production levels	Flaring natural gas – oil companies choose to burn the gas instead of reinjecting it into the ground or developing it as a resource; they argue it would be uneconomic
	Good quality oil with more efficient refining	'Bunkering' by locals also depletes precious reserves of gas and oil
	Availability of natural gas, which can be used effectively in the future to assist development	Oil leaks and spills – thousands of barrels are leaked into the Delta's precious wetlands each year from oil heads and pipelines (Some commentators argue that more oil is spilled from the delta's network of pipes and terminals each year than in the ecological disaster caused by the Deepwater Horizon leak in 2010 but less notice is taken because of the relative location)
	Joint venture operations between NNPC and foreign TNCs give Nigeria some control and security over an important national resource	Oil wells and pipelines are open to attacks from militant groups who damage facilities and risk security
		Much of the wealth from the area is 'leaked' as TNC profits or payments to government
Human welfare	Many Nigerians are employed in the petrochemical industry in the area, increasing incomes, spending and skill levels	Local tribes rarely benefit directly from the wealth, which has given rise to the militant activity
	Foreign investment and wealth can be used to help develop the area economically and socially in terms of infrastructure, health and education	Atmospheric and land pollution – families live among the oil fields, breathing in methane gas and coping with frequent oil leaks
		Leaked oil when heated by the sun can become highly inflammable
		Flares from burning natural gas are common and the most visible impact of the industry – surrounding farms are seen to be constantly covered in smoke
		Pipelines – dense lines of unsightly pipelines cover the ground along communication routes
		Contaminated water – pollution means that access to safe drinking water is a major problem facing local communities
		Traditional agriculture and fishing – oil operations affect both farming and fishing in the area. Staple crops, such as cassavas and yams, can still be grown but the soil is losing fertility. Locals who rely on fish for protein in their diet are unable to fish because of polluted water
		Villages and communities – whole villages and communities in the delta have been displaced as oil spills or forest fires mean they lose all productive land
		Destruction of forests – forest fires in rainforest and mangroves surrounding oil installations are fairly common and can be triggered by gas flares, explosions or by oil igniting. Fires can take weeks to extinguish and destroy much of the local farmland and natural forest.

- Nigeria's electricity sector is small compared to other countries with similar population size and standards of living. There is enormous potential for more electricity to be generated using natural gas and renewable sources such as HEP, solar power and biomass.

- The Nigerian government made flaring illegal in 2010, as it wastes energy and contributes to global warming. Shell committed to ending flaring in 2008, but environmental groups accuse them of reneging on this because of the expense.

- Shell has paid compensation to the Nigerian government for the damage caused by spills.

- An amnesty programme was agreed between the government and militants in 2009–10 which reduced attacks on oil facilities, but there has since been a resurgence of sabotage.

- The progress of Boko Haram, an Islamic extremist insurgent group in the north of Nigeria, has caused the withdrawal of the army, meaning there is less military protection of the delta region.

- The government is planning to transform the NNPC into a more profit-driven company that will seek more private financing and reinvest in infrastructure.

- Investment in infrastructure has established the West African Gas Pipeline (owned by a subsidiary of Chevron) which pipes natural gas to other West African states including Togo, Benin and Ghana.

- Nigeria continues to discuss a planned Trans-Saharan Gas Pipeline to carry natural gas from its oilfields to Algeria's Beni Saf gas export terminal so that they can supply gas to Europe.

Case study to illustrate how aspects of physical environment affect the availability of water: Water shortages in Iztapalapa, Mexico City

Figure 11.76 Map of Mexico City and location of the borough of Iztapalapa

Greater Mexico City had a GDP of US$411 billion in 2011, ranking it as the eighth richest city in the world. Mexico City is a mega-city and one of the most important financial centres in the Americas. It is located in the centre of Mexico covering an area of around 1,485 km². With a population of around 22 million, it is the largest metropolis in the western hemisphere. The city consists of sixteen boroughs, one of which is Iztapalapa, lying to the southeast of the central business district. Iztapalapo is one of the poorer districts and itself has a population of nearly two million. Before being subsumed into the greater Mexico City Metropolitan Area (MCMA) it was an independent settlement, so is sometimes known as 'the city within a city'.

Physical environment of Mexico City

Topography and drainage: Mexico City is located in a mile-high mountain-rimmed basin, called the Valley of Mexico, near the southern end of the plateau of Anáhuac. The heart of the city is at an altitude of 2,240 m (7,350 feet) above sea level, making it one of the loftiest cities in the world. It is built largely on land reclaimed by draining and filling an old lake. There are hardly any permanent rivers in the valley today so groundwater is the main source of water.

Climate: Mexico City has a subtropical highland climate, which is cool and dry. The high altitude determines the climate of the city, which experiences hot summers and mild winters. Temperatures vary around an annual average of 64°C depending on the altitude of the district. Much of the surrounding valley is a lake basin with no outlet and the city is prone to both flooding and drought. Most rainfall occurs in the summer and early autumn (May to October) with little or no precipitation during the

remainder of the year. The lower region of the valley to the north and east (including Iztapalapa) receives less rainfall (around 720 mm per annum) while other districts in the south experience as much as 820 mm and in the mountains around the city rainfall is up to 1,500 mm each year.

Geology and vegetation: Mexico City is surrounded by a ring ('doughnut' shaped) of porous rock forming an aquifer underneath the surrounding mountains. The mountains are topped by forests of pine and oak, though much of this has been cleared as the city has expanded onto the foothills.

Figure 11.77 Crippled by poor urban planning, a surge of population growth has forced thousands of people to populate the hills surrounding Mexico City

Background to water supply problems in Mexico City

Mexico City residents face huge water scarcity issues. Eight million residents in the country's capital suffer from inadequate water supply. The city's topography, climate and geology have caused water to be a major issue for the city since its foundation 700 years ago. Uncontrolled growth has made it one of the most densely populated conurbations in the world (Figure 11.77). Parts of the city benefit from reasonably heavy rainfall in summer but the ageing water system is buckling under the pressure to supply water to 22 million people. Every day at least one million people are affected by severe shortages of water. Both the water supply system and the sewer systems of Mexico City are managed by a single government body – the Sistema de Aguas de la Ciudad de México (SACM).

Water supply: The physical environment means that the city is reliant on the groundwater aquifer to supply the bulk of the municipal water. Concerns over the continued depletion of the aquifer, combined with the extra demand from a growing population, have meant that water is also pumped in from outside of the region by inter-basin transfers from rivers to the west of Mexico City. The Lerma River transfer project was completed in the mid-twentieth century but has now largely been sucked dry. The larger, more complex Cutzamala system was completed between 1970 and 1994.

The main sources of water are:

- **abstraction of groundwater** (73 per cent)
- **Cutzamala system** (18 per cent)
- **Lerma system** (6 per cent)
- **rivers and springs** (3 per cent).

The Cutzamala system is a feat of hydraulic engineering and supplies water from the Cutzamala River in the Balsas basin, nearly 100 miles to the southwest, to 11 boroughs via one of the largest water supply systems in the world. It lifts water more than 1,000 m in altitude by a series of pumps and storage reservoirs (see Figure 11.78). The Mazahua Indians who have lived around the Cutzamala for centuries now lack access to their own river water. This conflict over the resource has led to violent protests. Overexploiting the river to supply Mexico City residents has also led to Chapala Lake in the distant state of Jalisco drying up. Conagua (Mexico's national water commission) has recently tendered for a third transfer line to be connected from the system, in order to avoid reductions in supply when maintenance work is carried out on the network's ageing pipes.

Sanitation: Mexico City is served by a single combined sewer system, collecting municipal and industrial wastewater, and storm water. The ageing drainage infrastructure can no longer cope with wastewater so less than 10 per cent of the city's sewage is treated. The remainder flows in open canals and is often washed out into the streets by rains that flood the bowl-shaped city, overwhelming many homes in poor districts. Much of the city's excess rain and sewage is

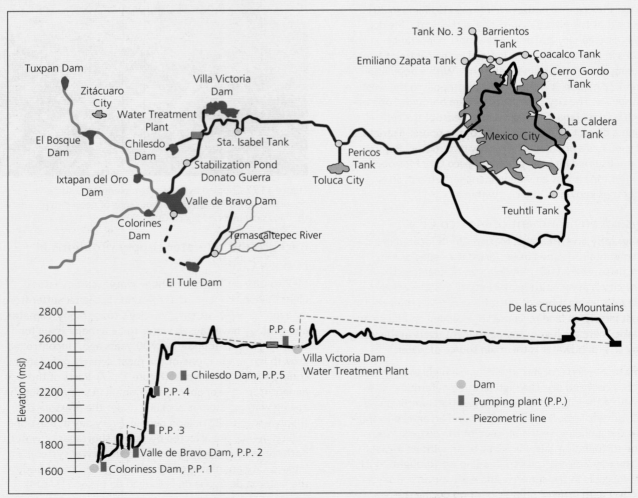

Figure 11.78 The Cutzamala inter-basin transfer supplying fresh water to Mexico City

fed through a network of tunnels and canals to Mezquital Valley, 50 miles to the north in the neighbouring state of Hidalgo. Here, the wastewater is used to irrigate fields where produce for human and livestock consumption is grown and sold back to the city. Increasing contamination of these waters has raised numerous health concerns.

Challenges

The city faces numerous environmental challenges relating to its water supply. The problems include:

- **severe shortages** of potable water supply to many boroughs (including Iztapalapa)
- **regular flooding** and a mix of sewage with rain water during periods of heavy rain
- aquifer overexploitation is causing **subsidence** and affecting buildings, some are tilting precariously; geologists claim that the city is sinking by as much as 75 mm each year
- **leaks** – losing about 1,000 litres of water per second (40 per cent of total water) from the system before it even reaches customers, not just a few big leaks, but thousands of small ones across miles of underground piping. In addition:
  - subsidence puts extra pressure on water distribution pipes
  - poor quality water can also corrode pipes
- **disparity of distribution** – those in higher income neighbourhoods get government subsidised tap water, while many living in poorer districts (beyond the reach of the city's water pipes) have to buy it at a higher cost
- back-up reservoirs are being depleted – authorities think they will run dry in a matter of years.

Impacts on the residents of Iztapalapa

The chronic water shortages are caused by poor infrastructure, political inaction (largely because of corruption and lack of continuity in elected representatives), poverty and population pressure. This has had a number of consequences for residents of Iztapalapa:

- Household taps are usually dry and even when water flows, people are reluctant to drink it because of its appearance (rusty, yellow in colour), taste (sulphuric or metallic) and texture (gritty). The city ranks number one in the world for gastro-intestinal infections (around 90 per cent of adults are infected with Helicobacter pylori).
- Water is delivered to poor districts by lorry tankers (called 'pipas') (Figure 11.79). Every day, long lines of tankers fill up at pumping stations, each one filling with 4,000 gallons. In Iztapalapa, there are 1,000 trucks distributing water to two million people, which is

Figure 11.79 Delivery of water to households in Iztapalapa

nowhere near enough to meet the needs of those people. They distribute water out of large hoses into buckets.

- Water is not free and residents pay up to 10 per cent of their household income for deliveries, meaning they cannot afford to buy other things.
- Efficiency is a problem and deliveries can take up to five days to arrive after they have been ordered. It is expensive, inefficient and customers are not satisfied but the alternative, really the only choice for many of the poorest, is self-service from a municipal tap.
- Mexicans consume more bottled water than any other nation. A study found that Mexicans used about 127 gallons of bottled water per person a year, more than four times the consumption in the USA. Rising incomes mean they can afford to buy more bottled water and it is now used for cooking and for bathing babies because of the unreliability of other supplies. Shops or street vendors on cycles selling water to meet people's daily needs are becoming increasingly common.
- People have to adapt their lifestyle to an erratic water supply – for example, using half a bucket of water to wash in and not being able to flush a toilet until two or three people have used it.

Management responses and solutions

- **Demand management:** Ramón Aguirre, Director of SACM, claims that residents need to ration their water much better. City residents are keen on long showers and washing their cars, homes and clothes well; the average Mexico City resident uses 300 litres of water per day compared to 180 per day in some European cities. On Easter Saturdays, residents traditionally have huge water fights, in which everyone from grandparents to young children join in hurling bucket loads over each other. A programme is planned to educate children that water is valuable and needs to be used wisely.

- **Supply management:**
  - **Deeper aquifer abstraction:** A new plan has suggested drawing drinking water from a mile-deep aquifer. The government is prepared to spend $40 million to pump and treat the deeper water, which they say could supply some of the city's population for as long as a century.
  - **Rainwater harvesting (domestic):** Some parts of the city receive high rainfall and local entrepreneurs have initiated social enterprises to install rainwater collection and storage systems in homes and businesses costing around $1,000 each. While it takes some pressure off the supply system, the cost is prohibitive to most of the city's residents. However, 'Isla Urbana', a non-profit NGO, is currently working to install rainwater harvesting systems in marginalised communities.
  - **Rainwater capture (ecological):** The city government have employed planners and environmentalists to seek ecological solutions. As the city is on the top of the plateau, the origin of most water is rain and not river surface water.

The ancient forests on top of the surrounding mountains have all but disappeared because of the urban expansion. Walls of gabions have been put in place to protect the remaining forest and prevent soil erosion in an attempt to re-establish the water cycle in the basin. New forest is also being planted on the watershed, so that when it rains the forest vegetation prevents rapid run-off. The water is held up and stored in the soil at the top and slowly infiltrates and percolates into the ground, filling the pore spaces in the aquifer, so increasing the recharge rate.

- **Other measures:**
  - fixing leaks in the whole system
  - increasing the use of recycled water, which can be used for bathing and recharging aquifers
  - authorities are overseeing the construction of new drainage tunnels to alleviate flooding
  - they have also pledged to finish a massive water treatment plant north of the city.

Fieldwork opportunities

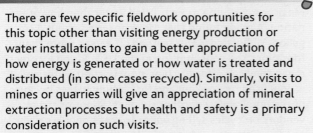

There are few specific fieldwork opportunities for this topic other than visiting energy production or water installations to gain a better appreciation of how energy is generated or how water is treated and distributed (in some cases recycled). Similarly, visits to mines or quarries will give an appreciation of mineral extraction processes but health and safety is a primary consideration on such visits.

Most power stations offer educational visits for students. The Dinorwig HEP installation inside Mynydd Gwefru (Electric Mountain) in Snowdonia is particularly interesting. For more information visit www.electricmountain.co.uk. It is also possible to visit renewable installations such as wind farms.

Primary data collection can be undertaken in river valleys containing reservoirs, both upstream and downstream of the reservoir, to establish impacts on discharge, sediment, bedload, etc. Likewise, the microclimate effects around reservoirs such as wind speed, humidity and effects on vegetation could be measured.

Review questions

1 Outline the stages necessary in the development and extraction of a resource.

2 Distinguish between the meaning of measured reserves, indicated reserves, inferred resources and possible resources.

3 Explain the advantages and disadvantages of using:

  a) different sources of water from rivers and lakes, reservoirs, aquifers and seas

  b) different sources of renewable energy

  c) open-pit mining and underground mining.

4 Examine the environmental impacts of resource exploitation of:

  a) water abstraction from aquifers

  b) fossil fuels

  c) open-pit mining for mineral ores.

5 How must today's policies, markets and technologies change to deliver tomorrow's sustainable energy?

6 Using a case study of either an energy, mineral ore or water resource issue, assess the impacts on the local or regional physical and human environments.

Further reading

General

'Earth's Natural Resources', John V. Walther, (published by Jones and Bartlett Learning), 2013

Energy

Malek Al-Chalabi, 'Peak oil', *Geography Review*, Volume 25, Number 2, November 2011, pp 33–35.

Sam Friggens, 'Should we subsidise renewables?' *Geography Review*, Volume 27, Number 3, February 2014, pp 30–31.

Peter Jones and Daphne Comfort, 'What's wrong with fracking?' *Geography Review*, Volume 27, Number 4, April 2014, pp 34–37.

A series of factsheets based on the annual World Energy Outlook, also published by the International Energy Agency www.worldenergyoutlook.org

2015 World Energy Issues Monitor (Executive Summary) – an annual report published by the World Energy Council www.worldenergy.org

World energy statistics published by Enerdata https://yearbook.enerdata.net/

UK Renewable Energy Roadmap: Update 2013 – Department of Energy and Climate Change

Nuclear Power in the United Kingdom – World Nuclear Association www.world-nuclear.org

Energy profile of Nigeria – updated 2013 www.eoearth.org/view/article/152513/

Randers, J. (2012) *2052 – A Global Forecast for the next Forty Years.* Chelsea Green Publishing Co.

Mineral ores

'World mineral statistics 2010-2014', British Geological Survey – www.bgs.ac.uk/mineralsUK/statistics/worldStatistics.html

D Rogich and G Matos, 'The Global Flows of Metals and Minerals', (2009) *US Geological Survey,* – http://pubs.usgs.gov/of/2008/1355/pdf/ofr2008-1355.pdf

S Lei and F Lan, 'The changing patterns of mineral market trade and their impact on China's economy', (2010) *Journal of Resources and Ecology* – www.jorae.cn/fileup/PDF/2010010111.pdf

'Trends in the mining and metals industry (Mining's contribution to sustainable development)', International Council on Mining and Metals (ICMM), (2012) – www.icmm.com/document/4441

Water

'The Human Right to Water and Sanitation', published by the United Nations – www.un.org/waterforlifedecade/pdf/human_right_to_water_and_sanitation_media_brief.pdf

Water in conflict – www.globalpolicy.org/the-dark-side-of-natural-resources-st/

The UN's inter-agency mechanism on all freshwater issues produces a number of topic focuses and factsheets, e.g. transboundary waters, water scarcity, water and climate change www.unwater.org/

'Future water: The government's water strategy for England' – published by DEFRA www.gov.uk/government/publications/future-water-the-government-s-water-strategy-for-england

Hollander, K. (2012) *Several Ways to Die in Mexico City*: Feral House.

Question practice

A-level questions

1. What is the difference between stock resources and flow resources?

    a) Stock resources include any energy resources that can be stored for transport and future use, such as coal and oil. Flow resources are those that cannot be stored easily and include gas, wind and solar power.

    b) Both are renewable resources but flow resources are continuously available, such as wind, solar and tidal power, while stock resources have to be managed otherwise they can be depleted, such as timber, crops and fish.

    c) Flow resources are those that can be naturally renewed or replaced within a relatively short time span, such as wind, water or timber. Stock resources are deposits with a fixed and finite supply, such as mineral ores or coal.

    d) Stock resources include mineral ores such as iron, copper and aluminium, which are not used to generate energy. Flow resources are those that are used to generate energy, such as gas, wind and solar power.   (1 mark)

2. What is meant by water stress?

    a) It is when annual water supplies fall to less than 500 m³/person on average.

    b) The demand for water exceeds the amount of water available during a certain period, or when poor quality of water restricts its availability for use.

    c) The population receives an adequate supply of water from nature but does not have the money to develop a supply infrastructure such as piped water.

    d) It is a water deficit in an area because the area receives less than 500 mm/year of rainfall.   (1 mark)

3. How is acid rain formed?

    a) It is formed by an increased use of catalytic converters in car exhaust systems releasing poisonous gases, such as carbon monoxide, into the atmosphere.

    b) Toxic waste materials from tailings ponds leach into groundwater or leak into rivers and enter the hydrological cycle.

    c) It is formed by burning fuelwood and other biomass fuels in the open, because they release toxic fumes into the atmosphere.

    d) Pollutants such as sulphur dioxide are released when burning fossil fuels in vehicles or power stations. The chemicals react with water in the atmosphere to produce acid solutions that are deposited when it rains.   (1 mark)

4. Which of the following best describes an 'indicated reserve' of a stock resource, such as a mineral ore?

    a) The mineral deposit has been surveyed extensively and the quantity, grade and physical characteristics are so well established that the tonnage and grade can be declared.

    b) The area has not yet been fully explored but a mineral deposit is reasonably expected to exist in the region because of the known geological conditions.

    c) The deposit has not been surveyed extensively but there is continuity with a proven reserve. Sufficient sampling has been done to estimate with confidence that the deposit is probably mineable.

    d) Limited geological sampling of the ore deposit and estimates of quantity and grade are based on an assumed continuity of an existing reserve. There is uncertainty and insufficient information to estimate tonnages or grades.   (1 mark)

5. The world average water availability is 6,236 m³/person. Using Table 11.24 and Figure 11.80, analyse the relationship between water availability and consumption by sector across different regions of the world. (6 marks)

Table 11.24 Water availability by region, 2013

Region	Average water availability (m³/person)
Arab World	500
Sub-Saharan Africa	1,000
Caribbean	2,466
Asia-Pacific	2,970
Europe	4,741
Latin America	7,200
North America (includes Mexico)	13,401

Source: FAO AQUASTAT, UNESCO

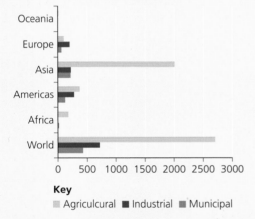

Key
Agriculcural ■ Industrial ■ Municipal

Figure 11.80 Total water withdrawal by sector (km³/yr)
Source: FAO AQUASTAT

6. With reference to Figure 11.81 and your own knowledge analyse the global production of crude oil and assess the role of OPEC in the geopolitics of energy supply and pricing. (9 marks)

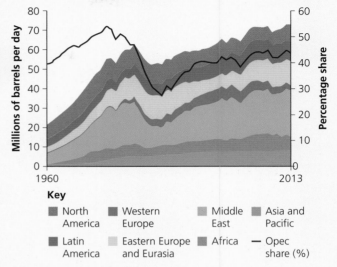

Key
■ North America ■ Western Europe ■ Middle East ■ Asia and Pacific
■ Latin America Eastern Europe and Eurasia ■ Africa — Opec share (%)

Figure 11.81 World crude oil production
Source: OPEC

7. Assess the extent to which ensuring mineral ore security is environmentally sustainable. (9 marks)

8. 'Strategies adopted to control water supply are more important for water security and sustainability than those used to manage water consumption.'

To what extent do you agree with this view?

(20 marks)

Part 3

Skills and Fieldwork Investigation

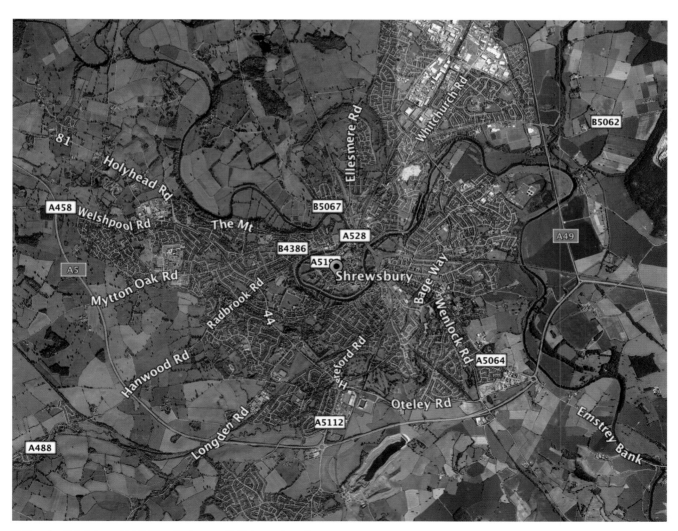

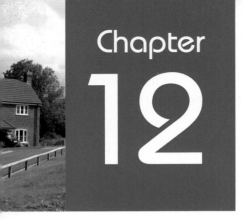

Chapter

12

Geographical skills and fieldwork

One of the AQA's three assessment objectives states that you should be able to:

'*Use a variety of relevant quantitative, qualitative and fieldwork skills to:*
- *investigate geographical questions and issues;*
- *interpret, analyse and evaluate data and evidence;*
- *construct arguments and draw conclusions.*'

These skills are assessed in the AS and A-level examination papers as well as in the Non-Examined Assessment (NEA) that is the fieldwork report.

Both the AQA AS and A-level specifications state that their intention is that geographical skills should be developed in an integrated way throughout the whole course.

During both AS and A-level courses, students should:
- understand the nature and use of different types of geographical information, including qualitative and quantitative data, primary and secondary data, images, factual text and discursive/creative material, digital data, numerical and spatial data and other forms of data, including crowd-sourced and 'big data'
- collect, analyse and interpret such information, and demonstrate the ability to understand and apply suitable analytical approaches for the different information types
- undertake informed and critical questioning of data sources, analytical methodologies, data reporting and presentation, including the ability to identify sources of error in data and to identify the misuse of data
- communicate and evaluate findings, draw well-evidenced conclusions informed by wider theory,
- construct extended written argument about geographical matters.

12.1 Qualitative skills and quantitative skills

Key terms

Qualitative data – These are non-numerical data that are used in a relatively unstructured and open-ended way. It is descriptive information, which often comes from interviews, focus groups or artistic depictions such as photographs.

Quantitative data – Quantitative data are numerical data such as metric-level measurements that are associated with the scientific and experimental approach and are criticised for not providing an in-depth description.

Coding – Coding is carried out when analysing answers to interviews. The coder (person who analyses the data) looks though all the answers to a question, develops a broad classification system based on the responses, and then uses a code to categorise responses.

Sampling – To make statistically valid inferences, when it is impossible to measure the whole population, by selecting a group which will be representative of that population. Such a group is known as a sample.

Statistical population – The whole from which a sample will be chosen for a research exercise.

Data used by geographers can be divided broadly into **qualitative data** and **quantitative** data.

Qualitative methods are ways of collecting data which are concerned with describing meaning, rather than with drawing statistical inferences. What qualitative methods (for example, in-depth interviews and participant observation) lose on reliability they gain in terms of validity. They provide a more in-depth and rich description of the subject under study. It is qualitative data that is most often used in cultural geography.

Quantitative methods (or geostatistics) are those which focus on numbers and frequencies rather than on meaning and experience. Quantitative methods (for example, till analysis, questionnaires and surveys) provide information which is easy to analyse statistically and whose reliability is known. Quantitative methods are associated with the scientific and experimental approach and are sometimes criticised for not providing an in-depth description.

Qualitative data

Interviews

One of the main ways of collecting qualitative data is through interview surveys. These can be detailed and flexible with open-ended questions and the opportunity for respondents to give their opinions without being limited to responses in option boxes. The interviewer has to be well prepared, with a specific aim for each interview, which matches the topic under consideration as well as the role of the interviewee. They can also be loosely structured so particularly interesting points can be pursued and you can adapt to follow the flow of the conversation.

Examples of where interviews would be appropriate include:
- studying the attitudes of residents to the development of a nearby housing estate
- finding out, from an entrepreneur, the reason for their choice of location of a new commercial development.

Once a series of interviews have been carried out then the technique of **coding** is used to put the qualitative data into a form in which it can be analysed using statistical methods. An example would be open-ended questionnaire data. The 'coder' (person who analyses the data) looks though all the answers to a question, develops a broad classification system based on the responses, and then uses a code to categorise responses. Often unusual or unexpected responses are left in an 'unclassified' category, and classes with very few answers may be aggregated to the nearest similar answer.

Quantitative data

One of the main skills that you are expected to use throughout fieldwork is the ability to apply geographical and the geospatial technologies (for example, GIS) that are used to collect, analyse and present geographical data. Alongside this are skills associated with descriptive statistics related to central tendency, dispersion and correlation.

Sampling

Sampling techniques are used to collect both qualitative and quantitative data.

In statistics, the entire pool of items under study is known as the **statistical population**. Sampling is a shortcut method for investigating this whole population. Data is gathered on a small part (a sample) of the whole parent population. The information obtained from the sample allows statisticians to develop hypotheses about the larger population.

Why sample?

Sampling is used when it is impossible, or simply not necessary, to collect large amounts of data. Collecting small amounts of carefully selected data will enable you to obtain a representative view of the feature as a whole. You cannot, for example, interview all the shoppers in a market town or all the inhabitants of a village, but you can look at a fraction of those populations and from that evidence indicate how the whole is likely to behave.

When you have established the need for a sample survey, you will have to decide on a method that will collect a large enough representative body of evidence. If, for example, you are interviewing the inhabitants of a village, you must ensure that your interviews cover all age ranges in the population.

Sampling considerations

Figure 12.1 (page 622) summarises the things you need to take into account when deciding whether or not to use a sampling survey.

Types of sampling

Three main types of sampling strategy:
- random
- systematic
- stratified

Within these types, you may then decide on a point, line or area method.

Random sampling

This is summarised in Figure 12.2 (page 622)

The advantages of random sampling are that it can be used with large sample populations and that it

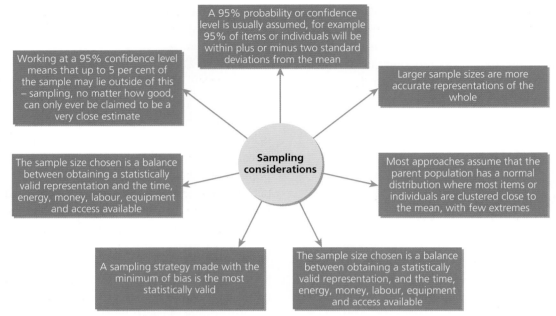

Figure 12.1 Sampling considerations

avoids **bias.** This means that each member of the population has the same chance of being selected. The disadvantages are firstly, that it can lead to poor representation of the overall parent population or area if large areas are not hit by the random numbers generated. This is made worse if the study area is very large. Secondly there may be practical constraints in terms of time available and access to certain parts of the study area.

Systematic sampling

Samples are chosen in a systematic, or regular way. This means they can be:

- evenly/regularly distributed in a spatial context, for example, every 10 m along a transect
- at equal/regular time intervals, for example, every hour or at the same time every day
- regularly numbered, for example, every tenth person or building.

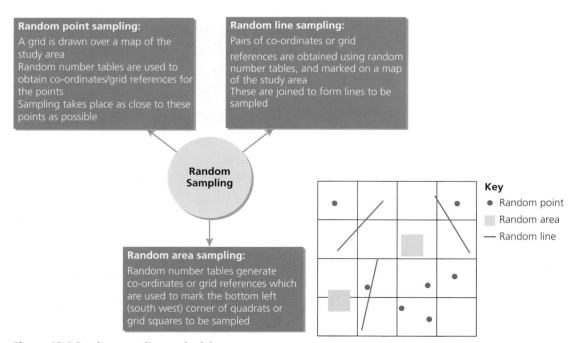

Figure 12.2 Random sampling methodology

Like random sampling, systematic sampling can be done either as a point, a line or an area. A grid can be used and the points can be at the intersections of the grid lines. If it is a transect survey then the points can be at regular intervals along that transect. Systematic line sampling could use the grid lines on a map or a series of transects (for example, a transect every 20 m along a beach). Systematic area sampling could be a series of grid squares on a map chosen to create a regular pattern.

The advantages of systematic sampling are that it is more straightforward than random sampling; a grid doesn't necessarily have to be used, sampling just has to be at uniform intervals and good coverage of the study area can be easily achieved. Disadvantages are that it is more biased, as not all members or points have an equal chance of being selected and so this can lead to over or under representation of a particular pattern.

Stratified sampling

This method is based on knowing something in advance about the population or area in question. The parent population is then grouped into subsets of known size. These make up different proportions of the total population and therefore sampling should be stratified to ensure that results are proportional to the size of the subset and representative of the whole. For example, if you are surveying a population with a view to examining lived experience of a place and you know its age distribution, your sample must reflect that distribution. Choice of sample within the subsets can still be either systematic or random.

The advantages of stratified sampling are:

- it can be used with random or systematic sampling, and with point, line or area techniques
- if the proportions of the subsets are known, it can generate results which are more representative of the whole population
- it is very flexible and applicable to many geographical enquiries
- correlations and comparisons can be made between subsets.

The disadvantages of stratified sampling are:

- the proportions of the subsets must be known and accurate if it is to work properly

- it can be hard to stratify questionnaire data collection; it may be hard to identify people's age or social background effectively.

Sample size

The size of sample usually depends upon the complexity of the survey being used. When using a questionnaire it is necessary to sample sufficient people to take into account the considerable variety introduced by the range of questions. Sample size can be restricted by practical difficulties and this may affect the reliability of results. Your aim should be to keep the sampling error as small as possible. You are not a professional sampler and cannot be expected to conduct hundreds of interviews, but on the other hand, sampling only 20–30 people in an energy conservation survey is not representative of the population as a whole.

Key questions

What sort of sample method would be most appropriate for the following investigations?

- **A survey of differing infiltration rates within a drainage basin**
- **A land use survey along a coast as part of an investigation into sustainable management**
- **Pebble size and roundness on a beach**
- **The cultural characteristics of a place**
- **The health of the population of a place.**

Justify your choices.

Ethical implications

You should have an awareness of ethical issues which are embedded in any study that involves the collection, analysis and representation of geographical information about human communities. The most common ethical dilemmas in human geography focus around participation, consent, safe-guarding/confidentiality of personal information as well as giving something back.

In physical geography the main ethical considerations are about consent/access to study sites and potential damage or pollution of study sites.

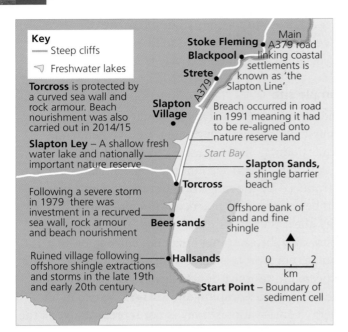

Figure 12.3 Sketch map of Start Bay

12.2 Specific skills

Core skills

There are certain core skills that we take for granted. These include you having sufficient literacy skills to be able to write and interpret written reports, as well as numeracy skills to measure quantitative data and make relevant calculations. Other core skills are outlined below.

Sketch maps

Any annotated map/diagram must feature the following elements:

- a title or location of the place in the map
- a key to all symbols
- annotations or labels explaining, elaborating or emphasising particular features. Annotations are usually more than 10 words whereas labels are less.
- an indication of scale
- a north pointer.

Using field sketches and photographs

In your personal investigation, field sketches and photographs are excellent ways to record exactly what you have seen. Field sketches enable you to pick out from the landscape the features that

you wish to identify and perhaps comment upon (Figure 12.3). They are particularly useful in physical geography investigations, for example, Figure 12.4 shows a field sketch of a glacial area. However, there is no reason why the technique cannot be used in an urban study. Sketching does not require artistic talent. It is far more important in geographical investigations to produce a clear drawing with good and useful labels.

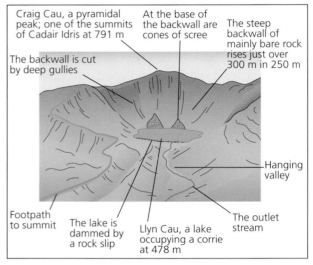

Figure 12.4 A field sketch of Llyn Cau, Cadair Idris

If you lack the confidence to produce sketches but need to show detail from the area under investigation, then taking photographs is just as valid. However, you must not use them as an attractive 'space-filler' in your report. Any photographs you include must be annotated or labelled. This can be done digitally or using an acetate overlay. A photograph is only useful when the reader is directed to the information it shows.

Ordnance survey maps

You must be able to interpret and use Ordnance Survey (OS) maps (at a variety of scales). Each grid square of an OS map contains a huge amount of information. This can be on the physical landscape and landforms of the area (using contour lines and map symbols) and/or the human landscape. OS maps can be used as a source of information or as a base map for display of fieldwork investigation findings.

You are expected to be able to work out location (grid references), use the scale and direction indicators, interpret contour patterns, describe settlement form and function as well as inferring land use.

Cross sections of physical landscapes can be constructed from OS maps. These could be used in fieldwork reports, though they will not be directly examined.

Geographical information systems (GIS)

Geographical information systems (GIS) are an integral part of twenty-first century geographical study. They are used extensively by environmental planners, government departments, public utility companies and commercial companies. GIS have the ability to store, retrieve, manipulate and analyse a wide range of spatially-related data. They can:

- help with questioning and understanding data
- enable multiple interrogation of complex data
- illustrate difficult abstract concepts in a dynamic visual way
- make use of three-dimensional representations
- provide opportunities for modelling and decision making.

GIS software is a set of computer tools for collecting, storing, processing and displaying sets of information linked to places on maps. Digital maps, satellite images or aerial photographs can link information. The maps are produced either as raster or vector data.

Raster maps are images made by a series of coloured dots on a screen (pixels), just like high quality digital photographs. These are mainly used as background maps to other information. Digital versions of OS maps are raster maps.

With vector maps, each feature (for example, roads, buildings or woods) is recorded using XY co-ordinates, enabling a GIS to link information from spreadsheets and databases to the maps. Vector data is stored in themed layers such as roads, water and settlements. These layers can be added to a base map when and if needed. GoogleMaps is an example of vector mapping in use.

Cartographic skills

Cartographic skills fall into two main categories. The first group of skills involves the reading and

interpretation of a variety of different types of maps. These include atlas maps and weather maps.

Atlas maps

An atlas is a collection of maps based on a single theme (for example, a road atlas) or a series of themes (physical, political, population, resources, etc.). Most academic atlases require you to understand and be able to use lines of latitude and longitude to identify and describe location. You need to be able to use the index and understand how colour and symbols are used to convey information.

Atlases contain huge amounts of information that can be useful when studying and comparing countries or regions. Note that it is important, in this rapidly changing world, that the atlas is up to date.

Weather maps

A **synoptic weather chart** (or map) is a summary of the current situation. In weather terms this means the pressure pattern, fronts, wind direction and speed and how they will change and evolve over the coming few days. Weather maps and synoptic charts are useful when you are planning fieldwork. They not only tell you what the weather is likely to be, but also tell you why the weather is like it is. You are likely to see these in two forms, as shown in Figure 12.5 and Figure 12.6 (page 626).

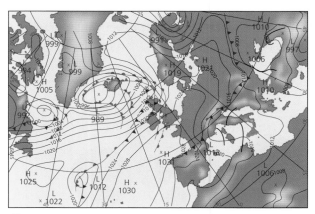

Key

▲▲▲ Cold front   ⬤⬤⬤ Warm front   ▲⬤▲⬤ Occluded front

—996— Isobar (with level of atmospheric pressure in millibars)

| L 1012 | Low pressure area with pressure in millibars | H 1030 | High pressure region with pressure in millibars |

Figure 12.5 A weather chart that shows surface pressure and some limited analysis

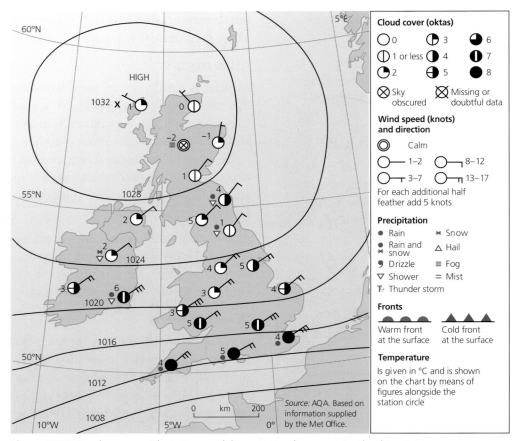

Figure 12.6 Weather map with station models: an anticyclone over Scotland in winter

Maps with located proportional symbols

These are maps that include symbols which are proportional in area or volume to the value they represent (Figure 12.7). Symbols of representative sizes, such as squares or circles, or even small graphs, such as bar graphs or pie graphs, can be placed on a map to show spatial differences.

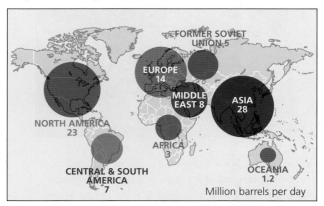

Figure 12.7 World petroleum consumption by region (1980–2012). *Source: US Energy Information Administration.*

It is important that you take great care in placing symbols on a map. It is essential to avoid too much overlap, but it must also be clear which area or place the symbol represents.

Maps showing movement – flow lines, desire lines and trip lines

Flow lines and desire lines are similar in that they both represent the volume of movement from place to place (Figure 12.8 and Figure 12.9). They are useful to show such features as:

- traffic movements along particular routes (for example, roads, railways and waterways)
- migration of populations
- movement of goods or commodities between different regions
- movements of shoppers.

In both methods the width of the line is proportional to the quantity of movement. A flow line represents the quantity of movement along an actual route, such as a train or bus route.

A desire line is drawn directly from the point of origin to the destination and takes no account of a specific route.

Trip lines can be drawn to show regular trips, for example where people shop; lines could be drawn from a town to nearby villages (Figure 12.10).

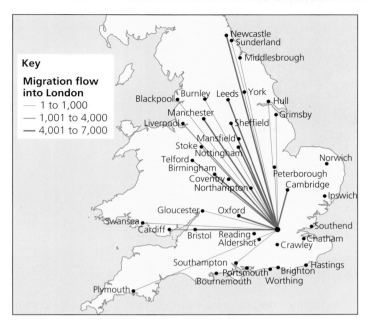

Figure 12.8 A desire line map showing net migration of people to London, 2009–2012

Maps showing spatial patterns – choropleth, isoline and dot maps

Choropleth maps

A choropleth is a map on which data values are represented by the density of shading within areas (Figure 12.11, page 628). The data are usually in a form that can be expressed in terms of area, such as population density per square kilometre. To produce such a map certain stages have to be followed:

The material has to be grouped into classes. Before you can do this you have to decide on the number and range of classes required to display your data clearly.

A range of shadings has to be devised to cover the range of the data. Darkest shades should represent the highest figures and vice versa. It is good practice

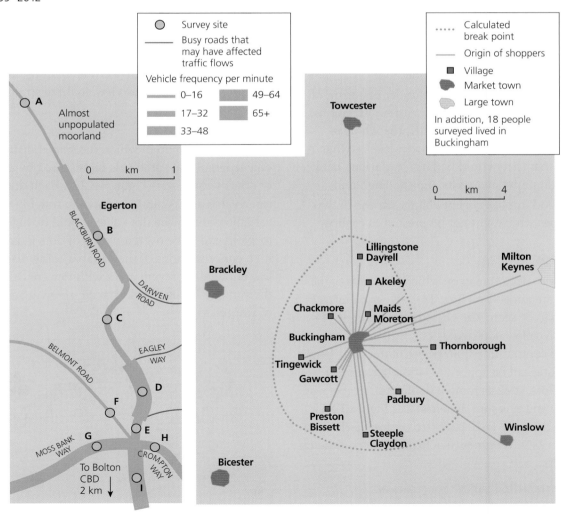

Figure 12.9 A flow line map showing traffic flows near Bolton

Figure 12.10 Trip line map showing the origins of shoppers on a market day in Buckingham

Source: AQA

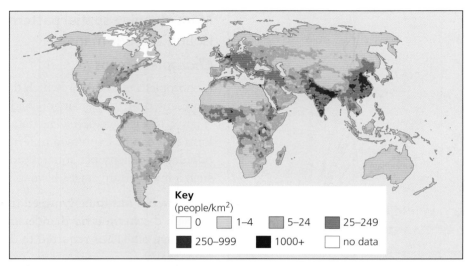

Figure 12.11 Global population density estimates, 2015

not to use the two extremes of black and white because black suggests a maximum value while white implies that there is nothing in the area.

Choropleth maps are fairly easy to construct and are visually effective as they give the reader a chance to see general patterns in an a real distribution. There are, however, a few limitations to the method:

- It assumes that the whole area under one form of shading has the same density, with no variations. For example, on maps of the UK, the whole of Scotland may be covered by one category, when it is obvious that there could be large variations between the central populated areas and the Highlands.
- The method implies abrupt changes at the drawn boundaries which will not be present in reality.

Isoline maps

It is possible to draw a map on which all points of the same value are joined by a line. This allows patterns in a distribution to be seen. The best-known example of such isolines (also called isopleths) is on Ordnance Survey maps where contour lines join places of the same height. This technique can be applied to a number of other physical factors, such as rainfall (**isohyets**), temperature (**isotherms**) and pressure (**isobars**) (see Figure 12.5, page 625 and Figure 12.6, page 626), as well as human factors, such as travel times (**isochrones**) for commuters and shoppers.

Some rules regarding isoline construction:

- They connect points of equal value.
- They represent continuous surfaces (like a ground surface). It doesn't suddenly disappear.

- Isolines do not cross or touch (with the exception of vertical gradients, like cliffs.)
- Values on one side of the isoline are higher or lower than values on the other side.
- The interval between isolines is the same over the entire map (unless otherwise specified).
- Isoline maps of the environment almost always represent an overhead view, looking straight down.

Dot maps

A dot map is a map in which the spatial distribution of a geographical variable is represented by a number of dots of equal size (Figure 12.12). Each dot has the same value and is plotted on a map roughly where that variable occurs. The dot value should be high enough to avoid excessive overcrowding of dots in areas with high concentrations of the variable being mapped and low enough to prevent areas with low concentrations of the variable having no dots at all, so giving a false sense of emptiness.

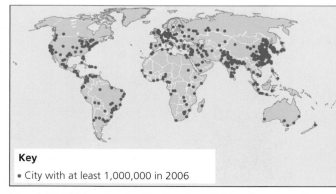

Key
• City with at least 1,000,000 in 2006

Figure 12.12 Map showing the location of cities of over 1 million people

Dot maps have two limitations:

- Large numbers of dots are difficult to count. Therefore, although dot maps are good at giving an impression of distribution, they are less valuable if you need a precise idea of the values they represent.
- There must be some accompanying information about the distribution of the variable or the map could be misleading.

Key questions

What sort of maps would be most appropriate for displaying the following?

- **The population distribution within a country**
- **The location of dairy farms in the UK**
- **The ethnicity of different cities in the UK**
- **The number of doctors per 1,000 of population for the UK**
- **The main global migrations of the past 30 years.**

Justify your choices.

Graphical skills

Line graphs

Simple and compound line graphs

Line graphs are appropriate when you want to show absolute changes in data. For example, they are suitable for showing changes in food production through time, population change or stream discharge. When several lines are plotted on the same graph, it is important to recognise whether it is a simple or a compound line graph:

- On a simple line graph, the line represents the actual values of whatever is being measured on the vertical axis
- On a compound line graph, the differences between the points on adjacent lines give the actual values. To show this, the areas between the lines are usually shaded or coloured and there is an accompanying key (Figure 12.13).

It is possible to show two sets of data on the same graph. The left hand vertical axis can be used for one scale and the right hand vertical axis for a different scale. This can often give a useful visual impression of the connection between two sets of data.

When using line graphs you should:

- plot the independent variable on the horizontal axis and the dependent variable on the vertical axis. If you are plotting data over time, time should always be plotted on the horizontal axis
- try to avoid awkward scales and remember that the scale you choose should enable you to plot the full range of data for each variable
- clearly label the axes
- use different symbols if you are plotting more than one line.

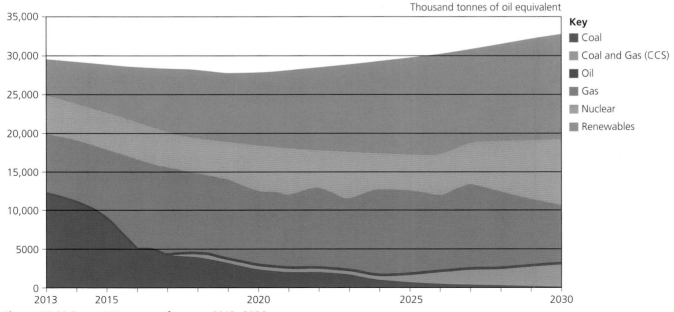

Figure 12.13 Future UK sources of energy, 2013–2030

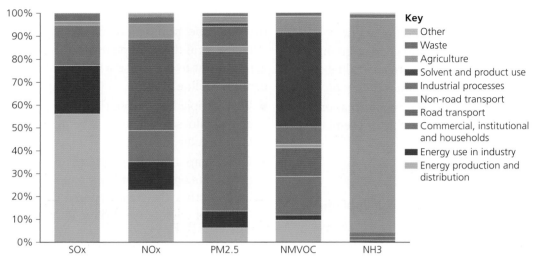

Figure 12.14 Share of emissions of main air pollutants by sector group for the 33 countries of the European Economic Area (EEA)
Source: European Environment Agency

Bar graphs – simple, comparative, compound and divergent

A bar graph (or chart) has vertical columns rising from a horizontal base (Figure 12.14). The height of each column is proportional to the value that it represents. The vertical scale can represent absolute data or figures as percentages of the whole.

Bar graphs are easy to understand. Values are obtained by reading off the height of the bar on the vertical axis. They show relative magnitudes very effectively.

Using an appropriate scale, it is also possible to show positive and negative values on the same graph – for example, profit and loss (Figure 12.15).

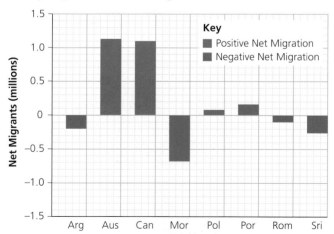

Figure 12.15 A gain-loss (divergent) bar chart showing net migration for selected countries.

Scatter graphs, and the use of best-fit line

Scattergraphs are used to investigate the relationship between two sets of data (Figure 12.16). They can be used simply to present data, but they are particularly useful in identifying patterns and trends in the relationship that might lead to further inquiry.

A general trend line (**best-fit**) can be added to the graph so that the relationship can be easily observed. This is the red line in Figure 12.16. If it runs from bottom left to top right, it indicates a positive relationship; if it runs from top left to bottom right, the relationship is negative.

Other features of scattergraphs include:

- they can be plotted on arithmetic, logarithmic or semi-logarithmic graph paper
- the independent variable goes on the horizontal axis and the dependent variable on the vertical axis
- it is possible for a correlation to emerge even when a relationship is only coincidental
- points lying some distance from the best-fit line are known as **residuals** (anomalies). These can be either positive or negative. Identification of residuals may enable you to make further investigations into other factors that could have influenced the two variables.

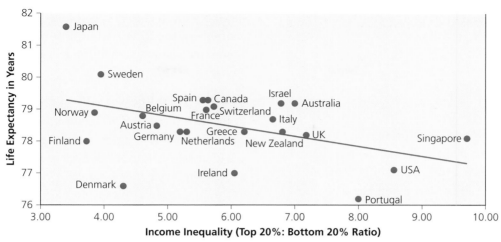

Figure 12.16 Scattergraph to show the relationship between income equality and life expectancy

Source: www.inequality.org

Pie charts and proportional divided circles

The pie chart is divided into segments according to the share of the total value represented by each segment (Figure 12.17). This is visually effective – the reader is able to see the relative contribution of each segment to the whole. On the other hand, it is difficult to assess percentages or make comparisons between different pie charts if there are lots of small segments.

When a number of pie charts are drawn proportional to the value each represents in total, they are called

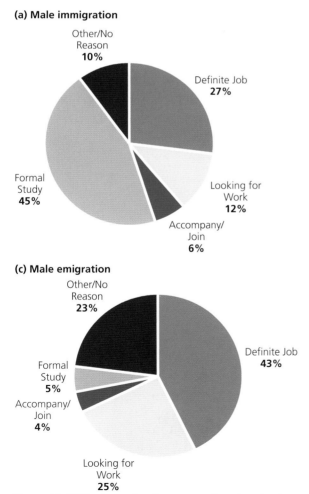

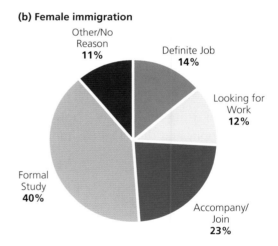

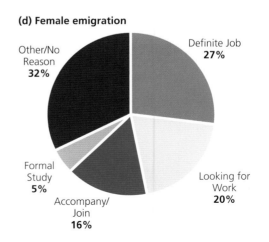

Figure 12.17 Pie charts showing reasons for migration by sex, 2011

Source: ONS

proportional divided circles. The construction of a proportional circle is as follows:

- Use the formula

$$r = \sqrt{\frac{V}{\pi}}$$

  where *V* is the value that you want the total pie chart to represent and *r* is the radius of the pie chart. Take the value for π as 3.142

- Draw a circle of radius *r* on graph paper

  It is important that you are able to state the scale of your proportional circle. The *area* represents the value, so the scale is in the form of:

  x units of data = 1 square unit on the graph paper

Triangular graphs

Triangular graphs are plotted on special paper in the form of an equilateral triangle (Figure 12.18 and Figure 12.19). Although this looks on the surface to be a method that has widespread application, it is only possible to use it for a whole figure that can be broken down into three components expressed as percentages. The triangular graph cannot therefore be used for absolute data or for any figures that cannot be broken down into three components.

The advantage of using this type of graph is that the varying proportions and their relative importance can be seen. It is also possible to see the dominant variable of the three. After plotting, clusters will sometimes emerge, enabling a classification of the items involved.

Graphs with logarithmic scales

A logarithmic graph is drawn in the same way as an arithmetic line graph except that the scales are divided into a number of cycles, each representing a tenfold increase in the range of values (Figure 12.20). If the first cycle ranges from 1 to 10, the second will extend from 10 to 100, the third from 100 to 1,000 and so on. You may start the scale at any exponent of 10, from as low as 0.0001 to as high as 1 million. The starting point depends on the range of data to be plotted.

Graph paper can be either fully logarithmic or semi-logarithmic (where one axis is on a log scale and the other is linear or arithmetic). Semi-logarithmic graphs are useful for plotting rates of change through time, where time appears on the linear axis. If the rate of change is increasing at a constant proportional rate (for example, doubling each time period) it will appear as a straight line.

Logarithmic graphs are good for showing rates of change – the steeper the line, the faster the rate. They also allow a wider range of data to be displayed.

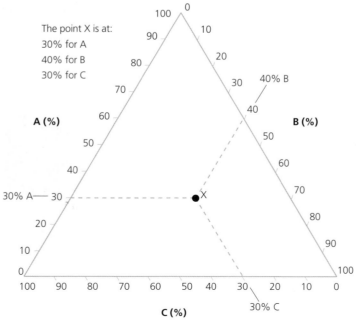

Figure 12.18 How to plot a point on a triangular graph

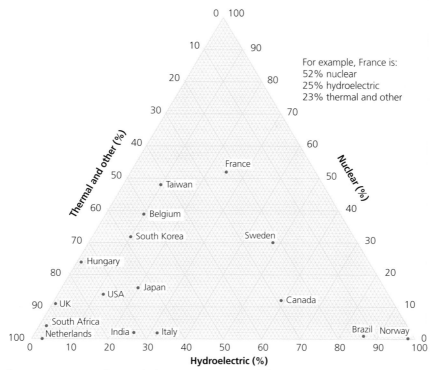

Figure 12.19 Triangular graph showing percentage of electricity produced by generating source for selected countries

Key question

Describe how the rates of human population growth change through time.

Remember that you cannot plot positive and negative values on the same graph and that the base line of the graph is never zero, as this is impossible to plot on such a scale.

Dispersion diagrams

Dispersion graphs are used to display the main patterns in the distribution of data. The graph shows each value plotted as an individual point against a vertical scale. It shows the range of the data and the

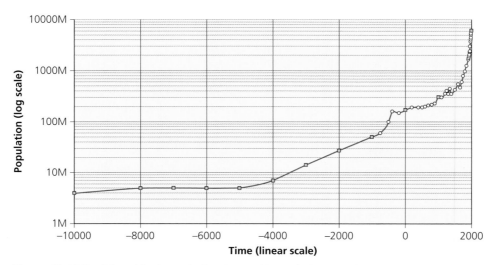

Figure 12.20 Semi-logarithmic graph showing world population growth.

Table 12.1 The lip orientation in degrees of 15 corries in the Glyders of Snowdonia and 15 on the Isle of Arran

Corrie number	1	2	3	4	5	6	7	8	9	10	11	12	13	14	15
Glyders	30	60	45	55	75	50	80	50	10	15	10	35	45	50	85
Arran	5	5	10	55	15	30	95	5	185	70	120	40	30	115	110

distribution of each piece of data within that range. It therefore enables comparison of the degree of bunching of two sets of data (Table 12.1). This is shown in Figure 12.21.

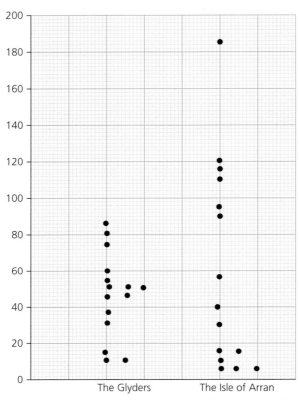

Figure 12.21 Dispersion diagram of corrie lip orientation for The Glyders (Snowdonia) and the Isle of Arran

Key questions

What sort of graphs would you choose to display the following?

- **The changing length of a glacier through time**
- **The percentage of global deaths from malaria**
- **The energy mix for electricity production in the UK from 1990 to 2015**
- **The top ten countries with technically recoverable shale-gas resources.**

Justify your choice.

Statistical skills

Measures of central tendency

There are three such measures: arithmetic mean, mode and median.

Arithmetic mean

The mean is calculated by adding up all the values in a data set and dividing the total sum by the number of values in the data set. So,

$$\bar{X} = \frac{\sum X}{N}$$

The arithmetic mean is of little value on its own and should be supported by reference to the standard deviation of the data set.

For the data set for corrie orientation (Table 12.1) the mean orientation for:

- The Glyders corries is: 695/15 = 46.34°. This is almost exactly NE, the direction away from the sun and warming southerly breezes.
- The Arran corries is: 890/15 = 59.34°. This shows a more easterly aspect, though still on the shady side of the mountain.

Mode

This is the value which occurs most frequently in a data set. It can only be identified if all the individual values are known.

Median

This is the middle value in a data set when the figures are arranged in rank order. There should be an equal number of values both above and below the median value. If the number of values in a data set is odd, then the median will be the $\frac{n+1}{2}$ item in the data set.

So, for example, if the total number of items in a data set is 15, the median will be the eighth value in the rank order of the data.

If the number of values in the data set is even, the median value is the mean of the middle two values.

Distribution of the data set

It is possible that each of these measures of central tendency could give the same result, but they are more likely to give different results. For them each to give the same result the distribution of a data set would have to be perfectly 'normal', and this is extremely unlikely when using real data. It is more likely that the distribution of the data set will be skewed towards the lower end of the distribution (positive skew) or towards the upper end of the distribution (negative skew). The more it is skewed, the greater the variation in the three measures of central tendency.

None of these measures gives a reliable picture of the distribution of the data set. It is possible for two different sets of data to have the same values for mean, mode and median. Measures of the dispersion or variability of the data should therefore also be provided.

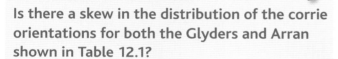

Key question

Is there a skew in the distribution of the corrie orientations for both the Glyders and Arran shown in Table 12.1?

Measures of dispersion

There are three measures of dispersion or variability: range, inter-quartile range and standard deviation.

Range

This is the difference between the highest value and the lowest value in a data set. It gives a simple indication of the spread of the data.

For the data set for corrie orientation (Table 12.1) the range of orientations for:
- the Glyders corries is: $85 - 10 = 75°$
- the Arran corries is: $185 - 5 = 180°$

Inter-quartile range

The inter-quartile range is calculated by ranking the data in order of size and dividing them into four equal groups or quartiles. The boundary between the first and second quartiles is known as the upper quartile and the boundary between the third and fourth quartiles is the lower quartile. They can be calculated as follows:

- The upper quartile (UQ) is the value that occurs at $\frac{(n+1)}{4}$ th position in the data set when arranged in rank order (from highest to lowest).

- The lower quartile (LQ) is the value that occurs at $\frac{3(n+1)}{4}$ th position in the data set.

- The difference between the upper and lower values is the inter-quartile range (IQR):

  $$IQR = UQ - LQ$$

The IQR indicates the spread of the middle 50 per cent of the data set about the *median* value, and thus gives a better indication of the degree to which the data are spread, or dispersed, on either side of the middle value.

For the data set for corrie orientation (Table 12.1) the results are shown in Table 12.2.

Table 12.2 The Upper quartile, lower quartiles and interquartile range for the lip orientation of 15 corries in the Glyders of Snowdonia and 15 on the Isle of Arran

Location of corries	Upper quartile	Lower quartile	Interquartile range
The Glyders	60	30	30
Isle of Arran	110	10	100

Standard deviation

The standard deviation of a range of data is a measurement of the degree of dispersion about the *mean* value of a data set. It is calculated as follows:

- The difference between each value in the data set and the mean value is worked out.
- Each difference is squared, to eliminate negative values.
- These squared differences are totalled.
- The total is divided by the number of values in the data set, to provide the **variance** of the data.
- The square root of the variance is calculated.

$$SD = \sqrt{\frac{\sum(x - \bar{x})^2}{n}}$$

The standard deviation is statistically important as it links the data set to the normal distribution. Figure 12.22 shows a **normal distribution**.

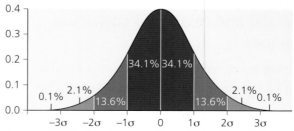

Figure 12.22 A plot of a normal distribution (or bell-shaped curve) where each band has a width of 1 standard deviation

Table 12.3 Calculating standard deviation

A	B	C		Arran corrie orientation Y	Y − Ȳ	(Y − Ȳ)²
Glyder corrie orientation x	x − x̄	(X − x̄)²				
30	−16.34	267.00		5	−54.34	2,952.84
60	12.66	186.60		5	−54.34	2,952.84
45	−1.34	1.80		10	−49.34	2,434.44
55	8.66	75.00		55	−4.34	18.84
75	28.66	821.40		15	−44.34	1,966.04
50	3.66	12.40		30	−29.34	860.84
80	33.66	1,133.00		95	35.66	1,271.64
50	3.66	12.40		5	−54.34	2,952.84
10	−36.34	1,320.60		185	125.66	15,790.44
15	−31.34	982.20		70	10.66	112.64
10	−36.34	1,320.60		120	60.66	3,679.64
35	−11.4	128.6		40	19.34	374.04
45	−1.34	1.80		30	−29.34	860.84
50	3.66	12.40		115	55.66	3,098.04
85	38.66	1,494.60		110	50.66	2,566.44
$\sum$x = 695	x̄ = 46.34	$\sum$(x − x̄)² = 7,773.4		$\sum$x = 890	x̄ = 59.34	$\sum$(x − x̄)² = 41,911.40
		σ = 22.76				σ = 52.86

A low standard deviation indicates that the data are clustered around the mean value and that dispersion is narrow. A high standard deviation indicates that the data are more widely spread and that dispersion is large. The standard deviation also allows comparison of the distribution of the values in a data set with a theoretical norm and is therefore of greater use than just the measures of central tendency.

Key question

The standard deviations for the two sets of corries are different. Using all the analysis so far, describe the two sets of data.

Inferential and relational statistical techniques

Comparisons are made between two sets of data to see whether there is a relationship between them.

Note that even if there is a relationship between two variables, this does not prove a causal link. In other words, the relationship does not prove that a change in one variable is responsible for a change in the other. For example, there may be a direct relationship between altitude and the amount of precipitation in a country such as the UK. These two variables (altitude and precipitation) are clearly linked, but a decrease in one does not automatically cause a decrease in the other – they are simply related to each other.

There are two main ways in which relationships can be shown:

- using scattergraphs
- measuring correlation using the Spearman rank correlation coefficient.

The Spearman rank correlation coefficient

This is used to measure the degree to which there is correlation between two sets of data (or variables). It provides a numerical value which summarises the degree of correlation, so it is an example of an objective indicator. Once it has been calculated, the numerical value has to be tested statistically to see how significant the result is.

The test can be used with any data set consisting of raw figures, percentages or indices which can be ranked. The formula for the calculation of the correlation coefficient is

$$r_s = 1 - \frac{6\sum d^2}{n^3 - n}$$

where d is the difference in ranking between the two sets of paired data and n is the number of sets of paired data.

The method of calculation is as follows:
- Rank one set of data from highest to lowest (highest value ranked 1, second highest 2 and so on).
- Rank the other set of data in the same way.
- Beware of tied ranks. In order to allocate a rank order for such values, calculate the 'average' rank they occupy. For example, if there are three values

which should all be placed at rank 5, add together the ranks 5, 6 and 7 and divide by three, giving an 'average' rank of 6 for each one. The next value in the sequence will be allocated rank 8.

- Calculate the difference in rank (d) for each set of paired data.
- Square each difference.
- Add the squared differences together and multiply by 6 (A).
- Calculate the value of $n^3 - n$ (B).
- Divide A by B, and take the result away from 1.

The answer should be a value between +1.0 (perfect positive correlation) and −1.0 (perfect negative correlation).

$$\sum(D - M)^2 = 562.5$$

$$R_s = 1 - \frac{6 \times 562.5}{13^3 - 13}$$

$$R_s = +0.55$$

Table 12.4a Calculating Spearman's rank correlation coefficient (R_s) for doctors/1,000 of population and confirmed malaria cases

Country	Doctors/1,000 population (D)	Rank (D)	Confirmed malaria cases/1,000 population (M)	Rank (M)	(D – M)	(D –M)²
Papua New Guinea	0.6	13	42.73	1	12	144
Solomon Islands	0.22	6	42.00	2	4	16
Cambodia	0.17	10	12.77	3	7	49
Vanuatu	0.12	11	8.92	4	7	49
Myanmar	0.61	4	5.99	5	−1	1
Laos	0.18	8.5	5.60	6	2.5	6.25
Indonesia	0.2	7	1.35	7	0	0
Timor-Leste	0.07	12	0.85	8	4	16
Thailand	0.39	5	0.49	9	−4	16
Vietnam	1.19	2	0.18	10	−8	64
Malaysia	1.2	1	0.10	11	−10	100
Philippines	1.15	3	0.06	12	−9	81
Micronesia	0.18	8.5	0.00	13	−4.5	20.25

Table 12.4b Calculating Spearman's rank correlation coefficient (Rs) for percentage of GDP spent on health care and confirmed malaria cases.

Country	Percentage of GDP spent on healthcare	Rank (D)	Confirmed malaria cases/1,000 population (M)	Rank (M)	(D – M)	(D – M)^2
Papua New Guinea	4.5	6	42.73	1	5	25
Solomon Islands	5.4	4	42.00	2	2	4
Cambodia	7.5	2	12.77	3	–1	1
Vanuatu	3.9	9	8.92	4	5	25
Myanmar	1.8	12	5.99	5	7	49
Laos	2.0	11	5.60	6	5	25
Indonesia	3.1	10	1.35	7	3	9
Timor-Leste	1.3	13	0.85	8	5	25
Thailand	4.6	5	0.49	9	–4	16
Vietnam	6.3	3	0.18	10	7	49
Malaysia	4.0	8	0.10	11	–3	9
Philippines	4.4	7	0.06	12	–5	25
Micronesia	12.6	1	0.00	13	12	144

$$\Sigma(D - M)^2 = 406$$

$$R_s = 1 - \frac{6 \times 406}{13^3 - 13}$$

$$R_s = -0.12$$

Some words of warning

- You should have at least 10 sets of paired data, as the test is unreliable if n is less than 10.
- You should have no more than 30 sets of paired data or the calculations become complex and prone to error.
- Too many tied ranks can interfere with the statistical validity of the exercise, although it is appreciated that there is little you can do about the 'real' data collected.
- Be careful about choosing the variables to compare – do not choose dubious or spurious sets of data.

Interpreting the results

When interpreting the results of the Spearman rank test, consider the following points:

What is the direction of the relationship? If the calculation produces a positive value, the relationship is positive, or direct. In other words, as one variable increases, so does the other. If the calculation produces a negative value, the relationship is negative, or inverse.

How statistically significant is the result? When comparing two sets of data, there is always a possibility that the relationship shown between them has

occurred by chance. The figures in the data sets may just happen to have been the right ones to bring about a correlation. It is therefore necessary to assess the statistical significance of the result. In the case of the Spearman rank test, the critical values for R_s must be consulted. These can be obtained from statistical tables, but Table 12.5 shows some examples.

According to statisticians, if there is a >5 per cent possibility of the relationship occurring by chance, the relationship is not significant. This is called the rejection level. The relationship could have occurred by chance more than five times in 100, and this is an unacceptable level of chance. If there is a <5 per cent possibility, the relationship is significant and therefore meaningful.

Table 12.5 Critical values for R_s

n	0.05 (5 per cent) significance level	0.01 (1 per cent) significance level
10	± 0.564	± 0.746
12	0.506	0.712
14	0.456	0.645
16	0.425	0.601
18	0.399	0.564
20	0.377	0.534
22	0.359	0.508
24	0.343	0.485
26	0.329	0.465
28	0.317	0.448
30	0.306	0.432

If there is a < 1 per cent possibility of the relationship occurring by chance, the relationship is very significant. In this case, the result could only have occurred by chance one in 100 times, and this is very unlikely.

How does this work? Having calculated a correlation coefficient, examine the critical values given in Table 12.5 (ignore the positive or negative sign). If your coefficient is greater than these values, the correlation is significant at that level. If your coefficient is smaller, the relationship is not significant at that level.

In our example, suppose we have calculated an R_s value of +0.55 from 13 sets of paired data; 0.50 is greater than the critical value at the 0.05 (5 per cent) level, but not that at the 0.01 (1 per cent) level. In this case, therefore, the relationship is significant at the 0.05 (5 per cent) level, but not at the 0.01 (1 per cent) level.

Key questions

How do the two results for R$_s$ calculated in Tables 12.4a and 12.4b support the statement below?

'The Spearman's rank correlation coefficient test only tells us whether there is a correlation. It does not imply causation.'

Chi-squared test

This technique is used to assess the degree to which there are differences between a set of collected (or observed) data and a theoretical (or expected) set of data, and the statistical significance of the differences.

The observed data (O) are those that have been collected either in the field or from secondary sources. The expected data (E) are those that would be expected according to the theoretical hypothesis being tested.

Normally, before the test is applied it is necessary to formulate a null hypothesis. In the example given here, the null hypothesis would be that there is no significant difference between the observed and expected data distribution. The alternative to this would be that there *is* a difference between the observed

and expected data, and that there is some factor responsible for this. An example is given below.

A group of students investigated the orientation of corries in the mountains of Snowdonia. The students wanted to investigate whether there was a preferred orientation.

They measured the orientation of 40 corries and placed them in one of four groups. This is shown in the Observed (O) column in Table 12.6. It looks like there is a preferred direction, but as this could be due to chance a chi-squared test was carried out.

They created a null hypothesis:

There is no significant difference between the observed orientation of the corries and the expected random orientation.

Table 12.6 was completed.

In the column headed Observed (O) are listed the numbers of corries in each of the categories of orientation. The total number of corries is 40. The column labelled Expected (E) contains the list of *expected* frequencies (E) in each of the categories, assuming that the corries are evenly orientated. In the column O − E, each of the expected frequencies is subtracted from the observed frequencies, and in the last column the result is squared. The relevant values are then inserted into the expression for chi-squared,

$$\chi^2 = \sum \frac{(O-E)^2}{E}$$

and the result is 10.8.

The aim of a chi-squared test, therefore, is to find out whether the observed pattern agrees with or differs from the theoretical (expected) pattern. This can be measured by comparing the calculated result of the test with its level of significance.

Table 12.6 Calculating chi-squared

Orientation	Observed (O)	Expected (E)	(O − E)	(O − E)^2	$\frac{(O-E)^2}{E}$
316° – 045°	16	10	6	36	3.6
046° – 135°	13	10	3	9	0.9
136° – 225°	6	10	−4	16	1.6
226° – 315°	5	10	−5	25	2.5
Total	40	40			χ^2 = 8.6

To do this the number of degrees of freedom must be determined using the formula $(n - 1)$, where n is the number of observations, in this case the number of cells which contain observed data (4). So, $4 - 1 = 3$. Table 12.7 gives the distribution of chi-squared values for these degrees of freedom.

Table 12.7 Critical values of chi-squared

Degrees of freedom	Significance level	
	0.05	0.01
1	3.84	6.64
2	5.99	9.21Z
3	7.82	11.34
4	9.49	12.28
5	11.07	15.09
6	12.59	16.81
7	14.07	18.48
8	15.51	20.09
9	16.92	21.67
10	18.31	23.21
11	19.68	24.72
12	21.03	26.22
13	22.36	27.69
14	23.68	29.14
15	25.00	30.58
16	26.30	32.00
17	27.59	33.41
18	28.87	34.80
19	30.14	36.19
20	31.41	37.57
21	32.67	38.93
22	33.92	40.29
23	35.17	41.64
24	36.42	42.98
25	37.65	44.31
26	38.88	45.64
27	40.11	46.96
28	41.34	48.28
29	42.56	49.59
30	43.77	50.89
40	55.76	63.69
50	67.51	76.15
60	79.08	88.38
70	90.53	100.43
80	101.88	112.33
90	112.15	124.12
100	124.34	135.81

Note: When there are A rows and B columns respectively, degrees of freedom = (A – 1) × (B – 1). If there is only one row, then there are (B – 1) degrees of freedom.

Then there are the levels of significance. There are two levels of significance: 95 per cent and 99 per cent (0.05 and 0.01 in Table 12.7). At 95 per cent there is a 1 in 20 probability that the pattern being considered occurred by chance, and at 99 per cent there is only a 1 in 100 probability that the pattern is chance. The levels of significance can be found in a book of statistical tables. They are also known as confidence levels.

If the calculated value is the same or greater than the values given in the table, then the null hypothesis can be rejected and the alternative hypothesis accepted.

In the case of our example the value of chi-squared is 8.6. This falls between the values for the 0.05 per cent and 0.01 per cent. Because the value is greater than that of the 0.05 per cent significance level then the null hypothesis can be rejected.

Some further points on this technique:
- The numbers of both observed and expected values must be large enough to ensure that the test is valid. Most experts state that there should be a minimum of five.
- The number produced by the calculation is itself meaningless. It is only of value for use in consulting statistical tables.
- As in a Spearman rank test, only significance (or confidence) levels of 95 per cent and 99 per cent should be considered when rejecting the null hypothesis. Levels of confidence greater than these simply allow the null hypothesis to be rejected with even greater confidence.
- It is strongly recommended that you do not apply the test to more than one set of observed data because the mathematics becomes too complex.

It is important to note that the chi-squared test should only be used as a support for your own ideas on the geographical significance of your study. All results, and the statistical analysis of them, should be related to the original hypothesis and/or the established theory in that aspect of the subject.

Your results may support established theory or your hypothesis, or they may not. If they do not, there may be some reason or factor you can identify that is responsible. This could lead to further studies. Above all, your investigation should make geographical sense. This is far more important than demonstrating your ability to use mathematics or statistics.

ICT skills

Use of electronic databases

There are a variety of electronic databases that can be accessed by geographers (Figure 12.23). Schools and colleges often have access to Digimap, an interactive map database that contains Ordnance Survey maps which themselves contain vast amounts of information.

When you use the internet as a research tool you must always ask the questions:

- who published the information (organisation/agency/individual)
- who wrote the information (expert/interested individual)
- the age of the material
- why the material exists (academic research/special interest groups)
- how one-sided (biased) are the arguments in the information.

Use of innovative sources of data such as crowd sourcing and 'big data'

Crowd sourcing

It is difficult to analyse individual people's behaviour and attitudes using traditional large-scale data sources such as population censuses and social surveys, because they tend to deal with large anonymous groups rather than individuals. They focus on easily measured attributes and characteristics of a population, rather than attitudes and behaviours. They also only

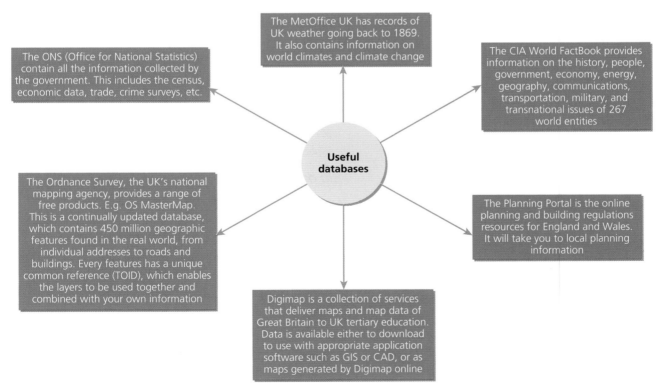

Figure 12.23 Useful databases

offer a snapshot rather than a continuously changing view. However, new data sources (such as that available through social networks) contain a wealth of information about people's geographical behaviour at an individual level; they have the potential to revolutionise our understanding of social phenomena. These sources are commonly referred to as geo-spatial crowd-sourced data (GSD) or **volunteered geographical information.**

GSD is increasingly being used to study individual people's daily behaviours and explore how these dynamics can potentially lead to civil unrest. It is also being used to look at the way people respond to natural disasters.

Since the start of the twenty-first century there has been a sequence of unfortunate natural disasters, including the devastating Indian Ocean earthquake and tsunami (26 December 2004), Hurricane Katrina (September 2005), the Wenchuan earthquake (May 2009), the Haitian earthquake (January 2010), the Tohoku earthquake and tsunami (March 2011), Hurricane Sandy (October 2012) and the Nepal earthquake (April 2015). These have all focused international attention on the immediate need for maps and geospatial data on the impacted areas and the critical role of geographically-distributed information-sharing communities in providing that information.

User-generated platforms such as Twitter, Facebook, Flickr and YouTube have large volumes of information with geographic footprints or geographic associations that can be used, even though the original contributors to these platforms may not have intended the information to be geographic or to have any specific geographic purpose.

Example of crowd-sourced data use – relief of Haitian earthquake (2010)

Two important questions that needed to be answered immediately after the Haitian earthquake hit on 12 January 2010 were: Who needs help? And where? Relief efforts had to get supplies and resources to the parts of the country most desperately in need, but it was difficult to know where to deploy resources because there was no systematic plan or data in place to help make such decisions. Even some of the most basic informational needs, like detailed roadmaps and locations of critical assets, were not available.

Figure 12.24 Google Earth image overlay showing the destruction of the earthquake in Haiti

People and organisations around the world realised that they didn't have to be physically present in Haiti to provide meaningful assistance to those who were. Information about opportunities to contribute spread quickly through a variety of online outlets, including blogs, emails, tweets and status updates.

OpenStreetMap (OSM) volunteers from around the world downloaded satellite images in order to trace and record the outlines of streets, buildings and other places of interest (Figure 12.24). These traces were uploaded into the OSM database and worked together with material from on-the-ground volunteers in Haiti who, using portable GPS devices, were able to upload additional information. In the few weeks after the disaster, there had been nearly 10,000 edits to the Port-au-Prince region and its immediate surroundings within OpenStreetMap by hundreds of people located worldwide. All of this information answered the two important questions and enabled help to be targeted where it was needed most.

Big data

Geographers have always used large data sets, for example, weather data or population census data. Because of expense and logistical problems this could only be collected periodically and analysed partially. There is now a deluge of data which can be held indefinitely in cloud-based servers. This is called 'big data' (Figure 12.25).

Analysing this data has, until recently, been limited in its use. One recent development, geographical information systems (GIS), has tools that enable spatial data to be collected, stored, archived, manipulated, analysed and presented (Figure 12.26). Even this is limited in scope.

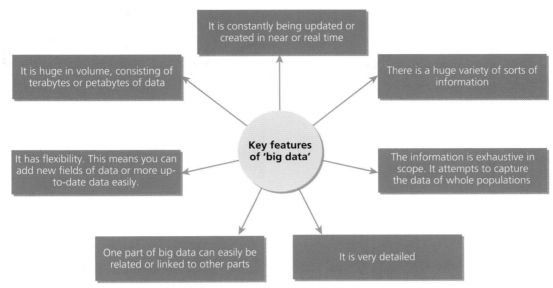

It is constantly being updated or created in near or real time

It is huge in volume, consisting of terabytes or petabytes of data

There is a huge variety of sorts of information

Key features of 'big data'

It has flexibility. This means you can add new fields of data or more up-to-date data easily.

The information is exhaustive in scope. It attempts to capture the data of whole populations

One part of big data can easily be related or linked to other parts

It is very detailed

Figure 12.25 Key features of 'big data'

Other uses of ICT

Digital technology is becoming ever more portable and flexible in its use. A mobile device is able to help you gather field information (via GPS information, photographs, video clips, annotated sketches, voice recorder, etc.).

Some of the more 'traditional' ICT skills (for example, Excel spreadsheets) enable you to analyse data, while word processing software along with other technology such as digital maps (Multimap, Memory-map, Google maps, Google Earth) allows you to present data in an interesting and meaningful way.

12.3 Fieldwork investigation

You are expected to undertake four days of fieldwork during your A-level course and two days for your AS level course. Ideally this should cover topics in both human and physical geography. At AS, the second half of Paper 2 will have a series of questions focusing on fieldwork investigation and specific geographical skills. It will assess your ability to apply your knowledge and skills to unseen information and resources, with reference to fieldwork that you have undertaken.

At A-level you are required to undertake an independent investigation that demonstrates the required fieldwork knowledge, skills and understanding. You may incorporate field data and/or evidence from field investigations collected on your own or as part of a group.

This independent investigation must:
- be based on a question or issue that you have developed relating to any aspect of the course content
- include data and evidence collected from field investigations
- include results of your own research and relevant secondary data

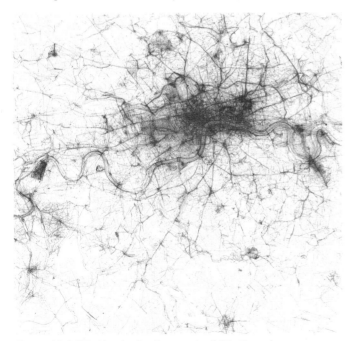

Figure 12.26 Tourist destinations made visible through mass user-generated data from the photograph network Flickr. Blue points on the map are pictures taken by locals (people who have taken pictures in this city dated over a range of a month or more). Red points are pictures taken by tourists (people who seem to be a local of a different city and who took pictures in this city for less than a month)

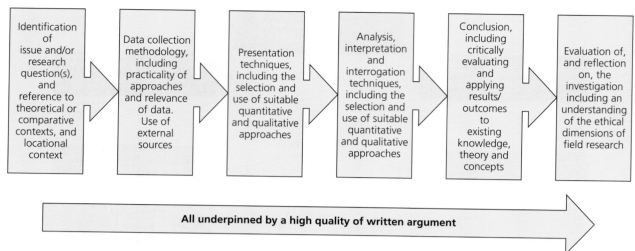

| Identification of issue and/or research question(s), and reference to theoretical or comparative contexts, and locational context | Data collection methodology, including practicality of approaches and relevance of data. Use of external sources | Presentation techniques, including the selection and use of suitable quantitative and qualitative approaches | Analysis, interpretation and interrogation techniques, including the selection and use of suitable quantitative and qualitative approaches | Conclusion, including critically evaluating and applying results/ outcomes to existing knowledge, theory and concepts | Evaluation of, and reflection on, the investigation including an understanding of the ethical dimensions of field research |

All underpinned by a high quality of written argument

Figure 12.27 The fieldwork pathway

- include presentation and analysis of collected data and conclusions drawn from this analysis.

At both AS and A-level you need to be able to:

- prepare for fieldwork, including background reading, drawing up aims and objectives for the enquiry, planning research in the field and from secondary sources, using data sampling techniques and carrying out health and safety procedures
- collect primary data in the field and use relevant secondary data sources
- process and present data using relevant graphical and cartographical techniques
- analyse data, including using statistical techniques where relevant
- draw conclusions related back to the original aims and objectives and link these conclusions to both the place studied and the general ideas forming the basis of the enquiry
- review the success, or otherwise, of all stages of the enquiry
- consider how the enquiry could be further developed.

AS fieldwork will be tested in Paper 2 Section B.

You **must** therefore participate in personal investigative work in the field to ensure you are familiar with such skills.

The fieldwork pathway
Choosing your fieldwork topic

Figure 12.27 shows the fieldwork pathway. Figure 12.28 shows some of the things you must consider when

choosing your fieldwork topic. Everybody's situation is unique; you must choose a topic where the aims can be met. Clearly, advice from your teachers will be invaluable. The main consideration is that preparation for this unit must involve enquiry work outside the classroom that includes data collection in the field. It might include, for example, data collected in specialist field study centres, work experience settings, internet research and use of libraries or archives.

Geographical information systems and fieldwork

GIS are an effective mapping tool and can be major elements in geographical fieldwork. Figure 12.29 summarises the key aspects of GIS and how they can be used in fieldwork.

A number of websites and associated programs have great potential for use in fieldwork. They include Multimap, Aegis, Digital Worlds, Quikmaps and Google Earth. Many allow you to use maps and photos together (Figure 12.30, page 646); others allow annotations to be added. Some types of software require a degree of training, whereas others are more straightforward to use. It all depends on your own level of ICT skills.

Global positioning systems (GPS) technology can also be useful in fieldwork. Data can be recorded at points along a transect or other sampling system, and the position of each point can be recorded at the same time. GPS-located data can then be fed into a GIS program, bringing data recording and mapping together.

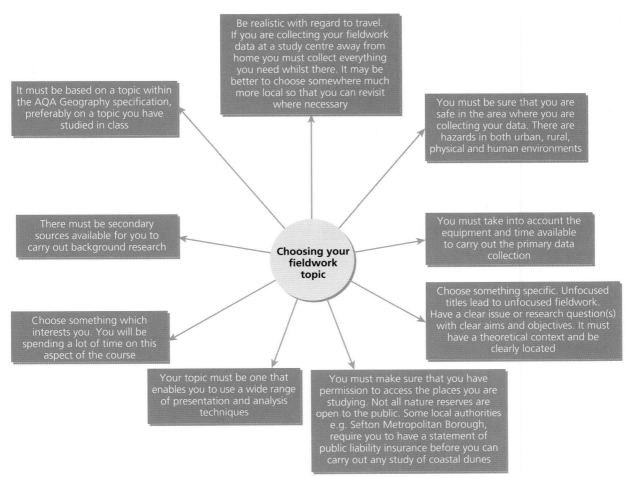

Figure 12.28 Choosing your fieldwork topic

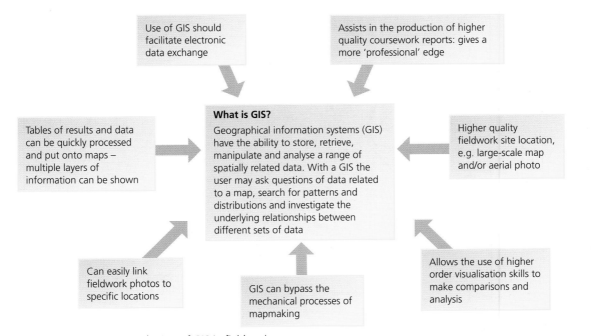

Figure 12.29 Maximising the use of GIS in fieldwork

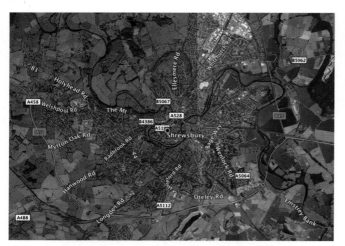

Figure 12.30 Using GIS software to overlay a satellite photo onto a map

Source: © Google; © Infoterra Ltd & Bluesky Image; © Getmapping plc

An example of fieldwork

The following suggestion for a piece of fieldwork is based on 'Changing places'.

Figure 12.31 shows the location of a planned housing development on the rural/urban fringe of Shrewsbury. Outline planning permission has been granted to build 15 houses on arable land on the western edge of the large village of Bayston Hill.

Issue/research question

In early 2012 plans to change Bayston Hill's status to a town were abandoned following concerns by residents. It is already the largest village in Shropshire, even larger than any of the county's market towns. The concerns expressed were related to the fact that many of the people in the village were concerned that the village would grow more quickly and lose its rural 'feel'. On the other hand, Shrewsbury has been described as being 'in the grip of an overwhelming housing shortage'. There is a great need for new houses.

The issue is: how can these two be reconciled?

Data collection

Secondary sources of data will include:

- the recent censuses for Bayston Hill to see how it has grown over time
- details of the housing development, which can be found through the local authority planning portal (Figure 12.32)
- Ordnance survey and other map collections including GIS.

Primary data collection could include carrying out surveys of the interested parties (landowner, developer,

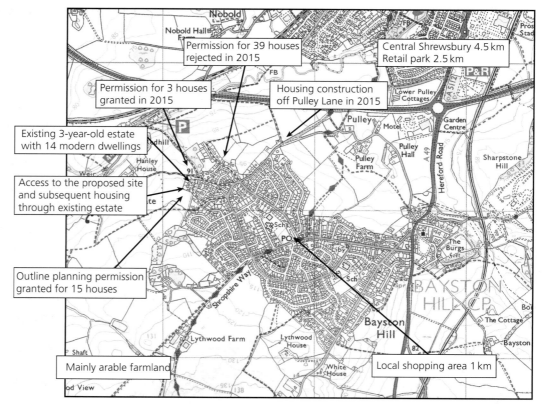

Figure 12.31 Location of the proposed housing development

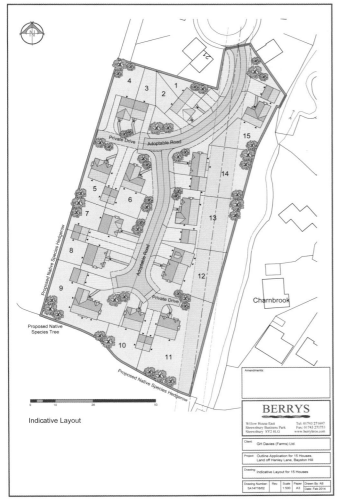

Figure 12.32 Plan of the proposed housing development

Figure 12.33 Access to proposed development to left of island and between existing houses

planning officer, local councillor and residents of the nearby estate).

Questionnaires about attitudes to the changes could be carried out with a larger number of the village's inhabitants. This could be in the form of an age-related stratified sample.

Traffic surveys could be carried out to look at concerns over access to the site by developers and subsequent occupiers of the houses (Figure 12.33).

Presentation and analysis

The interview answers could be coded, depending on what the responses were. Questionnaire responses can be easily summarised in a table and appropriate graphs drawn. The analysis will depend on what the researcher finds.

Conclusions

The conclusion to this report depends on what exactly the researcher finds. Any summary should include the benefits of having more housing in this area as opposed to the perceived damage that such developments will have on this place.

Further reading

The Royal Geographical Society (RGS) produces a comprehensive guide to fieldwork, *Fieldwork techniques*, which gives information and instructions for a range of techniques that can be used to carry out geographical fieldwork investigations in different locations and settings. www.rgs.org

The Barcelona Fieldwork centre has produced a series of geography resources which are available free. http://geographyfieldwork.com

Dr Bethan Davies of AntarcticGlaciers.org has written a comprehensive guide to A-level fieldwork called 'A-Level Geography Fieldwork Investigation'. Although some of this is really aimed at university-level fieldwork, it does give sound advice on how to go about designing and carrying out fieldwork.

The Field Studies Council (FSC) has a website dedicated to A-level and AS level fieldwork. This is constantly being updated to match the changing requirements of A-level and AS level. www.geography-fieldwork.org

The Geography Review published by Philip Allan for Hodder Education has a regular feature on geographical skills.

The Ordnance survey has a large range of resources based on maps. They range from simple map skills to complex use of GIS. www.ordnancesurvey.co.uk/mapzone/gis-zone

The Office for National Statistics (ONS) produces a range of resources including the census results. This is useful background material when completing a 'Place' study. www.ons.gov.uk

Question practice

AS level questions

The following questions are a sample what you might expect to meet in AS Paper 2, Part B.

1. a) What are the advantages of sampling when collecting data as part of a fieldwork investigation? (2 marks)

b) Describe how you have used one of the sampling methods given below as a part of your fieldwork investigation. Outline why you chose that particular method of sampling.

- Random sampling
- Systematic sampling
- Stratified sampling (6 marks)

2. On a clear, windless day in January, a group of students carried out an investigation into temperatures at different points on a transect from NE to SW through the town of Shrewsbury. Working in pairs they each measured the temperature at their designated location at the same time using the same type of thermometer. The Ordnance survey map, Figure 12.35, shows the locations of the data collection. Table 12.8 shows the data they produced.

a) Complete the scattergraph in Figure 12.34 by plotting the data marked (*) in the table. (2 marks)

b) Use your completed graph to test the hypothesis that 'temperatures will increase towards the town centre because there is more human activity.' (6 marks)

c) Use the information from the Ordnance survey map extract (Figure 12.35) to help you explain the data in Table 12.8 and to draw a conclusion about this exercise. (9 marks)

3. You have experienced geography fieldwork as part of the course. Use this experience to answer the following questions.

a) State the aim of your fieldwork investigation.

b) Outline the geographical concept, process or theory that underpinned your fieldwork enquiry; explain how you used that concept, process or theory to develop the aim(s) of your investigation. (6 marks)

c) Making specific reference to your results, outline your conclusions and explain how they helped you to develop your understanding of the concept beyond what you learnt in the classroom. (9 marks)

Table 12.8

Distance from town centre (km)		Temperature (°C)
NE	6.2	−4
	5.6	−4
	5.1	−2
	4.3	+1*
	3.6	+2
	2.6	+2
	2.1	+2
	1.3	+2*
	0.7	+2
	0.3	+4
	0.2	+4*
	0.4	0*
	0.7	0
	1.4	+2
	1.7	−1
	2.9	−4
	3.6	−4
SW	4.4	−4

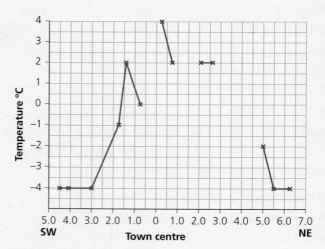

Figure 12.34 Scattergraph

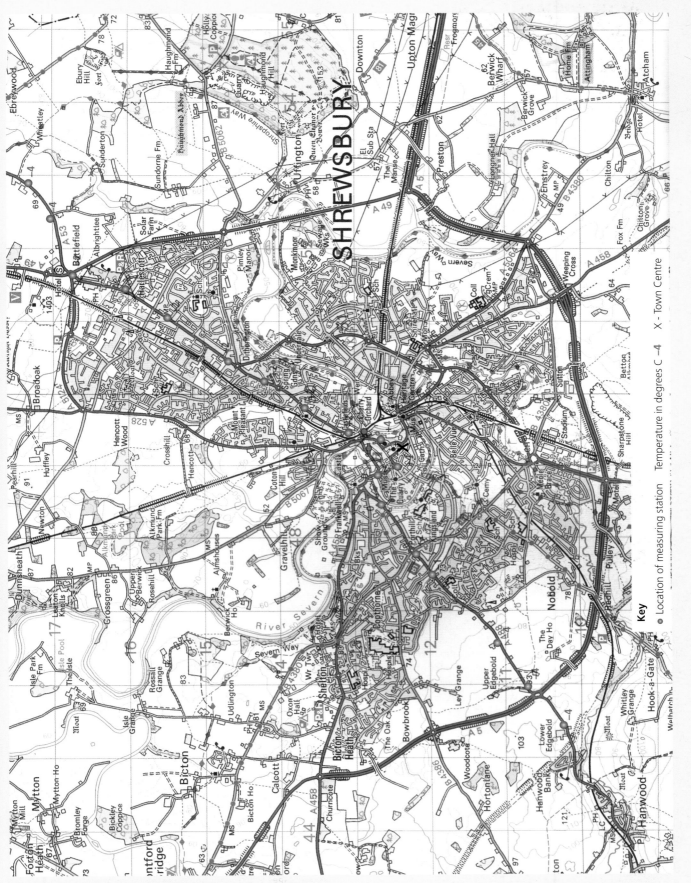

Figure 12.35 Ordnance survey map of Shrewsbury

Glossary

A

Abrasion – Where material carried by moving water or wind (or ice) hits exposed rock surfaces, thus wearing them away. Often referred to as a sandblasting or sandpapering effect.

Abstraction (water) – The removal of water available in the environment, either permanently or temporarily, from rivers, lakes, canals, reservoirs or from underground aquifers. **Over-abstraction** refers to the removal of too much water from a source upsetting the ecological and hydrological balance of that store.

Acid rain – Rain that has been made acidic by certain pollutants in the air.

Adaptation – Any alteration or adjustment in the structure or function of an organism or system which enables it to survive better in changing environmental conditions.
In the context of hazards, adaptation is the attempts by people or communities to live with hazard events. By adjusting their living conditions, people are able to reduce their levels of vulnerability. For example, they may avoid building on sites that are vulnerable to storm surges but stay within the same area.

Agents of change – These are the people who impact on a place whether through living, working or trying to improve that place. Examples would include residents, community groups, corporate bodies, multinational corporations, central and local government and the media.

Agglomeration – When companies in similar industries locate near to each other because of the benefits gained by sharing ideas and resources – called 'agglomeration economies'.

Agricultural productivity – The ratio of agricultural outputs to agricultural inputs.

Agriculture – The science or practice of farming including cultivation of the soil for the growing of crops and the rearing of animals to provide food, wool and other products.

Albedo – The reflectivity of a surface. It is the ratio between the amount of incoming insolation and the amount of energy reflected back into the atmosphere. Light surfaces reflect more than dark surfaces and so have a greater albedo.

Anaerobic (conditions) – Conditions where there is no air and little oxygen, such as waterlogging in soil.

Antarctic Convergence – A curve continuously encircling Antarctica where cold northward flowing Antarctic waters meet the relatively warmer waters of the sub-Antarctic.

Anthropogenic CO_2 – Carbon dioxide generated by human activity.

Anthropogenic factors – Those factors that are caused by human beings.

Appropriate technology – Technology that is suitable to the social and economic conditions of the geographic area in which it is to be used; it is environmentally sound and promotes self-sufficiency on the part of those using it.

Arid – The climate of an area that receives less than 250 mm of precipitation per year.

Aridity index – The ratio between precipitation (P) and potential evapotranspiration (PET).

Asthenosphere – That part of the mantle which lies directly beneath the lithosphere. In the original plate tectonic theory, convection currents within the asthenosphere generate the movement by which crustal plates are moved.

Asylum seeker – A person who has fled their country of origin and applies for asylum under the 1951 Convention on the grounds that they cannot return to their country of origin because of a well-founded fear of death or persecution. While they wait for a decision on their application to be concluded, they are known as an asylum seeker.

Atmospheric water – Water found in the atmosphere; mainly water vapour with some liquid water (cloud and rain droplets) and ice crystals.

Autotroph – An organism capable of synthesizing its own food from inorganic substances using light or chemical energy. Green plants, algae, and certain bacteria are autotrophs.

B

Backwash – The action of water receding back down the beach towards the sea.

Bankfull – The maximum discharge that a river channel is capable of carrying without flooding.

Base flow – This represents the normal day-to-day discharge of the river and is the consequence of slow-moving soil throughflow and groundwater seeping into the river channel.

Benthic – The ecological region at the lowest level of a body of water such as an ocean or a lake, including the sediment surface and sub-surface layers. The benthic environment comprises of a benthic community of organisms, known as **benthos**, which include worms, clams, crabs, sponges and other small organisms that live in the bottom sediments.

Bilateral (agreement) – An agreement on trade (or aid) that is negotiated between two countries or two groups of countries.

Biodiversity – The variation of life forms within a given ecosystem, biome, or for the entire Earth. Biodiversity is often used as a measure of the health of biological systems.

Biomapping – The mapping of emotions shown by people to certain places.

Biomass – The mass of living biological organisms in a given area or ecosystem at a given time. Biomass can refer to species biomass, which is the mass of one or more species, or to community biomass, which is the mass of all species in the community.

Biome – A major habitat category, based on distinct plant assemblages which depend on particular temperature and rainfall patterns, for example, tundra, temperate forest and rainforest.

Biosphere – The total sum of all living matter. The biological component of Earth systems (the others being atmosphere, hydrosphere and lithosphere).

Bottom up – When local people are consulted and supported in making decisions to undertake projects or developments that meet one or more of their specific needs.

BRIC – An acronym used to describe a group of four countries – Brazil, Russia, India and China – whose economies have grown very rapidly.

Brownfield site – Brownfield is a term used in urban planning to describe land previously used for industrial purposes or some commercial uses.

Bunkering – In the Niger Delta, this is a term used for the theft of oil, mainly by local people, from pipelines or other sources.

C

Capital flows – The movement of money for the purpose of investment, trade or to produce goods/provide services. Usually regarded as investment into a production operation.

Carbon sequestration – The capture of carbon dioxide (CO_2) from the atmosphere or capturing anthropogenic (human) CO_2 from large-scale stationary sources like power plants before it is released into the atmosphere. Once captured, the CO_2 gas (or the carbon portion of the CO_2) is put into long-term storage.

Carbon sink – A store of carbon that absorbs more carbon than it releases.

Carrying capacity – The maximum population size that an area or environment can sustain indefinitely.

Central Business District (CBD) – The centre of an urban area containing the major shops, offices and entertainment facilities.

Chalcopyrite – The most important ore of copper.

Channelling – Wind redirected down long straight canyon-like streets where there is less friction. These are sometimes referred to as urban canyons.

Chemical weathering – The processes leading to the decomposition or breaking down of rocks due to chemical reactions. This most often requires the presence of water and/or exposure to the air.

Climate – A region's long-term weather patterns. This is measured in terms of average precipitation (that is, the amount of annual rainfall, snow, etc.), maximum and minimum temperatures throughout the seasons, sunshine hours, humidity, the frequency of extreme weather and so on.

Climatic climax – A biological community of plants and animals which, through the process of ecological succession, has reached a state of dynamic equilibrium with its climate and soils.

Clone town – A term used to describe urban retail areas dominated by national, and in some cases, international chain stores.

Coastal sediment budget – The balance between sediment being added to and removed from the coastal system; that system being defined within each individual sediment cell.

Coding – Coding is carried out when analysing answers to interviews. The coder (person who analyses the data) looks through all the answers to a question, develops a broad classification system based on the responses and then uses a code to categorise responses.

Combined cycle technology – Power plants that use both a gas and a steam turbine together to produce more electricity from the same fuel than a traditional plant. The waste heat from the gas turbine is routed to the nearby steam turbine, which generates extra power.

Common markets – A group formed by countries in geographical proximity in which trade barriers for goods and services are eliminated. (This may eventually apply to removing any labour market restrictions, as in the EU.)

Community preparedness/risk sharing – This involves prearranged measures that aim to reduce the loss of life and property damage through public education and awareness programmes, evacuation procedures, and the provision of emergency medical, food and shelter supplies

Condensation – The process by which water vapour changes to liquid water.

Conglomerates – A collection of different companies or organisations which may be involved in different business activities but all report to one parent company – most transnational corporations are conglomerates.

Constructive waves – Waves with a low wave height, but with a long wavelength and low frequency of around 6–8/min. Their swash tends to be more powerful than their backwash and as a consequence beach material is built up.

Consumer – Organisms of an ecological food chain which receive energy by consuming other organisms.

Containerisation – A system of standardised transport that uses large standard-size steel containers to transport goods. The containers can be transferred between ships, trains and lorries enabling cheaper, efficient transport.

Continentality – The impact of increasing distance from the coast on the climate of an area.

Conventional oil and gas – Refers to petroleum, or crude oil, and raw natural gas extracted from the ground by conventional means and methods.

Cost-benefit analysis – A systematic analysis of the advantages and disadvantages likely to result from a development project, where an objective value is allocated to all economic, social and environmental aspects affected.

Counter-mapping – This describes a bottom-up process by which people produce their own maps, informed by their own local knowledge and understanding of places.

Counter-urbanisation – The movement of people from large urban areas into smaller urban areas or into rural areas, thereby leapfrogging the rural–urban fringe. It can mean daily commuting, but could also require lifestyle changes and the increased use of ICT.

Crude birth rate – The total number of live births per 1000 of a population per year – sometimes known as crude birth rate.

Crude death rate – The average number of deaths per 1000 of a population per year –sometimes known as crude death rate.

Cryospheric processes – Those processes that affect the total mass of ice at any scale from local patches of frozen ground to global ice amounts. They include accumulation (the build-up of ice mass) and ablation (the loss of ice mass).

Cryospheric water – The water locked up on the Earth's surface as ice.

Cultural diversity – The existence of a variety of cultural or ethnic groups within a society.

Customs unions – A trade bloc which allows free trade with no barriers between its member states but imposes a common external tariff to trading countries outside the bloc (for example, the European Union).

D

Decentralisation – The movement of population and industry from the urban centre to outlying areas. The term may encompass the processes of both suburbanisation and counter-urbanisation.

Deflation – Where wind removes dry, unconsolidated (loose) sand, silt and clay particles from the surface and transports them away.

Demographic dividend – The benefit a country gets when its working population outgrows its dependents such as children and the elderly. A boost in economic productivity results from growing numbers in the workforce relative to the number of dependents.

Demography (demographic) – Demography is the study of human population. A demographer is someone who studies the statistics and characteristics of populations.

De-multiplier effect – Opposite to the multiplier effect, when a withdrawal of income or investment leads to the closure of an activity, such as a factory. The loss of jobs and income for the workers means less money is spent in the economy usually giving rise to more job and income losses in a downward spiral effect.

Density dependent/independent factors – In ecology, these are limiting factors which restrict population growth. Density dependent factors are those caused by the size of the population itself, for example competition for limited food supplies. Independent factors are those which restrict population growth whatever the population size, for example severe environmental conditions.

Deposition – Occurs when the velocity of the wind decreases until it can no longer transport the grains it is carrying.

Deregulation – The process of reducing or eliminating state-imposed regulations on certain activities, such as removing constraints on foreign currency transactions in financial deregulation.

Dereliction – The state of buildings having been abandoned and become dilapidated.

Desert – An arid environment receiving very low levels of rainfall.

Desertification – There have and continue to be a range of definitions of desertification, and it is generally accepted that it is not simply the 'advance of deserts', but it can include sand dunes encroaching on land. However, a UNESCO source in 2015 defines it as 'the persistent degradation of dryland ecosystems by human activities and by climate change'.

Destructive waves – Waves with a high wave height with a steep form and high frequency (10–14/min). Their swash is generally stronger than their backwash, so more sediment is removed than is added.

Diaspora – A group of people with a similar heritage or homeland who have settled elsewhere in the world.

Disability-adjusted life years (DALYs) – A measure of morbidity within a society. They measure the number of years of healthy life lost by being in poor health or a state of disability.

Discharge – The amount of water in a river flowing past a particular point expressed as m^3/s (cumecs)

Disseminations – In geological terms, these are ore deposits that are scattered throughout a rock.

Drainage basin – This is an area of land drained by a river and its tributaries. It includes water found on the surface, in the soil and in near-surface geology.

Dry ports – An inland terminal directly connected by road or rail to a seaport, which operates as a hub for the shipment of sea cargo for export or receives imports to inland destinations.

E

Earthquake – As the crust of the Earth is constantly moving, there tends to be a slow build-up of stress within the rocks. When this pressure is released, parts of the surface experience, for a short period, an intense shaking motion. This is an earthquake.

Ecological footprint – A measure of the demand placed by humans on Earth's natural ecosystems; the total area of productive land and water required to produce the resources a population consumes and absorb the waste produced.

Economic inequality – The difference between levels of living standards, income, etc. across the whole economic distribution.

Economic migrant – A person who has voluntarily left their country of origin to seek, by lawful or unlawful means, employment in another country.

Economies of scale – The cost advantages that result from the larger size, output or scale of an operation as savings are made by spreading the costs or by rationalising operations.

Ecosystem – A system in which organisms interact with each other and the environment.

Edge city – A self-contained settlement which has emerged beyond the original city boundary and developed as a city in its own right.

Electrolysis (of water) – A technique used to drive an otherwise non-spontaneous chemical reaction by passing a direct electrical current (DC) through the material involved. In the case of water, electrolysis is used to decompose water into oxygen (O_2) and hydrogen gas (H_2).

Emigrant – A person leaving their native area or country in order to settle elsewhere.

Endogenous characteristics – In the context of place, this refers to the characteristics of the place itself. This would include aspects such as location, physical geography, land use and social and economic characteristics such as population size and employment rates.

Endoreic streams – Where rivers occupy drainage basins that are closed and do not flow out to the sea or other rivers, but instead end inland in lakes or swamps.

Enhanced greenhouse effect – The impact on the climate from the additional heat retained due to the increased amounts of carbon dioxide and other greenhouse gases that humans have released into the Earth's atmosphere since the Industrial Revolution.

Environmental sustainability – A state in which the demands placed on the environment can be met without reducing the quality of the environment for the future.

Ephemeral streams – Streams that flow intermittently in hot deserts areas following heavy thunderstorms.

Epidemiological transition – Describes changing patterns of population age distributions, mortality, fertility, life expectancy, and causes of death. It assumes that infectious diseases are replaced by chronic diseases over time due to expanded public health and sanitation.

Erosion – The wearing away of the Earth's surface by the mechanical action of processes of glaciers, wind, rivers, marine waves and wind.

Ethnography – A research method that explores what people do as well as say.

Eustatic change – A global change in sea level resulting from an actual fall or rise in the level of the sea itself.

Eutrophication – This occurs when excess fertilisers are washed off the land by rainwater into rivers and lakes. The resulting increase of nitrate or phosphate in the water encourages algal growth, which then forms a bloom over the water surface. This prevents sunlight reaching other water plants causing them to die.

Evaporation – The process by which liquid water changes to a gas. This requires energy, which is provided by the sun and aided by wind.

Evapotranspiration – The total output of water from the drainage basin directly back into the atmosphere.

Exfoliation – A process of mechanical weathering that results in the breaking, splitting or peeling-off of the outer rock layers. Also commonly known as 'onion skin weathering'.

Exogenous characteristics – This refers to the relationship of one place with other places and the external factors which affect this. The demographic, socio-economic and cultural characteristics of a place are shaped by shifting flows of people, resources, money and investment.

Exogenous streams – Rivers that originate external to the desert in adjacent highlands and more humid environments, flow from outside of the desert and pass through it.

Exploitation – The action of using natural resources to the fullest or for the most profitable use.

Exploration – The process of searching an area with the intention of finding and mapping natural resources.

F

Fatalism – A view of a hazard event that suggests that people cannot influence or shape the outcome, therefore nothing can be done to mitigate against it. People with such an attitude put in place limited or no preventative measures. In some parts of the world, the outcome of a hazard event is said to be 'God's will'.

Fauna – The animals of a particular region, habitat or geological period.

Fetch – Refers to the distance of open water over which a wind blows uninterrupted by major land obstacles. The length of fetch helps to determine the magnitude (size) and energy of the waves reaching the coast.

Fjord – Former glacial valleys drowned by rising sea levels.

Flora – The plants of a particular region, habitat or geological period.

Flow/transfer – A form of linkage between one store/component and another that involves movement of energy or mass.

Flow resources – Resources that are renewable and can be replaced. Examples include fresh water and timber. They are commonly expressed in terms of the annual rates at which they are regenerated.

Food chain – An arrangement of the organisms of an ecological community according to the order in which they eat each other, with each organism using the next lower organism in the food chain as a source of energy.

Food security – Food security exists when all people at all times have access to sufficient, safe, nutritious food to maintain a healthy and active life.

Food web – A scheme of feeding relationships, resembling a web, which unites the member species of a biological community.

Footloose – Used in geography to describe an activity, such as an industry, that can be placed at any location without having to consider factors such as raw materials or transport.

Fortress landscapes – This term refers to landscapes designed around security, protection, surveillance and exclusion.

Fragile environments – Those that are easily damaged or disturbed and then difficult to restore once destroyed.

Frequency (of hazards) – The distribution of a hazard through time.

G

Gentrification – This is the buying and renovating of properties often in more run-down areas by wealthier individuals.

Geopolitics – The study of international relations, as influenced by geographical factors.

Geo-sequestration – The technology of capturing greenhouse gas emissions from power stations and pumping them into underground reservoirs.

Glacier – A large mass of ice on land, moving downhill due to the influence of gravity.

Global commons – Resource domains or areas that lie outside the political reach of any one nation state.

Global governance – A movement of political integration aimed at negotiating responses to problems that affect more than one state or region.

Globalisation – A process by which national economies, societies and cultures have become increasingly integrated through the global network of trade, communication, transportation and immigration.

Glocalisation – A term used to describe products or services that are distributed globally but which are fashioned to appeal to the consumers in a local market.

Green belt – An area of open space and/or low-density land use around a town or city where further development is strictly controlled.

Greenfield site – An area of undeveloped land.

Greenhouse gas – Any gaseous compound in the atmosphere that is capable of absorbing infrared radiation, thereby trapping and holding heat in the atmosphere.

Groundwater – The water stored underground in the cracks and spaces in soil, sand and rock. It moves slowly through geologic formations called aquifers.

Groundwater flow – The slow movement of water through underlying rocks.

H

Habitat – An ecological or environmental area that is inhabited by a particular species of animal, plant, or other type of organism.

Hard engineering – Making a physical change to the coastal landscape using resistant materials, like concrete, boulders, wood and metal.

Health – Defined by the World Health Organisation (WHO) as a state of complete physical, mental and social well-being and not merely the absence of disease or infirmity.

Heliostats – Computer-controlled mirrors which keep the sun reflected on a target as the sun moves across the sky.

High-energy coast – A coastlines where strong, steady prevailing winds create high-energy waves and the rate of erosion is greater than the rate of deposition.

Homogenisation – The process of making things uniform or similar leading to places becoming indistinct from one another.

Horizons (soil) – Horizontal layers which are parallel to the soil crust. Each horizon has physical characteristics which differ from the layers above and beneath. Soil types are distinguishable by the composition of their horizon layers and each soil type usually has three or four horizons.

Hot spot – In certain places, a concentration of radioactive elements below the crust causes a hot spot to develop. From this, a plume of magma rises to eat into the plate above.

Hydrosphere – A discontinuous layer of water at or near the Earth's surface. It includes all liquid and frozen surface waters, groundwater held in soil and rock, and atmospheric water vapour.

I

Ice ages – A common term for periods when there were major cold phases known as glacials and ice sheets covered large areas of the world, interspersed with warmer interglacials. The last 2 million years, the Quaternary Period, contains the Pleistocene epoch lasting from about 1.8 million years to about 11,500 BP (before present). This was characterised by a series of episodes of glaciation, the last major glacial beginning around 120,000 BP reaching the Last Glacial Maximum about 18,000 BP. At that time about a third of the Earth's surface was covered by ice, compared with just a tenth today.

Ice sheet – a continental size mass of ice, covering at least 50,000 km² that is dome-shaped with flows of ice outward from the centre. Ice sheets can be up to 2,000 m thick today; those in the Quaternary glaciation could have been twice that thickness.

Ignition source – The means by which a wildfire can be started such as lightning strikes, falling power lines, cigarettes or agricultural burning which gets out of control.

Immigrant – A person moving into an area or country to which they are not native in order to settle there.

Improved drinking water – A source of water that by using intervening treatment is protected from outside contamination, particularly from faecal matter.

Index of Multiple Deprivation (IMD) – UK government qualitative study measuring deprivation across England, most recently carried out in 2015.

Indicator species – An organism whose presence, absence or abundance reflects a specific environmental condition. They can signal a change in the biological condition of a particular ecosystem, and thus may be used as a proxy to diagnose the health of an ecosystem.

Infiltration – The downward movement of water from the surface into soil.

Input – The addition of matter and/or energy into a system.

Insolation – The incoming solar radiation that reaches the Earth's surface.

Infant mortality rate – The number of children who die before their first birthday per 1,000 live births per year.

Integrated risk management – The process of considering the social, economic and political factors involved in risk analysis; determining the acceptability of damage/disruption and deciding on the actions to be taken to minimize damage/disruption.

Interception storage – The precipitation that falls on vegetation surfaces (canopy) or human-made cover and is temporarily stored on these surfaces. Intercepted water either can be evaporated directly to the atmosphere, absorbed by the canopy surfaces, or ultimately transmitted to the ground surface.

Interculturalism – The interaction and the exchange of ideas between different cultural groups.

International trade – The exchange of capital, goods and services across international borders. Inbound trade is defined as imports and outbound trade as exports.

Island arc – A chain of volcanic islands associated with an ocean trench where subduction is taking place. During subduction, the descending plate begins to melt and the magma created rises towards the surface to form explosive volcanoes. If these eruptions take place offshore, a line of volcanic islands forms.

Isostatic change – Local changes in sea level resulting from the land rising or falling relative to the sea.

L

Labour – Factor of production defined as the aggregate of all human physical and mental effort used to create goods or provide services.

Lag time – The time between the peak rainfall and peak discharge.

Lahars – A flow of wet material down the side of a volcano consisting of erupted ash and water. Found when heavy rainfall occurs after a volcanic eruption. Essentially they are volcanic mudflows. In the Philippines, if a typhoon occurs after a volcanic eruption, then lahars can result.

Land remediation – The removal of pollution or contaminants from the ground which enables areas of derelict former industrial land to be brought back into commercial use.

Lava – Molten rock (magma) flowing onto the surface. Acid lava solidifies very quickly, but basic lava (basaltic) tends to flow some distance before solidifying (for example, on the Hawaiian Islands).

Leaching – The loss of soluble substances and colloids from the top layer of soil by percolating precipitation. The materials lost are carried downward (eluviated) and are generally redeposited (illuviated) in a lower layer.

Leakages (economic) – Refers to a loss of income from an economic system. It most usually refers to the profits sent back to their base country by transnational corporations – also known as profit repatriation.

Leopold matrix – An environmental impact assessment method that uses a tabulated recording sheet to assess the impact on each environmental characteristic and then the importance of that impact to the environment as a whole.

Life expectancy (at birth) – The average number of years a person born in a particular year in a location is expected to live.

Lithosphere – The crust and the uppermost mantle; this constitutes the hard and rigid outer layer of the Earth. It is this layer which is split into a number of tectonic plates.

Liveability – Aspects of urban living which make life more comfortable and endurable for city dwellers. Liveability can mean different things to different people. It may include natural amenities such as parks and green space; cultural offerings, career opportunities, economic and political stability or basic safety. It is measured by the Global Liveability Ranking.

Locale – This is the place where something happens or is set, or that has particular events associated with it.

Location – 'Where' a place is, for example the co-ordinates on a map.

Longshore or littoral drift – Where waves approach the shore at an angle and swash and backwash then transport material along the coast in the direction of the prevailing wind and waves.

Low-energy coast – A coastline where wave energy is low and the rate of deposition often exceeds the rate of erosion of sediment.

M

Magnitude – The assessment of the size of the impact of a hazard event.

Marine processes – Are those that operate upon a coastline that are connected with the sea, such as waves, tides and longshore drift.

Maquiladora – A manufacturing operation (plant or factory) located in a free trade zone in Mexico. They import materials for assembly and then export the final product without any trade barriers.

Mass movement – The movement of material downhill under the influence of gravity, but may also be assisted by rainfall.

Media – Means of communication including television, film, photography, art, newspapers, books, songs, etc. These reach or influence people widely.

Megacity – A city or urban agglomeration (urban area incorporating several large towns or cities) with a population of more than 10 million people. According to the UN, London achieved megacity status in 2013. This classification included residents in the Greater London area.

Metacity – A conurbation with more than 20 million people.

Microclimate – The small-scale variations in temperature, precipitation, humidity, wind speed and evaporation that occur in a particular environment such as an urban area.

MINT – An acronym referring to the more recently emerging economies of Mexico, Indonesia, Nigeria and Turkey.

Mitigation – Includes any actions, strategies, measures or projects undertaken (by mankind) to offset the known detrimental impacts of a process.

Monoculture – The agricultural practice of continually producing or growing a single crop, plant, or livestock species or breed in a farming system over a period of time, usually over a wide area.

Morbidity – Morbidity relates to illness and disease. It can also be used to describe the incidence of a disease within society. Some diseases are so infectious that by law they must be reported. These include malaria, rubella and tuberculosis.

Mortality – Mortality relates to death. It can be measured by death rate, infant mortality, case mortality and attack rate.

Multilateral (agreement) – An agreement negotiated between more than two countries or groups of countries at the same time.

Multiplier (effect) – A situation where an initial injection of investment or capital into an economy (at any scale) in turn creates additional income by, for example, increasing employment, wages, spending, tax revenues, etc.; the overall increase in income arising from any new injection of spending or investment.

Mineral precipitation – The process in which dissolved minerals are released from water and form a deposit.

N

Naphtha – A generic term to describe a flammable liquid comprised of various hydrocarbon mixtures.

Natural change – The difference between birth rates and death rates. If birth rates are higher there is a natural increase in population; if death rates are higher, there is a natural decrease.

Natural hazards – Events which are perceived to be a threat to people, the built environment and the natural environment. They occur in the physical environments of the atmosphere, lithosphere and the hydrosphere.

Net migration change – The difference between the total number or average rate of immigrants and emigrants in an area or country over a given period of time. More immigrants than emigrants will give a positive net migration and vice versa.

Net primary production (NPP) – The rate at which an ecosystem accumulates energy or biomass, excluding the energy it uses for the process of respiration. This typically corresponds to the rate of photosynthesis minus respiration by the photosynthesisers.

(Net) replacement rate – The number of children each woman needs to have to maintain current population levels or give zero population growth by generation. It is a measured fertility rate. In richer developed countries the replacement rate is about 2.1 but ranges from 2.5 to 3.3 in less developed countries because of higher mortality. If fertility is above the replacement rate, population will grow and if it is below, population will decline.

Non-communicable disease – A medical condition of disease that is by definition non-infectious and non-transmissible among people.

Non-government organisations (NGOs) – Any non-profit, voluntary citizens' group with a common interest, which is organised on a local, national or international level. Sometimes referred to as a 'civil society organisation' (CSO).

Nutrient cycling – The movement of nutrients in the ecosystem between the three major stores of soil, biomass and litter.

O

Objective – Not influenced by personal feelings or opinions in considering and representing facts.

Oceanic water – The water contained in the Earth's oceans and seas but not including such inland seas as the Caspian Sea.

Outsourcing – A cost-saving strategy used by companies who arrange for goods or services to be produced or provided by other companies, usually at a location where costs are lower.

Overland flow – The tendency of water to flow horizontally across land surfaces when rainfall has exceeded the infiltration capacity of the soil and all surface stores are full to overflowing.

Overshoot – An ecological term referring to a point when the population and its associated consumption of resources exceed the long-term carrying capacity of its environment.

P

Particulate air pollution – A form of air pollution caused by the release of particles and noxious gases into the atmosphere. Emissions of particles can occur naturally but they are largely caused by the combustion of fossil fuels.

Peak discharge – the point on a flood hydrograph when river discharge is at its greatest.

Perception – With regard to hazards, this is the way in which an individual or a group views the threat of a hazard event. This will ultimately determine the course of action taken by individuals or the response they expect from governments and other organisations.

Perception of place – This is the way in which place is viewed or regarded by people. This can be influenced by media representation or personal experience.

Percolation – The downward movement of water within the rock under the soil surface. Rates vary depending on the nature of the rock.

Periglacial – Processes and landforms associated with the fringe of, or area near to, an ice sheet or glacier.

Photochemical – A form of air pollution that occurs mainly in cities and can be dangerous to health. Exhaust fumes become trapped by temperature inversions and, in the presence of sunlight, low-level ozone forms. It is associated with high-pressure weather systems.

Place – Defined as a location with meaning. Places can be meaningful to individuals in ways that are personal or subjective. Places can also be meaningful at a social or cultural level and these meanings may be shared by different groups of people.

Placelessness – Defined by the geographer Edward Relph as the loss of uniqueness of place in the cultural landscape so that one place looks like the next.

Placemaking – The deliberate shaping of an environment to facilitate social interaction and improve a community's quality of life.

Plagioclimax – The plant community that exists when human interference prevents the climatic climax vegetation being reached.

Plough pan – A compacted layer within the profile of a cultivated soil resulting from repeated compaction from ploughing.

Population density – The average number of people living in a specified area, usually expressed as the number of people per square kilometre.

Population distribution – The pattern of where people live. This can be considered at all scales from local to global, in an area or country.

Porphyry – A fine-grained hard igneous rock which contains large crystals and can contain hydrothermal ore deposits such as copper.

Potable (water) – Water that is suitable for drinking (it does not have to be completely pure but must not contain unacceptable levels of hazardous materials, nor look, taste or smell unpleasant).

Prediction – The ability to give warnings so that action can be taken to reduce the impact of hazard events. Improved monitoring, information and communications technology have meant that predicting hazards and issuing warnings have become more important in recent years.

Primary effects (hazard) – The effects of a hazard event that result directly from that event. For a volcanic eruption these could include lava and pyroclastic flows. In an earthquake, ground shaking and rupturing are primary effects.

Primary sector – The sector of the economy making direct use of natural resources, for example agriculture, mining, forestry and fishing.

Prisere – The succession of vegetational stages that occurs in passing from bare earth or water to a climax community.

Protectionism – A deliberate policy by government to impose restrictions on trade in goods and services with other countries – usually done with the intention of protecting home-based industries from foreign competition.

Pyramid of biomass – A diagrammatic representation of the amount of organic material, measured in grams of dry mass per m^2, found in a particular habitat at ascending trophic levels of a food chain.

Pyroclastic flows – Also known as nuées ardentes, formed from a mixture of hot gas (over 800°C) and tephra. After ejection from the volcano they can flow down the sides of a mountain at speeds of over 700 km/hr. Some vulcanologists apply the term nuées ardentes when the cloud is formed only from hot gas.

Pyrophytic vegetation – Pyrophytes are plants adapted to tolerate fire. Methods of survival include thick bark, tissue with a high moisture content and underground storage structures.

Q

Qualitative data – These are non-numerical data that are used in a relatively unstructured and open-ended way. It is descriptive information, which often comes from interviews, focus groups or artistic depictions such as photographs.

Quantitative data – Quantitative data is numerical data such as metric-level measurements that is associated with the scientific and experimental approach and is criticised for not providing an in-depth description; data that can be quantified and verified, and is amenable to statistical manipulation.

R

Radiative forcing – The difference between the incoming solar energy absorbed by the Earth and energy radiated back to space.

The Ramsar Convention – An international treaty for the conservation and sustainable use of wetlands. It is also known as the Convention on Wetlands. It is named after the city of Ramsar in Iran where the Convention was signed in 1971.

Raised beach – Areas of former wave-cut platforms and their beaches which are at a level higher than the present sea level.

Recharge (groundwater) – A hydrological process involving the downward movement of water by infiltration and percolation causing the replenishment of natural groundwater. It occurs naturally but can be done artificially through human intervention and management.

Refugee – In its broader context it means a person fleeing, for example, civil war or natural disaster but not necessarily fearing persecution as defined by the 1951 Refugee Convention. Legally however, a refugee is an asylum seeker whose application claim for asylum has been successful.

Reproductive age – The age at which women can give birth. In official demographic data it is usually considered to be between 15 and 44.

Reserves – That part of a resource that is available for use under existing economic and political conditions and with available technology.

Resilience – The sustained ability of individuals or communities to be able to utilize available resources to respond to, withstand and recover from the effects of natural hazard events. Communities that are resilient are able to minimize the effects of the event enabling them to return to normal life as soon as possible.

Resilience (ecological) – The amount of disturbance that an ecosystem can withstand without changing existing structures and processes.

Resource frontier – A newly-colonised region where resources have been discovered and are brought into production for the first time.

Resource peak – This marks the point in time when the maximum production rate of a resource occurs (at all scales from individual reserves to global) with production declining in subsequent years.

Retardants – Chemicals sprayed onto fires in order to slow them down. They are composed of nitrates, ammonia, phosphates and sulphates and thickening agents.

Retrofitting – In earthquake-prone areas buildings and other structures can be fitted with devices such as shock absorbers and cross-bracing to make them more earthquake-proof.

Ria – Former river valleys drowned by rising sea levels.

Riparian – This describes the land adjacent to a stream or river. In water conflicts and disputes it is used as a legal term to determine the rights of those living on the banks of a river to access the water for use.

Run-off – All the water that enters a river channel and eventually flows out of the drainage basin.

S

Sea-floor spreading – Movement of oceanic crustal plates away from divergent/constructive plate boundaries such as in the middle of the Atlantic Ocean.

Safe drinking water – Water that is safe for human consumption. The water must be free from harmful pollutants and bacteria that could make people ill.

Salinisation – The build-up of salts in soil, eventually to toxic levels for plants.

Saltation – A process where sand-sized particles are transported by bouncing and hopping along the surface.

Sampling – To make statistically valid inferences, when it is impossible to measure the whole population, by selecting a group which will be representative of that population. Such a group is known as a sample.

Saturated – This applies to any water store that has reached its maximum capacity.

Secondary effects (hazard) – These are the effects that result from the primary impact of the hazard event. In volcanic eruptions these include flooding (from melting ice caps and glaciers) and lahars. In an earthquake, tsunamis and fires (from ruptured gas pipes) are secondary effects.

Secondary sector – The sector of the economy involved in processing or manufacturing to produce finished capital or consumer goods.

Sediment – Any naturally-occurring material that has been broken down by the processes of erosion and weathering and has then been transported and subsequently deposited by the action of ice, wind or water.

Sediment cell – A distinct area of coastline separated from other areas by well-defined boundaries, such as headlands and stretches of deep water.

Semi-arid – The climate of an area that receives between 250 and 500 mm of precipitation per year.

Sense of place – This refers to the subjective and emotional attachment people have to a place. People develop a 'sense of place' through experience and knowledge of a particular area.

Seral stage – An individual stage within a sere, for example, colonisation or stabilisation.

Sere – The entire sequence of stages in a plant succession. Different seres are named after the starting point of the succession, for example, lithosere, hydrosere, psammosere and halosere.

Social segregation – This is when groups of people live apart from the larger population due to factors such as wealth, ethnicity, religion or age.

Soft engineering – Using natural systems for coastal defence, such as beaches, dunes and salt marshes, which can absorb and adjust to wave and tide energy.

Soil – The upper layer of earth in which plants grow, a black or dark brown material typically consisting of a mixture of organic remains, clay and rock particles.

Soil organic carbon (SOC) – The organic constituents in the soil: tissues from dead plants and animals, products produced as these decompose and the soil microbial biomass.

Statistical bias – The intentional or unintentional favouring of one group or outcome over other potential groups or outcomes in the statistical population.

Statistical population – The whole from which a sample will be chosen for a research exercise.

Stemflow – The portion of precipitation intercepted by the canopy that reaches the ground by flowing down stems, stalks or tree bole.

Stock resources – Non-renewable resources which can be permanently expended. Their quantity is expressed in absolute amounts rather than rates, for example, oil.

Store/component – A part of the system where energy/mass is stored or transformed.

Storm and rainfall event – An individual storm is defined as a rainfall period separated by dry intervals of at least 24 hours and an individual rainfall event is defined as a rainfall period separated by dry intervals of at least 4 hours (Hamilton and Rowe, 1949).

Storm flow – Discharge resulting from storm precipitation involving both overland flow, throughflow and groundwater flow.

Storm hydrograph – A graph of discharge of a river over the time period when the normal flow of the river is affected by a storm event.

Sub-aerial processes – Includes processes that slowly (usually) break down the coastline, weaken the underlying rocks and allow sudden movements or erosion to happen more easily. Material is broken down in situ, remaining in, or near, its original position. These may affect the shape of the coastline and include weathering, mass movement and run-off.

Sub-climax – The development of an ecological community to a stage short of the expected climax because of some factor, such as repeated fires in a forest, which arrests the normal succession.

Subduction – the process by which one plate is driven beneath another plate to form such features as ocean trenches at destructive/convergent plate boundaries.

Subjective – Based on or influenced by personal feelings, tastes or opinions.

Suburbanisation – The movement of people from living in the inner parts of a city to living on the outer edges. It has been facilitated by the development of transport networks and the increase in ownership of private cars. These have allowed people to commute to work.

Succession – The series of changes in an ecological community that occur over time.

Surface creep – Where saltating particles return to the surface and hit larger particles that are too heavy to hop; they slowly creep (slide or roll) along the surface from a combination of the push of the saltating grain and the movement of the wind.

Suspension – Transportation by wind where the smallest particles, generally less than 0.2 mm, are held in the air.

Sustainable city – A city which provides employment, a high standard of living, a clean, healthy environment and fair governance for all its residents.

Sustainable development – Development which recognises that the needs of the present have to be met but doing this without affecting the needs of future generations.

Sustainable Urban Drainage Systems (SUDS) – This is a relatively new approach to managing rainfall by using natural processes in the landscape to reduce flooding, control flooding and provide amenities for the community.

Swash – The rush of water up the beach after a wave breaks.

System – A system is a set of interrelated components working together towards some kind of process.

T

Tailings – Tailings consist of the slurry mix of crushed rock and effluents generated by the mechanical and chemical processes used to extract the desired mineral from the ore. (It produces a waste stream known as mine tailings.)

Tariffs – A tax or duty placed on imported goods with the intention of making them more expensive to consumers so that they do not sell at a lower price than home-based goods – a strategy of protectionism.

Temperature inversion – An atmospheric condition in which temperature, unusually, increases with height. As inversions are extremely stable conditions and do not allow convection, they trap pollution in the lower layer of the atmosphere.

Tephra – The solid matter ejected by a volcano into the air. It ranges from volcanic bombs (large) to ash (fine).

Terrestrial water – This consists of groundwater, soil moisture, lakes, wetlands and rivers.

Tertiary sector – The sector of the economy concerned with providing services.

Thermal fracture – The weathering of rock resulting from their rapid and repeated heating and cooling.

Throughfall – The portion of the precipitation that reaches the ground directly through gaps in the vegetation canopy and drips from leaves, twigs and stems. This occurs when the canopy-surface rainwater storage exceeds its storage capacity.

Throughflow – The movement of water downslope through the subsoil under the influence of gravity. It is particularly effective when underlying permeable rock prevents further downward movement.

Tides – The periodic rise and fall of the level of the sea in response to the gravitational pull of the sun and moon.

Top down – When the decision to undertake projects or developments is made by a central authority such as government with little or no consultation with the local people whom it will affect.

Topophilia – The affective bond between people and place or setting. (Tuan, 1974)

Topography – The relief and drainage of an area.

Total fertility rate – This is the average number of children born per woman in an area or country if all women lived to the end of their child-bearing years. It is considered to be a more direct and accurate measure of fertility than birth rate as it refers to the births per woman.

Total productive bio-capacity – All the food, water and energy resources produced by the Earth's natural systems annually to sustain us.

Transpiration – The loss of water from vegetation through pores (stomata) on their surfaces.

Transportation – The processes that move material from the site where erosion took place to the site of deposition.

Trophic level – An organism's position in the food chain. Level 1 is formed of autotrophs which produce their own food. Level 2 consume level 1 and level 3 consume level 2, etc.

Tsunami – Giant sea waves generated by shallow-focus underwater earthquakes, violent volcanic eruptions, underwater debris slides and landslides into the sea.

Tundra – A Lapland term for the climate and vegetation type found across extensive areas of northern Siberia, Scandinavia, Canada and Alaska. Tundra lies between the region of continual snow and ice of the Polar regions and the northern limit of tree growth (or the taiga environments) and at high altitudes above the tree line in the Alps, Rockies, Andes and Himalayas. Similar ecosystems are found around the ice-free fringes of Antarctica. It is characterised by a stony or marshy surface with mosses, lichens and other low-lying vegetation including shrubs and dwarf trees. Tundra has short summers but mean monthly temperatures do not exceed 10°C, which is just warm enough for snow and the surface layer of the permafrost (active layer) to melt. Currently tundra covers about 8 per cent of the land area of Earth, but during the glacials of the Quaternary glaciation tundra covered huge areas of central North America and Europe, including the British Isles.

U

Unconventional (oil and gas) reserves – Hydrocarbon reservoirs that have low permeability and porosity and so are difficult to exploit. They require enhanced recovery techniques such as fracture stimulation; for example, shale deposits, tar sands and heavy oils.

Under-employment – The practice in which a person is not doing work that makes full use of their skills and abilities.

United Nations – An international organisation founded in 1945 made up of 193 member states whose aim is to promote international peace and cooperation.

Urban growth – An increase in the number of urban dwellers. Classifications of urban dwellers depend on the census definitions of urban areas, which vary from country to country. They usually include one or more of the following criteria: population size, population density, average distance between buildings within a settlement, legal and/or administrative boundaries.

Urban heat island – The zone around and above an urban area which has higher temperatures than the surrounding rural areas.

Urbanisation – An increase in the proportion of a country's population that lives in towns and cities. The two main causes of urbanisation are natural population growth and migration into urban areas from rural areas.

Urban resilience – The capacity of individuals, communities, institutions, businesses and systems within a city to survive, adapt and grow no matter what kinds of chronic stresses and acute shocks they experience.

Urban morphology – The spatial structure and organisation of an urban area.

Urban policy – Strategies chosen by local or central government to manage the development of urban areas and reduce urban problems.

Urban resurgence – Urban resurgence refers to the regeneration, both economic and structural, of an urban area which has suffered a period of decline. This is often initiated by redevelopment schemes but is also due to wider social, economic and demographic processes.

Urban social exclusion – Economic and social problems faced by residents in areas of multiple deprivation.

Urban sprawl – The spread of an urban area into the surrounding countryside.

V

Venturi effect – The squeezing of wind into an increasingly narrow gap resulting in a pressure decrease and velocity increase.

W

Water balance – The balance between inputs (precipitation) and outputs (run-off, evapotranspiration, soil and groundwater storage) in a drainage basin.

Water stress – Water stress occurs when the demand for water exceeds the amount of water available during a certain period, or when poor quality of water restricts its availability for human use.

Water surplus (areas) – Areas where there is more than sufficient water available to meet human needs.

Wave-cut platform – A gently sloping (less than 5°), relatively smooth, marine platform caused by abrasion at the base of the cliff.

Wave refraction – When waves approach a coastline that is not a regular shape, they are refracted and become increasingly parallel to the coastline. The overall effect is that the wave energy becomes concentrated on the headland, causing greater erosion. The low-energy waves spill into the bay, resulting in beach deposition.

Weathering – The breakdown and/or decay of rock at or near the Earth's surface creating regolith that remains in situ until it is moved by later erosional processes. Weathering can be mechanical, biological/organic or chemical.

Well-being – The state of being comfortable, healthy, or happy.

World city – These are cities which have great influence on a global scale, because of their financial status and worldwide commercial power. Three cities which have traditionally sat at the top of the global hierarchy: New York, London and Tokyo are now being joined by the likes of Beijing, Shanghai and Mumbai. These cities house the headquarters of many transnational corporations (TNCs), are centres of world finance and provide international consumer services.

Z

Zonal soil – A soil which has experienced the maximum effect of climate and natural vegetation upon the parent rock, assuming there are no extremes of weathering, relief or drainage.

Index